Subsurface Sensing

WILEY SERIES IN MICROWAVE AND OPTICAL ENGINEERING

KAI CHANG, Editor
Texas A&M University

A complete list of the titles in this series appears at the end of this volume.

Subsurface Sensing

Edited by

AHMET S. TURK

A. KOKSAL HOCAOGLU

ALEXEY A. VERTIY

A JOHN WILEY & SONS, INC., PUBLICATION

Library of Congress Cataloging-in-Publication Data:

Turk, Ahmet S, 1977–
    Subsurface sensing / Ahmet S. Turk, A. Koksal Hocaoglu, and Alexey A. Vertiy.
        p. cm—(Wiley series in microwave and optical engineering)
    Includes bibliographical references and index.
    ISBN 978-0-470-13388-0
    1. Ground penetrating radar. 2. Nuclear Quadruple Resonance. 3. Metal detectors. I.
Hocaoglu, A. Koksal. 1967– II. Vertiy, Alexey, A., 1947– III. Title.
    TK6592.G7T87 2010
    621.36–dc22

                                                                    2009042936

Printed in The United States of America

ePDF ISBN: 9780470608562

10 9 8 7 6 5 4 3 2 1

For Betul, Selda, and Vera

Contents

Contributors

Steven Achal ITRES Research Ltd., Calgary, Alberta T2L 2K7, Canada

Serkan Aksoy Department of Electronics Engineering, Gebze Institute of Technology, Gebze, Kocaeli 41400, Turkey

Barry Allred USDA/ARS Soil Drainage Research Unit, Columbus, Ohio 43210

Hiroshi Asanuma Graduate School of Environmental Studies, Tohoku University, Sendai, Miyagi 980–8579, Japan

Pier Matteo Barone Physics Department, Università Roma Tre, Rome 00146, Italy

Erkul Basaran Department of Electronics Engineering, Gebze Institute of Technology, Gebze, Kocaeli 41400, Turkey

Ali Nezihi Bilge Institute of Graduate Studies in Science and Engineering, Yeditepe University, Kadikov, Istanbul 34755, Turkey

Esther Bloem Bioforsk, Norwegian Institute for Agricultural and Environmental Research, Soil and Environment Division, Frederik A. Dahls vei 20, Ås 1432, Norway

John Butnor Southern Institute of Forest Ecosytems Biology, USDA Forest Service, South Burlington, VT 05403

David Cleland School of Planning, Architecture and Civil Engineering, Queen's University of Belfast, Northern Ireland, UK

Dennis L. Corwin U.S. Salinity Laboratory, USDA-ARS, Riverside, CA 92507-4617

Lorenzo Crocco Institute for Electromagnetic Sensing of the Environment, National Research Council, Naples 80124, Italy

Alexander Grigorievich Denisov Academy of the Technological Sciences of Ukraine, Kiev, Ukraine

Andrea Di Matteo Physics Department, Università Roma Tre, Rome 00146, Italy

Roger Eigenberg Environmental Management Research Unit, U.S. Meat Animal Research Center, Clay Center, Nebraska 68933

Hartmut Ewald Institute of Electrical Engineering, Rostock University, Rostock D-18051, Germany

Hamid Farahani   Department of Biosystems Engineering, Clemson University Edisto Research and Education Center, Blackville, South Carolina 29634

Sergey Gavrilov   TÜBITAK Marmara Research Center, TUJRL, Gebze, Kocaeli 41470, Turkey

Mika Harbeck   TÜBITAK Marmara Research Center, Material Science Institute, Gebze, Kocaeli 41470, Turkey

A. Koksal Hocaoglu   TÜBITAK BILGEM Information Technologies Institute, Gebze, Kocaeli 41470, Turkey

Kurt H. Johnsen   USDA Forest Service, Southern Research Station, Research Triangle Park, North Carolina 27709

Susan S. Hubbard   Environmental Remediation Program, Lawrence Berkeley National Laboratory, Berkeley, California 94720

Sebastien Lambot   Department of Environmental Sciences and Land Use Planning, Université Catholique de Louvain, Louvain-la-Neuve 1348, Belgium

Elisabetta Mattei   Physics Department, Università Roma Tre, Rome 00146, Italy

John E. McFee   Explosives Detection Group, Defense Research and Development Canada, Suffield, Alberta T1A 8K6, Canada

Daniel McInnis   USDA Forest Service, Southern Research Station, Research Triangle Park, North Carolina 27709

Georgy V. Mozzhukhin   Radiophysics Department, Russian State University of Immanuel Kant, Kaliningrad, Russia

Ioan Nicolaescu   Military Technical Academy of Romania, Bucharest 050141, Romania

Caner Ozdemir   Department of Electrical and Electronics Engineering, Mersin University, Mersin 33343, Turkey

Engin Ozturk   Department of Electronics Engineering, Gebze Institute of Technology, Gebze, Kocaeli 41400, Turkey

Zafer Ziya Ozturk   Department of Physics, Gebze Institute of Technology, Gebze, Kocaeli 41400, Turkey

Elena Pettinelli   Physics Department, Universitá Roma Tre, Rome 00146, Italy

Karel Pospisil   Road Department, Centrum Dopravniho, Transport Research Centre, Brno 636 00, Czech Republic

Vincent E.A. Post   Department of Hydrology and Geo-Environmental Sciences, VU University Amsterdam De Boelalaan, 1081 HV Amsterdam, The Netherlands

Eduardo Proverbio   Department of Industrial Chemistry and Materials Engineering, University of Messina, Messina 98166, Italy

Stanislaw Radkowski   Faculty of Automotive and Constructive Machinery Engineering, Warsaw University of Technology, 02–524 Warsaw, Poland

Bulat Z. Rameev   Department of Physics, Gebze Institute of Technology, Gebze, Kocaeli 41400, Turkey

Alp Oral Salman TÜBITAK BILGEM, Information Technologies Institute, Gebze, Kocaeli 41470, Turkey

Lisa Samuelson Center for Longleaf Pine Ecosystems, School of Forestry and Wildlife Sciences, Auburn University, Auburn, Alabama 36849-5418

Motoyuki Sato Center for Northeast Asian Studies, Tohoku University, Sendai, Miyagi 980–8576, Japan

Evert C. Slob Department of Geotechnology, Delft University of Technology, Stevinweg 1, 2628 CN Delft, The Netherlands

Francesco Soldovieri Institute for Electromagnetic Sensing of the Environment, National Research Council, Naples 80124, Italy

Josef Stryk Road Department, Centrum Dopravniho, Transport Research Centre, Brno 636 00, Czech Republic

Ahmet S. Turk Department of Electronics and Telecommunication Engineering, Yildiz Technical University, Besiktas, Istanbul, Turkey

Alexey A. Vertiy TÜBITAK Marmara Research Center, International Laboratory for High Technologies, Gebze, Kocaeli 41470, Turkey

Piet van Genderen EEMCS/International Research Centre for Telecommunications and Radar, Delft University of Technology, 2600 GA Delft, The Netherlands

Bryan Woodbury Environmental Management Research Unit, U.S. Meat Animal Research Center, Clay Center, Nebraska 68933

Alexander G. Yarovoy EEMCS/International Research Centre for Telecommunications and Radar, Delft University of Technology, 2600 GA Delft, The Netherlands

Preface

Subsurface sensing is a topic with contributions from many diverse fields, including electrical engineering, civil engineering, geophysics, and mathematical statistics, to name a few. It involves detecting, locating, and identifying objects underneath a surface. For example, locating underground mines, detecting cracks in bridges, through-wall imaging, locating of victims in rubble, and detecting and identifying improvised explosive devices are some of the numerous applications in many scientific and engineering branches.

The major goal of this book is to provide a deep understanding of both the capabilities and limitations of the various sensor technologies used for detecting buried objects. It is designed primarily for researchers and engineers working in this field as well as for users of subsurface sensors. The book should be useful for self-study, as it is largely self-contained. It presents the theory along with the applications and existing technologies.

The text is designed as three parts. In the first part, ground-penetrating radar (GPR), electromagnetic induction (EMI), and microwave tomography are covered, as they are usually the main sensors used in subsurface sensing. Acoustic and seismic sensors, biochemical sensors, nuclear sensors, and optical sensors, generally used to improve the performance of the primary detector, are covered in the second part of the book. The final part deals mainly with application areas. Broad coverage of geophysical applications is provided. Applications in security, landmine detection, and transport and civil engineering are also discussed.

The scope is broad and truly multidisciplinary. A project of this kind inevitably involves making compromises in terms of both breadth and depth of coverage, against constraints of time and length of manuscript. We have expanded the content to include new applications along with recent results on more established topics. We have nearly doubled the content from our initial plan for the book, but the content is still not broad enough to cover this multidisciplinary topic in its entirety.

The book is organized as follows: The first chapter is introductory in nature; we describe the buried object detection problem and give a brief review of the technology. The second chapter covers the sensor types and their capabilities for different operations. Chapter 3 presents the fundamentals necessary to comprehend ground-penetrating radar. GPR system design, GPR hardware for ultrawide band impulse radar, and stepped-frequency continuous-wave radar are discussed. GPR antenna design and performance issues are both covered. GPR data-processing techniques to improve data interpretation to increase detection performance are reviewed. Imaging algorithms and numerical modeling of GPR are presented.

The next three chapters cover electromagnetic induction, microwave tomography, and acoustic–seismic sensors. Metal detection is a very broad technological field: from process control and system monitoring in the food industry to detecting landmines for military operations. These different requirements result in systems that vary in the sensor technology (e.g., fluxgate, coil) and in the complexity of the algorithm employed. Chapter 4 gives an overview of the inductive metal detection systems used mainly for humanitarian demining. How a simple coil can be used to develop an imaging system for landmine detection is discussed.

Chapter 5 presents several approaches to subsurface microwave tomography. Development of theoretical methods and practical realizations for a broad frequency range are covered. This chapter is also devoted to the description and application of linear and nonlinear inversion algorithms. Theoretical approaches are illustrated by experimental methods with a wide spectrum of applications. In particular, employment of subsurface tomography for the investigation of biological objects and studies in medicine, for nondestructive testing of materials, and for landmine detection are discussed. Most of these applications consider a multifrequency method. A number of the results suggest the development of the terahertz tomography method.

Chapter 6 reviews mechanical seismometers, electromagnetic sensors (geophones), piezoelectric sensors, capacitive sensors, and optical sensors. The principles behind mechanical seismometers, geophones, piezoelectric sensors, and capacitive sensors are described as well as the fundamentals of sensor installation. Multicomponent seismic monitoring techniques and limitations for earthquake and seismic monitoring are also discussed.

Chapter 7 is concerned with infrared and hyperspectral systems and other sensors, such as biological and chemical-based sensors. These sensors are generally used for specific application areas or to improve main detector performance. The operating principles, current capabilities, limitations, and improvement potentials of these sensors are described briefly.

Chapter 8 reviews the fusion techniques for multisensor systems. These techniques combine information from multiple sensors to achieve better accuracy. The theory behind these techniques is quite old and well established. The availability of sophisticated sensors and the existence of sophisticated information-processing systems makes it possible to employ multiple sensors in detection and identification systems. Diverse physical properties measured by these sensor technologies provide significant advantages over single-source data. This keeps multisensory data fusion an active area of research.

Chapter 9 deals with geophysical applications. Initially, an overview is given of research and methods for determining the electric and magnetic properties of soils and rocks; then related applications are discussed. An understanding of the electromagnetic response of fluid-filled porous materials is crucial in these applications. How to determine these properties and the implication for field measurements are the subjects of Section 9.2.

In many hydrogeophysical applications the most important tool is ground-penetrating radar, because the water content of the pore space is the most decisive parameter determining the effective electric permittivity. Several recently developed methods are discussed for water content determination from GPR measurements. An important distinction in GPR systems is the use of air-launched antennas or ground-coupled antennas. Surface water content, subsurface water content, and the detection of water table depth are discussed in detail. GPR data are also used for monitoring temporal changes in subsurface water content in the vadose zone, and attempts to combine GPR data full-waveform inversion with hydrological modeling are discussed. Finally, Section 9.3 ends with a hydrogeophysical case study in which low-frequency electromagnetic methods have been combined to map the distribution of subsurface salinity.

In contaminant remediation, hydrogeological and biochemical properties are most important because they influence the flow and transport of contaminants, their natural attenuation, and contaminant remediation efficacy. Section 9.4 details the mapping of hydrogeological parameters using hydrogeophysical methods and techniques. This involves determination of the distribution of the relevant properties and monitoring of the biochemical processes with geophysical techniques. The section concludes by looking at challenges and corresponding future directions for research and applications.

In recent years the importance of geophysical techniques has grown rapidly for agricultural applications. Section 9.5 provides a historical overview followed by an overview of geophysical techniques and methods used as related to different applications. The section gives detail information relating to several case studies that were carried out for a variety of purposes.

Archeology is another rapidly growing field of geophysical applications. This is the subject of Section 9.6, where a brief history is followed by a description of methods and techniques for specific archeological targets. This section contains several case studies showing a wide variety of methods used for archeological purposes. Magnetometry, electric resistivity tomography, microwave tomography, and ground-penetrating radar are discussed in these case studies.

Chapter 10 is devoted to applications of microwave and millimeter-wave imaging technologies for security purposes. Image reconstruction of objects in the case of plane-parallel dielectric layers and its practical application for detecting people behind walls are considered. Schemes of experimental devices and illustrations of the results are presented. A passive radar method in the millimeter-wave band which is very prospective for detecting concealed weapons, explosive materials, and other dangerous objects hidden under clothes is presented.

Among many application areas of subsurface sensing, landmine detection is probably the one that has attracted the greatest public attention in recent years. According to current estimations, at least 60 million (some estimate up to 200 million) undetected landmines spread around the world in countries on every continent have an enormous impact on economical and social developments in the areas affected. With current demining practice it will take centuries before these areas will be demined and the safety of the environment restored. Within the last two decades, considerable research and development effort has been expended to develop secure, reliable, and efficient high-tech means of landmine detection. Chapter 11 is a brief overview of the landmine problem and the most promising advanced technologies (such as EMI, GPR, electrooptical, and chemical sensors) for humanitarian demining.

Safe performance and improved quality control are in greater and greater demand for building and strategic civil structures (e.g., bridges, dams, power plants), while increasing age, deterioration, and corrosion damage reduce their reliability and structural performance. In this context, nondestructive inspection techniques (NDT) based on propagation and detection of acoustic waves, electromagnetic waves, and nuclear radiation plays a fundamental role in material testing and structural health monitoring. Chapter 12 is dedicated to the more recent uses of NDT in this field. Starting from the criticality of each different type of structure, an overview of the outcomes of as well as in laboratory applications is presented while evidencing the advantages and drawbacks of each technology.

Of the twelve chapters three have been written exclusively by invited contributors. Each chapter has been reviewed for technical correctness by the corresponding chapter editors: Turk for Chapters 1, 2, 3, and 12; Hocaoglu for Chapters 1 and 3; Vertiy for Chapters 1, 5, and 10; Bilge for Chapter 7; Slob for Chapter 9; Yarovoy and Sato for Chapter 11; Proverbio for Chapter 12. Our editorial work consisted of having each chapter reviewed to check for minor typographical errors, misspellings, and the like. The publisher has, where necessary, transformed the text into standard American English.

Color versions of selected figures can be found online at *ftp://ftp.wiley.com/public/sci_tech_med/subsurface_sensing*.

Finally, we want to express appreciation to all the contributors as well as to all the chapter editors. Clearly, without their efforts, this publishing project could not have been completed. We have gained in knowledge, and as editors we express our gratitude to all who have added expertise to the book.

AHMET S. TURK
A. KOKSAL HOCAOGLU
ALEXEY A. VERTIY

CHAPTER ONE

Introduction

In this introductory chapter we present an overview of subsurface sensing technologies and applications. We emphasize that subsurface sensing covers a wide range of disciplines, methods, and applications and present the main and auxiliary sensors commonly used in subsurface sensing. We emphasize the need for advanced signal and imaging techniques as well as sensor data fusion techniques. We also address a number of the barriers present in subsurface sensing and introduce common issues associated with the medium and certain properties of the soil and with temperature and humidity during data collection as they relate to the performance of sensing devices. We include both primary and secondary sources for subsurface sensing technologies and applications.

Subsurface detection and identification of buried geological and human-made structures currently represent an important research area and a progressive technological concept throughout the world. The scope of the problem is very complex and the basis of the topic depends on the wide area of demands: from military to commercial requirements, such as locating underground pipes, buried mines, and archeological or anthropological artifacts. The variety of unknown false and undesired targets under the ground complicates the object identification task. Moreover, the medium involved is usually lossy and inhomogeneous. The size, geometry, constitution (i.e., dielectric, metallic, explosive, etc.), and depth of the target object and the characteristics of the medium (i.e., wet soil, sand, etc.) are the principal parameters for detector designs. Unfortunately, there is not yet a single method or unique system that can manage to detect every type of object of any size, structure, depth, or soil. Therefore, a number of convenient sensor technologies must be considered regarding specific situations. These sensors may have common, similar, or different capabilities for the detection of desired targets.

Subsurface Sensing, First Edition. Edited by Ahmet S. Turk, A. Koksal Hocaoglu, and Alexey A. Vertiy.
© 2011 John Wiley & Sons, Inc. Published 2011 by John Wiley & Sons, Inc.

In this book we compile and present methods available to the growing number of potential users of subsurface sensing technologies, regardless of their educational or academic backgrounds, involved in physics, geophysics, civil engineering, archeology, or electrical engineering. A variety of sensor technologies and radar systems, based primarily on electromagnetic, acoustic, infrared, and chemical characteristics, are examined for their pros and cons as related to operating principles, strengths, limitations, and feasibilities. Furthermore, some convenient multisensor approaches to a wide range of applications are examined to reach the best detection performance in specific cases.

Subsurface sensing is a multidisciplinary research area combining expertise in wave physics, sensor engineering, image processing, and other areas, such as agricultural engineering and geophysics. It covers all areas of subsurface sensing technologies, such as radar, interferometer, ultrasonics, acoustics, microwaves, millimeter waves, submillimeter waves, infrared, and optics. Some examples of common applications are the detection, identification, and classification of objects, structures, and matter under and at surfaces, mapping pollution plumes underground, and nondestructive evaluation and testing of materials. Other sensing areas, such as seeing through walls, locating victims in buildings or rubble, personnel and vehicle surveillance, and intrusion detection and assessment are closely related and utilize similar sensing technology.

Detecting, locating, and identifying objects that are obscured beneath a covering medium all share the problem of distinguishing the effect of a dispersive, diffusive, and absorptive medium from the desired details of the subsurface structure and functionality. The problem is similar whether the wave probe is electromagnetic or acoustic, whether the medium is soil or concrete, or whether the target is a landmine or a pipeline. The properties of the medium are one of the leading items that determine the technology and technical specifications. For example, the effectiveness of inductive metal detectors to detect metal-cased antipersonnel mines is greatly reduced by metal clutter, such as metal fragments, spent ammunition, shrapnel, and cans in the soil. The presence of highly magnetic minerals may cause strong attenuation, which prevents the penetration of electromagnetic signals at certain frequency bands. Characterization of the medium, an important area for subsurface sensing, still constitutes a challenging task.

A second problem arises from varying properties of the medium. Subsurface inhomogeneities such as rocks, tree roots, and water packets are also a major concern and have the potential to cause false alarms. In such random inhomogeneous and highly cluttered environments, techniques to image objects quantitatively are limited. The use of a single sensor technology in such a complex environment is usually not reliable. Sensor or information fusion techniques are often used to overcome the shortcomings of available technologies. The fusion of more than one physical probe and the analysis of complementary information from multiple probes is an important avenue to progress in difficult subsurface sensing problems [1]. Combining different sensor inputs to optimize the information gathered is also a challenging area. Due to the lack of a theoretical basis, success in one problem is not easily applied to other domains. Therefore, sensor or information fusion is often applied on an ad hoc basis.

Ground-penetrating radar (GPR) is one of the best assessed and most recognized remote sensing tools. The use of GPR technology dates back to the late 1920s [2]. Advances in GPR data visualization and processing have increased its use significantly in a growing number of applications, from archeology to agriculture. In the 1990s it was particularly popular for use in the Humanitarian Demining Research and Development Program in the United States. This led to a marked increase in the number of GPR systems made available. GPR technology has reached a level of maturity but is still an active area of research, due to its limitations and varying effectiveness with the electrical properties of the medium. Imaging and visualization of GPR data sets are still one of the most active areas of research, as the raw GPR data are often difficult to discern and interpret. Advances in computer technology made it possible to process GPR data in real time. Today's GPR systems even allow for on-the-spot decision-making capabilities.

Metal detectors find numerous applications daily, such as screening people before allowing them access to airports, schools, and other critical buildings. They are also used as a primary sensor for many subsurface sensing applications. For example, in archeological explorations they are used to find metallic items of historical significance. In geological research they are used to detect metallic composition of soil or rock formations. Perhaps one of the most important applications for metal detectors is to locate mines or other explosive devices, such as unexploded ordnance. The main challenge for metal detectors in subsurface sensing applications is not just metallic debris in the areas surveyed; in addition to temperature and moisture conditions, the electromagnetic properties of soil also greatly influence the performance of metal detectors. It is for this reason that the electromagnetic parameters of the host medium are taken into account for object identification [3]. Inductive metal detectors are the type used most commonly when searching for antipersonnel landmines. Electromagnetic induction–based metal detectors are also popular for salinity monitoring of agricultural lands and soil water content measurements. Detecting all the subsurface characteristics of agricultural, archaeological, and other types of sites is not possible using a single remote-sensing instrument. This is also true for salinity monitoring. Metal detectors are thus used in combination with other sensors, such as GPR systems [4], as, once again, this requires advanced signal and sensor data fusion techniques to increase the effectiveness of subsurface instruments in varying soil conditions.

Tomographic imaging is a nondestructive technique used to detect critical deformities on roads, railways, highways, and bridge decks. There is also a huge interest in tomographic imaging methods to see through walls for military and security applications. Tomographic imaging techniques include optical tomography, electromagnetic tomography, electrical impedance tomography, and magnetic resonance electrical impedance tomography, to name a few. Tomographic imaging techniques have also been used to detect buried objects such as landmines. Some of these application areas are covered in Chapters 5 and 11.

Acoustic techniques are powerful tools for subsurface sensing, especially in the area of geological research, mainly because of the ability of acoustic signals to travel relatively long distances in a variety of rocks and sandstones, which in contrast to soils, show modest scattering and absorption [5]. In civil engineering, defects appearing in structures are determined by monitoring vibroacoustic signal parameters. In a recent study it was demonstrated that acoustic-to-seismic coupling was an effective and extremely accurate technique for the detection of buried land-mines in soils [6]. There is growing interest in using acoustic methods for buried object detection and nondestructive testing of structures [7].

Sensor technologies using biological and chemical methods, nuclear quadrupole resonance, x-ray imaging, and infrared and hyperspectral systems are the auxiliary sensor types generally used in specific application areas or to improve primary detector performance. These sensors are discussed to some extent in Chapter 7. Other chapters are devoted to specific application areas in security, transportation, and civil engineering and agriculture.

RELEVANT RESOURCES

The latest research on subsurface sensing is distributed throughout journals and conference proceedings from various fields. One dedicated journal is *Subsurface Sensing Technologies and Applications*. As subsurface sensing is a multidisciplinary research area, relevant information is spread among a number of journals: *IEEE Transactions on Geoscience and Remote Sensing*, *IEEE Transactions on Microwave Theory and Techniques*, *IEEE Transactions on Magnetics*, *IEEE Transactions on Antennas Propagation*, *IEEE Instrumentation and Measurement Technology*, *IEEE Sensors Journal*, *Journal of Engineering and Environmental Geophysics*, *Journal of Applied Geophysics*, *Journal of Applied Physics*, *International Journal of Infrared and Millimeter Waves*, *Journal of the Optical Society of America*, *Journal of the Acoustical Society of America*, *Journal of Geophysical Research*, *Journal of Materials Chemistry*, *Biosensors & Bioelectronics*, *Soil Science Society of America Journal*, and *Remote Sensing of Environment*, to name a few. A journal with an information fusion emphasis is *Information Fusion*. The *International Journal of Approximate Reasoning*, *Fuzzy Sets and Systems*, and *Transactions on Neural Networks* are some other sources of information on sensor data fusion.

REFERENCES

1. McKnight, S. W., Silevitch, M. B. 2003. *Subsurf. Sens. Technol. Appl.*, 4(4).

2. Goodman, D., Schneider, K., Piro, S., Nishimura, Y., Pantel, A. G. 2007. Advances in subsurface imaging for archeology. In *Remote Sensing in Archaeology*, J. Wiseman, F. El-Baz, Eds. Springer Science + Business Media, Berlin.

3. Das, Y. 2006. Effects of soil electromagnetic properties on metal detectors. *IEEE Trans. Geosci. Remote Sens.*, 44(6): 1444–1453.

4. Metternicht, G., Zinck, A. 2008. *Remote Sensing of Soil Salinization: Impact on Land Management*. CRC Press, Boca Raton, FL.

5. Flammer, I., Blum, A., Leiser, A., Germann, P. 2001. Acoustic assessment of flow patterns in unsaturated soil. *J. Appl. Geophys.*, 46: 115–128.

6. Korman, M. S., Sabatier, J. M. 2004. Nonlinear acoustic techniques for landmine detection. *J. Acoust. Soc. Am.*, 116(6): 3354–3369.

7. Donskoy, D. M. 2008. Nonlinear acoustic methods. In *Encyclopedia of Structural Health Monitoring*, C. Boller, F.-K. Chang, Y. Fujino, Eds. Wiley, Hoboken, NJ.

CHAPTER TWO

Sensor Types

2.1 INTRODUCTION

The main sensors for buried-object detection are based on electromagnetic, acoustic, seismic, and optical technologies. Electromagnetic sensors are divided into three groups: electromagnetic induction (EMI), ground-penetrating radar (GPR), and microwave tomography [1,14,38]. An EMI or metal detector uses low-frequency electromagnetic fields to induce eddy currents in the metal components of buried objects. It is very mature, popular, and relatively cheap sensor technology conducted in a wide range of environments for target detection. The main limitation of an EMI sensor is the possible metallic clutter of the medium, especially for objects of low conductivity or low metallic content. GPR is another widely used and established electromagnetic sensor type. Its operating principle is the reflection and backscattering of radio-frequency (RF) waves from buried metallic or dielectric objects. GPR detects all anomalies under the ground, even those that are nonmetallic. The most important problem is clutter due to unwanted objects in the soil, buried targets that may have constitutive parameters similar to those in the soil. Microwave and millimeter-wave tomographic systems, which operate based on electromagnetic inverse scattering phenomena, are employed for near-surface imaging and nondestructive testing with relatively high resolution.

Acoustic and seismic sensors benefit from the principles of sound or seismic wave reflection from an object [55]. Although they are not completely established sensor technologies, such detector systems can have low false-alarm rates. Nevertheless, deeply buried objects, rough surfaces, and vegetation- or

Subsurface Sensing, First Edition. Edited by Ahmet S. Turk, A. Koksal Hocaoglu, and Alexey A. Vertiy.
© 2011 John Wiley & Sons, Inc. Published 2011 by John Wiley & Sons, Inc.

mineral-covered or frozen ground are the most difficult for detection. Optical sensors use the infrared and hyper spectral bands to distinguish a target from temperature and light reflectance differences [63]. If the heat-absorbing capacity of an object differs greatly from that of the soil, the sensor performance will be highly satisfactory. An infrared detector can quickly scan wide areas from high altitudes. However, this is not a suitable sensor for the detection of deeply buried objects.

Beyond the primary sensor technologies mentioned above, a few assistant detector systems can be evaluated either for main sensor performance improvement or for specific applications such as explosive material detection. For example, electrical impedance tomography determines the electrical conductivity distribution of the surface, and x-ray sensors image buried objects using x-ray radiation [65,66]. Moreover, systems based on electrochemical, piezoelectric, nucleonic, and biological methods have proved useful for explosive vapor detection. Detailed investigations of detectors from both the main and auxiliary sensor groups are described below.

2.2   GROUND-PENETRATING RADAR

2.2.1   Overview

The terms *ground probing, subsurface detecting*, and *surface penetrating* refer to a wide range of electromagnetic methods designed for the detection and identification of buried artifacts or structures beneath the surface, generally termed *ground-penetrating radar* (GPR). GPR uses electromagnetic wave propagation and scattering principles to locate, quantitatively identify, and image variations in electrical and magnetic properties under the ground. GPR may be initiated from the Earth's surface, from a land vehicle, or from an aircraft and has high resolution in subsurface imaging, equal to that of any geophysical method, approaching the centimeter range under the right conditions. Since GPR senses subsurface electrical inhomogeneities, the detection performance of such subsurface features as depth, orientation, water density, size, and shape are related to the contrast in electrical and magnetic properties. The detection performance can be improved further by quantitative feature interpretation through modeling. GPR technology is largely application oriented; implementation of its hardware and software determines the overall system performance characteristics. The detection range can vary from a few centimeters to tens of meters. Metallic or dielectric structures can be detected. Identification of objects as small as in the centimetric range is possible. In principle, any metallic object and dielectric discontinuity can be detected by a GPR system. The target may be a long, thin, cylindrical, or spherical object or a planar soil layer and can be classified according to its physical geometry and electrical constitution. All these performances depend strongly on the soil properties and radar parameters, which are primarily operational frequency, transmitter power, system dynamic range, and electromagnetic wave polarization [1–6].

Two types of GPR systems, impulse and stepped frequency, are employed in most current applications. *Impulse GPR* uses time-domain electromagnetic pulse radiation and scattering principles. The shape of the impulse signal determines the frequency bandwidth. An object is detected by means of the amplitude and time delay of the pulse received. For *swept-* or *stepped-frequency GPR*, the operational band is clearly defined by the designer. The electromagnetic wave scattering problem is solved in the frequency domain. The target information is obtained from the amplitude and phase of the signal received [1,6].

GPR is not very different from conventional radar [18]. The principal differences are the short-range concept, near-field analysis, and lossy inhomogeneous propagation medium. The detection technique is usually based on a backscattered radiation signal from the underground object, but in specific cases, forward scattering can also be employed to yield target information. Since this is short-range radar, clutter is the main problem in target detection and identification. The sources of the clutter are antenna ringing, transmitter/receiver (T/R) coupling, ground surface reflection, and unwanted (spurious) backscattering from other buried structures or soil inhomogeneities. Preventive measures such as proper antenna matching, shielding enclosure design, and adaptive methods for processing the signal should be taken to improve radar performance [3,5].

The object-oriented design procedure and critical parameters for GPR systems are explained briefly in the following sections. More specifically, range and resolution limits, operating frequencies, suitable broadband antennas, proper continuous-wave or impulse signal-generator design, low-noise RF receiver units, signal processing techniques, detection and imaging performances, and application results are discussed in Chapter 3.

2.2.2 Principles of Operation and Design

Ground-penetrating radar has an enormously wide application area: from underground structure exploration to small-object detection just beneath the surface. The operational principle consists of the generation of impulse or radio-frequency (RF) power by the signal source, radiation of the electromagnetic wave by the transmitter antenna, characterization of the soil and air–earth interface, reception of the backscattered electromagnetic wave from the target object by the receiver antenna, and finally, sensing, digitizing, and processing the RF signal received by the receiver (Fig. 2.1a). Critical parameters such as soil type and target properties strongly affect detection performance. The main goal of the design is to provide application-oriented information and to illustrate the technical options available for an operator or designer. To operate GPR successfully, the following requirements must be satisfied:

1. RF or impulse signal generator of sufficiently high power to attain the penetration depth required
2. Suitable antenna to couple the electromagnetic wave into the ground efficiently with sufficient bandwidth, high gain, narrow beam, and low input reflection

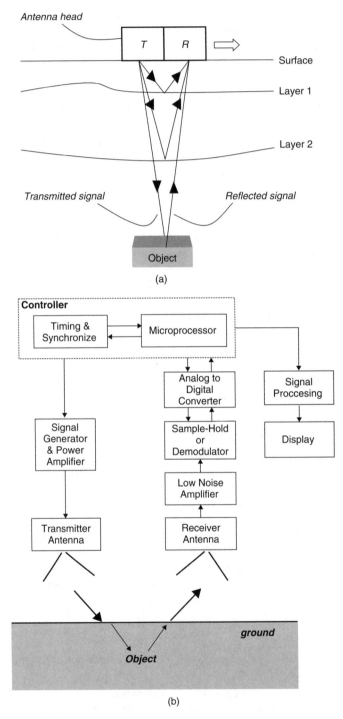

FIGURE 2.1   (a) GPR operation setup. (b) System block diagram.

3. Proper frequency bandwidth to obtain a sufficiently large scattered signal from the smallest buried object
4. Appropriate receiver hardware to achieve adequate signal-to-noise and signal-to-clutter ratios for the best target detection performance
5. Adaptive signal-processing techniques to distinguish and classify buried objects

Selection of the center frequency, bandwidth, and power of a GPR system are the key factors in design. The electromagnetic wave reflection from the ground surface and the attenuation under the ground increase dramatically with increasing frequency and soil conductivity. As a general rule, material of high conductivity will have a high attenuation value. Thus, GPR object probing in sand, gravel, dry soil, and fresh water is much easier than probing in clay, wet soils, mineral soils, and salty water. Since the ground attenuation and reflection parameters involve power loss and clutter, the lower frequencies are used to detect deeply buried objects. Nevertheless, the higher frequencies are required for better resolution and detailed echos to determine small objects [4]. Thus, the ultrawideband (UWB) GPR is generally preferred to benefit from both low and high frequencies. The UWB antenna design is a significant area in the sense of GPR performance. The transmitter and receiver antennas are designed to comply with the remaining hardware to radiate and couple the signal into the ground efficiently [3].

Detection range and horizontal–vertical resolutions are most critical design parameters for the best performance in applications. The detection range of GPR is calculated by the radar equation, which considers transmitter power, antenna gains, receiver sensitivity, target object or sublayer reflectivity, and propagation losses due to wave attenuation in the soil. Horizontal GPR resolution is based on the antenna beamwidth and illumination area. The thickness of the depth-range layers on a GPR screen is related directly to the vertical (range) resolution (see Fig. 2.3). For impulse radars, the bandwidth (BW) is inversely proportional to the range resolution (ΔR_z):

$$BW = \frac{c}{2\Delta R_z \sqrt{\varepsilon_r}} \tag{2.1}$$

where c is the free-space speed of the light and ε_r is the relative permittivity of the soil. For stepped-frequency radars with a Δf frequency step, the range resolution may be defined as

$$\Delta R_z = \frac{c}{2N \,\Delta f \sqrt{\varepsilon_r}} \tag{2.2}$$

A GPR system consists basically of transmitter, receiver, and controller units. The transmitter unit includes a signal generator, power amplifiers, and a transmitter antenna. The receiver unit includes a receiver antenna, a low-noise amplifier (LNA), a sample-and-hold detector or mixer demodulator, and an analog-to-digital converter (ADC). The controller unit includes a timer, a synchronizer of the T/R units, a microcontroller, and a data acquisition card. Finally, the raw data are processed to obtain clear subsurface imaging for the detection and identification of buried objects. The block diagram of a typical GPR is shown in Fig. 2.1b.

2.2.3 Signal Sources and Modulation

Many types of modulation techniques can be employed for GPR systems. The most frequently used GPR systems are impulse radar, which generates short pulses or impulses in the category of amplitude modulation (AM), and swept or stepped-frequency radar, which uses frequency modulation (FM) followed by a synthesized pulse. Impulse GPR is the most popular system commercially, preferred for a wide range of applications. The operational principle is based on impulse signal generation at the peak power and frequency bandwidth required. The signal sources generate monocycle or monopulse waveforms that can have pulse durations of a few hundred picoseconds to a few nanoseconds. The impulse waveforms are generally Gaussian shaped, and their frequency bandwidths range from a few hundred megahertz to a few gigahertz, depending on the application [8]. The main difference between the monocycle and monopulse waveforms is in their frequency-domain behavior. A monopulse signal has many dc or low-frequency components. Nevertheless, the monopulse waveform has some physical advantages for time-domain radar signal transmission and radiation [2].

Frequency-modulated continuous-wave (FMCW) and stepped-frequency continuous-wave (SFCW) GPR systems transmit sweeping or sequentially stepping RF carriers between the operating frequency bands continuously. The signals received are down-converted to intermediate frequency (IF), and I/Q (in phase/quadrature) data are stored for every step. Complex Fourier transformation is used to obtain the time-domain response of a reflected or backscattered target signal [11]. FMCW and SF GPRs are used when the targets of interest are shallow and operational frequencies of 1 GHz and above need to be maintained. They are easier to design than are wideband impulse radars. Such swept- and stepped-frequency radars have the advantage of selection and control of the operational frequency bandwidth. This utility provides better radiation efficiency with high-power generators and high-gain antennas, a lower noise level at narrower band, and consequently, wider dynamic range compared with impulse radar.

Impulse GPR There are two major concerns with the impulse generator used in GPR applications: (1) pulse shape and pulse width, and (2) pulse amplitude. In many cases, both of these parameters are highly critical. The pulse shape and width are broadband characteristics yielding to the detection of buried small objects. On the other hand, pulse amplitude refers to the system power related to the operational detection depth of the target in soil.

Various high-powered switching transistors [e.g., field-effect transistors (FETs)] can be used to acquire the typical GPR pulses of a few hundreds of picoseconds to a few nanoseconds in duration. Avalanche transistors are often used to generate peak power levels, from 50 W to several kilowatts. For the 3- to 10-GHz UWB, the selection of devices available is much smaller. Step-recovery diodes (SRDs) can generate 50- to 200-ps edges with amplitudes of several volts [4].

Similar design procedures are used in avalanche and SRD-mode monocycle impulse generators. Avalanche transistors or step-recovery diodes are operated in the breakdown region to obtain fast switching by discharging the stored energy into short transmission lines. These are output loads, which include the transmitter antenna input impedance. The time-domain impulse signal output and frequency-domain response of avalanche mode impulse generators are described in Chapter 3.

FMCW and SFCW GPR   An FMCW radar system transmits a continuously sweeping or stepping RF carrier signal controlled by a voltage-controlled oscillator (VCO) over a chosen frequency bandwidth. The operational frequency band is chosen principally on the basis of the maximum depth and minimum size of the buried object. The backscattered signal from the target object is mixed with a sample of the waveform transmitted and the results in a difference frequency, called an intermediate frequency (IF), derived from an I/Q mixer pair. The IF output contains the amplitude and phase data of the signal received. The phase information yields the range of the target since it is correlated with time delay. Only if there is a target on the scope will an IF signal be produced. If changing the transmitter frequency is a linear function of time, a target return will occur at a time T_r given by

$$T_r = \frac{2R}{u} \tag{2.3}$$

where R is the range in meters and u is the velocity of an electromagnetic wave on the propagation path (i.e., air, ground, or both) in meters per second.

Since FMCW radar measures essentially the phase of the IF signal, which is related directly to the target range, it requires a high degree of linearity of frequency sweep with time to avoid spectral widening of the IF, which relates to system range resolution degradation. In practice, the sweeping nonlinearity levels should remain below 0.1% [1]. For appropriate system design, the amplifier and antenna gain responses over the frequency band must also be linear. A time-domain equivalent to the impulse radar can be acquired by inverse Fourier transformation.

SFCW radar is a continuous-wave system that is sequentially repetitive at different frequencies, as shown in Fig. 2.2. This type of radar radiates a sequence of N carrier frequency steps, and the amplitude and phase of the signal are stored. The principle of down-converting to an IF signal is similar to that for FMCW radar. Complex Fourier transformation is used to obtain the time-domain response of a reflected or backscattered target signal. The range resolution is defined as

$$\Delta R = \frac{u}{2N\,\Delta f} \tag{2.4}$$

where ΔR is the range resolution and Δf is the frequency increment of every step. For both FMCW and SFCW radar, it is evident that the thermal noise level is

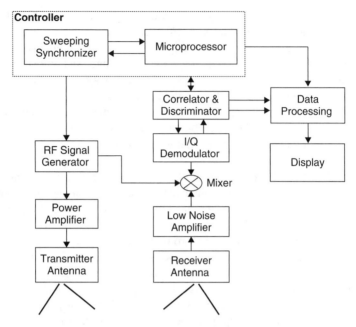

FIGURE 2.2   Swept- or stepped-frequency GPR.

much lower than that of the receiver of time-domain impulse radar. Nevertheless, discriminatiing the target backscattering signal from the ambiguous clutter at the receiver unit is a more difficult problem, due to the uselessness of the antialiasing filters. Therefore, some calibrating methods, such as iterative range gating, are employed primarily for such systems.

2.2.4   Antennas

The performance of the transmitter and receiver (T/R) antenna pair is related strongly to GPR performance since the shape of GPR RF signal radiated and received is determined by the antenna. Wideband antennas, which satisfy electrical characteristics such as high directivity gain, low input reflection level, suppressed side and back lobes, convenient polarization with respect to target shape, high T/R shielding, and linear phase response over the operational frequency band should be designed for both impulse and stepped-frequency GPRs to reach the largest dynamic range, best-focused illumination area, lowest T/R antenna coupling, reduced ringing, and uniformly shaped RF signal radiation. Moreover, some physical restrictions and lightweight conditions required especially by handheld prototypes should also be considered.

For these reasons, dipole, bowtie, planar/conical spiral, log-periodic array, rectangular/circular horn, and TEM horn antennas are preferred for most GPR applications. The dipole is the basic GPR antenna, due to its simple, easy-to-use, linearly polarized structure and lightweight design [9]. Some dielectric, capacitive, and resistive loadings can be implemented to improve dipole radiation characteristics and efficiency. However, the operational bandwidth and gain response of the dipoles are considered as somewhat poor and unsatisfactory. Thus, two types of planar antennas, bowtie and spiral, are more likely to be chosen, especially for high-performance UWB GPRs [1]. The Archimedean and logarithmic spirals are theoretically called frequency-independent antennas. In practice, the arm length and feed radius determine the lower and upper cutoff frequencies of the operational band, respectively [3]. The bowtie is a dipolelike wideband antenna used frequently for pulse radiation. It shows relatively higher gain and broader band performance than those of a dipole of similar length [10]. The bowtie and spiral antennas have similar electrical and physical characteristics. The main difference lies in the polarization response. For a circular object, the bowtie is more suitable since it produces linear polarization and has relatively good linear-phase response over the wideband. But if the target is cylindrical, circular polarized planar spirals can promise significant advantages toward avoiding possible false polarization on scanning.

Vehicle-mounted GPR systems have a physical advantage in antenna design, so it is advisable to use some appropriate three-dimensional antennas that can yield enhanced radiation performance. The ridged horn, log-periodic, conical spiral, discone, and TEM horn antennas are common types that can usually provide wider-band, lower-voltage standing-wave ratios (VSWRs) and higher gain and narrower beam width characteristics than can planar antennas. The horns are the most popular among current GPR designs. Rectangular or circular horn antennas are suitable for SF GPR systems, due to their higher band operations in the gigahertz region. Standard gain horns operate as transverse electric (TE)/transverse magnetic (TM) mode selective and can reach very high gain and low VSWR values with good linearity over the bandwidth ratio $2:1$ [12]. This ratio can be expanded to $10:1$ for octave gain and to $20:1$ for ridged horn models. The main restriction of the horn antenna is the lower cutoff frequency, since its length must be several wavelengths. Therefore, standard gain horns are generally not appropriate antennas for impulse GPR systems since the frequency band of impulse signals starts in the megahertz region. The log-periodic antenna is a well-known linear array structure that is composed of different sizes of dipoles to obtain broadband radiation characteristics. The lengths of the dipole elements are chosen appropriately for the frequency band desired. The lowest operating frequency and the bandwidth ratio of the antenna, which correspond to the longest dipole element and number of dipoles, respectively, usually determine its physical dimensions. Log-periodic arrays are preferred for SF-GPRs since they produce ringing signals on impulse radiation caused by elemental couplings.

The TEM horn is one of the most promising antenna types for impulse GPR systems because of its wider frequency band, higher directivity gain, narrower beamwidth, and lower back reflection characteristics than those of planar antennas [1,4]. It consists of a pair of triangular or circular slice–shaped conductors forming a V-dipole structure and characterized by L, d, α, and θ parameters, which correspond to the antenna length, feed point gap, conductor plate angle, and elevation angle, respectively (see Section 3.4.4). Essentially, the arm length of the TEM horn limits the lower cutoff frequency of the radiated pulse, the plate angle designates

TABLE 2.1 GPR Antenna Selection Chart

Antenna Model	Gain (dBi)	Bandwidth (f_{max}/f_{min})	VSWR	Polarization	Suitability
Standard gain horns	8–15	1.5–10	<1.5	Linear/circular	Vehicle-mounted SF GPR
Ridged horn	5–12	10–20	<2	Linear	Vehicle-mounted SF GPR
TEM horn	2–5	3–8	<3	Linear	Vehicle-mounted impulse GPR
PDTEM horn	3–10	20–25	<2	Linear	Vehicle-mounted impulse GPR
Loaded dipole	1–3	1.5–2	<1.5	Linear	Handheld impulse GPR
Log-periodic	5–8	6–15	<2	Linear	Vehicle-mounted SF GPR
Bowtie	2–3	2–5	<2	Quasilinear	Handheld impulse GPR
Planar spirals	2–3	2–5	<2	Circular	Handheld impulse/SF GPR
Conical-spiral	2–5	5–10	<2	Quasicircular	Vehicle-mounted impulse/SF GPR

the polarization sensitivity, and the plate elevation angle determines the structural impedance of the antenna with d and α. A conventional TEM horn antenna usually shows bandpass filter-like gain behavior over a large bandwidth. Therefore, dielectric-filling techniques are employed to improve the operational band, decreasing the lower cutoff frequency by increasing the electrical size [11]. Furthermore, the PDTEM horn, which uses a partial dielectric-loading approach, promises more satisfactory impulse radiation responses, due to its at least twice broadened bandwidth [7]. The typical broadband gain and input reflection characteristics of empty and loaded TEM horns are shown in Section 3.4.4.

Impedance matching is another critical point in avoiding ringing signals on a radiated pulse. Since most of the impulse signals are generated by avalanche mode transistors, which produce antenna impedance–dependent pulse shapes, the antenna impedance over the wide operational band should be stabilized by the matching circuit design. Moreover, a wideband fast-rise-time balanced-to-unbalanced (Balun) transformer is required to avoid structural imbalances [3]. Otherwise, such possible ringing effects may cause difficult clutter problems that can dramatically degrade the detection and identification performance of GPR. Table 2.1 exhibits a practical overview chart for GPR antenna selection.

2.2.5 Data Processing

The development of advanced processing algorithms for GPR data is very important for achieving a high performance level for the detection and identification of subsurface objects. The signal-processing methods are based on interpreting the backscattered signal parameters; these are the amplitude and phase of the signal received for continuous or stepped-frequency GPR and the time-domain shape of the signal received for impulse GPR. In both cases, the amplitude represents how much power returned back from the target. The time delay in impulsive oscillations scattered by the target and the change in phase between the transmitted and received signals indicates the depth range of the target for impulse GPR and stepped-frequency GPR, respectively. The phase of the electromagnetic wave changes when propagating over the path proportional to its effective wavelength. Thus, every wavelength distance corresponds to a 360° phase difference. But it is not easy to find the target range if the medium is inhomogeneous and cluttered. At this stage, basic preprocessing procedures such as background removal, bandpass filtering (BPF), time-varying gain (TVG), signal averaging, two-dimensional spatial filtering, and peak detection are applied primarily to the raw data. Beyond that, advanced signal-processing algorithms such as pattern recognation are generally needed to make adaptive correlations and right decisions to obtain better detection performance and clearer subsurface imaging [1]. The data collecting and signal-processing techniques for GPR are discussed in more detail in Chapters 3 and 11.

In the case of impulse radar, the pulse signals received are collected in the scanning direction. The A- and B-scan raw data series are obtained as shown in Fig. 2.3. The A-scan data represent the impulse signals received for any scanning

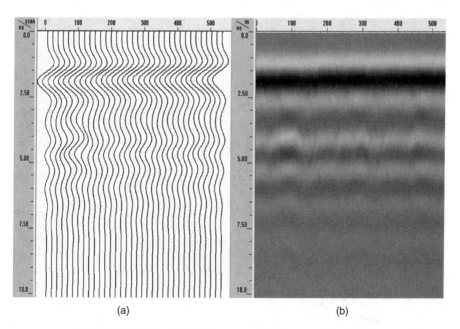

FIGURE 2.3   Typical (a) A-scan and (b) B-scan data of GPR.

step. The B-scan plot is simply a two-dimensional representation of consecutive A-scans in the scanning direction. Here, the x and y axes correspond to the scanning direction and the propagation time, respectively. Time also corresponds to the range, since the electromagnetic wave has a velocity of

$$u = \frac{3 \times 10^8}{\sqrt{\varepsilon_r \mu_r}} \qquad \text{m/s} \tag{2.5}$$

where ε_r and μ_r are the relative constitutional dielectric and magnetic coefficients of the propagation medium. Thus, if any anomaly occurs under the ground, the signal amplitude changes on a time scale that expresses the related depth range. When the soil structure is not well known, inhomogeneous, and there are many unwanted buried objects, the detection and discrimination of the target will be difficult. Most useful GPR signal-processing algorithms are the background-removal, differentiated power density analysis, and subband processing of the signal received. The main goal is to find the target emphasizing the anomalies under the ground. For background estimation and removal, an average A-scan data set is calculated from multiple A-scan data that do not contain any signal from the target, and this average is then taken as a reference. The other A-scan data in a B-scan plot are calculated relatively. The magnitude of the GPR data is computed after the background removal process has been completed. Further processing can be done separately in several frequency bands. The advantage of subband processing is that it allows one to take into account the different target responses and clutter behaviors in different frequency regions. For example, the average RCS values of the probable target objects can be calculated for every subband, and some correlation algorithms can be estimated to discriminate and classify the objects detected.

2.2.6   System Performance

The following measurement results are typical figures illustrating the subsurface detection performance of GPR for small objects buried under the ground. The field data were collected at GPR test fields of 3×3 m soil pools. A cylindrical dielectric object with a 5-cm radius is buried in the terrain at a depth of a few centimeters. The GPR sensor was held roughly 5 cm above the terrain. The scanning speed is 0.3 m/s in the sweeping direction, and this process is repeated in the walking direction with 4-cm spatial resolution.

A gray-scale B-scan plot of the GPR data received is shown in Fig. 2.4. The noise and clutter signals should be cleaned to increase the signal-to noise ratio (SNR) of the GPR system. Called *preprocessing*, this includes signal-processing techniques such as bandpass and spatial filtering, data averaging, background removal, and stacking. Then the cumulative energy distribution reflected from the target can be calculated to obtain a detection-warning signal, as shown in Fig. 2.5, with an A-scan sample plot shown in Fig. 2.6. A B-scan plot of other test data collected for a larger dielectric object buried at a shallow depth is also presented in Fig. 2.7, using a different color map.

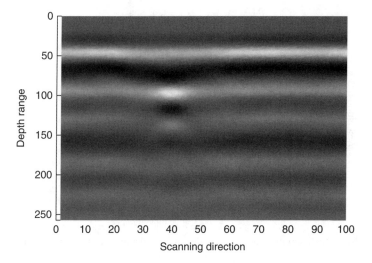

FIGURE 2.4 B-scan of impulse GPR raw data measurement.

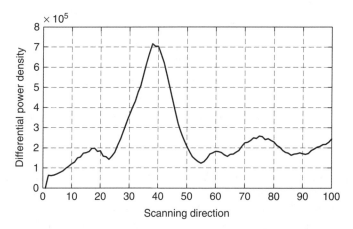

FIGURE 2.5 Background-removed cumulative energy data along the scanning direction.

2.3 ELECTROMAGNETIC INDUCTION DETECTOR

2.3.1 Overview

Electromagnetic induction (EMI), which is the basis for metal detectors, is one of several sensor modalities widely deployed for subsurface buried-object detection and identification. The basic technology is very mature. EMI sensors were used for landmine detection in World War I, and EMI metal detectors were developed further during World War II. Afterward, the use of EMI sensors was also directed at detecting conducting objects in other application areas, such as mineral exploration, nondestructive body testing, archeological investigation, and security enforcement.

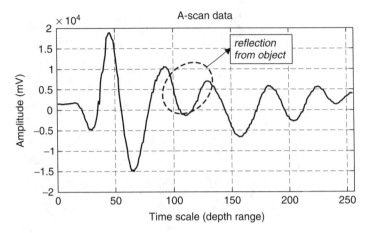

FIGURE 2.6   Demonstration of A-scan impulse data of the fortieth scanning line.

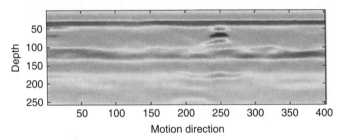

FIGURE 2.7   Grayscale color map imaging of B-scan data measured for a dielectric object buried at 5 cm.

The operating principle of the EMI sensor is to transmit a time-varying primary electromagnetic field, which induces currents in any metallic object that the primary field penetrates. When the primary field is turned off abruptly, eddy currents in the metallic object produce a secondary electromagnetic field that is measured by the receiver. The frequency range employed is generally limited to a few tens of kilohertz [13]. EMI sensors usually consist of a pair of transmitter and receiver coils, as illustrated in Fig. 2.8. An electrical current flowing in the transmitter coil of the wire produces the primary field, and the electromagnetic waveform is often either a broadband pulse or a continuous wideband. The field transmitted induces a secondary current in the earth (ground) as well as in any buried conducting objects. This secondary current is usually detected by sensing the voltage induced in the same or another coil of wire called the *receiver coil*. In the case of pulsed excitation, the transmitter waveform is quenched quickly and the receiving coil measures the decaying secondary field that has been induced in the earth (ground) and subsurface objects. In the case of wideband excitation, the receiving coil is placed within the magnetic cavity so that it senses only the weak secondary field radiated by the earth and buried objects [14–17].

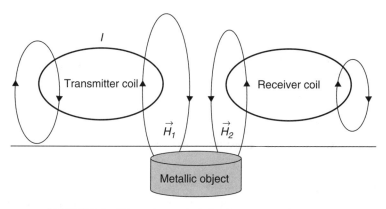

FIGURE 2.8 Electromagnetic induction system phenomena.

2.3.2 Operation Theory

The detection performance of the EMI sensor depends primarily on the system parameters: the primary magnetic field level created by the transmitter coil, the target range, the eddy current inducement capability of the target object, and the receiver coil probe characteristics. Assuming that the fields are time dependent and that the coil position is stationary ($\partial S/\partial t = 0$), the electrical current/magnetic field and the magnetic field/electrical voltage transformation ratios of the transmitter and receiver coils are calculated by the related Maxwell equations:

$$\nabla \times E = -\mu \frac{\partial H}{\partial t} \Rightarrow \oint_c E\,dl = -\mu \int_S \frac{\partial}{\partial t}(H\,dS)$$

$$\Rightarrow V_{AB} = i2\pi f \mu\, NHS \cos \varphi \qquad (2.6)$$

$$\nabla \times H = J + \varepsilon \frac{\partial E}{\partial t} \Rightarrow \oint_c H\,dl = \int_S J\,dS + \varepsilon \int_S \frac{\partial}{\partial t}(E\,dS)$$

$$\Rightarrow \oint_c H\,dl = I \qquad (2.7)$$

where E is the electric field, H the magnetic field, f the frequency, N the number of turns, S the coil surface area of the coil, J the current density, I the coil current, φ the angle between the magnetic field and the coil surface normal vectors, (ε, μ) the dielectric permittivity and magnetic permeability of the coil material, and V_{AB} the induced voltage (Fig. 2.9).

Equation (2.6) denotes that the coil probe sensitivity increases with the operational frequency, the number of coil turns, the magnetic permeability of the coil ferrite, and the effective area of the coil. The physical shapes of the coils are also important for sensitivity in terms of the coil's effective area and the field pattern. The field amplitude is inversely proportional to the range. Since EMI sensors use very low frequencies, the ground surface reflectivity and the soil conductivity losses

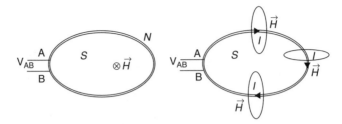

FIGURE 2.9   Operational theory of receiver and transmitter coils.

are very slight values. Nevertheless, the specific soil types that contain highly conductive and magnetic molecules or particulars can cause some attenuation, and remarkable clutter levels can occur. The eddy currents set up in an object, and hence the induced field, depend on the size, shape, and composition of the object. Because of this, different objects generally have different EMI responses. The basic issues are then how much information about an object can be inferred from its EMI response and whether that information can be used to distinguish between target and clutter. The objects that contain many more metallic elements will naturally produce more powerful eddy currents and secondary field on the receiver coil, but the shape of the metallic part in the sense of the angular position with respect to the primary electromagnetic field vector will be an important parameter for detection sensitivity as well.

2.3.3   System Performance

The system detection capability of buried metallic objects is generally a well-known parametric problem, as mentioned above. It is possible to detect any small amount of metallic or metal-like object under the dynamic range limitations of the system. The range equation is not as complicated as that of a GPR. For example, the soil attenuation and reflection loss parameters are not strong enough to affect system performance; even in most cases, the size of the metallic content of the object is the most effective system parameter. Experimental results [14] show that the measured voltages (the receiver voltage, after eliminating the transmitter–receiver coupling voltage) at the receiver coil are 3000, 450, and 50 mV for 15-, 8-, and 5-mm-diameter spherical aluminum objects, respectively. These figures decrease by half for iron objects of the same size.

   The concept of EMI sensor capability involves not only the detection of metal objects, but also the discrimination from clutter of targets of interest. The traditional detection technique is made based on the secondary field energy calculation method for the decision regarding the presence or absence of an object. In some physical applications, this approach can lead to excessively high false-alarm rates, due to the presence of the other metal pieces or granules in the soil. Several modifications to traditional EMI sensors are considered in order to facilitate target discrimination and classification. Wideband sensor operation is one of the most promising methods. In this way, the frequency-dependent response of the field induced by a buried conducting object is used to distinguish the target from the clutter [15]. Another way is to use a simple phenomenological model that describes the measured time-domain waveform as a weighted sum of decaying exponentials to provide an accurate discrimination model for EMI sensors. The measured EMI response can be modeled as a sum of decaying exponential signals whose characteristic decay rates are intrinsic to the target interrogated. Since the decay rates are not dependent on target–sensor orientation, decay-rate estimation has been proposed as an effective and robust method for target identification by Ho et al. [17]. In general, the decay rates associated with metallic objects are slower than that of the earth (ground), so there is more energy in the signal received when a metallic object is present under the surface of the earth. This simple phenomenology allows very basic signal processing to be employed; for example, either an energy detector or the overall amplitude of the signal in a given time frame may be used when the goal is to detect any metallic subsurface objects. Nevertheless, such processing can be the source of many false alarms in highly cluttered mediums.

2.4 MICROWAVE TOMOGRAPHY METHOD

2.4.1 Overview

Microwave tomography (MWT) is an electromagnetic sensor–based technique used to monitor the subsurface and inner contents of structures and to detect concealed objects. The operating principle and hardware units of a diffraction tomographic system are similar to those of stepped-frequency GPR. However, MWT systems generally use higher frequencies, up to sub THz and THz frequency bands (i.e., detection of hidden objects for security purposes [40]), and their detection and imaging process is based on solution of the inverse scattering problem. The operating principle exploits a sweeping microwave signal radiated toward the buried object, and this signal causes a wave diffracted from the target that is collected by a receiver antenna. Wideband antennas, mostly dielectric loaded horns and bowtie antennas, are employed for microwave and MM wave transceivers design. Then I/Q data signatures of the buried target are processed using image reconstruction algorithms, usually based on simplified models of electromagnetic scattering such as Born–Kirchhoff approximations. The diffraction tomographic sensor can yield high-resolution subsurface mapping and target imaging when high frequencies are exploited [45]. Nevertheless, for these cases the detection depth is less then 10 cm, due to the strong soil attenuations at high frequencies, and they are not yet ready for real-time detection, due to the long scanning time.

2.4.2 Background and Operation Theory

Tomographic imaging methods are often used for imaging of cross sections (slices) of inhomogeneous objects. These methods found practical application in medicine (x-rays, ultrasound, and nuclear magnetic resonance tomography); aerospace, chemical, and power technologies [eddy current tomography (ECT)]; geophysics [diffraction tomography (DT), multifrequency backscattering tomography]; and control of subsurface areas, through-wall detection, and inside-wall imaging (microwave/millimeter-wave tomography).

Since the 1980s, considerable research interest has focused on subsurface tomographic methods (STMs) for electromagnetic wave imaging of cylindrical objects placed in homogeneous material half-space. The first generation of STM, classic STM (CSTM) [20,22,37,43], was based on principles of DT [19,42]. These methods provide quasi-real-time qualitative reconstruction of the object function in terms of the two-dimensional polarization current density distribution induced in the object. The main aims of such methods are to develop explicit formulas to solve the imaging problem that are implemented in an efficient way via fast numerical algorithms such as fast Fourier transformation (FFT).

Limitations of DT stimulated the development of STM, which are different variations of CSTM. A generalized diffraction tomographic algorithm for subsurface imaging from multifrequency multimonostatic ground-penetrating radar (GPR) data was described by Deming and Devaney [25]. The algorithm is based on the Born approximation (BA) for vector electromagnetic scattering. The object contrast function, which is connected with the relative differences of the dielectric permittivities between the targets and the host medium, is estimated analytically by inverting a linear integral equation using regularization of the inverse scheme [46]. Soil attenuation and realistic near-field models for the transmitting and receiving antennas were included in the mathematical inversion [52]. A novel DT algorithm has been proposed by Cui and Chew for imaging two-dimensional dielectric cylinders buried in a lossy earth, where the air–earth interface is also taken into account [24]. This algorithm can be used by a low-frequency system (<100 MHz), which can penetrate much deeper than the ground-penetrating radar. Hansen and Johansen employed BA and plane-wave expansion of the dyadic Green's function for the interface problem to derive a three-dimensional DT inversion scheme for fixed-offset GPR [27]. In 2001, Vertiy et al. described a solution algorithm for the inverse problem in the framework of STM for the case of complex values of one of the space frequencies in a generalization of the two-dimensional Fourier transform of the object function [39]. This algorithm accounted for losses of the medium surrounding the object. The evanescent part of the scattered field plane-wave spectrum has also been incorporated into the algorithm. The algorithm was also tested with experimental multifrequency data obtained using a microwave setup that allowed reconstruction of three-dimensional isosurfaces (by stacking a set of two-dimensional reconstructions) of antipersonnel and antitank plastic mines embedded in sand. Application of STM for millimeter-wave imaging has been considered by Vertiy et al. [40]. This has led to improvements in subsurface object image quality.

During the past decade, imaging of inhomogeneous objects of arbitrary shape embedded in a planar layered background has been an important research area in subsurface sensing. Wiskin et al. suggested a method of imaging geophysical anomalies or material defects [41]. Solution of the direct problem required a Green's function for a stratified medium. In the inversion, a pair of Lippmann–Schwinger-like integral equations is solved simultaneously via the Galerkin procedure for the unknown total internal field and object function. To increase the applicability of the problem, investigators decided to employ a set of frequencies. Song and Liu presented a fast three-dimensional electromagnetic nonlinear inversion method in a multilayered medium via a novel scattering approximation [34]. In this method, the inverse problem is cast into a weighted least-squares problem and solved via a conjugate gradient scheme. Synthetic tomographic experiments demonstrated the three-dimensional EM imaging technique. The nonlinear inverse problem of reconstructing three-dimensional objects buried in layered media has also been considered by Li et al. [30]. Such a problem is solved iteratively via the conjugate gradient approach; within each iteration, the problem is linearized by BA and distorted BA. The effects of aperture size and noise on the inversion results were investigated. Numerical results showed that both methods can be used for determination of the object location and size. In 2005, Song and coauthors applied the contrast-source inversion (CSI) method to solving the three-dimensional electromagnetic nonlinear inverse scattering problem in multilayered media [35]. Numerical experiments showed that a reasonably accurate background model was important for solving the inverse scattering problem, especially in the case of high-contrast layering.

Proper extension of the contrast source extended the Born model, recently introduced in the two-dimensional scalar case, has been proposed to establish a novel full-wave inversion method for nonlinear vectorial three-dimensional scattering problems [50]. In particular, a suitable auxiliary function, embedding the parameters of the unknown targets, is introduced as a fundamental unknown of the inverse problem instead of the widely used contrast function. Numerical examples confirm the effectiveness and reliability of the inversion approach proposed, which compares favorably with other methods previously described in the literature.

In the solution of inverse scattering problems, support information is useful in reducing the degree of nonlinearity of the relationship among the data and the unknowns as well as the overall computational burden [51]. Moreover, by taking into account the spectral properties of the scattered fields, it can be helpful to reduce the effect of noise on data. The three-dimensional inversion strategy in which one performs a preliminary estimation of the support of the targets has made it possible to process the experimental data provided by the Institute Fresnel of Marseille, France.

Tomographic methods for imaging of objects buried in a layered medium have also been proposed. Methods based on DT use BA or RA (Rytov approximation) and a homogeneous Green's function, regardless of the medium being imaged. For example, Mast et al. presented results from a ground-penetrating radar imaging experiment on a concrete slab [31]. A multifrequency diffraction tomographic method and BA were used. The three-dimensional multilayer imaging was implemented by a plane-to-plane backward propagation method. Multiple-layer reconstruction is attained by applying boundary conditions at the layer interfaces. Each plane of the three-dimensional source distribution (parallel to the interface) is reconstructed independently. Three-dimensional rendering and planar slicing of this image provide a means of visualizing the results. Lehman considered the problem of noninvasively locating objects buried in a layered medium, such as landmines in the ground or objects concealed in a wall [28]. The resulting planar diffraction tomography algorithm incorporates evanescent field information to achieve "super-resolution." The algorithm is based on the extended Porter–Bojarski theory, BA, and generalization of the Fourier transform, which allows for complex spatial frequencies. The model examined is specialized to a reflection mode and a multimonostatic, wideband wave probing system. The algorithm can be extended to three dimensions by stacking a set of two-dimensional reconstructions.

Crocco and Soldovieri incorporated a nonhomogeneous (planar multilayer) Green's function in a linearized tomographic technique based on BA in order to reconstruct from GPR data a spatial map of the dielectric and/or conductive properties of an investigation domain located in the second layer of a three-layered medium [23]. In the case of TM polarization, it was considered a multifrequency, multiview, multistatic measurement configuration Data are collected over a linear domain with finite extent. The singular-value decomposition was introduced to determine a regularized solution and to solve the optimal measurement configuration problem. A combination of the diffraction tomographic Hilbert space inverse wave algorithm together with the planar multilayer Green's function for inverting a reflection mode multistatic data set was proposed by Lehman in 2005 [29]. In this approach, a free-space Green's function is used to propagate a field from a point within the medium to the receiver, and a multilayer Green's function is applied to propagate a field from a transmitter through the layers to a point within the medium. The distorted Born approximation for a forward-scattering model is realized.

Chaumet et al. developed a fast method for solving the three-dimensional inverse scattering problem based on the coupled dipole method [32] and applied it to complex background configurations such as buried objects in a layered medium [21]. The sample is illuminated with various angles of incidence, and the scattered far-field field is measured for each illumination, as in optical diffraction tomography. The method proposed is a full-vectorial inversion scheme. Although this method is not quantitative, the simulated experiment showed that it was possible to localize the objects and to discriminate between conducting and transparent objects. Two-dimensional tomographic analyses of the through-wall radar imaging (TWI) problem using a CSI-based imaging technique for layered media have been conducted by Song et al. [36]. Numerical experiments on imaging in the frequency range 100 to 600 MHz showed that object locations, shape, and their constitutive parameters can be reconstructed accurately through the use of multifrequency data with a suitable array aperture size. The TWI problem solution has also been considered by Vertiy and Gavrilov [38]. They found that advanced through-wall images of the human body can be obtained by introducing an effective dielectric constant instead of the wall and behind-wall media permittivities. As a result, the TWI problem is solved by STM and was also addressed by other groups deal with Born-based algorithms [47,48]. In particular, a solution approach presented by Soldovieri et al.

[49] is able to account for the observation diversity via a multiarray measurement technique.

2.4.3   System Design and Performance

Next we consider a multifrequency tomographic algorithm for imaging the cylindrical inhomogeneity of arbitrary cross section and electrical properties placed in a layer of a plane multilayered material half-space. The object function imaged is the two-dimensional normalized polarization current density distribution (the normalized polarization current) induced in the inhomogeneity when a TM-polarized plane wave impinges on the object. The problem is solved by a regularized version of the integral equation for the scattered electric field. This work is an extension of previous reports on object imaging by STM and DT where simple (or reducible to simple) problem configurations are assumed. A homogeneous material half-space containing a buried cylindrical inhomogeneity of arbitrary cross section and electrical properties was used in problems of subsurface imaging solved by STM [39,40]; a person was placed behind a wall in a TWI problem reduced to a subsurface imaging problem [38]; a two-dimensional region (object) located in a region filled by an immersion medium in an object imaging problem was reduced to an imaging problem of a two-dimensional weakly scattering object in a homogeneous background solved by a DT method. Obviously, a layered medium represents a more realistic model than a homogeneous two-dimensional half-space or a homogeneous background upon which inverse scattering methods described in prior reports are based. Hence, it is of practical importance and interest to extend the imaging methods of prior reports from such simple configurations to multilayered medium configurations [33,44].

The advancement attained compared to previous papers is that the one-dimensional Fourier transform of a two-dimensional Green's function for an arbitrary plane multilayered medium and total electric field inside this medium have been incorporated in the integral equation to be inverted, relating to Fourier transform of backscattered field data with the object function being sought. Also, it has resulted in a reconstruction technique for imaging inhomogeneity placed in a stratified environment. The reconstruction technique proposed is new and makes it possible to calculate two- or three-dimensional data arrays (by stacking two-dimensional arrays) for imaging of the inhomogeneity slices and isosurfaces being investigated. Practically, an explicit formula can be used for the calculation. The main contribution of this work is the development of a multifrequency tomographic method for imaging of a cylindrical inhomogeneity of arbitrary cross section and electrical properties placed in a layer of a plane multilayered material half-space. In particular, they elaborated both the procedure for transformation of a set of integral equations (for a set of frequencies) into a system of linear equations and a technique for its regularized solution determination.

Typical schemes of microwave tomography systems are shown in Figs. 2.10 and 2.11, with the corresponding performance results given in Figs. 2.12 and 2.13, respectively. It is obvious that MWT systems have a great advantage in reaching high-resolution subsurface imaging.

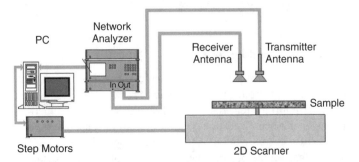

FIGURE 2.10 Setup of a typical microwave tomography system.

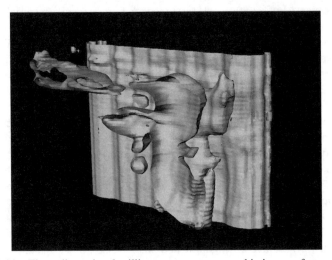

FIGURE 2.11 Three-dimensional millimeter-wave tomographic image of a concealed gun in a plastic box.

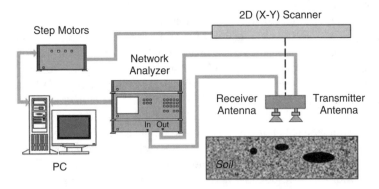

FIGURE 2.12 Setup of a typical subsurface microwave tomographic system.

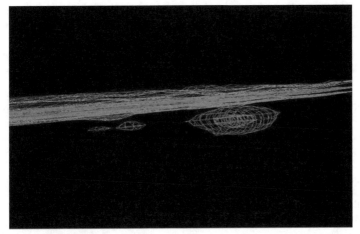

FIGURE 2.13 Subsurface microwave tomographic reconstructed images: (left) small stone; (middle) plastic antipersonnel mine; (right) plastic antitank mine.

2.5 ACOUSTIC AND SEISMIC SENSOR

2.5.1 Overview

Most technologies for the detection of shallow buried objects are electromagnetic methods. They measure the contrast in ferrous content, electrical conductivity, or dielectric constant between the object and the soil in which they are buried. Plastic objects have no or little such contrast because of their nonconductive constitutions. They have dielectric constants that are low, very close to the dielectric constant of dry soil. This makes them very difficult to detect by these technologies. Their acoustic compliance, on the other hand, makes them easier to detect by seismic and acoustic technologies. These methods search for buried objects by causing them to vibrate by introducing sound or seismic waves into the ground. As the materials with different properties vibrate differently, the pattern of ground motion leads to the detection and possibly the identification of the buried object.

Seismic sensors provide a means of sensing the mechanical properties of buried objects remotely. Seismic technologies are commonly employed in exploration for oil. The equipment used in such applications is designed to find layered geology or employ field techniques too expensive to apply. Geophones provide an inexpensive solution to record seismic data and thus sense buried objects. The main difficulty in this approach is the generation of clean Rayleigh waves. Also, the geophones have to contact the ground to perform the measurements. This is not desirable for detecting objects containing explosives.

Seismic motion can also be generated by an airborne acoustic wave. A loudspeaker above the ground can be used for this purpose. A phenomenon called *acoustic-to-seismic coupling* occurs when an airborne acoustic wave is incident at the ground surface. This term refers to the coupling of acoustic energy into the ground as seismic motion. Acoustic-to-seismic coupling causes particles to vibrate on the soil surface. More recent techniques use a laser Doppler vibrometer (LDV) to measure and study the spatial pattern of the particle velocity amplitude of the surface. A variety of sensors, such as radars, microphones, and ultrasonic devices, have also been tried to detect the vibrations and the backscattered sound.

There are three main approaches to buried-object detection using acoustics:

1. *Linear acoustic technique.* This technique has proven to be an extremely accurate technology for locating buried landmines [53]. In this technique, loudspeakers insonify broadband acoustical noise over the soil and a laser Doppler velocimeter (LDV) equipped with X-Y scanning mirrors is used to detect increased soil vibration across a scan region.

2. *Nonlinear acoustic technique.* This technique uses a single speaker to drive two tonal excitations to provide the airborne sound to produce A/S coupling. An accelerometer is used to measure the surface acceleration [54].

3. *Excitation of elastic waves and surface displacement measurement.* The main objective of this approach is to excite elastic waves in the soil and then measure the surface displacements [55]. It uses an electrodynamic shaker to excite elastic waves. As the waves propagate through a scan region, the surface displacements are measured using radar.

The acoustic and seismic techniques mentioned above are described briefly in the following sections. For more detailed information, refer to Chapter 6, which includes the sensor physics and performances of these methods.

2.5.2   Linear Acoustic Techniques

Acoustic–seismic systems proposed by Sabatier et al. have shown great promise in detecting low-metal mines [53,56]. In these systems, a loudspeaker located above the ground surface insonifies the target region at the surface. Acoustic energy is coupled into the ground, producing vibrations. The velocity of the ground surface resulting from these vibrations is measured with a Laser doppler vibrometer (LDV) at several positions over a rectangular grid. The output of the system at a single spatial position is a collection of complex velocities at a set of frequencies, typically within a range from about 100 to 300 Hz for large objects and within 100 to 1000 Hz for small ones. The data collected over a region form a three-dimensional complex image which can be regarded as a stack of two-dimensional images, referred to as *spectral images*. These images can be viewed as the displacement energy of the soil above the buried object at various frequencies. The analysis of these images leads to detection and identification of the buried object [57,58].

There are three different versions of the linear acoustic systems proposed by Sabatier et al.:

1. The first version uses a stop-and-stare data collection scheme [53,56]. The LDV collects data staring at a single spatial position and then stares at a position close to the preceding one. A region of interest is thus scanned by stopping

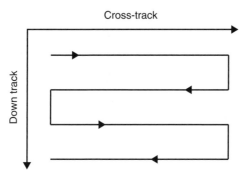

Cross-track

Down track

FIGURE 2.14   Sweeping scheme for laser measurement of the velocity of ground vibration.

and staring. The data are collected with some predefined spatial resolution, which obviously depends on the size of the object to be detected. As a rule, the spatial resolution is selected such that a minimum of three equally spaced positions is scanned along the major axis of the object to increase the detection rate. The scan time of this type of data collection scheme may be too slow for applications such as mine detection, where a faster scanning speed is always desirable.

2. A faster version of the linear acoustic system scans a region without stopping [59]. A laser beam sweeps over an area by moving back and forth continuously, as shown in Fig. 2.14. The spatial resolution in the downtrack can be adjusted by factoring in the size of the buried object. However, cross-track resolution depends on the imaging techniques employed as well as the sweeping speed. Unlike the case in previous systems, spectral images are not readily available. Transforming the time-domain data into spectral images is a critical step in the second approach. The scanning speed of an area by the sweeping beam approach is faster than that of the stop-and-stare approach.

3. A much faster scanning speed is achieved using an array of LDVs placed on a moving platform [60]. The area is scanned by moving the platform. The LDVs on the platform are positioned such that they point to uniquely spaced positions in the cross-track direction (Fig. 2.15). When the platform is moved in the downtrack direction, each LDV scans the region along the path. Since this system uses more than one LDV, a normalization problem is likely to occur due to the nonuniqueness of the LDVs.

2.5.3   Nonlinear Acoustic Techniques

A nonlinear acoustic technique proposed by Donskoy employs a dual-frequency (f_1 and f_2) acoustic signal transmitted toward a buried object [54]. It relies on reception of a signal generated on the soil-buried object interface. This signal has frequencies which differ from those of the signal radiated initially. The dual-frequency sound causes an object to radiate a signal at the intermodulation frequencies ($f_1 \pm f_2$). Thus, the presence of the difference frequency in the signal radiated leads to detection of the object. Some solid objects, such as bricks, steel disks, tree roots, and

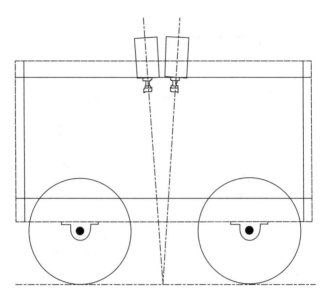

FIGURE 2.15   LDV array on a moving platform.

rocks, do not radiate this new signal with the difference frequency. This makes it easier to differentiate certain objects from debris.

The dynamic behavior of a buried object is strongly dependent on the stiffness of the soil and the buried object. The stiffness of solid objects such as rocks and shrapnel is usually much higher than that of the soil in which they are buried. Their burial of depth also affects their stiffness. The stiffness increases as the depth of the object increases. When the stiffness of the object is much higher than that of the soil, the nonlinear motion is not observable; rather, it is a linear motion. Therefore, the nonlinear acoustic technique is not sensitive to such solid objects and deeply buried objects. Donskoy explains this by modeling the buried object as a mass–spring system [54]. This model is discussed in Section 2.5.5.

2.5.4   Simultaneous Use of Elastic and Electromagnetic Waves

The detection system proposed by Scott et al. uses elastic and electromagnetic waves simultaneously [55]. Elastic surface waves are generated using an elastic wave transducer, which is an electrodynamic shaker in direct contact with the soil (Fig. 2.16). The shaker is excited by a differentiated Gaussian pulse with a center frequency of 400 Hz. The waves propagate through the scan region. A radar-based sensor is used to measure the surface displacements of the soil throughout the scan region. The displacements of the surface are measured as a function of time and position. The sensitivity of the radar is typically around 1 nm. As in the linear and nonlinear acoustic techniques, the technique relies on the flexible mechanical structure of the object to be detected. If the object is solid and inflexible, it will not exhibit resonant response. The surface displacements of the soil are then not

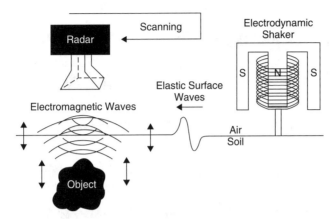

FIGURE 2.16   Use of elastic and electromagnetic waves for object detection.

likely to yield to the detection of the object. Flexible objects such as plastic pipes can be detected using this technology. The main advantage of this method is that it can be used even if the surface is covered by vegetation. Electrical arc and air acoustic sources have also been investigated.

2.5.5   Detection Algorithms

An important step in object detection by acoustic–seismic technology is to process the data acquired. The processing steps for detecting an object are dependent on the data acquired, sensor type, soil type, target type, and environmental conditions. For example, the data acquired can be in the time or the frequency domain, and this may require applying different processing steps to prepare for further processing. Figure 2.17 shows a sensor output in the time domain. Each row represents the time-domain signals acquired by sweeping a region on the ground along the scan direction. The location of the buried object is shown by the dashed line in the figure. A number of imaging techniques, such as short-time fast Fourier transform and the Yule–Walker method, can be applied to obtain spectral images. In some cases, on the other hand, the sensor readily provides the spectral images, as shown in Fig. 2.18. In the sample spectral images shown in the figure, the buried object is located at the lower left of the 8×12 grid with a spatial resolution of 2 cm. If multiple sensors are used to collect the signals at each row in Figs. 2.17 and 2.18 a processing step to equalize the channels may be necessary if the sensors are not unique. This step is skipped if a single sensor is used to scan the same region by stopping and staring, as shown in Fig. 2.14. It is therefore necessary to design detection algorithms specific to the detection technique. Some examples of automated methods for detecting and discriminating buried objects from clutter objects are discussed below.

Model-Based Detection Techniques   Mathematical models are often used for the detection and identification of objects. For buried objects, the depth and physical

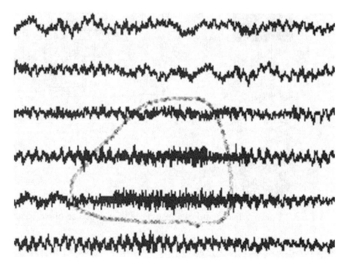

FIGURE 2.17 LDV measurements of an acoustic–seismic object detection system producing time-domain signals.

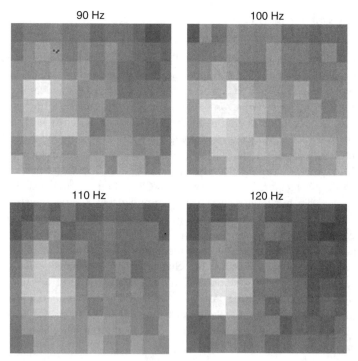

FIGURE 2.18 LDV measurements of an acoustic–seismic object detection system producing spectral images.

properties of the objects and the soil are the common parameters of the models. Unfortunately, some of these parameters, such as the depths of the objects, are not known, or sometimes their estimation is too time consuming and impractical. These problems render the model useless even if a perfect model is available. Also, it is obvious that a general model for any type of buried objects cannot be given. Therefore, each model must be specific to each problem.

A mathematical model for a buried mine using coupled damped harmonic oscillators was developed by Donskoy to interpret acoustic measurements taken on the soil surface [54]. On the basis of this model, it can be shown that the transfer function of a buried mine is in the following form:

$$H(s) = \frac{V(s)}{F(s)} = K \frac{s^3 + n_1 s^2 + n_2 s}{s^4 + p_1 s^3 + p_2 s^2 + p_3 s + p_4} \tag{2.8}$$

where V is the velocity of the vibration on the surface of the ground and F is the driving acoustic force at a particular spatial location [61]. The coefficients are somewhat complicated functions of the model parameters. They are very difficult to estimate and are not given here. Assuming that F(s) is nearly constant over all frequencies, the frequency spectrum of the velocity of the ground surface vibration is proportional to $|H(j\omega)|$ and is what is actually measured by the LDV:

$$|V(j\omega)| \approx \left| \frac{(j\omega)^3 + n_1(j\omega)^2 + n_2(j\omega)}{(j\omega)^4 + p_1(j\omega)^3 + p_2(j\omega)^2 + p_3(j\omega) + p_4} \right| \tag{2.9}$$

A sample Bode diagram of the transfer function in Eq. (2.8) is shown in Fig. 2.19. This system has four poles and three transmission zeros. As the depth of the mine increases, the leftmost peak shifts to the left and the rightmost peak shifts to the right.

It can easily be verified that an autoregressive-moving average (ARMA) process with order (4,4) has the same form as that shown in Eq. (2.8). This suggests that the acoustic measurements (i.e., the velocity of the vibration of the ground surface) have frequency spectra that can be modeled by an ARMA process with order (4,4). This leads to pole analysis for detection of buried objects [61].

Observe that the change in the phase is dramatic at the same frequency of the leftmost peak of the magnitude plot in the Bode diagram in Fig. 2.19. Another approach uses this property to extract features based on phase information [62]. It lowers false-alarm rates significantly at given detection probabilities.

Shape-Based Detection   A common approach to object detection is to use shape information [57,58]. For buried objects, the spectral images usually provide good shape information. In Fig. 2.20, for example, the shape of the object shown in the spectral images at various bands is quite round. In such cases, *roundness* can be used as a feature.

Shape-based detection algorithms usually have a preprocessing step that may include background subtraction, normalization, and thresholding. Connected component analysis usually follows this step. Features such as eccentricity, minor-axis

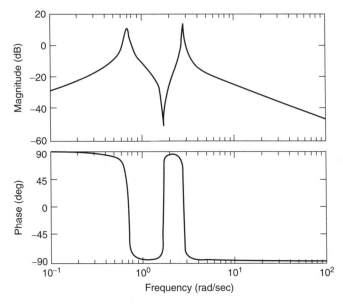

FIGURE 2.19 Sample Bode diagram of the transfer function in Eq. (2.8).

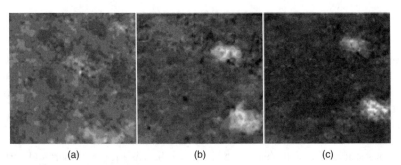

(a)   (b)   (c)

FIGURE 2.20 Infrared imagery taken at different times of day: (a) 8:00 A.M.; (b) 2:00 P.M.; (c) 3:00 P.M.

length, major-axis length, area, and compactness are useful for discriminating among objects [57]. Hocaoglu et al. used independent component analysis to extract additional features to enhance the detection rate [58]. Similar features can also be applicable to images representing the basis vectors of independent components.

2.6 OPTICAL DETECTORS (INFRARED AND HYPERSPECTRAL)

Detection of buried objects by optical sensing is regarded as one of the most promising of current techniques [63,64]. Its main advantage is that passive sensors are employed, which is especially important in military applications. A common practice is to use several sensors (cameras) sensitive at different wavelengths. The spectral reach reflection region (from 0.4 to 2.5 μm) is available only for daylight detection. Detection concepts using midwave infrared (3 to 5 μm) and longwave infrared (8 to 12 μm) are usually emphasized in a day/night system.

There are two primary approaches to optical detection; the sensor can detect either (1) the thermal signature of the buried object itself or (2) a disturbance of the surface soil caused by the process of burying an object. The first approach, which uses the thermal contrast governed by the alternation of temperature over day and night, is referred as the *thermal detection method*. As buried objects heat up at a different rate from that for soil, a thermal signature can be obtained to detect them, a technique known as the *nonthermal detection method*. The principal rationale behind the nonthermal detection method is that the spectral properties of the surface layer are different from those of the subsurface soil. The act of burying an object will bring some subsurface materials, such as the SiO_2 content of rocks and soils, to the surface. Thus, the spectral properties of the soil above the buried object will differ from the spectral properties of the soil around the same region. The contrast between the spectral properties will lead to the detection of a freshly buried object exploiting an 8- to 9.4-μm Reststrahlen feature, which refers to the absorption of energy as a function of silica content. As the silica content increases, the emissivity decreases more and shifts as wavelengths become longer.

Soil contains both large and small particles. In fact, many of the larger particles are usually coated with smaller particles. The smaller particles in the surface layer, on the other hand, tend to be cleaned from the larger particles by wind and rain. This weathering process over time causes the clean large particles to dominate the surface layer, thus changing the spectral property of the soil at the surface layer. When the soil is disturbed, the chemical composition of the surface will change. The weathering process over time will restore the surface to its original state. Therefore, this technique may not be suitable for detecting objects that were buried a long time ago.

Both methods have extreme variability in performance. Their performances depend heavily on environmental conditions such as humidity and temperature. Figure 2.20 shows images of two buried objects in the thermal band, obtained by a thermal camera with an operating wavelength of 7.5 to 13 μm. The objects are 10 cm deep and are circular with a diameter of 15 cm. The best contrast between the areas near the objects and the surrounding areas is obtained at the warmest time of the day. But the objects lose heat more quickly at night, thus show up as dark spots in the imagery instead of bright spots. Laser illumination or microwave radiation can be used to induce these differential temperature profiles.

2.7 BIOCHEMICAL SENSORS

2.7.1 Chemical Sensors and Biosensors

According to the IUPAC, a *chemical sensor* is "a device that transforms chemical information, ranging from the concentration of a specific sample component to total composition analysis, into an analytically useful signal" [67]. The definition of *biosensors* is analogous, but a condition is added saying that the biological recognition element must be "in direct spatial contact with a transducer element." Following this definition, an immunosensor should be distinguished from, for example, biochemical or biological assays, where the detector is generally a separate instrument, even though quite similar principles are employed. Besides, sensors systems rely on reagentless detection mechanisms.

All chemical or biochemical sensors contain at least two basic functional units: a receptor and a transducer. In the receptor the chemical information contained in the sample (i.e., the concentration of a specific sample component or sample composition) is converted into a form of energy that can be transformed by the transducer into a useful electrical or optical signal. For chemical sensors, organic and inorganic polymers, macromolecules, metals, metal oxides, or inorganic salts have been used as receptors, whereas antibodies and enzymes are the most common receptors for biosensors. Many different techniques are available to apply the sensitive material to the transducer, forming films and monolayer or multilayer systems. However, the receptor part must be such that it changes an inherent physical property while interacting with the analyte molecules. This physical property can be the work function, electric permittivity, mass, optical absorbance, electrical conductivity, electrical potential, and much more. Many transducers are available for convenient measurement of those physical properties and changes thereof.

Sensors are commonly classified according to the underlying transduction principle or transducer type (e.g., electromechanical, mass sensitive, electrical, electrochemical, optical) or the nature of the receptor (chemical vs. biochemical). Due to a manifold of possible combinations of receptor and transducer, a wide variety of sensors is imaginable. The user needs to choose those that best suit the specific measurement conditions (e.g., measurement medium, temperature), analyte type, concentration range expected, and possible interferants.

Sensitivity and selectivity of a sensor toward a certain analyte molecule are the result of the number and type of possible fundamental chemical interactions or, depending on the material used, reactions between receptor and analyte molecules. The receptor may be designed to maximize these interactions or the reaction rate. In this way, high sensitivity and selectivity for a certain analyte may be achieved. This applies especially to bioreceptors, owing to their ability to form only complexes with suitable counterparts.

The second approach is to use a set of partially selective sensors and special data evaluation methods. In general, most chemical sensors are never fully specific for a single chemical compound, and they always show some cross-sensitivity to interferants present in the sample or respond to chemical compounds having similar chemical properties. This only partial selectivity of chemical sensors is even taken advantage of in sensor array systems intended for the detection of a multitude of different analytes or to account for the influence of present interferants on the measurement result. A relatively limited number of unspecific sensors with overlapping but nevertheless varying sensitivities, responding to a range of volatile compounds, are combined in an array. When the sensors are exposed to a sample, each sensor will give an individual response. These different signals are then combined into a "pattern," which is, ideally, unique for a certain sample, such as a fingerprint. If the system measures a similar sample again, the sensor responses and the pattern will also be very similar. If the composition of the sample differs, sensor signals will differ as well and form a deviating pattern. To compare the patterns produced and to recognize similar patterns, mathematical pattern recognition tools are used [68,69]. Such a system can be trained to classify and quantify many different analytes using the same sensors, or to identify the presence of interferants.

This approach resembles the concept of an "electronic nose," mimicking the human nose, first proposed by Persaud and Dodd [70]. In its more than 25 years of existence, this approach has been extended to a variety of technologies and applications. Nowadays, chemical and biochemical sensor or sensor array instrumentation is used widely in many analytical measurement tasks [71–74]. Applications fields are as varied as environmental [75], medical diagnostic [76], food safety [77], and military and civil security applications, such as detection of explosives or warfare agents. The initial idea of odor sensing gave way to a more technical view of the technology as an analytical device for chemical compounds.

2.7.2 Transducer Technology

Next, the basic principles of the most common transducers used in explosives and mine detection with chemical and biochemical sensors are described. The sensing materials proposed by the various authors are described in Section 7.1. The small review includes only very common types of electromechanical transducers (acoustic transducers and the cantilever) and optical transducers measuring intrinsic or extrinsic optical properties of the attached receptor (fiber optic probes and surface plasmon resonance). Information on other transducer techniques is available elsewhere (e.g., [72,78]).

Acoustic Wave Transducers   In acoustic wave transducers, acoustic waves are generated in a piece of a piezoelectric crystal by an alternating electrical field applied to attached electrodes. Standing acoustic waves are created either in the material bulk or at the surface, depending on the electrode configuration, crystal orientation, and crystal material. Measurable acoustic wave characteristics are altered upon interaction between the analyte molecules and the receptor on the transducer surface. Many configurations are known: notably the bulk acoustic wave sensor (BAW) or quartz crystal microbalance (QCM) and the surface acoustic wave sensor (SAW) [79]. Both are very common in chemical sensing and biosensing in the gas and liquid phases [80,81]. Generally speaking, acoustic wave transducers are mass sensitive devices—hence the term *microbalance*.

To fabricate a QCM transducer, a crystal of piezoelectric quartz is cut along a certain crystallographic axis, normally the *AT-cut*, into thin plates of submillimeter thickness and is then covered with thin, circular metal electrodes on either side [82]. By an electrical field applied to the electrodes, standing acoustic waves are created in the material bulk in the section between the electrodes. A QCM transducer configuration for gas sensing, and the characteristics of the acoustic wave utilized,

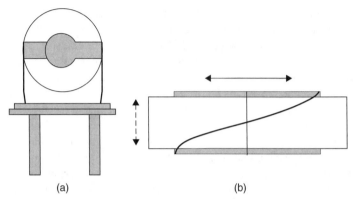

(a)   (b)

FIGURE 2.21   Principle of a QCM transducer: (a) Sketch of a QCM for gas sensing, where a crystal is suspended by electric contacts over a socket to allow free vibration; (b) the direction of wave propagation and particle movement of a bulk acoustic wave in a quartz plate (side view of the crystal) is indicated by the dashed and solid arrows, respectively. The relative amplitude of particle movement is indicated by the gray curve.

are depicted in Fig. 2.21. For the frequency excitation at resonant frequency, the QCM is part of an electronic oscillator circuit. Typical frequencies are in the range 10 to 30 MHz. Any mass load on the crystal decreases the wave frequency; ideally, according to Sauerbrey [83], the decrease is directly proportional to the mass load. The same relation holds in liquids, when possible changes in viscosity of the surrounding medium are negligible or suppressed [84].

For gas-sensing purposes the sensitive material is applied to both sides of the QCM, whereas in liquid media only one side is covered with the sensitive material and exposed to the ambient medium. Sorption of analyte molecules into the receptor layer or due to the interaction of bioreceptors with analyte during the sensing process changes the mass load on the crystal and thus the frequency of the acoustic wave. This shift is recorded by a frequency counter. The first use of QCM with a sensitive material as sorbent matrix as a sensor was presented by King [85]; the first QCM biosensor followed it almost a decade later [86].

Wohltjen and Dessy [87] presented the first SAW gas sensor in 1979. SAW sensors are based on a principle similar to that of the QCM. However, this device uses surface acoustic waves: specifically Rayleigh waves. In liquids a variation is used that employs shear horizontal waves (SH-SAWs). Typical frequencies are in the range 100 to 500 MHz. As sensitivity of acoustic wave transducers depends on the frequency, the higher frequencies make them more sensitive than the QCM. However, in practical use, the experimental gain in sensitivity remains below the theoretical value, due to increased noise level and viscoelastic effects in the sensitive layer.

A SAW device consists of two interdigital transducers (IDTs) of thin metal electrodes on a polished piezoelectric substrate [88]. The most common substrates for SAWs are ST-cut quartz or $LiNbO_3$. The geometry of the IDT fingers and the substrate type determine the wavelength and thus the frequency of the standing

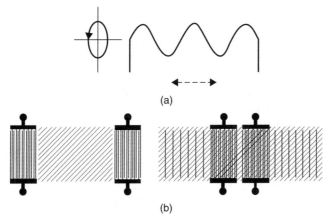

FIGURE 2.22 Principle of a SAW transducer using Rayleigh waves. (a) The direction of wave propagation and particle movement of the surface acoustic wave at the surface of the piezoelectric substrate is indicated by the dashed and solid arrows, respectively. (b) Delay line and resonator configuration of a SAW transducer. The hatched area indicates an area covered with sensitive material.

wave. An alternating electrical potential applied to one IDT causes the surface to expand and contract. This motion generates surface waves, which propagate across the substrate along some distance to the receiver IDT. At the receiver IDT the motion is converted back to an alternating voltage, which is read by a frequency counter (delay line setup; Fig. 2.22). In an alternative setup, where the two IDTs are placed close together and are surrounded by reflector gratings on the outside, a resonant standing wave is created which reaches from the resonator IDT to the receiver IDT (resonator setup).

The transducer surface is covered with a thin layer of the sensitive chemical or biochemical receptor. The propagation velocity of the acoustic wave in the delay line configuration or the frequency of the wave in resonance in the resonator configuration is highly influenced by the physical properties of the transducer surface, including the layer of sensitive material. The acoustic wave is influenced by contributions of changes in mass, layer stiffness, dielectric constant, and electrical conductivity during interaction with the analyte molecules, and depends on ambient temperature and pressure [89]. However, the relationship between frequency and mass loading can be reduced to something similar to Sauerbrey's equation mentioned above if effects other than mass loading are negligible (e.g., viscosity, dielectric constant of surrounding medium) or are held constant (e.g., temperature, pressure).

Cantilevers Microfabricated cantilevers are used primarily as force sensors in atomic force microscopy (AFM). The first cantilever sensor was reported by Wachter and Thundat [90]. Cantilevers are typically rectangular bars of micrometer size [91–93] attached at one side to a support, allowing free bending and vibration of a beam in different modes. They may be operated in air as well as in liquids.

Depending on the ambient, medium cantilever sensors are operated in either static or dynamic mode. In static mode, preferred in liquids, the sensitive layer is applied onto one cantilever surface and interaction of analyte molecules with the receptor produces a surface stress. The surface stress results in a static bending of the cantilever. In dynamic mode, the cantilever is oscillating at its resonance frequency, typically some hundred kilohertz. The cantilever is coated on its upper and lower surfaces with the sensitive material. Upon adsorption of mass on the cantilever, the resonance frequency is shifted to a lower value, which is recorded as a sensor signal. The static deflection and the vibration of the beam can be measured in a number of ways. The optical method is similar to that used in AFM. A laser beam is directed to the cantilever and reflected. The position of the reflected laser changes with the bending angle and is recorded with a diode array. For resistive readout a bimaterial cantilever beam with a piezoresistive layer is produced. Contraction of the piezoresistor during bending of the cantilever is indicated by changes in resistivity. Capacitive readout uses the cantilever beam as one part of a capacitor. The other part is either fixed at the side of the cantilever or below. The latter two methods are very compact, but the electrical signals recorded are very small. This requires amplification or acquisition on the cantilever transducers. However, production in CMOS technology [94] allows easy integration of all electronic circuitry required and of several cantilever beams or other transducers on a small sensor chip [95] to form very compact sensor arrays.

Optical Transducer   Optical sensors are based on the interaction of electromagnetic radiation with matter. Transducers are simple optical probes used to sense the influence of analyte on the characteristic properties of radiation, such as amplitude or intensity, frequency, phase, state of polarization, or more complex methods based on refractrometry, reflectometry, or interferometry [96]. Changes in optical properties occur directly, due to interaction of the receptor with the analyte molecules or indirectly as molecules bearing labels are sensed. In the field of bio- and chemosensors, a large number of optical detection principles have been published [97,98].

Optical fiber, perhaps one of the most exploited platforms [99], can be used in either extrinsic or intrinsic ways. In an extrinsic sensor, optical fiber is used to take light to and from the sensing element. In an intrinsic sensor, the fiber itself acts as a sensing element. Total internal reflection as it occurs within the fiber produces an evanescent field reaching out of the fiber surface. This field can be used as a probe, as propagation of the wave through the fiber changes if the electrical properties are varied within the reach of the field. Other use includes the excitement of fluorescence of molecules at the surface—to name a very common method.

First used as biosensors in 1983 [103], surface plasmon resonance (SPR) sensors are now one of the most commonly used optical sensor techniques for liquid sensing based on evanescent fields [100–102]. SPR systems are available commercially and very widely promoted for study of the biomolecular interactions. Surface plasmons are surface electromagnetic waves that propagate parallel to a metal–dielectric

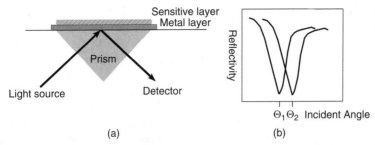

FIGURE 2.23 Principle of SPR: (a) Kretschmann configuration to produce SP in the metal layer; (b) shift in the angle of maximum absorbance caused by changes in the electric properties of the sensitive layer during interaction with the analyte.

interface. Since the wave is on the boundary of the metal and the external medium (air or water), these oscillations are very sensitive to any change in this boundary, such as the adsorption of molecules to the metal surface or the interaction of immobilized (bio)receptors with analyte molecules. Electromagnetic radiation is used to excite SP waves; however, special constructions (e.g., Kretschmann or Otto configuration) are necessary, as SP cannot be excited directly (Fig. 2.23). To make energy transfer possible, the wave vectors of the evanescent field and the plasmon must match. Commonly, the angle of the incoming radiation is changed during sensor measurement, but one could also use light of a different frequency. If the wave vectors match, maximum adsorption of the incoming radiation is going to occur, creating surface plasmons. The angle of incidence at this point depends on the refractive index of the medium on the outer surface of the metal layer. If the refractive index of the receptor layer changes due to interaction with the analyte molecules, the angle of maximum absorption also changes, resulting in a shift in the surface plasmon resonance angle (i.e., the sensor signal).

2.7.3 Electric and Electrochemical Transduction

All transduction principles summarized above, as well as electric transduction, rely on sorption or chemical interactions between a receptor and analyte molecules. Electric transducers simply measure changes in capacitance or electrical potential within the receptor due to sorption processes. In contrast to this, in electrochemical sensors a chemical reaction is involved in which the analyte molecule is consumed. Different transducer configurations are available for the gas and liquid phases in biochemical or chemical receptors [104–106].

In electrochemical biosensors that use enzymes, the product formation or disappearance of a reactant in a chemical reaction involving the analyte and catalyzed by the enzyme is measured. In fact, the most common biosensor is the glucose sensor using amperometry [107]. Determination of the chemical species can be carried out in an amperometric or potentiometric measurement technique using electrochemical cells or a special transducer such as a field-effect transducer. Amperometric transducers measure the electrical current resulting from the oxidation or reduction of an electroactive species formed, for example, in the enzyme reaction. The redox reaction takes place at an electrode, which is held at constant potential. The Clark oxygen electrode is the first amperometric biosensor. A current is produced in proportion to the oxygen concentration in solution. The electrical current is diffusion limited and a direct indicator of the concentration of the electroactive species. Potentiometric devices measure the potential of a working electrode compared to the reference electrode in an electrochemical cell. For potentiometric measurements, the relationship between the concentration and the potential is governed by the Nernst equation, being a logarithmic dependency of the potential on the concentration of the species in solution.

Chemiresistors measure the electric current through a sensitive material (e.g., a metal oxide or conducting polymer). Conductivity of the sensitive material is changed by reaction of the analyte with the sensitive material.

2.8 NUCLEAR SENSORS

2.8.1 Atomic Nucleus

An atom consists of a centrally located positively charged nucleus surrounded by negatively charged electrons revolving in certain physically defined orbits. The nucleus is much heavier than the electrons. The nucleus itself is made up of *neutrons* and *protons*, collectively called *nucleons*. The neutrons are electrically neutral and the protons are positively charged. The electrons are negative with the same magnitude of charge of protons. The masses of nucleons and electrons are often expressed in *atomic mass units* (1 amu $= 1.6606 \times 10^{-27}$ kg). The proton, neutron, and electron masses are 1.007277, 1.008665, and 0.0005486 amu, respectively. This shows that a neutron and a proton do have almost the same mass and that an electron is lighter by a factor of about 1840.

The number of protons (Z) is called the *atomic number*, and the total number (A) of nucleons in a nucleus is called the *mass number*. The number of neutrons is represented as N. If the nuclei have the same Z but differing N, they are called *isotopes*; if the same N but differing Z, *isotones*; and the same A but differing N, Z, or both, *isobars*. This is symbolized as

$$_Z^A X \qquad (2.10)$$

where X is an element. There are about 266 stable nuclides and 70 radioactive nuclides in nature. A radioactive nuclide is unstable and approaches a stable configuration by emitting radiation in the form of gamma energy or particles such as alpha, beta, or neutron, or by splitting into fragments.

2.8.2 Types of Nuclear Radiation

When nuclei come close enough together, they can interact with one another through the strong nuclear force, and reactions between the nuclei can occur. As in chemical reactions, nuclear reactions can either be exothermic (i.e., release energy) or endothermic (i.e., require energy). There are four major types of nuclear reaction:

1. *Fission*, in which a heavy nucleus absorbs a thermal neutron and splits into two small nuclei. The splitting of a *parent* nucleus produces two *daughter* nuclei:

$$n + {}^{235}U \rightarrow {}^{141}Ba + {}^{92}Kr + 3n$$

where n stands for "neutron" and it can be noted that 3n is produced from this fission.

2. *Fusion*, in which two small nuclei come together and fuse into one daughter nucleus with a release of energy:

$$p + p \rightarrow {}^{2}H + e^{+} + v + 0.42 \text{ MeV}$$

where p, H, and e^{+} stand for a proton, hydrogen, and a positron, respectively, and v stands for a neutrino.

3. *Neutron capture*, a reaction to produce radioactive isotopes, in which:

- The nuclear charge (Z, the atomic number) is unchanged.
- The nuclear mass (A, the number of protons + neutrons, the atomic mass) increases by one.
- The number of neutrons (N) increases by one (note that N always $= A - Z$):

$$ {}^{59}Co + {}^{1}_{0}n \rightarrow {}^{60}Co + \gamma \qquad (2.11) $$

This reaction can be shown as ${}^{59}Co(n, \gamma){}^{60}Co$.

4. *Radioactive decay*, in which nuclei "spontaneously" eject one or more particles and lose energy. The decay can take place with an energy to eject either α and β particles or γ-rays, becoming nuclei of lighter atoms.

The most common use of fission is in nuclear reactors, which provide power. Fusion, the source of life on this planet, in the form of the Sun, is much more interesting. The Sun has "burned" for about 5 billion years; the burning is actually nuclear fusion.

2.8.3   Principles of Radiation Detectors

The physical basis of radioisotope instruments is a radiation detector and an electronic unit to convert the output from the detector into a signal capable of operating a visual display or automatic control system. The primary importance of this type of measurement is to evaluate the interaction of radiation with matter. The most significant process is *ionization*, in which the energy of the particles (alpha or beta) or the photon, as it interacts or passes through matter, is used to eject electrons from their atomic orbits. This type of ionization process produces a pair of ions, a negatively charged electron and a positively charged atom. This process is important, because most detection principles are based on ionization. Detection techniques have been developed quantitatively to provide measurement of the strength of radioactive sources, the total energy being carried by the particles or photons emitted.

Ion Chambers and Proportional Counters   The most commonly used detector, the ionization chamber, is simple, robust, inexpensive, and reliable. It can be used to measure high and low radiation intensities. Filling gas ionized by either particle or electromagnetic radiation causes the electric current. These types of detectors are highly efficient for the detection of α and β particles, but their efficiency for high-energy electromagnetic radiation (γ, photons) is low when air at atmospheric pressure is used as a filling gas. However, the detection efficiency can be increased by using high-atomic-number filling gas at high pressure. Ionization chambers work with dc devices; a dc amplifier can be used to amplify the output current.

Ion chambers used to measure β particles have a thin metal window usually protected against damage by a coarse grid. Under very dirty and heavy conditions, dust or oil can be deposited on the detector window, and jets of air can be used to keep the window clean. Sealed ionization chambers have a very small temperature coefficient, so cooling facilities are required only if the temperature goes above $50°C$. In addition to this, high-pressure ion chambers are widely used for the detection of γ-radiation. Walls of ionization chambers used for thermal neutron measurements have two aluminum cylinders covered by fissile material ($^{238}U + ^{235}U$). When a thermal neutron collides with fissile material, fission results, and the fission products produced act like highly energetic ionizing particles. They ionize the filled gas in the ionization chamber.

Proportional counters are used in analytical applications to measure low-energy x-ray radiation. Since these are used for low radiation energy, their entrance gates are fitted with thin (0.1-mm) beryllium windows. Proportional counters filled with BF_3 for slow neutron detection have solid copper walls and are ideal for industrial and field applications. All proportional counters have a very small temperature coefficient.

Geiger–Muller Counters   If an applied voltage on a proportional counter is increased steadily, a stage is reached at which a single ionization event initiates the passage of continuous current. The chamber is said to be breaking down. Bombardment of the anode by the initial avalanche can then produce an ultraviolet photon, which in turn produces photoelectrons, which leads to further avalanches. The discharge becomes self-perpetuating and spreads all over the anode wire. Under certain conditions there is a discharge to grow and then to be terminated, or *quenched*. The chamber is then performing as a Geiger–Muller counter. Two factors are of special importance for the quenching process: the composition of the filling gas and the effect of the discharge spreading all over the anode, which is to create a space charge that reduces the number of avalanches.

The Geiger–Muller (G-M) detector is a radiation detector that operates in region V, the G-M region, shown in Fig. 2.24. G-M detectors produce larger pulses than do

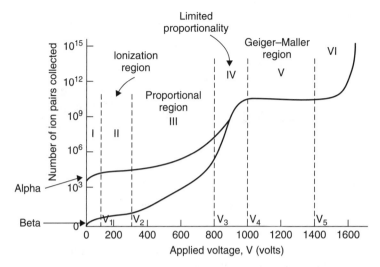

FIGURE 2.24 Gas ionization curve.

other types of detectors. However, discrimination is not possible, since the pulse height is independent of the type of radiation. Counting systems that use G-M detectors are not as complex as those using ion chambers or proportional counters.

Scintillation Counters Scintillation counters are used for the detection of electromagnetic radiation because their higher sensitivity enables lower-activity γ-sources to be used. Scintillation counters consist of two principal parts: a scintillator and a photomultiplier. In these, the secondary electrons arising from an initial ionizing event are detected through the photons of visible light (scintillation) which they generate in suitable materials. Light emission from a single ionizing event generally cannot be seen by the naked eye, and therefore the scintillator is optically coupled to a photomultiplier tube. Commercial photomultiplier tubes have maximum sensitivity in the region of 400 nm, so the choice of scintillators is limited to those that emit light near this wavelength. They must have properties such as chemical stability transparency and a high yield of light photons. There are five types of scintillators: scintillators of organic solid compounds such as antracene, scintillators of liquid organics such as toluene, scintillators of transparent plastic such as Plexiglas, scintillators of inorganic materials such as zinc sulfide and barium platinocyanide, and scintillators of thermoluminescent materials such as lithium fluoride.

The process of radiation detection using a scintillator is not really complete until a usable electrical signal has been made available. That is why a photomultiplier tube in optical contact with a scintillator should be used. An evacuated glass envelope has at one end an optically flat surface or window through which light passes from the scintillator. There is a thin film of photoelectric material such as cesium–antimony alloy on the inside surface of the window. Light photons that

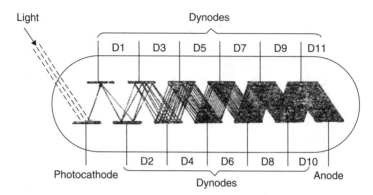

FIGURE 2.25   Photomultiplier tube.

fall on this from outside the tube generate photoelectrons that travel to electrodes. A schematic diagram of a photomultiplier tube is shown in Fig. 2.25.

Semiconductor Detectors   Semiconductor detectors are fabricated from solid materials in the form of a thin slab with metallic contacts on opposite sides to act as collecting electrodes. If such a device is made with absolutely pure silicon or germanium, it may acts in the same way as a gas-filled detector acts, in that ionizing events within the material can produce detectable pulses. But ideally pure materials are not available. Instead, materials with trace amount of impurities can be used. During the manufacturing process, very high purity silicon and germanium are prepared and very small amount of phosphorus, gallium, and arsenic are added. The mixture is formed into a large single crystal by melting and slow cooling, as uniformly as possible. Interaction of radiation with this single crystal produces highly energetic electrons which interact with other electrons, and electron holes (i.e., ion pairs) are produced. They become stable within 10^{-12} s. Applying an electric field to this accumulated charge leads to an electric signal.

High-resolution lithium-drifted silicon and germanium detectors are now being used in x-ray fluorescence analyzers. Ge(Li) is also used for neutron activation analysis.

REFERENCES

1. Daniels, D. J. 1996. *Surface Penetrating Radar*. IEE Radar, Sonar, Navigation and Avionics Series 6. IEE, London.

2. Turk, A. S., Hocaoglu, A. K. 2005. Buried object detection. In *Encyclopedia of RF and Microwave Engineering*, Vol. 1. Wiley-Interscience, Hoboken, NJ, pp. 541–559.

3. Turk, A. S., Sen, B. 2003. Ultra-wide-band antenna designs for GPR impulse radar systems. *IEEE EMC Symp. Rec.*, May.

4. Sahinkaya, D. S. A., Turk, A. S. UWB GPR for detection and identification of buried small objects. *Conference on Radar Sensor Technology VIII and Passive Millimeter*

Wave Imaging Technology VII, SPIE Proc., Vol. 5410, 174–184, April 14–15, Orlando, FL, 2004.

5. Sezgin, M., Kurugollu, F., Tasdelen, I., Ozturk, S., Real time detection of buried objects by using GPR. *Conference on Detection and Remediation Technologies for Mines and Minelike Targets IX*, SPIE Proc., Vol. 5415, 447–455, April 12–16, Orlando, FL, 2004.

6. Chan, L. C., Moffat, D. L., Peters, L. 1979. A characterization of subsurface radar targets. *Proc. IEEE*, 67(7):991–1000.

7. Turk, A. S. 2004. Ultra wide band TEM horn design for ground penetrating impulse radar system. *Microwave Opt. Technol. Lett.*, 41(6):333–336.

8. Lee, J. S., Nguyen, C. 2001. Novel low-cost ultra-wideband, ultra-short-pulse transmitter with MESFET impulse-shaping circuitry for reduced distortion and improved pulse repetition rate. *IEEE Microwave Wireless Component Lett.*, 11(5).

9. King, R. W. P., Smith, G. S. 1981. *Antennas in Matter*. MIT Press, Cambridge, MA.

10. Schlager, K. L., Smith, G. S., Maloney, J. G. 1994. Optimization of bow-tie antenna for pulse radiation. *IEEE Trans. Antennas Propag.*, 42(7):975–982.

11. Yaravoy, A. G., Schukin, A. D., Kaploun, I. V., Lightart, L. P. 2002. The dielectric wedge antenna. *IEEE Trans. Antennas Propag.*, 50(10):1460–1471.

12. Chumachenko, V. P., Turk, A. S. 2001. Radiation characteristics of wide-angle H-plane sectoral horn loaded with dielectric of multiangular shape. *Int. J. Electron.*, 88(1):91–101.

13. *RAND. 2003. Alternatives for Landmine Detection*. RAND Science and Technology Policy. RAND Corporation, Santa Monica, CA.

14. Yamazaki, S., Nakane, H., Tanaka, A. 2001. Basic analysis of a metal detector. In *Proceedings of the IEEE Instrumentation and Measurement Technology Conference*, Budapest, Hungary, May 21–23.

15. Gao, P., Collins, L., Geng, N., Carin, L., Keiswetter, D., Won, I. J. 2000. Classification of landmine-like metal targets using wideband electromagnetic induction. *IEEE Trans. Geosci. Remote Sens.*, 38(3):1352–1361.

16. Bell, T. H., Barrow, B. J., Miller, J. T. 2001. Subsurface discrimination using electromagnetic induction sensors. *IEEE Trans. Geosci. Remote Sens.*, 39(6):1286–1293.

17. Ho, K. C., Collins, L. M., Huettel, L. G., Gader, P. D. 2004. Discrimination mode processing for EMI and GPR sensors for hand-held land mine detection. *IEEE Trans. Geosci. Remote Sens.*, 42(1):249–263.

18. Skolnik, M. I. 1970. *Radar Handbook*. McGraw-Hill, New York.

19. Born, M., Wolf, E. 1999. *Principles of Optics*, 7th exp. ed. Cambridge University Press, Cambridge, UK.

20. Candy, J. V., Pichot, C. 1991. Active microwave imaging: a model-based approach. *IEEE Trans. Antennas Propag.*, 39:285–290.

21. Chaumet, P. C., Belkebir, K., Lencrerot, R. 2006. Three-dimensional optical imaging in layered media. *Opt. Express.*, 14:3415–3426.

22. Chommeloux, L., Pichot, C., Bolomey, J. 1986. Electromagnetic modeling for microwave imaging of cylindrical buried inhomogeneities. *IEEE Trans. Microwave Theory Tech.*, 34:1064–1076.

23. Crocco, L., Soldovieri, F. 2003. GPR prospecting in a layered medium via microwave tomography. *Ann. Geophys.*, 46:559–572.

24. Cui, T. J., Chew, W. C. 2000. Novel diffraction tomographic algorithm for imaging two-dimensional targets buried under a lossy earth. *IEEE Trans. Geosci. Remote Sens.*, 38:2033–2041.

25. Deming, R. W., Devaney, A. J. 1997. Diffraction tomography for multi-monostatic ground penetrating radar imaging. *Inverse Probl.* 13:29–45.

26. Detlefsen, J., Dallinger, A., Huber, S., Schelkshorn, S. 2005. Effective reconstruction approaches to millimeter-wave imaging of humans. In *Proceedings of the URSI General Assembly*, New Delhi, India, Oct. 23–29.

27. Hansen, T. B., Johansen, P. M. 2000. Inversion scheme for ground penetrating radar that takes into account the planar air–soil interface. *IEEE Trans. Geosci. Remote Sens.*, 38:496–506.

28. Lehman, S. K. 2002. Superresolution planar diffraction tomography through evanescent fields. *Int. J. Imag. Syst. Technol.*, 12:16–26.

29. Lehman, S. K. 2005. Hilbert space inverse wave imaging in a planar multilayer environment. *J. Acoust. Soc. Am.*, 117:2929–2936.

30. Li, F., Liu, Q. H., Song, L.-P. 2004. Three-dimensional reconstruction of objects buried in layered media using Born and distorted Born iterative methods. *IEEE Trans. Geosci. Remote Sens.* 1:107–111.

31. Mast, J. E., Johansson, E. M. Udpa, S. S., Han, H. C., Eds. 1994. Three-dimensional ground-penetrating radar imaging using multi-frequency diffraction tomography. *Proceedings of the SPIE Symposium, Advanced Microwave and Millimeter Wave Detectors Conference, San Diego, CA. Proc. SPIE*, 2275:196–204.

32. Purcell, E. M., Pennypacker, C. R. 1973. Scattering and absorption of light by nonspherical dielectric grains. *Astrophys. J.*, 186:705–714.

33. Vertiy, A., Gavrilov, S. 2006. Imaging of buried objects by tomography method using multifrequency regularization process. In *Proceedings of the 11th International Conference on Mathematical Methods in Electromagnetic Theory (MMET-06)*, Kharkiv, Ukraine, June 26–29, pp. 152–157

34. Song, L.-P., Liu, Q. H. 2004. Fast three-dimensional electromagnetic nonlinear inversion in layered media with a novel scattering approximation. *Inverse Probl.*, 20:S171–S194.

35. Song, L.-P., Liu, Q. H., Li, F., Zhang, Z. Q. 2005. Reconstruction of three-dimensional objects in layered media: numerical experiments. *IEEE Trans. Antennas Propag.*, 53:1556–1561.

36. Song, L.-P., Yu, C., Liu, Q. H. 2005. Through-wall imaging (TWI) by radar: 2-D tomographic results and analyses. *IEEE Trans. Geosci. Remote Sens.*, 43:2793–2798.

37. Tabbara, W., Duchène, B., Pichot, C., Lesselier, D., Chommeloux, L., Joachimowicz, N. 1988. Diffraction tomography: contribution to the analysis of some applications in microwaves and ultrasonics. *Inverse Probl.*, 4:305–331.

38. Vertiy, A. A., Gavrilov, S. P. 2005. Subsurface tomography application for through-wall imaging. In *Proceedings of the 9th International Conference on Electromagnetics in Advanced Applications (ICEAA–05) and 11th European Electromagnetic Structures Conference (EESC-05)*, Torino, Italy, pp. 223–226.

39. Vertiy, A. A., Gavrilov, S. P., Aksoy, S., Voynovskyy, I. V., Kudelya, A. M., Stepanyuk, V. N. 2001. Reconstruction of microwave images of the subsurface objects by diffraction tomography and stepped-frequency radar methods. *In Zarubejnaya Radioelektronika. Uspehi Sovremennoy Radioelektroniki*, pp. 17–52.

40. Vertiy, A. A., Gavrilov, S. P., Voynovskyy, I. V., Stepanyuk, V. N., Ozbek, S. 2002. The millimeter wave tomography application for the subsurface imaging. *Int. J. Infrared Millimeter Waves*, 23:1413–1444.

41. Wiskin, J. W., Borup, D. T., Johnson, S. A. 1997. Inverse scattering from arbitrary two-dimensional objects in stratified environments via a Green's operator. *J. Acoust. Soc. Am.*, 102:853–864.

42. Wolf, E., Consortini, A., Eds. 1996. Principles and development of diffraction tomography. *In Trends in Optics*. Academic Press, San Diego, CA, pp. 83–110.

43. Zorgati, R., Duchène, B., Lesselier, D., Pons, F. 1991. Eddy current testing of anomalies in conductive materials: 1. Qualitative imaging via diffraction tomography techniques. *IEEE Trans. Magn.*, 27:4416–4437.

44. Gavrilov, S. P., Vertiy, A. A. 2007. Imaging of layer inhomogeneity in stratified environment via tomographic reconstruction. *Electromagnetics*, 25(6):473–494.

45. Vertiy, A. A., Gavrilov, S. P. 1998. Modeling of microwave images of buried cylindrical objects. *Int. J. Infrared Millimeter Waves*, 19(9):1201–1220.

46. Leone, G., Soldovieri, F. 2003. Analysis of the distorted Born approximation for subsurface reconstruction: truncation and uncertainities effect. *IEEE Trans. Geosci. Remote Sens.*, 41:66–74.

47. Soldovieri, F., Solimene, R. Brancaccio, A., Pierri, R. 2007. Localization of the interfaces of a slab hidden behind a wall. *IEEE Trans. Geosci. Remote Sens.*, 45:2471–2482.

48. Soldovieri, F., Solimene, R. 2007. Through-wall imaging via a linear inverse scattering algorithm. *IEEE Geosci. Remote Sens. Lett.*, 4:513–517.

49. Soldovieri, F., Solimene, R., Prisco, G. 2008. A multiarray tomographic approach for through-wall imaging. *IEEE Trans. Geosci. Remote Sens.*, 46:1192–1199.

50. Catapano, I., Crocco, L., D'Urso, M., Isernia, T. 2006. A novel effective model for solving 3D nonlinear inverse scattering problems in lossy scenarios. *IEEE Geosci. Remote Sens. Lett.*, 3:302–306.

51. Catapano, I., Crocco, L., D'Urso, M., Isernia, T. 2009. Support-aided 3D microwave imaging: testing on the Fresnel 2008 database. *Inverse Probl.*, 25:024002.

52. Soldovieri, F., Persico, R., Leone, G. 2005. Effect of source and receiver radiation characteristics in subsurface prospecting within the distorted Born approximation. *Radio Sci.*, 40.

53. Sabatier, J. M., Xiang, N. 2001. An investigation of a system that uses acoustic-to-seismic coupling to detect buried anti-tank mines. *IEEE Trans. Geosci. Remote Sens.*, 39:1146–1154.

54. Donskoy, D. M. 1998. Nonlinear vibro-acoustic technique for landmine detection. *Proc. SPIE*, 3392:211–217.

55. Scott, W. R., Larson, G. D., Martin, J. S., Rogers, P. H. 2000. Seismic/electromagnetic system for landmine detection. *J. Acoust. Soc. Am.*, 107(5)(Pt. 2):28.

56. Sabatier, J. M., Hickey, C. J. 2000. Acoustic-to-seismic transfer function at the surface of a layered outdoor ground. *Proc. SPIE*, 4038:633–644.

57. Hocaoglu, A. K., Gader, P., Keller, J., Nelson, B. 2002. Anti-personnel landmine detection and discrimination using acoustic data. *Subsurf. Sens. Technol. Appl.*, 3(2):75–93.

58. Hocaoglu, A. K., Gader, P. 2002. Continuous processing of acoustics data for landmine detection. *In Detection and Remediation Technologies for Mines and Minelike Targets VII*. *Proc. SPIE*, 4742:654–664.

59. Xiang, N., Sabatier, J. M. 2003. Acoustic-to-seismic landmine detection using a continuously scanning laser Doppler vibrometer. *In Detection and Remediation Technologies for Mines and Minelike Targets VIII*. *Proc. SPIE*, 5089:591–595.

60. Burgett, R. D., Bradley, M. R., Duncan, M., Melton, J., Lal, A. K., Aranchuk, V., Hess, C. F., Sabatier, J. M., Xiang, N. 2003. Mobile mounted laser Doppler vibrometer array for acoustic landmine detection. In *Detection and Remediation Technologies for Mines and Minelike Targets VIII*. *Proc. SPIE*, 5089:665–672.

61. Hocaoglu, A. K., Gader, P. D., Ritter, G. X. 2003. Acoustic/seismic imaging using spectral estimation for landmine detection. In *Proceedings of the International Conference on Requirements and Technologies for the Detection, Removal and Neutralization of Landmines and UXO*, Brussels, Belgium, Sept.

62. Wang, T., Keller, J., Gader, P. D., Hocaoglu, A. K., Phase signatures in acoustic–seismic landmine detection, *Radio Science*. vol. 39, pp. RS4S02/1–13, 2004.

63. Salisbury, J., Walter, L., Vergo, N., D'Aria, D. 1991. *Infrared (2.1–2.5 mm) Spectra of Minerals*. Johns Hopkins University Press, Baltimore.

64. Hong, S., Miller, T. W., Tobin, H., Borchers, B., Hendrickx, J. M., Lensen, H. A., Schwering, P. B., Baertlein, B. A. 2001. Impact of soil water content on landmine detection using radar and thermal infrared sensors. *Proc. SPIE*, 4394:409–416.

65. Wexler, A. 1998. Electrical impedance imaging in two and three dimensions. *Clin. Phys. Physiol. Meas.* (*Suppl. A*):29–33.

66. Towe, B., Jacobs, A. 1981. X-ray backscatter imaging. *IEEE Trans. Biomed. Eng.*, 28:646–654.

67. Hulanicki, A., Geab, S., Ingman, F. 1991. Chemical sensors definition and classification. *Pure Appl. Chem.*, 63(9):1247–1250.

68. Gutierrez-Osuna, R. 2002. Pattern analysis for machine olfaction: a review. *IEEE Sens. J.*, 2(3):189.

69. Scott, S. M., James, D., Ali, Z. 2007. Data analysis for electronic nose systems. *Microchim. Acta*, 156:183–207.

70. Persaud, K., Dodd, G. 1982. Analysis of discrimination mechanisms in the mammalian olfactory system using a model nose. *Nature*, 299:352–355.

71. Rock, F., Barsan, N., Weimar, U. 2008. Electronic nose: current status and future trends. *Chem. Rev.*, 108:705–725.

72. Pearce, T. C., Schiffman, S. S., Nagle, H. T., Gardner, J. W. 2002. *Handbook of Machine Olfaction: Electronic Nose Technology*. Wiley-VCH, Weinheim, Germany.

73. Baltes, H., Fedder, G. K., Korvink, J. G., Eds. 1996–2003. *Sensors Update. Vols.* 1–13. Wiley-VCH, Weinheim, Germany.

74. Göpel, W., Gardner, J. W., Hesse, J., Eds. 2005. *Sensors Applications*, Vols. 1–5. Wiley-VCH, Weinheim, Germany.

75. Rodriguez-Mozaz, S., Lopez de Alda, M. J., Marco, M.-P., Barcelo, D. 2005. Biosensors for environmental monitoring: a global perspective. *Talanta*, 65:291–297.

76. Wang, Y., Xu, H., Zhang, J., Li, G. 2008. Electrochemical sensors for clinic analysis. *Sensors*, 8:2043–2081.

77. Logrieco, A., Arrigan, D. W. M., Brengel-Pesce, K., Siciliano, P., Tothill, I. 2005. DNA arrays, electronic noses and tongues, biosensors and receptors for rapid detection of toxigenic fungi and mycotoxins: a review. *Food Addit. Contam. A*, 22(4):335–344.

78. James, D., Scott, S. M., Ali, Z., O'Hare, W. T. 2005. Chemical sensors for electronic nose systems. *Microchim. Acta*, 149:1–17.

79. Janshoff, A., Galla, H.-G., Steinem, C. 2000. Piezoelectric mass-sensing devices as biosensors: an alternative to optical biosensors? *Angew. Chem. Int. Ed.*, 39:4004–4032.

80. Cooper, M. A., Singleton, V. T. 2007. A survey of the 2001 to 2005 quartz crystal microbalance biosensor literature: applications of acoustic physics to the analysis of biomolecular interactions. *J. Mol. Recognit.*, 20:154–184.

81. Länge, K., Rapp, B. E., Rapp, M. 2008. Surface acoustic wave biosensors: a review. *Anal. Bioanal. Chem.*, 391:1509–1519.

82. O'Sullivan, C. K., Guilbault, G. G. 1999. Commercial quartz crystal microbalances: theory and applications. *Biosens. Bioelectron.*, 14:663–670.

83. Sauerbrey, G. 1959. Use of vibrating quartz for thin film weighting and microweighting. *Phys.*, 155:206.

84. Martin, S. J., Granstaff, V. E., Frye, G. C. 1991. Characterization of a quartz crystal microbalance with simultaneous mass and liquid loading. *Anal. Chem.*, 63:2272–2281.

85. King, W. H., Jr. 1964. Piezoelectric sorption detector. *Anal. Chem.*, 36(9):1735–1739.

86. Shons, A. 1972. Immunoassay with coated piezoelectric crystals. *J. Biomed. Mater. Res.*, 6:565–570.

87. Wohltjen, H., Dessy, R. 1979. Surface acoustic wave probe for chemical analysis, Parts 1–3. *Anal. Chem.*, 51:1458.

88. Grate, J. W., Martin, S. J., White, R. M. 1993. Acoustic wave microsensors, Part I. *Anal. Chem.*, 65(21):940A–948A; Part II. *Anal. Chem.*, 65(22):987A–996A.

89. Ricco, A. J., Martin, S. J., Zipperian, T. E. 1985. Surface acoustic wave gas sensor based on film conductivity changes. *Sens. Actuat.*, 8:319–330.

90. Wachter, E. A., Thundat, T. 1995. Micromechanical sensors for chemical and physical measurements. *Rev. Sci. Instrum.*, 66:3662.

91. Goeders, K. M., Colton, J. S., Bottomley, L. A. 2008. Microcantilevers: sensing chemical interactions via mechanical motion. *Chem. Rev.*, 108:522–542.

92. Lang, H. P., Hegner, M., Gerber, Ch. 2006. Nanomechanical cantilever array sensors. In *Springer Handbook of Nanotechnology*, 2nd ed., B. Bhushan, Ed. Springer-Verlag, New York, pp. 443–459.

93. Fritz, J. 2008. Cantilever biosensors. *Analyst*, 133:855–863.

94. Lange, D., Brand, O., Baltes, H. 2002. *CMOS Cantilever Sensor Systems: Atomic-Force Microscopy and Gas Sensing Applications*. Springer-Verlag, Berlin.

95. Li, Y., Vancura, C., Barrettino, D., Graf, M., Hagleitner, C., Kummer, A., Zimmermann, M., Kirstein, K.-U., Hierlemann, A. 2007. Monolithic CMOS multi-transducer gas sensor microsystem for organic and inorganic analytes. *Sens. Actuat. B*, 26:431–440.

96. Gauglitz. G. 2005. Direct optical sensors: principles and selected applications. *Anal. Bioanal. Chem.*, 381:141–155.

97. McDonagh, C., Burke, C. S., MacCraith, B. D. 2008. Optical chemical sensors. *Chem. Rev.*, 108:400–422.

98. Borisov, S. M., Wolfbeis, O. S. 2008. Optical biosensors. *Chem. Rev.*, 108:423–461.

99. Leung, A., Shankar, P. M., Mutharasan, R. 2007. A review of fiber-optic biosensors. *Sens. Actuat*. B, 125:688–703.

100. Sharma, A. K., Jha, R., Gupta. B. D. 2007. Fiber-optic sensors based on surface plasmon resonance: a comprehensive review. *IEEE Sens*. J., 7(8):1118.

101. Homola, J. 2003. Present and future of surface plasmon resonance biosensors. *Anal. Bioanal. Chem*., 377(3):528–539.

102. Homola, J. 2008. Surface plasmon resonance sensors for detection of chemical and biological species. *Chem. Rev*., 108:462–493.

103. Liedberg, B., Nylander, C., Sundstrom, I. 1983. Surface plasmon resonance for gas detection and biosensing. *Sens. Actuat*. B, 4:299–304.

104. Grieshaber, D., MacKenzie, R., Voros, J., Reimhult, E. 2008. Electrochemical biosensors: sensor principles and architectures. *Sensors*, 8:1400–1458.

105. Stetter, J. R., Li, J. 2008. Amperometric gas sensors: A review. *Chem. Rev*., 108(2):352–366.

106. Janata, J. 2003. Electrochemical microsensors. *Proc. IEEE*, 91(6):864–869.

107. Wang, J. 2001. Glucose biosensors: 40 years of advances and challenges. *Electroanalysis*, 13(12):983.

CHAPTER THREE

Ground-Penetrating Radar

3.1 INTRODUCTION

In the last 20 years, ground-penetrating radar (GPR) has become a leading technology for the detection, identification, and imaging of subsurface artifacts, abnormalities, and structures such as pipes, mines, gaps, water channels, oil wells, tunnels, and roads. It has a very broad range of applications, including geophysics, hydrogeology, mineral mining, archeology, civil engineering, transportation, nondestructive testing, mine detection, and remote sensing [1–4].

GPR performance is associated with the electrical and magnetic properties of local soil and buried targets as well as with implementation of the GPR hardware and software. The central frequency and bandwidth of the GPR signal chosen are key factors in the detection of subsurface features. Conventional GPRs are usually designed for geophysical applications and use central frequencies below 1 GHz. The lower frequencies are preferred to detect something buried too deep, due to the dramatically increased attenuation of the soil with increasing frequency. Nevertheless, the higher frequencies are needed for better range resolution and detailed echos to determine small objects. Thus, GPR systems that transmit ultrawide band (UWB) impulse signals are proposed primarily to benefit from both low and high frequencies [5]. The impulse waveform is generally of Gaussian-shaped monocycle type in time with application-oriented pulse durations from a few nanoseconds to a few hundred picoseconds, which can correspond to a broadband spectrum from 100 MHz to 5 GHz [6].

It is also possible to use stepped-frequency (SF) GPR, which operates essentially like a network analyzer up to 8 GHz. The stepped-frequency technique offers some benefits compared to time-domain GPR systems. Most important, SF-GPR

Subsurface Sensing, First Edition. Edited by Ahmet S. Turk, A. Koksal Hocaoglu, and Alexey A. Vertiy.
© 2011 John Wiley & Sons, Inc. Published 2011 by John Wiley & Sons, Inc.

has a distinct advantage over conventional impulse GPR, where there is no effective control of the source frequency spectrum. Apart from increased resolution and increased depth of penetration, the signal spectrum received by SF-GPR offers the advantage of reading the real and phase parts, which can be made use of in analyzing subtle and complex inhomogeneities, particularly when carrying out tomographic inversion [7]. However, as the frequency-domain approach usually requires a larger measurement time, this technique is not suitable for a number of applications: namely, extensive public utility searches (pipes, wirings) under road networks, inspection of asphalt using specially equipped motor vehicles, and introspections of large structure engineering. The main reason is that state-of-the-art continuous-wave (CW) radar systems usually synthesize the waveform using a phase-loop-locked (PLL) oscillator or, alternatively, a direct digital synthesizer (DDS). Although PLLs are able to combine the high spectral purity of each synthesized frequency with the capability of scanning a very large bandwidth, they mostly suffer from long lock times because of the loop lowpass filter. The DDS is able to overcome this speed limit, but the available band may still not be sufficient for some high-resolution applications. Thus, the use of multiple synthesizer combinations is proposed to reach a high-speed ultrawideband (UWB) SF-GPR system [8,9].

Employment of bands higher than 8 GHz is technically available for SF-GPR designs. This approach is quite similar to that of microwave tomographic systems [14]. Thus, as expected, such SF-GPR designs could be preferred for the detection of shallow buried objects when it is necessary to obtain high-resolution subsurface images.

The target detection and identification performance of GPR depends significantly on the ability of the UWB antenna, which radiates uniformly shaped GPR impulse signals into the ground without distortion [10]. GPR antennas are operated close to the surface for efficient coupling of the energy into the ground. In this case, the convenient design of UWB transmitter and receiver (T/R) antennas is essential to radiate uniformly shaped GPR impulse signals into the ground and to receive with high efficiency pulses scattered from subsurface objects [1,2]. The antennas must have a flat, high-directivity gain, a narrow beam, and low sidelobe and input reflection levels over the operational frequency band to reach the largest dynamic range, best focused illumination area, lowest level of T/R antenna coupling, reduced ringing, and uniformly shaped impulse radiation. Furthermore, electromagnetic (EM) coupling effects between the transmitter and its receiver (or other receivers for array designs) and adaptive designs for EMI sensor (metal detector)–mounted operations should also be considered [11].

A brief explanation of the typical GPR system and its main system blocks, such as transmitter and receiver antennas, signal generator, high-speed receiver, analog-to-digital converter (ADC), system control, and data communication units was given in Section 2.2, especially for GPR users. The present chapter provides more detailed information, intended primarily for GPR designers. Thus, GPR systems are described in detail, including operational principles, block diagrams, hardware units, and adaptation techniques for application areas. The most promising signal-processing algorithms and useful imaging methods are also examined. To this end, two types of GPR designs, an UWB impulse GPR developed over a wide frequency range and a broadband SF-GPR system, are studied with all the spesifics.

The performance results of GPR systems used for popular applications are exhibited to give related information for better understanding of GPR capabilities. Moreover, other useful configurations and measurement techniques, such as borehole radar and multisensor approaches, are mentioned, particularly for specific applications.

3.2 GPR SYSTEM DESIGN

3.2.1 Design Procedure

GPR is a near-zone radar system used for the detection of underground inhomogeneities in constitutents such as objects, structures, and sublayers, based on electromagnetic wave-scattering principles. Thus, its operational effectiveness relies on meeting the following system requirements successfully:

- Efficient coupling of the electromagnetic signal radiation into the ground using appropriate transmitter and receiver antennas
- Adequate penetration range into the ground of the signal radiated, to reach the deepest target
- Receiving the strongest possible scattering signal from the smallest buried object
- Using a suitable frequency bandwidth to obtain the desired depth resolution and detector noise level

In addition to these performance requirements, there are other operational parameters and constraints that should be considered in the design, such as physical conditions and restrictions depending on the application area, properties of materials to be inspected, and planar resolution of subsurface imaging. Therefore, to determine the system requirements, we must take into consideration all the soil characteristics, target properties, and hardware design parameters.

The electrical properties of the medium (soil) are important parameters in GPR design. In concept the soil is quasihomogeneous, and its electrical behavior may be defined by its constitutional parameters: dielectric permittivity (ε), magnetic permeability (μ), and electrical conductivity (σ). Here, the relative dielectric and magnetic constants (ε_r and μ_r) designate the effective wavelength, and the soil conductivity determines the propagation loss. The variability of both material parameters and local geological conditions causes great difficulty in accurate prediction of propagation behavior. All of these parameters are generally the functions of frequency as well. For example, radio-frequency (RF) propagation and ground surface reflectivity losses increase sharply above 3 GHz, and in many cases, this fact physically limits the operational bandwidth. Nevertheless, when the buried objects decrease in size, the simulation results show that the radar cross-sectional (RCS) values are insufficient before the resonance frequency region (see Fig. 3.2). As is well known, the RCS of a target object represents the object's electromagnetic wave backscattering level. Since the RCS depends on the electrical sizes (measurements with respect to wavelength) of the object, and the effective wavelength is inversely proportional to the square root of the soil dielectric constant, the GPR frequency can be shifted downward in high-dielectric mediums (e.g., in wet soils). For example, the operating frequency must be at least 2 GHz to obtain sufficiently powerful backscattered signal from a 5-cm-diameter cylindrical target buried in dry soil. However, if this object were buried in wet soil, a frequency bandwidth of about 1 GHz would be enough for the backscattering level desired.

Due to the facts noted above regarding electromagnetic wave propagation and scattering principles, use of both lower and higher frequencies each has its own pros and cons. Thus, ultrawide band GPR systems yield significant performance advantages in detecting subsurface objects that are either deeply buried or small in size.

3.2.2 Radar Equation

The radar equation [Eq. (3.1)] determines the detection range for a buried target in different parametric cases, such as power, frequency, soil type, and antenna structures (see the configuration in Fig. 3.1):

$$P_r = \frac{P_t G_t G_r}{(4\pi)^3 R^4} \frac{RCS}{L_t} \lambda^2 \tag{3.1}$$

$$L_t(\text{dB}) = L_a(\text{dB}) + L_{r1}(\text{dB}) + L_{s1}(\text{dB}) + L_{s2}(\text{dB}) + L_{r2}(\text{dB}) \tag{3.2}$$

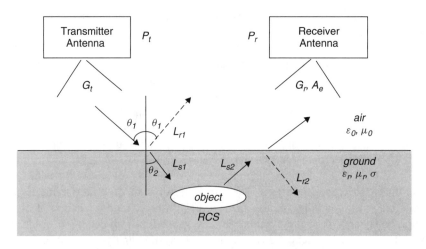

FIGURE 3.1 Propagation modeling of a GPR system.

where P_r = receiver power (W)

P_t = transmitter power (W)

G_t = transmitter antenna gain

G_r = receiver antenna gain

R = total range of the target (m)

A_e = receiver antenna effective area (m^2)

RCS = radar cross section of buried target object (m^2)

λ = wavelength in air (m)

L_t = total propagation loss (except free-space loss)

L_a = transmitter and receiver antenna efficiency loss

L_{r1} = surface reflectivity loss (air to ground)

L_{r2} = surface reflectivity loss (ground to air)

$L_{s1,2}$ = soil attenuation losses

The transmitter antenna gain (G_t) depends on the antenna type, geometry, and frequency. The gain at a given frequency band is taken for SF-GPR, while the peak gain of the radiated impulse signal or the average gain over the impulse band must be considered for impulse GPR. Typical gains can be from minus a few decibels over isotropic up to 10 dBi. The receiver antenna gain can be expressed in terms of the antenna aperture (A_e) and the efficiency (e) as

$$G_r = \frac{4\pi A_e}{\lambda^2} e \qquad (3.3)$$

where the antenna efficiency is generally related to the radiation losses due to some resistive loadings, the impedance mismatching at the feed point, and the structure of the antenna aperture. Such losses are typically counted in a few decibels.

The RCS value of the target depends essentially on the size, geometry, and structure of the object. For example, if the target size is too small with respect to the wavelength of the radar frequency, it is not expected to obtain a remarkable RCS. Moreover, metallic and highly conductive structures usually yield better RCS responses than do dielectrics (Fig. 3.2). The geometrical shape of the object is also important to reflect an incoming electromagnetic wave to the desired direction (back-reflection is taken into consideration primarily for monostatic radars). In addition, the mediums in which the objects are buried can significantly affect the RCS responses For example, if a medium is lossy (i.e., soil), the RCS usually degrades, due to the electromagnetic wave attenuation. Nevertheless, if the medium has a high permittivity value ($\varepsilon_r > 1$), it may even improve the RCS, especially for low frequencies, since the electrical size of the object increases due to a smaller effective wavelength in propagation. Model RCS formulas for canonical geometries are given at Table 3.1. Some numerical simulation plots calculated for cylindrical targets are presented in Fig. 3.2.

The reflectivity losses between the air–ground and ground–air layers are subject to the problem of electromagnetic wave scattering from a dielectric surface. Too

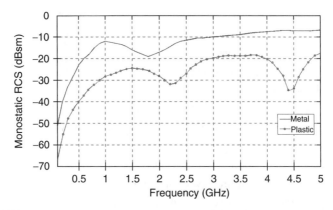

FIGURE 3.2   RCS behaviour of a cylindrical object (5 cm in diameter) in air.

TABLE 3.1 Radar Cross-Section Formulas of Basic Scatterer Geometries

Object	Aspect Direction	RCS (m²)	Shape Parameters
Sphere	Omnidirectional	πR^2	R: sphere radius (m); $R \gg \lambda$
Cylinder	Broadside	$\dfrac{2\pi a d^2}{\lambda}$	a: cylinder radius (m) d: cylinder length (m)
Flat plate	Normal	$\dfrac{4\pi w^2 h^2}{\lambda^2}$	w: plate width (m) h: plate length (m)
Corner reflector, dihedral	Symmetry axis	$\dfrac{8\pi A^2}{\lambda^2}$	A: reflector plate area (m²) λ: wavelength (m)
Corner reflector, triangular trihedral, rectangular trihedral	Symmetry axis	$\dfrac{4\pi L^4}{3\lambda^2}\ \dfrac{12\pi L^4}{\lambda^2}$	L: side length (m) λ: wavelength (m)

many parameters, such as antenna–ground surface and target–ground surface distances, the dielectric constant of the soil, the antenna radiation and target scattering waveshapes at the surface, and the tilt angles of antenna and object, are included in this formulation. In this case, some useful approximations are usually proposed for easy-to-calculate coherent estimations [1]. For example, the air–ground reflection loss is given by Eq. (3.4) presuming that the radiated wave on the surface is a uniformly distributed plane wave:

$$L_{r1} = 20 \log_{10} \frac{\sqrt{\varepsilon_0} \cos \theta_1 - \sqrt{\varepsilon_r} \cos \theta_2}{\sqrt{\varepsilon_0} \cos \theta_1 + \sqrt{\varepsilon_r} \cos \theta_2} \qquad \text{dB (see Fig. 3.1)} \qquad (3.4)$$

Another generic method, based on the transmission-line approach, uses the characteristic impedances of the air (Z_{air}) and the soil (Z_{soil}):

$$L_{r1} = 20 \log_{10} \frac{2Z_{soil}}{Z_{soil} + Z_{air}} \quad \text{dB} \qquad (3.5)$$

where

$$Z_{air} = 377 \ \Omega \qquad (3.6)$$

$$Z_{soil} = \sqrt{\frac{\mu_0 \mu_r}{\varepsilon_0 \varepsilon_r}} \frac{\cos_2^\delta + j \sin_2^\delta}{(1 + \tan^2 \delta)^{14}} \qquad (3.7)$$

Typical values for many soil materials: Z_{soil} is about 100 Ω; hence, $L_{r1} \approx 8$ to 10 dB and $L_{r2} \approx 5$ dB.

The soil attenuation losses can be formulated using the theory of electromagnetic wave propagation in a lossy medium. In this case, the electrical field expression for the monochromatic case is in the form

$$E(z, t) = E_0(\omega) e^{i\omega t - \gamma z} \qquad \omega = 2\pi f \qquad (3.8)$$

where the complex propagation constant is defined as

$$\gamma = \alpha + i\beta \qquad (3.9)$$

where

$$\alpha = \frac{\omega}{c} \sqrt{\frac{\varepsilon_r \mu_r}{2} \left(\sqrt{1 + \tan^2 \delta} - 1 \right)} \qquad (3.10)$$

$$\beta = \frac{\omega}{c} \sqrt{\frac{\varepsilon_r \mu_r}{2} \left(\sqrt{1 + \tan^2 \delta} + 1 \right)} \qquad (3.11)$$

$$\tan \delta = \frac{\sigma}{\omega \varepsilon_r \varepsilon_0} \qquad (3.12)$$

In this way, one can obtain the soil propagation path loss as

$$L_{s1,2} = e^{\alpha R'} \qquad (3.13)$$

or in nepers per meter (1 N = 8.686 dB),

$$L_{s1,2} = 8.686 \cdot 2\pi f R' \sqrt{\frac{\mu_0 \mu_r \varepsilon_0 \varepsilon_r}{2} \left(\sqrt{1 + \tan^2 \delta} - 1 \right)} \qquad (3.14)$$

TABLE 3.2 Estimated Propagation Losses at 100 MHz and 1 GHz for Various Mediums

Material	Relative Dielectric Constant	Conductivity (mS/m)	Loss at 100 MHz (dB/m)	Loss at 1 GHz (dB/m)
Air	1	0	0	0
Asphalt				
Dry	2–5	1–100	1–10	10–100
Wet	6–12	10–100	5–20	50–200
Clay	5–40	10–1000	5–100	50–1000
Concrete	4–20	1–100	2–20	20–200
Rock	4–10	0.1–10	0.01–1	0.1–10
Granite	5–8	0.01–10	0.2–10	2–100
Brick	8–12	0.01–1	0.2–2	2–20
Fresh water	81	0.01–10	0.1	1
Seawater	81	5000	100	1000
Snow	6–12	0.01–0.1	0.1–2	1–20
Sand: dry	3–6	0.001–1	0.01–2	0.1–20
Sandy soil (wet)	10–30	10–100	1–5	10–50
Loamy soil	3–30	0.1–100	0.1–50	1–500
Clayey soil	4–30	1–1000	1–200	10–2000

where f is the operating frequency and R' is the path length in the soil. Table 3.2 lists the RF propagation losses in some soils at 100 MHz and 1 GHz. For poor conductive materials ($\tan \delta \ll 1$), such as dry sand, one can write

$$\beta \simeq \frac{\omega}{c}\sqrt{\varepsilon_r \mu_r} \tag{3.15}$$

$$E(z, t) = E_0(\omega t - \beta z)e^{-\alpha z} \tag{3.16}$$

As a general rule, the system noise threshold determines the minimum detectable signal (MDS) level. Detection of the target is possible if the power backscattered from the object is sufficiently higher than both the noise and the clutter. The MDS can be calculated by using the system frequency bandwidth (BW), the receiver noise figure (NF), and the required signal-to-noise ratio (SNR):

$$\text{MDS} = kT_0 \cdot \text{BW} \cdot \text{NF} \cdot \text{SNR} \tag{3.17}$$

where k is Boltzmann's constant and T_0 is room temperature in kelvin ($kT_0 \simeq 4 \times 10^{-21}$ W/Hz). Obtaining a received power (P_r) that exceeds this value will not be sufficient for detection when clutter dominates the signal. The sources of clutter are primarily the undesired coupling signals from transmitter to receiver, the unwanted backscattering from other buried objects, and reflection from the ground surface. For this reason, the minimum SNR requirement of the system should be estimated accurately in the design, to guarantee the physical radar detection range. The SNR value should also be limited optimally to avoid using excessive RF power transmission because of its effect on the system's power budget.

3.2.3 Design Example

Suppose that we want to detect a dielectric cylindrical object 10 cm in diameter that is buried up to $R = 100$ cm. The system operating frequency chosen is 1 GHz. The soil is dry and loamy ($\varepsilon_r = 4$). The transmitter antenna has a 3-dBi gain, and the receiver antenna effective area is $A_e = 0.0012$ m^2. Using Eqs. (3.1) to (3.17), Table 3.2, and Fig. 3.2, we find that the total two-way reflection losses are 15 dB, the soil attenuation loss is 10 dB, the target has an RCS of -10 dBm^2, and the overall power loss (dynamic range) is calculated as about 94 dB. If we assume that the amplifier of the GPR receiver has a noise figure of 3 dB over 1 GHz of bandwidth, the MDS for the receiver characteristics specified yields 8×10^{-10} W or—61 dBm, assuming a signal-to-noise ratio (SNR) of 20. Under the circumstances, the peak power of the transmitter must be at least 33 dBm or approximately 2 W. For a 50-Ω impulse GPR system, this corresponds to 10 V of peak voltage of the transmitter output.

This radar budget equation is also adaptable for SF and FMCW GPR systems, which transmit sequentially repetitive frequency steps or a sweeping RF carrier signal controlled by a voltage-controlled oscillator over a chosen frequency bandwidth. The operational frequency band should be chosen based principally on the maximum depth and minimum size of the buried object. For both FMCW and SF radar, it is evident that the thermal noise level is much lower than that of the receiver of the time-domain impulse radar. Thus, their dynamic range is higher than that of impulse GPRs. Nevertheless, discrimination of the target backscattering signal from the ambiguous clutter at the receiver unit is a more difficult problem, due to the uselessness of simple antialiasing filters. Therefore, some calibrating methods, such as iterative range gating, are generally employed for such systems.

3.2.4 Radar Resolution

Resolution is defined as a radar system's capacity to discriminate individual targets in the subsurface, by either thickness or size. The resolution concept of the GPR is essentially divided into two parts:

1. Vertical (down-range or depth) resolution (ΔR_z)
2. Horizontal (cross-range or plain) resolution (Δx and Δy)

Vertical resolution (ΔR_z) is a critical performance parameter for GPR systems. The thickness of the depth-range layers on a GPR screen is related directly to the down-range resolution (see Fig. 3.8). For pulsed radars, the range resolution is connected with the pulse duration (τ) as $c\tau/2$. Since the impulse duration is roughly inversely proportional to the radar bandwidth (BW), ΔR_z can then be expressed as

$$\Delta R_z \simeq \frac{c}{2\text{BW} \cdot \sqrt{\varepsilon_r}} \tag{3.18}$$

where c is the free-space speed of the electromagnetic waves and ε_r is the relative permittivity of the soil. For example, if it is presumed that $\varepsilon_r = 9$ and $\Delta R_z = 2$ cm, the radar bandwidth required should be at least 2.5 GHz.

The horizontal resolution indicates the availability to distinguish two targets located one next to another at the same depth. It is based principally on the intersection area of the radiation projections (footprints) of T/R antennas. Thus, the antenna beamwidths, the height of the GPR head, and the depth of the buried object heavily determine the plain resolution of the GPR.

There are many studies in the literature to estimate the correlation between the radar resolution and the operational GPR parameters: primarily the frequency bandwidth and antenna types. For example, according to the experimental results of Rial et al. published in 2007 [12], the vertical resolution of a 1-GHz GPR system yields 20 cm in air ($\varepsilon_r = 1$). The measured horizontal resolutions are 20, 50, and 70 cm for GPR heights of 7, 91, and 147 cm, respectively. When we survey the literature, it should be noted that the vertical resolution estimations are usually compromised; however, the horizontal estimations can differ up to twofold, especially due to the use of different antenna types.

3.3   GPR HARDWARE

3.3.1   Ultrawideband Impulse Radar

Block Diagram   The impulse GPR usually has a wideband frequency spectrum since it is designed to detect both small objects buried at shallow depths and some huge layers at very great depths at differing soil conditions and climates. Such a system consists principally of a transmitter, a receiver, and controller units. The block diagram of a typical impulse GPR is shown in Fig. 3.3.

Impulse Generator   The GPR transmitter unit includes an impulse generator and a wideband transmitter antenna. The pulse generator produces a monocycle or monopulse-shaped impulse signal waveform which can have broad frequency spectrum characteristics, from a few MHz up to 5 GHz (see Fig. 3.5). The pulse width and pulse amplitude are two major concerns for proper design of the impulse generator. The narrower pulse widths imply more broadened operational frequency bands in order to detect buried small objects. The higher amplitudes yield more intensive target responses to going deeper for detection under the surface.

The shape of the pulse signal rigorously affects the operational frequency bandwidth of a GPR system. For example, a steplike pulse has a large amount of spectral energy at lower frequencies, with the spectrum falling off as an inverse frequency ($1/f$), whereas an impulsive-like pulse has a flat spectrum over the bandwidth. In every case, the high-frequency content of the pulse signal is limited by either the step rise time (T_s) or the impulse duration (T_d). For ideal impulse waveforms, the first null (f_0) in the spectrum occurs at $1/T_d$. The spectral power distribution of the impulse decreases rapidly beyond f_0 and typically follows an equation of the sinc function form. A monocycle signal can be analyzed as pulse AM modulation

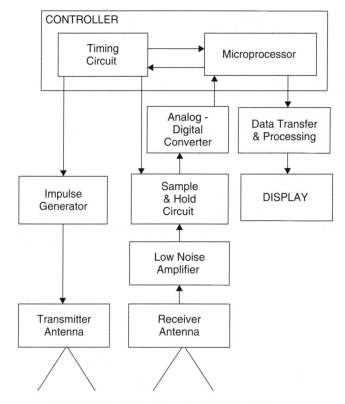

FIGURE 3.3 Typical impulse GPR block diagram.

of an RF sine-wave carrier. Thus, the spectrum of the monocycle impulse is shifted upward in the frequency domain with regard to the rise time and is centered on the carrier frequency, which corresponds to $1/T_d$. The generation of these various waveforms normally starts with a very fast rise time edge and a steplike pulse. Then one takes the first derivative of the step rise time to obtain an impulse signal. To obtain a monocycle, the first derivative of an impulse or the second derivative of a step pulse is taken. The UWB antenna and the wideband amplifier must be designed appropriately to avoid bandpass filtering, which can cause multicycling [5].

There are three common types of impulse generator technologies: high-speed switch MOSFETs, avalanche transistors, and step-recovery diodes (SRDs). Switch-mode MOSFETs are capable of generating peak voltages of some thousands of volts with a pulse width of a few nanoseconds. Avalanche transistors can produce a peak voltage of about a few hundred volts in the subnanosecond region. The SRD technology still has the best switching performance since SRDs are very fast. Although less than 100-ps pulse widths can be obtained, the available peak voltage is limited by a few tens of volts. Some suitable cascade configurations are also possible to increase the pulse power. The basic operational theory is similar for all the technology types.

The following is a monocycle impulse circuit design procedure, especially for high-frequency, high-amplitude GPR applications:

- A fast rise time and high-amplitude square-wave generator as a trigger
- An impulse generator with switch-mode transistors or diodes
- Wideband impedance matching using active FET circuits
- Pulse sharpening using high-speed Schottky diodes
- Pulse shaping using short-ended shunt stubs and *RC* tank circuits
- A wideband MMIC amplifier (if the signal power is insufficient)

The monocycle pulse generator shown in Fig. 3.4a is developed using short-ended transmission-line stubs realized by short-ended rigid cables, SRDs and a Schottky diode, a MESFET preamplifier, and a monolithic microwave integrated-circuit (MMIC) amplifier. At the input, a square-wave signal with a repetition frequency of up to 400 kHz is applied by the microprocessor in the GPR controller

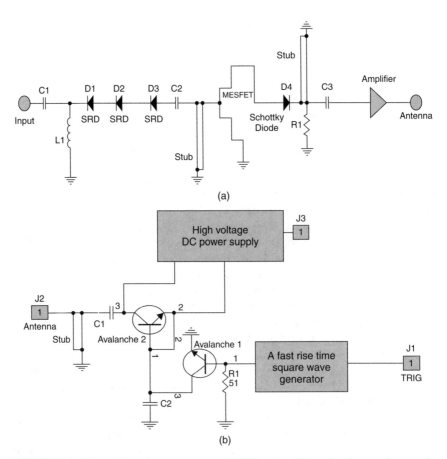

FIGURE 3.4 Typical impulse generators: (a) SRD mode; (b) avalanche transistor mode.

unit. The rise time of the signal is shortened with a comparator circuit. The signal proceeds to the impulse generator, composed of SRDs in active mode [13]. Operating in active mode enables the designer to shape the impulse by adjusting the dc bias current drawing from the dc source. The MESFET is used particularly for the circuit isolation and wideband impedance matching. Finally, the stubs, which are at the gate and output of the MESFET, and *RC* high-pass filters, are used to generate a very short impulse. It is noted that due to the MESFET buffer, the antenna cannot load back the circuit in an undesirable way. The output responses measured for this SRD design are presented in Fig. 3.5a and b, in the time and frequency domains,

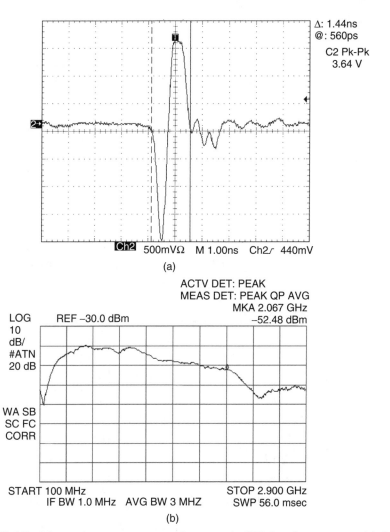

FIGURE 3.5   Measured output responses of monocycle SRD impulse generator: (a) time domain; (b) frequency domain.

respectively. The measured waveform has less than a 1.5-ns pulse duration in the time domain, and its frequency-domain band is between approximately 500 MHz and 2 GHz.

The design procedure of the avalanche mode monocycle impulse generator is similar to the SRD circuit. Here, as the primary difference, an avalanche transistor is used instead of short-recovery diodes. The transistor is operated in avalanche breakdown mode, used as a fast switch by discharging the stored energy into short transmission lines. These are output loads, which include the transmitter antenna equivalent impedance. A typical dual-cascade avalanche mode impulse generator design is shown in Fig. 3.4b. The output responses measured in this avalanche mode design are presented in Fig. 3.6a and b in the time and frequency domains, respectively.

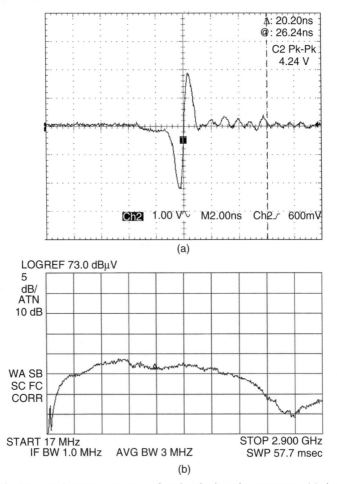

(a)

(b)

FIGURE 3.6 Measured output responses of avalanche impulse generator: (a) time-domain scope ($\times 10$ probe); (b) frequency domain.

Receiver Unit The impulse GPR receiver unit includes a wideband receiver antenna, a low-noise amplifier (LNA), a sample-and-hold (S&H) circuit, an analog-to-digital converter (ADC), and a data acquisition card. This unit operates so that:

- The RF impulse signals scattered from subsurface targets are received by the antenna.
- The signal received is amplified by the wideband LNA.
- The S&H circuit samples the RF signal with a specific rate and time shift, depending on the pulse repetition frequency (PRF), which is defined by the controller unit.
- The sampled data are digitalized by a 14- or 16-bit ADC and transferred by the data acquisition unit to the microprocessor for display and signal processing.

The receiving data are synchronized by the timing unit. System PRF is mostly adjustable between 10 kHz and 1 MHz. The most popular PRF value is 100 kHz. It means that hundreds of thousands of impulse signals are transmitted every second. The timing unit controls the triggering delays between the pulse generator and the S&H circuits to read the impulse stream data properly as they are received. Here the time-delay step value represents the depth resolution. The number of data per impulse stream are chosen as 256, 512, 1024, or 2048 for most GPR designs. A higher PRF means more data collection speed and thus, of course, better resolution. Nevertheless, the PRF is usually limited by about 100 kHz, due to the poor performance of impulse generators and power budget restrictions of handheld GPR systems.

A typical S&H receiver unit is illustrated in Fig. 3.7. It is designed for a frequency band from 100 MHz to 3 GHz, with less than a 3-dB noise figure. An RF signal is converted to a sampled signal with a trigger signal (100 to 400 kHz) in an S&H amplifier. High-speed switching and sampling requirements are accomplished with beam-led Schottky diodes. Schottky diodes have a switching speed in the picosecond range. The signal sampled is held by a capacitor and amplified by an operational amplifier to convert the digital signal. In this way, the impulse signal received is read by each triggering with a definite time shift determined by the GPR controller, like a digital oscilloscope.

It should be pointed out that jitter, which results in dynamic reading noise due to triggering signal shifting errors based on the send/receive trigger comparator circuit in the controller unit, is the most significant parameter defining the physical MDS level of the GPR system. Therefore, for high-dynamic-range impulse GPR

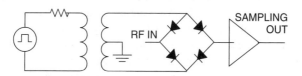

FIGURE 3.7   Sample-and-hold circuit.

design, it is strongly recommended that low-jitter-level comparators be used as well as RF circuit modeling, since the jitters are observed mostly in picoseconds that correspond to microwave bands in gigahertz. Performance test results of an impulse GPR system are shown in Fig. 3.8.

3.3.2   Stepped-Frequency Continuous-Wave Radar

Fundamentals and General Considerations   UWB radar can be realized either as a system transmitting signals with an instantaneous UWB spectrum or as a frequency-hopping system, which transmits at a single frequency at each moment of time, but over a period of time this frequency changes over the entire operational frequency band. A system with a narrow instantaneous spectrum is, in fact, simply a frequency-hopping system, where at any moment the system works with a narrowband signal whose frequency varies over time. There are two basic approaches to realizing frequency-domain systems: the stepped-frequency approach, where the frequency jumps from one fixed value to another, and the frequency sweep approach [or frequency-modulated continuous wave (FMCW)], where the frequency slowly varies, covering the entire operational band within a certain interval of time. In the first approach a system operates at a number of fixed frequencies well separated from each other. A typical example of such a system is a network analyzer. In the second approach the operational frequency of a system should vary linearly with time within a certain frequency band. A typical example of such a system is FMCW radar. System design according to both approaches mentioned above are usually analyzed in the frequency domain and treated as frequency-domain systems.

The frequency-domain approach has a solid background: a well-developed RF technology and a large choice of commercially available components. The frequency-domain approach typically leads to a higher signal-to-noise ratio (SNR), due to the higher and more uniform spectral density of the radiated signal, more accurate measured data, and, in principle, it allows the use of a much larger frequency bandwidth than does the time-domain approach [1,6]. On the other hand, the frequency-domain approach requires bulkier, more expensive equipment and a longer measurement time. It also involves very precise calibration of the entire radar, system, which is necessary for numerical synthesis of short pulses during postprocessing.

For ground-penetrating radars, the stepped-frequency approach has an additional (theoretical) advantage, as it allows for radar adaptation to the frequency dependence of the soil's dielectric permittivity and propagation losses. For example, for two types of soil, wet sand and wet clay, the difference between maximum penetration depths is as high as 5 m for frequencies around 400 MHz and less then 1 m for frequencies above 1.5 GHz. Knowing the properties of a soil is a very difficult problem because in real life there are no standard types of soils, but mixtures, and in addition they depend on the water content. Thus, practical implementation of the frequency dependence mentioned above is very challenging.

The first dedicated stepped-frequency continuous-wave (SFCW) GPR was developed by Koppenjan and Bashforth [15]. Further developments of SFCW GPRs are

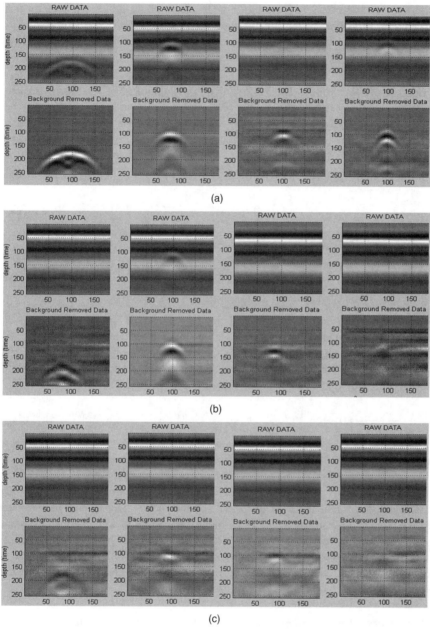

FIGURE 3.8 Subsurface detection performance of impulse GPR for the buried objects indicated (from left to right). (a) Sandy soil: (1) 25-cm-diameter metal at 15 cm, (2) 10-cm-diameter metal at 8 cm, (3) glass bottle at 5 cm, (4) 10-cm-diameter plastic at 5 cm. (b) Loamy soil: (1) 25-cm-diameter plastic at 15 cm, (2) 10-cm-diameter metal at 8 cm, (3) glass bottle at 5 cm, (4) 5-cm-diameter plastic at 5 cm. (c) Clayey soil: (1) 25-cm-diameter plastic at 15 cm, (2) 10-cm-diameter plastic at 5 cm, (3) glass bottle at 5 cm, (4) 5-cm-diameter plastic at 5 cm.

discussed elsewhere [9,16]. At the moment there are a few commercially produced SFCW GPRs, which are used primarily for landmine detection (such as
the AN/PSS-14 [17] and ALIS [18]), as for this application an extremely large
bandwidth is required.

Principle of Operation   The SFCW radar emits a number of successive frequencies. If they are distributed uniformly, the kth frequency is expressed by

$$f_k = f_1 + (k - 1)\Delta f \tag{3.19}$$

On a multidimensional space, the frequency can be written like a vector:

$$\overline{f} = [f_1 \quad f_2 \quad \cdots \quad f_N] \tag{3.20}$$

and the signals transmitted become

$$\overline{\dot{u(\overline{f}, t)}} = [Ae^{j2\pi f_1 t} \quad Ae^{j2\pi f_2 t} \quad \cdots \quad Ae^{j2\pi f_N t}] \tag{3.21}$$

where A is a constant. The signals defined by Eq. (3.21) propagate toward the
ground. Any change within the propagation medium as well as any object will
produce a return signal that will be captured by the reception antenna. Let us
suppose that only one object is within the unambiguous range; then the signal
backscattered by that object can be written

$$\overline{\dot{s(\overline{f}, t, t_i)}} = [r_1 Ae^{j2\pi f_1 (t - t_i)} \quad r_2 Ae^{j2\pi f_2 (t - t_i)} \quad \cdots \quad r_N Ae^{j2\pi f_N (t - t_i)}] \tag{3.22}$$

where r_k is the reflection coefficient for the kth frequency and t_i is the two-way
propagation delay to the object, $0 \le t_i \le 1/\Delta f$.

SFCW radar measures the phase delay between the signal transmitted in Eq.
(3.21) and the signal received in Eq. (3.22) and in this way locates the scatterer
that produced this signal. Due to its mode of operation, SFCW radar is phase
sensitive, so it is very important that the phase center of the antenna system be
stationary. The power spectral density of the echo received is sampled. Inverse
Fourier transform (IFT) could reconstruct the corresponding time-domain signal
from the frequency domain into the time domain.

Parameters   The maximum range that can be reconstructed depends on the interval between successive frequencies. Any distance greater than this will generally
exceed the principal reconstructed range interval. The resolution equals the recip-
rocal of the total frequency range covered by the samples. Proper tapering can
reduce the sidelobes in the synthesized profile that has been reconstructed. These
limitations refer to the IFT procedure. Other procedures for finding objects may be
used as well (i.e., subspace methods or other high-resolution methods).

The main parameters of any radar system are the unambiguous range and the down- and cross-range resolution. The unambiguous range (R_{un}) of SFCW radar is given by the formula [6]

$$R_{un} = \frac{v}{2\Delta f} \qquad (3.23)$$

where v is the propagation velocity of the electromagnetic fields in the medium and Δf is the frequency step. The down-range resolution (ΔR) depends on the bandwidth (B) of the radar and can be calculated using the formula [6]

$$\Delta R = \frac{v}{2B} \qquad (3.24)$$

where $B = N\Delta f$ (N is the number of frequencies). The resulting synthesized time-domain profile can be obtained via [6]

$$s_i = \frac{1}{N} \sum_{n=1}^{N} w_n S_n e^{j2\pi ni/N} \qquad (3.25)$$

where w_n is the weight assigned to the nth frequency sample and S_n is the sample for the nth frequency signal. Here s_i is the ith sample in the reconstructed signal. This corresponds to a time delay of

$$i\Delta t = i\frac{v}{2}\frac{1}{B} = \frac{iv}{2N\Delta f} \qquad (3.26)$$

It should be noted that two effects with opposite impact would affect the resolution in the synthesized profile:

1. In the soil the relative dielectric permittivity will be higher than 1. For example, the permittivity of dry sand is on the order of 3, so the wavelength in sand is 1.9 cm.

2. The synthesized profile is resolved by processing. If a straightforward IFFT had been applied to the data set, in the range profile one would have observed range sidelobe levels of some -13 dB. Generally this is not acceptable, as the high sidelobes generated on the air–ground interface may obscure the small echoes off shallow buried objects. Therefore, a tapering function is applied such that the sidelobes will be at some chosen (low) level. The effect on the main lobe of the IFFT response is, of course, that it is widened.

The cross-range resolution of the system is given by the antenna system footprint, which corresponds to the footprint of an antenna for a monostatic system. The footprint is obtained by overlapping the transmitting and receiving antenna footprints for bistatic radar systems. As the SFCW radar is a UWB system, the cross-range resolution can be achieved using either a physical array antenna system or by simulating it through processing (the synthetic aperture radar approach) [6].

Dynamic Range and Power Resolution   One of the major problems of ground-penetrating radar is the observability of weak scatterers (e.g., small objects) masked by a strong reflection (e.g., from an air–ground interface) or antenna coupling. To overcome this problem, radar should have a sufficiently large dynamic range. As an example, let us follow a rather elementary procedure to find an expression for the small-signal dynamic range of the receiver.

Suppose that the effective surface of the footprint of the antenna is A_a and the effective surface of the object is A_m. Assume both the object and the surface to be nonconducting. For example, let the soil be dry sand with a relative permittivity $\varepsilon_{r,s} = 2.6$, and let the object be a dielectric material with permittivity $\varepsilon_{r,m} = 4$. Then the reflection coefficient at the air–soil interface, if it is only soil (no object), is

$$r = -\frac{\sqrt{\varepsilon_{r,s}} - \sqrt{\varepsilon_{r,a}}}{\sqrt{\varepsilon_{r,s}} + \sqrt{\varepsilon_{r,a}}} \tag{3.27}$$

Here $\varepsilon_{r,a}$ is the dielectric permittivity of the medium. Since $\varepsilon_{r,a} = 1$ for an air medium, it follows that

$$r = -\frac{\sqrt{\varepsilon_{r,s}} - 1}{\sqrt{\varepsilon_{r,s}} + 1} \tag{3.28}$$

Now let S_t be the power density incident on the soil. Then the backscattered power will be $P_{r,1} = S_t A_a |r_s|^2$. If part of the soil is covered by another object, the total backscattered power will be

$$P_{r,2} = (A_a - A_m)|r_s|^2 S_t + A_m |r_m|^2 S_t \tag{3.29}$$

From this equation it can readily be seen that there will be no difference between observation with and without the object if the reflection coefficient of the object is identical to that of the soil. It can also be seen that a difference as small as $\Delta P = P_{r,2} - P_{r,1}$ should be within the power resolution of the radar to find the contrast between areas with and without the object. This small-signal power resolution (SSPR) can be computed from Eqs. (3.27) and (3.29) as [19]

$$SSPR = \frac{\Delta P}{P_{r,1}} = \frac{A_m}{A_a} \frac{|r_m|^2 - |r_s|^2}{|r_s|^2} \tag{3.30}$$

If the antenna is elevated above the ground by some 0.7 m and if the beamwidth of the antenna is 90°, with the values already given for the permittivities of the object and the soil and an object with a diameter of 5 cm, one would find a small-signal power resolution of −28.9 dB [19]. These considerations concern the properties of the receiver. Due to this large number, it is common practice when processing B- and C-scan images to subtract the background $P_{r,1}$ from each data point. Obviously, when the variation in the background itself is of the same order of magnitude as the variation due to the presence of an object, the distinction between the effects of surface roughness and the presence of a mine is difficult to detect. Signal processing accounting for coherent spatial integration and clutter decorrelation will then support this distinction.

Another property concerns the maximum signal level that may enter a receiver without saturating it. Here, also, variations in soil conditions dominate the requirement.

Ambiguity Function   SFCW radar is sensitive to range–Doppler coupling. The ambiguity function of the radar can be analyzed to evaluate the effect of Doppler due to antenna motion, in our case the possible motion of the antenna on the scanner system. The complex signal envelope can be described by

$$g(t) = \frac{1}{\sqrt{N}} \sum_{n=0}^{N-1} g_n(t - nT_R) \exp(-j2\pi n \, \Delta f \, t) \qquad (3.31)$$

where T_R is the pulse repetition period, Δf is the frequency step size, and

$$g_n(t) = \frac{1}{\sqrt{t_p}} \text{rect}\left(\frac{t}{t_p}\right) \qquad (3.32)$$

is the envelope of the individual pulse with unit energy.

Division by $\sqrt{N}$ will maintain the unit energy for the entire train. After that, the ambiguity function for this train of pulses can be derived as [20]

$$\chi(\tau, f_d)|_{p\geq 0} = \frac{1}{N} \exp(j2\pi p T_R f_d) \chi_c(\tau - pT_R, f_d + p\Delta f)$$

$$\cdot \sum_{m=0}^{N-1-p} \exp\{j2\pi m[\Delta f(\tau - pT_R) + T_R f_d]\} \qquad (3.33)$$

Here p is a parameter referring to the various phase- and amplitude-modified repetitions of the basic component ambiguity function χ_c. For negative values of p the expression would be symmetrical. The contour plot of this ambiguity function is shown in Fig. 3.9. It exhibits a number of Doppler- and delay-shifted kernel functions. The kernel at the origin (i.e., with delay and Doppler shift zero) is the relevant one for this application. The slope of the tilted line is $\frac{\Delta f}{T_R}$. From the dimensions of this figure it can then be shown that the antenna may move at a speed of a few meters per second without significant distortion of the range profile.

Effect of Phase Noise   Phase noise can limit the resolution of high-resolution radars [21]. The radar transmitter is based on frequency sources with a limited stability over time. Therefore, the signal transmitted, and obviously also the signal received, are distorted due to cumulative phase noise. Here cumulative phase noise is the phase change accumulated over the time delay τ between transmission and reception of an echo signal. The noise-free signal received would ideally have a phase $\phi(t) = 2\pi f_c t$, apart from an arbitrary initial phase. The actual reference frequency, however, is $f_c + \delta_f(t)$. The phase measured as a result of this will deviate

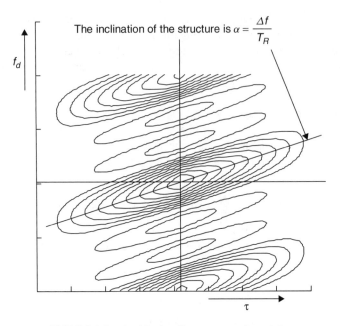

f_d

The inclination of the structure is $\alpha = \dfrac{\Delta f}{T_R}$

τ

FIGURE 3.9   Ambiguity diagram near the origin.

from the ideal value, leading to a noisy background disturbance. This background is expressed by the *phase noise–limited dynamic range* (PNDR), defined as the ratio between the maximum squared IDFT output and the variance of the noise induced by the phase noise at ranges away from the object:

$$\text{PNDR} = \frac{(NC_f)^2}{NC_f^2(1 - C_f^2)} \tag{3.34}$$

Here N is the number of frequencies and C_f is the characteristic function of the effective random frequency, with

$$C_f = \exp\left[-\frac{(2\pi\tau\sigma_f)^2}{2}\right] \tag{3.35}$$

σ_f is the standard deviation of the effective random frequency. Knowing that in our application $2\pi\tau\sigma_f \ll 1$, we can approximate C_f by

$$C_f = 1 - \frac{(2\pi\tau\sigma_f)^2}{2} \tag{3.36}$$

Consequently, it reduces to

$$\text{PNDR} = \frac{N}{(2\pi\tau\sigma_f)^2} \tag{3.37}$$

Design Example   The SFCW principle can be realized in various ways. To illustrate the design procedure and give an example of SFCW hardware, we describe briefly a multichannel SFCW developed in IRCTR for landmine detection [22].

Choosing the Operational Bandwidth   Given the application for ground penetration, the radar must provide low frequencies for deep penetration. An overview of the absorption of electromagnetic fields in various types of soil is provided by Daniels [1]. From this and related sources it can be concluded that the lower frequency should not be above 400 MHz. The resolution in air should be as small as a few centimeters to find mines laid at or near the surface of the soil. Adopting a resolution of 3 cm in air, the bandwidth of the radar should be on the order of 5 GHz. The reflection of the soil is a function of frequency: the higher the frequency, the weaker the reflection. Moreover, as shown in Figs. 3.9 and 3.10, the penetration is higher for low frequencies. Thus, the bandwidth of 5 GHz should be obtained in the lowest possible bands.

One problem in the design and manufacture of antennas is the ratio between the lowest and highest frequencies to be transmitted and received. It can be seen that the lower frequency is the lower bound of the frequency band; given a fixed bandwidth of 5 GHz, this ratio will be higher. For this reason the lower frequency should not be too low, and a compromise must be chosen among the various effects. Given these considerations, we have chosen 400 MHz as the lower frequency limit. The upper frequency counts for the cross-range resolution. After taking the procedure for synthesizing the range profile into account, we have selected an upper limit of almost 4.9 GHz.

Choosing the Frequency Step   The total range to be covered by the system is not very long, so the distance between the antennas and the soil is normally less than 1 m. In this application, the depth of the buried mines is up to 1 m below the

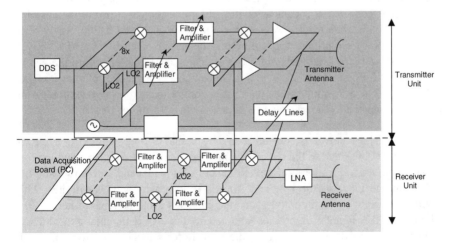

FIGURE 3.10   SFCW block diagram.

surface. Of course, other effects may come into play: Objects other than the surface and subsurface may reflect signal energy through the sidelobes of the antennas or the cabling between the transmitter or receiver, the antennas may produce internal reflections, and so on. In general, we must anticipate secondary reflections and reserve some room in the synthesized range profile to dimension the geometry such that aliased or irrelevant signals do not affect the area containing the mines.

For this reason we have chosen the unambiguous range R_{un} to be at least 4 m. In fact, we have chosen a frequency step size of 35 MHz, so

$$R_{un} = \frac{3 \times 10^8}{2} \cdot \frac{1}{35 \times 10^6} = 4.28 \text{ m}$$

By this calculation, one needs approximately $5000/35 = 142$ frequencies. In this situation, 128 steps are selected because of the ease of implementing a fast IFFT. So the bandwidth is calculated as $(128 - 1) \times 35 = 4445$ MHz. The lowest frequency being 400 MHz, the highest frequency will be 4845 MHz and the ratio between the highest and lowest frequencies is $12:1$. The corresponding resolution in air, without tapering, is 3.37 cm.

System Block Diagram   The system design (Fig. 3.10) comprises the following essential blocks.

Transmitter   The SFCW source is based on a direct digital synthesizer (DDS; manufacturing model STEL-9949). The DDS runs on a 1-GHz clock and steps through 115 to 360 MHz in eight steps at 35-MHz intervals. Each step is of duration 100 μs. The signal generated by the DDS is split into eight channels. Then the transmitter imparts bandwidth coverage to the eight channels by dividing the frequency range of each channel. The band spread of these channels ranges from 400 to 4845 MHz at intervals of 525 MHz each. Each channel then feeds into a power amplifier via a digital attenuator. The attenuator is digitally controlled by the control processor and is meant to control the quality of the spectrum transmitted across the entire bandwidth. A power combiner is located at the output of the individual power amplifiers for combining the power of each channel additively. Finally, the output of the transmitter is a single coaxial line leading to the antenna via a direction coupler. The coupler is meant to tap a portion of the power for calibration. The transmitter and signal generator are located away from the scanner and are connected to the transmitting antenna via a 15-m coaxial cable.

Receiver   The receiver is located away from the scanner via 15 m of coaxial cabling. It is connected to the receiving antenna via a direction coupler which feeds in an input from a calibration device. There is a wideband low-noise amplifier (LNA) at the input of the receiver. This is followed by a power divider which divides the input power into eight separate channels using bandpass filters (BPFs). Thereafter, the signal is amplified and fed to the I/Q demodulators, which are fed the corresponding signals from the transmitter's local oscillators (LOs), for downmixing. There are 16 such demodulators, the output of each being a dc signal.

The dc signals obtained at the output of an amplitude (I) and phase (Q) mixer for each channel quantify the phase delay due to the two-way propagation from the antenna to an antipersonell mine.

Antenna System   The antenna assembly consists of two Archimedean cavity-backed spirals of opposite rotation. Each spiral has two arms fed in antiphase through an ultrawideband balun, which in addition to balancing assures impedance matching between the coaxial cable and the antenna. The width of the arms is 10 mm and the gap between them is 5 mm. The balun is a very important element of the antenna system; actually, it limits the upper frequency to about 4.5 GHz [23]. The antenna transmitter and receiver signals are cross-polarized to reduce direct coupling and ground (surface) returns. The physical dimensions of the two antennas are given by the frequency range; the lower frequency will determine the size of the antenna. In our case the diameter of each spiral is about 34 cm, and due to this large value, part of the signals backscattered by the ground will be reflected back toward the ground, producing multiple bounces.

Timing and Control Unit   This unit is based on a microcontroller and controls the switching and frequency-stepping operation of the system. It interacts with the main computer and is independently programmable from the computer.

Data Acquisition System   The data acquisition system (DAS; manufacturing model DAP 5216A) is an eight-channel device. It can sample eight channels simultaneously at a rate of 150 kHz. The inputs are routed to the ADC via a changeover switch, which first samples the I-channel and then the Q-channel. The DAS incorporates switched capacitor antialiasing filters at its input so that the user can adjust the quality of frequency response of the antialiasing filters online. The entire system can be controlled from the main computer. The DAS is hosted inside the main computer and interfaces directly to MATLAB software on the main computer.

Computer and Display   This is a basic PC. It is the interactive user-controlled man–machine interface (MMI) to the SFCW-GPR system. It not only processes the data received from the SFCW-GPR off-line using MATLAB but also controls the entire radar system frequency-stepping and data acquistion timing control. This is achieved via a control processor, which is connected to the computer. It uses computer hard disks for data storage.

Radar Control Software   The SFCW radar is computer controlled, all the measurements are made automatically, and in addition, the results of the measurements in the shape of range profiles (A-scans) or B-scans are displayed on the computer screen, providing the user with "live" information. When the data are acquired, the transmission and reception antennas are fixed on the cradle of a computer-controlled scanner. Through the user-friendly interface of the radar control panel, the user can choose an operational mode as the calibration or normal mode. In the calibration mode the radar will measure the dc offset by setting the attenuators to the maximum values, and the offset will be subtracted from each A-scan.

In addition, a measurement with a direct connection between the transmitter and the receiver is made to generate a *reference profile*. This profile is used to check the functionality of the radar before starting a new set of measurements, together with dc offset correction for on-site automatic calibration. In the second mode of operation, the transmission and reception antennas are connected to the radar and it performs measurements, the data being displayed on the screen. The SFCW-GPR scope's interface has a menu with several options, such as "File," which allows the user to set the locations for the antenna calibration file, for the attenuation file, and for where the data are to be saved; "Setup," which allows the user to set the values of attenuators in order to equalize the spectrum of transmitted power; and "Technical Mode," which permits the user to chose the operational mode.

The radar operates on two antennas, one for transmission and one for reception. This will have no impact on the performance of the combiner/resolver, as these are buffered by amplifiers, the former by the RF amplifier and the latter by a wideband LNA. We ensure 60 dB of isolation between the antennas from the point of view of the reduction of AM noise. The design has not taken into account any limitations regarding the size and weight of the system, which should be considered in the case of industrial development.

Experimental Verification of the Parameters   A demonstrator based on a suggested design has been built. The radar transmits eight frequencies simultaneously, as shown in Fig. 3.11. This transmission scheme decreases the data acquisition time drastically. The radar works with circularly polarized electromagnetic waves, being able to detect metallic as well as nonmetallic targets, no matter what their

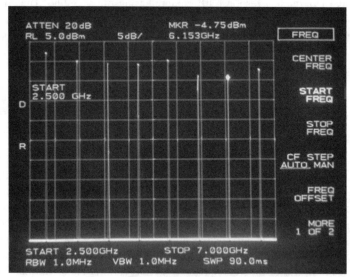

FIGURE 3.11   Measured spectrum of the radar (by stopping the frequency-hopping procedure).

shape is, and moreover, using a matched filter synthetic aperture radar (MF-SAR) procedure, their shape can be reconstructed.

The demonstrator has been tested, and it proved able to meet the performance expected [24,25]. Among the most important features of the radar, the following should be emphasized for practical operations:

- High-speed data acquisition, because eight signals are transmitted simultaneously
- Onboard signal processing and on-site calibration
- Friendly interface which allows the user to check the calibration of the system and to visualize "live" measurements (A and B scans)
- Software control of power transmitted
- Ability to detect metallic as well as nonmetallic objects
- Ability to picture the shape of the object
- Ability to provide data in the frequency as well in the time domain
- Possibility of focusing the image by changing the calibration reference plane

The radar is operational and can be used to investigate additional and more sophisticated scenarios.

Radar Calibration The radar calibration has to cope with several problems:

- The output power is not the same for all frequencies.
- The last mixer in the receiver suffers from nonzero dc-offset values, which must be corrected.
- The electronics and the cabling cause frequency-dependent amplitude and phase deviation.
- Delays within the transmitting and receiving antennas are frequency dependent.
- Another complicating factor is that there are multiple bounces of the echo signal off the ground. It bounces several times between the receiving antenna and the ground. To make a proper calibration, the first reflection should be isolated. Time-domain gating can do this.

Equalization of the Transmission Powers Power equalization is needed because the power provided by the transmitter is not the same for each frequency and because attenuation of the two Sucoflex cables (from transmitter cabinet to transmitting antenna and from receiving antenna to receiver cabinet) as well as the gain of the low-noise amplifier (LNA) is frequency dependent. To determine the attenuation needed for each frequency, the signal is measured with a direct connection, excluding the antenna system (transmitter cable after an attenuator connected to the input of the LNA). The power distributions before and after equalization, are pictured in Fig. 3.12. Comparing Fig. 3.12a and b, one can see that before equalization the power ripple is about 15 dB, whereas after it is only around 4 dB.

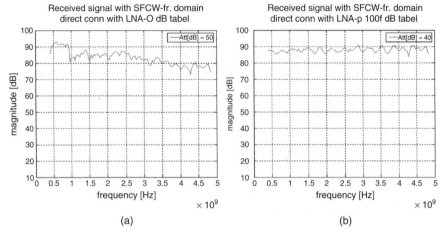

FIGURE 3.12 Power distribution diagrams (a) before and (b) after equalization. (From [23].)

Radar Resolution

Cross-Range Resolution An example of the cross-range resolution achieved is presented in Fig. 3.13. The separation between the two objects is 6 cm, and it can be seen that the two objects can be distinguished easily, with room for further reduction of the separation.

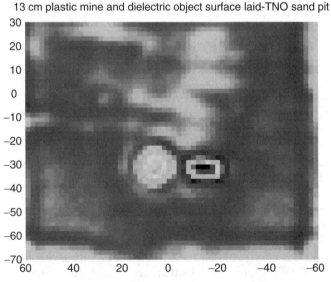

FIGURE 3.13 Horizontal cut at the level of the air–soil interface after background subtraction and SAR (two dielectric objects separated by 6 cm).

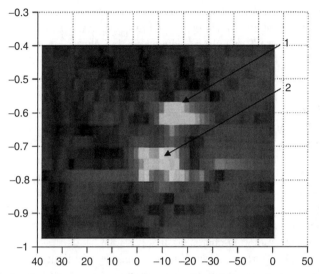

FIGURE 3.14   Vertical cross section through a three-dimensional data volume for two dielectric objects located 11 cm apart.

Down-Range Resolution   To check the down-range resolution of the radar, a C-scan has been made with two dielectric objects: a 3.4 × 15.7 cm dielectric object and a 13-cm-diameter plastic landmine, located 11 cm apart in the vertical plane. The results after synthetic aperture radar (MF-SAR) processing are presented in Fig. 3.14. It can be seen that the objects are clearly resolved. Actually, on a vertical axis, the separation between the two objects is covered by two pixels, so the down-range resolution is better than 5.5 cm. However, other measurements with a lower separation between objects should be made to measure the precise value of the down-range resolution [26].

Conclusion   This section gave a brief overview of SFCW radar fundamentals. We demonstrated how these fundamentals can be applied to the design of an operational SFCW-GPR using an experimental system setup and its performance results.

3.4   GPR ANTENNAS

3.4.1   General Design Requirements

The performance of a GPR depends exactly on proper design of the transmitter and receiver (T/R) antenna pair. In practice, GPRs use primarily broadband signals, which can vary from about 10 MHz to 5 GHz for impulse systems and from 1 to 8 GHz for stepped- or swept-frequency systems. Thus, wideband (even ultrawide band) T/R antennas are usually selected, especially for impulse GPR systems. It is desirable that the T/R antennas designed have flat, high directivity gain, narrow beamwidth, low sidelobe and input reflection levels over the operational frequency band to attain the largest dynamic range, the best-focused illumination area, the lowest T/R coupling, reduced ringing clutter, and uniformly shaped impulse radiation [1]. Electromagnetic coupling effects between the T/R antennas and adaptive designs for metal detector-mounted operations should be considered as well [2].

A useful reference for the general case of antennas in matter is given by King and Smith [27]. Further considerations in the selection of suitable antenna types are based on the types of targets and radar systems used. Dipole, bowtie, spiral, and TEM horn are the most popular ultrawide band (UWB) antenna models proposed for impulse GPR systems. For stepped-frequency GPR, wideband selective octave horns, ridged horns, log-periodic dipole arrays, discones, dielectric rod–inserted waveguides, cavity-backed planar spirals, conical spirals, and pencil beam dish antennas (only in certain specific cases) can also be employed, due to their high gain and wideband radiation characteristics [28–30].

Some physical requirements, such as compact size and light weight, can restrict the convenient antenna design of the GPR rigorously, especially for handheld models. Therefore, a loaded dipole is the basic GPR antenna, as it is a simple, easy to use, and linearly polarized structure. Nevertheless, the operational bandwidth and gain response of the dipoles are mostly poor and unsatisfied for high-performance wideband radars. Thus, two types of planar antennas, bowtie and spiral, are most likely to be chosen [1]. Nevertheless, vehicle-mounted GPR systems usually mitigate the physical restrictions, and then three-dimensional antennas can be put to use suitably for hyperwide or multiband operations. Such a GPR, which benefits from the entire frequency band up to 10 GHz, will accomplish significantly improved detection and identification performances both for large objects buried deeply and for very small targets at shallow depths under the ground. The TEM horn, which has mainly bowtie-originated characteristics, is an appropriate structure, due to its wider band, higher gain, and narrower beamwidth characteristics than those of typical planar antennas. Dielectric loading techniques can be applied to improve the gain pattern, increasing the antenna electrical size [30]. For example, the PDTEM horn was introduced by Turk as an efficient UWB impulse radiator over a $20:1$ frequency band [32].

Finally, impedance matching is another critical issue in increasing the antenna radiation efficiency for performance enhancement and reducing the time-domain ringing effects of the GPR signal. Ringing is the essential source of the clutter, in particular for impulse GPR systems, since most impulse signals are generated by switch-mode transistors that produce antenna impedance-dependent pulse shapes. Thus, the antenna impedance must be stabilized by a matching circuit design over the wide operational band. In addition, the antenna structure should be balanced to isolate it from structural imbalances using a wideband balun (balanced-to-unbalanced) transformer [2]. The essential antenna models' suitable for GPR operations are presented in the following sections, together with their performance results and comparisons.

3.4.2 Wideband Planar Antennas

The simplest GPR antenna is an electrically small dipole element. It is a linearly polarized antenna that has the far-field intensity proportional to its electrical length. Thus, the antenna gain and radiation efficiency can be very small at lower frequencies. To avoid performance degradation, resonance dipoles such as half-wave dipoles are usually preferred for impulse GPR. Although whole, the dipoles have a physical advantage, particularly for handheld GPR systems because of their small size, light weight, and easy-to-fabricate structures; they contain frequency-dependent input impedance and radiation gain characteristics. Some dielectric and absorber loading techniques are usually applied to improve their operational frequency bandwidth [27].

The bowtie is a dipolelike-characterized wideband antenna type frequently used for pulse transmission (Fig. 3.15a). It is basically a triangular (or circular/elliptical slice) plate version of a dipole structure that yields higher gain and wider-band performance than those of a simple dipole [28]. For a circular object, the bowtie is a suitable antenna type since it produces linear polarization and has relatively good linear-phase response over the wideband.

During detection of noncircular targets such as pipes and wires, polarization of the radiated electromagnetic wave gains great importance, due to the strong risk of performance degradation caused by cross-polarization positioning of the target object. Thus, circular polarized antennas (i.e., spirals) are usually considered for such application cases. The design of spiral antennas can be planar or conical versions (Fig. 3.15c and d). The spirals are, theoretically, frequency-independent antennas. But in practice, both types of spirals have band limitations by the feed radius and the arm length for the upper and lower frequencies, respectively. The cavity-backed version of planar spiral antennas has become very popular since the initial designs of handheld GPR systems [23].

The bowtie and spiral antennas have similar electrical and physical characteristics. The main difference is in the polarization responses. For example, the bowtie is a more convenient antenna type than wideband for circular-shaped object detection, due to its linear polarization and linear phase response. Contrastingly, spiral antennas are generally suitable for pipelike longitudinal objects. Frequency-domain responses of bowtie and spiral antennas are shown in Fig. 3.17.

3.4.3 Aperture Antennas

Standard gain horn, ridged horn, dielectric loaded waveguide, and dish antennas can be considered for stepped-frequency GPR systems. All of these models are very high-gain and narrow-beam band-selective antennas that are used from the L-band to the X-band (from 1 to 10 GHz). The typical gain of the standard horn antenna is around 15 dBi over a 2 : 1 frequency bandwidth. To increase the antenna directivity gain with a narrower beamwidth, a model using proper dielectric loadings is available [29]. For multiband GPR operations, a ridged horn that yields gain around 10 dBi over a 10 : 1 frequency bandwidth can be chosen (Fig. 3.15f). When the highest gain and narrowest beamwidth are principal requisites and the physical antenna size

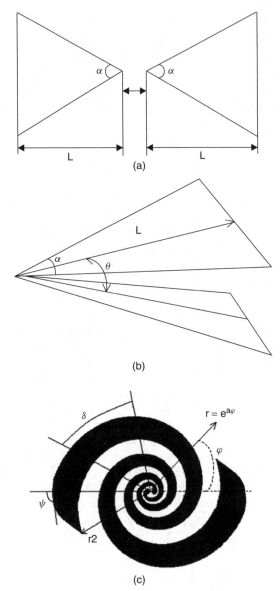

FIGURE 3.15 Typical GPR antennas: (a) triangular plate bowtie; (b) TEM horn; (c) two-armed logarithmic spiral; (d) conical spiral; (e) log-periodic antenna; (f) rectangular horn.

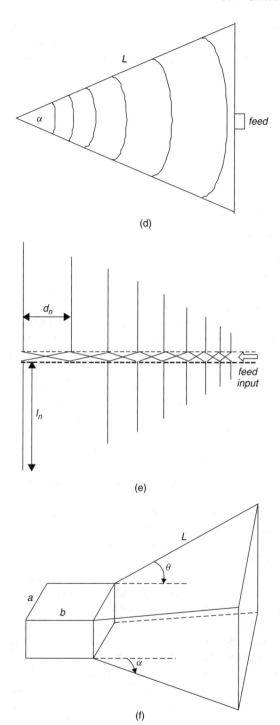

(d)

(e)

(f)

FIGURE 3.15   (*Continued*)

of the antenna is not restricted (e.g., in vehicle-mounted forward-looking systems), the dish antenna is an appropriate solution, due to its high gain up to 30 dBi and its pencil beam characteristics. The electrical size of the dish aperture is related directly to the directivity gain and HPBW [31]:

$$G = 20 \log \frac{D}{\lambda} + 9.94 \qquad \text{dBi} \tag{3.38}$$

$$\text{HPBW} = \frac{180\lambda}{\pi D} \qquad \text{deg} \tag{3.39}$$

where G is the antenna gain for 100% efficiency and D is the diameter of the circular dish.

3.4.4 TEM Horn Antennas

The TEM horn is one of the most promising antennas for impulse GPR systems because it has a wider frequency band, higher directivity gain, a narrower beamwidth, and a lower back-reflection than planar antennas [2]. Such a structure consists of a pair of triangular or circular slice–shaped conductors, forming a V-dipole structure. It is characterized by L, d, α, and θ parameters, which correspond to the arm length of the antenna, the feed point gap, the conductor plate angle, and the elevation angle, respectively (Fig. 3.15b). In general, the arm length of the TEM horn limits the lower cutoff frequency of the radiated pulse, the plate angle designates the polarization sensitivity, and the plate elevation angle determines the structural impedance of the antenna with d and α. A conventional TEM horn usually exhibits bandpass filterlike antenna gain behavior over a wide bandwidth. Thus, some practical dielectric-filling techniques can be used in GPR applications to improve the operational band, decreasing the lower cutoff frequency by increasing the electrical size [30]. In this case, the operational band of the antenna can be shifted downward in the frequency domain. However, the frequency bandwidth probably cannot be broadened.

A partial dielectric-loaded TEM (PDTEM) horn structure has been proposed by Turk to achieve low VSWR and high directivity gain over a bandwidth broadened at least twice [32]. It is obvious that such a UWB antenna can yield a much more satisfactory impulse radiation performance for GPR. The PDTEM structure is modeled as a microstrip line, and its geometry is designed to match the antenna output impedance to the feed source impedance (nominal, 50 Ω) by decreasing the segment characteristic impedances along the antenna line (see Fig. 3.19). For this purpose, using multiangular dielectric profile loading and arranging the segment widths of the plate wings properly, one can implement an efficient geometrical design for the TEM horn. This approach is quite effective for higher frequencies, where the electrical length of the antenna line is sufficiently long to transform the output impedance. Nevertheless, in the case of a lower-frequency band, the antenna output impedance rather than the characteristic impedance strongly determines the matching performance. Thus, an extensional resistive aperture loading may be required to keep VSWR levels reasonable at lower frequencies. Finally, the feed gap is coated by a piece of low-conductivity absorber to improve the overall VSWR performance.

The fusion of GPR with an EMI sensor (metal detector) is rapidly becoming one of the most popular subsurface detection techniques. The additional requirements for multisensor adaptive impulse GPR antennas are the following specifications: (1) a metal reduced antenna structure designed not to cause clutter for the EMI sensor, and (2) adaptive structural design to avoid highpass filterlike waveguide behavior of the GPR headbox wound by EMI sensor coils. For this purpose, a grating model Vivaldi-shaped version of the PDTEM horn (called PDVA) can be proposed for metal detector combined GPR systems [5]. The PDVA geometry design is similar to that of a PDTEM horn (Fig. 3.16b). The essential difference is the grating model antenna wing approach shown in Fig. 3.16a, which reduces the amount of metal in the structure to minimize the clutter in a metal detector while retaining the antenna characteristics [33]. Furthermore, since PDVA wings are bent more than PDTEM horn wings as a structural feature, the electromagnetic wave guidance of a coil-wound GPR headbox is much less effective for a PDVA horn than for a PDTEM horn. Typical performance results of various TEM horn models are presented in Fig. 3.18.

3.4.5   Array Configurations

As mentioned earlier, some physical restrictions may be disregarded in the case of semiportable vehicle-mounted systems, and thus hyperwide or multiband GPR operations can be implemented by using larger antennas. The principal basis for the antenna array approach is usually a requirement for wide-area scanning at high speed. In this situation, many T/R antenna pairs are used in multiple configurations, such as a few identical T/R antenna pairs operated in line-switch mode, or one illuminator transmitter antenna located above many sectoral receivers for beamforming. Beyond that, hyperwide and multiband performances can also be implemented by UWB antenna arrays, since a GPR system that benefits from the entire frequency band up to 10 GHz will accomplish significantly improved detection for both large objects buried deeply and small targets at shallow depths. The same phenomenon is also valid for open-air impulse radars.

In this situation, a combination of PDTEM horns can be proposed as an array to obtain hyperwideband antenna characteristics from 100 MHz to 10 GHz for multiband GPR (Table 3.3). The array consists of a PDVA and a TEM horn antenna, which are designed particularly for higher and lower band operations, respectively (Fig. 3.16c). A wideband 50- Ω two-way RF splitter is used for the antenna feed. Each element of the PDTEMA is designed geometrically to match the antenna output impedance to the nominal 50- Ω source impedance by decreasing the segment characteristic impedances along the antenna line [34]. The performance results of a TEM horn array (PDTEMA) are presented in Figs. 3.20 to 3.22.

3.4.6   Designs and Performance Results

Antenna designs suitable for wideband GPR operations are listed in Table 3.3. Their performance results are shown in Figs. 3.17 and 3.18. The plots in Fig. 3.17

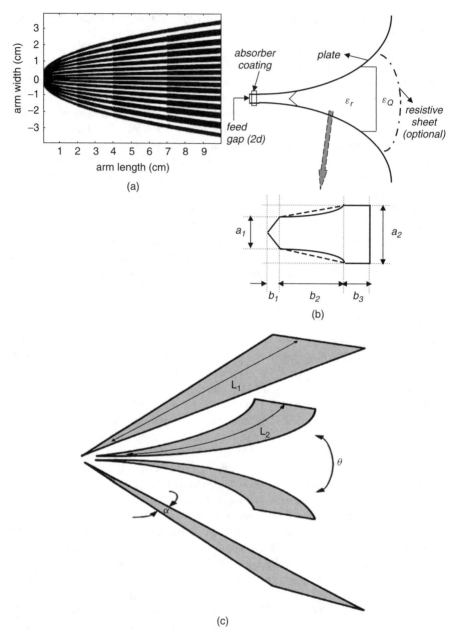

FIGURE 3.16 Multisensor adaptive TEM horn structures: (a) geometry of the antenna arm; (b) profile of dielectric loading; (c) array configuration.

TABLE 3.3 Description of GPR Antenna Models

Figure	Model	Physical Description
3.15c	Spiral	$r_1 = 0.5$ cm, $r_{2y} = 5.5$ cm, $r_{2d} = 7.3$ cm, $N = 4.5$ turns
3.15a	BT-10	$\alpha = 90°$, $\theta = 180°$, $d = 0.25$ cm, $L = 5$ cm
3.15b	TEM-10	$\alpha = 20°$, $\theta = 60°$, $d = 0.15$ cm, $L = 10$ cm, $\varepsilon_r = 1$ air-filled
3.15b	PDTEM-10	$\alpha = 20°$, $\theta = 60°$, $d = 0.15$ cm, $L = 10$ cm, $\varepsilon_r = 3$ dielectric loaded
3.16b	VA-10	$\alpha = 20°$, $\theta \in 0°$ to $160°$, $d = 0.4$ cm, $L = 10$ cm, $\varepsilon_r = 1$ air-filled
3.16b	PDVA-10	$\alpha = 20°$, $\theta \in 0°$ to $160°$, $d = 0.4$ cm, $L = 10$ cm, $\varepsilon_r = 3$ dielectric-loaded
3.16c	PDTEMA-45	$\alpha^1 = 20°$, $\theta^1 = 90°$, $d^1 = 0.25$ cm, $L^1 = 45$ cm, aperture: 10×15 cm $\alpha^2 = 20°$, $\theta^2 = f(l) \in 0°$ to $120°$, $d^2 = 0.2$ cm, $L^2 = 25$ cm, dielectric profile: $\varepsilon_r = 3.5$, $a_1 = 4$ cm, $a_2 = 13$ cm, $b_1 = 3$ cm, $b_2 = 9$ cm, $b_3 = 7$ cm, 5.5 cm thick

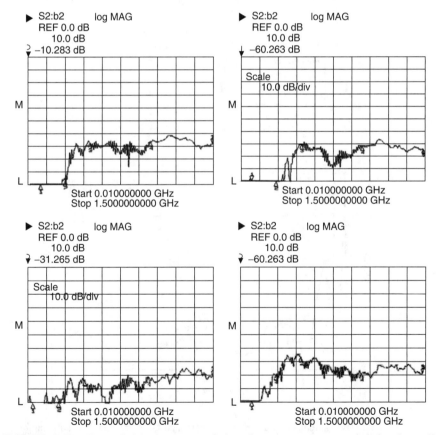

FIGURE 3.17 Transmission responses of the planar antennas measured for dry soil: (a) logarithmic spiral; (b) Archimedean spiral; (c) bowtie (BT-10): cross-polarization; (d) bowtie (BT-10): copolarization.

demonstrate the transmission performances of the planar GPR antennas, which are spirals and bowtie models, measured in soil for a 1.5-GHz operation band. It is clearly seen that bowtie and spiral antennas of similar size also have similar gain behaviors, except for cross-polarization.

The gain measurement results of bowtie, TEM horn, PDTEM, PDVA, and Vivaldi antennas (VAs) are given in Fig. 3.18a. The plots show that TEM10 and VA10 exhibit sufficient wideband characteristics from 500 MHz up to 6 GHz. However, the −3 and −10 dB antenna gain bands are limited in the high-frequency

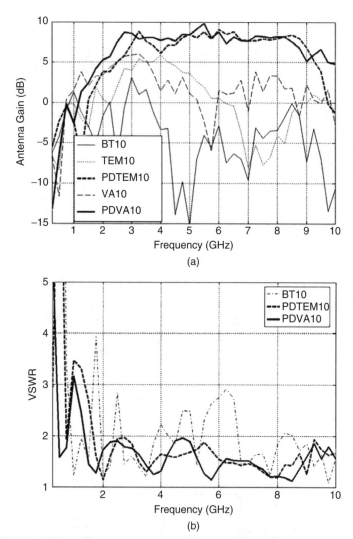

FIGURE 3.18   UWB antenna characteristics of designed GPR antennas: (a) antenna gains; (b) VSWR characteristics; (c) received impulse signal of PDTEM10.

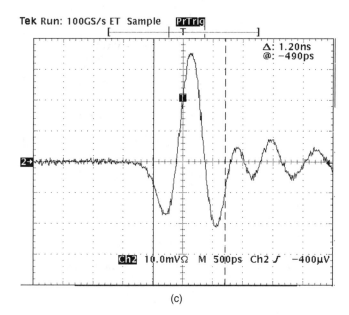

Tek Run: 100GS/s ET Sample   PriTrig

Δ: 1.20ns
@: −490ps

Ch2 10.0mVΩ   M 500ps Ch2 ∫ −400μV

(c)

FIGURE 3.18   (*Continued*)

region. Nevertheless, the gain band can be extended to higher frequencies using a partial dielectric loading technique. As a result, PDTEM10 and PDVA10 attain twofold broadened gain characteristics. Their −3 dB gain band is about 6 : 1 (1.5 to 9 GHz) and their −10 dB gain band is 25 : 1 (400 MHz to 10 GHz).

The VSWR is also a significantly critical parameter for GPR performance. The antenna input should be matched with the pulse generator to prevent corruption and ringing on the radiated signal and to avoid near-zone clutter. Otherwise, unwanted ringing signals can cause rigorous clutter for a GPR system, and the scattered signal coming from the subsurface object cannot be distinguished. Figure 3.18b shows that the input reflection levels of the PDTEM10 and PDVA10 are pretty satisfactory and a VSWR below 2 (band average 1.6) can be obtained over a 10 : 1 frequency band without aperture loading. When the antenna is put inside the GPR head, which acts as an absorber-filled cavity, the lower cutoff frequency of the VSWR band is extended to 400 MHz, and the PDVA10 can reach up to a 25 : 1 bandwidth ratio. Thus, an input impulse reflection level of about 0.15 can easily be achieved for a typical 3-GHz impulse GPR band. This level tends to 0.1 for full-band (10 GHz) transformation [33].

The TEM horn array model, which is illustrated in Fig. 3.16c and specified in Table 3.3, was measured from 50 MHz to 10 GHz (Fig 3.19) and compared with reference antennas. The gain measurement results for the PDTEMA-45, calibrated log-periodic array (LPA), and calibrated ridged horn (RH) antennas are plotted in Fig. 3.20. The LPA demonstrates relatively narrow bandpass filter-like behaviors similar to those of typical TEM horns. The PDTEM horn and PDVA designs can reach extended bandwidth ratios up to 25 : 1 as a result of the remarkable

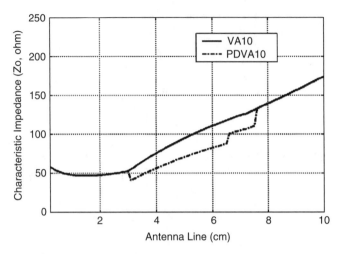

FIGURE 3.19 Characteristic impedance matching of TEM horns along the antenna line.

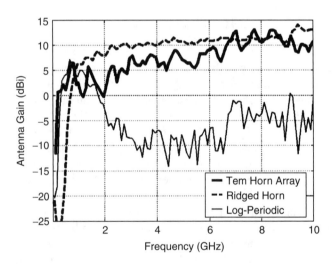

FIGURE 3.20 Antenna gain of a TEM horn array.

gain improvements at higher frequencies achieved by partial dielectric loading techniques [32,33]. A typical gain band of the RH is 20:1, similar that of loaded TEM horns. However, the PDTEMA attains at least twice that of broadened UWB gain characteristics. Its −10 dB gain band is more than 50:1 (150 MHz to 10 GHz), as shown in Fig. 3.20. The lower-band (1 GHz) and full-band (10 GHz) IFFT plots of impulses received are presented in Fig. 3.21. It is obvious that the PDTEMA can perform wideband radiation of log-periodic and ridged-horn antennas individually at similar physical sizes.

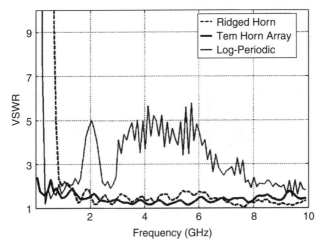

FIGURE 3.21 Input reflection characteristic of a TEM horn array.

The input reflection characteristic of the PDTEMA is rather good and a VSWR value below 2 with an average value of 1.5 is clearly attainable over the 50:1 frequency band, as shown in Fig. 3.22. Thus, input impulse reflection levels of about 0.15 can easily be achieved for 1-GHz subband and 10-GHz full-band transformations [34].

3.4.7 Analysis of Coupling Effects

The GPR system is affected explicitly by clutter on the receiver antenna induced by T/R coupling fields. In this situation, proper RF shielding enclosures should be designed to prevent coupling between T/R pairs and different receiver antennas to the extent possible. The electromagnetic (EM) coupling simulation of T/R antennas shown in Fig. 3.23 basically demonstrates how EM couplings can be so

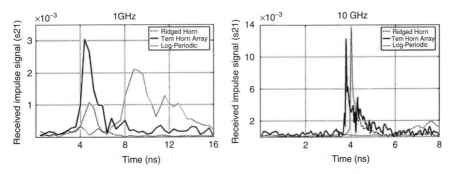

FIGURE 3.22 Impulse signals received for 1- and 10-GHz bands.

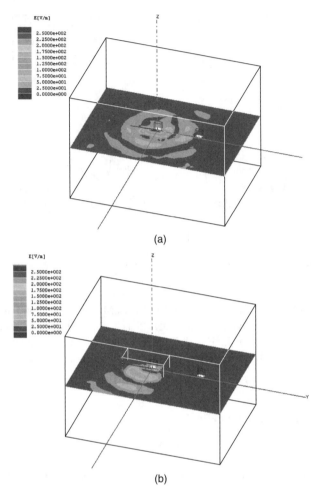

FIGURE 3.23 Field distribution at the aperture of a TEM horn antenna ($f = 750$ MHz): (a) without shielding; (b) with shielding.

effective and consequently be so problematic for GPR systems, since they are usually the main source of clutter, which strongly degrades the target detection and identification performance of a system.

3.5 SIGNAL-PROCESSING TECHNIQUES

In this section we introduce GPR data-processing techniques. The general objectives of these techniques are to improve either the performance of a detection algorithm or data interpretation for visual analysis. The latter is frequently involved in the algorithm development phase, as one makes a visual analysis of sensor data to see how much processing is necessary. For GPR, interpretation of the raw data is difficult even for a specialist. Improving the interpretation by processing raw GPR data is therefore an important step toward achieving the objectives. This preprocessing step is designed to accomplish one or more of the following objectives:

- Improving data interpretation
- Increasing detection performance
- Providing accurate spatial and depth interpretation

Many advanced signal-processing techniques have been developed for these objectives. Some of the commonly used approaches are summarized below to give the reader a quick tool for preprocessing the raw GPR data. Exploiting a priori knowledge about the objects to be detected and specifications of the sensor usually help to design this step. For example, depending on the size, shape, and depth of the objects to be detected, an adaquate signal-processing technique needs to be utilized. In other words, there is no unique scheme for data processing. Most commonly used preprocessing methods need to be customized for special cases, such as the purpose of the measurements, the GPR system itself, and measurement conditions. All these factors make the GPR sensor interesting to researchers, and it therefore remains an active area of research.

GPR sensors produce one-, two-, or three-dimensional data. GPR sensors get a downward view into the ground. Each antenna generates time- or frequency-domain data in vector form containing the response of the ground to GPR excitation. In the case of frequency-domain radar systems, the data are usually transformed to the time-domain equivalent. A horizontally arranged array of these sensors gets slices of downward views into the ground. A stack of these vertical slices is obtained by surveying a lane (i.e., scanning a lane by moving the sensors in a certain direction). This collection procedure forms a three-dimensional volume of data sets [35]. When only a single sensor is involved, two-dimensional data sets are formed. A three-dimensional data formation is illustrated in Fig. 3.24. The signal collected presents

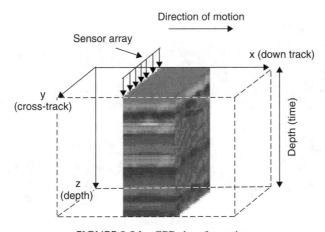

FIGURE 3.24 GPR data formation.

signal strength vs. time delay. The direction x is the down track, representing the scans along the travel direction of the sensor or array of sensors. The direction y is the cross track (along the antenna array). The horizontal axis z represents the time and is approximately related to depth.

For the preprocessing algorithms introduced below, unless otherwise noted, we use the vector notation $\mathbf{x}$ to denote a signal received at location i along the down track and j along the cross track. This one-dimensional signal of length M is called an *A-scan signal*. It can be represented mathematically as

$$\mathbf{x} = f(x, y, z)|_{x=i, y=j} \tag{3.40}$$

where f is the signal measured at position (x,y,z) and z varies from 1 to M.

The ensemble of A-scans in the down track forms two-dimensional matrices called *B-scan signals*, which can be represented as

$$\mathbf{X} = f(x, y, z)|_{y=j} \tag{3.41}$$

where x varies from 1 to N and z varies from 1 to M, or

$$\mathbf{X} = f(x, y, z)|_{x=i} \tag{3.42}$$

where y varies from 1 to K and z varies from 1 to M. The resulting images are of size $M \times N$ and $M \times K$, respectively.

A *C-scan signal* consists of three-dimensional data, as shown in Fig. 3.24 and is frequently represented by horizontal slices, each slice corresponding to a certain depth. The slices can be represented as

$$\mathbf{Y} = f(x, y, z)|_{z=k} \tag{3.43}$$

where x varies from 1 to N and y varies from 1 to K. Figure 3.25 shows the GPR returns in the form of B- and C-scan slices of the same object.

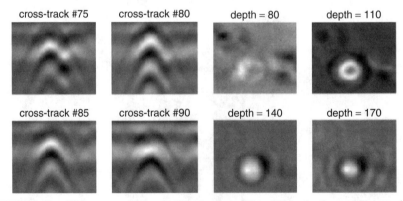

FIGURE 3.25 GPR returns with a depth index from 80 to 170 in the form of B-scan slices (the first two columns) and C-scan slices (the last two columns) of the same object.

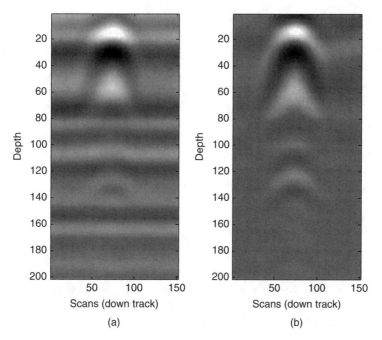

FIGURE 3.26 Raw GPR data; (b) after removing the average along the scan direction.

3.5.1 Background Modeling and Removal

GPR data sets collected in the field require some data manipulation to obtain more usable forms of information. Background subtraction is one such manipulation of the data. Background subtraction causes subsurface features of interest to be captured more easily. It is often used to reduce the effects of the air-to-ground reflection, which appears in the form of dominant horizontal lines, as shown in Fig. 3.26a. The strong reflection is due to high contrast between the dielectric constant of the ground and air.

The air–ground interface is a significant source of high-energy interference. This response is the main component of clutter noise. It dominates the data and obscures the target information. The responses of objects buried at shallow depths are strongly affected by the clutter noise. It is a potential source of false alarms for anomaly detection algorithms [36]. In this section we describe commonly used approaches to suppressing clutter noise.

Spatial Filtering Spatial filtering is a technique commonly used to model the background. To estimate the background, an average of the GPR data along the scan direction is computed. The resulting background estimation is strongly affected by high-energy anomalies due to the presence of buried objects. To avoid such problems, instead of taking all scans into account, only scans from the blank areas may be used to improve the quality of the estimate [37]. The background estimated is then subtracted from the raw GPR data.

Let x be a one-dimensional signal

$$x(i) = f(x, y, z)|_{y=j, z=k, x=i} \tag{3.44}$$

$i = 1, 2, \ldots, N$, where f is the GPR return measured at position (x,y,z), x varies from 1 to N, and z and y are fixed at depth k and cross-track position j, respectively. Assume that the soil content is homogeneous. Assume also that the ground surface is not coarse and thus the horizontal bands in the data remain evenly along the scan direction. These assumptions lead to a model with a constant signal corrupted by independent, identically distributed noise. Assuming further that the noise is additive, the measured signal takes the form

$$x(i) = f(x, y, z)|_{x=i, y=j, z=k} = \theta + N(i) \tag{3.45}$$

$i = 1, 2, \ldots, N$, where θ is constant for depth k and $N(i)$ is the measurement error for the ith sample. If the measurement noise over time is modeled by a zero-mean normal distribution $N(0, \sigma^2)$, the signal observed has mean θ and variance σ^2. Employing a finite number of observations, the sample mean defined in Eq. (3.46) is the maximum likelihood estimator for additive Gaussian noise:

$$\hat{\theta} = \frac{1}{N} \sum_{i=1}^{N} x(i) \tag{3.46}$$

Notice that $\hat{\theta}$ is an estimation of the horizontal band at depth k. It is then subtracted from the measured signal to enhance the raw data:

$$\tilde{x}(i) = x(i) - \hat{\theta} \tag{3.47}$$

$i = 1, 2, \ldots, N$, where $\tilde{x}(i)$ is the result of spatial filtering at depth k and cross-track position j.

Figure 3.26a shows raw GPR data. To achieve the best performance in estimating the background, samples from the blank area (i.e., with no objects) are used. The first 20 scans are adequate for this purpose. The vector $\hat{\theta} = [\hat{\theta}(1)\ \hat{\theta}(2)\ \cdots\ \hat{\theta}(L)]^T$, where $\hat{\theta}(k)$ is an estimation of the horizontal bands at depth k, is calculated and subtracted from each A-scan. The result of the background subtraction for these data is shown in Fig. 3.26b. Notice that due to the domination of the horizontal bands, the raw data do not reflect the features of interest at lower depths. The background subtraction, shown in in Fig. 3.26b, helps to bring up the features of interest to be captured more easily.

The method described above is based on consistency along the scan direction. If the soil content is inhomogeneous and/or the ground surface is coarse, the approach described above may fail and thus it may not effectively remove the effect of the air-to-ground reflection. One approach to overcoming this problem is to estimate the background adaptively [38]. For adaptive estimation of the background, the moving average of the scans is subtracted from the raw data:

$$\tilde{x}(i) = x(i) - \frac{1}{S} \sum_{s=i-S}^{i-1} f(x, y, z)|_{x=s, y=j, z=k} \tag{3.48}$$

$i = 1, 2, \ldots, N, k = 1, 2, \ldots, M$, where $S \geq 1$ and its size determines the length of the window for a moving mean.

Remember that we modeled the measurement noise using a zero-mean normal distribution $N(0, \sigma^2)$. If the observations $x(1), x(2), \ldots, x(N)$ form a random sample from a Laplace distribution with mean $\theta(k)$ and variance $2/\sigma^2$, the maximum likelihood estimate for $\theta(k)$ is the median of the samples. The background subtraction is then applied as follows:

$$\tilde{x}(i) = x(i) - \text{median} \left[x(i-S), x(i-S+1), \ldots, x(i-1) \right] \tag{3.49}$$

$i = 1, 2, \ldots, N, k = 1, 2, \ldots, M$, where $s \geq 1$ and its size determines the length of the window for a moving median.

A generalization of this approach is to use ordered weighted-average (OWA) filters. OWA operators can be applied to enhance the GPR returns as follows:

$$\tilde{x}(i) = x(i) - \sum_{h=-s}^{s} w_h x_{(h)} \tag{3.50}$$

where $\sum_{h=-s}^{s} w_h = 1$ and $x_{(h)}$ is the hth largest of the vector $[x(i-s)\, x(i-s+1)$ $\cdots\, x(i+s)]^{\mathrm{T}}$.

As stated earlier, the background estimation by spatial filtering can be carried out effectively when the medium is homogeneous. However, the presence of objects and variations in ground homogeneity can cause inaccurate background estimation. This results in distortion of the target's response. This is especially serious for small objects and those buried shallowly. For such cases, advanced signal-processing techniques such as PCA and ICA are utilized.

Principal Component Analysis Principal component analysis (PCA) is a technique often used to reduce multidimensional data sets. PCA involves calculation of the eigenvalue decomposition of a data matrix. This operation is often interpreted as revealing the internal structure of data and is used to remove outliers. It is for this reason that PCA is a technique commonly used to estimate the background of GPR data [39,40].

Consider an $M \times 1$ time-domain signal $\mathbf{x}_i$ received at location i along the down track. We can treat the vectors $\mathbf{x}_i$ as random quantities. The mean vector of the population is defined as

$$\mathbf{m_x} = E\{\mathbf{x}\} \tag{3.51}$$

where $E\{\cdot\}$ is the expectation operator (or expected value operator).

The covariance matrix $\mathbf{C_x}$ of the vector population $\mathbf{x}$ is a matrix of order $M \times M$ and is defined as

$$\mathbf{C_x} = E\{(\mathbf{x} - \mathbf{m_x})(\mathbf{x} - \mathbf{m_x})^{\mathrm{T}}\} \tag{3.52}$$

Having N observations, $i = 1, 2, \ldots, N$, one can form a data matrix $\mathbf{X} = [\mathbf{x}_1 \; \mathbf{x}_2 \; \cdots \; \mathbf{x}_N]$ of size $M \times N$. The mean vector $\mathbf{m_x}$ and the covariance matrix $\mathbf{C_x}$ can be approximated from the samples using

$$\mathbf{m_x} = \frac{1}{N} \sum_{i=1}^{N} \mathbf{x}_i \tag{3.53}$$

and

$$C_{\mathbf{x}} = \frac{1}{N-1} \sum_{i=1}^{N} (\mathbf{x}_i - \mathbf{m_x})(\mathbf{x}_i - \mathbf{m_x})^{\mathrm{T}} \tag{3.54}$$

Let $\mathbf{v}_i$ and λ_i, $i = 1, 2, \ldots, M$, be the eigenvectors and corresponding eigenvalues of the covariance matrix $\mathbf{C_x}$. Arrange the eigenvalues in descending order so that $\lambda_{(1)} \geq \lambda_{(2)} \geq \cdots \geq \lambda_{(M)}$. Let $\mathbf{v}_{(i)}$ be the eigenvector associated with the eigenvalue $\lambda_{(i)}$. The covariance matrix can be expressed as the sum of a signal covariance matrix and a noise covariance matrix

$$\mathbf{C_x} = \mathbf{S_x} + \mathbf{W_x} \tag{3.55}$$

in which $\mathbf{S_x} = \sum_{i=1}^{p} \lambda_i \mathbf{v}_{(i)} \mathbf{v}_{(i)}^{\mathrm{T}}$ and $\mathbf{W_x} = \sum_{i=p+1}^{M} \lambda_i \mathbf{v}_{(i)} \mathbf{v}_{(i)}^{\mathrm{T}}$.

The eigenvectors $\mathbf{v}_{(1)}, \ldots, \mathbf{v}_{(M)}$ are known as the *principal eigenvectors*. Retaining only the information in the signal subspace eigenvectors, that is, forming a lower-rank approximation to $\mathbf{C_x}$, effectively enhances the signal-to-noise ratio. Let $\mathbf{V}_1$ and $\mathbf{V}_2$ be matrices whose columns are formed from the eigenvectors $\mathbf{v}_{(i)}$ as

$$\mathbf{V}_1 = \lfloor \mathbf{v}_{(1)}, \mathbf{v}_{(2)}, \ldots, \mathbf{v}_{(p)} \rfloor \quad \text{and} \quad \mathbf{V}_2 = \lfloor \mathbf{v}_{(p+1)}, \mathbf{v}_{(p+2)}, \ldots, \mathbf{v}_{(M)} \rfloor$$

where $p < M$. Then $\mathbf{x}$ can be decomposed into the sum of three components:

$$\mathbf{x} = \mathbf{V}_1 \mathbf{V}_1^{\mathrm{T}}(\mathbf{x} - \mathbf{m}_x) + \mathbf{V}_2 \mathbf{V}_2^{\mathrm{T}}(\mathbf{x} - \mathbf{m}_x) + \mathbf{m}_x \tag{3.56}$$

The first term is due to the artifacts caused by buried objects. The rest are due to the background and noise. Then the visual quality of the GPR data can be achieved by removing the second and third terms. To decide on the value of p, the percentage of the energy to be preserved is calculated. For example, to preserve $\alpha_0 = 95\%$ of the energy of the signal, we choose the smallest p that satisfies the following:

$$\alpha = \frac{\sum_{i=1}^{p} \lambda_i}{\sum_{i=1}^{M} \lambda_i} \geq 0.95$$

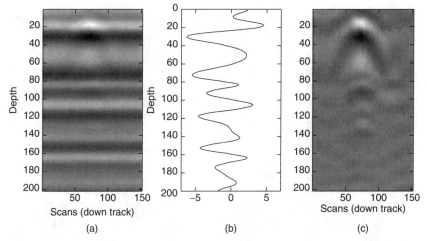

FIGURE 3.27   Raw GPR data; (b) corresponding mean vector; (c) GPR data after removal of the mean vector.

Figure 3.27a shows a raw GPR data. The corresponding mean vector $\mathbf{m_x}$ is shown in Fig. 3.27b. After subtracting the last term, $\mathbf{m_x}$, the raw GPR data reduce to $\mathbf{g} = \mathbf{x} - \mathbf{m_x}$ and are shown in Fig. 3.27c. The buried object is now visible. The second term, $\mathbf{V_2 V_2^T}(\mathbf{x} - \mathbf{m_x})$, is shown in Fig. 3.28 for $p = 1, 2, 5$, and 13. For $p = 1$ and 2, the artifacts due to the buried object are evident, as shown in the figure. On the other hand, the resulting image for $p = 5$ shows no evidence of the buried object. Therefore, $p \geq 5$ is a satisfactory choice to enhance the signature of the object.

Figure 3.29 shows the GPR signal after removing the sum of the second and third terms, which is equivalent to $\mathbf{V_1 V_1^T}(\mathbf{x} - \mathbf{m_x})$. Notice that the artifacts due to the buried object are corrupted for $p = 1$ and 2. These two values for p cause the elimination of about 50% of the energy in the signal. For $p = 5$ and 13, the results are similar and are close to the ideal. In the case of smaller objects, on the other hand, one would probably choose a larger value for α_0 to preserve the artifacts due to the buried objects.

3.5.2 Zero Offset Removal

Most signal-processing algorithms applied to GPR data are not invariant to dc offsets in the signal. Their existence in the signal may cause misinterpretation of the results and therefore should be removed. The zero-offset removal consists of estimating the mean of an A-scan and subtracting the estimated mean from the samples of this A-scan. It is well known that when a signal x has a Gaussian density function, the mean estimator

$$\hat{m} = \frac{1}{N} \sum_{i=1}^{N} x(i) \tag{3.57}$$

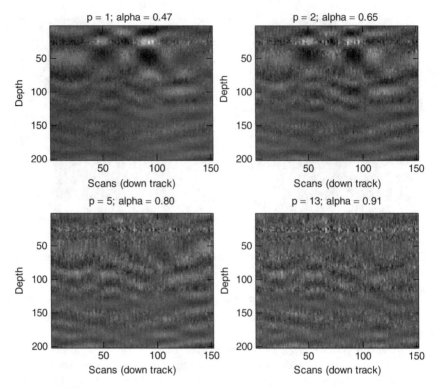

FIGURE 3.28   Estimated backgrounds for $p = 1, 2, 5$, and 13, respectively.

is optimal in the maximum likelihood sense. Even if the density function of the signal is not Gaussian but its density is symmetrical with respect to the mean, the same estimator provides a reasonable estimate to remove the dc offset [41].

3.5.3   Time-Varying Signal Gain

Radar signals are attenuated rapidly as they propagate into the ground. Signals at greater depths are weaker then those of shallower depths. Amplitudes at different depths can be equalized by applying a time-dependent gain function. Figure 3.30 shows a sample A-scan and a time gain function to compensate for losses at greater depths.

3.5.4   Ground-Bounce Localization

The air–ground interface is a significant source of high-energy interference. To reduce the effects of the air-to-ground reflection, background subtraction is used. However, if the ground surface is course, it is difficult to maintain the radar antenna at a fixed distance above the ground [42]. Then, estimating the background is a major challenge. In such cases it is necessary to adjust the position of the entire profile in the data window vertically before applying background subtraction.

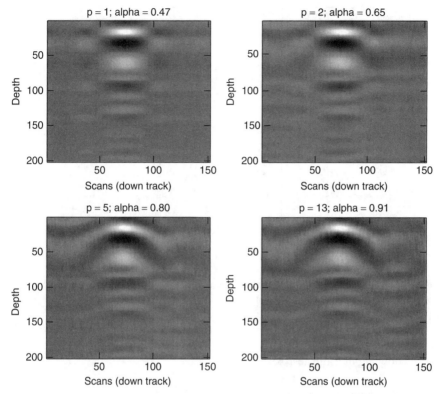

FIGURE 3.29 GPR signal after applying the background subtraction for $p = 1, 2, 5$, and 13, respectively.

Various methods are proposed for locating the ground bounce. The first negative and first positive peaks are considered a close approximation of the ground surface. They are the most common approaches to locating the ground bounce. The first break position, zero-amplitude point, and midamplitude point are among the other approaches [43]. Once the ground bounce is located, the early time samples of each signal, up to a few samples beyond the ground bounce, are discarded. Only data corresponding to regions below the ground surface are processed.

Consider the $M \times 1$ time-domain signal $\mathbf{x}_i$ received at location i along the down track. Assume that we identify the ground level as the signal's first positive peak. The first positive peak can be found as follows:

$$m^* = \frac{\arg \max \mathbf{x}_i(m)}{m} \qquad (3.58)$$

To align the position of the entire profile in the data window vertically, the GPR signal at location i along the down track is modified as follows:

$$\mathbf{s}_i(m) = \mathbf{x}_i(m + m^*) \qquad \text{for } m = 1, \ldots, M - m^* \qquad (3.59)$$

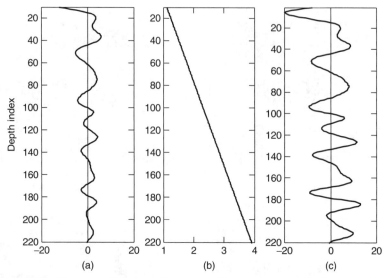

FIGURE 3.30 Concept of time-varying gain: (a) raw A-scan; (b) time-gain function; (c) result of applying time-varying gaining.

Experimental results indicate that using a calibrated time value in advance of the first positive peak of the direct wave gives the most consistently accurate results for shallow depth estimates [43].

3.5.5 Frequency Filtering

Both low- and high-frequency noise exist in GPR returns. Antenna–ground inter-actions are the main source of low-frequency noise. The high-frequency noise on a GPR profile is produced mainly by the communication equipment of the radar system. Both can be removed by using a bandpass filter. Another option is to use a lowpass and a bandpass filter in a cascade configuration. An important point to take into account is that the filter designed should exhibit a minimum phase response. A linear phase characteristic is necessary to preserve the shape of a given signal within the passband.

Infinite impulse response (IIR) filters cannot produce linear-phase characteristics. Only special forms of causal finite impulse response (FIR) filters can produce a linear phase [44]. For the filter to exhibit a minimum phase response, the impulse response of the filter is chosen symmetric about its midpoint. The choices of settings of bandwidths, slope, and so on, have effects on the size of the window. This step involves a trade-off: choosing the size of the window large enough to minimize smearing, yet small enough to allow reasonable implementation.

3.5.6 Wiener Filter

In this section we model clutter as a degradation process and apply reverse filtering to reduce clutter noise. Consider a degradation process modeled as a degradation function that, together with an additive noise term $\eta\,(n)$, operates on an input $x(n)$ to produce a degraded output $y(n)$. Let $x(n)$ and $y(n)$ be arbitrary, zero-mean, random sequences with size M. If we assume that $y(n)$ is the output of a deterministic linear system with impulse response $h(n)$, then

$$y(n) = \sum_{k} h(k)x(n-k) + \eta(n) \tag{3.60}$$

In the frequency domain, this model can be expressed as follows:

$$Y(u) = H(u)X(u) + N(u) \tag{3.61}$$

where Y, H, X, and N are the Fourier transforms of the output signal y, the degradation function h, the input signal x, and the additive noise $\eta(n)$, respectively. For the sake of discussion, this model assumes position invariance. The objective is to obtain an estimate $\hat{x}(n)$ of the original signal given $y(n)$ and some knowledge about H and $\eta(n)$, as shown in Fig. 3.31. We assume that the signal and the noise are uncorrelated.

The mean-square error between the original signal $x(n)$ and an estimate $\hat{x}(n)$ of the original signal is given by

$$e^2 = E\{(x - \hat{x})^2\} \tag{3.62}$$

where $E\{\cdot\}$ is the expected value operator. The minimum of the error function is given in the frequency domain by the expression

$$\hat{X}(u) = \frac{H^*(u)}{H^*(u)H(u) + S_\eta(u)/S_x(u)} Y(u) \tag{3.63}$$

where $H^*(u)$ is the complex conjugate of the degradation function $H(u)$, and $S_x(u)$ and $S_\eta(u)$ are the power spectra of the original signal and the noise, respectively. This result is known as the *Wiener filter* [44] and is commonly referred to as the *least-squares error filter* [45].

The power spectra of the original signal and the noise are seldom known. An approach used frequently when these quantities are not known or cannot be estimated is to approximate the term $S_\eta(u)/S_x(u)$ by a constant $1/K$. When the signal is very strong relative to the noise, this term goes to zero and the Wiener filter becomes the inverse filter for the degradation process.

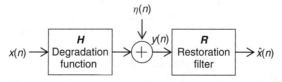

FIGURE 3.31 Model of clutter as a degradation function and the restoration process.

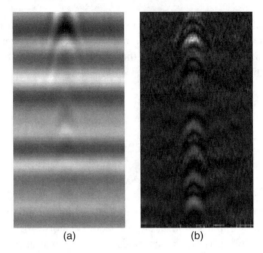

(a)   (b)

FIGURE 3.32   Effects of a Wiener filter: (a) raw GPR data; (b) result of the filter.

The data collected from the blank areas can be used to estimate the degradation process. This estimate stays accurate when the medium is homogeneous. Consider the raw GPR data shown in Fig. 3.32d. The Wiener filter applied to A-scans is shown in Fig. 3.32b. Notice that the horizontal bands are effectively removed in the blank areas. The resulting image is not attractive in terms of visual quality. However, when used as an input, it increases the performance anomaly detection algorithms.

3.6   IMAGING ALGORITHMS

3.6.1   The Problem of Focusing (Migrating)

Many subsurface sensing researchers across different disciplines have applied miscellaneous imaging algorithms in various fields, from mine detection to geophysics [1,46–48]. The usual ground-penetrating radar (GPR) system collects the reflectivity of the ground and that beneath objects when the radar is moving on top of the ground. A typical GPR image yields information as to the spatial position and reflectivity of a buried object. For the monostatic arrangement of the radar sensor, a single-point scatterer appears as a hyperbola in the space–time GPR image when the radar sensor moves over the surface, as demonstrated in Fig. 3.33. This type of hyperbolic image is sometimes adequate if the main goal is just to detect a pipe or similar object. However, size, depth, and electromagnetic (EM) reflectivity information regarding the buried object are also crucial in most GPR applications. If this is case, the hyperbolic diffraction (or dispersion) in the space–time GPR image should be transformed to a focused image that shows the object's true location and size together with its reflectivity. The process of reversing the hyperbolic diffraction or any other type of dispersion in getting a focused image is often called *migration*

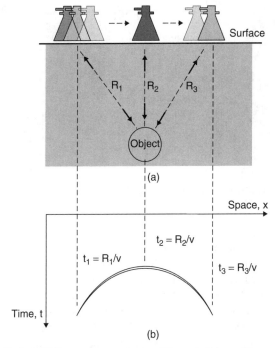

FIGURE 3.33 Typical GPR measurement setup (B-scan); (b) resulting space–time image that contains the well-known hyperbolic aliasing.

(or *focusing*). For this ultimate goal, many algorithms have been adopted by various researchers [49–62]. The Kirchhoff wave equation [49] and frequency- and wavenumber ($\omega - k$) -based [50,51] migration techniques are widely accepted and applied. The similarities between the acoustic and electromagnetic wave equations have led to the use of the same processing techniques for GPR image processing as in the case of acoustic imaging [51–55]. Among these, the wavenumber-domain focusing techniques were developed originally in seismic imaging applications [51] and have been widely implemented and adapted to modern synthetic aperture radar (SAR) imaging [56–59]. The wavenumber-domain algorithms have been developed by various groups in the SAR community, and called by different names, including seismic migration [56,57] and $\omega - k$ (or $f - k$) migration [60–62]. In this section, the fundamental and most significant focusing algorithms—hyperbolic summation, Kirchhoff migration, phase-shift migration, $\omega - k$ (Stolt) migration, and SAR-based focusing—are covered briefly, and some other techniques are mentioned as well.

3.6.2 Exploding Source Model

Before going into the details of various popular migration methods, we explain briefly the *exploding source model* (ESM), which is used widely in most of these algorithms. Claerbout [63] claimed this clever idea: of thinking as if the scattered

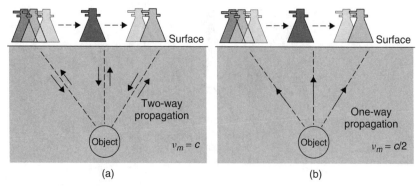

FIGURE 3.34 Geometry for (a) B-scan GPR data collection scheme; (b) utilizing the exploding source model.

field collected at the radar transmitter originates from the sources at the targets. A reference time of $t = 0$ is selected for convenience when the sources "explode" and send EM waves to the sensors at the surface. Implementation of the model is illustrated in Fig. 3.34. The real data collection scheme is shown in Fig. 3.34a, where, in fact, two-way propagation between the radar and the object takes place. In the exploding source model; however, the scattered wave is assumed originally to be radiated from a source on the scatterer. Therefore, one-way propagation is assumed from the scatterer to the radar sensor as depicted in Fig. 3.34b. Since the travel time in this case would be half of the original problem, compensation should be made for the velocity of the EM wave by dividing it by 2. Therefore, the velocity of the propagation in the medium in ESM is taken as $v_m = c/2$.

Migration using ESM is basically realized by employing the following two procedures:

1. Extrapolation of the received signal back to exploding source points.
2. Realizing the migrated image by displaying the backward-extrapolated EM wave at $t = 0$. Therefore, the main idea is to get a focused image of the wave at the scatterer at the instant of exploding.

3.6.3 Hyperbolic (Diffraction) Summation

For GPR problems, the radar collects the scattered or backscattered EM signal from subsurface objects together with many cluttering effects, caused mainly by the air-to-ground interface and inhomogeneities within the ground. The phase of the signal received is proportional to the trip distance that the EM wave possesses for homogeneous mediums. Therefore, for the monostatic configurations, the backscattered signal from a single point scatterer experiences different round-trip distances as the radar moves over the surface. For each spatial point, the frequency diversity of the backscattered signal can be used to get a one-dimensional range profile by taking the inverse Fourier transform (IFT) of the frequency-diverse data. Putting all range profiles side by side produces a two-dimensional B-scan GPR image in the space–time (or space–depth) domain. As explained earlier, a single point scatterer shows up as a parabolic hyperbola in the space–time GPR image due to different trip distances as the radar scans over the ground. The real object is, in fact, located at the apex of this hyperbola.

For a point scatterer located at (x_0, z_0), the parabolic hyperbola in the GPR image is characterized by the following equation when the radar is moving on a straight path along the X-axis.

$$r = \sqrt{z_0^2 + (X - x_0)^2} \tag{3.64}$$

Here X represents the synthetic aperture vector and r gives the depth of the hyperbola. Assuming that a B-scan GPR image is obtained by the summation of a finite number of hyperbolas that corresponds to different points on the object(s) below the surface, the following procedure can be applied to focus the defocused image [64]:

1. For each pixel point (x_i, z_i) in the two-dimensional original B-scan GPR image, find the corresponding hyperbolic template using Eq. (3.64) and trace the pixels under this template.

2. Record the image data for the pixels under this template. This step of the procedure provides one-dimensional field data, E_s, whose length, N, is the same as the total number of sampling points in X.

3. Then take the root-mean-square (rms) value of the total energy contained in these one-dimensional complex data as follows:

$$\{\text{rms at } (x_i, z_i)\} = \sqrt{\frac{|E_s|^2 |E_s^*|^2}{N}} \frac{1}{\sqrt{N}} \sum_{i=1}^{N} |E_{s,i}|^2 \tag{3.65}$$

Here the summation runs over the elements of vector $|E_s|^2$.

4. The rms value calculated is recorded in the new GPR image at the point (x_i, z_i). This procedure is repeated for all pixels in the original GPR image.

The hyperbolic summation (HS), also known as *diffraction summation*, is illustrated through the example shown in Fig. 3.33. In this example, the measured B-scan subsurface backscattered electric field data are collected by the help of a stepped-frequency continuous-wave radar (SFCWR) setup. The data measured are taken from a subsurface scene where a thick iron pipe 16.5 cm in diameter and 47 cm in length was buried flat 15 cm below the ground. The ground medium is full of dry and homogeneous sand whose relative dielectric constant is about 2.4 for C-band frequencies. The classical B-scan GPR image is obtained by Fourier-transforming the frequency-domain backscattered field as shown in Fig. 3.35a. The reflection from the ground surface can be seen in the B-scan GPR image around $z = 0$ m for the entire synthetic aperture along X. As is obvious from the figure, the image contains the hyperbolic image pattern. After applying the HS methodology explained above, the focused GPR image is obtained as depicted in Fig. 3.35b.

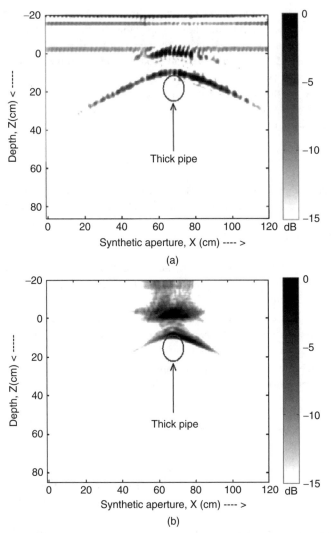

FIGURE 3.35   Measured B-scan GPR images for the monostatic case: (a) hyperbolic image of a buried pipe; (b) focused image after applying the method proposed.

3.6.4 Kirchhoff Migration

Kirchhoff migration or *reverse-time wave-equation migration* is equivalent to a hyperbolic summation method with some correcting factors included in the solution [65]. In fact, Kirchhoff migration is the solution of the following scalar wave equation for the wave function $\varphi(x, z, t)$ in the propagating medium:

$$\left(\frac{\partial^2}{\partial x^2} + \frac{\partial^2}{\partial z^2} - \frac{1}{v_m^2} \frac{\partial^2}{\partial t^2} \right) \varphi(x, z, t) = 0 \tag{3.66}$$

Here v_m is the velocity inside the propagating medium and equal to $c/2$ by utilizing the "exploding source" concept. The differential equation above can be solved by using the *Kirchhoff integral theorem* [66]. The solution for the far-field approximation is the following:

$$P(x, z, t) = \frac{1}{2\pi} = \int_{-\infty}^{\infty} \left[\frac{\cos\theta}{v_m R} \frac{\partial}{\partial t} P\left(x, z, t - \frac{r}{v_m}\right) \right] dx \qquad (3.67)$$

In this equation, θ is the angle of the incident wave to the depth axis (z) and $r = [(x - x_m)^2 + z^2]^{1/2}$ is the path from the target point at (x, z) to the observation point at $(x_m, 0)$.

In the Kirchhoff migration method, the following correction factors are considered, in contrast to hyperbolic summation:

1. Compensation for spherical spreading is taken into account. A $1/\sqrt{v_m r}$ correction factor for two-dimensional propagation and a $1/v_m r$ correction factor for three-dimensional propagation are included.

2. The directivity factor $\cos\theta$ is also considered. This factor corrects the diffraction amplitudes.

3. The phases and amplitudes are also corrected. The phase is corrected by $\pi/2$ and $\pi/4$ for two- and three-dimensional propagation cases, respectively. The amplitude of the wave is corrected proportionately to the square of the frequency for the two-dimensional propagation case and to the frequency for the three-dimensional propagation case. See the literature for further details on Kirchhoff migration [66–68].

3.6.5 Phase-Shift Migration

The phase-shift migration (PSM) method was introduced by Gazdag [50]. This method utilizes the ESM concept as well [69]. In brief, the algorithm iteratively performs a phase shift to migrate the wave field to an exploding time of $t = 0$.

The goal in the PSM algorithm is also to evaluate the wave field at $t = 0$ by extrapolating the downward (z-directed) EM wave with a phase factor of $\exp(jk_z z)$.

The PSM algorithm can be summarized briefly via the following steps:

1. A two-dimensional measured raw data set in the (ω, k_x) domain is multiplied by a phase-shift factor; C along the depth axis (or z-axis) is given as

$$C = e^{jk_z \Delta z} \qquad (3.68)$$

This factor can also be written in terms of wavenumber along the data-collecting axis (k_x) as

$$C = \exp(jk_z \Delta z) = \exp(j\sqrt{k^2 - k_x^2}\Delta z) = \exp\left(jk\sqrt{1 - \left(\frac{k_x}{k}\right)^2}\Delta z\right) \qquad (3.69)$$

Putting $k = \omega/v_m$ (where ω is the angular frequency and v_m is the speed of the wave inside the propagation medium) into Eq. (3.68), one can get

$$C = \exp\left(j\frac{\omega}{v_m}\Delta z\sqrt{1 - \left(\frac{v_m k_x}{\omega}\right)^2}\right) \tag{3.70}$$

In Eq. (3.70) the incremental depth parameter Δz is chosen as $\Delta z = v_m \Delta t$. Here Δt is the time-sampling interval of the input data. After this modification, the final form of the phase-shift factor is equal to

$$C = \exp\left(j\omega\Delta t\sqrt{1 - \left(\frac{v_m k_x}{\omega}\right)^2}\right) \tag{3.71}$$

The two-dimensional measured raw data set $E_s(\omega, k_x)$ is multiplied by C along the depth axis for the steps of Δt.

2. By utilizing the ESM concept, the imaging task is performed as taking the inverse Fourier transform (IFT) of $E_s(\omega, k_x)$ after selecting the time variable as $\Delta t = 0$. Therefore, only one FT operation is required at one point (when $\Delta t = 0$) for the focused image. After updating the factor C for every value of Δt and ω, we have the data in the three-dimensional (ω, k_x, k_z) domain as

$$E_s'(\omega, k_x, k_z) = CE_s(\omega, k_x) \tag{3.72}$$

The new data set, $E_s'(\omega, k_x, k_z)$, is summed up along the frequency axis and indexed for different values of Δt as

$$E_s(k_x, k_z, t) = \sum_w E_s'(\omega, k_x, k_z) \tag{3.73}$$

3. Setting $\Delta t = 0$ and taking the two-dimensional IFT along with respect to k_x and k_z, the focused image in the (x,z) domain can be calculated as

$$E_s(x, z) = \text{IFT}\{E_s(k_x, k_z)\} \tag{3.74}$$

3.6.6   Frequency–Wavenumber (Stolt) Migration

Frequency–wavenumber $\omega - k$ migration, also known as *Stolt migration*, is also based on the ESM concept and the scalar wave equation [51]. In terms of the computation time, it is faster than the focusing algorithms presented previously. The Stolt algorithm works well for constant-velocity propagation mediums. The solution to $\omega - k$ migration is in fact identical to that of Kirchhoff migration [62]. A brief explanation of the algorithm is presented below.

The algorithm begins with a three-dimensional scalar equation for a wavefunction $\varphi(x, y, z, t)$ inside a constant-velocity propagation medium:

$$\left(\frac{\partial^2}{\partial x^2} + \frac{\partial^2}{\partial y^2} + \frac{\partial^2}{\partial z^2} - \frac{1}{v_m^2} \frac{\partial^2}{\partial t^2} \right) \varphi(x, y, z, t) = 0 \tag{3.75}$$

In the Fourier space, the frequency and spatial wavenumbers obey the relationship

$$k_x^2 + k_y^2 + k_z^2 = k^2 = \frac{\omega^2}{v_m^2} \tag{3.76}$$

Stratton [69] demonstrated that any given wavefunction can be written as a summation of an infinite number of plane-wave functions $E(k_x, k_y, \omega)$ as

$$\varphi(x, y, z, t) = \left(\frac{1}{2\pi}\right)^{3/2} \int\!\!\!\int\!\!\!\int_{-\infty}^{\infty} E(k_x, k_y, \omega) e^{-j(k_x x + k_y y + k_z z - \omega t)} \, dk_x \, dk_y \, d\omega \tag{3.77}$$

For the GPR operation, of course, the field is assumed to be measured on the $z = 0$ plane. Therefore, Eq. (3.77) reduces to a Fourier transform pair, where $e(x,y,t)$ is the field measured on the z-axis:

$$\varphi(x, y, 0, t) \overset{\Delta}{=} e(x, y, t) = \left(\frac{1}{2\pi}\right)^{3/2} \int\!\!\!\int\!\!\!\int_{-\infty}^{\infty} E(k_x, k_y, \omega) e^{-j(k_x x + k_y y - \omega t)} \, dk_x \, dk_y \, d\omega \tag{3.78}$$

It is very important to note that Eq. (3.78) designates a three-dimensional forward Fourier transform relationship between $e(x, y, t)$ and $E(k_x, k_y, \omega)$ for the negative values of the time variable (t). Then the inverse Fourier transform can also be defined in the following way:

$$E(k_x, k_y, \omega) = \left(\frac{1}{2\pi}\right)^{3/2} \int\!\!\!\int\!\!\!\int_{-\infty}^{\infty} e(x, y, t) e^{j(k_x x + k_y y - \omega t)} \, dx \, dy \, dt \tag{3.79}$$

Afterward, we can use the ESM to focus the image by setting $t = 0$ in Eq. (3.78) and using

$$E(k_x, k_y, \omega) = e^{jk_z z} E(k_x, k_y, \omega, z = 0) \tag{3.80}$$

one can get

$$e(x, y, z, 0) = \left(\frac{1}{2\pi}\right)^{3/2} \int\!\!\!\int\!\!\!\int_{-\infty}^{\infty} E(k_x, k_y, \omega) e^{-j(k_x x + k_y y + k_z z)} \, dk_x \, dk_y \, d\omega \tag{3.81}$$

Equation (3.81) offers a focused image. However, to be able to use the advantages of the fast Fourier transform (FFT), the data in the (k_x, k_y, ω) domain should be transformed to the (k_x, k_y, k_z) domain. Therefore, a mapping procedure is required from the ω domain to the k_z domain. The relationship between ω and k_z and $d\omega$ and dk_z can easily be obtained from Eq. (3.76) as

$$\omega = v_m(k_x^2 + k_y^2 + k_z^2)^{1/2} \qquad (3.82a)$$

$$d\omega = \frac{v_m^2 k_z}{\omega} dk_z \qquad (3.82b)$$

Substituting these equations into Eq. (3.81), it is straightforward to obtain

$$e(x, y, z) = \left(\frac{1}{2\pi}\right)^{3/2} \int\int\int_{-\infty}^{\infty} \frac{v_m^2 k_z}{\omega} E_n(k_x, k_y, k_z) e^{-j(k_x x + k_y y + k_z z)} dk_x \, dk_y \, dk_z$$

$$(3.83)$$

Here $E_n(k_x, k_y, k_z)$ is the mapped version of the original data $E(k_x, k_y, \omega)$ from a three-dimensional (k_x, k_y, ω) grid to a (k_x, k_y, k_z) grid. To be able to use the FFT routine, an interpolation scheme should also be applied since the mapped data will not be on a linear grid after the transformation. Equation (3.83) proposes a very well focused subsurface image of the possible scatterers. The image can be obtained very rapidly thanks to the use of FFT. It is also notable to state that the mapped data are scaled by the factor of $v_m^2 k_z/\omega$. This scaling is sometimes called the *Jacobian transformation from* ω *to* k_z. Equation (3.83) is, of course, valid for the three-dimensional problem. In most subsurface imaging problems, however, the raw data are collected in two-dimensions. Therefore, Eq. (3.83) can easily be reduced to a two-dimensional problem in the space–depth domain:

$$e(x, z) = \left(\frac{1}{2\pi}\right) \int\int_{-\infty}^{\infty} \frac{v_m^2 k_z}{\omega} E_n(k_x, k_z) e^{-j(k_x x + k_z z)} dk_x \, dk_z \qquad (3.84)$$

3.6.7   SAR-Based Focusing

The geometrical similarities between the strip map synthetic aperture radar (SAR) problem and the B-scan GPR problem has led to the use of SAR-based focusing in subsurface imaging problems [60–62,70–72]. The illustration geometries for SAR and the B-scan GPR are shown in Fig. 3.36. Detailed formulation for strip map SAR imaging is given by Soumekh [71]. Dual algorithms for the GPR problem are presented by several researchers [62,70]. Here we present a $\omega - k$ migration domain SAR imaging algorithm based on the plane-wave decomposition of spherical wavefronts. The geometrical layout of a typical B-scan GPR problem is shown in Fig. 3.36b as the two-dimensional scattered electric field $E_s(x, \omega)$ is recorded

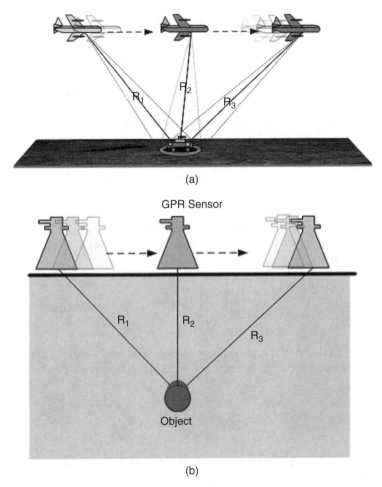

(a)

GPR Sensor

(b)

FIGURE 3.36 Geometrics for (a) an SAR problem and (b) a B-scan GPR problem.

for different synthetic aperture points and frequencies. Assuming that the propagation medium is homogeneous, the frequency-domain backscattered field from a point scatterer at distance r from the antenna will have the form

$$E_s(\omega) = \rho e^{-j2\omega r/v_m} \tag{3.85}$$

where $\omega = 2\pi f$ is the angular frequency, ρ the strength of the scattered field from the point target, and v_m the velocity of the wave. The number "2" in the exponential is taken into account for the two-way propagation between the radar and the scatterer. This static measurement at a single spatial point is simply an A-scan. In the case of a B-scan setup, however, two-dimensional GPR data are obtained by collecting a series of A-scan measurements along the synthetic aperture axis, say X. For an arbitrary measurement point x_n on the synthetic aperture path, the distance r from the point target at (x_i, z_i) to the sensor is equal to

$$r = \sqrt{(x_n - x_i)^2 + z_i^2} \qquad n = 1, 2, \ldots, N \tag{3.86}$$

where N specifies the total number of A-scan measurements. If there exist M point scatterers within the subsurface environment, the total two-dimensional scattered field is the sum of all scattered fields from each point scatterer along the B-scan path:

$$E_s(x, \omega) = \sum_{i=1}^{M} \rho_i e^{-j2(\omega/v_m)\sqrt{(x-x_i)^2 + z_i^2}} \tag{3.87}$$

Here ρ_i stands for the scattered field strength from the ith point scatterer. Taking the one-dimensional FT of Eq. (3.87) along the X-direction gives the field in the spatial frequency (k_x) domain as

$$E_s(k_x, \omega) = \sum_{i=1}^{M} \rho_i \int_{-\infty}^{\infty} e^{-j2(\omega/v_m)\sqrt{(x-x_i)^2 + z_i^2}} e^{jk_x x} \, dx \tag{3.88}$$

By utilizing the ESM, we again replace v_m with $c/2$. Then, applying the *principle of stationary phase* [73], the integral above can be solved approximately as

$$E_s(k_x, \omega) \simeq \frac{e^{-j\pi/4}}{\sqrt{4k^2 - k_x^2}} \sum_{i=1}^{M} \rho_i e^{-j\left(k_x x_i + \sqrt{4k^2 - k_x^2} z_i\right)} \tag{3.89}$$

where the ratio $e^{-j\pi/4}/\sqrt{4k^2 - k_x^2}$ is the complex amplitude term and has a constant phase. Therefore, it can be neglected for image-displaying purposes. Thus, Eq. (3.89) can be normalized to give

$$\overline{E}_s(k_x, \omega) = \sum_{i=1}^{M} \rho_i e^{-j\left(k_x x_i + \sqrt{4k^2 - k_x^2} z_i\right)} \tag{3.90}$$

Here $\sqrt{4k^2 - k_x^2}$ is simply the wavenumber in the z-domain: namely, k_z. With this construct we can map the data from the (k_x, ω) domain to the (k_x, k_z) domain by applying the nonlinear transformation of $k_z = \sqrt{4k^2 - k_x^2}$. Since the mapped data will no longer be on a uniformly spaced grid, the data should also be interpolated. At the end of these mapping and interpolation processes, the GPR data will be in the form

$$\tilde{E}_s(k_x, k_z) = \sum_{i=1}^{M} \rho_i e^{-j(k_x x_i + k_z z_i)} \tag{3.91}$$

where $\tilde{E}_s$ represents mapped and interpolated data on an equally spaced rectangular grid of k_x and k_z so that two-dimensional inverse fast Fourier transform (IFFT) can be used. If the two-dimensional IFT of Eq. (3.91) is taken with respect to k_x and k_z, we obtain

$$e_s(x, z) = \sum_{i=1}^{M} \rho_i \int\!\!\!\int_{-\infty}^{\infty} e^{-j(k_x x_i + k_z z_i)} e^{j(k_x x + k_z z)} \, dk_x \, dk_z \qquad (3.92)$$

which results in

$$e_s(x, z) = \sum_{i=1}^{M} \rho_i \delta(x - x_i, z - z_i) \qquad (3.93)$$

In this equation, $\delta(x, z)$ is the two-dimensional impulse (Dirac delta) function, which pinpoints perfectly the locations of M point scatterers. In reality, the data are collected within a certain bandwidth of frequencies and a synthetic aperture length, which means that the limits of the integrals in Eq. (3.92) are finite in reality. Therefore, the impulse functions degrade to a sinc (sinus cardinalis) function for practical applications. A flowchart representation of the algorithm steps is given in Fig. 3.37 for a clear understanding of the SAR-based focusing algorithm. To summarize:

1. Collect the scattered field data either in the time domain to have $E_s(x, t)$ or in the frequency domain to have $E_s(x, \omega)$.

2. Take the two-dimensional FT of $E_s(x, t)$ or one-dimensional FT of $E_s(x, \omega)$ to transform the data onto the wavenumber–frequency domain as $E_s(k_x, \omega)$ and normalize to get $\overline{E}_s(k_x, \omega)$.

3. Map the data from the $k_x - \omega$ domain to the $k_x - k_z$ domain and interpolate to have the data on a uniformly spaced rectangular grid as $\tilde{E}_s(k_x, k_z)$.

4. Take the two-dimensional IFFT of $\tilde{E}_s(k_x, k_z)$ to get the final focused image in Cartesian coordinates as $e_s(x, z)$.

An example of SAR-based $\omega - k$ migration imaging is demonstrated in Fig. 3.38. Two metal pipes were buried flat around $z = 30$ cm and $z = 40$ cm in a dry and homogeneous sand environment, as shown in Fig. 3.38a. Using a stepped-frequency continuous-wave radar (SFCWR) system, B-scan measurements were taken along a straight path while the frequency was altered from 4.0 GHz to 7.1 GHz. A raw GPR image obtained through one-dimensional IFT of the measured spatial–frequency data is shown in Fig. 3.38b. Obviously, the image is distorted in the synthetic aperture domain due to the well-known hyperbolic behavior. Since the pipes are close to each other, the tails of these hyperbolas interact. The focused GPR image after applying the SAR-based $\omega - k$ migration algorithm is shown in Fig. 3.38c.

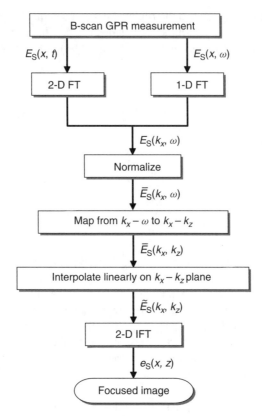

FIGURE 3.37 Flowchart representation of a SAR-based $\omega - k$ migration algorithm.

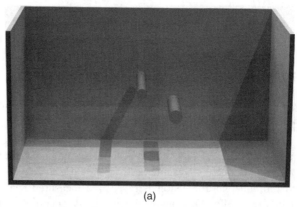

(a)

FIGURE 3.38 (a) Configuration of two closely placed pipes buried in a sand medium; (b) classical B-scan GPR image; (c) focused image after a SAR-based $\omega - k$ migration algorithm.

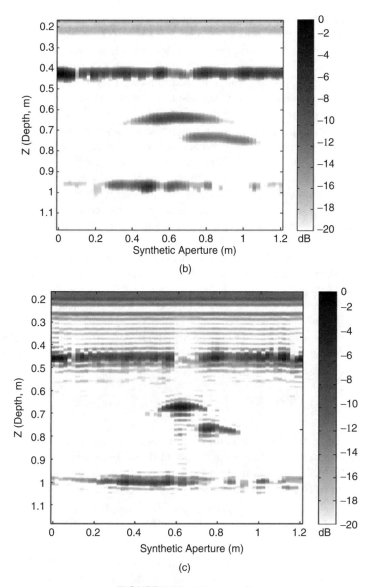

FIGURE 3.38 (*Continued*)

3.6.8 Other Methods

In addition to the common migration methods for imaging the subsurface environment mentioned above, many other methods have been introduced and studied by different GPR researchers. Fisher et al. [74] applied reverse-time migration algorithm for GPR profiles. Capineri et al. [75] employed a Hough transformation technique to the B-scan GPR data to obtain better-resolved images of pipes. Leuschen and Plumb [53] and Morrow and Van Genderen [76] implemented back-propagation techniques based on finite-difference time-domain (FDTD) reverse-time migration methods to solve the focusing problem. Ozdemir et al. applied a hyperbolic summation technique to focus the hyperbolic dispersion in B-scan GPR images [64]. Among these techniques, the Hough transform method and back-propagation algorithm are the most famous and widely used.

The first use of the *Hough transform*, a widely accepted image-processing tool, was by Kaneko [77]. In the algorithm, hyperbolic curves and/or straight lines in the space–time image are first searched for and detected. Generally, a Hough transformer transforms a straight line in the image to a single point in the polar coordinate system. In GPR applications, the space–time image is full of cluttering artifacts behind the hyperbolic curves. Before applying the Hough transform, therefore, an edge-detection procedure should be employed to find the hyperbolic lines. After detecting a hyperbolic curve, the first derivative (the slope) is used for the initial estimation of the target's spatial (x) location and the velocity of the wave. For an estimation of the target's depth (z) information, the hyperbolic curve is transformed to Hough space. Then the horizontal line that best focuses the target point is found [77–79] with the help of a search scheme. At the end of this procedure, the target is focused successfully around its true location.

Another popular approach to focusing the GPR image is to use the *back-propagation algorithm* [76,80,81]. The algorithm starts with the assumption that the signal measured, $E_s(x, z_0)$, is simply the convolution between the target's reflectivity, $\rho(x_0, z_0)$, and the propagation function, $h(x_0, z_0)$:

$$E_s(x, z_0) = \int_{-\infty}^{\infty} \rho(x_0, z_0) h(x_0 - x, z_0) \, dx_0 \tag{3.94}$$

Here the x-axis is the B-scan axis of length D. A detailed definition of the propagation function $h(x_0, z_0)$ for lossless and lossy medium cases was given by Baysol and Kosloff [52]. At the end of the algorithm, the target's reflectivity can be calculated by inspection as follows, thanks to the Fourier theory:

$$\rho(x_0, z_0) = \int_{-D/2}^{D/2} E_s(x, z_0) h^*(x_0 - x, z_0) \, dx \tag{3.95}$$

Other methods, such as the finite-difference time-domain (FDTD)-based time-domain inversing technique [82–84], spectral domain approaches [85,86], and reverse-time migration algorithms [53,87], are also described in the GPR literature.

3.7   NUMERICAL MODELING OF GPR

3.7.1   Introduction

Ground-penetrating radar (GPR) can be used for the detection of buried objects for many applications in civil engineering: detection of mines and unexploded explosives, evaluation of archeological structures, mining, nondestructive testing and diagnostics, pavement and site characterization, and utility mapping in cities. GPR systems use wideband signals for better resolution and to gain greater ability to detect underground objects. Under real conditions, underground targets are embedded in an inhomogeneous medium, with different scatterers having complex geometries and material structures. This situation makes the analytical solution of GPR problems very difficult, especially since objects buried near the surface can markedly alter the character of electromagnetic wave echoes returned. The basic model for analysis of a GPR signal is evaluation of the radar range equation [101]. In addition, there are various numerical methods, such as the transmission-line method, finite-element method, and moment method. In this area, the finite-difference time-domain (FDTD) method appears to be more popular and suitable because of its ability and the flexibility to analyze the complex transient scattering from a buried object or the transient radiation from a complex source in the presence of a boundary [114].

The FDTD method is based on replacing analytical time and spatial partial differential operators in Maxwell's equations with numerical derivatives extracted from Taylor series expansion, generally in the sense of central differences [169]. These derivatives force the electric and magnetic field components to be shifted in time, therefore also in space. To evaluate this requirement, Yee's unit cell is proposed for proper location of the field components. The stability of the FDTD algorithm is satisfied by the Courant stability condition. To use limited computational sources, a wide class of absorbing boundary conditions is applied in the FDTD algorithms, the most popular and widely used being the perfectly matched layer (PML). The FDTD method also has a lot of different types with important application steps in its rich literature [156]. There are several fundamental advantages of the FDTD method for GPR simulations:

- There is a possibility of direct time-domain solution of Maxwell's equations in three-dimensional space without requiring extraction of the Green's function for different problems.
- It is a well-understood method for providing solutions for narrow- and broadband sources.
- In the sense of materials and geometry, it can be used for complex objects and soil properties.
- It has the ability to model full GPR system, including antennas.
- The computational costs for computing time and CPU power are reasonably acceptable.

In the sense of the advantages given above, the FDTD method has been used extensively to model complex subsurface radar and GPR problems. Moreover, it should be noted that the success of GPR applications is strongly connected to computer simulations for understanding of the complex transient electromagnetic wave propagation phenomenon. The other important point in numerical GPR modeling is to understand the electromagnetic properties of buried objects and soil. The soil's properties, especially are complicated and cannot be modeled easily.

Researchers have shown that a constant real permittivity and a model of simple two-pole conductivity would be suitable enough for adequate FDTD modeling of the soil [149].

3.7.2 Historical Background

The historical background of time-domain modeling of the GPR problem by the FDTD method is very rich, and a lot of work is reported in the literature. Much of this deals with two principal subjects: fundamental GPR and GPR antenna modeling by the FDTD method.

Studies on Fundamental GPR Modeling by FDTD   One of the first works on time-domain FDTD modeling of GPR problems involved subsurface radar based on the response of a point source in two-dimensional inhomogeneous media. Rebars were used as the buried objects. The singularity of the point source was treated by finer meshing of the problem region near the point source. The Liao absorbing boundary condition was employed to terminate the FDTD problem space [135,136]. Two-dimensional FDTD analysis of TM (transverse magnetic) scattering from perfectly conducting circular cylinders located above (may be evaluated as a type of rough surface) and below a dielectric surface were investigated. Specifically, three wave models with a total field/scattered field formulation were used for extraction of the scattering field from the object. Mur's second-order absorbing boundary condition was used to terminate the layered FDTD problem space. Using Green's functions, the far field of the scattered field was calculated. The magnitude and phase of the far-field scattered patterns were compared with a method of moments solution, and excellent agreement was observed [165].

A three-dimensional FDTD solution including scattered field formulation for GPR modeling with a wire detector has also been presented. Mur's first- and second-order absorbing boundary conditions were applied for problem space termination. Results were obtained for various canonical targets [130]. A subsurface radar model of the GPR was investigated employing the pure scattered field formulation in the three-dimensional FDTD algorithms. The dielectric soil was considered as a horizontally aligned buried perfect electric conductor box. Berenger's PML for layered media was employed as the absorbing boundary condition [88]. To obtain the scattered signal, subtraction with differentiating techniques was applied to a raw GPR signal with respect to the position of the radar [139]. Simulation of GPR electromagnetic wave scattering in a partially conducting layered medium has also been modeled by the FDTD method for simple buried geometric objects [93].

Bourgeois and Smith have applied the three-dimensional FDTD method to another subsurface radar problem with a separated aperture sensor for the detection of metal plates, a Plexiglas block, and a Stycast block. Ramped sinusoidal signals were used for the wave excitation and lossy material was used for soil modeling [91]. The authors have extended their work to nonmetallic mine (dielectric block) detection [92] with a generalized perfectly matched layer to terminate the FDTD problem space [103].

More realistic three-dimensional FDTD GPR simulation in the sense of two arbitrarily polarized transmitters and one receiver (small dipoles) by pulse excitation has been investigated in order to cancel direct coupling between the receiver and the transmitter, or between the receiver and the ground surface. The ground is modeled as a lossless homogeneous dielectric [140] or an inhomogeneous heterogeneous layer with surface roughness [141]. To achieve the maximum scattered fields from the buried objects several techniques have been used: subtraction, isolation between the receiver and transmitters using absorbers [108], time-domain separation of the pulses, and time windowing [109]. Later, the optimization of antenna separation [142] and suitable frequency-band selection for the GPR over the lossy and heterogeneous grounds with the isolation of antennas [143] were investigated. Single or multiple dielectrics and the conducting targets (rectangular prisms and cylindrical disks) were buried and also investigated. Berenger's PML for the layered medium was used to terminate the infinite FDTD problem space.

Undesired and dominant direct coupling effects from transmitter GPR antennas to receiver GPR antennas that have conductive shields can be isolated with an absorber modeled by a PML-type absorbing boundary condition. Additional advantages of the absorbers are reduction of the coupling of the extra noise and the increasing directivity of the antennas [144].

In the case of deep-sounding GPR FDTD simulations ($\sim$2500 m), an extrapolation method is used in the FDTD algorithm, since classical FDTD simulation requires a long computational time. To this end, two different FDTD simulations are performed. One-dimensional FDTD simulation takes soil attenuation into account and provides for two-way propagation losses using simple analytical calculations. Three-dimensional FDTD simulation aims to model the antenna region. Thus, the total transfer function can be extracted from the coupling of one- and three-dimensional FDTD simulations in a short computational time [151].

The effects of GPR frequency variation have also been investigated and it has been shown that frequency variation may not change the detectability of buried objects, especially in soil regions that lack extreme humidity [145]. As an extension of this work, the authors have worked on eliminating direct coupling between the transmitter and the receiver antennas by proposing a different configuration of the receiver and two transmitter antennas [110], and later by setting the antennas $180°$ out of phase over randomly heterogeneous ground models [111]. The receiver antenna is located in the middle of the transmitter antennas. Four different transmitter and receiver antenna configurations were evaluated. Thus, optimization of transmitter–receiver antenna separation in GPR applications was demonstrated [112].

Detailed FDTD modeling of GPR problems requires more computational power. These requirements are generally overcome by using powerful single and/or parallel processing computers with a developed mathematical model [95]. Using a current sheet source, the oblique incident wave (the side facing the field) was excited in FDTD GPR simulation, and the possibility of detection of buried cylinder was investigated using resonances. The aim was to minimize the undesired clutter effects with improved target discrimination capabilities for buried metallic objects [154] and multiple-subsurface metallic objects [155].

Another interesting work deals with a field-programmable gate array (FPGA) implementation of the pseudo-two-dimensional GPR FDTD algorithm. It was claimed that FPGA hardware implementation increases the computational speed of the pseudo-two-dimensional GPR FDTD algorithm about 24-fold compared to software implemented on a 3.0-GHz personnel computer. To obtain this acceleration, data quantization using a fixed-point algorithm, careful memory design, pipelining, and parallelism was employed. The appropriate bit width was chosen for the best ratio of relative error to cost. This and future planned implementations of the three-dimensional GPR FDTD algorithm open new possibilities for faster calculations in real-time applications [100].

A hybrid method combining alternating direct implicit FDTD (ADI-FDTD) and the method of moments in the time domain (MoMTD) has been proposed to simulate such GPR problems as the detection of cracks in a marble block. The main aim of hybridization based on Huygens' principle is to use both the arbitrary material-modeling ability of the FDTD and the capability of thin wire structure modeling of the MoMTD. A V thin-wire antenna was located close to the inhomogeneous media and analyzed using the hybridization technique [89].

GPR simulation in inhomogeneous media containing conductive cylindrical objects in free space, buried in the ground, and buried in a trench with Gaussian pulse excitation has been investigated using the FDTD method [152]. An FDTD software tool called GprMax has been developed for the realistic target modeling of a wide class of GPR problems for a better understanding of wave propagation. Two- and three-dimensional GPR solutions are possible with dispersive and lossy material modeling. Antenna feeding features using transmission lines are incorporated in the FDTD algorithm [106].

A hybrid method combining the method of moments (MoM) and the FDTD method has been proposed to simulate the two-dimensional GPR problem. The arbitrary thin-wire antenna modeling ability of the MoM to use the electric field integral equation, and the ability of arbitrary object treatment of the FDTD method inside inhomogeneous media, were used together. First, the antenna field was calculated by the MoM and applied to Huygens' surface. Then the field was converted to the time domain as a source of the FDTD solution. This procedure was applied in an inverse manner for the evaluation of antenna coupling currents up to the convergence [104].

In some GPR simulations, fine geometric modeling may be needed. To perform this requirement, subgrids are used in the main FDTD grids. But this will cost a lot in terms of computational time and resources. To overcome this problem, the two-dimensional alternating direct implicit FDTD (ADI-FDTD) method has been employed in the subgrid region of the FDTD method. Two communication schemes have been described between the ADI-FDTD and FDTD methods at the boundary of two grids. The recursive integration PML was used to truncate the GPR problem space. The merits of the computational effectiveness of the ADI-FDTD method have been shown for an overresolved problem [102].

Quantification regarding how much information is omitted from two-dimensional FDTD GPR modeling as contrasted with three-dimensional FDTD GPR modeling has been studied in a bridge deck deterioration evaluation problem. Since use of the two-dimensional model in computationally large GPR models is required for fast results, the quantification has been analyzed in relation to the timing of scattering features, propagation loss, and so on. [171].

An energy-density spectrum of weak GPR scatterers such as plastic mines has been investigated by the FDTD method for the derivation of an estimation procedure using measured GPR data. The FDTD method was used to synthesize the electromagnetic signatures of weak three-dimensional scatterers such as buried landmines. Theoretical signatures of the FDTD method were compared with a measured signature using a metric correlation coefficient. The work showed that landmine detection with clutter discrimination is possible using knowledge of the energy-density spectrum for extracting spectral characteristics [115].

Dispersive Modeling   Three-dimensional FDTD GPR modeling including the transmitter and receiver antennas and antenna feeds with the dispersive character of the earth has been evaluated. Shielded and resistively loaded bowtie antennas with differentiated Gaussian pulse excitation were used and the soil was modeled as a lossy dispersive medium. The Debye model was considered to evaluate the earth dispersion using the recursive convolution technique. Metallic, and air- and water-filled plastic cylindrical pipes were buried. An MUR-type boundary condition with a subtraction technique was used to model the infinite problem space [90].

Another dispersive and conductive-loss soil modeling of three-dimensional FDTD GPR simulation was analyzed by a Debye soil model as a special case of an Nth-order Lorentz soil model [158]. The piecewise linear recursive convolution technique with PML extension for the dispersive medium was applied to the FDTD algorithm. Results were given for plastic and metallic buried pipe with various moisture contents of the soil medium. This formulation also facilitates parallelization of the code [99]. Detailed three-dimensional FDTD analysis for the effect of dispersive, inhomogeneous, and conductive soil modeling was also given for multiterm Lorentz and/or Debye models. Electric dipoles were used as transmitter and receiver antennas. Metal and plastic pipes were buried underground [159].

Another FDTD study aimed at producing dispersive and lossy soil modeling of a GPR simulator with bowtie antennas for buried air void, air, and round metal cylinders was handled as a fully three-dimensional problem [170]. A more accurate three-dimensional FDTD GPR model for both flat and rough realistic dispersive soil ground with a Gaussian variation has been proposed using an experimentally measured transmitted field as the model's excitation. This model simulates GPR systems with a transmitter parabolic reflector and a multistatic receiver array. The simulated results were validated with the results measured, showing a good agreement [119]. As a verification tool, the FDTD method for GPR applications at the higher microwave frequencies (X and Ku bands) in a dispersive environment was utilized, with good agreement of the experimental results [120].

Later, three-dimensional FDTD GPR consisting of a pair of resistor-loaded bowtie antennas was proposed for a dispersive two-term Debye soil model with static conductivity. Target signatures and target characteristics of buried metal and plastic pipes were obtained in order to distinguish the various buried materials [164]. The success of three-dimensional FDTD GPR simulations for cavity detection and tunnel inspection has been shown for dispersive FDTD earth modeling in a Debye medium [132].

Rough Surface Modeling   FDTD simulation of GPR problems for an irregular ground surface have been investigated in a three-dimensional environment with scattered field formalism [161]. A two-dimensional FDTD method was used to analyze the delay and amplitude characteristics of scattered GPR waves with realistic dispersive soil surface roughness. Monte Carlo simulations and cross-correlation functions for the metallic and nonmetallic mines were used to generate clutter statistics and signal variations. Using this procedure, the signals from nonmetallic mines could be extracted from the ground surface clutter signals for shallows buried objects [150].

The clutter response of realistic dispersive soils for GPR use was investigated in another study of rough surface modeling by the FDTD method. First, the three-dimensional and high-resolution FDTD solution surrounding an antenna region was employed; then the rough-resolution two-dimensional FDTD solution was used to simulate scattering by the rough dispersive ground. The dispersion of the soil is evaluated in the FDTD algorithm using Z transforms, and PML was used to terminate the FDTD problem space. A collaborative study was undertaken using multistatic linear array radar [167].

To enhance GPR images to eliminate undesired effects of heterogeneous surfaces, a redatuming process through downward and upward continuations based on the full-wave equation was applied to an FDTD-generated GPR radar profile. The technique was utilized at several urban public transportation sites, such as reinforced bridge decking containing cemented aggregate with rebars. The redatuming technique first shifts the reference datum (i.e., collected data surface) from the original reference to an imaginary reference at greater depth relating to the later GPR response, then shifting the datum back to the original reference. The results show that this technique is an effective way to suppress undesired diffractions and interferences that come from geometrical and material heterogeneity [125].

Imaging and Inversion Modeling   FDTD implementations of the reverse-time migration algorithm with matched filter definitions for mono- and bistatic antenna configurations suitable for GPR applications have been developed to obtain buried-object images from the measured back-propagating data for the monostatic case and from the convolution of incident and back-propagated fields for the bistatic case [129].

An iterative inverse optimization algorithm has been proposed for reconstruction of buried multiple-target permittivities. The forward-scattering data were calculated by a 2.5-dimensional GPR FDTD simulation. The convergence rates and initialization value of the inverse algorithm were discussed. Successfully reconstructed images were obtained for a single target and for two targets buried at the same or different depths [107]. GPR FDTD simulation for of the imaging buried objects by the improved back-propagation algorithm has been proposed with two-dimensional and three-dimensional imaging results. Real GPR data were used for imaging [126].

Three-dimensional GPR FDTD work has been performed for the identification of antipersonnel landmines (modeled as buried circular dielectric disks) by validating the convolutional models, and the target size and target depth determined (notably by Roth et al. [153]). Imaging of the dielectric structure for a buried object has been developed for the use of inversion algorithms based on a conjugate gradient search for the minimum of an error functional calculated from the difference between the two-dimensional FDTD data measured and predicted. The perturbation method was used for functional calculation by inserting error into the model in reverse time and correlating the back-propagated field with the incident field [157].

A Fourier transform–based three-dimensional FDTD numerical GPR imaging algorithm of buried arbitrary shapes has been proposed. Fourier transform in a selected frequency was used to detect the specification of the target shape by using data over target scanning. After the time-domain data saved were converted to the frequency domain for each point of scanning, the amplitude distribution for imaging was obtained. A plane wave, a dipole, and a Hertzian dipole were used as different sources of simulation, but it was reported that similar imaging results were obtained. The targets were modeled as an air hole, a conductive material, and a surface crack. The results show that the frequency selected for imaging was of great importance for the reconstructed shape quality of buried objects [105].

The FDTD method was used to solve a forward problem in order to analyze an inverse GPR problem to test combining the matched filter–based reverse-time and particle swarm optimization algorithms to characterize concrete structures. The air gap, conductor, and the water were investigated as buried objects [162].

Higher-Order Modeling Some studies of GPR simulations by higher-order FDTD modeling have been undertaken. Two-dimensional GPR FDTD simulation based on the lossy wave equation has been investigated using a newly proposed composite absorbing boundary condition. The spatial derivatives of the wave equation were discretized by the fourth-order finite difference formulation for better modeling of the waves. Various geophysical synthetic examples were investigated [96]. Another three-dimensional FDTD fourth-order scheme in space has been developed for GPR simulations [94].

GPR Antenna Modeling by FDTD Antennas are a critical part of GPR systems, as they greatly affect GPR detection of buried objects. Due to wideband signal use in GPR systems, wideband antenna design is a critical step in successful detection. Increasing the directivity of GPR antennas to ensure better resolution is also a very important design issue. Lightweight, low-cost, portable antennas with better radiation resistance and higher gain are the features of a GPR system most often requested. Impedance matching of a GPR antenna between the antenna feed and the ground surface is another important design issue. In the following sections we discuss important studies of GPR antenna design by the FDTD method.

Dipole Antennas   A linear dipole antenna loaded by resistors has been investigated for use with three-dimensional FDTD GPR simulations. It was observed that the current waveforms on the antenna relating to the antenna pattern depend strongly on the soil parameters, loaded resistors, and the height of the antenna above the ground [113].

Resistively loaded V dipoles modeled as thin triangular-shaped conductive sheets were used to detect cylindrical antipersonnel mines using the GPR fully modeled by the three-dimensional FDTD method. V dipoles are thin and tapered linearly. Critical parametric studies, such as the height of the dipole antenna above the ground, the soil, the pulse, and the antenna parameters, were investigated parametrically. Experimentally validated FDTD results showed that V dipole antennas greatly reduced the clutter and make it much easier to distinguish buried targets [137].

A FDTD-MoM hybrid technique for complex GPR antenna (dipole, skewed dipole, and bowtie) modeling in the presence of heterogeneous ground has been proposed. The ability of the MoM in free-space antenna modeling and the ability of the FDTD method in heterogeneous ground modeling were combined using the equivalence principle. Thus, a less expensive hybrid type of GPR FDTD antennas modeling was developed by Huang et al. [116].

The characterization of horizontal dipole antennas on a half-space boundary in GPR applications has been investigated by considering the interference of space and lateral waves. In the FDTD model of these types of antennas, as validation of the experiments clearly showed interference from the space and lateral waves. This made possible a clearer understanding of the radiation mechanisms of GPR antennas [146].

Near-field horizontal dipole radiation mechanisms are examined by two- and three-dimensional FDTD simulations for better understanding of GPR antenna behavior. The effects of ground electrical properties, antenna height, and observation distance on near-field antenna patterns extending to far-field patterns have been investigated. Electric and magnetic near-field radiation patterns have been produced for a variety of scenarios. The experimental results have been used to validate the simulated results, with good agreement [147].

A dielectric embedded dipole antenna based on a butterfly radiator has been proposed to stabilize the performance of GPR antennas for various ground types. The antenna was shielded by a metal case to achieve low coupling. A theoretical model of this antenna has been analyzed by the FDTD method. FDTD simulations of a butterfly antenna showed the propagation of a clean pulse with late-time ringing. Two approaches, full and partial filling of the shielding case, have been investigated. Partial filling caused input impedance variations to increase, depending on operational frequency, whereas full filling caused the antenna transfer function to widen. The antenna performance was optimized via the absorber thickness, radiator position, and other characteristics [117].

Dielectric Rod Antennas   Three-dimensional GPR FDTD modeling of an ultra-wide band dielectric rod antenna has been proposed. The advantages of a dielectric rod antenna are localized illumination and weak antenna–ground interaction.

Therefore, a dielectric rod antenna may be preferable, especially for the mine detection of GPR applications. Simulations showed that by changing the antenna height above the ground surface, it was possible to control the spot size of the antenna and to minimize surface clutter [148].

Another ultrawide band GPR dielectric rod antenna for detecting shallow targets such as antipersonnel mines has been investigated by three-dimensional FDTD simulations. The lowest hybrid mode of a circular dielectric waveguide with tapered permittivity design was excited. Low antenna clutter and weak antenna–ground interactions with the broadband behavior of a relatively frequency dependent antenna have been observed by Chen et al. [97]. It was shown that by controlling antenna height, the illumination spot size could be controlled to decrease the surface clutter considerably [98].

BowTie Antennas The effect of partly or fully ferrite-coated covering on the cavity conduction surface of a resistively loaded bowtie antenna for GPR applications has been analyzed by a fully three-dimensional FDTD method. The results showed that remarkable improvement in the antenna characteristics was not ensured by the ferrite coating in a given antenna geometry [138].

An efficient, lightweight, low-cost GPR bowtie antenna that has tapered capacitive loading implemented on the circular end by constructing slots that increase linearly with width is investigated by detailed taper modeling of the FDTD method. This greatly increased the radiation efficiency of the antenna and the maximum antenna amplitude of the input pulse was approximately 47% higher than that if a conventional bowtie antenna, with the effective elimination of open-end reflections. The late-time ringing of the antenna could also be suppressed by using microwave absorbers [127].

GPR FDTD modeling of a fully polarimetric horn-fed bowtie antenna has been developed for detection of unexploded ordnance by realistic antenna analysis. The effects of resistive card design, reflection coefficients, and antenna patterns have been investigated for better understanding of antenna radiation characteristics [122,123]. GPR FDTD modeling of an ultrawide band dual-polarized dielectric-loaded horn-fed bowtie antenna has been investigated. Fully polarimetric simulations with parametric study of the effect of a resistive taper an R-card termination have been investigated. It was shown that a linear taper was better than an exponential taper for short taper lengths and that suitable overlap between a PEC and an R-card improved the transitions and reduced the diffractions at the end. R-card termination also reduced antenna ringing with the broadband behavior of the antenna. These results made at possible to optimize the antenna in the sense of the feeding cables, wave launchers, dielectric loading, and the resistive film loading [124].

A detailed parametric study analyzing resistor-loaded bowtie antennas for such antenna parameters as the flare angle, antenna length, and height above the simple ground in the GPR system was undertaken to achieve low clutter and better target discrimination [163]. A modified bowtie antenna based on an ultrawideband radiator with metallic shielding (a simple rectangular box) was analyzed by the FDTD method to obtain better radiation efficiency with reduced dimensions and to suppress late-time ringing, especially in short-range GPR applications [128]. An adaptive antenna application of the two-dimensional GPR FDTD method was proposed as a slotted bowtie antenna. By varying the position of the straight slot (or by shunting the antenna slot into different positions), the antenna could be adapted (the main pulse amplitude and transfer function varied) to the ground region with different permittivities [168].

Modeling of resistively loaded surface GPR antennas such as bowtie antennas has been performed by the three-dimensional FDTD method in Cartesian coordinates. Specifically, the antenna panels were modeled either as perfect electrical conductors (PECs) or as having a more realistic model, such as a Wu-King type of conductivity profile. The simulations showed that PEC modeling of the antennas was quite sensitive to the environment and geometry of the antenna panels, but that Wu-King antenna modeling was remarkably more robust to the environment and geometry of the antenna panels [121].

A bowtie antenna with exponential discrete resistor-loaded GPR FDTD simulations has been compared to a conventional resistor-loaded bowtie antenna. The results showed that an exponential discrete resistor-loaded antenna could terminate the reflections at the end and was less sensitive to the GPR environment and showed more broadband character [133].

An FDTD numerical model of bowtie antennas as receivers and transmitters has been developed for GPR-coupled antenna simulations, especially for propagation through concrete structures. Development of the antenna model was supported by free-space measurements. Thus, the design of a good GPR antenna is focused on real-world GPR applications [118].

Horn Antennas   A novel hybrid method based on the plane-wave spectra interactions has been proposed to model realistic ultrawide band GPR antennas such as dipoles and TEM horn antennas. In this method, the problem space is divided into two independent subregions: one for an antenna and the other for an air–ground interface with buried objects. Coupling between the regions is achieved by plane-wave spectrum interactions. The scattering field from the ground surface and the buried objects is calculated using the FDTD method. The fundamental advantage of this method is to reduce the computational requirements [134].

A modified TEM horn antenna with discrete exponential resistive loading has been modeled by a three-dimensional FDTD algorithm for GPR applications. A TEM horn antenna is shielded (covered) with a rectangular conducting box having an open end in order to prevent undesired signal to the antenna from the air and from other electronic devices. A wave absorber attached to the inner box wall is used to decrease undesired multiple reflections between the antenna and the box. It was reported that the dynamic range for the GPR system could be increased in this way [131].

A three-dimensional GPR FDTD ultrawide band dielectric horn antenna with two PEC conical launcher plates feeding a two-wire line has been investigated. A detailed analysis of the design curves of the surge impedances for the various geometries and dielectric loading with impact variations of the dielectric constant was given. The attractiveness of the FDTD modeling for the dielectric horn antenna was noted [166].

Although many of the studies cited above are based on GPR FDTD simulations in Cartesian coordinates, cylindrical coordinates were also used for the computation of transient electromagnetic waves with a buried cylindrical disk in inhomogeneous, dispersive, and conductive media [160].

3.7.3 GPR Modeling by the FDTD Method

The time-domain modeling of the GPR by the two-dimensional FDTD method is investigated next. In the classical concept of buried-object detection, generally transmitter and receiver antennas are located very closed to the ground surface. This does not provide a way to analyze the effects of the incident angle for buried-object detection in GPR applications. To analyze the incident angle effect for the detection abilities of GPR, a horn antenna is used as a transmitter and the scattered electromagnetic field variation is calculated in the time domain by the FDTD method. A Berenger-type PML is used to terminate the FDTD problem space [88]. But the PML for the layered media (air–ground) is also extended by Gürel and Oğuz [109]. The reflected electromagnetic waveforms are observed in the time domain and converted to the frequency domain by fast fourier transform (FFT). An air–ground interface is considered in two-dimensional space. Air gap and dielectric rectangular prisms are buried under the ground and their detectability is analyzed for different incident angles.

Incident Wave Source in the Time Domain An H_z -polarized differentiated Gaussian pulse is used to excite the electromagnetic waves by the transmitter antenna as

$$f(t) = e^{-(t-t_0)^2/\tau^2} \sin(2\pi f_0 t) \tag{3.96}$$

where $t_0 = 45 \times \Delta t s$, $\tau = 10^{-9.3}$ s, and the central frequency $f_0 = 1.8$ GHz. The time and frequency responses of the incident time-domain signal are given in Fig. 3.39.

Horn Antenna Design The horn antenna is designed for excitation of the given time-domain source. Although increasing the antenna directivity is an important issue, for the better results the small modification is also necessary in the FDTD algorithm [156]. According to that, the geometrical details of the used horn antenna are given in Fig. 3.40a. In this step, the length L is 1.81λ, the aperture length A is 0.774λ, the aperture angle of α is $5.9°$, the waveguide width of a is 0.4λ, and the distance between the source point and the waveguide wall w is 0.2λ.

Waveguide feeding to the horn antenna is excited between the first (dominant mode) and second modes having cutoff frequencies at $f_{first} = 1.25 \times 10^9$ Hz and $f_{second} = 2.5 \times 10^9$ Hz. To understand how much of the source waveform energy can be propagated from the horn antenna, the relation between the source waveform

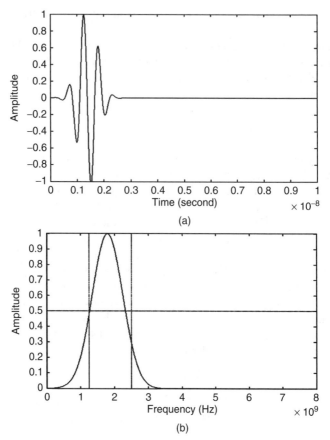

FIGURE 3.39   Normalized time- and (b) frequency-domain representations of the incident wave source.

spectrum and the first and second cutoff frequencies of the horn feed waveguide are shown in Fig. 3.39b as vertical lines. Moreover, between the cutoff frequencies of the horn feed waveguide, half of the source signal amplitude is shown with a horizontal line. It can be concluded that a large part of the source signal energy is propagated by using the proposed horn antenna in the dominant mode of the horn feed waveguide.

The FDTD calculated radiation pattern of the horn antenna in polar coordinates is shown in Fig. 3.40b. The field distributions of the horn antenna for the pulse excitation are shown in Fig. 3.40c, where the number of the FDTD time step is $n = 450$. The results show that the horn antenna directivity is acceptable for buried-object detection.

Numerical Details   In the applied two-dimensional FDTD algorithm, the unit size of the spatial FDTD grids is chosen as $\Delta x = \Delta y = \lambda/30 = 0.01$ m, where λ is the wavelength. Accordingly, the unit time step $\Delta t = 2.1213 \times 10^{-11}$ s is

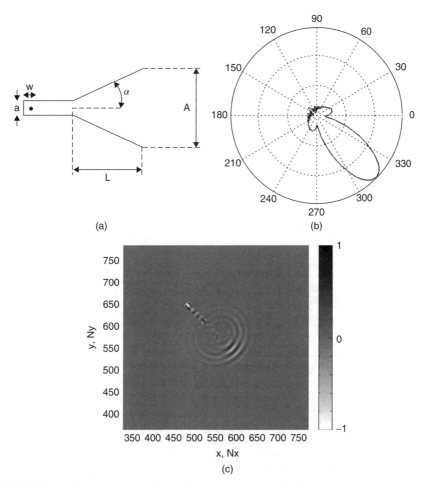

(a)   (b)

(c)

FIGURE 3.40   Geometrical details of a horn antenna; (b) radiation pattern of a horn antenna normalized in polar coordinates; (c) H_z field distribution of the horn antenna for the pulse excitation ($n = 450$).

calculated using the Courant stability condition. Under these circumstances, the electrical dimensions of the problem are $24\lambda \times 32\lambda$, of which the real physical dimension without the PML region is 7.2 m $\times$ 5.6 m for an $30°$ incident angle, and $18\lambda \times 45\lambda$, of which the real physical dimension is 5.4 m $\times$ 13.5 m for an $60°$ incident angle. The total number of unit cells is 960×1200 for a $30°$ incident angle and 800×1590 for a $60°$ incident angle, adapted for different incident angles. The thickness of the PML is 4λ. The height of the $\overline{ON}$ line from the ground (Fig. 3.41) is 13λ for a $30°$ incident angle and 7.75λ for a $60°$ incident angle. The number of time steps for the E_x field calculations without buried objects for $30°$ and $60°$ incident angles is taken as $n = 1820$. The total number of time steps for buried object simulation is $n = 4800$ for a $30°$ incident angle and $n = 6300$ for a $60°$ incident angle.

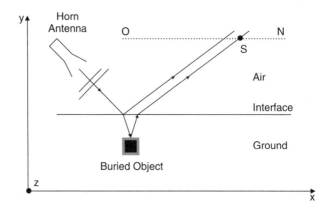

FIGURE 3.41   Geometrical details of the GPR problem investigated.

The total depth of the soil layer is 7λ. All the objects are buried at the same depth from the air–soil interface, 5λ, and they have also the same size, $1.2\lambda \times 1.5\lambda$. They are located for direct impingement of the electromagnetic waves of the horn antenna, meaning that the air gap and the dielectric object are horizontally 22λ far from the PML layer. But the horizontal distance between two rectangular dielectric objects is 1.6λ. The relative dielectric permittivities of the objects and the soil are taken as $\varepsilon_r^{\text{soil}} = 6.8$ and $\varepsilon_r^{\text{object}} = 12$, respectively, and $\varepsilon_r^{\text{air}} = \varepsilon_r^{\text{air gap}} = 1$ for the air and the air gap.

Scattered Field from Buried Objects   Three different buried objects are used to investigate the scattering characteristics of transient waves by two-dimensional FDTD GPR modeling. To understand the GPR performance for closer and more difficult real problems such as plastic mine detection, the objects chosen are an empty cavity (air gap), a dielectric rectangular prism, and two rectangular prisms (rather than perfect electric materials). Different incident angles (i.e., $30°$, $60°$) are used for the detailed analysis. The geometry of the problem investigated is shown in Fig. 3.41.

An H_z -polarized electromagnetic wave is used to illuminate the air–ground interface and the buried objects. The results are given in three different manners. First, two-dimensional electric field distribution is shown at various time steps (snapshots). Second, the time response of the electric field is given at the symmetric point S of the antenna above the soil surface. Third, the scattered total electric field distributions along and above the soil surface (the line $\overline{ON}$) are saved in the time domain. Using FFT, the spatial spectrum of the scattered data with the maximum amplitude evaluation is calculated. All the results are obtained with and without buried objects and compared with the evaluations.

In the manner of the explanation above, the results without buried objects are shown in Fig. 3.42. The E_x electric field distribution of a $30°$ incident angle and the H_z magnetic field distribution of a $60°$ incident angle are shown in Fig. 3.42a and b, respectively. The time-domain scattered fields of $30°$ and $60°$ incident angles at the symmetric point S of the transmitter are shown in Fig. 3.42c and d, respectively.

The results for an air gap are shown in Fig. 3.43. The E_x electric field distributions of the $30°$ and $60°$ incident angles are shown in Fig. 3.43a and b, respectively. The time-domain scattered fields of the $30°$ and $60°$ incident angles at the symmetric point S of the transmitter are shown in Fig. 3.43c and d, respectively. Moreover, the scattered total electric field distributions of $30°$ and $60°$ incident angles along the $\overline{ON}$ axis are shown in Fig. 3.43e and f, respectively.

The results for a rectangular dielectric object are shown in Fig. 3.44. The E_x electric field distributions of $30°$ and $60°$ incident angles are shown in Fig. 3.44a and b, respectively. The time-domain scattered fields of the $30°$ and $60°$ incident angles at the symmetric point S of the transmitter are shown in Fig. 3.44c and d. Moreover, the scattered total electric field distributions of the $30°$ and $60°$ incident angles along the $\overline{ON}$ axis are shown in Fig. 3.44e and f.

The results for two rectangular dielectric objects are shown in Fig. 3.45. The E_x electric field distributions of the $30°$ and $60°$ incident angles are shown in Fig. 3.45a and b, respectively. The time-domain scattered fields of the $30°$ and $60°$ incident angles at the symmetric point S of the transmitter are shown in Fig. 3.45c and d. Moreover, the scattered total electric field distributions of the $30°$ and $60°$ incident angles along the $\overline{ON}$ axis are shown in Fig. 3.45e and f.

3.7.4 Results and Discussion

In this section a detailed historical background regarding FDTD modeling of GPR simulations has been given. The studies are collected in two main groups: fundamental GPR and GPR antenna modeling by FDTD. In the first group, research on direct two- and three-dimensional FDTD modeling of GPR is described classified in terms of dispersive, rough surface, imaging, inversion, and higher-order modeling. In the second group, special attention is given to GPR antenna modeling by two- and three-dimensional FDTD methods. Modeling progress on dipole, dielectric rod, bowtie, and horn antennas is given. Later, time-domain simulation of GPR in two-dimensional Cartesian coordinates is presented by the FDTD method. The differentiated Gaussian pulse is used to excite H_z -polarized electromagnetic fields, including detailed design consideration of the horn antenna with the radiation pattern in polar coordinates and field distributions in the entire problem space.

Soil and buried objects are modeled as a pure dielectric medium. Thus, investigations about modeling of difficult and more realistic problems such as buried plastic mine detection are considered. In fact, time-domain scattering from three types of buried objects is investigated, such as an empty cavity (air gap), a dielectric material of rectangular prism shape, and two dielectric rectangular prism-shaped materials. In classical GPR problems, since the receiver and transmitter are generally located very close to the air–soil interface, the incident angle effects cannot be analyzed in detail. In this work, the incident field is excited for two different incident angles ($30°$ and $60°$). In that case, results are given in the manner of two-dimensional total field distributions in a whole problem space: the time-domain

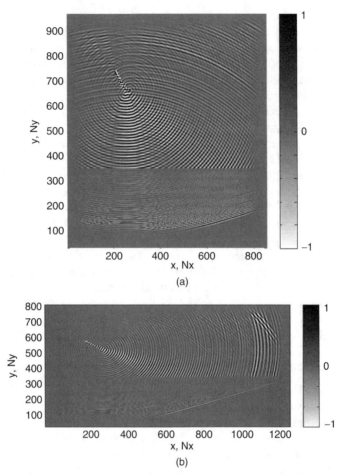

FIGURE 3.42 E_x electric field distribution of the $30°$ incident without a buried object; (b) H_z magnetic field distribution of the $60°$ incident without a buried object; (c) time-domain scattered field of the $30°$ incident at the symmetric point S of the transmitter without a buried object; (d) time-domain scattered field of the $60°$ incident at the symmetric point S of the transmitter without a buried object.

signal of the scattered field collected at the symmetric point of the transmitter and the scattered field distributions along the line of the $\overline{ON}$ axis, parallel to and above the air–soil interface.

The results show that the two-dimensional FDTD method is capable of modeling the GPR problem properly in the time domain and all the buried objects can be clearly seen in the two-dimensional field distributions on the entire problem space compared with a case with no buried objects. In time-domain field results at the symmetric point of the horn antenna, transient reflections from the air–soil interface and from the buried objects can be seen, especially at a $30°$ angle to the incident wave. For a $60°$ incident wave angle, time-domain field results are

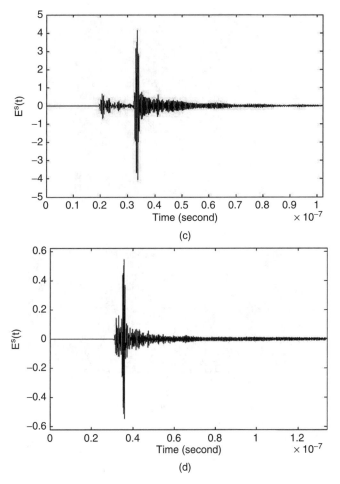

FIGURE 3.42 (*Continued*)

not very clear for the discrimination of reflections from buried objects and from the air–ground surface, although details of buried objects are clear enough in field distributions. This may happen because of the different reflection and transmission properties of the air–soil surface according to the incident angle, which is an argument of the reflection and transmission coefficients.

Moreover, the data collected from the scattered field distributions along the line of the $\overline{ON}$ axis parallel to the air–ground interface are very useful for the detection of buried objects with pure dielectric modeling. Although clear differences in scattered field distribution for one and two buried dielectric materials are present, very close scattered field distributions are observed for one buried object and the air gap. In that case, the discrimination is possible using the time-domain data of the scattered field at the symmetric point for a transmitter horn antenna.

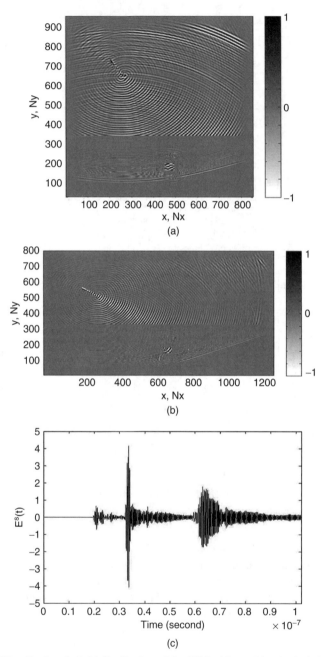

(a)

(b)

(c)

FIGURE 3.43 E_x electric field distribution of the $30°$ incident with a buried air gap; (b) E_x electric field distribution of the $60°$ incident with a buried air gap; (c) time-domain scattered field of the $30°$ incident at the symmetric point S of the transmitter with a buried air gap; (d) time-domain scattered field of the $60°$ incident at the symmetric point S of the transmitter with a buried air gap; (e) scattered total electric field distribution of the $30°$ incident along the $\overline{ON}$ axis with a buried air gap; (f) scattered total electric field distribution of the $60°$ incident along the $\overline{ON}$ axis with a buried air gap.

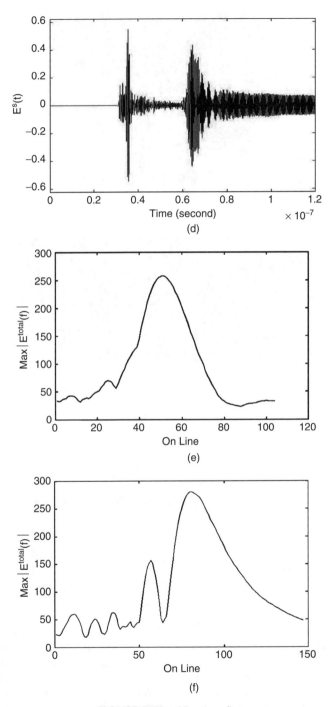

FIGURE 3.43 (*Continued*)

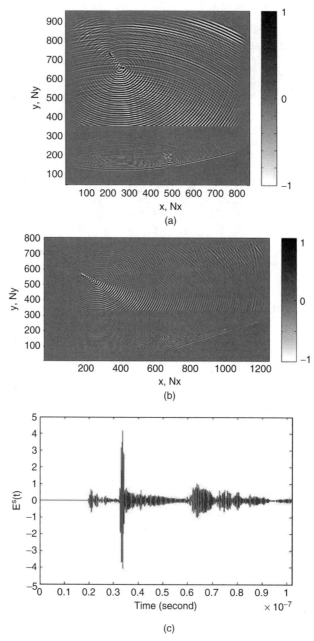

FIGURE 3.44 E_x electric field distribution of the $30°$ incident with a rectangular dielectric object; (b) E_x electric field distribution of the $60°$ incident with a rectangular dielectric object; (c) time-domain scattered field of the $30°$ incident at the symmetric point S of the transmitter with a rectangular dielectric object; (d) time domain scattered field of the $60°$ incident at the symmetric point S of the transmitter with a rectangular dielectric object; (e) scattered total electric field distribution of the $30°$ incident along the $\overline{ON}$ axis with a rectangular dielectric object; (f) scattered total electric field distribution of the $60°$ incident along the $\overline{ON}$ axis with a rectangular dielectric object.

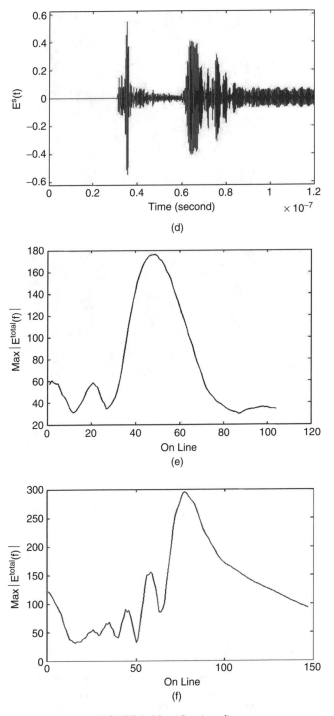

FIGURE 3.44 (*Continued*)

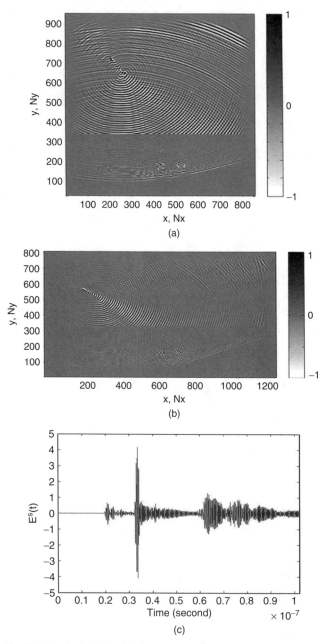

FIGURE 3.45 E_x electric field distribution of the $30°$ incident with two rectangular dielectric objects; (b) E_x electric field distribution of the $60°$ incident with two rectangular dielectric objects; (c) time-domain scattered field of the $30°$ incident at the symmetric point of the transmitter with two rectangular dielectric objects; (d) time-domain scattered field of the $60°$ incident at the symmetric point of the transmitter with two rectangular dielectric objects; (e) scattered total electric field distribution of the $30°$ incident along the $\overline{ON}$ axis with two rectangular dielectric objects; (f) scattered total electric field distribution of the $60°$ incident along the $\overline{ON}$ axis with two rectangular dielectric objects.

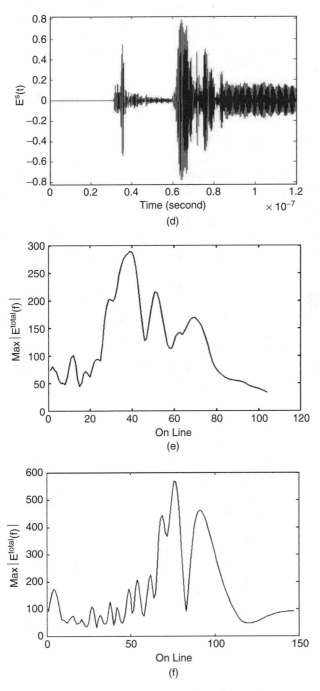

FIGURE 3.45   (*Continued*)

To investigate the incident angle effect with better FDTD modeling of the GPR problem, three-dimensional FDTD simulations are proposed as future work. This can provide a chance to analyze the incident effect in the sense of two horizontal planes and yield more information about the transient scattering phenomena related to different aspects of GPR problems.

3.8 DETECTION AND CLASSIFICATION ALGORITHMS

Ground-penetrating radar (GPR) is a device commonly used to detect buried objects. In this section we review fundamental detection and classification approaches for landmine detection. The algorithms discussed in this section are applicable to the detection and classification of buried objects encountered in other fields, such as civil engineering and archeology. The most common strategy for detection and classification algorithms is depicted in Fig. 3.46. Given data collected over a region of ground, the detection and discrimination algorithm has two functions: determining regions of interest in the data corresponding to potential mines (detection), and classifying each region as a mine or not a mine (discrimination). A prescreening algorithm usually follows the preprocessing step to identify (detect) potential targets. The general nature of the prescreening algorithms is their low computational complexity. Once potential targets are identified, further investigation is carried out by extracting features from the region of interest (ROI). Feature extraction can be applied in two steps, as shown in Fig. 3.46. The first sets of features may be low in terms of computational complexity and are used to eliminate or reduce false detections. Some ROIs

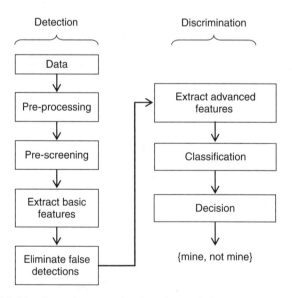

FIGURE 3.46 General strategy for detection and discrimination algorithms.

are eliminated in this step and thus do not go through the rest of the process. Elimination of false alarms in this step helps to reduce the computational effort. The rest of the ROIs are investigated further by extracting additional, usually more complex features.

Preprocessing algorithms are applied to the sensor data to enhance the signal-to-noise ratio and/or to highlight significant features of the data. It is important to note that detection and feature extraction algorithms perform better if a suitable set of preprocessing algorithms are applied. The hidden Markov model (HMM) approach of Gader et al. [172] uses a preprocessing algorithm to accentuate the edges in the diagonal and antidiagonal directions. The HMM algorithm uses the edge information to detect the hyperbolic signature produced from a landmine. On the other hand, the same preprocessing algorithm does not lead to outstanding performance for the Choquet morphological shared-weight neural networks (CMSNNs) presented in the same work [172]. The CMSSN algorithm does not rely on the edge information and uses a different set of preprocessing algorithms. Some of the commonly used preprocessing algorithms for GPR data are reviewed in Section 3.5.

At present, the most common techniques for mine detection are GPR-based landmine detection systems. Over the past two decades, many signal-processing algorithms have been proposed for GPR data. Typical landmine detection approaches for these systems are based on abrupt change detection [173]. GPR returns form a relatively smooth background for blank areas (i.e., locations where no landmine or other buried objects are present) compared to nonblank areas. Abrupt change detection algorithms use this phenomenon to locate buried objects. However, the rough ground interface, subsurface inhomogeneties such as tree roots, small pieces of metal and rocks in the ground, and discontinuities in soil texture reduce the performance of GPR. Clutter dominates the data and obscures the mine information. This is especially important for antipersonnel mines, as they are buried at shallow depths. Detecting landmines with relatively low energy is the main challenge. Most detection algorithms employ a preprocessing algorithm to reduce clutter. Statistical and parametric clutter modeling techniques have been proposed [174–179]. Other approaches include adaptive background subtraction [177] and background modeling using time-varying linear prediction [178] and its improvement [179]. In the first part of this section we review the linear prediction, PCA, and ICA-based algorithms. These algorithms are based on the idea of detecting abrupt changes. In the rest of this section, mathematical morphology–based approaches are discussed. These are trainable methods that perform feature extraction and classification simultaneously.

3.8.1 Linear Prediction–Based Detection

To detect and localize landmines in B-scans, methods based on abrupt change detection have been proposed by several authors [180–183]. Abrupt changes in B-scans can be sought not only along the time (vertical) axis but also along the spatial (horizontal) axis. Consider two data subsets $\mathbf{x}_1$ and $\mathbf{x}_2$ collected sequentially. Let $D(\mathbf{x}_1, \mathbf{x}_2)$ be a dissimilarity measure between the two sets. An abrupt change occurs if the dissimilarity is larger than a threshold. The detection performance is highly dependent on the choice of dissimilarity measure employed. A commonly used dissimilarity measure is $|\mathbf{x}_1 - \mathbf{x}_2|^T|\mathbf{x}_1 - \mathbf{x}_2|$, which can be regarded as the energy of the difference between the two signals. Tuning the threshold is a classic problem in most detection algorithms. When a fixed threshold does not give satisfactory results, the threshold is made adaptive to improving the performance [181].

Modeling the GPR signal by the linear prediction technique is one approach to detecting landmines by anomaly detection. This approach has been explored by several authors. Wu et al. [183] proposed removing the ground bounce response by modeling it as a shifted and scaled version of an adaptively estimated reference ground bounce. Ho et al. [181] utilized a one-sided linear prediction technique in the frequency domain to model the background and clutter reflection for a handheld GPR system. Torrione et al. [184] implemented the filtering approach in multiple dimensions to exploit spatial correlation information from a vehicular system providing multichannel data. Chan et al. [182] proposed a generalized version of two-sided linear prediction and employed it to perform processing in spatial domain.

In this section we give the methodology and motivation for linear prediction–based detection. Our modeling is based on detecting changes along the track and we use time-domain data, but they can easily be extended to anomaly detection along the time axis or to frequency-domain data. Suppose that $x(n)$ is the measured GPR response at depth k and down-track location n. From a system modeling point of view, the block diagram is as shown in Fig. 3.47. The goal here is to develop an adaptive model of the GPR data by a finite impulse response (FIR) filter. If the filter parameters are adjusted to minimize the error between $x(n)$ and the output $y(n)$, the filter parameters and/or modeling error $e(n)$ can be used to detect the anomaly.

Suppose that the filter has $M + 1$ coefficients, with $h(0)$ set to zero. Then $y(n)$ satisfies that

$$y(n) = \sum_{k=1}^{M} h(k)x(n - k) \qquad (3.97)$$

where $h(\cdot)$ are the filter parameters and $x(n)$ is the measured GPR response. We interpret $y(n)$ as a linear prediction of $x(n)$ from the previous samples $x(n - 1)$,

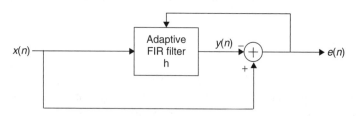

FIGURE 3.47   Linear prediction-based detection.

$x(n-2), \ldots, x(n-M)$. The error between the desired output $x(n)$ and the filter output $y(n)$ is given by

$$
e(n)^2 = [x(n) - y(n)]^2 = \left[x(n) - \sum_{k=1}^{M} h(k)x(n-k) \right]^2 \qquad (3.98)
$$

In this formulation, there are M unknown filter parameters. Thus, a minimum of M equations is required to estimate the filter parameters. These equations can be obtained along either the horizontal axis or the time axis. The latter is developed here to formulate the problem. Let $\mathbf{x}(n)$ be a column vector containing GPR returns at position n. The time-varying filter coefficients are adjusted to minimize the energy of the error vector $\mathbf{e}(n)$, defined by

$$
\mathbf{e}(n) = \mathbf{x}(n) - [x(n-M) \ \mathbf{x}(n-M+1) \quad \cdots \quad \mathbf{x}(n-1)]\mathbf{h}_n
$$
$$
= \mathbf{x}(n) - \mathbf{X}_n\mathbf{h}_n \qquad (3.99)
$$

where $\mathbf{h}_n = [h_n(M) \ h_n(M-1) \quad \cdots \quad h_n(1)]^{\mathrm{T}}$ are the filter parameters at time n and the matrix $\mathbf{X}_n$ is the collection of M past GPR returns.

The estimate of $\mathbf{h}_n$, denoted by $\hat{\mathbf{h}}_n$, is computed using standard least squares as

$$
\hat{\mathbf{h}}_n = (\mathbf{X}_n^{\mathrm{T}}\mathbf{X}_n)^{-1}\mathbf{X}_n^{\mathrm{T}}\mathbf{x}(n) \qquad (3.100)
$$

The filter output then becomes

$$
\mathbf{y}(n) = \mathbf{X}_n\hat{\mathbf{h}}_n \qquad (3.101)
$$

Figure 3.48 shows the GPR return of an AP mine buried at a 2-in. depth. The energy of the GPR return is shown in Fig. 3.48c. The mine is located at position 150. The prediction error for $M = 5$ and its energy are shown in Fig. 3.48b and d. The energy of the raw GPR return and the energy of the prediction error are, for comparison, normalized by their maximums. Notice that the energy of the raw GPR return cannot be used to detect this object for this case. However, the energy of the prediction error has a peak at the location of the mine. A constant threshold can be applied to detect this landmine.

The parameter M is highly critical for detection performance. An ROC curve can be generated using different values of M. Then an optimal value for M can be chosen based on this ROC curve. Further improvements can be achieved by setting some of the values of $\mathbf{h}_n$ to zero. One can give an offset and exclude P of the previous samples. For example, $\mathbf{h}_n$ can be set equal to $\mathbf{h}_n = [h_n(M) \quad \cdots \quad h_n$ $(P+1) \ 0 \quad \cdots \quad 0]^{\mathrm{T}}$, where $P < M$, and only the parameters that are not set to zero can be learned. This is equivalent to having a guard region to avoid possible target pixels in the computation when position n corresponds to the background. This leads to another interpretation: One can randomly select from the previous samples and exclude them from the formulation. A generalization of this approach

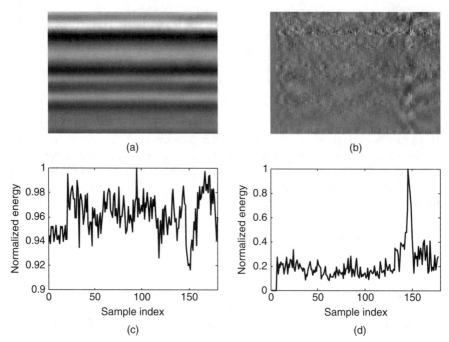

FIGURE 3.48   The concept of linear prediction-based detection: (a) B-scan of GPR data; (b) prediction error; (c) energy of the raw GPR data; (d) energy of the prediction error.

is a two-sided window. This results in a noncausal FIR filter. Then the filter output becomes

$$y(n) = \sum_{k=-M}^{M} h(k)x(n-k) \qquad (3.102)$$

Consider the three cases shown in Fig. 3.49. In case I, the shaded samples [i.e., $x(n-k)$, where $k = 6, 5, 4, -4, -5$, and -6] are taken into account and the rest of the samples are excluded by presetting the corresponding filter parameters to zero. Figure 3.49 shows the energy of the prediction errors for the three cases. These results clearly indicate that the domain of the observation window h has a strong effect on the performance of the linear prediction approach. The domain can be learned by a search algorithm such as the genetic algorithm.

3.8.2   L-Estimator–Based Detection

In this section we model the GPR signal by linear combinations of order statistics filters (L-filters). The main advantage of L-filters is that for a known noise distribution it is possible to choose the filter weights in such a way that it becomes the optimal filter in the mean-square-error sense. Mean filters, trimmed mean filters, and ranked order filters, including the median filter, can all be expressed as L-filters.

	$h(-6)$	$h(-5)$	$h(-4)$	$h(-3)$	$h(-2)$	$h(-1)$	**$h(0)$**	$h(1)$	$h(2)$	$h(3)$	$h(4)$	$h(5)$	$h(6)$
Case-I				0	0	0	**0**	0	0	0			
Case-II	0		0		0	0	**0**	0	0		0		0
Case-III			0	0		0	**0**	0		0	0		

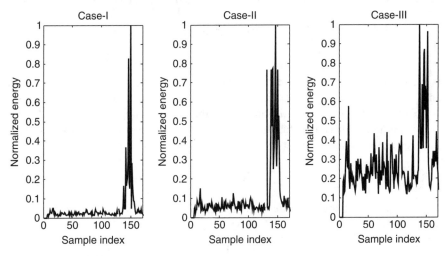

FIGURE 3.49 Normalized energy vs. sample index, cases I to III.

L-filters consist of a nonlinear operation, although they can express linear filters such as the mean filters. The output of the filter is obtained as a weighted sum of the ordered data values in the moving window. Let $y(n)$ be the output of the filter at time n. Suppose that the filter has M coefficients and we implement a *forward predictor* based on the L-estimators. Then $y(n)$ satisfies

$$y(n) = \sum_{k=1}^{M} h_k x_{(n-k)} \tag{3.103}$$

where we arrange M previous samples [i.e., $x(n-1), x(n-2), \ldots, x(n-M)$] in descending order so that $x_{(n-M)} \leq x_{(n-M+1)} \leq \cdots \leq x_{(n-1)}$.

We interpret $y(n)$ as a prediction of $x(n)$ from the previous samples $x(n-1)$, $x(n-2), \ldots, x(n-M)$. The prediction error [i.e., the difference between the measured signal at time of n and the model output $y(n)$] is given by

$$e(n) = x(n) - y(n) = x(n) - \sum_{k=1}^{M} h_k x_{(n-k)} \tag{3.104}$$

We wish to find the weights that minimize the mean-squared error (MSE),

$$E\{|e(n)|^2\} = \mathbf{h}^T \mathbf{R} \mathbf{h} - 2\mathbf{h}^T \mathbf{r} + E\{x^2(n)\} \tag{3.105}$$

where $\mathbf{h} = [h(1)\ h(2)\ \cdots\ h(M)]^T$ are the filter parameters,

$$R = \begin{bmatrix} E\{x_{(n-M)}x_{(n-M)}\} & E\{x_{(n-M)}x_{(n-M+1)}\} & \cdots & E\{x_{(n-M)}x_{(n-1)}\} \\ E\{x_{(n-M-1)}x_{(n-M)}\} & E\{x_{(n-M+1)}x_{(n-M+1)}\} & \cdots & E\{x_{(n-M-1)}x_{(n-1)}\} \\ \cdots & \cdots & \cdots & \cdots \\ E\{x_{(n-1)}x_{(n-M)}\} & E\{x_{(n-1)}x_{(n-M+1)}\} & \cdots & E\{x_{(n-1)}x_{(n-1)}\} \end{bmatrix}$$

$$(3.106)$$

is the correlation matrix of the order statistics $x_{(n-M)} \le x_{(n-M+1)} \le \cdots \le x_{(n-1)}$, and $\mathbf{r} = [E\{x_n x_{(n-M)}\}\ E\{x_n x_{(n-M+1)}\}\ \cdots\ E\{x_n x_{(n-1)}\}]^T$ is the cross-correlation between the current sample and the past M samples.

To minimize MSE as a function of $\mathbf{h}$, we take the partial derivative of MSE with respect to $\mathbf{h}$ and then set it equal to 0. Solving yields

$$\mathbf{h} = \mathbf{R}^{-1}\mathbf{r} \qquad (3.107)$$

If we require that the estimate be unbiased [i.e., $E\{y(n)\} = E\{x(n)\}$], $\mathbf{h}$ must satisfy

$$E\{x(n)\} = \sum_{k=1}^{M} h_k E\{x_{(n-k)}\}$$

$$= \mathbf{h}^T\mathbf{b} \qquad (3.108)$$

where $\mathbf{b} = [E\{x_{(n-M)}\}\ E\{x_{(n-M+1)}\}\ \cdots\ E\{x_{(n-1)}\}]^T$. We end up with a least-squares problem with linear constraints. The optimal coefficients are

$$\mathbf{h} = \mathbf{R}^{-1}\mathbf{r} + \frac{E\{x(n)\} - \mathbf{b}^T\mathbf{R}^{-1}\mathbf{r}}{\mathbf{b}^T\mathbf{R}^{-1}\mathbf{b}}\mathbf{R}^{-1}\mathbf{b} \qquad (3.109)$$

for the unbiasedness condition (see Vertiy and Gavrilov [14] for details).

The concept of applying the L-filters is illustrated in Fig. 3.50. The filter can be applied in a forward-predictor fashion or in a noncausal fashion. In the latter case, Eq. (3.103) can easily be reformulated; for example,

$$y(n) = \sum_{k=1}^{L} h_k x_{(k)} \qquad (3.110)$$

where $x_{(i)}$ is the ith smallest of the set $\{x(n-M), x(n-M+1), \ldots, x(n-P), x(n+P), \ldots, x(n+M))\}, 0 < P \le M$, and L is the number of elements in the set.

In some cases we are interested not only in locating the mine along the track direction, but also in its depth. Both order statistics and linear filters described in the previous section can be used to detect anomalies along the track and the time axis. For frequency-domain data, this corresponds to subband processing of GPR

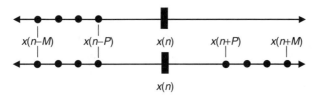

FIGURE 3.50 Window setup for forward prediction (top) and a noncausal filter (bottom).

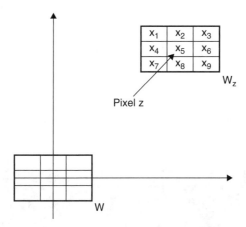

FIGURE 3.51 Forming the observation window by translation.

data. To detect anomalies not only along the track but also in the time axis, we apply the filter to the pixel values in a window translated in the time axis and along the track.

Consider the image domain shown in Fig. 3.51. Let Ψ be a filter to operate on the window $W = \{w_1, w_2, \ldots, w_n\}$, consisting of n pixels. The window is translated to a pixel $\mathbf{z}$ to form the observation window W_z. The filter is then applied to the GPR returns in this window. The process of applying the anomaly filter Ψ to the GPR data is shown in Fig. 3.52. First, an observation window is formed, as explained earlier. The GPR returns at each row of the observation window are then sorted. The optimal weights are estimated based on the observation vectors from 1 to m, where m denotes the number of rows in the observation window. The L-filter is then applied to each row separately. Finally, the prediction errors are computed for each row. The average of the prediction errors is used as a confidence value to indicate the presence of an anomaly. After estimating the filter parameters, it is not necessary to compute the prediction errors of all the rows used in estimating the filter coefficients. Only a few rows about the center of the window may be used to find the average prediction error.

The results of applying the anomaly detectors based on the linear predictor covered in the preceding section and the L-estimators are shown in Fig. 3.53. The results presented are for the raw GPR data shown in Fig. 3.48. The two filters, with

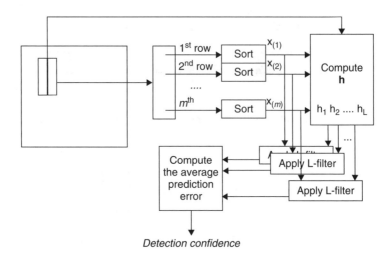

FIGURE 3.52 Anomaly detection based on *L*-estimators.

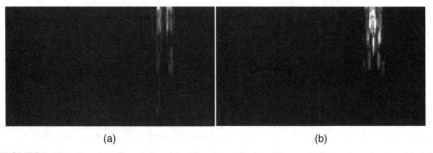

(a)   (b)

FIGURE 3.53 Results of applying (a) a linear prediction filter and (b) an *L*-estimator (right) for the raw GPR data shown in Fig. 3.48.

the same window size, are applied to several GPR data containing the signatures of antipersonnel and antitank mines and other clutter objects. The results in Fig. 3.53 show the typical performances of these two filters. Notice that unlike the result obtained by the L-estimator, the linear predictor causes two modalities, and it has no clear indication of depth information for the landmine.

3.8.3 Principal Component–Based Detection

Principal component analysis (PCA) is a widely used transform in applications ranging from spectral decorrelation to dimensionality reduction. PCA involves calculation of the eigenvalue decomposition of a data matrix. This operation is often interpreted as revealing the internal structure of the data and used to remove the outliers. In Chapter 3 we showed how to apply this technique to enhance the visual quality of B-scans. In this section we show how to apply this technique for landmine detection.

PCA techniques have previously been applied to GPR data analysis [186,187] for detection of mines on preprocessed data using cross track–depth scans. This approach is based on the observation that the clutter and noise either remain relatively constant from scans to scans or contain less energy than that of the mine scans. The algorithms proposed use clutter signatures as the differentiating features. Therefore, common features of clutter scans must be captured in a training phase. In the test phase, the new scans that display significantly distinct features are considered to contain mines. The use of anomaly detection using PCA is a key to the success of the handheld stand-off mine detection system (HSTAMIDS) developed for the U.S. government [188]. The PCA technique developed in this section has a different approach. It does not require a training mode to learn the principal components of the background from the background clutter samples. The idea behind this algorithm is similar to the use of eigenfaces for face detection [189]. It generates a mine detection alarm if the current sample does not project well to the subspace spanned by the principal components of the past few samples.

Consider a GPR return $\mathbf{x}_i$ received at sample index i. Form a matrix Γ_i from the past M past GPR returns:

$$\Gamma_i = [\mathbf{x}_{i-M} \quad \mathbf{x}_{i-M+1} \quad \cdots \quad \mathbf{x}_{i-1}] \tag{3.111}$$

The average of the training set is defined by

$$\Psi_i = \frac{1}{M} \sum_{k=1}^{M} \mathbf{x}_{i-k} \tag{3.112}$$

Each vector sample $\mathbf{x}_{i-k}$ differs from the average by the vector

$$\Phi_i^k = \mathbf{x}_{i-k} - \Psi_i \tag{3.113}$$

The covariance matrix $\mathbf{C}_i$ of the vector population Γ_i is

$$\mathbf{C}_i = \frac{1}{M} \mathbf{A}_i \mathbf{A}_i^{\mathrm{T}} \tag{3.114}$$

where the matrix $\mathbf{A}_i = [\Phi_i^1 \quad \Phi_i^2 \quad \cdots \quad \Phi_i^M]$. Let $\mathbf{v}_k$ and $\lambda_k, k = 1, 2, \ldots, N$, be the eigenvectors and corresponding eigenvalues of the covariance matrix $\mathbf{C}_i$. Arrange the eigenvalues in descending order so that $\lambda_{(1)} \geq \lambda_{(2)} \geq \cdots \geq \lambda_{(N)}$. Let $\mathbf{v}_{(k)}$ be the eigenvector associated with the eigenvalue $\lambda_{(k)}$. Let $\mathbf{V}_p$ be a matrix whose columns are formed from the eigenvectors $\mathbf{v}_{(k)}$ as follows:

$$\mathbf{V}_P = \lfloor \mathbf{v}_{(1)}, \mathbf{v}_{(2)}, \ldots, \mathbf{v}_{(P)} \rfloor \tag{3.115}$$

where $P \ll N$.

Given the current GPR return $\mathbf{x}_i$ received at sample index i, one can project it to the subspace Ω spanned $\mathbf{V}_p$.

$$\mathbf{y}_i = \mathbf{V}_P^T (\mathbf{x}_i - \Psi_i) \tag{3.116}$$

This removes noise while preserving signals that contribute to large variations from scan to scan. The similarity between the current sample and the past M samples can be measured by a distance measure. We can use the common Euclidian distance,

$$\varepsilon_i^k = (\mathbf{y}_i - \mathbf{y}_i^k)^T (\mathbf{y}_i - \mathbf{y}_i^k) \tag{3.117}$$

or the Mahalanobis distance,

$$\varepsilon_i^k = (\mathbf{y}_i - \mathbf{y}_i^k)^T \mathbf{D}_P^{-1} (\mathbf{y}_i - \mathbf{y}_i^k) \tag{3.118}$$

$k = 1, \ldots, M$, where

$$\mathbf{y}_i^k = \mathbf{V}_P^T (\mathbf{x}_{i-k} - \Psi_i) \tag{3.119}$$

and

$$\mathbf{D}_P = \begin{bmatrix} \lambda_{(1)} & 0 & \cdots & 0 \\ 0 & \lambda_{(2)} & \cdots & 0 \\ \cdots & \cdots & \cdots & 0 \\ 0 & 0 & 0 & \lambda_{(P)} \end{bmatrix} \tag{3.120}$$

The value of ε_i^k is related to how the current sample $\mathbf{x}_i$ is similar to the sample $\mathbf{x}_{i-k}$ in the transformed space. Let ε_i denote a vector that contains a collection of these distances:

$$\varepsilon_i = \lfloor \varepsilon_i^M \quad \varepsilon_i^{M-1} \quad \cdots \quad \varepsilon_i^1 \rfloor \tag{3.121}$$

The vector element having the minimum value (i.e., $\varepsilon_i^{\min} = \min\{\varepsilon_i^M, \varepsilon_i^{M-1}, \ldots, \varepsilon_i^1\}$), gives the worst-case scenario for the similarity between the current sample and the past M samples, as we use the most similar sample from the past M samples to detect the anomaly. We can also use the maximum and median values of the vector ε_i to detect the anomaly. If ε_i is above a threshold, this is in indication of an abrupt change. Figure 3.54 shows raw GPR data and the corresponding similarity values between the current sample and the preceding 16 samples. The position of the landmine is indicated by two vertical lines. The values of $\varepsilon_i^{\min}$, $\varepsilon_i^{\text{median}}$, and $\varepsilon_i^{\max}$ for each scan location i are normalized by their maximum values, for the sake of comparison. Notice that the minimum distance is capable of detecting the initial location of a landmine. On the other hand, the maximum distance ends up with two peaks, corresponding to blank–landmine and landmine–blank transitions. The median distance shows a result similar to that for the maximum distance for this example, but it is more robust in general.

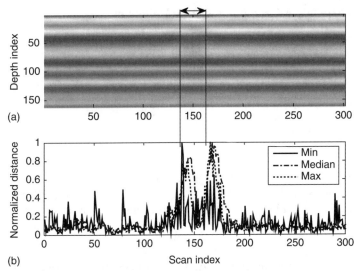

(a)

(b)

FIGURE 3.54 The concept of abrupt change detection using PCA. (a) The GPR return; the boundaries of the landmine are marked with vertical lines. (b) Scan index versus the normalized minimum, maximum, and median distances between the current sample and the past 16 samples.

These minimum and maximum distance values can also be computed as $\varepsilon_i^{\min,r} = \min\{\varepsilon_i^M, \varepsilon_i^{M-1}, \ldots, \varepsilon_i^r\}$ and $\varepsilon_i^{\max,r} = \max\{\varepsilon_i^M, \varepsilon_i^{M-1}, \ldots, \varepsilon_i^r\}$, where $1 < r \le M$. This formulation uses a guard area with r samples to make sure that when we compute the similarity of GPR returns for the landmine and blank areas, we do not compare the landmine returns by themselves. The width of the guard is chosen large enough to cover half the largest landmine signature in the scan direction. Figure 3.55 shows $\varepsilon_i^{\min,12}$ and $\varepsilon_i^{\max,12}$ for the GPR data in Fig. 3.54. To compute the distances, we used the past 20 samples. It is easy to see that the first peak corresponds to the landmine location and can be used to generate a mine detection alarm. The second peak corresponds to the landmine–blank area transition and may be eliminated by postprocessing.

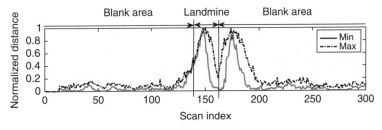

FIGURE 3.55 Scan index versus $\varepsilon_i^{\min,12}$ and $\varepsilon_i^{\max,12}$ for the GPR data in Fig. 3.54.

3.8.4   Independent Component–Based Detection

Independent component analysis (ICA) is a method of finding underlying (i.e., statistically independent and non-Gaussian) components from multidimensional statistical data. It was introduced in the early 1980s by J. Herault, C. Juttent, and B. Ans. Due to its generality, the ICA model has been applied to many problem domains, such as blind–source separation, feature extraction, and image compression [190–196].

The ICA model is

$$\mathbf{x} = \mathbf{As} \tag{3.122}$$

where $\mathbf{x}$ represents the measured data, the components of $\mathbf{s}$ are unknown independent components, and the matrix $\mathbf{A}$ is an unknown matrix of coefficients [191]. ICA estimation algorithms estimate $\mathbf{A}$, and then an estimate of $\mathbf{s}$ is given by

$$\mathbf{s} = \mathbf{Wx} \tag{3.123}$$

where $\mathbf{W}$ is the pseudoinverse of $\mathbf{A}$. ICA assumes that the components of the weight vector $\mathbf{s}$ are statistically independent. In other words, the assumption is that information on the value of s_i does not give any information on the value of s_j, and vice versa. If the independent components are Gaussian, their joint density is completely symmetric. Unfortunately, this makes it impossible to estimate the matrix $\mathbf{A}$ since the joint density does not contain any information on the directions of the columns of $\mathbf{A}$. Therefore, non-Gaussianity plays an important role in ICA.

There are several ways to measure non-Gaussianity, one of which is *negentropy*, based on the information-theoretic quantity of (differential) entropy [191]. Hyvärinen formulates the ICA problem as minimization of mutual information between the transformed variables s_i and a family of contrast (objective) functions. Here we use the FastICA algorithm developed by Hyvärinen to find the basis functions [190]. Consider the $M \times 1$ time-domain signal $\mathbf{x}_i$ received at location i along the down track. This signal can be represented by a linear mixture of signals radiated from objects and clutter. The problem can then be formulated as a blind source separation problem. Assume that the signal can be represented by a linear combination of L signals:

$$\mathbf{x} = \sum_{k=1}^{L} \mathbf{A}_k s_k \tag{3.124}$$

where $\mathbf{A}_k$ is the kth column of mixing matrix $\mathbf{A}$.

Unlike PCA, this formulation does not guide directly on components to be excluded from the signal to reduce clutter. In other words, automatic selection of subspace components corresponding to clutter is not natural for ICA. This step usually requires close examination of independent components. Assume that the components are arranged as follows:

$$\mathbf{x}^{\text{clutter}} = \sum_{k=1}^{L_1} \mathbf{A}_k s_k \tag{3.125}$$

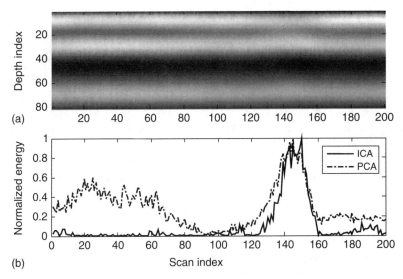

FIGURE 3.56 (a) Raw GPR data with an AP mines (b) results of applying ICA and PCA.

where $L_1 < L$. Then the rest of the components represent the object:

$$\mathbf{x}^{\text{object}} = \mathbf{x} - \mathbf{x}^{\text{clutter}} \qquad (3.126)$$

Let us consider the GPR data in Fig. 3.56a. An antipersonnel mine is present at position 150. The number of independent components to be estimated is assumed to be 5. Before proceeding with the calculation of basis, ICA uses centering, whitening, and dimensionality reduction algorithms as preprocessing steps to simplify and reduce the complexity of the problem. We used PCA for whitening. ICA computation is performed using the Fastica algorithm. In this implementation, the deflation approach to ICA computation is used. After ICA computation we obtain the transformations **A** and **W**. Two independent components are selected by visual examination. Transformation 26 is applied to the mean of the data removed. The energy of the transformed signal and whitened signals are shown in the figure. For the sake of comparison, the energy values are normalized by their maximum values. For this example, ICA finds a more informative basis set than PCA to generate a mine detection alarm at a reduced false-alarm rate. Although the results are attractive, this technique is not suitable for learning the clutter in the field of operation, as convergence is not guaranteed. This issue for landmine detection has been reported by several authors, and different approaches to overcoming this problem have been discussed [192–196].

3.8.5 Mathematical Morphology for Detecting Landmines

In this section we review the literature on the use of mathematical morphology to detect landmines. Mathematical morphology is an important discipline and has been used widely to analyze the geometrical and topological properties of objects in digital images and extract signal characteristics. Morphological operations are based on morphological transformations of signals by sets [185]. They are non-linear, translation-invariant transformations that involve probing an image with structuring elements. The performance of the morphological operations depends solely on selection of the value and shape of the structuring element. However, the images encountered in many automatic target recognition (ATR) applications are extremely complicated. Selecting the structuring elements is not an easy task for such applications.

Morphological shared-weight neural networks (MSNNs) combine the morphological algorithm and neural networks [197]. The hit-and-miss transform is one of the morphological operations invariant to gray-level shifts. Due to this property, the morphological algorithm used in MSNN is the gray-level hit-and-miss transform. It is used to extract features from two- or three-dimensional data. The size and gray values of the structuring elements in the hit-and-miss structure greatly affect the features extracted. Optimizing the gray values of the structuring elements and the network weights is performed using the standard back-propagation algorithm.

Morphological shared-weight neural networks are an extension of shared-weight neural networks (SSNNs) [198], which have been employed in many different areas, such as discriminating textures and the handwritten character recognition problem [199–202]. SSNNs use linear correlational filters for feature extraction. Linear correlational or matched filters have been used extensively for automatic target recognition. The idea behind these filters is to obtain a high confidence value at the location of the target to be detected and low confidence values at other locations. A major drawback of these approaches is that they are poor in terms of detecting targets that are not similar to those in the training set.

The many types of neural networks that have been developed are widely used for classification tasks. However, extracting effective features is difficult. SSNNs and MSNNs are heterogeneous neural networks, and as they learn feature extraction and classification simultaneously, they are attractive. Choquet morphological shared-weight neural networks (CMSNNs) have a network topology similar to those of SSNNs and MSNNs. CMSNNs use a pair of Choquet integral–based morphological operators as a neural node [203]. These nodes provide not only the morphological hit-and-miss transformation but also the tophat transform and morphological gradient. The gray-scale hit-and-miss transform is used to measure how a shape h fits under f using an erosion operator and how a shape m fits above f using a dilation operator. The hit-and-miss transform is defined by

$$f \otimes (h, m) = (f \ominus h) - (f \oplus m^*) \tag{3.127}$$

where

$$(f \ominus g)(x) = \min\{f(z) - g_x(z) \; : \; z \in D[g_x]\} \tag{3.128}$$

$$(f \oplus g)(x) = \max\{f(z) - g_x^*(z) \; : \; z \in D[g_x^*]\} \tag{3.129}$$

and $g_x(z) = g(z - x)$, $g^*(z) = -g(-z)$, and $D[g]$ is the domain of g. Equation (3.128) and (3.129) are the erosion and dilation operators, respectively. These operators are the most elementary operators of mathematical morphology. The Choquet integral is a generalization of the standard erosion and dilation operators with flat structuring elements. The sensitivity of standard morphological operators to noise is well known. On the other hand, the Choquet integrals define robust morphological operators. This property, along with the combination of the gray-scale mathematical morphology with the Choquet integral, provides a potential use of this approach to real-world problems in information fusion and object detection.

A CMSNN has two parts. The first part is used to extract features from the inputs by using one or more Choquet integral–based-hit-and-miss transforms (CHMTs). The second part is a neural network classifier. Each CHMT produces a feature map, which can also be viewed as an output image. Each of these feature maps is used as an input to a neural network classifier. The classifier has p outputs, where p represents the number of classes, such as *antipersonnel mine, antitank mine*, and *clutter object*. It is desired that the CMSNN output at node i is 1 and all others are zero when the input belongs to class i. The basic architecture is shown in Fig. 3.57. The architecture of CHMTs is described in Chapter 8.

The applicability of CMSNNs in landmine detection problems using a very significant data set acquired from a variety of geographical locations and environmental conditions has been investigated and the performance compared to that of MSNNs [204]. The experiments showed that CMSNNs perform better than MSNNs and that domain reduction (domain learning) has the potential to increase

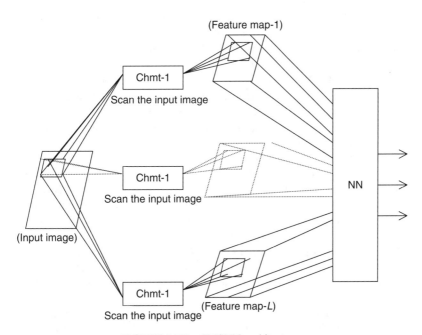

FIGURE 3.57 CMSNN architecture.

detection performance. The most attractive feature of CMSNNs is their ability to learn features. Landmine signatures are extremely complicated and change all the time, due to soil properties and weather conditions. Mines at different depths and orientations are still a major problem, and their signatures change by these factors. Discovering effective features for all these situations is difficult without an automated process. CMSNNs' ability to optimize the structuring elements of morphological filters was demonstrated on a GPR data set collected from off-road from dirt and gravel lanes. For such a complicated and changing image, a detection rate of 96% has been achieved by CMSNNs at 0.03 false alarm per square meter. The performance of CMSNNs was compared to that of three other algorithms: the continuous HMM (cHMM) model, the discrete HMM (dHMM) model, and the fuzzy ATR [205,206]. Although CMSNN outperforms MSNN, the performance of the CMSNN is not better than that of these three algorithms. Fusing their outputs did not achieve a significant improvement.

REFERENCES

1. Daniels, D. J. 1996. *Surface Penetrating Radar*. IEE Radar, Sonar, Navigation and Avionics Series 6. IEE, London.
2. Turk, A. S., Hocaoglu, A. K. 2005. Buried object detection. In *Encyclopedia of RF and Microwave Engineering*, Vol. 1. Wiley-Interscience, Hoboken, NJ, pp. 541–559.
3. Borchert, O., Aliman, M., Glasmachers, A. 2007. Directional borehole radar calibration. In *Proceedings of the 4th International Workshop on Advanced Ground Penetrating Radar (IWAGPR 2007)*, Naples, Italy, pp. 19–23.
4. RAND. 2003. *Alternatives for Landmine Detection*. RAND Science and Technology Policy. RAND Corporation, Santa Monica, CA.
5. Sahinkaya, D. S. A., Turk, A.S. 2004. UWB GPR for detection and identification of buried small objects. *Proc. SPIE*.
6. Taylor, D. 2001. *Ultra-wide Band Technology*. CRC Press, Boca Raton, FL.
7. Kong, F. N., By, T. L. 1993. Theory and performance of a GPR system which uses step frequency signals. *J. Appl. Geophys.*, 33:453–445.
8. Parrini, F., Pieraccini, M., Atzeni, C. 2004. A high-speed continuous wave GPR. In *Proceedings of the 10th International Conference on Ground Penetrating Radar*, Delft, The Netherlands, pp. 183–186.
9. Stickley, G. F., Noon, D. A., Cheriakov, M., Longstaff, I. D. 2000. Gated stepped frequency ground penetrating radar. *J. Appl. Geophys.*, 43(2–4):259–269. Special Issue on Ground Penetrating Radar (GPR 98), C. T. Allen, R. G. Plumb, Eds.
10. Chan, L. C., Moffat, D. L., Peters, L. 1979. A characterization of subsurface radar targets. *Proc. IEEE*, 67(7):991–1000.
11. Turk, A. S., Sahinkaya, D. A., Sezgin, M., Nazli, H. 2007. Investigation of convenient antenna designs for ultra-wide band GPR systems. In *Proceedings of the 4th International Workshop on Advanced Ground Penetrating Radar (IWAGPR 2007)*, Naples, Italy, pp. 192–196.
12. Rial, F. I., Pareira, M., Lorenzo, H., Arias, P., Novo, A. 2007. Vertical and horizontal resolution of GPR bow-tie antennas. In *Proceedings of the 4th International Workshop on Advanced Ground Penetrating Radar*, Naples, Italy, pp. 187–191.

13. Lee, J. S., Nguyen, C. 2001. Novel low-cost ultra-wideband, ultra-short-pulse transmitter with MESFET impulse-shaping circuitry for reduced distortion and improved pulse repetition rate. *IEEE Microwave Wireless Component Lett.*, 11(5).

14. Vertiy, A. A., Gavrilov, S. P. 1998. Modeling of microwave images of buried cylindrical objects. *Int. J. Infrared Millimeter Waves*, 19(9):1201–1220.

15. Koppenjan, S. K., Bashforth, M. B. 1993. The Department of Energy's ground penetrating radar, an FM-CW system. In *Underground and Obscured Object Imaging and Detection*, N. K. Del Grande, I. Cindrich, P. B. Johnson, Eds. *Proc. SPIE*, 1942:44–55.

16. Noon, D. A. Longstaff, I. D. Yelf, R. J. 1994. Advances in the development of step frequency ground penetrating radar. In *Proceedings of the 5th International Conference on Ground Penetrating Radar (GPR 94)*, Kitchener, Ontario, Canada, pp. 117–132.

17. Hatchard, C. 2003. A combined MD/GPR detector: the HSTAMIDS system. Presented at the International Conference on Requirements and Technologies for the Detection, Removal and Neutralization of Landmines and UXO (EUDEM2-SCOT 2003). Vrije Universiteit Brussel, Brussels, Belgium, Sept. 15–18.

18. Sato, M., Fujiwara, J., Takahashi, K. 1999. The development of the hand-held dual sensor ALIS. In *Detection and Remediation Technologies for Mines and Minelike Targets IV. Proc. SPIE*, 6553:1–10.

19. Van Genderen, P., Nicolaescu, I. 2003. System description of a stepped frequency continuous wave radar for humanitarian demining. In *Second International Workshop on Advanced Ground Penetrating Radar*, Delft, The Netherlands, pp. 9–15.

20. Rihaczek, A. W. 1985. *Principles of High-Resolution Radar*. Peninsula Publishing, Newport Beach, CA.

21. van Genderen, P. 2001. The effect of phase noise in a stepped frequency continuous wave ground penetrating radar. In *CIE International Conference on Radar IEEE*, Beijing, China, pp. 581–584.

22. van Genderen, P., Hakkaart, P., van Heijenoort, J., Hermans, G. P. 2001. A multifrequency radar for detecting landmines: design aspects and electrical performance. In *31st European Microwave Conference*, London, pp. 249–252.

23. van Genderen, P., Nicolaescu, I., Zijderveld, J. 2003. Some experience with the use of spiral antennas for GPR landmine detection. In *Proceedings of the International Conference on Radar 2003*, Adelaide, Australia, Sept., pp. 219–223.

24. Nicolaescu, I. 2003. *Parameters Analysis of Stepped-Frequency Continuous-Wave Ground-Penetrating Radar Used for Landmines Detection*. Research Report IRCTR-S-007-03. Publisher: Delft University of Technology, Delft, The Netherlands.

25. Nicolaescu, I. 2003. *Stepped Frequency Continuous Wave Radar Used for Landmines Detection*. Research Report IRCTR-S-004-03. Delft, The Netherlands.

26. Nicolaescu, I., van Genderen, P., van Heijenoort, J. 2003. Range resolution and calibration of an ultra wideband stepped frequency continuous wave ground penetrating radar. In *Conference Proceedings, International Radar Symposium 2003*, Dresden, Germany, pp. 301–306.

27. King, R. W. P., Smith, G. S. 1981. *Antennas in Matter*, MIT Press, Cambridge, MA.

28. Schlager, K. L., Smith, G. S., Maloney, J. G. 1994. Optimization of bow-tie antenna for pulse radiation. *IEEE Trans. Antennas Propag.*, 42(7):975–982.

29. Chumachenko, V. P., Turk, A. S., 2001. Radiation characteristics of wide-angle H-plane sectoral horn loaded with dielectric of multiangular shape. *Int. J. Electron.*, 88(1):91–101.

30. Yaravoy, A. G., Schukin, A. D., Kaploun, I. V., Lightart, L. P. 2002. The dielectric wedge antenna. *IEEE Trans. Antennas Propag.*, 50(10):1460–1471.

31. Skolnik, M. I. 1970. *Radar Handbook*. McGraw-Hill, New York.

32. Turk, A. S. 2004. Ultra-wide band TEM horn design for ground penetrating impulse radar system. *Microwave Opt. Technol. Lett.*, 41(6):333–336.

33. Turk, A. S. 2006. Ultra-wide band Vivaldi antenna design for multi-sensor adaptive ground-penetrating impulse radar. *Microwave Opt. Technol. Lett.*, 48(5):834–839.

34. Turk, A. S., Nazli, H. 2008. Hyper-wide band TEM horn array design for multi band ground-penetrating impulse radar. *Microwave Opt. Technol. Lett.*, 50(1):76–81.

35. Gader, P. D., Khabou, M. A., Koldobsky, A. 2000. Morphological regularization neural networks. *Pattern Recognition*, 33(6):935–944.

36. Torrione, P., Collins, L. 2006. Ground response tracking for improved landmine detection in ground penetrating radar data. In *Geoscience and Remote Sensing Symposium, 2006 (IGARSS 2006)*, July 31–Aug. 4, pp. 153–156.

37. Caldecott, R., Young, J. D., Hall, J. P., Terzuoli, A. J. 1985. *An Underground Obstacle Detection and Mapping System*. Technical Report EL-3984. Electroscience Laboratory, The Ohio State University, Columbus, OH, and the Electric Power Research Institute, Palo Alto. CA.

38. Brunzell, H. 1999. Detection of shallowly buried objects using impulse radar. *IEEE Trans. Geosci. Remote Sens.*, 37:875–886.

39. Karlsen, B., Larsen, J., Sorensen, H. B. D., Jakobsen, K. B. 2001. Comparison of PCA and ICA based clutter reduction in GPR systems for anti-personal landmine detection. In *Proceedings of the 11th IEEE Signal Processing Workshop*, pp. 146–149.

40. Yu, S., Mehra, R. K., Witten, T. R. 1999. Automatic mine detection based on ground penetrating radar. In *Detection and Remediation Technologies for Mines and Minelike Targets IV*, Orlando, FL. *Proc. SPIE*, pp. 961–972.

41. Duflos, E., Hervy, P., Nivelle, F., Perrin, S., Vanheeghe, P. 1999. Time-frequency analysis of ground penetrating radar signals for mines detection applications. In *Proceedings of the IEEE International Conference on Systems, Man, and Cybernetics*, Tokyo, Vol. 1, pp. 520–525.

42. Gader, P., Lee, W. H., Wilson, J. N. 2004. Detecting landmines with ground-penetrating radar using feature-based rules, order statistics, and adaptive whitening. *IEEE Trans. Geosci. Remote Sens.*, 42(11):2522–2534.

43. Yelf, R. 2004. *10th International Conference on Ground Penetrating Radar*, Delft, The Netherlands, June 21–24.

44. Oppenheim, A. V., Schafer, R. W., Buck, J. R. 1999. *Discrete-Time Signal Processing*. Prentice Hall, Upper Saddle River, NJ.

45. Manolakis, D. G., Ingle, V. K., Kogon, S. M. 2000. *Statistical and Adaptive Signal Processing*. McGraw-Hill, Singapore.

46. Peters, L., Jr., Daniels, D. J., Young, J. D. 1994. Ground penetrating radar as a subsurface environmental sensing tool. *Proc IEEE*, 82(12):1802–1822.

47. Vitebskiy, S., Carin, L., Ressler, M. A., Le, F. H. 1997. Ultrawide-band, short pulse ground-penetrating radar: simulation and measurement. *IEEE Trans. Geosci. Remote Sens.*, 35:762–772.

48. Carin, L., Geng, N., McClure, M., Sichina, J., Nguyen, L. 1999. Ultra-wide-band synthetic-aperture radar for mine-field detection. *IEEE Trans. Antennas Propag.*, 41:18–33.

49. Schneider, W. A. 1978. Integral formulation for migration in two and three dimensions. *Geophysics*, 43:49–76.

50. Gazdag, J. 1978. Wave equation migration with the phase-shift method. *Geophysics*, 43:1342–1351.

51. Stolt, R. H. 1978. Migration by Fourier transformation. *Geophysics*, 43:23–48.

52. Baysal, E., Kosloff, D. D., Sherwood, J. W. C. 1983. Reverse time migration. *Geophysics*, 48:1514–1524.

53. Leuschen, C. J., Plumb, R. G. 2001. A matched-filter-based reverse-time migration algorithm for ground-penetrating radar data. *IEEE Trans. Geosci. Remote Sens.*, 39:929–936.

54. Gu, K., Wang, G., Li, J. 2004. Migration based SAR imaging for ground penetrating radar systems. *IEE Proc. Radar Sonar Navig.*, 151(5):317–325.

55. Song, J., Liu, Q. H., Torrione, P., Collins, L. 2006. Two-dimensional and three-dimensional NUFFT migration method for landmine detection using ground-penetrating radar. *IEEE Trans. Geosci. Remote Sens.*, 44(6):1462–1469.

56. Cafforio, C., Prati, C., Rocca, F. 1991. Full resolution focusing of Seasat SAR images in the frequency–wave number domain. *J. Robot. Syst.*, 12:491–510.

57. Cafforio, C., Prati, C., Rocca, F. 1991. SAR data focusing using seismic migration techniques. *IEEE Trans. Aerosp. Electron. Syst.*, 27:194–207.

58. Milman, A. S. 1993. SAR imaging using the w-k migration. *Int. J. Remote Sens.*, 14:1965–1979.

59. Callow, H. J., Hayes, M. P., Gough, P. T. 2002. Wavenumber domain reconstruction of SAR/SAS imagery using single transmitter and multiple-receiver geometry. *Electron. Lett.*, 38:336–337.

60. Gunawardena, A., Longstaff, D. 1998. Wave equation formulation of synthetic aperture radar (SAR) algorithms in the time-space domain. *IEEE Trans. Geosci. Remote Sens.*, 36:1995–1999.

61. Anxue, Z., Yansheng, J., Wenbing, W., Cheng, W. 2000. Experimental studies on GPR velocity estimation and imaging method using migration in frequency–wavenumber domain. In *Proceedings of ISAPE*, Beijing, pp. 468–473.

62. Gilmore, C., Jeffrey, I., LoVetri, J. 2006. Derivation and comparison of SAR and frequency–wavenumber migration within a common inverse scalar wave problem formulation. *IEEE Trans. Geosci. Remote Sens.*, 44:1454–1461.

63. Claerbout, J. F. 1985. *Imaging the Earth's Interior*. Blackwell Scientific, Oxford, UK.

64. Ozdemir, C., Demirci, S., Yigit, E., Kavak, A. 2007. A hyperbolic summation method to focus B-scan ground penetrating radar images: an experimental study with a stepped frequency system. *Microwave Opt. Technol. Lett.*, 49(3):671–676.

65. Yilmaz, O. 1987. *Seismic Data Processing*. Society of Exploration Geophysicists, Tulsa, OK.

66. Jones, D. S. 1964. *The Theory of Electromagnetism*. Pergamon Press, Elmsford, NY.

67. Morse, P. M., Feshbach, H. 1953. *Methods of Theoretical Physics*, Vol. 1, Sect. 7.3. McGraw-Hill, New York.

68. Lecomte, I., Hamran, S. E., Gelius, L. J. 2005. Improving Kirchhoff migration with repeated local plane-wave imaging? A SAR-inspired signal-processing approach in prestack depth imaging. *Geophys. Prospect.*, 53:767–785.

69. Stratton, J. A. 1941. *Electromagnetic Theory*. McGraw-Hill, New York.

70. Yigit, E., Demirci, S., Ozdemir, C., Kavak, A. 2007. A synthetic aperture radar-based focusing algorithm for B-scan ground penetrating radar imagery. *Microwave Opt. Technol. Lett.*, 49:2534–2540.

71. Soumekh, M. 1992. A system model and inversion for synthetic aperture radar imaging. *IEEE Trans. Image Process.*, 1:64–76.

72. Kovalenko, V., Yarovoy, A., Ligthart, L. P. 2006. A SAR-based algorithm for imaging of landmines with GPR. In *International Workshop on Imaging Systems and Technology* (IST 2006), Minori, Italy, pp. 65–70.

73. Chew, W. C. 1995. *Waves and Fields in Inhomogeneous Media*, 2nd ed. IEEE Press, New York.

74. Fisher, E., McMechan, G. A., Annan, A. P., Cosway, S. W. 1992. Examples of reverse-time migration of single-channel ground-penetrating radar profiles. *Geophysics*, 57:577–586.

75. Capineri, L., Grande, P., Temple, J. A. G. 1998. Advanced image-processing technique for real-time interpretation of ground-penetrating radar images. *Int. J. Imag. Syst. Technol.*, 9:51–59.

76. Morrow, I. L., Van Genderen, P. A. 2001. 2D polarimetric backpropagation algorithm for ground-penetrating radar applications. *Microwave Opt. Technol. Lett.*, 28:1–4.

77. Kaneko, T. 1990. Radar image processing for locating underground linear objects. In *IAPR Workshop on Machine Vision Applications*, Tokyo, pp. 35–38.

78. Dell'Acqua, A., Sarti, A., Tubaro, S., Zanzi, L. 2004. Detection of linear objects in GPR data. *Signal Process.*, 84(4):785–799.

79. Carlotto, M. J. 2002. Detecting buried mines in ground penetrating radar using a Hough transform approach. *Proc. SPIE*, 4741:251–261.

80. Kagalenko, M. B., Weedon, W. H. 1996. Comparison of backpropagation and synthetic aperture imaging algorithms for processing GPR data. *IEEE-APS Soc. Int. Symp. Dig.*, 3:2179–2182.

81. Kocak, T., Draper, M. 2006. A back-propagation neural network landmine detector using the delta-technique and S-statistic. *Neural Process. Lett.*, 23(1):47–54.

82. Guangyou, F. 2001. FDTD and optimization approach to time-domain inversing problem for underground multiple objects. *Microwave Opt. Technol. Lett.*, 31:384–387.

83. Gurel, L., Oguz, U. 2000. Three-dimensional FDTD modeling of a ground-penetrating radar. *IEEE Trans. Geosci. Remote Sens.*, 38(4):1513–1521.

84. Bourgeois, J. M., Smith, G. S. 1996. A fully three-dimensional simulation of a ground-penetrating radar: FDTD theory compared with experiment. *IEEE Trans. Geosci. Remote Sens.*, 34:36–44.

85. Bolomey, J., Lesselier, D., Pichot, C., Tabbara, W. 1981. Spectral and time domain approaches to some inverse scattering problems. *IEEE Trans Antennas Propag.*, 29:206–212.

86. Soldovieri, F., Persico, R., Leone, G. 2004. Frequency diversity in a linear inversion algorithm for GPR prospecting. In *Proceedings of the Tenth International Conference Ground Penetrating Radar (GPR 2004)*, Vol. 1, pp. 8–90.

87. Zhou, H., Sato, M., Liu, H. 2005. Migration velocity analysis and prestack migration of common-transmitter GPR data. *IEEE Trans. Geosci. Remote Sens.*, 43(1):86–91.

88. Berenger, J. P. 1994. A perfectly matched layer for the absorption of electromagnetic waves. *J. Comput. Phys.*, 114:185–200.

89. Bretones, A. R., Martin, R. G., Rubio, R. G., Garcia, S. G., Pantoja, M. F. 2004. On the simulation of a GPR using an ADI-FDTD/MoMTD hybrid method. In *10th International Conference on Ground Penetrating Radar*, June, 21–24, pp. 13–15.

90. Bourgeois, J. M., Smith, G. S. 1996. A fully three-dimensional simulation of ground-penetrating radar: FDTD theory compared with experiment. *IEEE Trans. Remote Sens.*, 34(1):36–44.

91. Bourgeois, J. M., Smith, G. S. 1997. A complete electromagnetic simulation of a ground penetrating radar for mine detection: theory and experiment. In *IEEE AP-S International Symposium*, Vol. 2, pp. 986–989.

92. Bourgeois, J. M., Smith, G. S. 1998. A complete electromagnetic simulation of the separated-aperture sensor for detecting buried land mines. *IEEE Trans. Antennas Propag.*, 46:1419–1426.

93. Calhoun, J. 1997. A finite difference time domain (FDTD) simulation of electromagnetic wave propagation and scattering in a partially conducting layered earth. In *IEEE International Geoscience and Remote Sensing Symposium (IGARSS)*, Aug. 3–8, pp. 922–924.

94. Cassidy, N. J., Murdie, R. E. 2000. The application of mathematical modelling in the interpretation of near-surface ground penetrating radar sections. In *Proceedings of the 8th International Conference on Ground Penetrating Radar*, May 23–26, pp. 842–847.

95. Cassidy, N. J., Tuckwell, G. W. 2002. Mathematical modelling of ground penetrating radar: parallel computing applications. In *Proceedings of the 9th International Conference on Ground Penetrating Radar*, Apr. 29–May 2, pp. 514–519.

96. Chen, H. W., Huang, T. M. 1998. Finite-difference time-domain simulation of GPR data. *J. Appl. Geophys.*, 40:139–163.

97. Chen, H. C., Rao, K. R., Lee, R. 2001. A tapered-permittivity rod antenna for ground penetrating radar applications. *J. Appl. Geophys.*, 47:309–316.

98. Chen, H. C., Rao, K. R., Lee, R. 2003. A new ultrawide-bandwidth dielectric-rod antenna for ground-penetrating radar applications. *IEEE Trans. Antennas Propag.*, 51(3):371–377.

99. Chew, W. C., Teixeira, F. L., Straka, M., Oristaglio, M. L., Wang, T. 1997. Parallel 3D PML-FDTD simulation of GPR on dispersive, inhomogeneous and conductive media. In *IEEE AP-S International Symposium*, July 13–18, pp. 380–383.

100. Chew, W., Kosmas, P., Leeser, M., Rappaport, C. 2004. An FPGA implementation of the two-dimensional finite-difference time-domain (FDTD) algorithm. In *12th International Symposium on Field-Programmable Gate Arrays (FPGA 2004)*, Feb. 22–24, pp. 213–222.

101. Daniels, D. J. 2004. *Ground Penetrating Radar*, 2nd ed. IEE, London.

102. Diamanti, N., Giannopoulos, A. 2007. An investigation into the implementation of ADI-FDTD subgrids in FDTD GPR modeling. In *Proceedings of the 4th International Workshop on Advanced Ground Penetrating Radar*, June 27–29, pp. 122–126.

103. Fang, J., Wu, Z. 1995. Generalized perfectly matched layer: an extension of Berenger's perfectly matched layer boundary condition. *IEEE Microwave Guided Wave Lett.*, 5(12):451–453.

104. Farnoosh, N., Shoory, A., Moini, R., Sadeghi, H. H. 2007. A hybrid MOMFD-FDTD ground penetrating radar modeling technique to detect multiple buried objects. In *9th International Symposium on Signal Processing and Its Applications*, Feb. 12–15, pp. 1–4.

105. Ghasemi, F. S. A., Abrishamian, M. S. 2007. A novel method for FDTD numerical GPR imaging of arbitrary shapes based on Fourier transform. *Non-Destructive Test. Eval. Int.*, 40:140–146.

106. Giannopoulos, A. 2005. Modelling ground penetrating radar by GprMax. *Constr. Build. Mater.*, 19:755–762.

107. Guangyou, F. 2001. FDTD and optimization approach to time-domain inversing problem for underground multiple objects. *Microwave Opt. Technol. Lett.*, 31(5):384–387.

108. Gürel, L. Oğuz, U. 1999. Employing PML absorbers in the design and simulation of ground penetrating radars. In *IEEE AP-S International Symposium*, July 11–16, Vol. 3, pp. 1890–1893.

109. Gürel, L., Oğuz, U. 2000. Three dimensional FDTD modeling of a ground-penetrating radar. *IEEE Trans. Geosci. Remote Sens.*, 38(4):1513–1521.

110. Gürel, U., Oğuz, L. 2002. Transmitter-receiver–transmitter configurations of ground penetrating radar. *Radio Sci.*, 37(3):5–1 to 5–7.

111. Gürel, U., Oğuz, L. 2002. Transmitter–receiver–transmitter configured ground-penetrating radars over randomly heterogeneous ground models. *Radio Sci.*, 37(6):6–1 to 6–9.

112. Gürel, L., Oğuz, U. 2003. Optimization of the transmitter–receiver separation in the ground-penetrating radar. *IEEE Trans. Antennas Propag.*, 51(3):362–370.

113. Guangyou, F., Zhongzhi, Z. 1997. The calculation of a SP-GPR antenna near lossy media interface by FD-TD method. In *Asia Pacific Microwave Conference*, Dec. 2–5, Vol. 3, pp. 1193–1196.

114. Hee, Y., Uno, T., Adachi, S. 1994. FDTD analysis of two dimensional transient scattering of cylindrical wave via buried conducting objects. *Electron. Commun. Jpn. I*, 77–55:93–102.

115. Hoa, K. C., Carin, L., Gader, P. D., Wilson, J. N. 2008. An investigation of using the spectral characteristics from ground penetrating radar for landmine/clutter discrimination. *IEEE Trans. Geosci. Remote Sens.*, 46(4):1177–1191.

116. Huang, Z., Demarest, K. R., Plumb, R. G. 1999. An FDTD-MoM hybrid technique for modeling complex antennas in the presence of heterogeneous grounds. *IEEE Trans. Geosci. Remote Sens.*, 37(6):2692–2698.

117. Kirana, Y. A., Yaravoy, A. G., Lighhart, L. P. 2003. Optimization of the dielectric embedded dipole antenna. In *IEEE AP-S International Symposium*, June 22–27, Vol. 3, pp. 729–732.

118. Klysz, G., Ferrieres, X., Balayssac, J. P., Laurens, S. 2006. Simulation of direct wave propagation by numerical FDTD for a GPR coupled antenna. *Non-Destructive Test. Eval. Int.*, 39:338–347.

119. Kosmas, P., Wang, Y., Rappaport, C. M. 2002. Three-dimensional FDTD model for GPR detection of objects buried in realistic dispersive soil. *Proc. SPIE*, 4742:330–338.

120. Kumar, S. B., Jacob, J., Mathew, K. T. 2003. Experimental and theoretical analysis of different buried object systems using time domain reflected signal analysis. In *2nd International Workshop on Advanced Ground Penetrating Radar*, May 14–16, pp. 76–81.

121. Lampe, B., Holliger, K. 2004. Modeling of resistively loaded surface GPR antennas. In *10th International Conference on Ground Penetrating Radar*, June 21–24, pp. 25–28.

122. Lee, K. H., Chen, C. C., Lee, R., O'Neill, K. 2002. A numerical study of the effects of realistic GPR antennas on the scattering characteristics from unexploded ordnances. In *IEEE International Geoscience and Remote Sensing Symposium (IGARSS)*, June 24–28, Vol. 3, pp. 1572–1574.

123. Lee, K. H., Venkatarayalu, N. V., Chen, C. C., Teixeira, F. L., Lee, R. 2002. Numerical modeling development for characterizing complex GPR problems. In *9th International Conference on Ground Penetrating Radar*, Apr. 29–May 2. *Proc. SPIE*, 4758:652–656.

124. Lee, K. H., Chen, C. C., Teixeira, F. L., Lee, R. 2004. Modeling and investigation of a geometrically complex UWB GPR antenna using FDTD. *IEEE Trans. Antennas Propag.*, 52(8):1983–1991.

125. Liu, L., He, K., Xie, X., Du, J. 2004. Image enhancement with wave equation redatuming-application to GPR data collected at public transportation sites. *J. Geophys. Eng.*, 4:139–147.

126. Lei, W., Liu, L., Huang, C., Su, Y. 2004. Subsurface imaging of buried objects from FDTD modeled scattered field. In *3rd International Conference on Computational Electromagnetics and Its Applications*, Nov. 1–4, pp. 516–520.

127. Lestari, A. A., Yarayov, A. G., Lighthard, L. P. 2000. Capacitively tapered bowtie antenna. Presented at the *Millennium Conference on Antennas and Propagation*, Apr. 9–14.

128. Lestari, A. A., Kirana, Y. A., Suksmono, A. B., Kurniawan, A., Bharata, A., Yarovoy, A. G., Ligthart, A. B. 2004. Compact UWB radiator for short-range GPR applications. In *10th International Conference on Ground Penetrating Radar*, June 21–24, pp. 141–144.

129. Leuschen, C. J., Plumb, R. G. 2001. A matched-filter-based reverse-time migration algorithm for ground-penetrating radar data. *IEEE Trans. Geosci. Remote Sens.*, 39(5):929–936.

130. Luneau, P., Delisle, G. Y. 1996. Underground target probing using FDTD. In *IEEE AP-S International Symposium*, July 21–26, Vol. 3, pp. 1664–1667.

131. Liu, L., Su, Y., Mao, J. 2008. FDTD analysis of ground penetrating radar antennas with shields and absorbers. *Front. Electr. Electron. Eng. China*, 3(1):90–95.

132. Liu, S., Zeng, Z., Deng, L. 2007. FDTD simulation for ground penetrating radar in urban applications. *J. Geophys. Eng.*, 4:262–267.

133. Liu, L. L., Su, Y., Huang, C. L., Mao, J. J. 2005. Study about the radiation characteristics of bow-tie antennas with discrete resistor-loaded. In *Asia Pacific Microwave Conference*, Dec. 4–7, Vol. 4, p. 3.

134. Martel, C., Philippakis, M., Daniels, D. J., Underhill, M. 2001. Modelling the performance of realistic ultra-wide band ground penetrating radar (GPR) antennas. In *IEE 11th Antennas and Propagation Conference*, Apr. 17–20, Vol. 2, pp. 655–659.

135. Moghaddam, M., Chew, W. C., Anderson, B., Yannakakis, E., Liu, Q. H. 1991. Computation of transient electromagnetic waves in inhomogeneous media. *Radio Sci.*, 26(1):265–273.

136. Moghaddam, E. J., Yannakakis, E., Chew, W. C., Randall, C. 1991. Modeling of the subsurface interface radar. *J. Electromagn. Waves Appl.*, 5(1):17–39.

137. Montaya, T. P., Smith, G. S. 1999. Land mine detection using a ground-penetrating radar based on resistively loaded vee dipoles. *IEEE Trans. Antennas Propag.*, 47(12):1795–1806.

138. Nishioka, Y., Maeshima, O., Uno, T., Adachi, S. 1999. FDTD analysis of resistor-loaded bow-tie antennas covered with ferrite-coated conducting cavity for subsurface radar. *IEEE Trans. Antennas Propag.*, 47(6):970–977.

139. Oğuz, U., Gürel, L. 1997. Subsurface scattering calculations via the 3D FDTD method employing PML-ABC for layered models. In *IEEE AP-S International Symposium*, July 13–18, Vol. 3, pp. 1920–1923.

140. Oğuz, U., Gürel, L. 2000. Three dimensional FDTD modeling of a ground-penetrating radar. In *IEEE AP-S International Symposium*, Vol. 4, pp. 1990–1993.

141. Oğuz, U., Gürel, L. 2001. Simulation of ground-penetrating radars over lossy and heterogeneous grounds. *IEEE Trans. Geosci. Remote Sens.*, 39(6):1190–1197.

142. Oğuz, U., Gürel, L. 2001. Simulation of TRT - configured ground penetrating radars over heterogeneous grounds. In *IEEE AP-S International Symposium*, July 8–13, Vol. 3, pp. 757–760.

143. Oğuz, U., Gürel, L. 2001. On the frequency-band selection for ground-penetrating radars operating over lossy and heterogeneous grounds. In *IEEE AP-S International Symposium*, July 8–13, Vol. 3, pp. 761–764.

144. Oğuz, U., Gürel, L. 2001. Modeling of ground-penetrating-radar antennas with shields and simulated absorbers. *IEEE Trans. Antennas Propag.*, 49(11):1560–1567.

145. Oğuz, U., Gürel, L. 2002. Frequency responses of ground-penetrating radars operating over highly lossy grounds. *IEEE Trans. Geosci. Remote Sens.*, 40(6):1385–1393.

146. Radzevicius, S. J., Daniels, J. J. 2000. GPR H-plane antenna patterns for a horizontal dipole on a half-space interface. In *8th International Conference on Ground Penetrating Radar*, May 23–26, pp. 712–717.

147. Radzevicius, S. J., Chen, C. C., Jr., Peters, L., Daniels, J. J. 2003. Near-field dipole radiation dynamics through FDTD modeling. *J. Appl. Geophys.*, 52:75–91.

148. Rao, K. R., Lee, R., Chen, C. C. 2000. Numerical modeling of an ultra-wide band-width dielectric rod antenna for ground penetrating radar applications. In *IEEE AP-S International Symposium*, July 16–21, Vol. 4, pp. 620–623.

149. Rappaport, C. M., Weedon, W. H. 1996. Efficient modeling of electromagnetic characteristic of soil for FDTD ground penetrating radar simulation. In *IEEE AP-S International Symposium*, July 21–26, Vol. 1, pp. 620–623.

150. Rappaport, C., Shenawee, M. E. 2000. Modeling GPR signal degradation from random rough ground surface. In *IEEE International Geoscience and Remote Sensing Symposium*, (*IGARSS*), July 24–28, Vol. 7, pp. 3108–3110.

151. Reineix, A., Martinat, B., Berthelier, J. J., Ney, R. 2001. FDTD method for the theoretical analysis of the Netlander GPR. In *Conference on the Geophysical Detection of Subsurface Water on Mars*, Aug. 6–10, pp. 88–89.

152. Roshchupkina, L. A. V., Pochanin, G. P. 2005. FDTD simulation of videopulse scattering by elliptic objects in different media. In *5th International Conference on Antenna Theory and Techniques*, May 24–27, pp. 361–363.

153. Roth, F., Genderen, P. V., Verhaegen, M. 2004. Radar scattering models for the identification of buried low-metal landmines. In *10th International Conference on Ground Penetrating Radar*, June 21–24, pp. 689–692.

154. Shubitidze, F., O'Neil, K., Shamatava, I., Sun, K., Paulse, K. D. 2003. Investigation of side looking EM field scattering from a buried metallic object to support UXO discrimination. In *IEEE AP-S International Symposium*, June 22–27, Vol. 2, pp. 223–226.

155. Shubitidze, F., O'Neil, K., Shamatava, I., Sun, K., Paulse, K. D. 2003. Analysis of GPR scattering by multiple subsurface metallic objects to improve UXO discrimination. In *IEEE International Geoscience and Remote Sensing Symposium*, *(IGARSS)*, July 21–25, Vol. 7, pp. 4163–4165.

156. Taflove, A., Hagness, S. 2005. *Computational Electrodynamics: The Finite Difference Time Domain Method*. Artech House, Norwood, MA.

157. Takeuchi, S. K., Kim, H. J. 2007. A full waveform inversion algorithm for interpreting crosshole radar data. In *4th International Workshop on Advanced Ground Penetrating Radar*, June 27–28, pp. 169–174.

158. Teixeira, F. L., Chew, W. C., Straka, M., Oristaglio, M. L., Wang, T. 1997. 3D PML-FDTD simulation of ground penetrating radar on dispersive earth media. In *IEEE International Geoscience and Remote Sensing Symposium (IGARSS)*, Aug. 3–8, Vol. 2, pp. 945–947.

159. Teixeira, F. L., Chew, W. C., Straka, M., Oristaglio, M. L., Wang, T. 1998. Finite-difference time-domain simulation of ground penetrating radar on dispersive, inhomogeneous, and conductive soils. *IEEE Trans. Geosci. Remote Sens.*, 36(6):1928–1937.

160. Teixeira, F. L., Chew, W. C. 2000. Finite-difference computation of transient electromagnetic waves for cylindrical geometries in complex media. *IEEE Trans. Geosci. Remote Sens.*, 38(4):1530–1543.

161. Tjuatja, S., Fung, A. K., Wu, S. H., Zhou, P., Li, Z. J. 1997. Remote sensing of buried objects: an analysis using FD-TD simulation. In *IEEE International Geoscience and Remote Sensing Symposium (IGARSS)*, Aug. 3–8, pp. 1144–1146.

162. Travassos, X. L., Vieira, D. A. G., Ida, N., Vollaire, C., Nicolas, A. 2008. Inverse algorithms for the GPR assessment of concrete structures. *IEEE Trans. Magn.*, 44(6):994–997.

163. Uduwawala, D., Norgren, M., Fuks, P., Gunawardena, A. W. 2004. A deep parametric study of resistor-loaded bow-tie antennas for ground penetrating radar applications. First International Conference on Industrial and Information Systems. *IEEE Trans. Geosci. Remote Sens.*, 42(4):732–742.

164. Uduwawala, D., Gunawardena, A. W. 2006. A fully three-dimensional simulation of a ground-penetrating radar over lossy and dispersive grounds. In *First International Conference on Industrial and Information Systems (ICIIS)*, Aug. 8–11, pp. 143–146.

165. Wong, P. B., Tyler, L., Baron, J. E., Gurrola, E. M., Simpson, R. A. 1996. A three wave approach to surface scattering with applications to remote sensing of geophysical surfaces. *IEEE Trans. Antennas Propag.*, 44(4):504–514.

166. Venkatarayalu, N. V., Chen, C. C., Teixeira, F. L., Lee, R. 2004. Numerical modeling of ultrawide-band dielectric horn antennas using FDTD. *IEEE Trans. Antennas Propag.*, 52(5):1318–1323.

167. Yang, B., Rappaport, C. 2001. Response of realistic soil for GPR applications with 2-D FDTD. *IEEE Trans. Geosci. Remote Sens.*, 39(6):1198–1205.

168. Yaravoy, A. 2004. Adaptive bow-tie antenna for ground penetrating radar. In *10th International Conference on Ground Penetrating Radar*, June 21–24, pp. 121–124.

169. Yee, K. 1966. Numerical solution of initial boundary value problems involving Maxwell's equations in isotropic media. *IEEE Trans. Antennas Propag.*, 14(8):303–307.

170. Zhan, Y., Liang, C., Fang, G. 2000. Development of a 3-D simulator using the FDTD method. In *5th International Symposium on Antennas, Propagation and Electromagnetic Theory (ISAPE)*, Aug. 15–18, pp. 620–623.

171. Zhan, Y., Belli, K., Fascetti, S. W., Rappaport, C. 2008. Effectiveness of 2D FDTD ground penetrating radar modeling for bridge deck deterioration evaluated by 3D FDTD. In *IEEE International Geoscience and Remote Sensing Symposium (IGARSS)*, July 6–11, p. 4.

172. Gader, P. D., Hocaoglu, A. K., Mystkowski, M., Zhao, Y. 2000. Hidden Markov models and morphological neural networks for GPR-based land mine detection. In *Detection and Remediation Technologies for Mines and Minelike Targets V. Proc. SPIE*, 4038:1096–1107.

173. Potin, D., Vanheeghe, P., Duflos, E., Davy, M. 2006. An abrupt change detection algorithm for buried landmines localization. *IEEE Trans. Geosci. Remote Sens.*, 44(2):260–272.

174. Brunzell, H. 1999. Detection of shallowly buried objects using impulse radar. *IEEE Trans. Geosci. Remote Sens.*, 37:875–886.

175. Dogaru, T., Carin, L. 1998. Time-domain sensing of targets buried under a rough air–ground interface. *IEEE Trans. Antennas Propag.*, 46:360–372.

176. van der Merwe, A., Gupta, I. J. 2000. A novel signal processing technique for clutter reduction in GPR measurements of small, shallow land mines. *IEEE Trans. Geosci. Remote Sens.*, 38(6):2627–2637.

177. Wu, R., Clement, A., Li, J., Larsson, E. G., Bradley, M., Habersat, J., Maksymonko, G. 2001. Adaptive ground bounce removal. *Electron. Lett.*, 37(20):1250–1252.

178. Ho, K. C., Gader, P. D. 2002. A linear prediction land mine detection algorithm for hand held ground penetrating radar. *IEEE Trans. Geosci. Remote Sens.*, 40(6):1374–1384.

179. Ho, K. C., Gader, P. D., Wilson, J. N. 2004. Improving landmine detection using frequency domain features from ground penetrating radar. In *Proceedings of the IEEE International Conference on Geoscience and Remote Sensing Symposium*, Vol. 3, pp. 1617–1620.

180. Potin, D., Vanheeghe, P., Duflos, E., Davy, M. 2006. An abrupt change detection algorithm for buried landmines localization. *IEEE Trans. Geosci. Remote Sens.*, 44(2):260–272.

181. Ho, K., Gader, P. 2002. A linear prediction land mine detection algorithm for hand held ground penetrating radar. *IEEE Trans. Geosci. Remote Sens.*, 40(6):1374–1384.

182. Chan, T. C. T., H. C. So, Ho, K. C. 2008. Generalized two-sided linear prediction approach for land mine detection. *Signal Process.*, 88(4):1053–1060.

183. Wu, R., Clement, A., Li, J., Larsson, E. G., Bradley, M., Habersat, J., Maksymonko, G. 2001. Adaptive ground bounce removal. *Electron. Lett.*, 37(20):1250–1252.

184. Torrione, P. A., Collins, L. M., Clodfelter, F., Lulich, D., Patrikar, A., Howard, P., Weaver, R., Rosen, E. 2006. Constrained filter optimization for subsurface landmine detection. In *Detection and Remediation Technologies for Mines and Minelike Targets XI*, R. S. Harmon, J. T. Broach, J. H. Holloway, Jr., Eds. *Proc. SPIE*, 6217.

185. Astola, J., Kuosmanen, P. 1997. *Fundamentals of Nonlinear Digital Filtering*. CRC Press, Boca Raton, FL.

186. Yu, S. H., Witten, T. R. 1999. Automatic mine detection based on ground penetrating radar. In *Detection and Remediation Technologies for Mines and Minelike Targets IV*. *Proc. SPIE*, 3710:961–972.

187. Changa, S.-S., Ruanea, M. F. 2003. Feature extraction of ground penetrating radar for mine detection. *Proc. SPIE*, 5089:1201–1209.

188. Bartosz, E. E., Dejong, K., Duvoisin, H., Solomon, G. Z., Steinway, W., Warren, A. 2004. Nonlinear processing of radar data for landmine detection. In *Detection and Remediation Technologies for Mines and Minelike Targets IX. Proc. SPIE*, 5415:892–895.

189. Turk, M., Pentland, A. 1991. Eigenfaces for recognition. *J. Cogn. Neurosci.*, 3(1):71–86.

190. Hyvärinen, A., Karhunen, J., Oja, E. 2001. *Independent Component Analysis*. Wiley, Hoboken, NJ.

191. Hyvärinen, A. 1999. Fast and robust fixed-point algorithms for independent component analysis. *IEEE Trans. Neural Networks*, 10:626–634.

192. Karlsen, B., Larsen, J., Sorensen, H. B., Jakobsen, K. B. 2001. Comparison of PCA and ICA based clutter reduction in GPR systems for anti-personnel landmine detection. In *Proceedings of the 11th IEEE Signal Processing Workshop on Statistical Signal Processing*, pp. 146–149.

193. Karlsen, B., Sørensen, H. B., Larsen, J., Jakobsen, K. B. 2003. GPR detection of buried symmetrically shaped mine-like objects using selective independent component analysis. In *Proceedings of the 2003 Detection and Remediation Technologies for Mines and Mine-like Targets (AeroSense 2003)*. *Proc. SPIE*, 5089:375–386.

194. Palit, P. P., Agarwal, S. 2002. Independent component analysis for GPR based hand held mine detection. *Proc. SPIE*, 4742:367–377.

195. Hocaoglu, A. K., Gader, P. D. 2003. Continuous processing of acoustic data for landmine detection. In *Detection and Remediation Technologies for Mines and Minelike Targets VII. Proc. SPIE*, 4742:654–664.

196. Collins, L. M., Torrione, P. A., Munshi, V., Throckmorton, C. S., Zhu, Q., Clodfelter, F., Frasier, S. 2002. Algorithms for landmine detection using the NIITEK ground penetrating radar. In *Detection and Remediation Technologies for Mines and Minelike Targets VI: 2002 International Symposium on Aerospace/Defense Sensing and Controls*, Orlando, FL, Apr. pp. 709–718.

197. Won, Y., Gader, P. D., Coffield, P. 1997. Morphological shared-weight networks with applications to automatic target recognition. *IEEE Trans. Neural Networks*, 8:1195–1203.

198. Le Cun, Y., Boser, B., Denker, J., Henderson, D., Howard, R., Hubbard, W., Jackel, L. 1989. Backpropagation applied to handwritten zip code recognition. *Neural Comput.*, 1(4):541–551.

199. Le Cun, Y., Mattan, O., Boser, B., Denker, J., Henderson, D., Howard, R., Hubbard, W., Jackel, L., Baird, H. 1990. Handwritten zip code recognition with multilayer

networks. In *Proceedings of the International Conference on Pattern Recognition*, Atlantic City, NJ.

200. Gader, P., Miramonti, J., Won, Y., Coffield, P. 1995. Segmentation free shared weight networks for automatic vehicle detection. *Neural Networks*, 8(9):1457–1473.

201. Theera-Umpon, N., Khabou, M. A., Gader, P. D., Keller, J., Shi, H., Li, H. 1998. Detection and classification of MSTAR objects via morphological shared-weight neural networks. Presented at the SPIE Conference on Algorithms for SAR Imagery V, Orlando, FL, Apr.

202. Khabou, M. A., Gader, P. D., Shi, H. 1999. Entropy optimized morphological shared-weight neural networks. *Opt. Eng.*, 38:263–273.

203. Hocaoglu, A. K., Gader, P. D. 1999. Choquet integral–based morphological operators. In *Proceedings of the SPIE Conference on Nonlinear Image Processing IX*, San Jose, CA, Jan., pp. 46–56.

204. Hocaoglu, A. K. 2000. Choquet Integral Based–Morphological Operators with Applications to Object Detection and Information Fusion. Ph.D. dissertation, University of Missouri–Columbia.

205. Gader, P. D., Hocaoglu, A. K., Mystkowski, M., Zhao, Y. 2000. Hidden Markov models and morphological neural networks for GPR-based landmine detection. In *Proceedings of the SPIE Conference on Detection and Remediation Technologies for Mines and Minelike Targets IV*, Orlando, FL, Apr.

206. Gader, P. D., Frigui, H., Nelson, B., Vaillette, G., Keller, J. M. 1999. New results in fuzzy set based detection of landmines with GPR. In *Proceedings of the Conference on Detection and Remediation Technologies for Mines and Minelike Targets IV*, Orlando FL, pp. 1075–1084.

CHAPTER FOUR

Electromagnetic Induction

4.1 INTRODUCTION TO METAL DETECTORS

Metal detectors are widely used in research and industry for a variety of applications: for example, geophysical prospecting, nondestructive testing (NDT) of materials, in the food industries, for distance and proximity sensing, in security systems, and for landmine detection. It is useful to distinguish between magnetic and inductive metal detectors (MDs) or electromagnetic induction sensors (EMIs). Saxby (1868) [1] first described the use of a simple magnetic metal detector: a magnet needle for searching defects [e.g., cavities in cast iron (e.g., gun barrels)]. Years later, with the introduction of the Hall probe (1879) [2], the Förster probe (1937), and the fluxgate or saturation probe [3], the use of magnetic leakage flux sensor techniques (e.g., for NDT problems) increased rapidly.

A *magnetic metal detector* (e.g., the magnetometer or gradiometer) can sense or detect ferrous metal pieces by the measured distortion of the (static) Earth magnetic field in three dimensions caused by buried ferrous metal pieces (the Earth's magnetic flux density is in the microtesla range). This type of sensor is a passive sensor because it has no active excitation system. An active magnetic metal detection probe generates its own magnetic excitation field. Passive probes are currently the most widely used. With highly sensitive magnetometers or gradiometers (which have nano- and picotesla resolution), it is possible to use this technique for geophysical and archaeological surveying by measuring the spatial distribution of the magnetic field. One of the main applications of magnetic metal detectors is the detection of buried unexploded ordnance (UXO).

Subsurface Sensing, First Edition. Edited by Ahmet S. Turk, A. Koksal Hocaoglu, and Alexey A. Vertiy.
© 2011 John Wiley & Sons, Inc. Published 2011 by John Wiley & Sons, Inc.

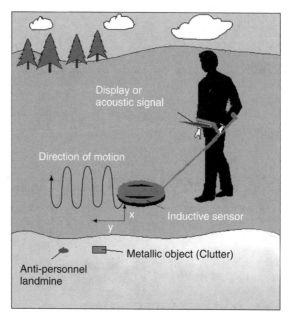

FIGURE 4.1 Principle of inductive handheld metal detector for humanitarian demining.

In the *inductive metal detector*, discovered in 1872 by Hughes [4], a coil of wire carrying an alternating current is affected by any conductive (metallic) material if it is penetrated by the alternating magnetic field of the coil. This effect is related to the inducted eddy currents in the conductive (metallic) material. Inductive sensors have a wider area of applications than magnetic sensors because they are not limited to ferrous metallic objects. They are used in research, geophysical prospecting, process control in industry (noncontact recording and inspection of material properties and nondestructive testing), the food industry, civil engineering and offshore technologies (e.g., pipeline inspections, suction dredging), and for landmine detection. The inductive metal detector is the device most used to search for antipersonnel landmines (Fig. 4.1). The handheld system allows "de-miners" to scan (sweep) areas in front of them to detect buried metal pieces in locations where no other technologies are applicable, due to the topology and condition of the terrain. The high sensitivity, the simple construction of MDs, the easy-to-use operation, and the cost of the system are important advantages compared with other demining technologies (e.g., GPR, acoustic, infrared, and laser vibrometry sensors). The main focus of this chapter is on inductive metal detectors as applied to humanitarian demining.

The basic construction of an inductive metal detector is very simple; it is just a coil through which an ac or power impulse current flows. However, the cause-and-effect chain is very complex, which makes it difficult to interpret the sensor signal received. Figure 4.2 illustrates the principle of an inductive metal detection probe. Inductive metal detectors are based on the interaction of the (primary)

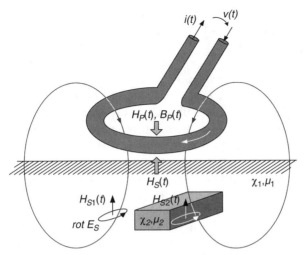

FIGURE 4.2 Principle of inductive metal detection: electromagnetic induction probe for metal detection.

varying magnetic field with a electrical conductive metal piece. The varying primary magnetic field $H_P(t)$ in a piece of metal cause eddy currents whose (opposite secondary; Lenz's law) resulting magnetic fields $H_S(t)$ interact with the primary field such that the probe impedance, or inducted secondary voltage in the coil, is changed. The eddy currents are the result of the electric field strength E induced in the electrical conductive piece (or the soil) due to the time-varying primary magnetic field $H_P(t)$. The physical description of EMI probes is based on Maxwell's equations [Eqs. (4.1) and (4.2)] and the terms in Eqs. (4.3) to (4.5). The magnetic flux density **B** (or magnetic field strength **H**) caused by the excitation current is given by *Ampère's law* (Maxwell's first equation), and the induced electric field strength **E** is given by the electromagnetic induction law, known as *Faraday's law* (Maxwell's second equation):

$$\text{rot } \mathbf{H} = \mathbf{J} + \frac{\partial \mathbf{D}}{\partial t} \tag{4.1}$$

$$\text{rot } \mathbf{E} = -\frac{\partial \mathbf{B}}{\partial t} \tag{4.2}$$

and the equations of the relationships between field components:

$$\mathbf{B} = \mu \mathbf{H} \tag{4.3}$$

$$\mathbf{J} = \kappa \mathbf{E} \tag{4.4}$$

$$\mathbf{D} = \varepsilon \mathbf{E} \tag{4.5}$$

with the constraints on the vector fields div $\mathbf{B} = 0$ and div $\mathbf{D} = Q$.

Even in the case of nonconductive or low electrically conductive but magnetically permeable objects, the primary magnetic field will be "disturbed" (i.e., a spatial change in the magnetic flux density in the coil), which is measurable as a change of impedance or a change in the induced (secondary) coil voltage.

In general, the sensor signal received cannot assign clearly defined properties to objects detected because the general solution of Maxwell's equations is ambiguous. For example, it is not very difficult to demonstrate for an inductive sensor that you can find more than one arrangement of one or more metal pieces which induce in the sensor coil the same change in the amplitude and phase. This is related to the *inverse problem* of electromagnetism: how to decide the form and material properties of a disturbance by using only magnetic values (e.g., induced coil voltage) measured from a plane to a constant distance from the disturbance (i.e., the disturbing entity). In recent years a few new approaches to solving the electromagnetic inverse problem have been developed using scattering theory. With a priori information and carefully chosen assumptions (e.g., for perfect conductive objects), solutions were obtained using ground-penetrating radar (GPR) systems and the using magnetic sensors. A few of these new approaches have been adopted for use in inductive metal detector problems [5].

Although the sensitivity of MD is high, and the distinction between ferrous and nonferrous metallic objects is simple (e.g., by phase-shift detection), other important information is needed regarding the depth and form of the metallic object that cannot be extracted from simple evaluation of the signal. For example, a few metal detectors employed in treasure hunting (e.g., coins) use an internal lookup table of amplitudes and phase values of a limited number of "treasures" to discriminate between coins and silver or gold nuggets. This strategy works in a "cooperative" soil; however, it is useless for humanitarian demining because there is a major problem with the high false-alarm rate caused by the huge amount and variety of metallic objects (clutter and debris) and the influence of the soil due to varying electrical and magnetic properties.

Soil can be classified into two types: cooperative and uncooperative. *Cooperative soil* has either no influence on the operation of metal detectors or the influence is constant and can be compensated electronically. *Uncooperative soil* generates signals in the metal detector, due to its components and properties [such as stones: their in-homogeneity, spatially distributed electrical and magnetic properties, and dispersive behavior (described in Section 4.3)]. This soil influence increases the false-alarm rate, so that MDs generate audio signals where no buried metal object or landmine exists.

In the following section, we provide an overview of the state-of-the-art technology of inductive metal detection for humanitarian demining. In recent years many publications have come from national and international research projects regarding metal detection, and a lot of active patents exist. All existing patents are based on special coil arrangements, special hardware components, and new soil compensation techniques in conjunction with new signal-processing procedures to reach better classifying systems for objects detected. Nevertheless, fundamental questions regarding MD technology have not really been solved: What is the size, geometric form, orientation, and depth of buried metallic objects detected in soil?

4.2 INDUCTIVE METAL DETECTORS: TYPES OF PROBES, EXCITATION, AND COIL ARRANGEMENTS

Inductive sensors can be classified based on the following criteria:

- *Type of excitation:* harmonic (continuous) or impulsive
- *Type of inductive probe:* parameter probe (one coil) or transformer probe (two coils)
- *Coil arrangement:* absolute, differential, or multidifferential probe
- *Geometry of the coils:* circular, rectangular, or oval
- *Alignment of the coils:* coplanar, noncoplanar, or orthogonal position
- *Arrangement and number of the coils:* monostatic, bistatic, and multistatic measurement arrangements

The last classification criterion is often used for radar systems, to determine how many spatially arranged transmitters and receivers (antennas) work together and how they stay (static) or are moved for receiver signal acquisition. For metal detector systems, monostatic coils (one transmitter and one receiver in a fixed position) are the most common arrangements, but in recent years bistatic and multistatic arrangements have been introduced, with improved signal-processing algorithms for solving the inverse problems of MDs, such as depth and size of an object detected.

EMI theory is used to solve the electromagnetic inverse problem for the low-frequency range and for a transducer that has a very small geometric size (the diameter of the coil is much smaller than the wavelength of the electromagnetic magnetic field used; $d \ll \lambda$) rather than the equivalent wavelength of the electromagnetic excitation current used. Different solutions exist for magnetic (ferrous) metal detectors and for radar data (GPR) based on scattering theory [5].

It is useful to distinguish between two general types of MD probes: parameter and transformer. In Fig. 4.3 are shown two different types of inductive probe and their equivalent electric circuit diagrams. In a *parameter probe* the ac through the coil generates a (primary) varying magnetic field H which interacts with the conductive piece; the magnetic field of the eddy current induced in the piece is opposite that in the primary field. The coil impedance changes as a result of the superposition of both fields. The parameter probe has one coil only, and the change in the impedance Z has to be measured and indicated. In general, this type of metal detector has more disadvantages than advantages: It has the simplest metal detector construction (low cost), but all the environmental influences (e.g., the operating temperature) have a direct effect on the detection results.

The type of probe most used in MDs is the *transformer probe*, which consists of a primary coil (inductance L_p or transmitter coil, Tx) and a secondary

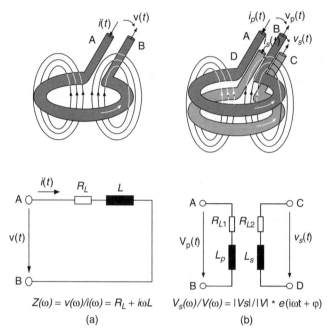

$$Z(\omega) = v(\omega)/i(\omega) = R_L + i\omega L \qquad V_s(\omega)/V(\omega) = |Vs|/|V| * e^{(i\omega t + \varphi)}$$

(a)   (b)

FIGURE 4.3   (a) Parameter probe, one coil, and (b) transformer probe, two or more coils, with their equivalent electric circuit diagrams.

coil (inductance L_s or receiver coil, Rx), considered an air-coupled transformer (sometimes with a coupled core). The amplitude and phase of the secondary voltage $V_S(t)$ depend on the air–ground core rather than the mutual inductance M of the coils, which is determined by the ground under the coils (soil properties) and the buried metal pieces. With a transformer probe, any change in the operating temperature of the sensor does not influence the measurement values much, because the change in parameter is similar on both sides (i.e., in the primary and secondary coils). Figure 4.4 shows a typical arrangement of secondary coils in a transformer probe.

An absolute probe is the simplest form of MD probe. A *differential probe* or *balanced probe* exists in different embodiments: the differential or double-D probe (middle in Fig. 4.4), the multidifferential probe, and the coaxial differential probe. The reason for introducing a multidifferential probe for MDs is that the dependency of the signal induced on the orientation (angle) between the coil moving direction and the target is reduced (see Section 4.2.2). The alignment of the coils is mostly coplanar, but could be noncoplanar or orthogonal. The number of secondary coils and the geometry of the coils (i.e., circular, rectangular, oval, etc.) are other relevant design features. In practice, a differential coil has a residual voltage that can be balanced electronically or by using additional coils/circuits.

The most popular MD has coplanar coils in a monostatic arrangement: one fixed-position Tx coil and the Rx coil. It is also possible to arrange the coils as

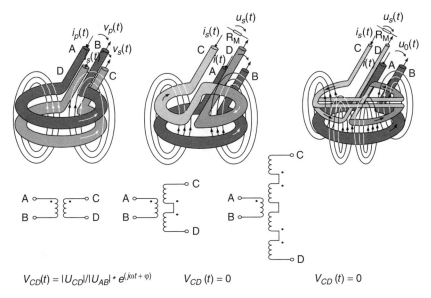

$$V_{CD}(t) = |U_{CD}|/|U_{AB}| * e^{(j\omega t + \varphi)} \qquad V_{CD}(t) = 0 \qquad V_{CD}(t) = 0$$

FIGURE 4.4 Absolute, differential, and multidifferential probes of a metal detector.

noncoplanar (e.g., by a fixed angle or orthogonal) in case the z-component of the eddy current induced in the soil or the metallic object would need to be measured. From the point of EMI theory, the eddy currents will have a z-component only in an inhomogeneous material [6] (see Section 4.3).

Normally, a noncoplanar arrangement reduces the amount of the voltage induced in the secondary coil, which then has to be compensated by higher amplification in the electronic section. This must be optimized as to the signal-to-noise ratio obtained or necessary for the threshold of audio signal generation. As is well known, the signal-to-noise ratio must be maximized to obtain a low false-alarm rate in conjunction with high sensitivity.

The wire used in the coils consists of single copper wire, Litz wire (Litzendraht wire), or multilayer printed-circuit-board coil or coaxial wire. The Tx and Rx coils could have a different design because of the electrical properties required (i.e., the resistance R of the coil and the inductance L of the coil). A few metal detectors use additional coils (or windings, e.g., short-connected) for automatic calibration or balance procedures.

The diameter of the coils determines the sensing range of an MD based on the limited magnetic field distribution of the coil. For a single wire loop circular coil of radius R, the magnetic flux density $B(x)$ in the centerline ($r = 0$ cm) as a function of the distance x is given by

$$B(x) = \frac{\mu_0 I R^2}{\sqrt[3]{R^3 + x^2}} \tag{4.6}$$

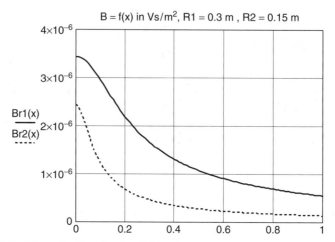

FIGURE 4.5 Magnetic flux density $B(x)$ of a circularly coil vs. the relative distance $x = X/R$.

The magnetic flux density $B(x)$ in the centerline for two different diameter values of R ($R_1 = 30$ cm, $R_2 = 15$ cm, $NI = 1$ A) is shown in Fig. 4.5. For a fixed distance x, the larger coil has a higher flux density, which in turn demonstrates a higher eddy current density in a metallic object (and soil), which explains the higher sensitivity (or wider sensing range).

A decrease in the magnetic flux density $B(x)$ is dependent on the diameter. A greater searching coil diameter enables a greater range for the magnetic flux. In general, the standard sensing range of most MDs in air for stable long-time operation is three to five times the diameter of the searching coil. In practice, two major factors limit the coil diameter: the manageability of MD handheld systems, and the lower sensitivity regarding the huge coil diameter of the smallest metallic objects (e.g., due to more soil noise).

Due to the electrical conductivity and magnetic permeability of soil, the sensing range of an MD in soil is always lower than in air because the eddy currents (primary excitation) shield the magnetic field and reduce the sensing range and sensitivity resulting from this skin effect. The electric conductivity and magnetic permeability (magnetic susceptibility $\chi = \mu_r - 1$) of soil usually demonstrates a wide scale variability [7–10] (see Section 4.3), as follows:

Variation of electrical conductivity: $\chi = 10^2$ to 10^{-6} S/m

Variation of magnetic susceptibility: $\chi = \mu_r - 1 = 200$ to 6000×10^{-5} SI

The electric conductivity and magnetic properties of the soil are dependent on the content of the soil and actual environmental conditions (e.g., rain, temperature). However, the electric conductivity of the soil is always lower than that of buried metallic objects, by about the factor of 10^3 to 10^4. The porosity and granularity of the components, the bulk density of the upper layer of the soil, and the weather conditions influence operation of the MD in the field. The influence of electromagnetic soil properties is described in Section 4.3.

4.2.1 Continuous-Wave Metal Detectors

Continuous-wave (CW) metal detectors exist as single-, dual-, or multifrequency sensors in the frequency range from a few hundred hertz up to 100 kHz. Single-frequency sensors work in the frequency range of a few kilohertz. The most popular professional metal detectors for landmine de-mining are dual- or multifrequency sensors. The influence of the soil (i.e., spatial change in the conductivity and permeability of the soil) generally necessitates *soil compensation* for *safe detection* in operation. For a sufficient soil compensation strategy for MDs, more than one measurement frequency (more than two features) is required. Multifrequency probe technology is also denoted as *broadband electromagnetic induction spectroscopy*, because the focus of the internal signal processing is spectrum (matching) analysis. The main part (block diagram) of the standard CW metal detector is shown in Fig. 4.6.

The signal from the harmonic generator (oscillator) needs to be boosted in the amplifier. The amplifier output, in turn, loads the excitation coil (the primary, or transmitter coil). The induced voltage in the secondary coil (the receiver coil) should be amplified, filtered, and phase-sensitive-rectified, for which a signal shifted 90° from the harmonic generator is required. The output signals of a phase-sensitive rectifier are the real and imaginary parts of the secondary voltage received. The signal-processing part must display and classify the results (e.g., to generate an audio signal in case of the presence of a metallic piece). The signal-processing part can be realized very simply using analog circuits for a level- and/or phase-sensitive detector, or by a digital signal processor unit, which makes complex signal classification and feature extraction algorithms possible. The real and imaginary parts of the secondary voltages received for each frequency are used for the detection and classification procedure: If n frequencies are used, $2n$ features can be handled.

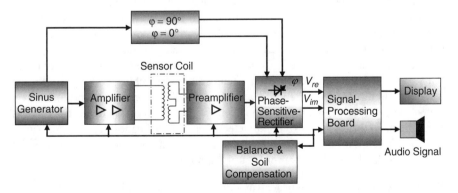

FIGURE 4.6 Standard CW metal detector.

For example, one component (the real or imaginary part of the secondary voltage) of one frequency can be used for internal *soil compensation*. With the features of the other frequencies from the secondary voltage, $2n - 1$ features are available to detect and classify metallic objects. The simplest form of classification is a phase- and level-sensitive detection, which has to be distinguishable from the threshold level for audio signal generation: the threshold level appointed by the level of the internal signal, which generates the audio signal (the volume and the frequency as well the pitch) for the operator in case the metallic target comes closer to the searching coil. The correct choice of this threshold is an important task in any MD field operation. The balance status of the MD can be controlled manually or automatically.

A simple solution for discriminating among various unknown buried objects can be the use of an internal *lookup table*, where for each measurement frequency and a fixed number of well-known collectible metal objects (e.g., coins), $2n - 1$ features are stored. The lookup table could be realized as a trainable system for the operator in the field. Depending on the version of coil arrangements used (absolute or differential coils), compensation of the secondary voltage (residual voltage) and soil compensation (balance status) could be accomplished using hardware or software.

In dual- or multifrequency excitation, the signal-processing part controls the oscillator, amplifier, filter, and compensation segments for each frequency using time-division multiplexing technology. From the point of view of systems theory, the transmitter coil can be fed by a mixed signal from all the frequencies used at the same time (a synthesized sum signal, e.g., $f_1 + f_2$). However, the "system," a coil above a conductive half-space (soil) with an internal metallic piece, is not a linear system. Hence, the highest frequencies will be shielded by the other (lower) frequency components for deep penetration in the soil and by the metal pieces, so that overall detectability and sensitivity will be decreased, even for the use of adapted amplitude for the set of mixed frequencies.

One of the oldest and simplest versions of a monofrequency metal detector is the *free-running* or *coupled oscillator* with internal audio signal generation. The beat frequency oscillator (BFO) is a special type of LC-oscillator (a searching coil is part of one resonant circuit), in which a carrier wave oscillator is coupled with a heterodyne oscillator so that the intermediate frequency is in the audio range. When a metallic object is present under the searching coil, the difference in the resonance circuits (frequencies, BFO) generates the audio signal.

The major disadvantages of this design/construction are the poor long-term stability, the influence of the operational temperature (searching coil and electronic oscillator circuit) and power supply, any type of soil compensation, and their interdependence. Nevertheless, the design is very simple (low cost) and the sensitivity is pretty high, so that this type of metal detector has an application for "treasure hunting" for private purposes. However, these MDs are not employed in commercial systems for landmine detection or security applications.

Signal Generation   The physical model of signal generation in an MD caused by the presence of a metallic object in air was noted in Section 4.1 and shown in

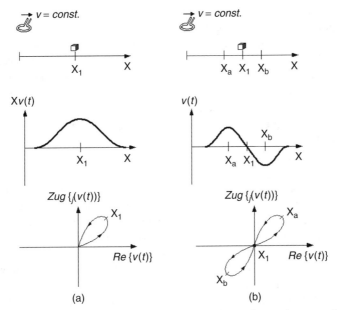

FIGURE 4.7 Output signals of a CW metal detector (secondary voltage vs. distance x), coil moving above a metallic object: (a) absolute probe; (b) differential probe.

Fig. 4.2. The secondary voltage $V(x)$ induced as a function of position x for absolute and differential probes when an MD is moving horizontally above a metallic object (at a constant distance/depth x/h) is shown in Fig. 4.7. It is a one-dimensional signal (a one-dimensional scan) where the motion induction effect is neglected because the moving speed of a handheld MD probe is not expected to be very high.

For an absolute probe the secondary voltage is highest when the metallic object is (standing) in the center of the probe (position X_1). The maximum amplitude depends on the distance between the coil and the metallic object, as well as from the electrical and magnetic properties. If the differential coil is moving over a metallic object, the voltage induced increases to a (positive) maximum if the object stands in the center of the (right) differential coil (position X_a). If the metallic object stands between the secondary coils (center of the primary coil, position X_1), the voltage induced is zero because the secondary coils have the same magnetic field distribution. In addition, if the object is under the left differential coil, a (negative) maximum value results (position X_b).

The form of this output signal depends on the geometry and orientation of the metallic object: the pitch of the coil and the distance, and the depth and the path taken by the metallic object as it crosses the differential coil (the angle between the direction of movement and the secondary coil orientation). If a metallic object crosses the probe between the secondary coils (coil rotated $90°$ as shown in Fig. 4.7), no signal will be generated in the coils. That is one reason for introducing multidifferential coils: reducing the angle dependency of the differential coil.

The position of the metallic object could be determined very easily in both cases: (1) with an absolute probe it is the maximum of the coil voltage, and (2) with a differential coil the object is on the point where the voltage is zero between the two maxima. Searching the maxima in the one-dimensional scan respectively the lateral position x of the buried metallic object is called *pinpointing*. The pinpointing procedure is carried out by the de-miner alternately for directions x and y. However, the depth (the z-component of the buried object) cannot be determined easily since the magnitude of the secondary voltage is a complex function of the size and depth and is influenced by the electric and magnetic soil parameters.

Another type of data representation for MD is the plot of the secondary voltage (the real and imaginary parts) in the complex plane. If the probe is moving above a metallic object, each value measured generally yields a point in the complex plane. By moving the probe over a metallic object, a typical loop for an absolute probe (single loop) and a differential probe (double loop) can be drawn (see Fig. 4.7). Another measurement frequency that is used is generated by another loop, so that, for example, for a two-frequency CW metal detector ($f_1 = 2.4$ kHz, $f_2 = 19.2$ kHz), one sweep (a one-dimensional scan) of the probe obtains two loops, which are characteristic for each metal object in a fixed (constant) depth, as shown in Fig. 4.8. Each frequency chosen generates other eddy currents in the object (different influence) so that each loop is "characteristic" for a buried metallic object. But this "pattern" will not be unique for any object, and the form of the loop will change with the depth of the object, in conjunction with the electrical and magnetic properties of the soil.

In general, any type of soil reduces the sensitivity and detectability as well as object discrimination. The homogeneous and bulk density of the soil, and the

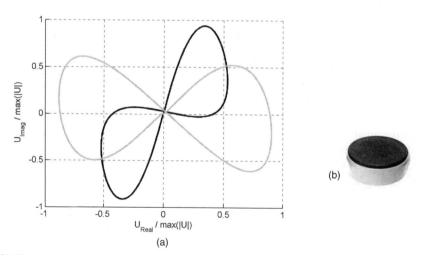

FIGURE 4.8 (a) Complex plane representation of the output signals of a two-frequency CW metal detector above a metallic object in air (M2B landmine, 5 cm depth/distance) (a) bold: $f_1 = 2.4$ kHz (b) grey: $f_2 = 19.2$ kHz; (b) M2B antipersonnel landmine. (From [8].)

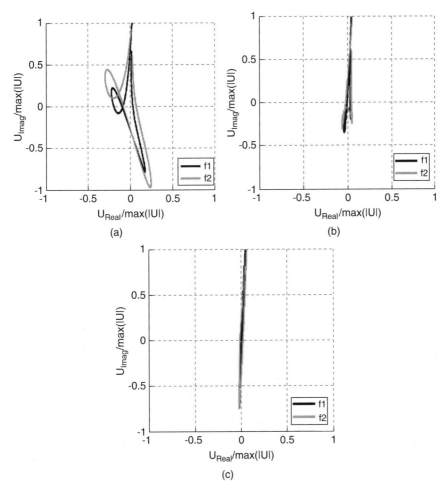

FIGURE 4.9 Output signals of a two-frequency CW metal detector above a metallic object (M2B landmine) for different depths: (a) 2 cm; (b) 5 cm; (c) 10 cm.

spatial distribution of their electrical and magnetic properties, affect MD operation. In uncooperative soil, it could influence an MD dramatically (Fig. 4.9), so that due to the current soil condition, a small metallic object (e.g., the antipersonnel landmine M2B) is undetectable [8,9]. This condition or status has to be detected or measured by the MD itself during field operation. All MDs have a signal-processing procedure for reducing this soil influence. The soil compensation has to be realized well enough to provide safe field operation for detecting objects in a harsh environment, and it has to be done before the field operation begins.

In humanitarian de-mining, the influence of the MD has to be characterized during field operations as described in operation manuals [11,12]. The methods proposed describe determination of the effect of soil on the detection performance of MD. The fixed-depth and equivalent detection depth tests, as well as the ground reference height (GRH), are determined first by each field operation using a metallic test object (e.g., a small metallic ball). The greater the GRH, the greater the influence of the soil. From a practical point of view, the equivalent detection depth and GRH are empirical measurements of how noisy or how uncooperative a soil is [11].

Signal Processing and Operational Modes   The philosophy of signal processing for CW metal detectors is as follows: The absolute value of the secondary voltage is not very interesting because this value depends on the soil properties, coil used, and other factors. The voltage change due to moving the coil above the area suspected to have buried metal objects provides all the necessary information. Thus, in the first step of any MD operation, the secondary voltage has to be compensated or balanced manually or automatically while the probe stands over metal-free ground. Subtraction of the ongoing or residual secondary voltage realized from this position sets all displayed values to zero. This is the simple type, or *static soil compensation*; for *dynamic soil compensation*, the balance procedure will start automatically or continuously during MD operation if the probe does not move [13].

After the balance procedure has been completed, the probe will display any change in the voltage induced only while the probe is moving over the ground. The simplest signal-processing version for a detection system is the level detector for the amount of secondary voltage. Shown in Fig. 4.10 is a typical group of signals obtained by a CW metal detector that crosses buried nonferrous and ferrous metallic objects (plot for one frequency only; an absolute probe). With the fixed threshold chosen (level L_1) the MD will generate audio signals only if the magnitude of the complex voltage is higher than level L_1. No distinction is possible between the different types of materials (nonferrous and ferrous metal or the influence of soil). With phase- and level-sensitive detection, different materials (object X, or the influence of soil) could be detected, and known inference signals (e.g., known debris objects) could be suppressed: For that the secondary voltage has to rotate with a fixed angle ϕ_1 so that after the rotation, object 1 can be discriminated by

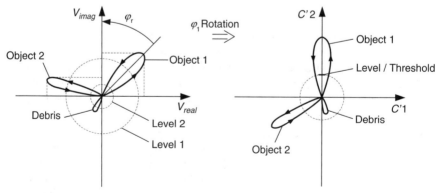

FIGURE 4.10   CW metal detector: level- and phase/level-sensitive detection, absolute probe.

the level or threshold chosen for the imaginary part, and object 2, by the real part of the voltage (see Fig. 4.10).

When for the two components of the original signal, the real part of the voltage, $v(\omega_1) = c_1$, and the imaginary part, $v(\omega_1) = c_2$, the new components c_1' and c_2' can be expressed by rotation of angle ϕ_1 of the coordinates as

$$c_1' = c_1 \cos \phi_1 - c_2 \sin \phi_1$$
$$c_2' = c_1 \sin \phi_1 - c_2 \cos \phi_1$$

(4.7)

Additionally, individual scaling of each component is possible. This method can be generalized, and the basic idea for this signal-processing method (called the vector mode) for disturbing signal suppression in eddy current testing was described by Libby [14]. With n features of the signal, $n - 1$ unwanted signals could be suppressed by $n - 1$ sequential phase rotations ϕ_{n-1}. Each measurement frequency used generates two features, the real and imaginary parts of the coil voltage, so that, theoretically, $2n - 1$ unwanted signals could be suppressed. The suppression of unwanted signals to the target signal will be optimal if their signal components are orthogonal to each other. In practice, the angle is less than $90°$ and it is difficult to find a set of frequencies where the angle between the components of all disturbing and target signals is maximal and close to $90°$ (an independent variable).

With the phase rotation ϕ_i of each unwanted signal (or signal component), which has to be determined theoretically or by experiment, the influence of the mapped target signal component will be minimal if the system is linear. Most systems are nonlinear, so in the second common signal-processing procedure, the difference mode, the vector mode is enhanced. The difference mode gives better results for the suppression of disturbing signals, especially if the signals have the form of loops and the form characterizes the objects to be detected. In the difference mode, the signal components $c_1(t) \ldots c_N(t)$ from N measurement frequencies, which are influenced by the disturbing signals, will be rotated and scaled *individually* and allocated by the *difference* of two signal components, so that the new set of difference components is, for example,

$$d_1 = c_1' - c_4'$$
$$d_2 = c_2' - c_3'$$

(4.8)

where c_1', c_2' and c_3', c_4' are the scaled and rotated components of the signals from the first and second measurement frequencies used, as shown in Fig. 4.11. The suppression of disturbing signals will be zero or minimized, and the mapped target signal will be maximized (multiparameter optimization). The strategy of finding automatically the optimum angle for the phase rotation and scaling values depends on the problem to be solved. Various approaches have been described [15].

With a priori information about the distributor (e.g., the influence of the soil) and the target signal (the signature of one antipersonnel landmine of type X), different strategies are possible for optimized soil compensation of the MD. Regarding metal detection (level detection) and image reconstruction of space-resolved captured MD

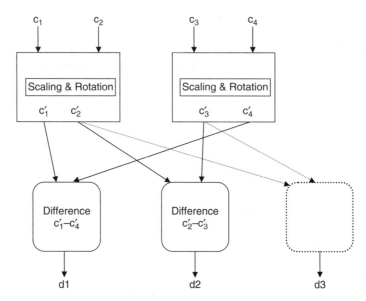

FIGURE 4.11   CW metal detector signal processing using the difference mode.

data (see Section 4.5.1), a soil compensation algorithm may be preferable, where the influence of the soil of the known specific target signal will be minimized. Another strategy proposed for the target or landmine recognition problem is the multivariable *feature-preservation soil compensation* for reducing the false-alarm rate of MDs, where the soil compensation algorithm is optimized with respect to the maximum signal components of the target features (see Section 4.5) [8].

After the phase rotation, scaling, allocation, and mapped target signal, a threshold level has to be chosen to generate an audio signal in case a buried metallic object is detected. When moving or sweeping the metal detector over metal-free ground, no audio signal should be generated by the detector. In field operation and in natural soil conditions, the spatial inhomogeneous distribution of the soil generated by moving the probe in the secondary coil, a change of voltage produces soil noise. In the ideal threshold, no audio signal is generated by the soil noise, which has to be ensured by the soil compensation of the MD.

The amount of a real or imaginary part of the secondary voltage (or in the case of the differential receiver coil, the unbalanced voltage) from each frequency can be used for the simple soil compensation (vector or difference mode), so that for the soil compensation strategy of the MD, the phase rotation, scaling, allocation, and mapped target signal could be used as shown in Fig. 4.16 (see Section 4.3.1 for optimized soil compensation in the difference mode).

MD operation is distinguished by two different operational modes, static and dynamic. The *static mode* is commonly used by CW detectors. After static soil compensation (balance status), the phase-rectified induced voltage (i.e., the real and imaginary parts) passes through a low-pass filter to reduce noise, and then the signals go to the level or phase detector and/or to visualization of the voltage received on a display [$v(x)$ or complex XOY-plane]. The cut-off frequency is determined by the highest-frequency section of the signal, which comes from the highest-speed components of the searching coil as it moves above a metallic object. Any dc value (e.g., drift, the influence of the soil) will be displayed. In the d*ynamic mode*, no dc values of real or imaginary voltage pass a bandpass or high-pass filter, which has principally a tunable low cut-off frequency.

In this context, static and dynamic mode operations should be distinguished from the static and dynamic soil compensation modes mentioned earlier. The dynamic soil compensation mode is used primarily by PI metal detectors. If the searching coil does not move over the ground (is kept in a fixed position), so that the PI output signal does not change, after a defined time period the ground compensation procedure begins again automatically and the current position (ground) will be compensated even when the coil remains over a buried metallic object. When the searching coil moves, a buried metal object will be detected; otherwise, the probe will be compensated in the next time period. Care must be taken to ensure that the soil compensation procedure of the PI probe is carried out over metal-free ground. Both compensation modes have advantages and disadvantages, so the method chosen is based on the experience and attitude of the deminer.

4.2.2 Pulsed Induction Metal Detectors

Pulsed induction metal detection is another fundamental inductive metal detector design. A short, powerful impulse from the magnetic field H, generated by a strong pulse-shaped current in the primary coil, is used as system excitation: a coil above a conductive half-space. The excitation current pulses used can be unipolar or bipolar. The secondary voltage received or induced is recorded vs. the time, as shown in Fig. 4.12. The amplitude and decay of the voltage impulse received within a few microseconds depends on the coil parameter, electrical and magnetic properties of the ground or soil, and buried objects. Any conductive materials change the decay rate of the secondary voltage induced at the switch-on and switch-off slopes, due to the secondary field of eddy currents induced in metallic objects. The switch-off slope is used primarily for PI signal processing, due to the higher rising rate of this slope. For a nonferrous metallic object, deceleration of the decay of the induced voltage is lower than in air because the eddy currents or their secondary magnetic field affect the damping. A material with different conductivity causes other decay of the induced voltage (the dashed curve in Fig. 4.12). In a ferrous metal object the inducted voltage is higher in the first few microseconds and the decay seems to be less, due to the permeability of the material. Later, the decay of the voltage is greater because of the lower induced eddy currents in the low-conductive material.

The general signal-processing method used in the time domain is the integration of induced voltage over time. To distinguish between different states of the PI (standing over metal-free ground or a buried metal object), integration of the induced secondary voltage should be carried out during several time periods: for example, the integrals I_{BA} and I_{CB} for two starting times T_A and T_B and the time duration ($T_{D1} = T_B - T_A$ and $T_{D2} = T_C - T_B$). In recent years, many patents

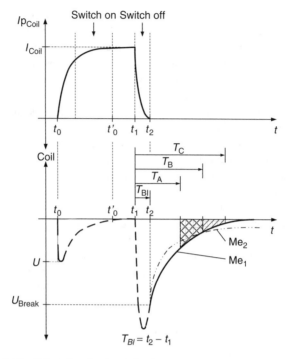

FIGURE 4.12 Principle of pulse induction metal detector, transformer probe.

propose different versions of time-domain signal processing in conjunction with special coil arrangements [13] to improve the discriminating power of PI metal detectors.

The coil of a PI probe can be a parameter or transformer coil as well. A typical block diagram of a PI metal detector, which uses bipolar pulses, is shown in Fig. 4.13. For the parameter type, an electronic switcher must switch alternately between two states: the transmitting (Tx) and receiving (Rx) modes. By means

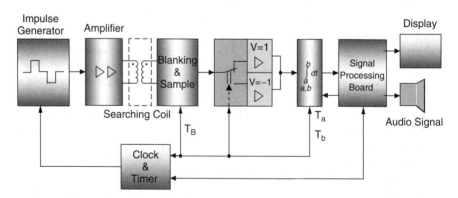

FIGURE 4.13 Pulse induction metal detector, transformer probe, and bipolar pulse.

of the transformer coil, the electrical properties of the Tx and Rx coils (and their construction) can be optimized in favor of their function. The blanking-out circuit assures that the preamplifier input (and that of other components) does not get the high voltage induced from the switch-on and switch-off slopes. Immediately after the delay T_{BL}, the blanking-out circuit opens a path for the induced secondary voltage to enter the low-noise broadband amplifier. In the next stage of signal processing (i.e., in the time-domain signal-processing method), the signal will be integrated with the value I_{BA} over the time period T_{D1} (see Fig. 4.12).

The repeating frequency of the PI probe is dependent on and limited by the parameters of the coil itself (and also the dynamic range of the power amplifier). The diameter, number, and type of winding determine the time constant $\tau (\tau = L/R)$ of the coil, so that the slope rate of the current through the coil is limited as well. In general, the PI metal detector's signal processing integrates the secondary voltage induced over the time (or the time domain), which is a real value. From this value (one feature) and their changes alone, the soil compensation and discrimination of objects must be realized.

The principal advantages of PI metal detectors are the high working stability and the option for a very large searching coil, which implies great penetration depths and a wide sensing range (see Fig. 4.10). The design of the detector is simple (electronic design and a coil) compared to that of other types of detectors. The biggest disadvantage of PI metal detectors is that this type of detector cannot discriminate easily between different type of metals: nonferrous versus ferrous. In some system configurations, the PI probe is combined with other magnetic probes, such as magnetometers and gradiometers. However, magnetometers react to magnetic metals only and thus not stainless steel objects. A few patents offer different coil arrangements and special signal procedures for solving this problem, but if the size and depths of metallic objects detected are different and unknown, in general it is not possible to determine the type of metal.

Signal Generation   Most PI probes use bipolar pulses (higher intensity) and process the induced voltage of the secondary coil from the switch-off slope (higher slope rate). All other parts of the signal have to be blanked out so that the integrator receives the signal induced after time T_{BL}. The first decay of the induced voltage (up to $t_2 = 0$ to 10 to 25 μs) is influenced by properties of the soil from the surface and objects that are closer to the surface or the coil. This time slot corresponds to a point in the frequency domain found by a CW metal detector to have a high cutoff frequency. At a later time ($t_{A,B,C} > 15$ to 50 μs), the decay of the curve is determined by the influence on the depth of metallic objects and soil properties. This fact is used by PI probes for the discrimination of metallic objects (PI signal processing) as well as for soil compensation. The voltage induced in the receiver coil is recorded. All signal processing should be implemented in the time domain. Use of a (variable) time domain for the signal-processing step makes possible integration of the secondary voltage induced over the time, which is proportional to the rate of decay. All information regarding existing electrically conductive pieces around the perimeter of the searching coil is contained in the decreasing curve of the secondary voltage. Discrimination between different types of metals is highly difficult and generally impossible. It is possible only for a well-known and constant testing condition such as the distance to a metallic object of the same size and form.

To make a metal detector sufficiently sensitive in the time domain and to contribute to the reduction of noise (integration over time), a narrowband filter processes a signal before integration. Choosing the blanking-time period T_{BL} (after the switch-off slope) and the time delay (the starting point T_A for the integrator) and time domain for the integrations (variable time windows, $T_{D1} = T_B - T_A$ and $T_{D2} = T_C - T_B$) in conjunction with a narrow filter with adaptive center frequency technology gives the PI probe high working stability and sensitivity.

In a ground compensation procedure for PI probe position x_1, the value of integration of the induced secondary voltage $I_{BA} = \int V_{TB-TA}(x_1)\, dt$ and the difference in the integral values of Eq. (4.9) should be zero:

$$\text{PI}_{\text{output}} = I_{CB} - I_{BA} = \int V_{TC-TB}(x_1)\, dt - \int V_{TB-TA}(x_1)\, dt \geq 0 \qquad (4.9)$$

The value of the $\text{PI}_{\text{output}} = I_{CB} - I_{BA}$ as a function of position x for the absolute and differential probe if the PI is moving horizontally above a metallic object (along a constant distance or depth h) will generate a similar curve, as shown in Fig. 4.7 for the amount of secondary voltage $v(t)$ of a CW metal detector. Each point represents the difference in the decay (or the difference $I_{CB} - I_{BA}$) of the switch-off slope between the point where the probe was ground-compensated and the current position.

In practice, the switch-on rising level of the magnetic field is limited by the rising current in the primary coil (the maximal switch-on rising level is $1/\tau$ with $\tau = L/R$), which is fixed by time constant τ ($\tau = L/R$, where R represents the resistor of the wire of the coil, with the losses in the primary coil and the excitation circuit, and L is the inductance of the coil). The switch-off slope has a lower time constant τ, so that in practice the switch-off slopes of unipolar or bipolar pulses are used, with blanking out of all other induced voltages in the secondary coil. The switch-off slopes of two subsequent bipolar pulses are also used.

Signal Processing and Operational Modes   All signal-processing procedures (e.g., the integration over time of the induced secondary voltage) have to be in the time domain, whereas all signal operations for CW metal detectors are in the frequency domain. The self-induced voltage from the switch-off slope of the PI probe in air is short ($v_{\text{coil}}(t) = L\, di/dt$) and the time-dependent characteristic (decay) is determined by internal electrical values of the coil only. The decay of the secondary voltage is dependent on the conductive material which is penetrated by the (primary) magnetic field impulse. The induced eddy currents in conductive half-space under the coil, as well as all metallic objects, decelerate the decay of the secondary voltage (the time-dependent magnetic field induces an electrical field strength that drives a current—the eddy current—in conductive materials). In materials with higher conductivity, at first ($T_{D1} = T_B - T_A$) there is more damping.

Then, a few microseconds later, it delays the decay of the secondary voltage because of higher induced eddy currents, so that a metallic object can be recognized by another rate of decay or the integrated voltage by a fixed time period. The decay rate is proportional to the conductivity and the size parameter of the object (or ground) below the coil.

The philosophy of signal processing for PI metal detectors is to supervise and analyze the decay of the induced secondary voltage after the switch-off slope of the primary voltage. To distinguish among the various conditions, the most used signal-processing technology is to integrate the signal received in a fixed or variable time domain. The absolute value of the decay is not (very) interesting, but its change can be if the probe is moving over the ground. The change in decay in different time domains due to moving the PI probe above a suspected area gives all information needed about the soil and buried metal objects, so the operational procedures in the field of the PI metal detector are quite similar to those of CW detectors. The first step in PI operation is to compensate or balance the probe manually or automatically while the probe stands over metal-free ground. This can be realized if the difference between the two integrals, $PI_{output} = I_{CB} - I_{BA}$, over the time of the induced secondary voltages [by fixed chosen time domains, $T_C - T_B$ and $T_B - T_A$; see Eq. (4.9)] is zero. In this case, all displayed values are zero from this position ($x = x_1$) (see Fig. 4.12). To reach this simple type of static soil compensation or the balance status, it is sufficient to change the starting point (time T_A or time period) of the voltage integration (or the integral of induced secondary voltage from two sequential bipolar pulses) only.

Few PI metal detectors have implemented dynamic soil compensation operation. The balance procedure starts automatically during the MD operation if the PI probe does not move for a short time (i.e., the PI searching coil stays still over the ground). Because of this, the deminer must consider in the field operation that the PI probe can be compensated (or balanced) on a metal-free ground like the CW metal detector. After the balance procedure and the compensation, the probe will display a change in the value of the integral only while the probe is moving over the ground. The PI_{output} signal as a function of position x for the absolute and differential probes if the PI is moving horizontal above a metallic object (in a constant distance or depth h) will generate a curve similar to that shown in Fig. 4.7 for the amount of secondary voltage $v(t)$ of the CW metal detector. Each point represents the difference in the decay [or the difference in the integral values $(I_{CB} - I_{BA})$ of the switch-off slope between the condition where the probe is ground compensated] and the current position.

The simplest signal-processing version for a PI system is a level detector to determine the value of the integral $PI_{output} = I_{CB} - I_{BA}$. (This value is equivalent to the change in decay of the slope from the point over the ground where the PI probe has been compensated.) By using a variable time delay T_{BL} (or T_A) for the start of the integration after the switch-off slope and by using a multiple integration system (with a variable and multiple time domain T_{Dn}) for each of the sequential bipolar secondary induced voltage slopes, a quite complex signal-processing system is formed in which the PI probe has more features to use to help distinguish among the different situations (e.g., influence of soil or of a buried metal object).

From the point of view of the EMI theory, the CW and the PI metal detector have the same sensing range (for the same coil diameters). However, in practice, the PI detectors can sense metal objects at a lower depth caused by the higher intensity of the short magnetic impulse. On the other hand, a PI metal detector can cause interference with other PI probes or devices in a distance of a few meters, due to the high-intensity primary magnetic impulse and the broadband characteristic of the preamplifier in the receiver path. CW metal detectors have narrowband amplifiers, which usually guarantees higher interference resistance.

4.3   INFLUENCE OF THE SOIL PROPERTIES

Knowledge of electrical and magnetic soil properties is the basis for all EMI survey and antipersonnel demining projects. The influence of inductive metal detectors on soil is one of the important factors that limit the success performance and reliability of MD operations in the field. In recent years a huge number of projects have explored soils to determine the characteristics of their electromagnetic properties for the prediction of the working range (e.g., the frequency range) of EMI sensors, their spatial distribution, and mathematical descriptions of their properties for MD modeling, particularly their frequency dependency [9–12]. Soil is a dispersive medium with regard to its electromagnetic properties. It is a complex system of matter consisting of a variety of components, such as organic and inorganic particulates, liquids, and gases. The homogeneity, bulk density, soil humidity, irregular allocation of stones in the upper soil layer, and the granularity and porosity of the constituents of solid matter in the soil determine the electrical, magnetic, and dielectric properties in a complex manner. From the point of view of EMI theory, the inductive metal sensor is a wire loop above a conducting half-plane, so that buried metal objects and all the electrical and magnetic properties of the soil (i.e., electrical conductivity, magnetic susceptibility, and dielectric permittivity) influence the MD signal. In general, many types of soil decrease the level of detection and discrimination of buried metallic objects because of the electric field induced in the soil (i.e., the eddy currents), which shields objects from the primary magnetic field of the MD.

All soil or paleosoil types consist of three different fractions of inorganic particles: sand, clay, and silt. It is useful to classify the types of soil as to the content of the three fractions of particles, as shown by the grain-size triangle in Fig. 4.14. For example, loam is a soil in which all three particle fractions are distributed uniformly. The actual formation of the soil determines its structure and properties. For the electrical, dielectric, and magnetic properties it is important where soil particles come from and which mineral contents are dissolved as well as the weathering status of the soil. The former material (i.e., stones, called *parent material*), such as vulcanites, basalts, or granites, and the environmental situation define the electromagnetic properties of a paleosoil.

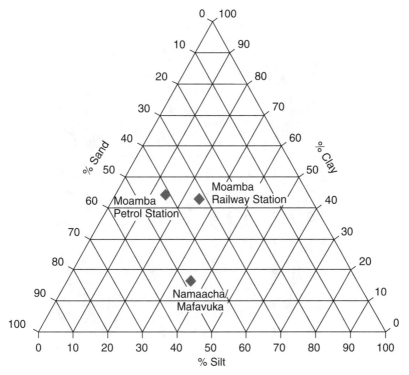

FIGURE 4.14 Grain-size triangle showing the texture of top soils on three different sites in southern Mozambique. (From [7], courtesy of GGA, Hannover, Germany.)

From the point of view of antipersonnel landmine detection using inductive metal detectors, the soil is classified in two major groups:

1. *Cooperative soil.* Sand or gravel as well as wet sand are cooperative soil, because the influence of the MD is constant due to the electrical conductivity and can be compensated for easily electronically.

2. *Uncooperative soil.* Uncooperative soil is mostly laterite with pyrite fractions, which disturbs all inductive metal detector operation. Due to its magnetic properties (ferrimagnetisms) and frequency dependency (dispersive media) as well as high spatial variation, it will generate a signal in the MD where no metallic object is buried (i.e., a false alarm). The spatial distribution of the magnetic susceptibility of an area with an uncooperative soil is shown in Fig. 4.15.

In the worst case, every few decimeters a new calibration (ground compensation) may be necessary, so that mine detection becomes impossible. Magnetic susceptibility and spatial distribution are the pivotal properties that influence MD operation in the field. The electromagnetic properties can be determined through geophysical measurements, however, and this soil map or geological map can be used to plan de-mining operations.

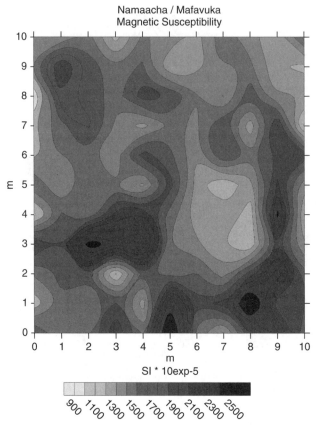

FIGURE 4.15 Spatial distribution of the magnetic susceptibility in the top soil: 10×10 m site in southern Mozambique (see Fig. 4.14). (From [7], courtesy of GAA, Hannover, Germany.)

Hence, the absolute values of properties are not that important—their spatial variability and frequency dependency (in the case of magnetic permeability) are more remarkable. In general, for all three properties—electrical conductivity, magnetic permeability (or magnetic susceptibility $\chi = \mu_r - 1$), and electrical permittivity—the soil is a dispersive medium. In the case of time-dependent excitation, additional losses in the media (soil) are caused by electrical and magnetic polarization effects, which could be described using special forms of time-dependent material properties [10], or in special cases for sinusoidal excitation in complex material properties [12]:

$$\chi(\omega) = \chi' + j\chi'' \tag{4.10}$$

$$\varepsilon(\omega) = \varepsilon' - j\varepsilon'' \tag{4.11}$$

$$\mu(\omega) = \mu' - j\mu'' \tag{4.12}$$

The real part characterizes the *effective property*, which is mostly constant for the low-frequency range, and the imaginary part is related to additional losses. The description of frequency dependency is approximated through relaxation processes and mechanisms in time-varying electric and magnetic fields by using one or more relaxation time constants, depending on the model of relaxation chosen (e.g., the Debye or Cole–Cole model [12]). These high-frequency properties have a greater influence on metal detector systems, which are based on ground-penetrating radar (GPR) and ultrawideband (UWB) systems, than does standard inductive MD.

For frequencies below 100 kHz and a searching coil diameter of less than 1 m, the influence of dielectric properties can generally be neglected or reduced through the coil construction. Even in salt water, which quite high in conductivity and electrical permittivity, there is no problem in compensating for this (constant) influence of the detector's operation.

The electrical conductivity and magnetic properties of soil depend on the content of the soil and on actual environmental conditions (e.g., rain, temperature), but the electrical conductivity of the soil is always lower, by about a factor of 10^3 to 10^4, than that of buried metallic objects. The porosity, granularity of the soil components, bulk density of the upper layer of the soil, and actual weather situation, such as temperature and moisture, influence the MD operation in the field and have to be compensated for electronically before operation in the field.

The spatial variability of the magnetic susceptibility is more important for MD operation than is their absolute value. The heterogeneity is caused by an irregular allocation of stones in the upper soil layer as well as by the variation of the bulk density. For uncooperative soils, typical correlation lengths measured on a paleosoil (e.g., in Mozambique) are between 0.5 and 1.0 m (see Fig. 4.15), which complicate dramatically safe demining of buried landmines using MD. The correlation lengths give the deminer information as to at which spatial distance an MD will acquire unbalanced status.

4.3.1 Soil Compensation for CW Metal Detectors

The standard solution for the soil compensation of mono- and multifrequency CW sensors uses one or more measured components for compensation. The eddy currents in the soil increase the losses of the coil, so that their real and imaginary parts will change. Each metal detector should compensate the soil influence before beginning the search operation mode. For an absolute probe, a simple soil compensation procedure could be as follows: subtraction of the induced secondary (unbalanced) voltage if the coil remains fixed over metal-free ground (here, any change in the distance of the coil to the ground will influence the result). For a differential coil, if the coil is kept in a noncoplanar position to the ground (e.g., 30 to 60°) so that the level of the induced secondary voltage is not zero (nonsymmetric field distribution under the coil). Because of the constant electrical and magnetic properties of the soil, the amount can be compensated by determining the necessary phase rotation for the signal components and mapped signal, as mentioned in Section 4.2 (difference mode). Most CW metal detectors use just one component (the imaginary or real part from one measurement frequency) to compensate for the influence of soil.

In the case of cooperative soil, (nondispersive soil regarding the magnetic susceptibility and their spatial distribution) this version of compensation works well because the disturber signal has a constant influence, which comes from the electrical conductivity of the soil. In uncooperative soil, the spatial distribution of the magnetic properties generates signals in an MD that overlap the target signal, so that for each features of the signal the phase rotation and scaling, combined with other features, has to be optimized for a minimum influence of the MD signal (Fig. 4.16). This can be achieved if all capture features of the MD signal are used and the soil's influence is known. Two strategies are available to reduce the false-alarm rate: mine/soil-sensitive compensation (MSSC) and feature-preservation soil compensation (FPSC), in conjunction with a data-based pattern recognition system [8] (see Section 4.5.3).

4.3.2   Soil Compensation for PI Metal Detectors

As mentioned in Section 4.2, the signal-processing method for PI metal detectors is implemented in the time domain as the integration of the induced voltage over time. One aspect of state-of-the-art soil compensation methods employing a PI metal detector is shown in Fig. 4.12: the difference of two integrals $(I_{CB} - I_{BA})$ over the interval of induced secondary voltages by fixed chosen time domains $T_{D2} = T_C - T_B$ and $T_{D1} = T_B - T_A$. For efficient soil compensation, the same strategy as that employed by CW metal detectors can be used: a *multiple integration system*, with N integrals I_N of the secondary voltage over time, where the integration time periods can be overlapped so that the parameter space expands. Each separate integral I_N over the voltage gives a new parameter (a new feature), which can be combined for soil compensation. The problem here is to find time slots for N integrations of two sequential switch-off pulses, where the difference of two or more values (integrations) is mostly zero, which is equivalent to the status of static soil compensation. In conjunction with a time-dependent tunable broad bandpass filter, the PI probe can reach very high stability and sensitivity in field operations.

Another approach to PI metal detectors is dynamic soil compensation. That is, if the PI$_{output}$ signal does not change in a fixed time span (the PI probe is not moving), the internal ground compensation procedure again starts automatically, and the current position over the ground will be *ground compensated* even when the coil is fixed over a buried metallic object. At present, optimal soil compensation is the key factor for any successful survey of a metal detector used for landmine detection. For field operations on cooperative soil, the state-of-the-art technology of the existing MD is sufficient. Due to the high spatial distribution of the soil properties in uncooperative soil, particularly magnetic susceptibility, the false-alarm rate increases dramatically. The soil compensation must be adaptive for the demining procedure. One way proposed to solve these problems and to increase the reliability of MD could be the introduction of FPSC for CW detectors as well as multiple pulse integration techniques for PI detectors.

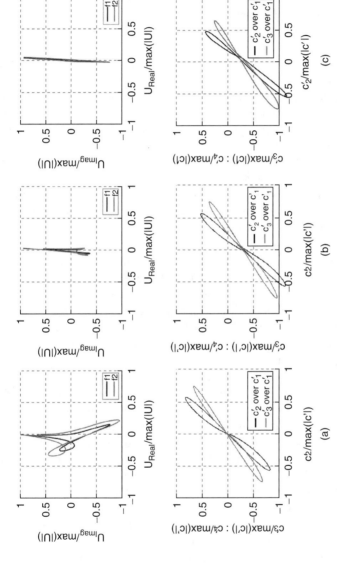

FIGURE 4.16 Soil-uncompensated loops (4.9) and soil-compensated loops (feature-preservation soil compensation) at 2-, 6-, and 10-cm depths in uncooperative soil, M2B mine. (From [8].)

4.4   MODELING INDUCTIVE METAL DETECTORS

Although inductive sensors, usually as an absolute or transformer-coupled air or ferrite core coil type, can be built very simply, accurate mathematical modeling proves to be quite complex even for the simplest calibration arrangements. The modeling of all inductive sensors mathematically is based on Maxwell's equations of the electrodynamics (as mentioned in Section 4.1). Maxwell's equations are the basis for any mathematical modeling of inductive sensors. There exist a number of different analytical and numerical methods for solving the various types of electromagnetic fields which are relevant for MD problems. In general, a partial differential equation with boundary conditions has to be solved, and the type of equation (elliptic, hyperbolic, and parabolic) is dependent on the EM problem with steady-state or transient equations (e.g., for harmonic solution). The partial differential equation can be homogeneous or inhomogeneous, depending on the excitation used. To model CW and PI metal detector problems, the stationary Helmholtz equation must be solved.

For nondestructive testing as well as (magnetic-inductive) geophysical prospection problems, the first approach to modeling a metal detector is a simple transformer model that uses the electrical (equivalent compact) elements of resistance R, self-inductance L, and mutual inductance M. Later, the first analytical solutions of the impedance or induced coil voltage of the MD probe for simplest symmetric coil arrangement are presented. The coil arrangement and analytical solution and its simplification are different for geophysical and NDT problems. These are the relations between the coil diameter and the form and size of test objects as well as the relationship between the electrical, magnetic, and dielectric properties of the materials. For example, the analytical solution for a metallic sphere above a coil over a great distance, that is, a magnetic dipole, can be used to describe the effect of the induced coil voltage if the eddy currents in the sphere are neglected [16,17].

At present, for most applications of inductive sensors, there are standard solutions whereby selection of the probe coils and parameterization of the devices are carried out using empirical values. With the help of a mathematical modeling approach and in addition to visualization of the measured fields, it is possible to optimize all constructive and physical parameters of inductive sensors for various settings of tasks (e.g., in the field of NDT, metal detection in environmental systems, and the food industry, as well as demining problems of antipersonnel landmines). To realize this approach, engineers use different numerical procedures and software packages.

In this section, the boundaries of the analytical method are discussed and possible modern numerical procedures for computer-aided drafting of inductive sensors for different metal detection applications are shown. For each application the coil model is generally chosen according to the setting of a task, and the coil form and core, as well as the operating frequency, are determined empirically. If required, some experimental investigations can also be used for optimization. If an experimental optimization approach is used, it is often very difficult, sometimes impossible, to

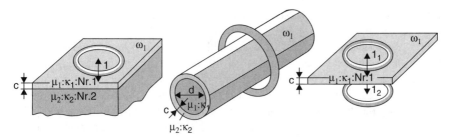

FIGURE 4.17 Typical probe coil arrangements in nondestructive testing (analytical solutions).

vary only one influence parameter. However, with the aid of mathematical modeling, the probe coil impedance effects expected can be estimated, and both the measuring frequency and the reactance coil geometry, including the field profile, can be optimized according to the problem. The boundary value problem (BVP) is usually solvable only for the simplest of coil arrangements and linear material properties: for example, a rotationally symmetrical air coil over a conductive half-space as shown in Fig. 4.17 (pickup coil) or for the tube probe coil represented by a uniform infinitely long core [18,19].

One of the first analytical solutions for modeling an inductive sensor arrangement (such as that which is typically used in NDT), was developed by Sobolev [18] and by Dodd and Deeds [19]. These initial approaches involved a symmetrical probe coil arrangement that was computed from the layer thickness, conductivity, and remote-sensing distance. An analytical expression for the impedance change in a coil carrying a low-frequency alternating current due to the presence of a small conducting sphere from the geophysical point of view was given by Wait [16] and later by Hugo and Burke [20,21]. Core and shielded probe coils prove that it is difficult to find an analytical solution even for very simple problems encountered in defect detection (i.e., the calculation of the probe coil impedance for a crack and the behavior of the sensors at edges). Such practical arrangements are important for probe design and accessible only with the help of a numerical (discrete) model approach [22,23].

4.4.1 Inductive CW Metal Detectors

In general, analytical procedures cannot solve nonsymmetric problems such as that of an MD coil moving over a buried metallic object. A short overview of a few different analytical approaches to the symmetric electromagnetic induction problem regarding metal detector problems is given by Bruschini [9]. By using a harmonic excitation signal $J(\omega)$ and the vector potential A (with the relation of $B = \mathrm{rot} A$) as well as linear materials properties (electrical conductivity κ and magnetic permeability μ), a solution in the frequency domain can be given by the (complex) stationary Helmholtz equation:

$$\Delta A(\omega) + k^2 A(\omega) = -\mu J(\omega) \tag{4.13}$$

where k is the wavenumber and the influence of the displacement current density D is neglected.

The numerical procedures involved in the modeling of inductive sensors are based on the discretization (Eq. 4.9) of the field area to be computed. One can divide the numerical procedures into five groups: (1) the finite-difference method (FDM), (2) the finite-element method (FEM), (3) the boundary element method (BEM), (4) the finite-integration technique (FIT), and (5) hybrid methods [37]. Initially, the FDM and FEM were applied to NDT problems by Ida and Lord [22], with the boundary integral method and finite-integration theory being employed successfully by Fawzi et al. [23]. The volume integral method can be used for eddy current coil arrangements with a ferrite core for solving NDT problems [24]. The solution approaches may be classified as for planar problems (i.e., for humanitarian landmine detection) and rotational problems, which are typically considered in the food industry and security areas.

For the rotationally differential coil arrangement shown in Fig. 4.18 (suction dredger and a saltwater–sand mixture in a tube of diameter up to 900 mm), a metal piece must be detected in a space such as a tube or cylinder (infinitely long). The investigation is carried out for a specific sensor and is involved in finding the optimal setup by utilizing the test frequencies, the geometry of the coil arrangement with respect to the default inside diameter of the volume flow, the range, and the material properties of the metal parts. The investigations were performed using the FIT simulated with the MAFIA software package [25,32]. The magnetic field generated by the primary coil at first induces equal voltages in the symmetric secondary coils. Without electrically or magnetically conductive metallic parts, this setup results in a zero-voltage difference. If a metallic part appears within the volume of the coil, a voltage difference arises from the resulting distortion of this symmetry. Figure 4.19 shows the discretization (mesh) used for the problem, and Fig. 4.20 demonstrates the effect on the amplitude of the induced voltage difference (differential coil) for one frequency as a function of the relative distance d/D of the secondary coils (D = coil diameter).

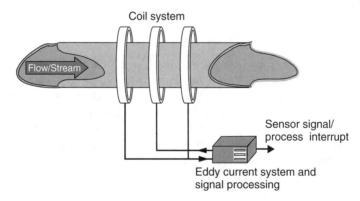

FIGURE 4.18 Differential coil arrangement for metal detection (water–sand stream).

FIGURE 4.19 Three-dimensional FIT modeling of the differential coil arrangement in $r-z$ coordinates: metal piece in the plane by $\phi = 0$ and $r = 0$.

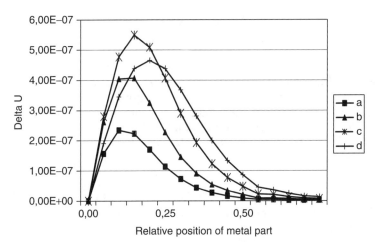

FIGURE 4.20 Amplitude of voltage difference vs. the relative position p: (a to d) different relative distances (d/D) of the secondary coils.

A distinctive feature in this modeling was a consideration of the influence of the high dielectric property of the saltwater−sand mixtures in the tube ($\varepsilon_r = 81$). The concentration of sand can change from 0% up to 40% (percentage of mass). In case a metal piece is detected, the sensor signals recorded could also be used for an internal flow control system (cross-correlation of the two secondary signals). With a two-frequency system (four features) it is possible to distinguish between the change of a stone−sand−water mixture and that of a metal piece (the inference signal is suppressed). A planar differential coil arrangement for the MD problem

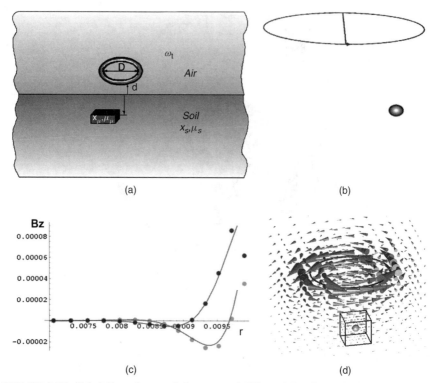

(a)   (b)

(c)   (d)

FIGURE 4.21   Modeling of a metal detector and differential coil over a metallic object in air. (a) Planar coil arrangement for metal detection. (b) Planar coil arrangement for metal detection: a ball of copper in air (three-dimensional problem). (c) Comparing the numerical results with the analytical solution for the distribution of the flux density component Bz in the metal pieces. (d) Distribution of the electric field: imaginary part of electric field strength. (From [32].)

(see Fig. 4.1) is shown in Fig. 4.21a. It is quite similar to the inverse NDT crack detection problem (detecting an air gap in a conducting half-space $\leftrightarrow$ detecting a metal piece in air).

In recent years there have been a number of investigations of the influence of all the parameters on the results of inductive metal detectors used in humanitarian demining because depending on the properties of the soil, there has been the problem of a high false-alarm rate affecting the detection capability [8,9, 26–31]. From the point of view of mathematical modeling, for MDs in general the three-dimensional problem must be solved. One of the greatest problems in using numerical approaches to modeling MD problems is the significant difference in the geometric sizes of objects (e.g., the ignition pin in an antipersonnel landmine is a few millimeters in size, whereas the single winding of the coils is in the decimeter range). To consider the eddy currents induced in small metal pieces, the discretization chosen must be sufficient for the needed approximation of the distribution of the eddy currents in a piece (the skin effect).

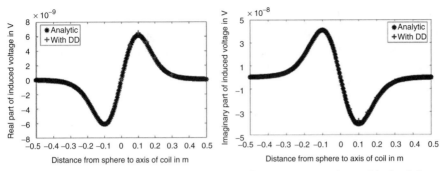

FIGURE 4.22 Comparing a simulated signature of MD: a copper sphere with $d = 2.8$ cm and $d = 20$ cm under the coils. (From [20,31].)

To achieve an efficient numerical solution, a domain decomposition method can be used to divide the calculation domain into subdomains. The MD problem, the three-dimensional distribution of the imaginary part of the electric field E for a coil above a piece of metal in air, and the flux density component Bz in the metal pieces by using the domain decomposition techniques are shown in Fig. 4.21 [31]. The domain decomposition techniques subdivide the calculation domain into sub-domains with different discretizations. A finer mesh is chosen around a metal object (e.g., a landmine) and a coarser one for the rest. Both subdomains are coupled with the help of interface conditions. Compared with conventional numerical methods, for the same total number of grid points, better accuracy can be achieved using the domain decomposition method with Lagrange multipliers, as has been shown for three-dimensional time-harmonic problems [31]. In Fig. 4.21 a circular transmitting coil and a (double D–shaped) differential receiving coil, both with a diameter of 30 cm located 20 cm above a copper sphere, are chosen for validation of the model problem.

The simulated signature of the MD in Fig. 4.22 shows analytical and numerical solutions (difference coil: diameter $d = 30$ cm, copper sphere: $d = 2.8$ cm, 20 cm under the coils, frequency 2.4 kHz). The good agreement among the solutions is clearly seen. The only difference is caused by the neglected eddy current in a metallic object in the semianalytical solution. Other approaches to solving the electromagnetic induction problem [e.g., for a perfect electrical conductor (PEC) or a (nonconductive) magnetic permeable piece (magnetic dipole) in soil/earth] cannot be used as references for the advanced soil compensation algorithm as well as for generating signatures of the metal detectors for the database (see Section 4.5), because their accuracy is not sufficient at this time [8]. Furthermore, the magnetic and electrical properties of soil are functions of the frequency. Different software packages [32–35] can be used to calculate the coil impedance due to the presence of a small metallic piece (e.g., a landmine). Landmine classification problems require very high computational accuracy for the coil impedance (or induced voltage) [8].

4.4.2 Inductive PI Metal Detectors

For modeling inductive PI metal detectors, the relevant Maxwell equation, the instationary Helmholtz equation, has to be solved in the time domain, so that, for example, the magnetic vector potential A ($B = \text{rot } A$ and by negligence of the displacement current density D) is given by

$$\sigma \, \frac{\partial A}{\partial t} \, \frac{\text{rot } 1}{\mu \text{ rot } A} = J(t) \tag{4.14}$$

This equation can be solved with a number of well-known numerical procedures, such as the finite-difference time domain (FDTD) or finite-element time domain (FETD) techniques (Section 4.4), but these techniques can require a great deal of computation time [36]. Different analytical and semianalytical approaches for modeling of inductive PI metal detectors have long been present for geophysical problems [17,28,30], where the solution is given primarily for a symmetric arrangement of coil and metallic sphere.

Another approach could be the use of superposition methods for linear systems, where the excitation signal will be transformed from the time domain into the frequency domain for each. If the switch-on and switch-off slopes of a PI detector as a function of $v(t)$ are available, then using the Fourier transformation of $v(t)$, the required magnitude and phase of the harmonic frequencies (the required number of frequencies) can be determined. The induced voltage of the PI detector (the time response) can be approximated by the superposition of all (harmonic) responses in air as shown in Fig. 4.23. The difference in the peak amplitude between the measured and simulated coil voltage values is caused by the blanked-out circuit of the PI metal detector that is used.

The procedure of solving the stationary Helmholtz equation for N frequencies and superposing the results is faster than the use of time-step integration methods for the numerical models. The eddy currents induced in the metallic object have to be considered for accurate modeling of the signature of the landmine. The superposition method is valid for linear systems only (e.g., a PI detector over a nonferrous metal in air). For a PI detector over the ground and/or above buried ferromagnetic objects, this assumption is not satisfied.

4.4.3 Influence of Soil Properties on Metal Detectors

Modeling the influence of soil properties on MDs is very difficult. On the one hand, conductive, dielectric, and magnetic soils are frequency dependent in general (see Section 4.3) and their spatial distribution is inhomogenous. On the other hand, an analytical solution is difficult to simplify (e.g., the symmetric coil arrangement, so that if the detector is moved above a soil surface with a buried metal object, the necessary signature could not be modeled analytically as a three-dimensional problem. The magnetic properties of the soil have the greatest influence on the induced voltage of the MD. Analytical approaches to a time-domain response for a buried object in the soil generally employ one or two important relaxation time

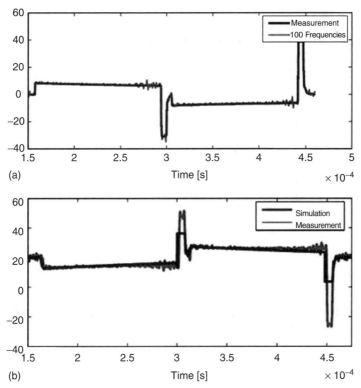

FIGURE 4.23 Modeling of a PI detector coil (in air) using superposition methods: (a) measured and Fourier-transformed excitation signal; (b) Results of the simulated and measured PI response signal in air. $N = 100$ harmonic frequencies.

constants to approximate the frequency-dependent magnetic susceptibility [28–30]. Numerical modeling of the influence of soil on a PI metal detector (e.g., determination of the ground penetration high [11]) can be approached using the superposition method: For uncooperative soils, a frequency-dependent magnetic susceptibility for each needed frequency f_i can be employed, so that the Helmholtz equation of the vector potential A has to be solved in the form.

$$\sigma \, \frac{\partial A(f_i)}{\partial t} \, \frac{\text{rot } 1}{\mu(f_i)} \, \text{rot } A(f_i) = J(f_i) \tag{4.15}$$

The coil voltage induced can be approximated as the sum of all voltages, $v_{\text{coil}}(t) \sim \sum v(f_i)$, because due to the frequency-dependent magnetic susceptibility, the problem is in general nonlinear (magnetic permeability and its frequency dependency).

The results of the influence of the frequency-dependent magnetic susceptibility of the induced voltage of a PI detector are shown in Fig. 4.24. Included are the difference in induced PI voltages in air and over the ground for the susceptibility of a magnetic soil with frequency-dependent and constant magnetic properties. The

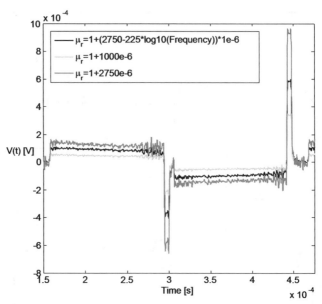

FIGURE 4.24   PI metal detector: influence of the frequency-dependent magnetic susceptibility of the difference in induced voltage in air and over the ground, $\kappa = 0.01$ S/m,
$\varepsilon_r = 1$.

frequency-dependent magnetic susceptibility of the soil, $\mu(f_i)$, is determined and
approximated for each harmonic excitation frequency using experimental soil data
from the geophysical investigation of uncooperative soil [7]. With this approach
it is possible, using measured magnetic soil properties from the test region, to
simulate the ground reference height (GRH) and equivalent detection depth before
field operations are begun [11].

The mathematical modeling of inductive sensors can be calculated by analytical
and numerical methods. The impedance of air-core coils is computable assuming analytical models based on linear electrical and magnetic material properties.
Preferably, inductive sensors with bore materials and nonlinear material characteristics are modeled using numerical procedures only. At present, the FEM, BEM,
and finite integration theory represent the strongest procedures for the calculation
of two- and three-dimensional magnetic field problems. Materials with nonlinear
properties can be computed, and an arbitrary form of the exciter signal response
can be approved.

Analytical and numerical approximation procedures for modeling both types of
MD problems are available. Very high accuracy is required for realistic CW and
PI metal detector modeling: for example, to calculate the signature of a landmine,
which can be realized by domain decomposition techniques (submesh techniques)
for the numerical methods (nonsymmetric, three-dimensional model). Modeling
the influence of the soil is another approach, which can be solved analytically and
numerically as well for the frequency-dependent magnetic property of the soil by

using a complex magnetic susceptibility model with one or more different magnetic relaxation times.

4.5 ADVANCED SIGNAL-PROCESSING AND PATTERN RECOGNITION SYSTEMS FOR METAL DETECTION

All CW and PI metal detectors are very sensitive; they can detect very small pieces of metal. Their differing detection powers are determined primarily using the analyzing signal method as well as the soil compensation strategy. One of the main disadvantages of all standard MDs is the fact that they cannot distinguish between one and two or more buried metallic objects under the searching coil, called a *multitarget situation*. Even the differential coil and multidifferential will not detect each metallic piece separately; in practice, the pinpointing procedure might not be feasible, which could indicate detection of a multitarget situation.

In the following section, new advanced signal-processing methods using time- and high spatially resolved metal detector data are described: identification of multitarget situation, a feature-preservation soil compensation strategy, and a data-based pattern recognition system. From the point of view of signal processing, one-, two-, and three-dimensional signal-processing algorithms will be able to extract or detect much more information from the MD data. The first example is the use of the time- and highly spatially resolved data acquisition as well as advanced data analysis.

4.5.1 Advanced Signal Processing Using Time- and Space-Resolved MD Data

When a differential CW metal detector probe is moved over ground with a buried metallic object, the continuously recorded one-dimensional signal [complex coil voltage $v_{coil}(x)$] has the chronological sequence shown in Fig. 4.7. The presence of a metallic object can be extracted from this signal by means of simple level and phase detection. By using a two-frequency MD probe and moving the probe in two directions (scanning in the x and y directions), an image can be generated (in false color) for each of the four recorded voltages, as shown in Fig. 4.25. This two-dimensional data set is called a *footprint* of the buried metallic object. It is very difficult to distinguish between the different signatures represented by the footprints.

By using high spatially resolved data sets in conjunction with an image-processing algorithm, more information about the buried object can be extracted. It is easy to recognize from the image where the buried object is localized, and it is simple to automate the pinpointing procedure: midway between the maxima (differential coil) in two dimensions, so that the exact localization of a metallic object in the XOY plane is possible. In the case of an absolute probe, the maximum is the pinpoint itself. Furthermore, it is possible to tell whether the scan has captured the signature completely. The disadvantage of these raw data images is that no information about the metallic object (i.e., the number of objects, size, form, and depth) is visible without additionally signal processing. For imaging

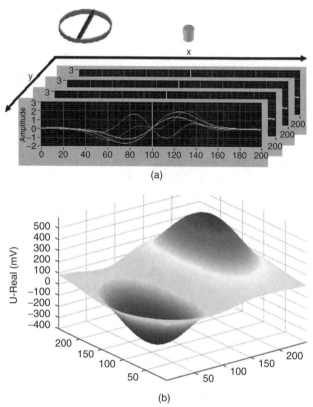

(a)

(b)

FIGURE 4.25   (a) High spatially resolved one- and two-dimensional data sets of the MD (scanned probe). (b) Typical footprint of a buried metal piece (real part of the secondary differential voltage, $f_1 = 2.4$ kHz).

sensors, a well-known technique is the deconvolution method [37], which could be applied to the inductive sensor.

When the sensor element (e.g., differential eddy current coil) is larger than the inhomogeneity (buried metal piece), the sensor signal, $P(x, y)$, is also dependent on the aperture size of the sensor element itself. In general, the sensor signal or picture $P(x, y)$ measured [e.g., the magnetic field distribution $B(x, y)$] is produced by convolving the aperture of the sensor $AP(x, y)$ and the aperture of the "defect," so that the original "image" can be reconstructed using the deconvolution method [9,37].

From the raw two-dimensional data set of the MD or the picture P [P(x, y), footprint] and by using the aperture function AP(x, y) of the sensor, an image of the original scene or geometry O(x, y) can be reconstructed. In the spatial domain, every picture P can be considered as a convolution of the aperture function AP of the sensor and the original (geometric) scene O:

$$P(x, y) = O(x, y) * AP(x, y) \tag{4.16}$$

with $*$ a convolution operator in the spatial domain, the original scene can be calculated as

$$O = P * AP^{-1} \qquad (4.17)$$

The notation in the frequency domain is

$$F\{O\} = \frac{F\{P\}}{F\{AP\}} \qquad (4.18)$$

where $F\{P\}$ is the Fourier-transformed picture [the raw two-dimensional sensor data $S(x, y)$] and $F\{AP\}$ is the Fourier-transformed aperture function of the sensor.

The aperture function AP of the sensor is also called the point spectrum function (PSF). In practice, the measured sensor signal data [the picture $P(x, y)$] is sometimes noisy, so that in Eq. (4.16) an additional term $N(x, y)$ for the noise should be introduced. Figure 4.26 shows the raw eddy-current images P of an MD (the real part of the complex secondary coil voltage) from a scan over a wire copper bent and the reconstructed image O using the deconvolution method.

The aperture function $AP(x, y)$ of the metal detector can be determined theoretically or experientially [7,9]. The result of the reconstruction using the deconvolution method is dependent on the quality of the raw data (i.e., noise, etc.), the aperture function used, and the method of noise cancellation (e.g., the Wiener filter) [9]. For deconvolution of the two-dimensional CW metal detector data on the real or imaginary part of voltages, the amount or phase data (images) could be used.

The sensor aperture function is a function of the depth chosen for an infinitely small metallic object, so that several aperture functions $AP(x, y, z_i)$ for different depths z_i should be created for the reconstruction procedure (three-dimensional data sets). The best results of the deconvolution reconstruction gives the $AP(x, y, z_i)$ values, where z_i is equivalent to the depth of the buried object: With this information, a prediction of the depth is possible.

Another example of the use of image processing of two-dimensional metal detector data is the recognition of a multitarget situation, as shown in Fig. 4.27. In the raw data image $P(x, y)$ or footprints no multitarget situation is visible, but in the reconstructed image $O(x, y)$ three metal balls are recognizable, and in the reconstructed phase image, the type of metal can be distinguished: ferrous and nonferrous metal pieces. This feature can be realized only by analysis of one-dimensional MD data, the phase of the loops. A multitarget situation in the xy-plane could be detected using this method. But the signatures of single objects interact depending on their distance from each other, so that their specific (one-dimensional) signature will be influenced. In general, in a multitarget situation, a landmine is not recognizable as a landmine of type X, so such MD data in general cannot be used for further pattern recognition procedures.

Furthermore, any layered objects in the z-direction cannot be detected with an MD. The metallic objects that are nearest the metal detector probe or the surface shield the other objects, due to their induced eddy currents, so that the primary

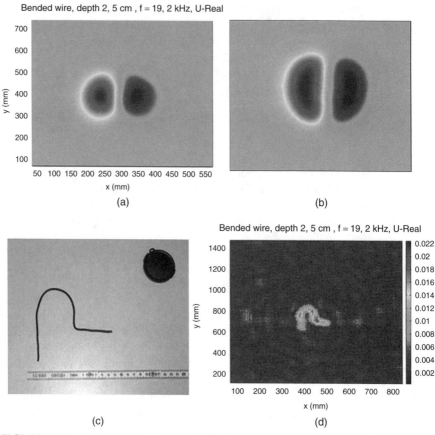

FIGURE 4.26 Image reconstruction of MD data: (a) raw image P (real part of the secondary differential voltage); (b) the sensor aperture function; (c) original scene of bent wire; (d) reconstructed image O from the MD data P (real part of the voltage).

field cannot penetrate the objects that are at greater depths. This type of multitarget situation cannot be discovered using a reconstructed image $O(x, y)$.

For the geometric alignment of metallic objects in the xy-plane (e.g., the three detonators used in the TMA-4 antitank mine), using a highly spatially resolved data acquisition system and deconvolution method it is possible to detect and recognize such a multi-ignition mine with high reliability, as shown in Fig. 4.28 (MD data from the landmine test field in Benkovac, Croatia, uncooperative soil). The lateral distance between the three metallic objects (i.e., the detonators) can be determined and matched against the original. The blurred signature of the captured two-dimensional metal detector data (e.g., in Fig. 4.28b, left) suggest a buried object in uncooperative soil. A clear signature, similar to the PSF itself, of a buried object indicates a single- or multitarget situation in cooperative soil; multitarget situations are observable in raw two-dimensional data sets only if the lateral distance between

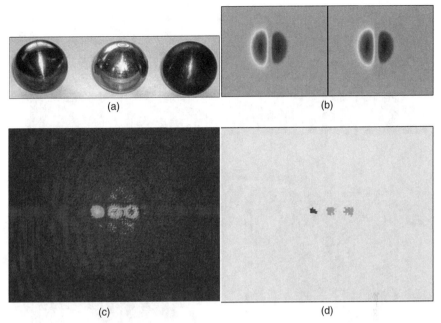

FIGURE 4.27   Image reconstruction of MD data in a multitarget situation at a 5-cm depth: (a) three metallic objects (ball $d = 28$ mm; steel, aluminum, and copper); (b) raw image data P, real and imaginary part of the voltage, $f_2 = 19.2$ kHz. Reconstructed images of the original scene O: (c) amount; (d) phase.

the metallic objects is sufficient in relation to the diameter of the searching coil. Any uncooperative soil generates blurred signatures, which are already observable in the raw two-dimensional data sets. With the next step, the deconvolution method for the reconstructed images $O(x, y)$, a single- or multitarget situation is detectable. In the single-target case, the pinpointing procedure extracts the optimal one-dimensional signature of the buried object, and following a signal-processing method for object recognition (e.g., a landmine or clutter), level, phase-sensitive detectors or other pattern recognition methods can be applied.

For uncooperative soil, which is recognizable by the blurred signatures of the space-resolved two-dimensional raw data of the metal detector, the influence of the soil has to be compensated in the two-dimensional raw data sets using an adapted soil compensation procedure before the deconvolution method can be used. The deconvolution of soil-uncompensated two-dimensional data sets generates artifacts in the images, which are not related to the original scene, as shown in Fig. 4.29. Moreover, it is not possible to determine whether or not a signature is generated by a metallic object and whether the signature has been captured completely. The deconvolution of soil-compensated MD data sets provides better results (see Fig. 4.26b, right).

All advanced signal-processing methods described for reconstruction of the original image O using the deconvolution method can be applied to a PI metal detector

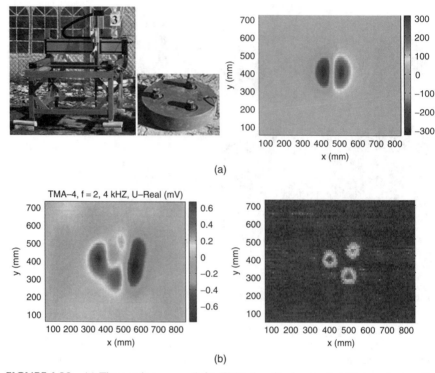

(a)

(b)

FIGURE 4.28 (a) Three-axis scanner (left), TMA-4 antitank mine (middle), and a two-frequency CW metal detector at the Benkovac test field, Croatia (right). (b) Raw data, real part, $f_1 = 2.4$ kHz (left) and the resulting deconvolution using soil-compensated data, real part of the voltage, $f_1 = 2.4$ kHz (right).

as well. As regards the captured space-resolved PI data, for each scanning point, a PI_{output} signal, which is the difference in the integrals over several scanning times [see Eq. (4.9)], is generated by the presence of a metallic object, an image quite similar to that from the CW detector scan, shown in Fig. 4.26. The original scene here is the same three balls that are shown in Fig. 4.27.

An automatic pinpointing procedure followed by signal-processing methods for object recognition can be employed as well. The presence and position of a metallic object and multitarget situation can be recognized using the deconvolution method. The lower signal-to-noise ratio of the PI images for this example (Fig. 4.30) compared to the CW data (see Fig. 4.27) is caused by the internal signal processing of the commercial PI metal detector used (broadband characteristic of the PI), not by the intrinsic power of the PI detector. In uncooperative soil, all raw data have to be soil compensated before deconvolution can be begun.

4.5.2 Advanced Soil Compensation for MD

There is a dramatic decrease in detection probability in soils with magnetic properties ($\mu_r > 1$) or inhomogeneous conductivity, because of the dominant soil signal,

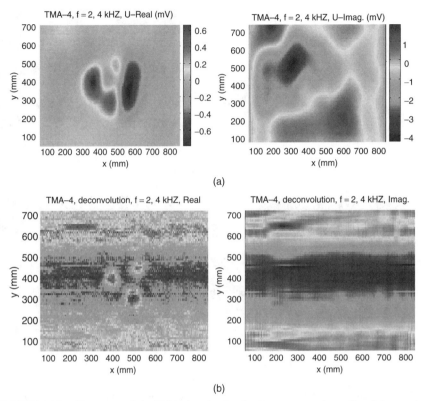

FIGURE 4.29 Signatures of the TMA-4 antitank mine in uncooperative soil and the results of the deconvolution using two-dimensional raw data sets; no soil compensation of the data sets. (a) $f_1 = 2.4$ kHz, real and imaginary part of the voltage; (b) results of the deconvolution: real part (left) and imaginary part (right).

especially for of mines with a low metal content. Compensation for the influence of the soil of an MD is the key factor for any safe and successful field operation. However, if the types of mines in the field are known, this a priori information can be used to optimize the soil compensation. With this information about the buried objects (e.g., landmines) with respect to their signatures, two methods or strategies can be introduced for the soil compensation adapted to the MD:

1. *Mine–soil sensitive compensation* (MSSC). The first strategy optimized the soil compensation with respect to visualization of the two-dimensional raw data from the MD: for example, recognition of a multitarget situation or determination of the status of uncooperative soil and estimation of the capability of the detection.

2. *Feature-preservation soil compensation* (FPSC). The second strategy, FPSC, optimized the mine–soil signal with respect to the best or highest variances of the features, which is needed for further pattern recognition procedures.

MDs provide n linear independent parameters so that up to $n - 1$ parameter can be used to suppress $n - 1$ disturbance values. Ground compensation algorithms

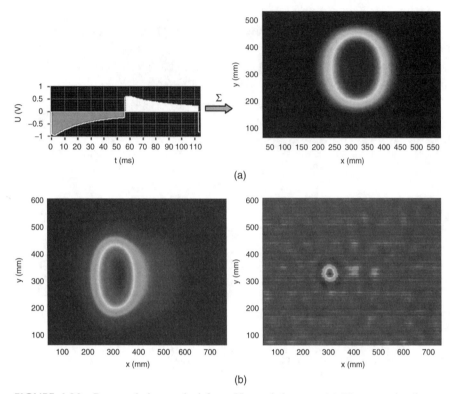

(a)

(b)

FIGURE 4.30 Deconvolution method for a PI metal detector: (a) PI output signal, raw data (left) and the signature of the PI data (right); (b) PI raw data image of three balls (left) and the results of the deconvolution (right).

from commercial metal detectors minimize the soil signal without reference to the different target object signatures. The proposed MSSC method used, for example, a mixture of four raw data components (real and imaginary parts from two frequencies) to one component with a maximum soil-to-mine signal ratio. This is possible with a priori information about the types of mines in the field, usually not more than one or two. Soil compensation is solved with a database of mine signatures (measured mine signatures under lab conditions) and space-resolved two-dimensional metal detector data. Figure 4.31 shows the influence of uncooperative soil (laterite) on the raw images (footprints) and the result of the MSSC algorithm (also used in Fig. 4.16).

Therefore, a database is used that provides signatures from mines in air. These signatures are scanned positions referenced under laboratory terms in various mine depths and orientations. Using this database, the soil effect is canceled in terms of the mine-to-soil signal ratio (MSR), defined by

$$MSR_i = \frac{\mathbf{c}_i \cdot \mathbf{p}_{mine}}{\mathbf{c}_i \cdot \mathbf{p}_{soil}} \qquad (4.19)$$

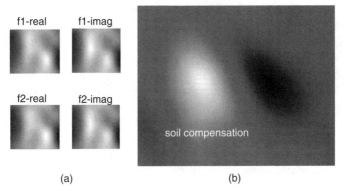

FIGURE 4.31   (a) M2A mine in laterite soil (depth: 10 cm) four raw data sets; (b) soil–compensated data using the MSSC algorithm. (From [8].)

The local soil is scanned on multiple points, moving the detector over a metal-free area. In the first step, the signals from the soil scan (each point with n raw data components $p = \{p_1 \cdots p_n\}$) are analyzed by the well-known principal component analysis (PCA) method, which gives the main component vectors, $c_1 \cdots c_n$, representing the influence of soil, where c_1 represents the greatest soil influence and c_n the least. In most cases the second principal component is nearly free from soil influences. These "soil-free" components are used in the second step for optimization of the MSR and thus may distinguish between two goals, MSSC and FPSC. By the MSSC algorithm, comparing the (desired) soil reduction with the (unwanted) reduction in the mine signal, the subset of component vectors c_i is used, which leads to the best MSR. The FPSC algorithm will be used only if the mine signal is strong enough (high metal content or shallow buried object), so that it is not important to obtain the maximum MSR. The soil-compensated component c_i is the dot product of the compensation vector c_i and the parameter vector p with $p = \{U_{\text{real-f1}}, U_{\text{imag.-f1}}, U_{\text{real-f2}}, U_{\text{imag.-f2}}\}$ by use of a dual-frequency detector.

Calculating the MSR for every mine at all depths in the database also gives a good approximation of the maximum detection depth (MSR > 0 dB) of every mine in the present soil with the actual soil compensation settings. The shape of the metal piece (the type of mine) cannot be recognized from either the raw data or from the reconstructed (deconvolution) image if the metal parts are small (weak signal strength) or not distributed spatially in the mine. In using the FPSC algorithm, the features (loops) from the pin pointing signature can be used for an object classification algorithm [9].

4.5.3   Pattern Recognition for Metal Detectors

Well-known methods of pattern recognition, such as phase and level detection of one-dimensional data, classical methods of cluster analysis of multifrequency data sets, and an artificial neural network (ANN), including a support vector machine (SVM), as well as fuzzy logic methods (fuzzy clustering and rule-based systems) can be considered for MDs. The important influence on all MD operation is the soil, so that for uncooperative soil, metal detector data must be soil compensated before their use in pattern recognitions systems. By CW metal detectors the patterns are the loops in the complex plane for each frequency by the presence of a metallic object, and the features are their real and imaginary values (see Section 4.2). To analyze or extract the main features and recognize the pattern automatically, the well-known pattern recognition methods mentioned above can be employed: for example, classical methods, ANN (including SVM), fuzzy-based systems (fuzzy clustering or rule-based systems), and data-based systems. Simple data-based systems can be realized in the form of lookup tables, which could be installed as learning or trainable systems, which are sometimes installed by treasure-hunting systems.

For a classification of the signature (a mine of type A or not a mine of type A), amplitude, phase, or geometrical features from the phase loop can be used. The latter is promising only for objects that give a strong signal. But with the aim also to classify small and/or deep-lying objects (low signal-to-noise ratio), phase and amplitude are the most stable features. Nevertheless, the mine signatures can be strongly distorted by the soil influence with increasing depth and are not usable for feature extraction. However, for better classification results, a compromise between the best possible soil compensation and the number and quality of parameters left for feature extraction has to be chosen. Therefore, soil compensation is carried out by calculating $n - 1$ weighted differences of different pairs from the n raw data components. The correct combination of raw data components is selected with the help of the signature database so that the soil influence is minimized and the $n - 1$ compensated output signals are free from redundancy. Figure 4.32 shows the structure of an advanced signal-processing method for MDs using a priori information for landmine detection. In addition, the influence of uncooperative soil and the presence of multiple metal objects are almost completely neglected by object classification methods. Under practical conditions these problems can be solved with image processing of both local-distributed sensor data and multiparameter sensor data and the use of a priori information for pattern recognition.

The strategy proposed for CW metal detector signal processing is as follows:

1. Capturing high-space-resolved MD data (data acquisition for $n \times n$ points, using a handheld system or three-axis scanner).
    Results: Two-dimensional raw data: eddy current images (footprints)
    1. Is it cooperative or uncooperative soil? (If uncooperative soil, soil compensation is needed!)
    2. Is the signature clear and scanned completely?
    3. In cooperative soil: pinpointing (the xy-position of the target)
2. Uncooperative soil → advanced soil compensation algorithm using a priori information for visualization and estimation of the capability of detection (MSSC).
    Criteria: maximized mine–soil signal.
    Results: 1. Is the signature scanned completely?
            2. Pinpointing (the xy-position of the target)

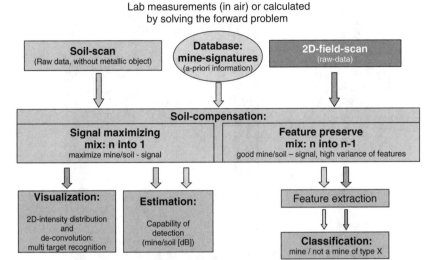

• The Database consist of signatures from all types of mines expected to be in the field.

FIGURE 4.32 Advanced signal-processing method for an MD proposed using a priori information for landmine detection. (From [40].)

3. Deconvolution of (soil-compensated) two-dimensional images

    Results: 1. Recognition of single- or multiple-target solution

               2. Prediction of the capability of the detection (signal-to-noise ratio)

               3. Prediction of the depth of the buried object

4. Single-target situation: advanced soil compensation algorithm using a priori information for pattern recognition (FPSC)

    Criteria: the best mine–soil signal by the highest variance of features

    Result: 1. Pinpointing (the *xy*-position) and extracting the optimal one-dimensional MD signals for the pattern recognition procedure

5. Pattern recognition using a data-based system with the optimal CW metal detector one-dimensional data (soil-compensated and optimal pinpointed loops)

    Results: 1. Estimation of the capability of detection of a landmine of type X

               2. Classification of a type X landmine or clutter

The algorithm proposed with respect to this signal-processing strategy can be used as an analog for PI metal detectors. For PI detectors it is important to increase the number of independent features of the MD output signal. More features can be generated using multiple time-domain integration systems, so that the reliability of the classification of PI metal detector data is improved.

All methods and algorithms described (i.e., soil compensation and pattern recognition algorithms) use a high-resolution three-axis scanning system and an off-line signal-processing system to acquire data. But these methods are also applicable

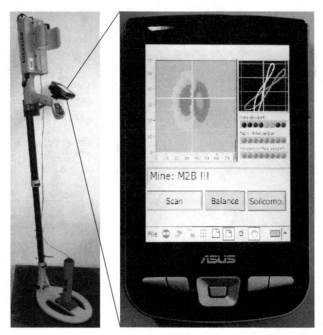

FIGURE 4.33 Modification of a standard metal detector (Foerster Minex) with a PDA as a user interface and an ultrasonic transducer for the space-resolved data acquisition system. (From [40].)

to a handheld MD in real time. Therefore, a standard CW two-frequency metal detector (e.g., the Foerster Minex 2FD-4500) can be modified (Fig. 4.33). The detector provides four raw data outputs (real and imaginary parts of the complex coil voltage at 2.4 and 19.2 kHz) that are connected to a mini-laptop for data acquisition and processing. The laptop is also connected to an ultrasonic space-resolved data acquisition system. A PDA is used as a graphical user interface, as shown in Fig. 4.33. The resolution achieved by the setup described is better than 2 mm, and the accuracy over the entire sensing area of 1 m^2 is below 5 mm.

Many approaches have been tested to acquire space-resolved data using a handheld detector. Differential GPS is often used by detection systems searching for unexploded ordinance (UXO), but it cannot provide the resolution required for landmine detection. Solutions based on acceleration sensors are only applicable to a short path, because of the drift sensitivity caused by the two integrations from acceleration over speed to the path. There are also solutions with optical sensors based on PIV [8], correlation, or spatial filtering [39]. Because the detector-mounted CCD sensor tracks the soil surface structure, no external device is needed, but the high dynamic range of light and shadow in grass or the moving shadows from trees precludes accuracy in the millimeter range. The transmitter for the space-resolved data acquisition system is mounted in the middle of the MD differential coil, sending pulses to three or four external receivers on the ground near the deminer. The position is determined in three dimensions by triangulation from the pulse time of flight. Knowledge of the third dimension is important, because the amplitude of the electromagnetic response caused from an object under the coil is influenced strongly by the sensor height. For a handheld metal detector, an ultrasonic three-dimensional reference system provides the space resolution needed for the acquisition of mine signatures.

In practice, it is difficult to align the external receiver device parallel to the ground because the terrain is not level in most cases. Already a slightly tangential deviation leads to a sensor height misinterpreted by up to some centimeters. But under the assumption that a well-trained deminer moves the sensor at a constant height averaged over the acquisition time, displacements of the external device can be corrected afterward by a three-dimensional-coordinate transformation which minimizes the average z-variation.

Additional artifacts, typical for a metal detector moved manually, are due to fast movements and outliers from the average sensor height. Such faulty measurements are easy to identify and filter out from the data set. Other approaches of advanced signal processing for MD data used an inverse data algorithm (solving the inverse problem of the MD) for estimation of the parameters of buried metallic objects (e.g., depth, or the form or type of landmine from the voltage received) [38].

A major problem of mine clearance using metal detectors is the high false-alarm rate, which is caused primarily by soil inhomogeneity and harmless metal parts (clutter). The key to significant lowering of the false-alarm rate is the use of a priori information about the types of mines in the field and the use of space-resolved multivariate MD data. The visualization of local-distributed sensor data gives an overview about the distribution of the metal pieces detected. The classification—type X mine or clutter—of a signature, also in the presence of uncooperative soil, is possible with the use of feature-maintaining soil compensation.

From the point of view of the theory of systems, it could be beneficial to combine or fuse different physical sensors to take advantage of each sensor for better prediction or safety measurement, in this case the detection and discrimination of buried metal objects. For example, with GPR, rocks and cavities in soil (air or nonconductive material) could be determined easily and in conjunction with data from the inductive metal detector [29,13].

Finally, a major problem for all kinds of MDs and handheld sensor fusion systems is the need for spatially resolved data acquisition. In addition, handheld systems must be manageable by an operator in the field. Each fused sensor system must have the same intrinsic power as that of a single sensor. This is one of the greatest challenges for sensor design: integrating multiple sensors (e.g., EM field sensitive) into a single handheld system. The sensitivity of the fused system is usually less than that of each individual sensor.

4.6 CONCLUSIONS

Metal detectors are widely used in industry and for humanitarian demining, especially for antipersonnel landmine detection. The influence of the soil on metal detectors increases the false-alarm rate dramatically. Compensating for this influence is the key factor in any field operation. With space-resolved acquisition of MD data, a priori information about buried objects, and advanced one-, two-, and three-dimensional signal-processing algorithms (e.g., the deconvolution method and feature-preservation soil compensation), more information about buried and detected metallic objects can be obtained. The a priori information is the signature of a buried metallic object in air. In conjunction with a data-based pattern recognition system, the information content and reliability of MD data will increase. Fast and precise solutions for the inverse problem of electromagnetism can provide strong support for new pattern recognition systems for handheld CW and PI metal detectors, which in turn will lead to higher security in humanitarian demining.

REFERENCES

1. Saxby, S. M. 1868. Letter to the editor. *The Standard London*.
2. Hall, E. H. 1879. On a new action of the magnet on electric currents. *Am. J. Math.*, 2:287–292.
3. Förster, F. 1937. *Z. Metallkd.*, 29:109–115.
4. Hughes, D. E. 1879. Induction balance and experimental researches therewith. *London, Edinburgh and Dublin Philosophical Magazine and Journal of Science*, Fifth Series, 8(46):50–57.
5. Potthast, R., et al. 2003. A "range test" for determining scatterers with unknown physical properties. *Inverse Probl.*, 19(3):533–547.
6. Hannakam, L. 1972. Eddy currents in a conducting half space by using a arbitrary form of the excitation coil [in German]. *Arch. Elektrotech.*, 54:251–261.
7. Igel, J., Preetz, H. 2006. *Report About the Analysis of Soil Samples from the Test Lanes in Ispra*. Prepared for the Project Network HUMIN/MD, GGA Report, Archive 125 794; and for Mozambique, GGA Report, Archive 0126165.
8. Krüger, H. 2009. Imaging and Classification of Inductive Metal Detector Data. Ph.D. dissertation, University of Rostock.
9. Bruschini, C. 2002. A Multidisciplinary Analysis of Frequency Domain Metal Detectors for Humanitarian Demining. Ph.D. dissertation, Vrije Universiteit Brussel.
10. Wtorek, J. 2003. *Electrical and Magnetic Properties of Soil*. EUDEM 2 Final Report. EUDEM 2, The EU in Humanitarian Demining: State of the Art on HD Technologies, Products, Services and Practices in Europe, Oct. 31.
11. CEN Workshop. 2008. *Humanitarian Mine Action: Test and Evaluation*, Part 2, *Soil Characterization for Metal Detector and Ground Penetrating Radar Performance*. Agreement Paper CWA 14747-2, Dec.
12. Benson, D. 1999. *Soil Properties and GPR Detection of Landmines: A Basis for Forecasting and Evaluation of GPR Performance*. Report GAL-972-1145. Department of National Defence, Canada, Oct.
13. Gaudin, C., Sigrist, Ch., Bruschini, C. 2003. *Metal Detectors for Humanitarian Demining: A Patent Search and Analysis*. EUDEM 2 Report. EUDEM2, The EU in Humanitarian Demining: State of the Art on HD Technologies, Products, Services and Practices in Europe, Nov.

14. Libby, H. L. 1971. *Introduction to Electromagnetic Nondestructive Test Methods*. Wiley-Interscience, New York.

15. Sword, C. K., Simaan, M. 1985. Estimation of mixing parameters for cancellation of discretized eddy current signals using time and frequency domain techniques. *J. Nondestruct. Eval.*, 5(1):27–35.

16. Wait, J. R. 1951. A conducting sphere in a time varying magnetic field. *Geophysics*, 16:666–672.

17. Wait, J. R. 1969. Quasi-static transient response of a conducting permeable sphere. *Geophysics*, 34(5) 789–792.

18. Sobolev, V. S. 1963. K teorii metoda nakladnoi katuski kontrole vichrevymi tokami. *Izv. Akad. Nauk. SSSR*, 2:78–88.

19. Dodd, C. V., Deeds, W. E. 1967. *Analytical Solution to Eddy-Current Coil Problems*. Report ORNL-TM-1987. Oak Ridge National Laboratory, Oak Ridge, TN.

20. Hugo, G. R., Burke, S. K. 1988. Impedance changes in a coil due to a nearby conducting sphere. *J. Appl. Phys.*, 21:33–38.

21. Burke, S. K. 1985. A perturbation method for calculating coil impedance in eddy-current testing. *J. Phys. D*, 18:1745–1760.

22. Ida, N., Lord, W. 1983. Simulating electromagnetic NDT probe fields. *IEEE Comput. Graph. Appl.*, 5:21–28.

23. Fawzi, T. H., Ali, K. F., Burke, P. E. 1983. Boundary integral solution equation analysis of induction devices with rotational symmetry. *IEEE Trans. Magn.*, 19(1):34–36.

24. Sabbagh, H. A., Sabbagh, L. D. 1988. An eddy-current model and algorithm for three dimensional nondestructive evaluation of advanced composites. *IEEE Trans. Magn.*, 24(6):3201–3212.

25. Ewald, H., Wolter, A. 2003. Optimization of inductive sensors using mathematical modeling. In *Sensors and Their Applications XII*. Sensor Series. IOP Publishing, Philadelphia, pp. 407–413.

26. Druyts, P., Das, Y., Craeye, Ch., Acheroy, M. 2006. Effects of the soil on the metal detector signature of a buried mine. In *Proceedings of the SPIE Defence and Security Symposium*, Orlando, FL, April.

27. Das, Y. 2005. Electromagnetic induction response of a target buried in conductive and magnetic soil. In *Proceedings of the SPIE Conference on Detection and Remediation Technologies for Mines and Minelike Targets X*, Orlando, FL, June. *Proc. SPIE*, 5794:263–274.

28. Das, Y. 2006. Time-domain response of a metal detector to a target buried in soil with frequency-dependent magnetic susceptibility. In *Proceedings of the SPIE Conference on Detection and Remediation Technologies for Mines and Minelike' Targets XI. Proc. SPIE*, 6217:621701.

29. Furuta, K., Ishikawa, J. 2009. *Anti-personnel Landmine Detection for Humanitarian Demining*. Springer, London.

30. Kaczkowski, P., Rissberger, E. J. 1995. *Pulsed Electromagnetic Induction (PEMI): Scientific and Technical Report*. Report SFIM-AEC-ET-CR-95092. Alliant Techsystems Inc., Mukilteo, WA.

31. Pertersen, S., van Rienen, U. 2008. A domain decomposition method for the computation of land mine signatures. *Int. Compumag. Soc. Newsl.*, 15(1):3–12.

32. MAFIA. 2002. *Manual*. MAFIA, Darmstadt, Germany.

33. EM-Studio 2006. *Manual*. EM-Studio, Darmstadt, Germany.

34. COMSOL. 2006. *COMSOL Multiphysics 3.2 Manual*. COMSOL Multiphysics, Göttingen, Germany.

35. Integrated Software. 2006. *FARADAY Manual*. Integrated software, Winnipeg, Manitoba, Canada.

36. Sadiku, M. N. O. 2009. *Numericals Techniques in Electromagnetics*, 3rd ed. CRC Press, Taylor & Francis Group, New York.

37. Mook, G., Lange, R. 1995. Non-destructive inspection of CFRP using Eddy current technique. In *Proceedings of ICCE/2*, New Orleans, LA, Aug. 21–24, pp. 519–520.

38. Lange, J., Hanstein, T., Helwig, S. L. 2006. Inversion of pulse induction and continuous wave metal detector data. In *URSI, Antennas and Propagation Society International Symposium*, July 9–14, p. 1167.

39. Krüger, H., Ewald, H. 2007. Image processing and pattern recognition of metal detector data. In *IEEE-ICONIC 2007, 3rd International Conference of Near-Field Characterization and Imaging*, St. Louis, MO, June 27–29, pp. 285–289.

40. Krüger, H., Ewald, H. 2008. Handheld metal detector with online visualization and classification for the humanitarian mine clearance. In *Conference Proceedings of IEEE Sensors 2008*, pp. 415–418.

CHAPTER FIVE

Microwave Tomography

5.1 OVERVIEW

Tomographic imaging methods are used for imaging of cross sections (slices) of inhomogeneous objects. These methods are found in practical applications in medicine (e.g., x-rays, ultrasonic, and nuclear magnetic resonance tomography), aerospace, chemical and power technologies (e.g., eddy current tomography), geophysics (e.g., diffraction tomography, multifrequency backscattering tomography), control of subsurface areas, through-wall detection, and inside-wall imaging (e.g., microwave tomography, millimeter-wave tomography).

The subsurface tomography method (STM) for imaging of objects placed in homogeneous material half-space has been considered [1,7,8]. Modeling and experiments on reconstruction of object shape and size for different values of medium electromagnetic parameters produced a reasonable performance. A generalized diffraction tomographic algorithm for subsurface imaging from multifrequency multimonostatic ground-penetrating radar data has been described by Deming and Devaney [2]. The algorithm is based on the Born approximation for vector electromagnetic scattering. The object function is expressed as a frequency-independent value. This assumption allows coupling of data measured at each frequency and thus incorporation of more information into the mathematical inversions. An object function is estimated by inverting the coupled equations analytically using the regularized pseudoinverse operator.

Subsurface Sensing, First Edition. Edited by Ahmet S. Turk, A. Koksal Hocaoglu, and Alexey A. Vertiy.
© 2011 John Wiley & Sons, Inc. Published 2011 by John Wiley & Sons, Inc.

Wiskin et al. [10] suggest an interesting method of geophysical anomalies or material defect imaging. For solution of the direct problem, they used a stratified Green's function. In the inversion, a pair of Lippmann–Schwinger-like integral equations is solved simultaneously via the Galerkin procedure for the unknown total internal field and object function. The inversion is an iterative optimization procedure requiring the solution of many forward problems. To increase the well-posedness of the problem, multiple frequencies have been employed.

Solving the through-wall imaging (TWI) problem [143] by a microwave tomographic technique has been considered [95]. A cross-sectional restoration of studied objects is based on the tomographic integral equation. Solution of this equation provides the possibility of finding an image function representing normalized polarization current distribution in the probing cross section. It is shown that advanced through-wall images of a human body can be obtained by introducing an effective dielectric constant instead of the wall and behind-wall media permittivity. This procedure substantially simplifies the through-wall-imaging problem. As a result, it is reduced to an imaging problem solved by STM. Using this method, images of a person placed behind a wall made from a variety of materials have been constructed .

The tomographic algorithm for imaging of cylindrical inhomogeneity of arbitrary cross section and electrical properties placed in a layer of a plane multilayered material half space is also considered. The electromagnetic response of the inhomogeneity for plane-wave excitation is estimated, the integral formulation for the scattered field is used, and a new approach to obtaining images of objects is presented. To reconstruct an object image, the scattered field is measured at various frequencies. Theoretical investigation of the inverse scattering problem has been considered in papers by Soldovieri and Persico et al. [11–17,20].

5.2   ELECTROMAGNETIC TOMOGRAPHY

5.2.1   Microwave Tomographic Approach to the Most Common Half-Space Geometry

GPR is a well-assessed diagnostic tool used to sound the subsurface and inner surfaces of structures based on the capability of electromagnetic waves to penetrate an optically dense medium. From such an interaction a backscattered field arises, which carries information regarding buried or embedded objects located in the zone illuminated by the transmitting antennas [18,21]. Therefore, retrieving information about buried objects—presence, location, geometry, and electromagnetic properties—begins with backscattered field measurements.

Usually, a GPR system operates within a multibistatic configuration using a system made up of transmitting (TX) and receiving (RX) antennas spaced by a fixed offset. For each location of the TX/RX antenna system, the transmitting antenna radiates a probing electromagnetic wave inside the medium to be investigated. This electromagnetic wave propagates in the medium, and when it impinges on a target, whose electromagnetic properties are different from those of the host medium, the backscattered field arises and is collected by the receiving antenna. By repeating this measurement along the air–structure interface (accessible/hidden medium), a radargram is obtained (i.e., a B-scan). The radargram accounts for a section of the zone investigated, where one axis coincides with the direction of movement of the antenna system and the other axis accounts for the two-way travel-time necessary for the electromagnetic wave to propagate from the TX antenna to the RX antenna through both direct coupling and scattering by a buried object.

A radargram is difficult to interpret to achieve clear and accurate information about the scene investigated, in terms of the presence, location, and geometry of buried objects. Such a direct "interpretation" is usually based on the a priori information available on the investigated scenario and on the expertise of the end user. In general, such an interpretation is very challenging, and this affects the overall reliability and accuracy of the GPR survey. This is shown clearly in Figs. 5.1 and 5.2, where the true geometry of a scene (in terms of electromagnetic properties and location of the "objects") is compared with its not easily interpretable radar image collected under a multibistatic configuration. Notably, even if the direct coupling between the TX and RX has been eliminated, the correct interpretation of such an image is not possible. Because of these drawbacks, it has been necessary to develop and analyze automatic processing approaches that are able to give clearer and more stable and interpretable reconstructed images compared to those provided by the starting raw data.

In recent years, a class of solution approaches based on radio-frequency (RF)/microwave tomography has gained increasing interest. Microwave tomography has become an increasingly popular interpretational tool for GPR applications. In fact, the possibility of recasting data processing as an inverse scattering problem [19,23] leads to an improvement in the interpretation of the simpler radargrams [18]. In addition, the adoption of more accurate models of the electromagnetic scattering phenomenon can help us to understand crucial aspects of a specific problem at a much deeper interpretational level. In addition, the theoretical investigation of the inverse scattering problem enables us to evaluate

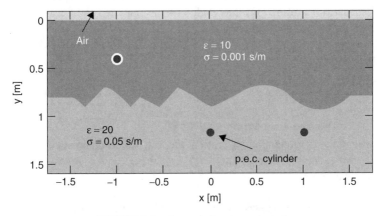

FIGURE 5.1 Scene being investigated.

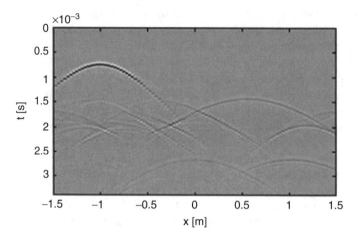

FIGURE 5.2   Raw data after elimination of direct coupling (B-scan).

reconstruction performance in terms of examples of spatial resolution achievable in a reconstructed image, and to provide guidelines on spatial and frequency sampling to be adopted in the survey criteria [20,25,144,146].

All the considerations noted above will be explained by means of the classical inverse scattering problem in GPR prospecting [16,20,27–29], which involves detecting, localizing, and determining the geometry and electromagnetic properties of targets buried in a lower half-space, starting from knowledge (measurements) of the scattered field collected at the interface between upper and lower half-spaces.

5.2.2 Theory: Formulation of the Problem for Penetrable Objects

In a microwave tomographic approach, the inverse scattering problem at hand is formulated from a mathematical point of view as the determination of "anomalies" in the electromagnetic properties with respect to the known properties of a background scenario, starting from knowledge of the scattered field collected under a fixed measurement configuration.

In formulating such an inverse problem, it is necessary to a priori define various quantities or factors, such as:

1. The background scenario in terms of its geometry and electromagnetic properties

2. The measurement configuration adopted

3. The type of embedded or buried objects (electromagnetically penetrable or not penetrable, elongated or not, etc.) under investigation, when applicable

The geometry of the background scenario (with respect to the targets considered as anomalies to be determined) is defined on the basis of various requirements, such as the nature of the realistic scene under investigation; the geometry (elongated or localized) of the objects to be determined; the necessity to keep the complexity and the computational cost of the inverse problem as low as possible. In general, one is concerned with one-, two-, or three-dimensional geometry and with homogeneous media or layered scenarios, and in all cases we have to define the electromagnetic properties of the background scenario in terms of dielectric permittivity, magnetic permeability, electrical conductivity, and their dependence on the work frequency. Accurate determination of the electromagnetic properties of the scenario is itself a challenging problem, and such a topic is discussed elsewhere in the book; here it is worth stressing how accurate knowledge of the electromagnetic properties of the background scenario is a key factor in ensuring the overall reliability of the results achieved under the microwave tomography–based approach [15].

The choice of measurement configuration depends on different constraints dictated by realistic observation conditions and by the need to have a sufficient amount of independent data so that an inverse scattering problem can be solved with a sufficient degree of accuracy and reliability. In general, different measurement diversities can be exploited.

- Illumination diversity (multiview configuration): when the zone investigated is probed by transmitting antennas located at different positions (in this case, the object investigated is "seen" under different views, and this is particularly useful in establishing the geometry of an object.)
- Observation diversity (multistatic configuration): when for each location of the transmitting antenna, the scattered field is collected by receiving antennas at different locations
- Frequency diversity (multifrequency configuration): when the transmitting and receiving antennas work in a frequency band

The importance of exploiting these measurement diversities when solving the theoretical difficulties related to an inverse scattering problem, especially for a nonlinear inverse problem, is shown in a thorough and detailed way below when we deal with a simplified and exact model of electromagnetic scattering. Exploitation of the foregoing diversities is still very limited in most applications when commercial GPR systems are used: In these applications, the measurement configuration adopted is usually a multibistatic (or multimonostatic one), where the transmitting and receiving antennas are moved by keeping the offset fixed between them. Such a configuration is the one used in this part of the chapter unless specified otherwise.

The third factor to be accounted for in the definition of an inverse problem is concerned with the a priori information about the electromagnetic properties of objects (e.g., electromagnetically penetrable or impenetrable objects, dielectric or metallic objects). In fact, such information leads to the choice of the (exact or simplified) model of the electromagnetic scattering at the basis of the solution approach. This point will be analyzed in detail when we focus our attention on approximate models of the electromagnetic scattering that allow us to simplify the inverse problem, such as the Born approximation for penetrable or dielectric objects [19] and the physical optics approximation for metallic objects [149–151].

Mathematical Basis of the EM Scattering Phenomenon and Difficulties Encountered in the Inversion

To simplify the discussion and to deal with the problem of main interest for applications, we focus attention on a two-dimensional geometry where the targets are buried in a half-space and the GPR measurements are collected along the air (upper medium)/soil (lower medium) interface (Fig. 5.3) and in the frequency domain. The background inhomogeneous scenario is built up of two homogeneous half-spaces (air and soil) separated by a flat interface at $z = 0$. The upper half-space (where the TX/RX antenna system moves) is made up of air with dielectric permittivity ε_0. The lower half-space has a relative dielectric permittivity ε_b and an electrical conductivity σ_b. The magnetic permeability is the same everywhere and is equal to e free-space permeability.

The incident field source is assumed to be a time-harmonic [time dependence $\exp(j2\pi f t)$] filamentary y-directed electric current of nonfinite extent and invariant along the y-axis. The source radiates at the single frequency f. The multifrequency case of a source radiating at different frequencies in the band $[f_{min}, f_{max}]$ is dealt with below.

In particular, a multimonostatic measurement configuration is assumed in which the TX antenna coincides with the RX antenna and the antenna system is moved at the air–soil interface with a measurement point x_s that ranges over the observation domain $\Gamma = [-x_M, x_M]$. The relevant targets are also assumed to be of nonfinite extent and invariant along the y-axis, and their cross section is enclosed in the rectangular investigation domain $D = [-a, a] X [z_{min}, z_{min} + 2b]$. When the objects under investigation are penetrable by electromagnetic waves, we assume as problem unknowns the relative dielectric permittivity $\varepsilon_r(x, z)$ and the conductivity $\sigma(x, z)$ distributions inside the investigation domain D. Accordingly, the inverse problem

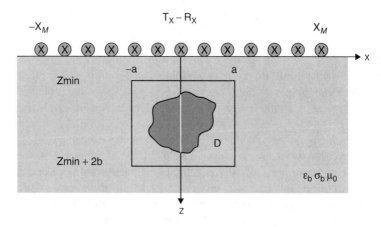

FIGURE 5.3   Geometry of the problem.

is recast in terms of the contrast function [16,33,145,146], defined as

$$\chi(\mathbf{r}') = \frac{\varepsilon_{eq}(\mathbf{r}') - \varepsilon_{eqb}}{\varepsilon_{eqb}} \tag{5.1}$$

where $\varepsilon_{eq}(x', z') = \varepsilon_0 \varepsilon_r(x', z') - j\sigma(x', z')/2\pi f$ and $\varepsilon_{eqb} = \varepsilon_0 \varepsilon_b - j\sigma_b/(2\pi f)$ are the equivalent complex dielectric permittivity of the targets and of the lower medium (soil), respectively; $\mathbf{r}' = x'\hat{x} + z'\hat{z}$.

The problem at hand is then formulated as the determination of the unknown contrast function $\chi(\mathbf{r}')$ starting from knowledge of the electric scattered field $E_s(\mathbf{r})$ collected at different abscissas $\mathbf{r} = x_s\hat{x}$ along the air–soil interface.

Now, it is necessary to give some definitions. The scattered field $E_s(\mathbf{r})$ is defined as the difference between the total field and the unperturbed field. The total field arises for two contributions: the first one is given by the field reflected by the air–soil interface and the field that accounts for direct coupling between the TX and RX antennas; the second contribution is the desired field backscattered by the buried objects. The unperturbed field coincides with the field arising from the background scenario when buried objects are absent: In this case it coincides with the field reflected by the air–soil interface plus the one arising from the mutual coupling between TX and RX antennas.

Formally, the problem of achieving the unknown electromagnetic properties of the target probed can be expressed through a pair of integral equations:

$$E_s^v(\mathbf{r}) = k_b^2 \int_D E(\mathbf{r}')\chi(\mathbf{r}')g_e(\mathbf{r}, \mathbf{r}') \, d\mathbf{r}' \qquad \mathbf{r} \notin D, \quad \mathbf{r} \in \Gamma \tag{5.2}$$

$$E(\mathbf{r}) = E_{inc}^v(\mathbf{r}) + k_b^2 \int_D E(\mathbf{r}')\chi(\mathbf{r}')g_i(\mathbf{r}, \mathbf{r}') \, d\mathbf{r}' \qquad \mathbf{r} \in D \tag{5.3}$$

where $k_b = 2\pi f \sqrt{\varepsilon_{eqb}\mu_0}$ is the wavenumber in the lower medium. Equation (5.2) accounts for the field scattered by objects at the measurement points and recorded by the receiving antennas, and thus accounts for the datum of the problem. The scattered field can be seen as the field radiated by an equivalent "source," related to the product of the unknown contrast function and of the total field $E(\mathbf{r})$ (see Fig. 5.4).

The total field, defined by Eq. (5.3), can be considered as the sum of two contributions. The first is the incident or background field E_{inc}, the field in the domain being investigated, D, when objects are absent (Fig. 5.5); the second term, defined as an integral over D, accounts for mutual interactions between objects in D.

From a physical viewpoint, we can say that when "excited" by the background field E_{inc} radiated by the transmitting antenna, the information on the targets, embedded in the contrast function χ, is transformed through Eq. (5.3) into the contrast source $J(\mathbf{r}') = j2\pi f \varepsilon_{eqb} E(\mathbf{r}')\chi(\mathbf{r}')$. Such a current $J(\mathbf{r}')$, which carries out information about the object, is then transformed in the scattered field E_s, which is "recorded" by the receiving antenna through Eq. (5.2).

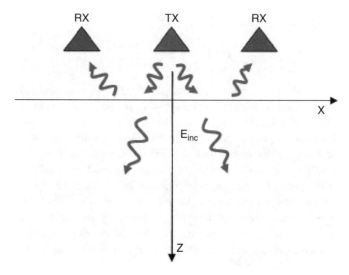

FIGURE 5.4 Background field.

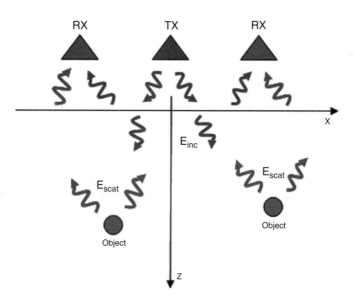

FIGURE 5.5 Total field.

The functions $g_e(\mathbf{r}, \mathbf{r}')$ and $g_i(\mathbf{r}, \mathbf{r}')$ are the external and internal Sommerfeld–Green's functions, respectively [19]. In particular, the external Green's function $g_e(\mathbf{r}, \mathbf{r}')$ accounts for the field generated at the point $\mathbf{r} = (x, z)$ in the upper half-space (the air) by the elementary source buried in the ground and located at the generic point $\mathbf{r}' = (x', z')$. Such a function is expressed via its plane-wave spectrum as [69]

$$g_e(\mathbf{r}, \mathbf{r}') = -\frac{j}{4\pi} \int_{-\infty}^{+\infty} \frac{2}{w_1 + w_0} \exp[-ju(x - x')] \exp[-jw_1 z'] \exp[+jw_0 z] \, du$$

(5.4)

where $w_1 = \sqrt{k_b^2 - u^2}$ and $w_0 = \sqrt{k_0^2 - u^2}$; the imaginary parts of the functions w_1 and w_0 are assumed less than or equal to zero so that the field does not amplify when it propagates far away by the source.

The internal Green's function $g_i(\mathbf{r}, \mathbf{r}')$ expresses the field generated at point $\mathbf{r}$ (in the lower half-space) by an elementary source placed at $\mathbf{r}'$ located in the same half-space. It can be expressed as

$$g_i(\mathbf{r}, \mathbf{r}') = g_0(\mathbf{r}, \mathbf{r}') - \frac{j}{4\pi} \int_{-\infty}^{+\infty} \frac{w_1 - w_0}{w_1 + w_0} \exp[-ju(x - x')] \exp[-jw_1(z + z')] \frac{du}{w_1}$$

(5.5)

where

$$g_0(\mathbf{r}, \mathbf{r}') = -\frac{j}{4} H_0^{(2)}(k_b |\mathbf{r} - \mathbf{r}'|)$$

(5.6)

and $H_0^{(2)}$ is the zeroth-order second kind of Hankel function. It can be noted that Eq. (5.5) is the sum of two contributions. The first, $g_0(\mathbf{r}, \mathbf{r}')$, which we term *homogeneous*, is the Green's function of an elementary source embedded in a homogeneous background with the same features in the lower half-space, and represents "direct" interaction between the source and observation points. The second, which we term *inhomogeneous*, takes into account the effect of the discontinuity in the dielectric constant of the background at the air–soil interface. The term $(w_1 - w_0)/(w_1 + w_0)$ in Eq. (5.5) can, in fact, be interpreted as a reflection coefficient in the spectral domain. Hence, it encodes the interaction between each of the plane waves constituting the expansion in Eq. (5.5) and the air–soil interface. Figure 5.6 depicts the effect of these two terms.

Equations (5.2) and (5.3), through the use of Eqs. (5.4) to (5.6), can be rewritten in "operatorial" form by the introduction of a pair of integral operators:

$$E_s = \mathcal{A}_e(\chi E)$$

(5.7)

$$E = E_{\text{inc}} + \mathcal{A}_i(\chi E)$$

(5.8)

The operators $\mathcal{A}_i$ and $\mathcal{A}_e$ are the internal and external radiation operators, respectively:

$$\mathcal{A}_e : \chi E \in X(L_D^\infty) \times T(L_D^2) \to E_s \in S(L_\Gamma^2)$$

(5.9)

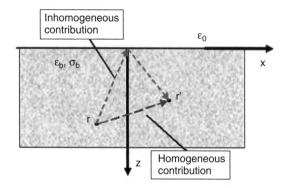

FIGURE 5.6 The two contributions of the internal Green's function.

$$\mathcal{A}_e(\chi E^v) = k_b^2 \int_D g_e(\mathbf{r}, \mathbf{r}')\chi(\mathbf{r}')E(\mathbf{r}')\, d\mathbf{r}' \qquad \mathbf{r} \notin D, \quad \mathbf{r} \in \Gamma \tag{5.10}$$

$$\mathcal{A}_i : \chi E \in X(L_D^\infty) \times T(L_D^2) \to (E - E_{\text{inc}}) \in T(L_D^2) \tag{5.11}$$

$$\mathcal{A}_i(\chi E^v) = k_b^2 \int_D g_i(\mathbf{r}, \mathbf{r}')\chi(\mathbf{r}')E^v(\mathbf{r}')\, d\mathbf{r}' \qquad \mathbf{r} \in D \tag{5.12}$$

wherein L_D^∞ is the subspace of all measurable bounded functions defined over the investigation domain D but for a zero measure set, $X(L_D^\infty)$ is the subspace of all contrast functions defined over the target region D, $T(L_D^2)$ is the subset of all square-integrable functions defined over D representing the scattered field inside the investigation domain, and $S(L_\Gamma^2)$ is the subset of all square-integrable functions defined over Γ, representing the scattered field measured over the observation domain Γ. These operators are linear in the product χE, or, equivalently, bilinear in E and χ: that is, they are linear in one of the two arguments when the other one is kept fixed.

The problem of retrieving the contrast function starting from the scattered field is ill-posed and nonlinear [16,23,119]. In fact, since $\mathcal{A}_e$ is a compact operator, it follows that Eq. (5.9) can be approximated, for any given accuracy, with a finite number of equations. Due to the latter property, a loss of information arises when transforming the induced current $J_{\text{eq}} = j\omega\varepsilon_b\chi E$ (through $\mathcal{A}_e$) into the scattered field. This means that according to the noise level on data, it is not possible to retrieve a nonfinite number of parameters of the contrast function (i.e., the contrast function can be retrieved only with a finite accuracy).

By considering the formal inversion of Eqs. (5.7) and (5.8), we obtain

$$E_s^v = \mathcal{A}_e[\chi(I - \mathcal{A}_i\chi)^{-1}E_{\text{inc}}] \tag{5.13}$$

From such an equation it follows immediately that a nonlinear relationship connects the unknown contrast function to the scattered field data. Therefore, not only a finite amount of information on the unknown target is transferred into the scattered field data, but this information is nonlinearly embedded in the data.

Up to now we have considered in the formulation the case of monochromatic data (i.e., at single frequency); when a multifrequency configuration is adopted, almost the same reasoning continues to hold. In particular, for the data space, the new data can be assumed as a vector $\mathbf{E}_s = [E_s(r, f_1), \ldots, E_s(r, f_i), \ldots, E_s(r, f_N)]$ made up of the a scattered field collected at the various work frequencies.

For the contrast function, we have the difficulty that when the frequency changes, the contrast function also changes, so that, formally, we have a new unknown at each frequency. On the other hand, if the dependence of the contrast on the frequency is known, this would allow us to fix as an unknown the contrast at one of the work frequencies and to determine the contrast at the other frequencies through a known relationship [136,154]. Finally, it is worth noting that when simplified models of electromagnetic scattering are adopted, where the aim is a "qualitative" reconstruction of the contrast function, it is possible to neglect the frequency dependence of the contrast function.

Reconstruction algorithms are subdivided into two main classes. The first is concerned with algorithms based on an approximate model of electromagnetic scattering (linear inversion algorithms). The second class tackles the problem in its complexity (nonlinear inversion algorithms). The drawbacks and advantages of these two classes of reconstruction algorithms should be considered.

As stated above, to face an inverse scattering problem requires that the nonlinear relation in Eq. (5.13) is inverted in order to achieve the solution. This causes many challenging mathematical difficulties. First, we recall that the approaches to quantitative reconstruction of the contrast function are cast mainly as the minimization of a cost functional that accounts for the distance in the data space between the measured scattered field data and the theoretical data predicted according to the model of electromagnetic scattering. This requires thorough attention to the question of local minima [24,119,120,125] where the local deterministic minimization schemes may be trapped, thus leading to a false solution of the overall diagnostic procedure. One might think to adopt a global minimization scheme, but its use is made almost impossible for applications related to inverse scattering, due to the large number of unknown parameters to be searched for and by the fact that the computational cost of such an approach grows exponentially with the number of unknowns.

Furthermore, accurate quantitative reconstruction, especially in the absence of a priori information about the scene investigated, requires precise knowledge of the incident field, which, in turn, requires accurate characterization of the behavior of the antenna in the presence of a layered scenario as well as the reliable electromagnetic characterization of the host medium. Finally, the computational cost of nonlinear inversion schemes should be accounted for when it is necessary to diagnose large investigation domains in a reasonable time.

The difficulties cited above can be mitigated by the adoption of simplified models of electromagnetic scattering, such as the Born approximation or physical optics approximation. In these cases we are concerned with a linear inversion problem and this has some practical and theoretical advantages, such as the absence of local minima for the cost function; the possibility to ensure the stability of the solution and to analyze the reconstruction performance of solution approaches thanks to well-assessed tools in the inverse problem literature. As a drawback of these approaches, no quantitative reconstruction of the electromagnetic properties of the targets is possible, since the underlying linear approximation is only suitable to localize the buried objects and provide information on their geometry.

5.2.3   Linear Inversion Algorithms

Born Approximation for Penetrable Objects   The advantages of approaches arising from simplified models of electromagnetic scattering have been briefly sketched above. Here we present the most common inversion approach, based on a simplified model of electromagnetic scattering that exploits the Born approximation [30–35, 119], which approximates the total field in Eq. (5.3) with the background field so that

$$E \approx E_{\text{inc}} \tag{5.14}$$

This assumption is based on the hypothesis that if the electromagnetic properties of the object are similar to those of the host medium, the presence of the scatterer does not introduces a significant perturbation of the total field induced in the object, so that it can be approximated with the background field.

The Born approximation was developed to solve the forward problem, which consists in the determination of the field scattered from a known object under the incidence of a known field [74]. For the forward problem, the Born approximation provides very accurate results under the hypothesis of a weak scatterer: an object whose dielectric permittivity is slightly different from that of the host medium and whose extent is small in terms of the probing wavelength [74].

To gain more insight into this point, by starting from Eq. (5.13), which relates the contrast function to the scattered field, and by assuming that $\|\mathcal{A}_i \chi\| < 1$, the transition operator, $(\mathcal{I} - \mathcal{A}_i \chi)^{-1}$, relating the incident field to the total field induced inside the scatterer, can be expanded into its Neumann series around the origin (i.e., the value of the dielectric permittivity in the background); as,

$$(\mathcal{I} - \mathcal{A}_i \chi)^{-1} = \mathcal{I} + \mathcal{A}_i \chi + (\mathcal{A}_i \chi)^2 + \cdots + (\mathcal{A}_i \chi)^n + \cdots \tag{5.15}$$

The series expansion on the right-hand side of Eq. (5.15) is denoted by the Born series.

Comparing Eq. (5.14) with (5.3) and using the expansion in Eq. (5.15), it can be noted that the Born model amounts to approximating the transition operator with the first term of its Neumann expansion. This implies assuming that the multiple

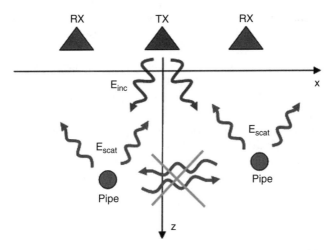

FIGURE 5.7 Electromagnetic scattering under the Born model.

scattering effects between the different scatterers and between parts of the same scatterer, which is the main cause of the nonlinearity of the problem, are negligible (Fig. 5.7). Under the Born approximation, electromagnetic scattering arises as each object stands alone in the background scenario (i.e., in the presence of more objects, the scattered field is given as the sum of the fields scattered by the single objects).

If such an assumption is verified, the inverse scattering problem changes to a linear problem, as the contrast function and the scattered field are related directly through the (linear) radiation operator $\mathcal{A}_e$:

$$E_s = \mathcal{A}_e(\chi E_{inc}) = \mathcal{A}(\chi) \tag{5.16}$$

When a general inverse scattering problem is tackled, the hypothesis of a weak scatterer can be relaxed. As a matter of fact, while this affects the "quantitative" reconstruction of the dielectric permittivity of the object. In any case, the adoption of a Born model inversion scheme also allows us to detect, localize, and determine the geometry of the object in the case of nonweak scattering objects, as already shown by a large number of numerical and realistic experiments [146,147]. Using an explicit expression of Eq. (5.16), the relationship between the unknown contrast function and the scattered field data is provided by the following linear integral equation [16,146,147]:

$$E_s(\mathbf{r}) = \mathcal{A}(\chi) = k_b^2 \int_D E_{inc}(\mathbf{r}')\chi(\mathbf{r}')g_e(\mathbf{r}, \mathbf{r}') \, d\mathbf{r}' \qquad r \in \Gamma \tag{5.17}$$

where the kernel of the integral equation is given by the product of the external Green's function and the incident field.

Linear integral equation (5.17) can be solved by a variety of methods, such as the minimization of a cost function via a conjugate gradient–based method or by explicit inversion of the relation (5.17).

Next, we focus our attention on inversion via the singular value decomposition (SVD) of the compact operator $\mathcal{A}(\chi)$ that connects the contrast function to the scattered field. The SVD of operator $\mathcal{A}$ in Eq. (5.16) provides the singular system $\{\sigma_n, u_n, v_n\}_{n=0}^{\infty}$ [26]. The set $\{\sigma_n\}_{n=0}^{\infty}$ denotes the sequence of the singular values ordered in a nonincreasing sequence accumulating to zero. The set $\{u_n\}_{n=0}^{\infty}$ is the orthonormal basis for the space of the visible objects (i.e., objects that could be retrieved by error-free data), and the set $\{v_n\}_{n=0}^{\infty}$ is the orthonormal basis for the closure of the range of the operator (i.e., the space of noise-free data within the Born approximation model).

The singular system satisfies the following relations, similar to those of eigenvectors:

$$\mathcal{A}u_n = \sigma_n v_n$$

$$\mathcal{A}^+ v_n = \sigma_n u_n \qquad (5.18)$$

where $\mathcal{A}^+$ denotes the adjoint operator of A [26,108]. By Eq. (5.18) it follows that

$$E_s = \sum_{n=0}^{\infty} \sigma_n < \chi, u_n > v_n \qquad (5.19)$$

where as long as the index n increases, the singular values become smaller and smaller. Thus, the contribution of the nth term in the summation (5.19) to the scattered field is weaker and weaker and can be overwhelmed by the possible noise on data. This means that these data components have to be filtered out during inversion.

From a different point of view, the formal solution can be expressed as

$$\chi = \sum_{n=0}^{\infty} \frac{1}{\sigma_n} < E_s, v_n > u_n \qquad (5.20)$$

The decay to zero of the singular values of $\mathcal{A}$ implies that the problem of inversion of the operator $\mathcal{A}$ is ill-posed since it does not satisfy the Hadamard condition for well-posedness concerning the continuous dependence of the solution on data. In other words, even small errors in the data can lead to a quite unreliable solution (i.e., the solution is strongly sensitive to small data perturbations). This problem is exacerbated by the fact that the radiation operator $\mathcal{A}$ has an exponential asymptotic decay of the singular values [128,148], so that beyond a given singular value index, the information provided by the scattered field data does not increase in a significant way.

To face the ill-posedness of a problem, different regularization schemes can be adopted [26,108] consisting of windowing the singular values in Eq. (5.20) by different regularization schemes according to the behavior of the singular values. However, due to the limited dynamic of the singular values above the asymptotic exponential decay, a suitable regularized solution is one obtained by means of the truncated SVD expansion [36]:

$$\chi = \sum_{n=0}^{N} \frac{1}{\sigma_n} < E_s, v_n > u_n \tag{5.21}$$

By restricting the solution space to that spanned by the first $N + 1$ singular functions, this regularization scheme does not amplify the effect of errors on data, so that the solution is made stable. The index N is chosen with regard to the degree of regularization one wants to apply in the inversion in order to obtain a stable solution and its choice is made on the basis of the SNR. It is evident that N represents a trade-off between the contrasting needs of ensuring the accuracy (increase in N) and stability (decrease in N) of the solution.

The SVD tool provides interesting information on the class of retrievable functions $\{u_n\}_{n=0}^{N}$ that dictates the features of the unknown that it is possible to reconstruct. Such information can be given in a synthetic form by considering some merit figures, such as the regularized reconstruction of the Dirac pulse function and the spectral content. Regularized reconstruction of the Dirac pulse allows us to discuss the resolution limits of the inversion algorithm and to outline the dependence of the resolution on both the measurement configuration and the location of the generic point of the investigation domain D. A theoretical and numerical investigation of the resolution limits has been the subject of many papers for measurement domains located in the far, Fresnel, and near zones and for homogeneous and layered scenarios [145].

The main conclusions of this investigation can be summarized as follows:

1. The horizontal resolution improves as long as the extent of the measurement domain increases and/or the investigation point comes closer and closer to the measurement domain. In other words, the resolution limits improve when the investigated targeted point is viewed under a large angle by the measurement domain.

2. The in-depth resolution is limited as long as the work frequency band increases and/or the investigation point comes closer and closer to the measurement domain. The effect of the extent of the measurement domain plays a secondary role compared to that of the frequency band.

It is clear from these considerations how regularized reconstruction of the Dirac pulse makes it possible to investigate in a local way (i.e., by accounting explicitly for the dependence of the location of the generic point in the investigation domain D) the reconstruction performances of the inversion algorithm.

The other merit figure used to evaluate the reconstruction performance of the linear inversion algorithm is the allowable spectral content sp, a function of the spectral variables η and ς, given by

$$sp(\eta, \varsigma) = \sum_{n=0}^{N} |\hat{\hat{u}}_n(\eta, \varsigma)| \tag{5.22}$$

where

$$\hat{\hat{u}}_n(\eta, \varsigma) = \iint_D u_n(x, z) \, \exp[-j(\eta x + \varsigma z)] \, dx \, dz \tag{5.23}$$

Therefore, by *spectral content* we mean the sum of the moduli of the Fourier transform of the singular functions (in the unknowns' space) retained in the TSVD expansion of Eq. (5.21). $\hat{\hat{u}}_n(\eta, \varsigma)$ is given as the two-dimensional Fourier transform of the nth singular function in the space of the unknowns. The relevance of this quantity is that it provides a visualization of the filtering properties of the truncated SVD regularized operator. In fact, since the solution is given by a linear combination of the singular functions above the chosen threshold, the only spatial frequencies that can, hopefully (even if not certainly), be retrieved are those where there is a meaningful spatial spectral content.

To give a practical example of the appearance of these two merit figures, that is, the regularized reconstruction of the Dirac pulse and the spectral content, we consider the numerical case of an investigation domain D whose extent ranges in depth from 0.1 to 1.5 m with a semihorizontal extent $a = 0.75$ m. The lower half-space has a relative dielectric permittivity $\varepsilon_b = 9$ and an electrical conductivity $\sigma_b = 0.01$ S/m. The measurement domain is located at the air–soil interface and has an extent of 1.5 m with a spatial step of 3 cm between the 51 measurement points. The work frequency band ranges from 300 to 900 MHz, with a frequency step of 30 MHz.

The numerical implementation of the solution algorithm requires the discretization of Eq. (5.16); this task is pursued by resorting to the method of moments (MoM) [133]. In particular, the unknown contrast function is expanded with 51 Fourier harmonics along the horizontal axis and 41 rectangular pulses along the depth [16]. The inversion of the linear operator $\mathcal{A}$ is performed via the TSVD scheme and in the summation of Eq. (5.21), singular values larger than 0.03 times the maximum value are retained (-30 dB).

Figure 5.8 depicts the regularized reconstruction of a Dirac pulse located at $x = 0$ and at various depths: 0.3, 0.6, 0.9, and 1.2 m. Note how at the increase in depth of the pulse, both the horizontal and in-depth resolutions worsen according to the reasoning above. This means that when an object is not located in the far zone with respect to the measurement domain, the achievable reconstruction is spatially variable.

The spectral contents are represented in Fig. 5.9; the support where the spectral content is significantly different from zero is compared with an "ideal area" achieved using the tools of diffraction tomography discussed later in the chapter. To evaluate an ideal spectral area, we assume two hypotheses: there is an infinitely

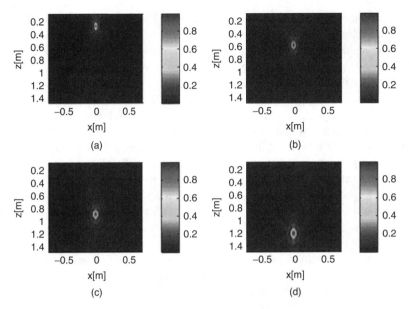

FIGURE 5.8 Regularized reconstruction of a point-target at various depths.

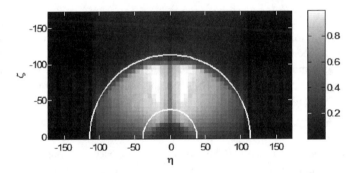

FIGURE 5.9 Allowable spectral content compared to the ideal case.

long observation domain, Σ, of a lossless soil, and the objects are not buried shallowly. According to these assumptions, the stationary-phase arguments can be exploited in an evaluation of the integral at the basis of the Born model [16,145]; this allows us to recognize a Fourier-like relationship between the unknown contrast function and the scattered field data; in particular, for the monostatic configuration at hand, a linear algebraic relationship between the spatial spectrum of the contrast and the spatial spectrum of the scattered field is established. In particular, the ideal spectral area is given as one limited by two semicircumferences (in the semiplane $\varsigma \geq 0$), whose radii are $2k_{b\min} = 4\pi f_{\min}\sqrt{\varepsilon_b\varepsilon_0\mu_0}$ and $2k_{b\max} = 4\pi f_{\max}\sqrt{\varepsilon_b\varepsilon_0\mu_0}$, respectively.

Some conclusions can be achieved by the examination of such a figure. First, a filtering effect arises due to the limitations introduced by the measurement configuration, which is a reflection configuration, and by the finiteness of the work frequency band. In particular, we observe a low-pass filtering effect for spatial variations of the unknown contrast function along the x-axis (direction of movement of the antennas) and a bandpass filtering effect for spatial variations of the unknown contrast function along the z-axis (depth direction). With bandpass filtering, only fast variations of the unknown contrast function along the depth are retrievable by the linear inversion algorithm under such a reflection configuration.

In addition, unless we go down to zero frequency, it is not possible to retrieve the constant term of the contrast function [corresponding to point $(0,0)$ in the spectral plane (η, ς)]; this means that even in the case of a negligible model error, that is, when the Born approximation is almost fulfilled, the linear inversion algorithm is not able to perform a quantitative reconstruction of homogeneous objects. When we consider the further difficulty of a limitation in the extent of the observation domain, an additional filtering effect is observed, mainly within the spectral zones (with respect to the ideal cases) lateral and closer to the η-axis.

Analysis of the spectral content and of the ideal retrievable area in the spectral plane allows us to establish guidelines regarding the resolution limits in dependence on the measurement configuration. The depth resolution is given as $\Delta z = \pi/2(k_{b,\max} - k_{b,\min}) = c_0/(4B\sqrt{\varepsilon_b})$, where $B = f_{\max} - f_{\min}$ and $c_0 = 1/\sqrt{\varepsilon_0\mu_0}$. The depth resolution improves as the relative dielectric permittivity of the soil increases.

All the foregoing considerations allow us to justify the reconstruction results presented in the following paragraph. The reconstruction result will be given as the spatial map of the modulus of the contrast retrieved; regions where the modulus of the contrast function is significantly different from zero account for the location and geometry of buried objects. Thus, in the following, when we refer to the tomographic reconstruction result we mean the modulus of the contrast function retrieved normalized with respect to its maximum.

Numerical Reconstructions   The first set of test cases is concerned with data computed on the basis of the Born model, so no model error is considered for the data. This set of test cases has a twofold aim, the first being to present the features of the reconstruction results for different extents of the same type of object: a rectangular target 0.4 m along the x-axis whose upper side is located at 0.6 m. The objects have different extents along the depth (i.e., 0.3, 0.2, 0.1, and 0.05 m; Fig. 5.10). All the reconstruction results point to the behavior expected above, so we observe that the linear inversion algorithm is able to reconstruct the two sides of the target parallel to the movement direction of the antennas, while no information is retrieved for the inner region of the target. The second aim of these test cases is to point out how the reconstruction changes at varying depth extents of the target. We can observe that the result deteriorates as the depth extent of the target decreases and an artifact appears between the two reconstructed spots when the situation shown in Fig. 5.10d is reached; in this case, the algorithm is not able to accurately reconstruct the two sides of the target, due to the limited resolution arising from the finiteness of the work frequency band.

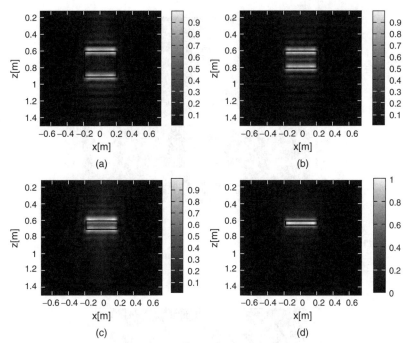

FIGURE 5.10 Tomographic reconstructions for different extents of a target along the depth. When such an extent becomes small, the approach is not able to discriminate the two sides of the target. Moreover, the "constant" part of the object is not retrieved, in agreement with the spectral content analysis.

The second set of test cases is concerned with the adoption of an inversion algorithm when a significant model error on the data is present. In particular, we consider two different causes of model error, the first one concerned with the presence of a single object whose dielectric permittivity is very different from that of the host medium. The data have been computed utilizing the finite-difference time domain (FDTD) GPRMAX code [153]. The time-domain data computed are then Fourier-transformed to pass to the frequency domain.

Various single-object cases are considered, assuming the same geometry of the object: a circle of radius of 0.1 m whose center is located at 0.6 m. Three different values of the relative dielectric permittivity filling the object have been considered: 9.1 (low model error), 4, and 16. Figure 5.11 depicts the reconstruction for a relative dielectric permittivity of 9.1; the result permits us to localize and accurately reconstruct the shape of the upper and lower parts of the circle. The situation is quite different for the other two cases, with relative dielectric permittivities of 4 and 16 (significant model error; Figs. 5.12 and 5.13). For both cases, the inversion algorithm is able to localize and reconstruct the upper part of the circle, whereas errors arise in reconstruction of the deeper part of the object.

The other cause of possible failure of the linear inversion algorithm is due to the mutual interactions arising when more objects are present in the investigational

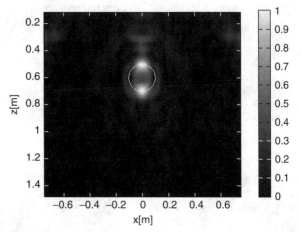

FIGURE 5.11 The linear inversion algorithm is able to point out the shallower and deeper parts of the target.

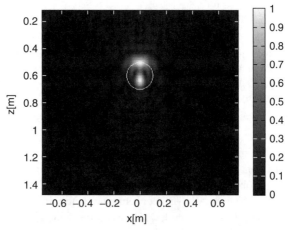

FIGURE 5.12 Tomographic reconstruction of an object with a relative dielectric permittivity of 4. The deeper side of the object is reconstructed at an incorrect depth, and a focusing effect arises.

domain. The negative effect of the mutual interactions between objects arises as the presence of artifacts (ghost spots) that may be of a level comparable to that of the spots accounting for true objects. In this framework, the first reconstruction is concerned with two object targets that have the same dimensions as the cases above and with a relative dielectric permittivity of 10; the two objects have centers spaced at 0.4 m. Figure 5.14 depicts the reconstruction for such a case, and no significant effect of the mutual interactions is visible. The second test refers to the same situation but when the relative dielectric permittivity of the targets is 12; in

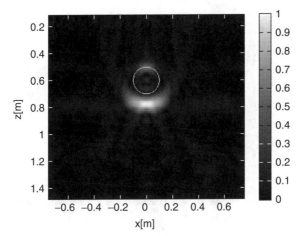

FIGURE 5.13 Tomographic reconstruction for an object with a relative dielectric permittivity of 16. The deeper side of the object is reconstructed at an incorrect depth and a defocusing effect arises.

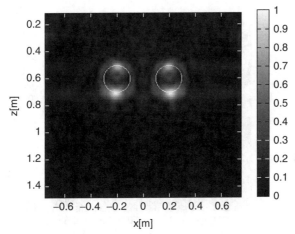

FIGURE 5.14 Tomographic reconstruction for two objects with a relative dielectric permittivity of 10 and a distance between centers of 0.4 m. No artifact appears.

this case the mutual interaction between the two objects starts to become significant and its effect on the image retrieved arises at a spot located midway between the two objects (Fig. 5.15). The effect of the mutual interactions is more significant when the two objects are at a shorter distance (0.3 m between their centers); in this case a more visible effect of the mutual interactions arises, as shown in Fig. 5.16.

Kirchhoff Approximation for Nonpenetrable Objects   A different electromagnetic scattering model has to be adopted when strong scattering from large metallic

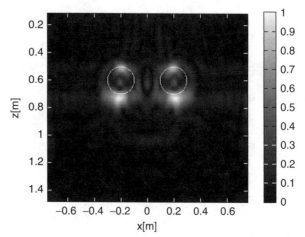

FIGURE 5.15 Tomographic reconstruction for two objects with a relative dielectric permittivity equal to 12 and a distance between centers of 0.4 m. An artifact appears in the middle between the two objects.

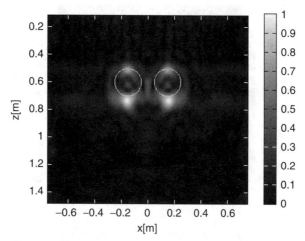

FIGURE 5.16 Tomographic reconstruction for two objects with a relative dielectric permittivity of 12 and a distance between centers of 0.3 m. The artifact becomes more significant.

objects is considered (Fig 5.17). In fact, in this case, the electromagnetic waves are not able to penetrate the object, and thus the only information carried out by the scattered field is concerned with the shape of the object. In this case, the phenomenon of scattering can be explained in the following way. The field impinging on the objects creates a current density over the surface of the object; this current radiates the scattered field that carries out the information about the support of the surface currents induced.

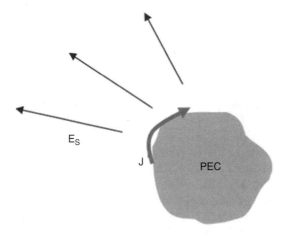

FIGURE 5.17 Electromagnetic scattering by a metallic object.

Also in this case, electromagnetic scattering is defined by two equations, which can be written compactly as

$$E_s = \mathcal{A}(J) \tag{5.24}$$

$$J = J_i + \mathcal{A}_i(J) \tag{5.25}$$

Equation (5.24) accounts for the scattered field as the field radiated by the current density induced on the surface of the object; analogous to the Born model, such an equation is fixed once we have fixed the reference scenario and the measurement configuration. Equation (5.25) accounts for the unknown current density J and is given as the sum of two contributions. The first is the physical optics current $\vec{J}_i = 2\hat{n} \times \vec{H}_i$, where $\hat{n}$ is the outward normal at the object's surface and $\vec{H}_i$ is the incident magnetic field (i.e., the magnetic field in the absence of the object). The second term of the equation accounts for the mutual interactions between different parts of the object and creates the nonlinearity of the inverse problem.

Also in this case, it is possible to simplify the inverse problem, thanks to adoption of the Kirchhoff approximation [150,151]. As is well known, this approximation is exploited in the forward model when the aim is to evaluate the field scattered by a known metallic object under the incidence of a known field (the forward model). In particular, the hypothesis at the basis of the Kirchhoff approximation is that the object is convex and has a curvature radius very much larger than that of the probing wavelength; in other words, the object has to behave locally as a planar object. Under the Kirchhoff approximation, the surface current density is approximated as

$$\vec{J} = \begin{cases} \vec{J}_{\text{opt}} = 2\hat{n} \times \vec{H}_i & \text{over the illuminated side} \\ 0 & \text{over the shadowed side} \end{cases} \tag{5.26}$$

and Eq. (5.24) can be rewritten as

$$E_s = \mathcal{A}(\vec{J}_{opt}) \tag{5.27}$$

It has been shown [150], that the Kirchhoff-based algorithm is valid beyond the hypotheses commonly adopted for the approximation to hold and that the inversion algorithm also works well for objects with a radius of curvature of the same order as the probing wavelength.

The discussion above can be exemplified for the shape reconstruction problem for a PEC object embedded in a homogeneous space in two-dimensional geometry. The host medium has dielectric permittivity $\varepsilon_b \varepsilon_0$ and magnetic permeability μ_0, where ε_b is its relative dielectric permittivity and ε_0 and μ_0 are the free-space dielectric and magnetic permittivities. The incident field source is a time-harmonic (time dependence $e^{j\omega t}$, with $\omega = 2\pi f$), filamentary, y-directed electric current (TM polarization) invariant along the y-axis. The radiation is multifrequency, with the frequency f varying in the band $[f_{min}, f_{max}]$. A multimonostatic measurement configuration is assumed, where the source and the observation points coincide. The abscissa x_s of the measurement points, which lays on the axis $z = 0$, varies in the interval $[-x_M, x_M]$. The object being searched for is assumed to be invariant along the y-axis with a cross section belonging to the rectangular investigation domain $D = [-a, a] \times [z_{min}, z_{min} + 2b]$.

The problem can be formulated as the determination of the unknown contour (indicated as Γ) starting from multifrequency and multimomonostatic measurements of the electric scattered field over a rectilinear domain [151]. Under the Kirchhoff approximation, the scattered field at position x_s can be written as

$$\vec{E}_s(x_s, \omega) = \mathcal{A}(\vec{J}_{opt}) = \frac{-\omega\mu_0}{4} \int_\Gamma H_0^{(2)}(kR) \vec{J}_{opt}(\gamma, \omega)\, d\gamma \tag{5.28}$$

where $k = 2\pi f \sqrt{\varepsilon_b \varepsilon_0 \mu_0}$ is the wavenumber of the host medium, R the distance between the source point $(x_s, 0)$ and a generic point (x, z) of the object's contour, $H_0^{(2)}(\cdot)$ the appropriate external Green's function for the homogeneous scenario (i.e., the Hankel function of zeroth order and second kind), and $\vec{J}_{opt}$ the surface-induced current density under the Kirchhoff model. Under this approximation the current density is assumed to be different from zero only on the illuminated side of an object; this means that the reconstruction procedure can provide information only on the illuminated side Γ_i of the object. Moreover, the illuminated region Γ_i depends on the antenna's location, x_s.

In Eq. (5.28) the current density $\vec{J}_{opt}$ over the illuminated side can be written as

$$\vec{J}_{opt}(\gamma, \omega) = 2\hat{n}x H_i = \frac{2}{j\omega\mu}\left(n_x \frac{\partial E_i}{\partial x} + n_z \frac{\partial E_i}{\partial z}\right)\hat{i}_y \tag{5.29}$$

where E_i and H_i are the incident electric and magnetic fields (i.e., fields in the absence of the scattering objects), respectively, and $\hat{n}$ is the outward-directed normal to the object's contour. Accordingly, the inverse problem for the two-dimensional geometry considered and the polarization chosen is scalar. Now, by exploiting the expression of the electric incident field for a filamentary source, we obtain, after simple passages,

$$E_s(x_s, \omega) = \frac{\omega \mu k}{8j} \int_{\Gamma_i} H_0^{(2)}(kR) H_1^{(2)}(kR)[\hat{n} \cdot \hat{R}] \, d\gamma \qquad (5.30)$$

where

$$\hat{R} = \left[\frac{x_s - x}{\sqrt{(x_s - x)^2 + z^2}}, \frac{-z}{\sqrt{(x_s - x)^2 + z^2}} \right] \qquad (5.31)$$

Now, if we assume that $kR \gg 1$ (i.e., the object is some wavelengths distant from the measurement line), we can exploit the asymptotic expression for the Hankel functions,

$$E_s(x_s, \omega) = \frac{\omega \mu k}{8j} \int_{\Gamma_i} \sqrt{\frac{2}{\pi k R}} \exp\left[-j(kR - \frac{\pi}{4})\right]$$

$$\sqrt{\frac{2}{\pi k R}} \exp\left[-j(kR - \frac{\pi}{4} - \frac{\pi}{2})\right] [\hat{n} \cdot \hat{R}] \, d\gamma \qquad (5.32)$$

which after some simple manipulations can be rewritten for a fixed x_s as

$$E_s(x_s, \omega) = \frac{j\omega\mu}{4\pi} \int_{\Gamma_i} \frac{1}{R} \exp(-j2kR)[\hat{n} \cdot \hat{R}] \, d\gamma \qquad (5.33)$$

The assumption $kR \gg 1$ also allows us to evaluate the integral in Eq. (5.32) by using a stationary-phase approximation [54]. It is easy to show that the stationary point accounts for the object's point where $\hat{n} \cdot \hat{R} = 1$. In other words, (1) only the object's points on the illuminated side of the contour that are located around the geometrical optics (GO) reflection point contribute to the scattered field in x_s; and (2) at this point, $\hat{n} \cdot \hat{R} = 1$. So a negligible error is made in evaluation of the integral in (5.33) if the factor $\hat{n} \cdot \hat{R}$ is set outside the integral equal to 1 and if the integration domain is restricted over a subset of Γ_i around the GO reflection point for a fixed position x_s. Accordingly, an approximate one-to-one mapping can be established between the scattered field in x_s and the associated subset of Γ_i contributing to it.

Pursuing the same argument, for other source–receiver positions x_s running along the observation domain from $-x_M$ to x_M, an approximate one-to-one mapping can also be assumed between the union Γ_u of the various Γ_i and the scattered field over the entire observation domain. So we can extend the integral (5.33) to points belonging to Γ_u as

$$E_s(x_s, \omega) = \frac{j\omega\mu}{4\pi} \int_{\Gamma_u} \frac{1}{R} \exp(-j2kR)\, d\gamma \qquad \text{for } x_s \in (-x_M, x_M) \qquad (5.34)$$

where the contour Γ_u is now independent from the observation variable x_s.

Now we are able to formulate the problem as being linear inverse by introducing the single layer distribution $\delta_{\Gamma_u}(x, z)$, whose support coincides with the curve Γ_u according to the strategy presented by Liseno et al. [152]. In this way, the scattered field can be rewritten as

$$E_s(x_s, \omega) = \frac{j\omega\mu}{4\pi} \iint_D \frac{1}{R} \exp(-j2kR)\delta_{\Gamma_u}(x, z)\, dx\, dz \qquad (5.35)$$

and the problem at hand now becomes the inversion of the linear integral operator $\mathcal{L}$:

$$\mathcal{L}(\gamma) = \frac{j\omega\mu}{4\pi} \iint_D \frac{1}{R} \exp(-j2kR)\delta_{\Gamma_u}(x, z)\, dx\, dz \qquad (5.36)$$

connecting the unknown distribution $\delta_{\Gamma_u}(x, z)$ to the scattered field data $E_s(x_s, \omega)$.

According to the Liseno et al. guidelines [152], the unknown contour of the object that is defined as the "support" of a distribution is determined by a two-step solution strategy. The first step is the analog of the one presented for the Born model: a regularized inversion of the integral relation of Eq. (5.36) by means of the truncated singular value decomposition (TSVD). The second step is concerned with determining the unknown contour of the object as the locus of the points where the TSVD retrieved version of the distribution is larger than a fixed threshold. The threshold is chosen according to the parameters of the measurement configuration and on the signal-to-noise ratio [152]. Such a step aims at curtailing spurious artifacts due to the noise and to the regularization scheme, whose reconstruction properties are very similar to those introduced for the Born model. Through the image retrieved, this threshold procedure allows to eliminate artifacts that might be incorrectly interpreted as actual scatterers. More details regarding the solution strategy are available elsewhere (e.g., [151,152]).

As proof of the feasibility of the shape reconstruction approach, next we present the reconstruction result for the same investigation domain and measurement configuration at that considered for the Born model case. The object under investigation is a PEC cylinder whose center is at a depth of 0.6 m and whose radius is 0.1 m. Figure 5.18 depicts the reconstruction after the TSVD inversion and threshold procedure (0.3 value). The reconstruction makes it possible to retrieve information about the illuminated side of the target while no information is achieved for the shadowed side of the object.

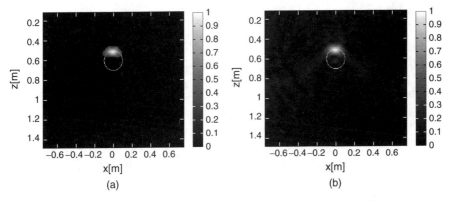

FIGURE 5.18   Shape reconstruction (a) after and (b) before the threshold procedure.

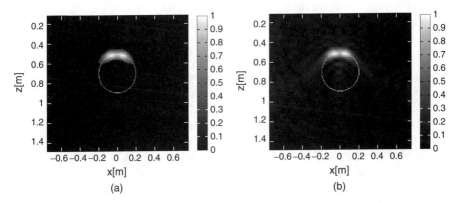

FIGURE 5.19   Shape reconstruction (a) after and (b) before the threshold procedure.

For the sake of comparison, we also report reconstruction after the TSVD inversion and when no threshold procedure is adopted; it is evident how the threshold procedure allows erasure of the undesired oscillating behavior due to the finiteness of the work frequency band. In addition, we report the reconstruction result for a cylinder of radius 0.2 m and depth of center at 0.7 m, which permits us to obtain good information about the location and shape of the illuminated side (Fig. 5.19). We also report on a comparison of the reconstruction results before and after application of the threshold procedure.

Finally, we point out how adoption of the solution approach for targets embedded in a homogeneous medium can also be used successfully with half-space geometry. For this test case, a scenario is assumed in which the upper medium is air, and for the lower medium (soil), $\varepsilon_b = 9$ and the conductivity is equal to 0.01 S/m. The measurement line is 2 m long and is located at the air–soil interface (at $z = 0$). The scattered field is collected at 51 equally spaced points spatial step 4 cm; 21 work frequencies with a step of 30 MHz are considered in a band ranging

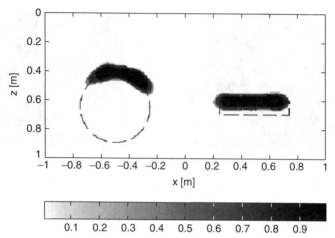

FIGURE 5.20 Normalized modulus of the TSVD reconstructed function with a threshold of 0.21 for $\varepsilon_b = 9$.

from $f_{min} = 100$ MHz to $f_{max} = 700$ MHz. There are two buried cylindrical PEC objects: the first has a circular cross section with radius 0.25 m and center at $(-0.5, 0.75)$ m; the second has a rectangular cross section of extent 0.5×0.1 m whose center is at $(0.5, 0.75)$ m. Figure 5.20 depicts the reconstruction after the TSVD inversion and the threshold procedure; the illuminated sides of the two objects are well localized and reconstructed.

To point out the effect of the relative dielectric permittivity of the lower medium on the reconstruction, we present the reconstruction results when $\varepsilon_b = 4$ and 16, with the soil conductivity still 0.01 S/m (Figs. 5.20 to 5.22). An improvement in the

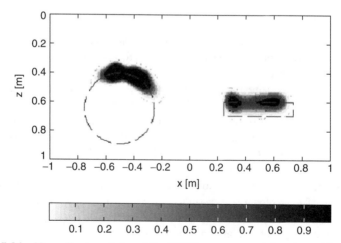

FIGURE 5.21 Normalized modulus of the TSVD reconstructed function with a threshold of 0.21 for $\varepsilon_b = 4$.

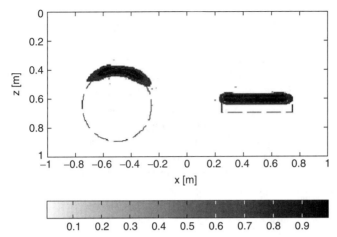

FIGURE 5.22 Normalized modulus of the TSVD reconstructed function with a threshold of 0.21 for $\varepsilon_b = 16$.

depth-resolution limits with an increase in ε_b can readily be appreciated due to the pulse behavior of the unknown along this direction. This can be explained by stating that the reconstruction performances of the Born and Kirchhoff models, meant to represent the class of retrievable unknowns by means of the TSVD scheme, are very similar, so that the depth resolution improves with an increase in the relative dielectric permittivity of the host medium.

5.3 MULTIFREQUENCY TOMOGRAPHIC METHOD

The reconstruction of microwave images of a subsurface object's cross sections by the methods of diffraction tomography and stepped-frequency radar is considered. In both reconstruction algorithm methods the apparatus of Fourier transformation is used. The images obtained from experimental data were compared using the methods considered. It is shown that good images of dielectric and metallic objects of different shape, which are located underground, can be obtained using these methods. The diffraction tomographic method is preferable in the case of imaging of a complex cross section of the object being studied.

The results of electromagnetic simulation on imaging of subsurface objects are also given in this section. Objects investigated are dielectric cylinders embedded in a lossy dielectric homogeneous half-space. It was supposed that objects had weak contrast, so for image reconstruction of objects, the plane-wave spectrum of a backscattered field is used. It was found that using only the evanescent part of the scattered field spectrum allows us to obtain images of the objects investigated.

The results of image processing of subsurface objects and detection with the acquisition of experimental data about a scattered field by the system, which radiates the pulse of electromagnetic waves and which then receives the returned signal, are presented. A short rise time for a sounding pulse (<50 ps) and, accordingly, the broad band of a test signal, have made it possible to use GPR and diffraction tomography to obtain images.

One of the most important problems in tomography microwave imaging of buried objects is caused by interface surface reflection. In previous chapters we showed the dramatic decrease of surface reflectivity by using Brewster angle as incident one. One more approach may be proposed to eliminate reflection from the surface: Leaky wave dielectric antenna composed of dielectric waveguide oriented parallel to interface surface in vicinity to interface border may transmit and receive electromagnetic waves through the border with minimum reflectivity. Incident angle in this case will be defined by dielectric permittivity and dielectric waveguide electromagnetic wave phase velocity.

The version of the subsurface tomography considered here is oriented toward real and practical applications in remote sensing, industry, medicine, etc.

5.3.1 Introduction

Microwave tomography is a technique for obtaining high-resolution images of objects cross sections [37,39]. Multifrequency diffraction tomography [40] and time-domain diffraction tomography [41] may be represented as frequency- and time-domain inversions, respectively, of a three-dimensional inhomogeneous medium. These methods are used to form the space images of different structures of subsurface regions by impulse radar systems.

For detection and identification of buried inhomogeneities using electromagnetic waves, Chommeloux et al. [1] proposed an original method. It is based on an integral representation of induced polarization current distribution in the object obtained from backscattered field measurement of amplitude and phase in the air by a probe, which is moved parallel to the air–medium interface. For object reconstruction, the Fourier relation between the diffracted field and the normalized polarization current was employed.

This approach was used by Vertiy and Gavilov [43] for image processing. It is shown that expansion in terms of the plane waves of the field scattered by the object located near a plane boundary of a dielectric half-space includes nondamping uniform and nonhomogeneous plane waves. Use of these various parts of the scattered field spectrum allows us to obtain different images of the object investigated. A comprehensive review of publications on the foundation of stepped-frequency ground penetrating radar (SFGPR), together with past and current achievements, has been given by Noon [44]. Use of SFGPR for the detection and localization of metallic or dielectric objects hidden in building structures is also discussed.

In the present work, the results of investigation in the area of microwave imaging of buried objects using diffraction tomography and stepped-frequency radar are given. In particular, the solution of the inverse problem for the case of complex values of one of the spatial frequencies in Fourier transformation of the imaging function (normalized polarized current) is considered in the framework of the tomographic method. From this, equations relating the imaging functions used in tomography and stepped-frequency radar methods have been obtained. Possibilities for use of the tomographic method are shown by modeling and experiments for the imaging of complex objects formed by several targets distributed in the a plane of a perpendicular surface of a medium containing the object. Resolution of the experimental setup in the plane parallel to the surface of the medium containing the object and deep to the object was compared by modeling and by experimental investigation. The results of the microwave images of undersurface objects (dielectric plates and landmines embedded in sand) are presented.

The results of electromagnetic modeling and imaging of subsurface objects by using the evanescent part of the plane-wave spectrum of a scattered field are also given.

5.3.2 Theoretical Consideration of the Diffraction Tomographic Approach

In this section we consider briefly image processing. For image reconstruction of objects, the plane-wave spectrum of a scattered field is used. It means that the scattered field $\psi(x, y_1)$ at line $y = y_1$ (the one-dimensional case, Fig. 5.23) is represented in the Fourier integral form [46]

$$\psi(x, y_1) = \int_{-\infty}^{\infty} \hat{\psi}(v, y_1) \, \exp(2\pi i x v) \, dv \tag{5.37}$$

where $\hat{\psi}(v, y_1)$ is the Fourier image of $\psi(x, y_1)$ and is defined as

$$\hat{\psi}(v, y_1) = \varphi(v) \, \exp[-i y_1 \sqrt{k^2 - (2\pi v)^2}] = \varphi(v) \, \exp(-i \gamma_1 y_1) \tag{5.38}$$

$$\gamma_1 = \sqrt{k^2 - 2\pi v^2} \tag{5.39}$$

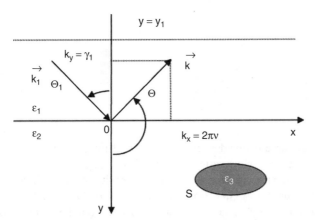

FIGURE 5.23   Region under consideration in the xy-plane and projections of a plane-wave vector $\vec{k}$ from the plane-wave spectrum of a scattered field on the axes of reference.

Here $k = \omega/c$ is the wavenumber of a plane wave in free space, ω the angular frequency, and c the velocity of light. Function $\varphi(\nu)$ is the angular spectrum of the scattered field. Variable ν is the space frequency in Eq. (5.38) and also defines a direction of propagation of the plane wave in an expansion of the scattered field in terms of the plane waves (Fig. 5.23).

The following relation connects value ν and angle θ:

$$2\pi\nu = \frac{\omega}{c}\sin\theta \tag{5.40}$$

The function $\varphi(\nu)$ may be written

$$\varphi(\nu) = c_1(\nu)c_2(\nu) \tag{5.41}$$

where

$$c_1(\nu) = \frac{ik_2^2 T}{\gamma_1 + \gamma_2} \tag{5.42}$$

T is the Fresnel transmittance of the boundary between medium 1 and medium 2 with dielectric permittivities $\varepsilon_1 = \varepsilon_0$ (air) and $\varepsilon_2 = \varepsilon_{r2}\varepsilon_0$, respectively; ε_0 is the dielectric permittivity of the vacuum; ε_{r2} is the relative dielectric permittivity of medium 2; $\gamma_2 = \sqrt{k_2^2 - (2\pi\nu)^2}$; $k_2 = \omega^2\varepsilon_2\mu_0 + i\omega\mu_0\sigma_2$, where k_2 and σ_2 are the wavenumber and conductivity of medium 2, respectively; and μ_0 is the magnetic permeability of the vacuum.

The function $c_2(\nu)$ may be written in integral form,

$$c_2(\nu) = \iint\limits_{S} K(x', y')\exp[-2\pi i(\alpha x' + \beta y')]\,dx'\,dy' \tag{5.43}$$

where

$$\alpha = \nu - \frac{\omega}{c}\frac{1}{2\pi}\sin\theta_1 \tag{5.44}$$

$$-2\pi\beta = \sqrt{\left[\left(\frac{\omega}{c}\right)^2\varepsilon_{r2} - (2\pi\nu)^2\right] + i\frac{\omega}{c}120\psi_2 + \frac{\omega}{c}\sqrt{(\varepsilon_{r2} - \sin^2\theta_1) + i\frac{c}{\omega}120\psi_2}}$$

$$\tag{5.45}$$

where θ_1 is an angle of incidence; S denotes that integration is over cross section S of the object under investigation; and function $K(x', y')$ represents the unknown normalized polarization current. Note that the results of the work by Chommeloux et al. [1] were used to obtain Eqs. (5.42) to (5.45). From (5.44) and (5.45) we can see that α is a real value for all ν and β is a complex value if $\sigma_2 \neq 0$ or if $\sigma_2 = 0$; $2\pi\nu > \omega/c\sqrt{\varepsilon_{r2}}$. If β is a complex value, the curve where the parameter ν changed from $\nu_{min} < 0$ to $\nu_{max} > 0$ can be set in complex α, β- space. Figure 5.24

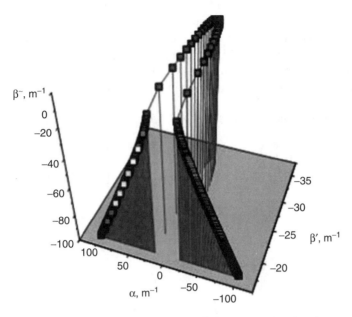

FIGURE 5.24 Parametric curve in complex α, β-space when changing the parameter ν.

shows this curve by the square points for the case $-100 < \nu < 100$ m^{-1}, $\varepsilon_{r2} = 4$, $\sigma_2 = 0$, $f = \omega/2\pi = 2.591$ GHz, and $\theta_1 = 10°$. In Fig. 5.25 the set of parametric curve projections calculated for different values of f is shown at the complex β-plane. Curves 1 to 7 were obtained for $f = 4.0 - (N-1)\ \Delta f$ in gigahertz. The number N assumed the values 1, 5, 10, 15, 20, 25, 30 and (for curves 1 to 7, respectively); $\Delta f = 0.045455$ GHz, $\varepsilon_{r2} = 4.0$, $\sigma_2 = 5.0 \times 10^{-2}$ S/m, $-30.0 <$

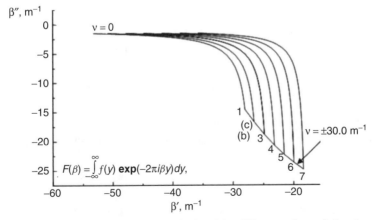

FIGURE 5.25 Set of parametric curves calculated for different values of f and projected on a complex β-plane.

$v < 30.0 \text{ m}^{-1}$, and $\theta_1 = 10°$. One can see that at $v \cong 0$, the functions $\beta''(\beta') \cong$ const for all frequencies f. But at the ends of the interval $-30.0 < v < 30.01/\text{m}$, a rapid change of values of β'' at increasing β' is observed and the curves are different for all frequencies f. It follows from this that an expansion of the variation intervals of f and v leads to an increase in regions using values of α, β', and β'' in complex α, β-space.

Let us now consider the function $c_2(v)$, which can be rewritten in the form

$$c_2(v) = \int_{-\infty}^{\infty} \left\{ \int_{-\infty}^{\infty} K(x, y) \exp[-2\pi i \alpha(v)x] \, dx \right\} \exp[-2\pi i \beta(v)y] \, dy$$

$$= \int_{-\infty}^{\infty} p[\alpha(v), y] \exp[-2\pi i \beta(v)y] \, dy \tag{5.46}$$

Let us also consider the integral

$$F(\beta) = \int_{-\infty}^{\infty} f(y) \exp(-2\pi i \beta y) \, dy \tag{5.47}$$

where $f(y)$ is an arbitrary function. The function $F(\beta)$ is defined for those complex values of β for which the integral exists. If the integral in Eq. (5.47) exists for values of $2\pi\beta$ in the strip $A < 2\pi \text{Im}(\beta) < B$, then, in accordance with Davies [47], the function $f(y)$ can be defined by the inversion integral as

$$f(y) = \int_{\frac{-i\gamma}{2\pi} - \infty}^{\frac{-i\gamma}{2\pi} + \infty} F(\beta) \exp(2\pi i \beta y) \, d\beta \qquad A < \gamma < B \tag{5.48}$$

From this result, the function $p(\alpha, y)$ in Eq. (5.46) can be written in the form of a contour integral in the complex β-plane along a contour L:

$$p(\alpha, y) = \int_{L} c_2(\alpha, \beta) \exp(2\pi i \beta y) \, d\beta \tag{5.49}$$

The contour is a direct line in the plane $\alpha = \text{const}$ of complex α, β- space parallel to axis β' at $\beta'' = -\gamma/2\pi$. The polarization current distribution $K(x,y)$ is defined as

$$K(x, y) = \int_{-\infty}^{\infty} \int_{\frac{-i\gamma}{2\pi} - \infty}^{\frac{-i\gamma}{2\pi} + \infty} c_2(\alpha, \beta) \exp[2\pi i (\alpha x + \beta y)] \, d\beta \, d\alpha \tag{5.50}$$

The Function $K(x,y)$ is calculated for the dielectric half-space with ε_2. If the variables x and y are taken in region S, the function $K(x, y) \simeq K(x', y')$ will give the polarization current distribution of the object under investigation. We take the image function in the form $|K(x, y)|$. Let $y = y_0$ be constant in Eq. (5.50). We can find the function $K(x, y_0)$ at different values of z, since in three-dimensions the scattered field is also changed along this axis. So we can obtain the polarization current distribution at an object in the x,y plane at constant depth y_0 measuring the scattered field $\psi(x, z, y_1)$ at this plane:

$$K(x, z) = \int_{-\infty}^{\infty} \int_{\frac{-i\gamma}{2\pi}-\infty}^{\frac{-i\gamma}{2\pi}+\infty} \hat{\psi}(\alpha, \beta, z, y_1)c(\alpha, \beta, y_1) \exp[2\pi i(\alpha x + \beta y_0)] \, d\beta \, d\alpha \quad (5.51)$$

where $c(\alpha, \beta, y) = \exp[i\gamma_1(\alpha, \beta)y_1]/c_1(\alpha, \beta)$. The function $K(x,y)$ or $K(x,z)$ depends on the frequency f and can be calculated for a set of frequencies $f_1, f_2, \ldots, f_N$ over the band Δf. In this case the image functions are defined as $|K(x, y)| = \left| \sum_{i=1}^{N} K_{f_i}(x, y) \right|$ or $|K(x, z)| = \left| \sum_{i=1}^{N} K_{f_i}(x, z) \right|$. One can see similar results in Fig. 5.26. Three- (Fig. 5.26) and two-dimensional (Fig. 5.27) images $|K(x, y)|$ of the object cross section (contour line for the two-dimensional case) were reconstructed. The object is a recognized standard target for calibration. The six targets all differ in size and are denoted by the numbers 1 to 6. The image was obtained at the following parameters of the inverse problem under consideration; dielectric permittivity $\varepsilon_{r2} = 5, \sigma_2 = 0.0$ S/m (left); $\varepsilon_{r2} = 5, \sigma_2 = 1.0 \times 10^{-2}$ S/m(right); $y_1 = -70.0$; $\theta_1 \simeq 0$. The set of frequencies $f_i(i = 1, 2, \ldots, 17)$ was defined by the formula $f_i = 4.0 - (i - 1) \times 0.03$ GHz. The Parameter v was changed in the range of values $-12 \leq v \leq 12$ m^{-1} and $\Delta v = 0.12$ m^{-1}. The Function $|K(x, y)|$ was

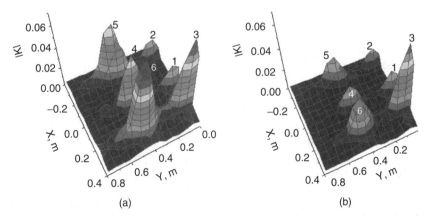

(a)　　　　　　　　　　　(b)

FIGURE 5.26 Three-dimensional images $|K(x, y)|$ of the object cross section under simulation.

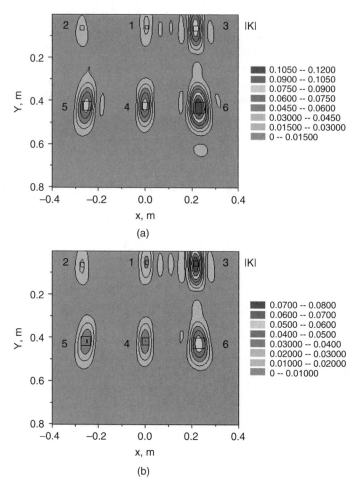

FIGURE 5.27 Two-dimensional images $|K(x, y)|$ of the object cross section under simulation.

calculated with steps $\Delta x = \Delta y = 0.01$ m in the regions $-0.4 \leq x \leq 0.4$ m^{-1} and $-0.4 \leq y \leq 0.4$ m^{-1}.

In the direct problem, the scattered field in the Born approximation was calculated by the method described by Vertiy and Gavilov [43] at the parameter values used in the inverse problem and at relative dielectric permittivity of the targets $\varepsilon_{r3} = 4.0$ and conductivity $\sigma_3 = 5.0 \times 10^{-3}$ S/m. The parameter ν was changed in the range of values $-20 \leq \nu \leq 20$ m^{-1} and $\Delta \nu = 0.20$ m^{-1} in this case.

The inverse Fourier transform of the space spectrum of the scattered field in the direct problem and Fourier transform of the scattered field in the inverse problem are made with the space-sampling step $\Delta x = 0.002$ m in the space coordinate range $-0.4 \leq x \leq 0.4$ m. The frequency- and space-sampling steps are taken so as to avoid problems of aliasing in calculation of the scattered field and the

Fourier transform of the scattered field and the polarization current distribution [1]. Figures 5.26 and 5.27 allow one to see the influence of electrodynamical and geometrical parameters of the inverse problem on the resolution attained by the diffraction tomography method and the image quality of a complex object. For example, in Figs. 5.26a and 5.27a, targets 1, 2, and 3 are less visible than targets 4, 5, and 6, although objects 1, 2, and 3 are located closer to the boundary of the unit of media at $y = 0$. This is because the losses equal zero in medium 2 in this case, since the targets 4, 5, and 6 are larger than targets 1, 2, and 3. This shows some mixing effect of losses and target size. Therefore, targets 4, 5, and 6 scatter a stronger electromagnetic field. Losses (Fig. 5.26b and 5.27b) in medium 2 essentially relax the normalized polarization currents in the region of targets 4, 5, and 6 because of the decreasing magnitude of an exciting electromagnetic field. In the field of targets 1, 2, and 3, values of polarizing currents vary insignificantly. Thus, target 3, which is smaller than target 6, is visibly better. Modeling showed that the object cross-sectional image could be obtained without distortion if the object size $S \leq 0.5\ S_1$, where S_1 is the size of the space-sampling region.

5.3.3 Impulse Synthesis Using the Stepped-Frequency Procedure

Let us suppose that three linear reflecting targets are placed under the surface at depths of $h_j (j = 1, 2, 3)$ and have limit size A along the x-axis ($A \simeq 17\lambda_0$, $\lambda_0 \cong 0.06$ m) (Fig. 5.28). Complex reflectivities $\dot{R}_j$ of these targets are

$$\dot{R}_j = |\dot{R}_j|\cos\varphi_j + i\sin\varphi_j \qquad (5.52)$$

where $|\dot{R}_1| = 0.8$, $\varphi_1 = 0.2$; $|\dot{R}_2| = 0.5$, $\varphi_2 = 0.5$; and $|\dot{R}_3| = 0.3$, $\varphi_3 = 0.9$. Values of h_j are $h_1 = 0.05$ m, $h_2 = 0.4$ m, $h_3 = 1$ m. The irradiating and receiving

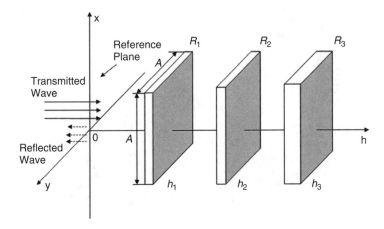

FIGURE 5.28 Reflector scheme.

antennas of stepped-frequency radar [44] are over the surface under investigation. The sample reflects the field at constant frequencies $f_n(n = 0, 2 \ldots, 16)$ from the frequency range 4.6 to 5.4 GHz with a step of 0.05 GHz.

The time-domain response of the reflecting medium is defined by the IDFT method at moments t_k [48]:

$$t_k = \frac{k}{N'\Delta f} \qquad k = 0, 1, \ldots, N' - 1 \tag{5.53}$$

where k is an integer changing from 0 to $N' - 1$; $N' = 216$ is the total number of frequency points; $\Delta f = 0.05$ GHz is a frequency step between f_n and f_{n+1}. The time-domain function $\tilde{R}(t_k)$ has the following form:

$$\tilde{R}(t_k) = \Delta f \sum_{n=0}^{N'-1} [R(n\,\Delta f) + iI(n\,\Delta f)] \exp\frac{i2\pi nk}{N'} \qquad k = 0, 1, \ldots, N' - 1 \tag{5.54}$$

where $R(f)$ and $I(f)$ are the real and imaginary parts of the complex signal in the frequency domain [44]. Figure 5.29 shows two- and three-dimensional images of $|\tilde{R}| = f(x, h)$ for given reflectors. The variable h is defined by

$$t = \frac{2h\sqrt{\varepsilon}}{c} \tag{5.55}$$

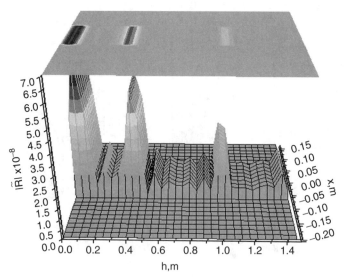

FIGURE 5.29   Two- and three-dimensional images of $|\tilde{R}| = f(x, h)$.

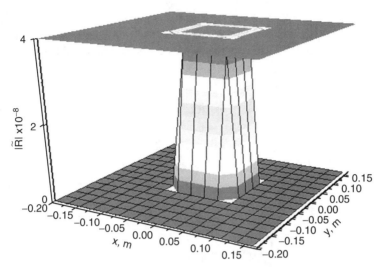

FIGURE 5.30   Image of a two-dimensional reflector placed at a distance h_3 from the medium surface.

where ε is the relative dielectric permittivity of the medium containing the reflecting objects; c is the velocity of light and is taken for values h_k:

$$h_k = \frac{t_k c}{2\sqrt{\varepsilon}} \tag{5.56}$$

We can also find the image function $|\tilde{R}| = f(x, y, h)$ of two-dimensional targets by using Eq. (5.54). Figure 5.30 presents an image of the reflecting object when variable $h = h_3$ is a constant value and the object has a square shape of $A \times A$ square meters. It is clear that stepped-frequency radar also allows us to obtain cross-sectional images of subsurface objects in the planes of the unit, parallel to boundary of media 1 and 2.

5.3.4   Comparison of the Two Methods Under Consideration

The diffraction tomographic and stepped-frequency radar methods use the same data from the scattered field, so it is interesting to compare the microwave images obtained using the two methods and the connection between the functions describing an object. Let us consider imaging the object crosssection in a plane perpendicular to the surface. In this case the field being measured is characterized by the function $\psi(x, f)$. In the stepped-frequency radar method the image function is $|\tilde{R}| = |\hat{\psi}(x, t)|$, where $\hat{\psi}(x, t)$ is the inverse Fourier transform of $\psi(x, f)$ with respect to f, and the depth $h = h(t)$. The function $\hat{\psi}(x, t)$ is written in the form

$$\hat{\psi}(x, t) = \int_0^\infty \psi(x, f) \, \exp(2\pi i f t) \, df \tag{5.57}$$

The function $|\hat{\psi}(x_j, h)|$ is the envelope of the complex zero-phase synthesized profile [44] at the x_j coordinate. This function can also be obtained by the tomographic approach considered. For this purpose we write the scattered field at line y_1 in the form

$$\psi(x, f) = \int\limits_{-\infty}^{\infty} \hat{\psi}(v, f) \, \exp(2\pi i x v) \, dv \tag{5.58}$$

and then

$$\psi(x, f) = \int\limits_{-\infty}^{\infty} \frac{ik_2^2 T \exp(-i\gamma_1 y_1)}{\gamma_1 + \gamma_2} \left\{ \int\limits_{-\infty}^{\infty} \int\limits_{-\infty}^{\infty} K(x', y') \, \exp[-2\pi i(\alpha' x + \beta y')] \right.$$

$$\left. dx' \, dy' \right\} \exp(2\pi i x v) \, dv \tag{5.59}$$

All values in Eq. (5.59) were defined earlier, but here it is assumed that α and β are real functions of v. Let us now take the inverse Fourier transform of this equation for f. On the left side we obtain the function $\hat{\psi}(x, t)$. Changing the order of integration on the right, we can write Eq. (5.59) as

$$\hat{\psi}(x, t) = \int\limits_{-\infty}^{\infty} \left\{ \int\limits_{0}^{\infty} \frac{ik_2^2(f)T(f) \, \exp[-i\gamma_1(v, f)y_1]}{\gamma_1(v, f) + \gamma_2(v, f)} \tilde{K}[\alpha(v, f), \beta(v, f)] \right.$$

$$\left. \exp(2\pi i t f) \, df \right\} \exp(2\pi i x v) \, dv \tag{5.60}$$

where $\tilde{K}[\alpha(v, f), \beta(v, f)]$ is the Fourier transform of the function $K(x', y')$ sought for with respect to x' and y'. Thus, the function $|\hat{\psi}(x, t)|$, used in the stepped-frequency radar method as the image function, is connected with the polarization current distribution $K(x', y')$ in the region of the object cross section by the modulus of the integral in formula (5.60). It should be observed here that the image obtained by the stepped-frequency method can be considered as some integral transformation of the object polarization current distribution. That is, using this method we obtain the original map. It follows from this that the images obtained by the two methods we are considering here can differ substantially.

As an example, Fig. 5.31 includes images of object 6 (see Figs. 5.26a and 5.27a) obtained using the diffraction tomographic method, the stepped-frequency method, and Eq. (5.60). The parameters of the problem in this calculation are the same as those in the earlier object computation. The difference between images reconstructed by the tomographic and stepped-frequency radar methods can be seen by examination of Fig. 5.31. We also note that the images in Fig. 5.31 are very close

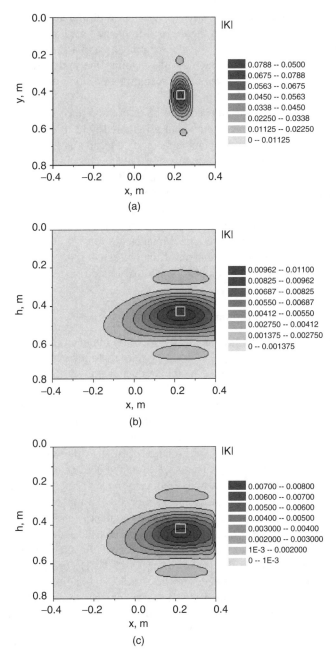

FIGURE 5.31   Images of object 6 (see Figs. 5.26a and 5.27a) using the tomographic method, the stepped-frequency method, and Eq. (5.60), respectively.

to those of the original structure, with the exception of the area where $x \simeq 0.4$ m. A possible explanation is that in the stepped-frequency method of construction of images, any conditions on the size of the scanning area in a space are not imposed. In the diffraction tomographic method of obtaining the undistorted image of an object, it is necessary that the size of the scanning area along the x-axis and y-axis exceeds the size of the object represented.

In general, the simulation showed that the diffraction tomographic method gave a better outcome for the same data about a scattered field. It can be explained by the fact that the diffraction tomographic method gives an image of the object polarization current distribution. This distribution outside the object must be equal to zero. The stepped-frequency radar image is a set of synthesized profiles (distributed in depth) that are collected as antennas move along the surface of the medium. In this case the synthesized profiles (or object images) can be obtained at each point where there is a scattered field. So if the scattered field is not only a reflected (as in geometric optics) field, the object images are reconstructed by the stepped-frequency radar method with distortions. Simulation using stepped-frequency radar showed that the image of the complex object (exhibited considerable deterioration.

5.3.5   Experimental Setup and Results

The experimental results presented below were obtained by using the setup described by Vertiy et al. [49]. The setup permits us to reconstruct cross-sectional images of various buried objects in the frequency range 1.25 to 5 GHz using tomographic and stepped-frequency methods. In the experiment we used data of a backscattered field at 32 frequencies in the frequency range 2.5 to 4.438 GHz with a step of $\Delta f \simeq 0.063$ GHz. Reconstruction of a buried object by the diffraction tomographic method was carried out in the x,y,z-system coordinates at a constant value $y = y_0$ in the x,y-plane (parallel to the interface) and in the x,y-plane (at depth) at $z = z_0$. The x,y,h-system coordinates (h is depth) was used for reconstruction by the stepped-frequency radar method. The backscattered field data were collected from an area of 0.3×0.3 m^2 with space steps of $\Delta x = \Delta z (\Delta y) \simeq 0.01$ m. The steps in the frequency and space coordinates are taken in accordance with the simulation results. The distance between the surface of the medium (dry sand) and the ends of the dielectric plates of the antenna system is 0.07 m. Electrical vector $\vec{E}$ of the electromagnetic field is in the x,y- or x,h-plane and is parallel to the x-axis. The angle of incidence of the electromagnetic wave is about $10°$. Dielectric objects (nonmetallic landmines) are made from materials having a relative dielectric permittivity close to that of the surrounding medium. The objects generally have $\varepsilon_{r3} \simeq 2.9$ [50]. Two marble ($\varepsilon_{r3} \simeq 9$) plates with cross sections of $\lambda_0/4 \times \lambda_0/10 \times 1.5\lambda_0 (\lambda_0 = 0.1$ m) were used for checking the setup resolution and comparing images reconstructed by the two methods. The relative permittivity of sand was taken as $\varepsilon_{r2} \simeq 4.0$. This value of ε_{r2} gave the best-reconstructed images in our experiments. It was assumed that the conductivity is $\sigma_2 \simeq 0$. Under these conditions, the positions observed for the subsurface object cross sections correspond to that of the actual subsurface object.

The Fourier transform of a scattered field with respect to ν was calculated in the range $\nu \approx -10 \div 10$ m^{-1} with a step of $\Delta\nu = 2.0$ m^{-1}. This ν range includes some part of the scattered-field plane-wave spectrum for $f < 3.0$ GHz. It was made with the purpose of improving the object image. It was shown earlier that extension of the range of values of the spatial frequency ν used results in an increase in the complex Fourier space of the spectrum region used in the algorithm of the inverse problem. Our experiments showed that this range of ν is optimal. Figure 5.32 shows images of the marble plates' cross sections obtained experimentally.

The images in Fig. 5.32a and b and Fig. 5.32c and d were obtained by diffraction tomography and stepped-frequency radar, respectively. The plates

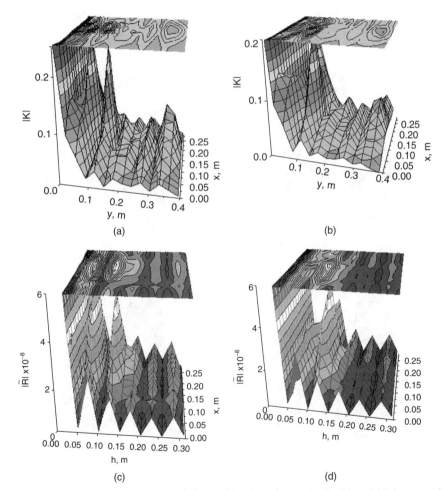

(a)   (b)

(c)   (d)

FIGURE 5.32 (a) and (c) Images of the marble plates into a depth; (b) and (d) images of the marble plates distributed along the x-axis.

are located at different depths in planes parallel to the sand surface. The space between the sand surface and the surface of the first plate is about $0.68\lambda_0$. The distance between the two objects (white contours) shown in Fig. 5.32a and c is about $0.68\lambda_0$. The distance between two objects (white contours) as shown in Fig. 5.32a and c is about $0.5\lambda_0$. Figure 5.32b and d illustrate the case when one of the objects is shifted by approximately $0.75\lambda_0$ along the x-axis and moved toward the other plate, so that the distance between the two objects is about $0.25\lambda_0$. From these pictures it can be seen that the diffraction tomographic method gives better images and has good resolution in both the cross-range direction and in the depth. The images obtained by the stepped-frequency radar method are larger in depth and have a low degree of resolution in cross section.

Figure. 5.33 shows cross sections of plates distributed along z-axes. In this case the images of the objects studied are discernibly worse. It can be explained by the

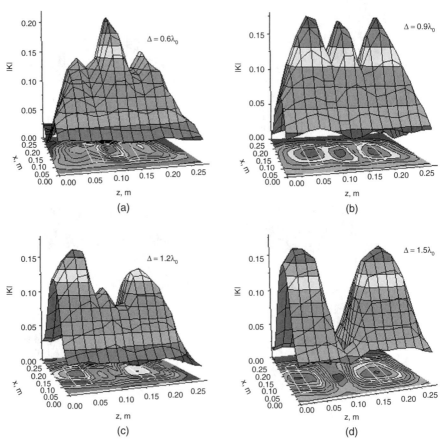

FIGURE 5.33 Images of the marble plates in the x, z, y_0-plane at $y_0 = 1.05\lambda_0$ (obtained from the experiment using the scheme shown in Fig. 5.34).

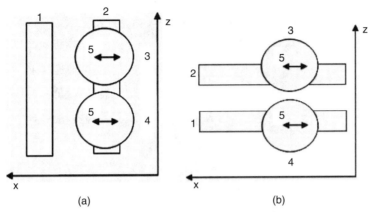

FIGURE 5.34 (a) Positioning schemes of a beam's cross sections (3 and 4, for the transmitting and receiving antennas, respectively) with respect to the marble plates (1 and 2). (b) The vector $\vec{E}$ orientation is indicated by line 5.

fact that when scanning across objects along z-axes, vector $\vec{E}$ of electromagnetic field is directed along the plates. This situation produces strong scattering and leads to interference of scattered fields between plates. The polarization current distribution (the image function) in this case has a complex structure and can be observed in the experiment (Fig. 5.34b).

Typical images of cross sections of a dielectric landmine with diameter d_m of $0.8\lambda_0$ and height h_m of $0.35\lambda_0$, which is embedded in sand at a depth y_0 of about $1.7\lambda_0$, are shown in Fig. 5.35. Figure 5.35a illustrates the polarization current distribution in depth under the sand surface at a value $z = z_0 = \text{const}$ and changing of coordinates x,y from 0 to $30\lambda_0$. The y-axis is directed down and the x-axis is set to the right. The surface polarization current distribution in the x,y-plane is shown at the top of part (a). Figure 5.35b displays cross sections of this target in the x,y-plane with a depth step $\Delta y \equiv \Delta H = 0.1\lambda_0$. The black circular area shows the target shape at real depth. Images shown in Fig. 5.35 confirm the good capabilities of the tomographic method. The images can be improved through the antenna system properties in the inverse problem algorithm and at filtering (deconvolution) of the images obtained, consisting of convolution of the antenna pattern with the target point spread function [52].

Additional three-dimensional images of buried plastic antipersonnel mines have also been reconstructed. The algorithm described has been used for calculation of the normalized polarization currents under the ground surface, depending on coordinates x,y, and z. Axis y is directed vertically into the ground, the xz-plane parallel to the ground surface. Three-dimensional images of mines embedded in sand are presented in Fig. 5.36. A reconstructed image of an antipersonnel plastic mine is shown in Fig. 5.36a. Figure 5.36b illustrates a plastic antitank mine placed under a sand surface (the outlook is from the depth). The objects are clearly visible under the sand surface.

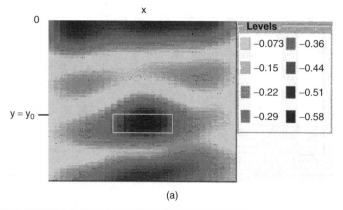

(a)

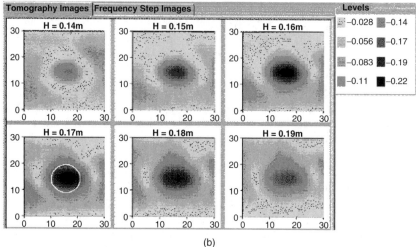

(b)

FIGURE 5.35   (a) Reconstructed image of an antipersonnel plastic mine placed under the surface of sand; (b) the target cross sections in the x, z-planes.

5.3.6   Imaging Subsurface Objects Using the Evanescent Part of the Plane-Wave Spectrum of a Scattered Field

Spectral partitioning in diffraction tomography has been considered in several studies [43,53,57]. The filtered back-propagation [53] algorithm of diffraction tomography is generalized to include evanescent wave components of the data in the reconstruction process. It was indicated that significant improvement in resolution could be obtained by incorporation of evanescent waves in the reconstruction.

Part of the nonevanescent spectrum of plane waves may be obtained by measurement of the scattered field at an arbitrary distance from the dividing line of dielectric media (e.g., air medium). The evanescent spectrum of the plane waves corresponds to nonhomogeneous plane waves in the spectrum of the scattered field with increase in distance from the interface. We supposed also that part of the

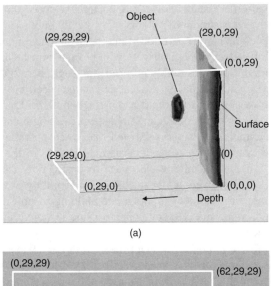

(a)

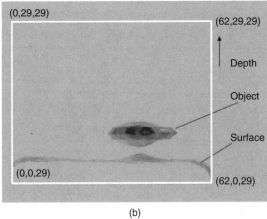

(b)

FIGURE 5.36 Reconstructed images of (a) an antipersonel plastic mine and (b) an antitank nonmetallic mine placed under the surface of sand.

spectrum associated with inhomogeneous plane waves and surface waves propagating along the boundary of two dielectrics (lateral waves and evanescent waves [54]) can be measured by means of a diffraction grating placed near the boundary. Microwave images of cylindrical dielectric objects embedded in a dielectric homogeneous lossy half-space can be reconstructed from the spectrum of plane waves of the backscattered field.

The problem of recovering the two-dimensional permittivity of objects buried in a lossy soil from measurement of a scattered field in the region above the interface between the air and soil has also been considered by pierri et al. [55]. The permittivity profile is represented as a superposition of a finite number of basis functions and is reconstructed using a linear model of the scattering. To improve the first-order approximation, a second-order model is considered. The improvement is achieved because the second-order model allows us to enlarge the spectral region to be taken into account during reconstruction. Authors of a recently published paper [56] show that the object spectrum can be partitioned into resolvable and nonresolvable parts, based on the cutoff between the propagating and evanescent fields. To reduce this effect, a beamforming on transmittal approach was suggested to direct the energy into either the propagating or the evanescent part of the spectrum. It was found that incorporating the evanescent fields into reconstruction algorithms is difficult, due to the spectral smearing of the object and field spectra. Using both a simulation and a laboratory experiment, the authors showed that when operating in a near-field environment, the evanescent power is approximately 28% of the power returned. The tomographic method of microwave image reconstruction has important practical applications in, for example, the medicine and military areas. On the other hand, the algorithm for this method allows the building of object images directly from evanescent or nonevanescent components of the scattered field using filtering in the Fourier space.

Reconstruction by a tomographic algorithm of the microwave images of cylindrical buried inhomogeneities has been described by Chommeloux et al. [1]. This algorithm is based on the integral representation of a backscattered electromagnetic field for a plane wave incident on a lossy half-space containing a cylindrical object of arbitrary cross section and properties. Distribution of the induced current distribution in the object of the image function is obtained from the backscattered field measurements in amplitude and phase. It was shown that there is a Fourier relation between the backscattered field and the normalized polarization current. Reconstruction of the object images was carried out taking into account the propagating components of the scattered field. Evanescent waves were neglected.

The results of modeling are represented below. Images of embedded underground objects using the tomographic method have been reconstructed from evanescent components of the scattered field only. The scattered field was found in the solution of the direct problem using the equations mentioned above. The Born approximation was used for the scattered field in the calculations. Solving the direct problem has a special interest, as it allows us to study the possibilities of the approach for obtaining the current function $K(x,y)$ of a cylindrical object of arbitrary shape. On the other hand, weakly scattering objects are usually objects of practical investigation.

Within the Born approximation, $\psi(x', y') \ll E_z^t(x', y')$(exciting field), and the expression for the normalized polarization current $K(x',y')$ is reduced to

$$K(x', y') = \frac{k_3^2(x', y')}{k_2^2} - 1 \qquad (5.61)$$

where $k_3(x', y')$ is the wavenumber for the object medium and k_2 is the wavenumber for medium 2 and does not depend on the incident and diffracted fields. If the electrodynamic parameters of the object under investigation do not depend on the coordinates x' and y', then according to Eq. (5.61), the function $K(x',y')$ also does not depend on this coordinates and may deleted from the integral in Eq. (5.43).

Then the expression for the Fourier image is of the form

$$\hat{\psi}(v, y_1) = C(v, y_1) K \iint_S \exp[-2\pi i(\alpha x' + \beta y')]\, dx'\, dy' \tag{5.62}$$

where K takes on a constant value in region S according to (5.61) and is equal to zero out of region S. Region S has a rectangular cross section with borders along the x-axis: $x_1 = a, x_2 = b, (b > a)$, and along the y-axis: $y_1 = c_1, y_2 = d(d > c_1)$. After analytical integration, the following expression was obtained for the function $\hat{\psi}(v, y_1)$ describing the Fourier image of the scattered field:

$$\hat{\psi}(v, y_1) = \frac{1}{4\pi^2 \alpha \beta} C(v, y_1) K I_1 I_2 \tag{5.63}$$

where

$$I_1 = (\sin 2\pi\alpha b - \sin 2\pi\alpha a) + i(\cos 2\pi\alpha b - \cos 2\pi\alpha a) \tag{5.64}$$

$$I_2 = (\sin 2\pi\beta\, d - \sin 2\pi\beta c_1) + i(\cos 2\pi\beta\, d - \cos 2\pi\beta c_1) \tag{5.65}$$

Equation (5.63) allows us to study the scattered field in a Born approximation using the inverse problem solution in (5.50) or (5.51) to obtain the function $K(x,y)$ imaging the object if the parameters of the problem are close to the experimental results. In this case the $K(x,y)$ function obtained differs from the one given initially, having a constant value in the given rectangular region and values of zero outside this region. It follows that in the inverse problem the integrand function is known only in some region of the plane of variables α and β as a result of the angle of incidence variation and the frequency of the plane wave, illuminating the boundary of mediums.

In a numerical experiment, the imaging function $\tilde{K}(x, y) \sim |K(x, y)|$ at a normal angle of incident $\theta \simeq 0$ was reconstructed in the frequency band $\Delta f = 4.0\text{--}3.52$ GHz at 17 frequencies with the function step $\Delta f = 0.03$ GHz. Relative dielectric permittivities and conductivities of the media are $\varepsilon_{r1} = 1, \varepsilon_{r2} = 5, \sigma_2 = 0.0, 1.0 \times 10^{-2}, 3.0 \times 10^{-2}, 5.0 \times 10^{-2}$ S/m, and $\varepsilon_{r3} = 4$ (object). The relative magnetic permeabilities of the media were taken as 1.0. The electric field amplitude of the incident plane wave is 1V/m and the incidence $\theta_1 \simeq 0$. The scattered field $\psi(x, y_1)$ has been calculated using formula (5.58) in the bounded limits of integration by $v = -30 \div 30$ m^{-1} at values of $x = -4\lambda_0 \div 4\lambda_0 (\lambda_0 = 0.1$ m) with constant step $\Delta x = 0.002$ m. These values of v are also used in the inverse problem. Images were reconstructed in the field $4\lambda_0 \times 4\lambda_0$ at values of $x = -2\lambda_0 \div 2\lambda_0$ and $y \simeq 0 \div 4\lambda_0$. The magnitude β assumes complex values; therefore, solution of the inverse problem in the form (5.50) is used. In our calculation we also varied the dimensions of the objects investigated (cylinders of rectangular cross section) as to the depth of their occurrence under the surface of media division (the object is located symmetrical to the x-axis). Figure 5.37 shows the spectra $|\hat{\psi}(v, y_1)|^2$ to v.

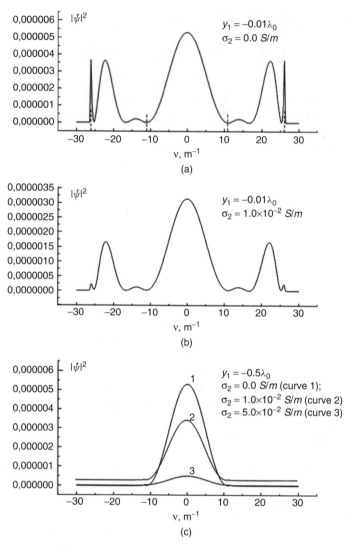

FIGURE 5.37 Spectra $|\hat{\psi}(\nu, y_1)|^2$ of a scattered field.

The curves in Fig. 5.37a and b were calculated for $y_1 = -0.01\lambda_0$ and different values of the conductivity: $\sigma_2 = 0.0$ S/m (Fig. 5.37a) and 1.0×10^{-2} S/m (Fig. 5.37b). Figure 5.37c illustrates a case for $y_1 = -0.5\lambda_0$ and $\sigma_2 = 0.0$ S/m for (curve 1, $\sigma_2 = 1.0 \times 10^{-2}$ S/m for curve 2, and $\sigma_2 = 5.0 \times 10^{-2}$ S/m for curve 3. The calculation is made for the frequency of the electromagnetic field, $f = 3.52$ GHz. The object is a dielectric cylinder of square cross section and dimension $S = 0.6\lambda_0 \times 0.6\lambda_0$ which is located at a depth $h = 1.3\lambda_0 (x = 0)$. The dashed lines in Fig. 5.37a show values of ν when $|\nu| = f/c$ and $|\nu| = f\sqrt{\varepsilon_{r2}}/c$. From these outcomes it follows that the significant part of a spectrum of plane waves is in a range

of values of $|v| > f/c$. It is part of the plane-wave spectrum of a scattered field asso-
ciated with evanescent waves. One can see that this part of the plane-wave spectrum
of a scattered field contributes to the scattered field at $f/c \le |v| \le f\sqrt{\varepsilon_{r2}}/c$. The
spectrum damps rapidly at $|v| > f\sqrt{\varepsilon_{r2}}/c$. The propagating and evanescent parts
of the backscattered spectrum are also decreased if the conductivity of medium 2
is increased. The evanescent portion of the backscattered field can be neglected
if the distance between the surface of medium 2 and the probing line is more
than $\lambda_0/2$. Figure 5.38 shows a modulus of the backscattered field on probing line
$(y = y_1)$. Figure 5.38a illustrates a case for $\sigma_2 = 0.0$ S/m and $y_1 = -0.01\lambda_0$ for
curve 1, $y_1 = -0.5\lambda_0$ for curve 2, and $y_1 = \lambda_0$ for curve 3. In Fig. 5.38b one can
the see same dependencies, but in this case, $y_1 = -0.01\lambda_0$ and the conductivity of
medium 2 is changed: $\sigma_2 = 0.0$ S/m for curve 1, $\sigma_2 = 1.0 \times 10^{-2}$ S/m for curve

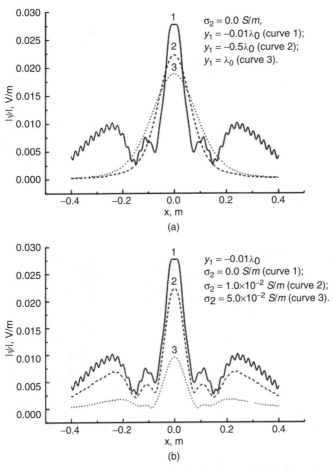

FIGURE 5.38 Distributions of the modulus of a backscattered field along the probing line
$y = y_1$.

2, and $\sigma_2 = 5.0 \times 10^{-2}$ S/m for curve 3. The essential distinction between curve 1 and curves 2 and 3 is shown in Fig. 5.38a. Thus, when coming to the surface of a dielectric half-space containing an inhomogeneity, the structure of the scattered field is changed because of the presence of evanescent waves in its spectrum. The increase in losses in medium 2, and also of the distance y_1, reduces the contribution of this part of the spectrum to the scattered field.

Use of the evanescent waves of the scattered field spectrum is of great interest for reconstruction of an image of the object embedded in a dielectric half-space. For example, Fig. 5.39 shows the reconstructed images of a dielectric cylinder studied

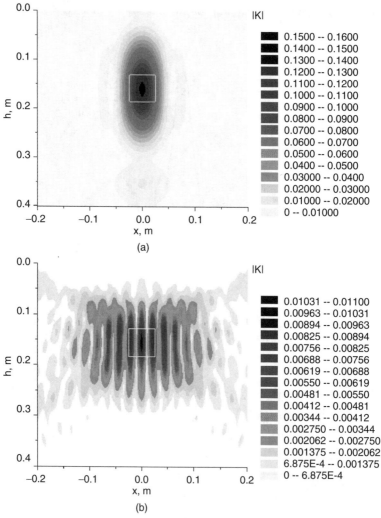

FIGURE 5.39 Images of a subsurface object used in the reconstructions of (a) $0.0 < |v| < 11.0$ m^{-1} (b) $13.8 < |v| < 30.0$ m^{-1} (c) $0.0 < |v| < 30.0$ m^{-1}.

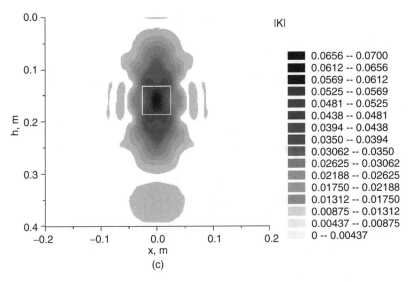

|K|

■	0.0656 -- 0.0700
■	0.0612 -- 0.0656
■	0.0569 -- 0.0612
■	0.0525 -- 0.0569
■	0.0481 -- 0.0525
■	0.0438 -- 0.0481
■	0.0394 -- 0.0438
■	0.0350 -- 0.0394
■	0.03062 -- 0.0350
■	0.02625 -- 0.03062
■	0.02188 -- 0.02625
■	0.01750 -- 0.02188
■	0.01312 -- 0.01750
■	0.00875 -- 0.01312
■	0.00437 -- 0.00875
	0 -- 0.00437

(c)

FIGURE 5.39   (*Continued*)

by the method described above. Data about the scattered field were calculated at $y_1 = -0.01\lambda_0$ and $\sigma_2 = 0.0$. In Fig. 5.39a one can see the subsurface object image obtained using scattered field values of $0.0 < |v| < 11.0$ m^{-1}, corresponding only to the propagating waves in the field spectrum. The position and cross-sectional shape of the object are shown by the black contour. We can conclude that in this case the position of the image corresponds to the position of the object, but the cross-sectional shapes are different. The size of the image is larger in the h-direction than in the x-direction. The structure of the image cross section is smooth. Figure 5.39b illustrates another example, when $13.8 < |v| < 30.0$ m^{-1}. In this case, evanescent waves of the field spectrum (at all frequencies from 3.52 to 4.0 GHz) are used in image reconstruction. Now the sizes (at level 0.8 from maximum) and the cross-sectional shape for image and object are close. Because of evanescent waves interface, the image has a periodical structure along the x-axis, and the structure of object image is smooth along the y-axis. The best image reconstruction results are obtained when the full range of $0.0 < |v| < 30.0$ m^{-1} is used. The object image for this case is shown in Fig. 5.39c.

5.3.7   Microwave Tomographic Technologies

Microwave Measurement Unit for Subsurface Tomography   The principle of a microwave tomographic setup was illustrated in Section 5.3.2. In our experiments, different equipment was used to measure amplitude and phase distribution of a scattered electromagnetic field, including vector analyzers at different frequency ranges and specially designed electronic schemes. Of course, the essential part of a tomography system is the antenna unit, which we consider below.

The creation of a portable microwave measurement unit is an important part of a project, for several reasons. First, the use of subsurface tomography processing imposes specific requirements on the technical specifications of underlying data collection hardware, which, unfortunately, cannot be satisfied using existing instruments if time of image reconstruction and convenience of work are of concern. These requirements are:

- A wide frequency range of operation, at least twice the initial frequency
- The ability to measure transmission coefficient on thousands of frequencies per second
- A high level of microwave frequency stability
- Built-in hardware means for compensation of a huge first reflection from an air–medium interface
- A wide dynamic range and linearity to be able to recover a useful weak signal from the target in the presence of huge background signals from the interface

The objective of this project, the creation of a handheld multisensor system, imposes additional crucial design requirements:

- Electromagnetic compatibility with other sensors
- Operation with a compact antenna system, preferably one with a single antenna (so in the reflection mode), but maintaining high sensitivity to signals from the target
- Illumination of minimum required microwave power
- Minimum weight, size, and power consumption

Fortunately, advances in the semiconductor industry and the availability of a wide range of integrated circuits with complex built-in functions provide the means to satisfy all the requirements.

As an example, a recently designed prototype of a complete microwave tomographic system is shown in Fig. 5.40, and a possible functional scheme of a vector measurement unit is shown in Fig. 5.41. A wide-frequency-range microwave quadrature transceiver operates in stepped-frequency mode with an agile digitally controlled oscillator. On one side it is connected to a personal computer (PC) using a standard serial interface, RS232 or USB, and on the other side, it is connected to an antenna. Conditionally, the setup may be divided into a high-frequency (HF) part, a low-frequency (LF) part, a digital part, and a power converter. The microcontroller (MCU) is the part of the unit that communicates through interface circuitry with the PC, controls all the execution modules, and performs measurement sequences, initial raw data processing, and many other tasks.

Let us trace signal propagation in a typical measurement sequence. Initially, the MCU sets a new frequency by sending a frequency code to the HF generator. From the output of the generator, microwave power is divided over two channels. Half of the power propagates through an output amplifier/modulator and diplexer to the

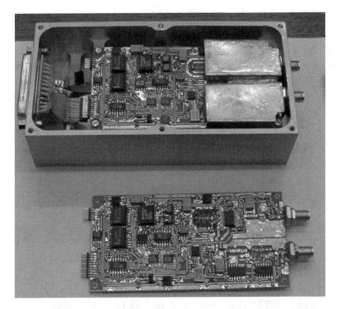

FIGURE 5.40   Prototype of a microwave vector measurement unit.

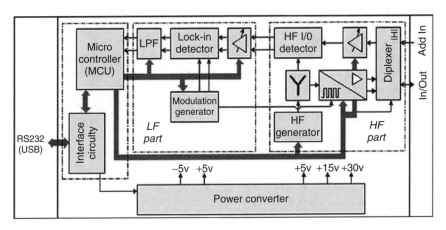

FIGURE 5.41   Functional scheme of a measurement unit.

output of the unit and excites an incident wave in the external antenna. The second half of the power is used as a reference signal for the quadrature (I/Q) detector in the receiving part. an incident wave, illuminated by the antenna, reaches the surface and propagates inside a medium. Part of the initial microwave power, reflected by the surface and objects under the surface, comes back and is received by the same antenna. Part of the unit, conditionally termed the *diplexer*, isn't actually a diplexer in the usual sense. The main task of this part is to rectify reflected signal from the total signal presented on a single unit's input and output. Several approaches will be tried, starting with the simple use of a circulator and proceeding up to special active compensation circuitry, which will be able not only to suppress the signal transmitted at the input to the receiver, but will also be able to compensate significantly for the constant reflected signal from the surface. This will be achieved by generating a special auxiliary microwave signal, individually on each frequency and controlled precisely by amplitude and phase. This signal, being added at the receiver input, to signal coming from the antenna, will compensate for the constant input signal and make it possible to apply additional amplification without saturating the receiver, thus increasing the ratio of useful signal to background signals. The design of this diplexer, which operates equally well at all frequencies from the band utilized, is a real advantage of this system.

So input signal from the diplexer passes through a controllable variable gain amplifier (VGA) and comes to an I/Q detector, which converts microwave signal to two in-phase and quadrature low-frequency signals on the modulation frequency. These signals are amplified additionally in a second VGA and passed to a lock-in detector followed by a lowpass filter (LPF). This set of modules converts I/Q signals on a modulation frequency to dc I/Q signals, each representing an average value of the I and Q components of the initial microwave input signal. This double conversion scheme eliminates many problems, such as drift characteristics, excessive flicker noise, and very narrow bandwidth, thus providing an excellent signal/noise ratio. The cutoff frequency of the LPF is changeable as well. The MCU sets its value depending on the speed of data collection.

Output dc I/Q signals from the LPF are passed to the MCU, which digitizes them, begins digital processing, and stores them in memory for further transmittal to the PC. This ends the cycle of measurement on a single frequency. After that, the MCU sets a new frequency from a predefined set and begins the next cycle. When a certain set of data has been collected, the MCU, after possible additional group processing, packs the data into a transmission frame and sends it to the PC. The MCU will have sufficient intermediate storage capacity and use a robust transmission protocol that guarantees the absence of data lost at the transmission stage.

An important part of measurement unit development is the creation of system and control software. During detailed functional scheme development and component selection, close attention was paid to power saving. The principal technical specifications for the measurement unit are summarized in Table 5.1.

Microwave Up–Down Converting Module The microwave module (Fig. 5.42) converts the frequency range 2 to 4 GHz in a single sideband (SSB) of 36.8 to 38.8 GHz. For this function, the microwave module includes a local 34.8-GHz oscillator, up-down converters, a signal splitter, ferrite circulators, and a waveguide-mounted rejection filter of 30.8 to 32.8 GHz sideband and carrier 34.8 GHz. Microwave power from the local oscillator comes in a signal splitter. Separated signals are passed through individual isolators and feed two balanced mixers, which are used as up- and down-converters. All isolators in the microwave module are

TABLE 5.1 Technical Specifications of the Measurement System

Parameter	Value	Unit	Notes
Operating frequency	2000–4000	MHz	
Microwave radiating power	≤100	mW	Pick value in active mode
Dynamic range	90	dB	
Data collection rate	0–2500	points/s	New frequencies per second
PC interface	Serial		RS232 or USB
Power supply (dc: 12–24 V)	≤15	W	Pick value in active mode
Weight of the unit	≤1.5	kg	
Dimensions	≤160 × 100 × 60	mm	

used for any feedback suppression. Signal with a frequency of 2 to 4 GHz from the output of the microwave measurement unit is applied to an up-converter and mixes with a 34.8-GHz signal from the local generator. Mixing products pass through the rejection filter, which has a rejection band at 30 to 35 GHz. (The filter has been described in our earlier publications: e.g., [43].) As a result, at the output the microwave signal has a frequency range of only 36.8 to 38.8 GHz, shaped by

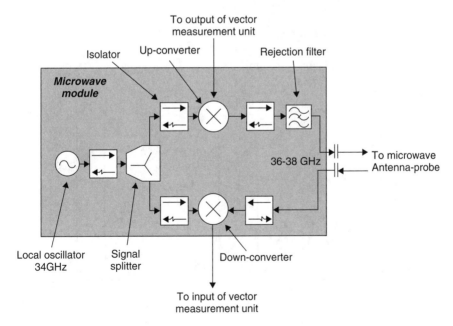

FIGURE 5.42   Up–down converting module.

the way in which the output signal is applied to the microwave antenna system. The back signal from the antenna arrives at the down-converter and mixes with signal from the local generator. The signal produced by the mixer is returned to the vector-measuring unit input. This scheme has been used for measurements in the millimeter waveband.

Antennas in a Tomographic Setup: Characteristics of Dielectric Rod In this section, the design and characteristics of a dielectric rod antenna (DRA) are considered. These characteristics illustrate the properties of antennas used in our tomographic experiments. We studied the characteristics of a complete range of dielectric rod antennas, to help us understand antenna properties so as to define ways to improve them. According to the plan, a complete range of measurements for dielectric rod antenna characteristics were conducted by matching with a generator (Fig. 5.43). The SWR (standing wave ratio) (or in other notation, return loss parameters S_{11} and S_{22}) performance is shown in Fig. 5.44. Vector network analyzer R4-38 (Russia), scalar network analyzer HP 8757D, (United States), and sweep oscillator HP 8350B (United States) have been employed for the measurements.

Before measurement the experimental setup was calibrated accordingly to the usual practice. Another example of the antenna matching the generator is given

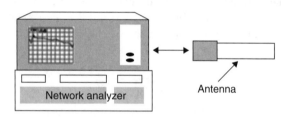

FIGURE 5.43   SWR parameter measurements.

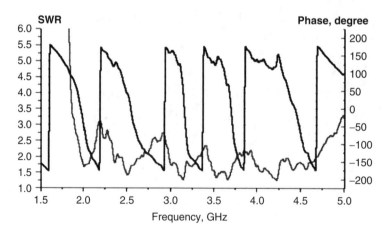

FIGURE 5.44   SWR parameter of the antenna investigated.

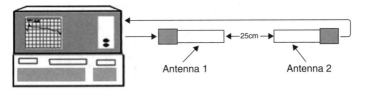

FIGURE 5.45 Measurements of direct antenna transmission.

as a measurement of the straight transfer ratio (parameters S_{12} and S_{21}) from antenna to antenna at different interpolarizations (Fig. 5.45), for copolarized antennas (Fig. 5.46) and for cross-polarized antennas (Fig. 5.47). Antenna characteristics for operating through the reflecting surface of a metal plane have been studied as

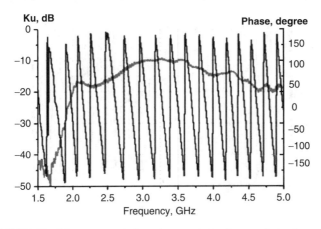

FIGURE 5.46 Direct transmission characteristics for copolarized antennas.

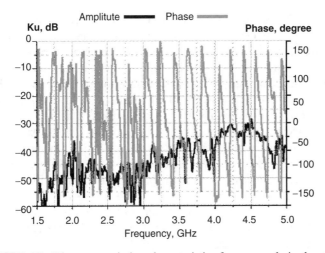

FIGURE 5.47 Direct transmission characteristics for cross-polarized antennas.

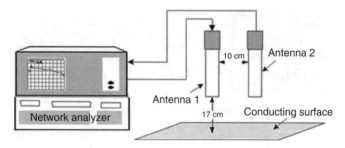

FIGURE 5.48 Antenna characteristic calibration by a metallic plate.

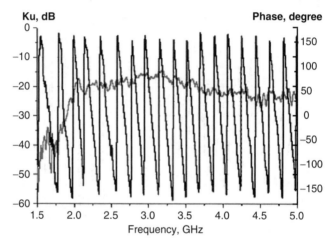

FIGURE 5.49 Transmission characteristics for copolarized antennas over a metallic plate.

well (Fig. 5.48). The measured transfer ratio for co-polarized antennas is presented in Fig. 5.49. A reflected signal for cross-polarized antennas is shown in Fig. 5.50. For copolarized antennas directed into free space, the characteristics of the signal passed are shown in Fig. 5.51. On the basis of measurements and experimental investigations, different approaches to the realization of effective wideband excitation of dielectric rod antennas may be considered. Later we show some new solutions, such as combining a DRA and a log-periodic antenna.

For correct measurement of a scattered field in tomographic experiments, it is very important to control near-field antenna characteristics. One possible measurement scheme is shown in Fig. 5.52, where we have used a dipole antenna as a probe (Fig. 5.53a). The antenna's field amplitude and phase distributions are shown in Figs. 5.54 and 5.55, respectively.

A possible scheme for a dielectric antenna system used for near-field measurements [58] is shown in Fig. 5.56a. In our opinion, within the range of antennas available, dielectric antennas could be a good choice for practical tomography, due primarily to the fact that DRAs have a smaller aperture in the Fresnel zone than that

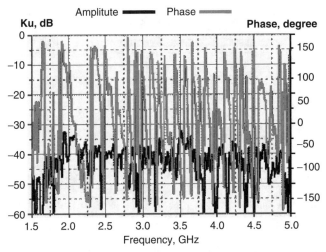

FIGURE 5.50 Transmission characteristics for cross-polarized antennas over a metallic plate.

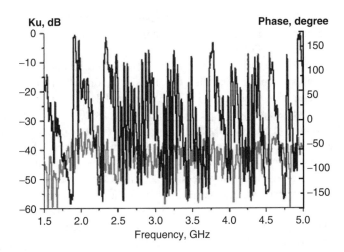

FIGURE 5.51 Transmission characteristics for copolarized antennas in free space.

of other antennas. It is also very important in tomography that the receiving and transmitting antennas be placed at a much smaller distance from each other than the operating wavelength. The usual coupling level should not exceed −40 dB.

The setup of the transmitting and receiving antennas includes a waveguide-to-coaxial adapter, The adapter circular cross-section diameter is 64 mm, the dielectric cones have a total length of 350 mm, and the operating frequency range is 2.5 to 5.0 GHz. The microwave transmission losses for a dielectric antenna pair may be estimated as approximately −10 dB under nonuniformity for an amplitude–frequency characteristic of —2 dB, the SWR being 2.0. In our practice we used different types of DRAs with circular and rectangular cross sections and

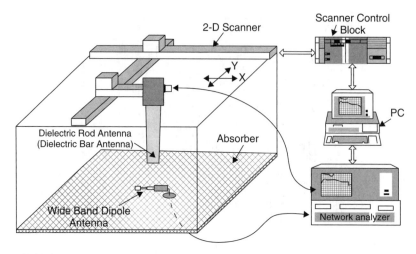

FIGURE 5.52 DRA near-field amplitude and phase distribution measurement.

tapering angles. We found that DRAs may easily be combined with other types of wideband antennas, such as a log-periodic antenna, which may be placed on the surface of a rectangular DRA.

Measurement of microwave scattering by different objects at various distances from the antenna system are important to an understanding of the efficiency of an experimental setup that includes a DRA. Experimental data on the measurement of a scattered field using a metal plane with dimensions 20 × 15 cm are presented in Fig. 5.57. A measured field scattered by a VS-50 antipersonnel mine is shown in Fig. 5.58. One of the main reasons for proposing the use of DRAs [58] for near-field measurements is the improved spatial resolution of the images obtained. We used dielectric rods of varying sizes that satisfied the fundamental mode of the corresponding dielectric waveguide (circular or rectangular).

An experimental setup for subsurface imaging using a DRA is presented in Fig. 5.59. The antenna system is located over a sand surface. The back side of a plastic VS-50 antipersonnel mine in sand at a depth of 5 cm is shown in Fig. 5.60a. The photograph makes the scanning conditions clear. A reconstructed image of the mine using the standard frequency-step method is shown in Fig. 5.60b. Use of the tomographic algorithm demonstrates in Fig. 5.60c the perfect reconstructed image of the buried mine.

It has been shown that the use of dielectric rod antennas in tomographic experiments covers quite wide frequency ranges, up to 135 to 150 GHz. However, we should also note difficulties with DRAs for perfect matching in wideband. That is why DRAs are generally used in transmission mode using a transmitter–receiver antenna pair. DRA technologies are opening unique possibilities in case you need to create acombined system that contains, for example, tomographic (or GPR) and NQR [or metal detector (MD)] sensors in one device. DRAs do not disturb NQR or MD coils and may easily be combined. On the other hand, DRAs are negligibly

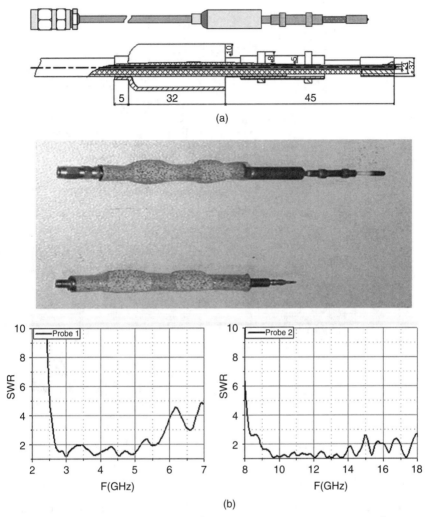

FIGURE 5.53 (a) Probe antenna for near-field antenna pattern measurement; probe antenna characteristics: operating frequency range 2.5 to 5 GHz, SWR<2. (b) Probe photograph and characteristics: probe 1, 2.5 to 5 GHz (left), probe 2, 8.7 to 12 GHz (right).

disturbed by coils. In Fig. 5.61 one can see a DRA combined with NQR and MD coils and the tomographic images obtained. Even if an NQR coil is placed on the surface of a DRA, satisfactory results can be obtained.

Wideband Bowtie Antenna Another type of wideband antenna frequently used in our tomographic experiments were backed bowtie antennas (Fig. 5.62). Examples of this type of antenna are shown in Fig. 5.63. This antenna set can be used to cover a wide frequency range, from 1 to 10 GHz (Fig. 5.64). One of main advantages of

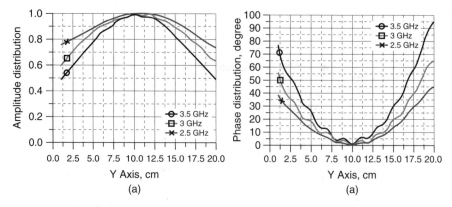

FIGURE 5.54   Antenna (a) amplitude and (b) phase distributions.

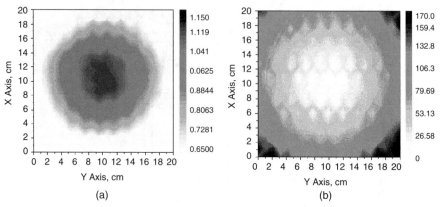

FIGURE 5.55   Two-dimensional image of antenna distributions at 3.5 GHz: (a) amplitude; (b) phase.

this antenna type is small size and light weight, which are also very important in practice, especially for tomographic systems that use arrays.

The basic specifications of the backed bowtie antenna are as follows:

Frequency range: 2600 to 5900 MHz

VSWR ratio: 1.4 : 1 (average), 2.0 : 1 (max.)

Input impedance (nominal): 75Ω

Cross-polarization rejection: >20 dB

Connector type: SMA male

Height (Z): 4.1 cm

Width (Y): 5.9 cm

Length (X): 3.9 cm

Weight: 0.2 kg

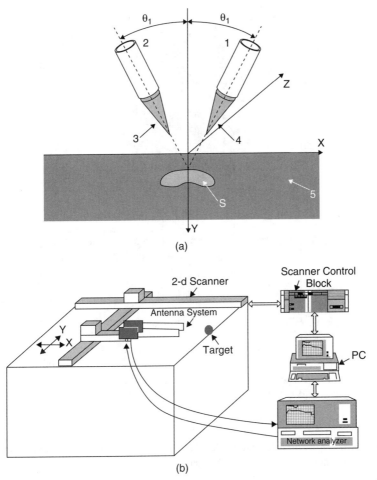

FIGURE 5.56 Schemes of (a) a DRA system used for subsurface measurements; (b) an experimental setup used for target scattered field measurements.

As one can see, this antenna type provides quite acceptable characteristics for use in wideband tomographic systems (Fig. 5.65).

5.3.8 Three-Dimensional Tomographic Imaging

Imaging of three-dimensional bodies' structure using microwaves and millimeter waves is a widely investigated topic at present. The imaging allows one to obtain important information about the shape, size, and localization of an object. Some image information about a body—for example, information related to the electric properties of the body—cannot be obtained by techniques that use different physical phenomena, including x-rays, γ-rays, positron electron annihilation reaction, nuclear magnetic resonance, ultrasound, electrons, and ions.

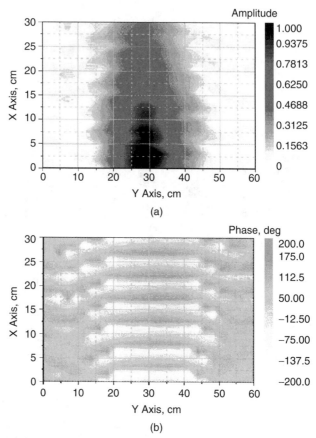

FIGURE 5.57   Measured scattered field of the metallic plate 20×15 cm in free space: (a) amplitude distribution; (b) phase distribution.

Some numerical methods and experimental techniques for microwave [62] and millimeter-wave [7,64] imaging of three-dimensional imhomogeneous bodies have been described. Joackimowicg et al. [62] reduced the reconstruction method to tomographic reconstruction of the body under investigation and provided different experimental results on dielectric rods and isolated animal organs. A spatial iterative algorithm for electromagnetic imaging based on Newton Kantorovich's procedure has been proposed by Detlefsen et al. [63]. The complex permittivity of an inhomogeneous lossy dielectric object of arbitrary shape is an object in the reconstruction. The reconstruction technique includes an integral representation of the electric field and the moment method. A general holographic imaging approach was considered by Gheen et al. [64]. This approach is characterized by the requirement of computer-aided reconstruction of the three-dimensional distribution of scattering centers from the scattered field. A wideband millimeter-wave imaging system for rapid inspection of personnel for concealed explosives, handguns, or other threats was described by Vertiy et al. [7].

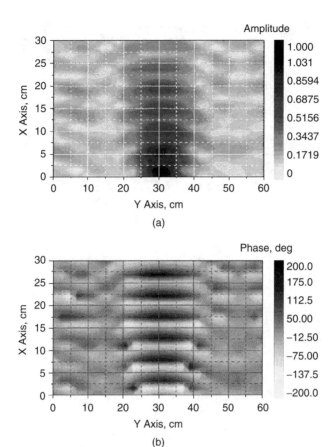

FIGURE 5.58 Measured scattered field of a plastic VS-50 antipersonnel mine in free space:
(a) amplitude distribution; (b) phase distribution.

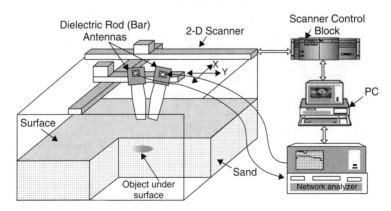

FIGURE 5.59 Experimental setup for subsurface imaging.

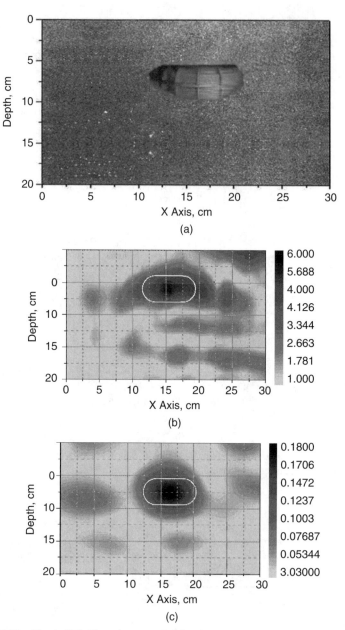

FIGURE 5.60 Plastic VS-50 antipersonnel mine in sand at a depth 5 cm: (a) photo of the mine in sand; (b) reconstructed mine image using the frequency-step algorithm, (c) reconstructed mine image using the tomographic algorithm.

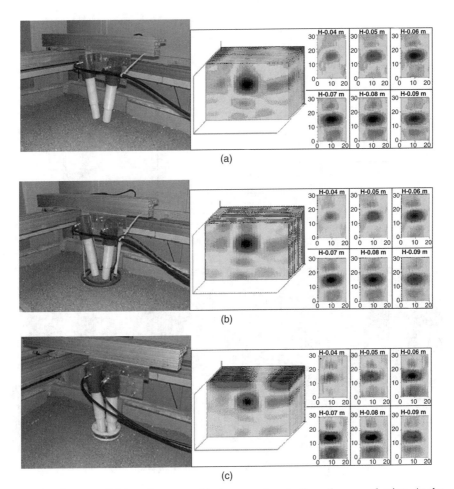

FIGURE 5.61 (a) DRA and tomographic images of a plastic antipersonnel mine. At the left side of the image are shown horizontal slices from the soil. (b) DRA combined with an MD coil and tomographic images of the same mine. (c) DRA combined with an NQR coil and the same mine image.

Below we consider a numerical three-dimensional imaging method that makes it possible to reconstruct a normalized electric current for NDT, biomedical, target recognation, and other applications. The method is based on the use of backscattering phenomena inside and outside the target investigated. The Fourier transform technique is used to obtain a solution to the integral equation derived for the normalized electric currents induced inside. A reconstruction algorithm similar to the tomographic algorithms can be extended to three-dimensional objects [8,9]. Experiments on imaging of three-dimensional objects by the method described were conducted using microwave and millimeter-wave setups developed at the International Laboratory for High Technologies (ILHT) at Tubitak-MRC, Turkey.

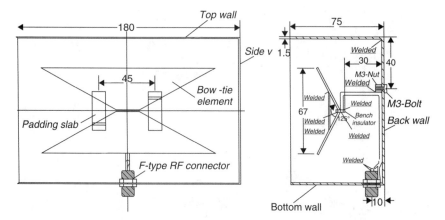

FIGURE 5.62 Front and side views of a backed bowtie antenna.

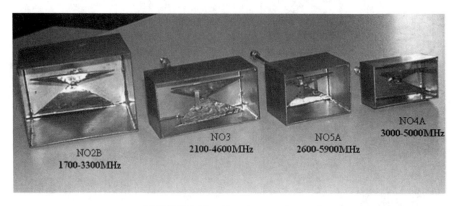

FIGURE 5.63 Bowtie antenna set.

Problem Definition Let's consider a three-dimesional object of volume V illuminated at angle θ_i by a plane wave $\vec{E}^i$ polarized linearly along the z-axis (Fig. 5.66). The object is characterized by dielectric permittivity $\varepsilon_{ob}(x, y, z)$ and conductivity $\sigma_{ob}(x, y, z)$. The external medium is uniform and is characterized by dielectric permittivity ε and conductivity σ. The permeability is that of the vacuum $\mu = \mu_0$ in each medium. Let $\vec{E}$ represent the total electric field and $\vec{E}^S$ the scattered field generated by the equivalent electric currents (polarization currents) radiating in the external medium so that $\vec{E} = \vec{E}^i + \vec{E}^S$. The equivalent electric currents are defined as

$$\vec{J}_{eq}(x, y, z) = [k_{ob}^2(x, y, z) - k^2]\vec{E}(x, y, z) \tag{5.66}$$

where k_{ob} is the wavenumber for the object medium,

$$k_{ob}(x, y, z) = \sqrt{\omega^2 \mu_0 \varepsilon_{ob}(x, y, z) + i \omega \mu_0 \sigma_{ob}(x, y, z)} \tag{5.67}$$

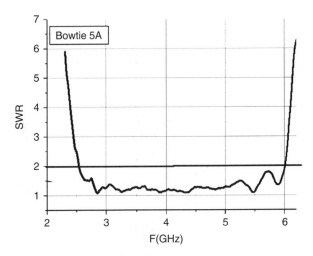

FIGURE 5.64 Typical SWR characteristics for a backed bowtie antenna.

and k is the wavenumber for the external medium,

$$k = \sqrt{\omega^2 \mu_0 \varepsilon + i \omega \mu_0 \sigma} \qquad (5.68)$$

at time dependence $e^{-i\omega t}$.

The scattered field $\vec{E}^S$ inside and outside the object can be calculated by using its integral representation

$$\vec{E}^S(x, y, z) = \frac{1}{k^2} (\text{grad div} + k^2)$$

$$\iiint_V \vec{J}_{eq}(x', y', z') G(x, y, z; x', y', z') \, dx' \, dy' \, dz' \quad (5.69)$$

where the function $G(x, y, z; x', y', z')$ is the Green's function for the three-dimensional case:

$$G(x, y, z; x', y', z') = \frac{e^{ik\sqrt{(x-x')^2+(y-y')^2+(z-z')^2}}}{4\pi\sqrt{(x-x')^2 + (y-y')^2 + (z-z')^2}} \qquad (5.70)$$

The Green's function in (5.70) can also be written in integral form:

$$G(x, y, z; x', y', z') = \frac{i}{2} \int\limits_{-\infty}^{\infty} \int\limits_{-\infty}^{\infty} \frac{e^{2\pi i v_1 (x-x')} e^{i\gamma|y-y'|} e^{2\pi i v_3 (z-z')}}{\gamma} \, dv_1 \, dv_3 \qquad (5.71)$$

where $\gamma = 2\pi v_2 = \sqrt{k^2 - 4\pi^2(v_1^2 + v_3^2)}$. The Green's function can be ieterpreted as the electric field created at a point (x, y) by a point source of current situated at a

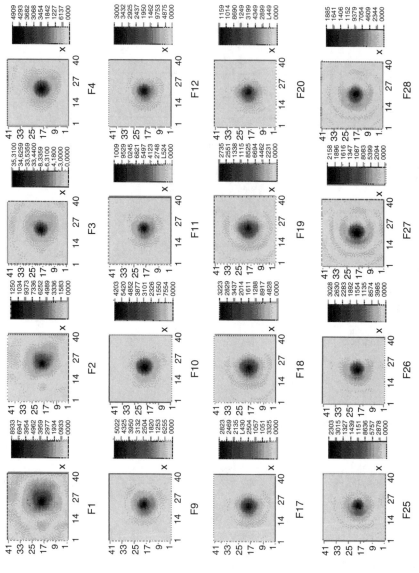

FIGURE 5.65 Near-field distribution for a backed bowtie antenna at different frequencies between 2.5 and 6 GHz.

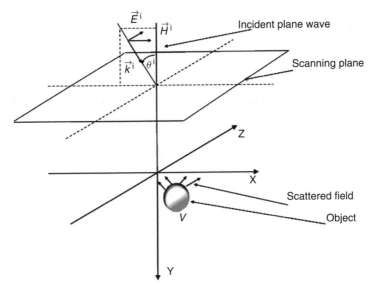

FIGURE 5.66 Geometry of the problem.

point (x', y'). The expression in Eq. (5.70) is the expansion of the Green's function in a plane-wave spectrum. The values ν_1, ν_2, and ν_3 are the space frequencies of the plane waves from the spectrum.

Assume that the depolarization is negligible. Then $\vec{E}^S$ can be approximated by

$$\vec{E}^S(x, y, z) = \iiint_V \vec{J}_{eq}(x', y', z')G(x, y, z; x', y', z') \, dx' \, dy' \, dz' \qquad (5.72)$$

with $\vec{E}^S$ as $\vec{E}^i$ parallel to the z-axis. Using as an expression for the normalized polarization current [5]

$$K(x, y, z) = \left[\frac{k_{ob}^2(x, y, z)}{k^2} - 1 \right]\left[1 + \frac{\psi(x, y, z)}{E_z^i(\theta^i, x, y)} \right] \qquad (5.73)$$

where $\psi(x, y, z)$ and $E_z^i(\theta^i, x, y)$ are z-components of the scattered and incident fields, recpectively, Eq. (5.72) can be written as

$$\psi(x, y, z) = k^2 \iiint_V E_z^i(\theta^i, x', y')K(x', y', z')G(x, y, z; x', y', z') \, dx' \, dy' \, dz'$$

$$(5.74)$$

Let the object be illuminated by a plane wave with an electric field of 1 V/m. In this case the incident field takes the form

$$E_z^i(\theta^i, x, y) = e^{ik(x \sin \theta^i + y \cos \theta^i)}$$

(5.75)

Let us define the Fourier transform $\hat{\psi}(\nu_1, y, \nu_3)$ of the scattered field $\psi(x, y, z)$ on a scanning plane at $y = \text{const}$ and the three-dimensional Fourier transform $\hat{K}(\alpha, \beta, \chi)$ of $K(x,y,z)$ as

$$\hat{\psi}(\nu_1, y, \nu_3) = \int\limits_{-\infty}^{\infty} \int\limits_{-\infty}^{\infty} \psi(x, y, z)e^{-2\pi i(\nu_1 x + \nu_3 z)} \, dx \, dz$$

(5.76)

$$\hat{K}(\alpha, \beta, \chi) = \int\limits_{-\infty}^{\infty} \int\limits_{-\infty}^{\infty} \int\limits_{-\infty}^{\infty} K(x, y, z)e^{-2\pi i(\alpha x + \beta y + \chi z)} \, dx \, dy \, dz$$

(5.77)

Taking the Fourier transform of (5.74) the along x and z directions, one can obtain the equation

$$\hat{\psi}(\nu_1, y, \nu_3) = \frac{k^2 i e^{-i\gamma y}}{2\gamma} \int\limits_{-\infty}^{\infty} \int\limits_{-\infty}^{\infty} \int\limits_{-\infty}^{\infty} K(x', y', z')e^{-2\pi i(\alpha x' + \beta y' + \chi z')} \, dx' \, dy' \, dz'$$

$$= \frac{k^2 i e^{-i\gamma y}}{2\gamma} \hat{K}(\alpha, \beta, \chi)$$

(5.78)

where

$$\alpha = \nu_1 - \frac{k}{2\pi} \sin \theta^i \qquad \beta = -\frac{1}{2\pi}(\gamma + k \cos \theta^i) \qquad \chi = \nu_3$$

(5.79)

Equation (5.78) shows that the Fourier transform of the normalized polarization current can be obtained from the scattered field measured. The object function $K(x,y,z)$ is determined by taking the inverse Fourier transform of Eq. (5.78) for ranges of ν_1, ν_2, and ν_3 at α, β, and χ defined in Eq. (5.79). Conducting measurements of the scattered field in the range of frequencies, it is possible to the find corresponding set of polarization currents induced in the object. The general (sought) object function is defined as the modulus of the polarization current sum that is determined.

Experimental Results To check the reconstruction algorithm considered, experiments were conducted on the microwave and millimeter-wave bands. The microwave experimental system contains:

- A desktop or portable computer equipped with a data acquisition board (DAQ board)

*Tomography program described here has been supported by the State Planning Organization of the Turkish Prime Ministry within a (5075519) project.

- A small, lightweight vector measuring unit (see Fig. 5.40 for a 2 to 4-GHz frequency range) or vector network analyzer
- A microwave or millimeter-wave antenna probe
- A scanner control block

The transmitting antenna radiates continuous-wave (CW) signals with a frequency-step mode. Frequencies f_j of the incident wave were taken in the bandwith $\Delta f = 2.0$ GHz from $f_1 = 2.0$ GHz to $f_{32} = 4.0$ GHz. The waves scattered by objects are recorded in the probing space. The return signal from the receiving antenna is registered by a microwave measurement unit. When using a perfectly matched antenna (see, e.g., Fig. 5.64), we prefer to connect one antenna with a transmitter and receiver by a directional coupler. It also allows us to improve the spatial resolution of a tomographic system. The transmitting and receiving antennas form an antenna block that can be moved by the mechanical scanner in front of the target area along direct lines on a plane lying parallel to the scanning surface. Signal-processing software based on Eqs. (5.78) and (5.79) makes it possible to reconstruct images of isosurfaces or different cross sections of the object under test which are normal or parallel to the scanning plane (vertical or horizontal slices). The spatial frequencies ν_k were taken in the bandwith $\Delta \nu = 20$ m^{-1} from $\nu_1 = -10$ m^{-1} to $\nu_{21} = 10$ m^{-1} for image reconstruction.

The tomographic system geometry is shown in Fig. 5.67. One can see protective dielectric layer (1) (geometrical thickness: $0.5\lambda_0$, $\lambda_0 = 0.1$ m), the position of the person (2), located on the surface of a sheet (3) made from foam plastic, and an absorber layer (4). The thickness of the free-space layer between the protective cover and the person was chosen as $\lambda_0 = 0.1$ m.

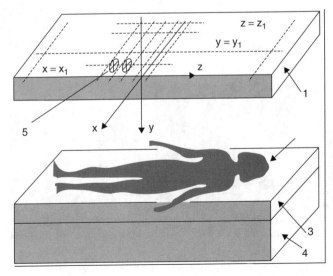

FIGURE 5.67   Experiment setup for human body imaging.

Before experiments with a real person, an experiment was conducted on imaging a head model. A photo of the model (phantom), which consists of a section of plastic (PVC) pipe and a foam plastic plate with four notches filled with water, is shown in Fig. 5.68c. The notches were cut out in the form of a cross. The plate diameter is $2.4\lambda_0$ ($\lambda_0 = 0.1$ m), the notch horizontal size is $0.9\lambda_0 \times 0.3\lambda_0$, and the depth of a notch is $0.15\lambda_0$. The straight-line distance between two nodes is $0.4\lambda_0$. Figure 5.68b illustrates a reconstructed horizontal slice of the notch cross.

Figure 5.67 also illustrates a womans body in the plane yOz; a probing plane $y = y_1$ above the protective layer, the transmitting and receiving antennas (5) at each scanning step (as the backscattered field is being measured, the transmitting and receiving antennas can be replaced by one antenna), and the scanning lines $z = z_1$ and $x = x_1$ (dashed lines) in the plane $y = y_1$. This plane is located at the coordinate $y \equiv y_1 = -0.15\lambda_0$. The maximum scanning area is $14.0\lambda_0 \times 20.0\lambda_0$ along the x and z axes, respectively. A typical scanning step is about 0.1 to $0.3\lambda_0$.

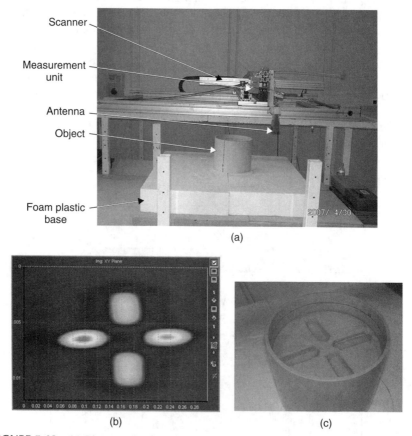

FIGURE 5.68 (a) Photograph of a microwave tomographic setup with a phantom object; (b) tomographic image of a phantom cross section; (c) a photograph of the phantom model.

The microwave setup is shown in Fig. 5.68a. As the test object we have used a phantom, as described by Gofre et al. [65], manufactured from foam plastic with four cavities filled with water. The tomographic slice demonstrates a good match with the model geometry (Fig. 5.68b).

A photo and tomographic image of a chicken leg are shown in Fig. 5.69. The frequency range used is 5 to 9.5 GHz. The image of bones inside the leg is clear.

As the next step in a practical application of this technology, we measured a human body structure. The isosurface image of a woman's body (from the knees up) is presented in Fig. 5.70. The scanning area is $6.912\lambda_0 \times 8.927\lambda_0$, and the scanning steps are $0.108\lambda_0$ and $0.113\lambda_0$ along the smaller and larger scanning areas. Figure 5.71 illustrates the corresponding horizontal slices. Three-dimensional imaging results at a frequency range of 2.2 to 3.7 GHz with a slice size of 0.691 m $\times$ 0.893 m are presented in Fig. 5.72. As one can see, images obtained by three-dimensional tomographic technology may be useful in medical applications (also see Fig. 5.73).

The algorithm based on Eq. (5.50) has been tested for a millimeter-wave tomographic system. The system contains a vector network analyzer used for the frequency range 10 MHz to 325 GHz, a two-dimensional scanner to support scanning with steps in the range from several millimeters to 0.25 mm ($\approx 0.033\lambda_0$, $\lambda_0 = 7.5$ mm), a desktop (or portable) computer equipped with a data acquisition board, a scanner control block, and an antenna system for probing. The antenna system consists of two dielectric rods of rectangular cross section which are fixed inside standard metal waveguides with cross sections of 7.1×3.6 mm^2. The shape of the ends of the dielectric rods is optimized to decrease reflections from them in

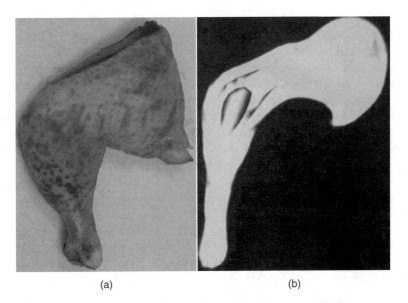

(a)   (b)

FIGURE 5.69   Chicken leg: (a) photograph; (b) tomographic image.

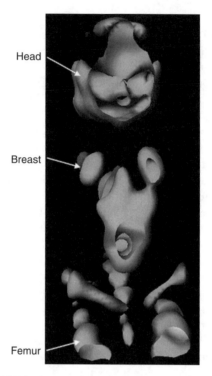

Head

Breast

Femur

FIGURE 5.70 Image of an isosurface of a woman's body.

the waveguides and in the air. The antenna system design allows us to decline the antennas in relation to the normal to the medium surface and to change the antenna height above the medium surface.

In the process of measurement the probe is placed near the plane interface between the air and the surface of sample being investigated, and is then shifted over the surface at lines directly parallel to it. One dielectric rod antenna creates an electromagnetic field around the space and another transforms into a waveguide mode field the backscattered electromagnetic field that appears in the space. Changing the backscattered field after calibration creates a complex signal at the output of the antenna system. The real and imaginary parts of this signal are saved in each point of scanning for selected frequencies in the operating frequency band. It is also a possible regime in which only one dielectric rod is used as a transmitting and receiving antenna.

Data measured over the scanning area can be collected at 32, 64, and 128 frequencies in the operating band at a constant step frequency. A photograph of the setup is presented in Fig. 5.74 where some parts of the tomograph and the sample being investigated are denoted. Some results on the imaging of objects are shown in Figs. 5.75 and 5.76. Figure 5.75 illustrates reconstructed a horizontal slice ($17.33\lambda_0 \times 17.33\lambda_0$) of a composite material plate containing a metal object

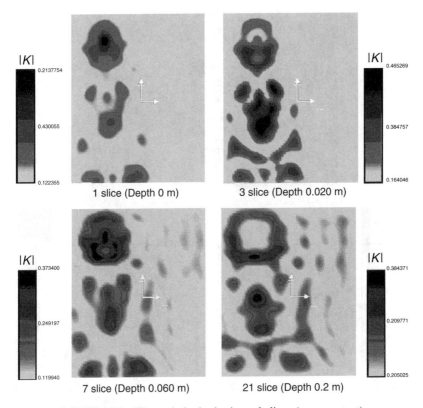

FIGURE 5.71 Woman's body: horizontal slices ($y = $ constant).

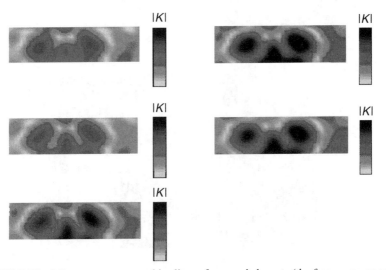

FIGURE 5.72 Microwave tomographic slices of women's breasts (the frequency range and scanning parameters are the same as in Fig. 5.70).

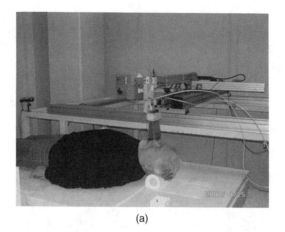

(a)

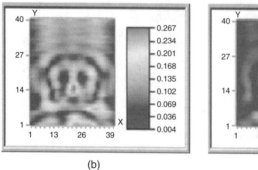

(b)

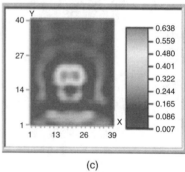

(c)

FIGURE 5.73 (a) Experimental tomographic setup; (b) and (c) tomographic imaging slices of a human head (the protective plastic cover has been removed).

Sub-terahertz Subsurface Tomography System

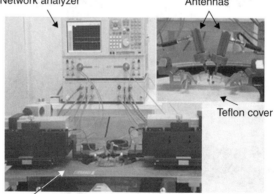

FIGURE 5.74 Photo of the millimeter and submillimeter wave tomography setup operating in between 40 GHz–325 GHz.

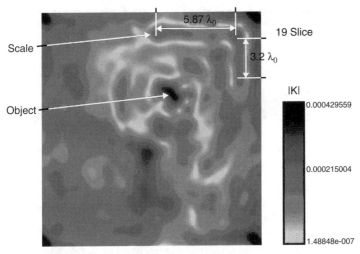

FIGURE 5.75 Damaged composite material plate (130×130 mm^2, nineteenth slice, 18 mm in depth).

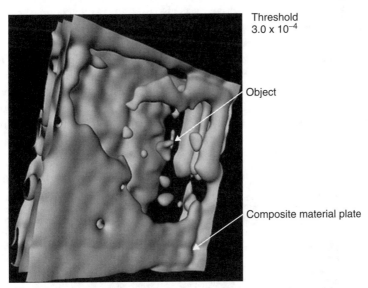

FIGURE 5.76 Damaged composite material plate (130×130 mm^2, isosurface) and object (isosurface).

of complex shape at depth $2.4\lambda_0$. Figure 5.76 illustrates an isosurface image of the damaged composite plate.

Let us consider a few more examples of reconstructed tomographic images in the millimeter-wave band. Next, experiments are conducted with a demountable Teflon cylinder with a Teflon cover and a cardboard letter *A* (Fig. 5.77). This cylinder,

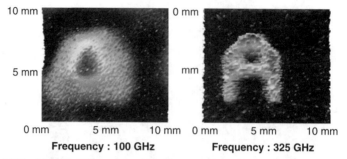

FIGURE 5.77   (a) 100 GHz tomography image of letter A; (b) Same letter A image at 325 GHz.

of diameter $31.6\lambda_0$ and thickness $7.4\lambda_0$ consists of three parts. One of them is a Teflon cylinder of diameter $14\lambda_0$ used as a platform on which is placed an object letter *A*. The thickness of the disk is $0.72\lambda_0$. The objects investigated may be of different shapes and sizes. The objects, which had the shapes of different letters, were cut from cardboard of thickness $0.053\lambda_0$. Image resolution was dramatically improved at 325 GHz [88].

The tomographic system described was also used for the detection of extraneous materials in postal envelopes. It used one dielectric rod antenna as a probe. In this case an envelope is located very near the antenna tip. The antenna probe is matched so that the reflected signal is zero on a clean area of this envelope before scanning. As a result, the antenna probe is mismatched at scanning above extraneous hidden materials in the envelope. The mismatched complex signal is used for processing by the tomographic algorithm upon the reconstruction of slices, on which one can see the sizes and position of hidden material (yogurt powder) inside of envelope. A reconstructed image of a horizontal slice of size $14\lambda_0 \times 5.4\lambda_0$ of an envelope is presented in Fig. 5.78.

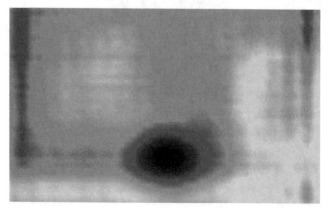

FIGURE 5.78   Reconstructed slice of a post envelope image with a small amount of yogurt powder.

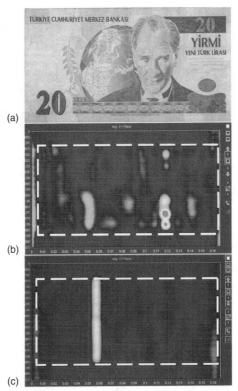

FIGURE 5.79   (a) 20 YTL Turkish currency; (b) tomographic image of counterfeit 20 YTL; (c) tomographic image of original 20 YTL hidden under a plastic layer.

Another interesting application of the proposed technology is related to the control and differentiation of original and counterfeit money. The experimental setup was the same as that for hidden letter imaging. The paper currency under test is placed under the Teflon layer in a way similar to that made with the letter *A* (Fig. 5.79). The tomographic method proposed clearly shows highly promising results. It is possible not only to detect paper currency, but it is also possible to distinguish if it is original or counterfeit. Other measurements also demonstrate the possibility to create a database for paper currency from different countries.

Millimeter-wave band tomography may also have another important application: distinguishing people by their handprints. The print images of two people are shown in Fig. 5.80. No doubt, increasing the operating frequency to 100 GHz, for example, will increase the resolution of tomographic images and improve the performance of the method.

5.3.9 Conclusions

In this section, a numerical method, system setup, and experimental results for microwave and millimeter-wave imaging of three-dimensional inhomogeneous

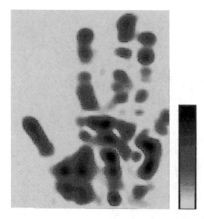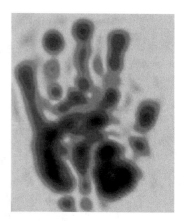

FIGURE 5.80   Millimeter-wave band tomographic handsprints of two different persons.

bodies, cross sections, and isosurfaces were described. This method is based on integral presentation of a scattered electric field for plane-wave excitation at various frequencies. Application fields are defined primarily by advantages and disadvantages compared with two-dimensional tomographic methods: for example, the subsurface tomography method (STM). The algorithm considered makes it possible to solve a very wide class of imaging problems. Medium losses may also be taken into account using this method. The results obtained by the algorithm and by STM are in good agreement. STM may have an advantage, due to the small time required for slice restoration. This allows us to obtain three-dimensional images of buried or hidden objects in quasi-real time.

Further, we described possible improvements in the algorithm involving the development of procedures that will decrease the calculating time for slice reconstruction and the development of tools that will allow us to take into account the characteristics of antenna system in a real device. The experimental results showed the effectiveness of the method in the microwave and millimeter-wave regions. The results obtained may find applications in geophysics, medicine, and nondestructive testing.

5.4   DIFFRACTION MULTIVIEW TOMOGRAPHIC METHOD IN THE MICROWAVE AND MILLIMETER-WAVE BANDS

5.4.1   Multiview Tomographic Method in the Millimeter-Wave Band

Various tomographic methods have been used widely in medicine, radio astronomy, the physics of the Earth's atmosphere, electronic microscopy, magnetic resonance spectroscopy, and other fields of science and technology [81,88]. There are many publications on applied and theoretical tomography [66,67], most devoted to the problems of computer tomography (CT). In CT methods, a finite number of linear integrals (projections) can be employed to reconstruct the internal structure of a body using sensing radiation (e.g., x-rays, ultrasonic or radio waves). These methods have the important practical application in x-ray diagnostics of imaging crosssections of the human body. There are for example, important applications of ultrahigh-frequency radio waves (e.g., microwaves): among them nondestructive testing of materials used in industrial products [68], surface and subsurface sensing, microwave imaging of inhomogeneous bodies in medicine [70,71], and others applications [38,72]. However, there are certain difficulties in the CT application of methods of obtaining tomographic images using microwaves. The main difficulty results from diffraction effects due to reflection and refraction on the boundaries of an imaged body whose dimensions are comparable to the wavelength of the probing radiation.

To obtain a microwave tomographic image of diffraction (scattering of electromagnetic wave by a sample), diffraction tomographic methods are used [73,74]. If the object scatters the electromagnetic wave weakly, and the scattered field is described by Born or Rytov approximations of the first order [67,74], the Fourier diffraction projection theorem [74], which is the mathematical basis of first-order diffraction tomography, can be employed. In the case of strong scattering, methods of high-order diffraction tomography or other methods that do not employ the Fourier diffraction projection theorem to solve inverse scattering problems are used to obtain a tomographic image [75–77,112]. There are also many unsolved theoretical problems connected with the scattering of quasi-optical beams by electromagnetic structures. In this section, methods of CT and first-order diffraction tomography are considered.

Fourier Reconstruction Method of CT Imaging   There are a number of methods for implementation of tomographic imaging, of which the Fourier reconstruction method is the most basic. Tomographic imaging using Fourier reconstruction is based on the Fourier transform, which provides an elegant and complete mathematical approach. The key to the use of this method is the projection theorem, which relates the one-dimensional Fourier transform of the spatial distribution projection of an object (a central section or slices of the two-dimensional Fourier transform of the distribution). The theorem is derived by considering the geometry for data collection shown in Fig. 5.81. Let the function $f(x,y)$ be defined in the u, v coordinate system. Using the rotation transformation, we can obtain

$$f(x, y) = f[(u \cos \theta - v \sin \theta), (u \sin \theta + v \cos \theta)] \tag{5.80}$$

The projection of $f(x,y)$ on the u-axis corresponding to the angle θ is a one-dimensional function of the single variable v, represented by

$$p(u, \theta) = \int_{-\infty}^{\infty} f[(u \cos \theta - v \sin \theta), (u \sin \theta + v \cos \theta)] \, dv \tag{5.81}$$

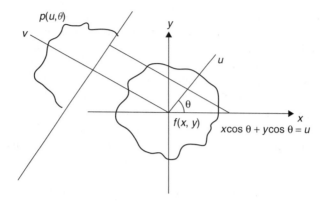

FIGURE 5.81   Geometry of data collection for tomographic imaging.

The one-dimensional Fourier transform of $p(u, \theta)$ with respect to variable u is

$$p(\omega, \theta) = \int_{-\infty}^{\infty} p(u, \theta) \exp(-iu\omega) \, du \qquad (5.82)$$

The Fourier transform of function $f(x,y)$, which is expressed in polar coordinates, is

$$F(\omega, \theta) = \int_{-\infty}^{\infty} \int_{-\infty}^{\infty} f(x, y) \exp[-i(\omega x \cos\theta + \omega y \sin\theta)] \, dx \, dy \qquad (5.83)$$

where ω and θ are the radial and angular coordinates in the frequency-domain plane. If we write Eq. (5.83) in the rotated coordinate system, we see that

$$p(\omega, \theta) = F(\omega, \theta) \qquad (5.84)$$

The function $f(x,y)$ can be writtten

$$f(x, y) = \frac{1}{4\pi^2} \int_{-\infty}^{\infty} \int_{-\infty}^{\infty} p(\omega, \theta) \exp[i(\omega x \cos\theta + \omega y \sin\theta)]\omega \, d\omega_x \, d\omega_y \qquad (5.85)$$

If the function $f(x,y)$ does not depend on the angle θ, it has the form of a Fourier–Bessel transform:

$$f(\sqrt{x^2 + y^2}) = \frac{1}{2\pi} \int_{0}^{\infty} \omega J_0 \left(\omega\sqrt{x^2 + y^2}\right) \left[2 \int_{0}^{a} p(u) \cos(\omega u) \, du \right] d\omega \qquad (5.86)$$

where J_0 is a Bessel function of order $m = 0$.

In our formulation we suppose that a cylindrical dielectric sample of radius a has losses related with the absorption of electromagnetic energy. This absorption is characterized by the function $f(x,y)$, which does not depend on θ, and it is a constant in the circular region of the radius a. We also suppose that we can measure the intensity of an electromagnetic wave propagating through the sample from the transmitter antenna to the receiver antenna. In the coordinate system (u,v), the transmission coefficient T will be obtained as

$$T(u) = \exp\left[-2 \int_v k''(u, v)\, dv \right] \tag{5.87}$$

where k'' is the imaginary part of the complex wavenumber k [78]. It follows from

$$-\tfrac{1}{2}\ln(T) = \int_{-\infty}^{\infty} k''(u, v)\, dv = p(u) \tag{5.88}$$

So we have the projection function $k''(x,y)$ on the u-axis, which does not depend on θ. As k'' is constant in the circle of radius a, we may obtain

$$p(u) = 2\sqrt{a^2 - u^2}\; \frac{\omega'}{c}\; \sqrt{\varepsilon'}\; \frac{\tan \Delta}{2} \tag{5.89}$$

where ω' is the frequency of electromagnetic irradiation, c the velocity of light, ε' the dielectric permittivity and $\tan \Delta$ the loss tangent.

In a second case, we studied a pipe-type cylindrical sample that has inner radius b and outer radius a. In our calculations, the radii are $a = 25$ mm, $b = 10$ mm, and $\omega' = 2\pi \times 33$ GHz, $\sqrt{\varepsilon'} = 1.5$, $\tan\Delta = 1.0 \times 10^{-3}$. The results of the calculations in Eq. (5.86) are shown in Figs. 5.82 and 5.83. Figure 5.84 illustrates the reconstruction of the cross section of a circular ($a = 10$ mm) dielectric cylinder with small absorption which is not located at the center. The radius a of the cylinder is 10 mm, and the center of the cylinder has coordinates $x_0 = 10$ mm and $y_0 = 10$ mm.

Imaging by First-Order Diffraction Tomography The basic equation of diffraction tomography can be obtained as a result of the inverse problem solution of a plane electromagnetic wave scattering from an object under investigation [67,74]. We consider a scattering object, for example in Shwartz space (R^2), characterized by refractive index $n(\vec{r}) = 1 + n_\delta(\vec{r})$, where f is equal to zero outside the refracting object. An incident harmonic [time dependence $\exp(-i\omega t)$] and plane wave $U_I(\vec{r}) = \exp[ik(\vec{\theta}.\vec{r})]$ is illuminating the object; here $\vec{\theta}$ is a unit vector pointing in the direction of wave propagation, and $k = \omega/c$, where ω is the angular radiation frequency and c is the velocity of light. In the case of direct scattering, the total

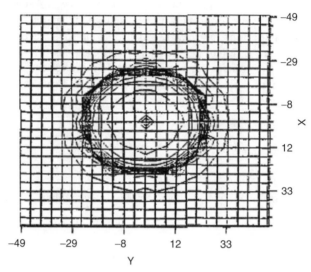

FIGURE 5.82 Reconstruction of the cross section of a circular ($a = 25$ mm) dielectric cylinder with a small amount of absorption (the axial-symmetry function of coordinates) carried out by the Fourier–Bessel transformation.

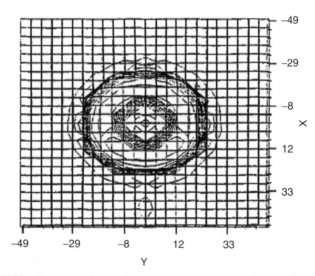

FIGURE 5.83 Reconstruction of the cross section of a pipe-type circular ($a = 25$ mm; $b = 10$ cm) dielectric cylinder with a small amount of absorption (the axial-symmetry function of coordinates) carried out by the Fourier–Bessel transformation.

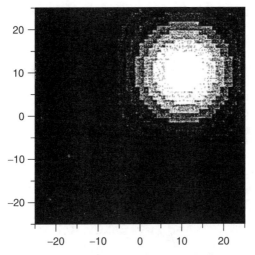

FIGURE 5.84 Reconstruction of the cross section of a circular ($a = 10$ mm) dielectric cylinder with a small amount of absorption which is not located at the center.

field $U = U_I + U_S$ [where $U_S(\vec{r})$ is a scattered wave] satisfies the given wave equation,

$$\Delta u + k^2 (1 + f)^2 u = 0 \tag{5.90}$$

and the boundary condition at infinity (Δ is the Laplacian). The scattered field U_S can be found using Eq. (5.90) in the first-order Born approximation. So for the scattered field, $U_S^{(1)}(\vec{r})$ can be written in integral form:

$$U_S^{(1)}(\vec{r}) = \int G(\vec{r} - \vec{r}') Q(\vec{r}') U_I(\vec{r}') \, d\vec{r}' \tag{5.91}$$

where $G(\vec{r} - \vec{r}') = (i/4) H_0^{(1)}(k|\vec{r} - \vec{r}'|)$ is the Green's function, which is a solution of the differential equation

$$\Delta G + k^2 G = -\delta(\vec{r} - \vec{r}') \tag{5.92}$$

Here $\delta(\vec{r} - \vec{r}')$ is Dirac's δ-function with an $(\vec{r} - \vec{r}')$ argument, $H_0^{(1)}$ the Hankel function of the first type of zeroth order, and $Q(\vec{r}) = k^2(2f + f^2)$. According to Eq. (5.92), the Green's function is the radiation field of a two-dimensional point source; than the expression (5.91), which is the scattered wave field, can be considered as a superposition of the fields of all two-dimensional point sources located in the region where $f \neq 0$. An integral representation of Eq. (5.91) of the scattered field $U_S^{(1)}(\vec{r})$ is permitted under the following condition:

$$U_S \ll U_I \tag{5.93}$$

Equation (5.90) can also be solved in the framework of a Rytov approximation of first order. In this approximation the total U_R is written

$$U_R = U_I \exp(k\Phi_R) \tag{5.94}$$

where Φ_R is a function assuming that $|\nabla\Phi_R|^2 \ll 1$, with ∇ the gradient operator. The value $w_R = U_I\Phi_R$ satisfies the expression

$$\Delta w_R + k^2 w_R = -\frac{1}{k}Q(\vec{r})U_I \tag{5.95}$$

and consequently, w_R has the form of the integral in Eq. (5.91):

$$U_I\Phi_R = \frac{1}{k}\int G(\vec{r}-\vec{r}')Q(\vec{r}')U_I(\vec{r}')\,d\vec{r}' \tag{5.96}$$

or

$$U_I(k\Phi_R) = \int G(\vec{r}-\vec{r}')Q(\vec{r}')U_I(\vec{r}')\,d\vec{r}' \tag{5.97}$$

A comparison of the integral expressions (5.91) and (5.97) shows that the following relations between $U_S^{(1)}$ and Φ_R (under the conditions of Born and Rytov approximations of the first order) can be applied to the field U:

$$U_S^{(1)} = U_I(k\Phi_R) \tag{5.98}$$

In the inverse scattering problem, the function f should be found with the known scattered field U_S. Solution of such a problem using Eqs. (5.91) and (5.97) allows us to obtain the primary equation of diffraction tomography. Then, if we know $U_S^{(1)}$ or $U_I(k\Phi_R)$, we can find the functions $Q(\vec{r})$ and $f(\vec{r})$ by calculations or experimentally. If the Rytov approximation is used, it is necessary to find $Q(\vec{r})$ from Eq. (5.97) by the known $k\Phi_R$ on a straight line N (Fig. 5.85) when the following function is known:

$$g(\xi, \vec{\theta}) = k\Phi_R(\xi\vec{\theta}_\perp + \eta\vec{\theta}) \tag{5.99}$$

where η is fixed and is greater than OB the largest cross-sectional dimension of the object under investigation), $\vec{\theta}_\perp$ and $\vec{\theta}$ are orts having the directions of the rectangular axis $O\xi$, $O\eta$. The scalars ξ and η are rectangular Cartesian coordinates of the vector $O\vec{u} = \xi\vec{\theta}_\perp + \eta\vec{\theta}$ in the $\vec{\theta}_\perp, \vec{\theta}$ system and define the lengths of vectors $\xi\vec{\theta}_\perp$ and $\eta\vec{\theta}$, respectively. The coordinate system ξ, η is connected to the propagation direction of the incident wave. If the direction of the ort is changed (rotating the coordinate system ξ, η relatively to the moving Cartesian rectangular system of coordinate x,y, where the original function is defined), the direction of the wave vector of this wave always coincides with the direction of the ort θ. According to Natterer [67],

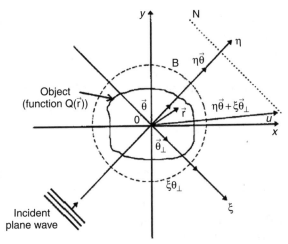

FIGURE 5.85 Data collection in the space domain for obtaining a tomographic image of an object characterized by a function using the Fourier method.

the following relation between Fourier images of functions $Q(\vec{r})$ and $g(\xi, \vec{\theta})$ (in the frequency domaini Fig. 5.86) is valid at fixed $\vec{\theta}_\perp$ and $\vec{\theta}$:

$$\hat{Q}(\vec{K}_c) = -\left(\frac{2}{\pi}\right)^{\frac{1}{2}} i \frac{\sqrt{k^2 - K_\xi^2}}{k^2} \exp\left[i\eta\left(k - \sqrt{k^2 - K_\xi^2}\right)\right] \hat{g}(K_\xi, \vec{\theta}) \qquad (5.100)$$

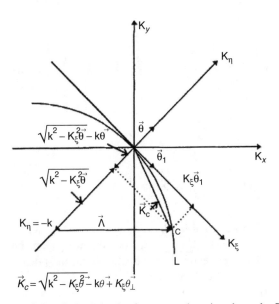

FIGURE 5.86 Part of the circle L in the frequency domain where the Fourier image of the function is obtained using Eq. (5.100).

where the vector $\vec{K}_c = K_\xi \vec{\theta}_\perp + \left(\sqrt{k^2 - K_\xi^2} - k\right)\vec{\theta}$ is defined in the $\vec{\theta}_\perp, \vec{\theta}$ vector system, K_ξ is the projection of the vector $\vec{K}_c$ on the axis OK_ξ, η is the length of the vector $\eta\vec{\theta}$ in the ξ, η coordinate system $k = \omega/c$, ω is the radiation angular frequency, and c is the velocity of light.

In Eq. (5.100), Fourier transformations are determined by the formula

$$\hat{f}(\gamma) = (2\pi)^{-n/2} \int_{R^n} \exp(ix\gamma)f(x)\,dx \qquad (5.101)$$

where $f \in L_1(R^n)$ and R^n is n-dimensional Shwartz space. The function $\hat{g}(K_\xi, \vec{\theta})$ is the result of integration of the function $g(\xi, \vec{\theta})$ by ξ in R^1 at fixed $\vec{\theta}$, and the function $\hat{Q}(\vec{K}_c)$ is the result of the integration of function $Q(\vec{r}')$ (where $\vec{r}' = \xi'\vec{\theta}_\perp + \eta'\vec{\theta}$) by ξ' and η' in R^2 at fixed vector $\vec{K}_{\bar{n}}$ and $Q(\vec{r}') = 0$ if $\vec{r}'$ does not belong to the region of the object under investigation.

When $\vec{K}_\xi$ changes from $-k$ to k, the end of the vector (Fig. 5.86) describes a semicircle with a center $-k\vec{\theta}$ which is located at the center of the coordinates. By changing the vector direction $\vec{\theta}$ and having data of $g(\xi, \vec{\theta})$, the Fourier image $\hat{Q}(\vec{K}_c)$ of the object under investigation is obtained in the frequency domain and in a circle with radius not less than k. Thus, if a finite resolution is possible or we neglect space frequencies higher than k, Eq. (5.100) is solved for an inverse scattering problem under the conditions of the Rytov approximation.

Expression (5.100) is a main equation of diffraction tomography. A similar ratio for the Born approximation in Eq. (5.91) was obtained by the convolution method [74] and has the following form for an incident wave propagating along the positive direction of the y-axis:

$$\hat{Q}(\vec{K}_c) = 2i\frac{\sqrt{k^2 - K_\xi^2}}{k^2} \exp\left(-i\sqrt{k^2 - K_\xi^2}\,\eta\right)\hat{g}(K_\xi, \eta) \qquad (5.102)$$

where

$$\hat{g}(K_\xi, \eta) = \int U_S(x, y = \eta)\exp(-iK_\xi x)\,dx\,dy \qquad (5.103)$$

$$\hat{Q}(\vec{K}_c) = \iint Q(\vec{r})\exp[-i(\vec{K}_c \cdot \vec{r})]\,dx\,dy \qquad (5.104)$$

The function $U_S^{(1)}(x,y)$ defined by the ratio (5.91) is supposed to be known at the straight line $y = \eta = \text{const} > 0$ in the x,y coordinate system. From a comparison of Eq. (5.100) (where the vector $\eta\vec{\theta}$ is directed along the positive direction of the y-axis) and Eq. (5.102), it can be seen that when the requirement in Eq. (5.98) is fulfilled, these expressions agree with the sign of U_S. Therefore, the same image of the object being investigated will result following inverse Fourier transformation.

Object Function Calculation Method Let the incident wave propagate along the positive direction of the η-axis of the rectangular Cartesian coordinate system ξ, θ, where the phase and amplitude of electromagnetic fields U and U_I and the straight line $\eta = \text{const} > OB$ are measured. Let the center of another rectangular Cartesian coordinate system x,y connected with the object coincide with the center of the ξ, η coordinate system and with the center of a circle O having radius OB (see Fig. 5.85). The object being investigated is characterized by the function $Q(\vec{r})$, which differs from zero in a certain region inside the circle O. Then, with the agreement of (5.100) and (5.102), and rotating the ξ, η coordinate system relative to the x,y coordinate system (also using the data measured for U and U_I), the Fourier image of the function in the frequency domain in a circle with radius not less than k can be found. An analogous result can be obtained if the system of data collection remains immovable and the object under investigation is rotating relative to the center O (Fig. 5.87), and hence the x,y coordinate system rotates relative to the ξ, η coordinate system.

When numerical methods of function investigation are used for inverse Fourier transformation, it is necessary to know the Fourier image of $\hat{Q}(\vec{K}_c)$ of an object at specific points in the ξ, η plane (these points are shown in Fig. 5.89 for the case when the Simpson formula is used for calculation of double definite integrals). It can be seen that it is possible to find positions when each of the points shown will be located on the circle L at values $K_\xi > 0$ of the vector $\vec{K}_c$ projection on the K_ξ-axis by changing the angle θ.

K_ξ and the angle θ, which are necessary for calculation of the Fourier image $\hat{Q}(\vec{K}_c)$ of the object, are obtained in the following manner. The equation of the

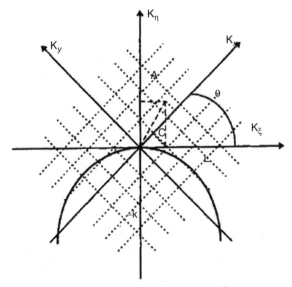

FIGURE 5.87 Scheme for finding of the Fourier image $\hat{Q}(\vec{K}_c)$ of the function $Q(\vec{r})$ when the object under investigation is rotated about the data measurement system.

circle in the K_ξ, K_η coordinate system, where selected points in the K_x, K_y coordinate system must be located, has the form

$$(K_\eta + k)^2 + K_\xi^2 = k^2 \tag{5.105}$$

Using the formulas for transformation from the K_x, K_y system to the K_ξ, K_η system,

$$K_\xi = K_x \cos\theta - K_y \sin\theta \tag{5.106}$$

$$K_\eta = K_x \sin\theta + K_y \cos\theta \tag{5.107}$$

the following equation may be found:

$$\frac{K_x^2 + K_y^2}{2k} = -K_\eta \tag{5.108}$$

Substituting K_η in Eq. (5.105), we can find K_ξ as

$$K_\xi = \pm\sqrt{K_x^2 + K_y^2}\sqrt{1 - \frac{K_x^2 + K_y^2}{4k^2}} \tag{5.109}$$

We use $K_\xi \geq 0$; thus, positive values are chosen for the root in Eq. (5.109). Using Eq. (5.109), it is followed that under the condition

$$\sqrt{K_x^2 + K_y^2} > 2k \tag{5.110}$$

the values of K_ξ are imaginary complex values. Equation (5.109) provides real values of K_ξ for all points K_x and K_y from the region under consideration, since in the K_x, K_y coordinate system, the region with a radius not less than k is considered. From Eqs. (5.106) and (5.107), an expression for $\tan\theta$ can be written

$$\tan\theta = \frac{K_\chi K_\eta - K_y K_\xi}{K_\chi K_\xi + K_y K_\eta} \tag{5.111}$$

K_ξ is defined by Eq. (5.109) and K_η is defined by the expression

$$K_\eta = \sqrt{k^2 - K_\xi^2} - k \tag{5.112}$$

Values of the K_x and K_y coordinates in Eqs. (5.109) and (5.111) are supposed to be known and defined by the given integration method when the inverse Fourier transformation is carried out. To diminish calculation errors for the function, the integration region is divided into regions using the inverse Fourier transformation. A scanner allowed us to obtain data about a scattered field with constant steps $\Delta\theta = 30°$ in the region $\theta(0 \leq \theta \leq 2\pi)$. For the Fourier image calculation $\hat{Q}(\vec{K}_c)$ at defined points using the Simpson formula, interpolation of the data is necessary.

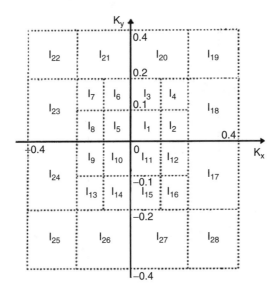

FIGURE 5.88   Integration region in the frequency domain.

Figure 5.88 shows the region of integration (a rectangular region with dimensions 0.8×0.8 mm^{-2}) in the K_x, K_y coordinate system, that was used in calculations of the function $Q(\vec{r})$ by the inverse Fourier transformation. Near the coordinate center, the region is divided into $S_1 = 16$ square cells (0.1×0.1 mm^{-2}). The rest of the region is divided into $S_2 = 12$ square cells with cell dimensions 0.2×0.2 mm^{-2}. The total number of integration cells is $S = S_1 + S_2 = 28$. Simpson's formula was used for each cell, which made it possible to find integral I:

$$I = \int_a^b \int_c^d f(x, y) \, dx \, dy = \frac{(b-a)(d-c)}{36} [f(x_0, y_0) + f(x_2, y_0) + f(x_0, y_2)$$
$$+ f(x_2, y_2)] + 4[f(x_1, y_0) + f(x_0, y_1) + f(x_2, y_1)$$
$$+ f(x_1, y_2)] + 16 f(x_1, y_1) \tag{5.113}$$

where $x_0 = a$, $x_1 = (a+b)/2$, $x_2 = b$, $y_0 = c$, $y_1 = (c+d)/2$, and $y_2 = d$, knowing the values of the complex function f of real variables in nine points of a cell (Fig. 5.89). Figure 5.89 shows the integration region I_l in the K_x, K_y coordinate system and nine points where it is necessary to know the Fourier image $\hat{Q}(\vec{K}_c)$ of the function $Q(\vec{r})$ according to Eq. (5.113).

The total integral giving a value of $Q(\vec{r})$ at a point (x,y) of the x,y coordinate system is obtained as a result of the summation of $N_1 = 16$ integrals of $I_l(l = 1, \ldots, 16)$ or $N_1 + N = 28$ integrals $I_l(l = 1, \ldots, 28)$. The interpolation procedure is shown in Fig. 5.90, where we see a layout of junctions when the function $g(\xi, \theta)$ of two variables ξ and θ is interpolated. It is supposed that the function values in

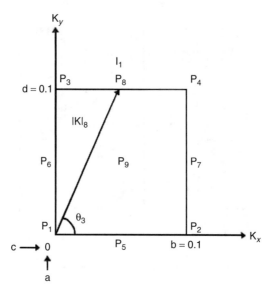

FIGURE 5.89   Integration region I_l in a K_x, K_y coordinate system and the points that are necessary to know in the function, according to Eq. (5.113).

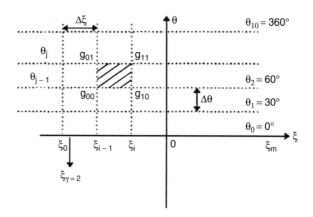

FIGURE 5.90   Interpolation scheme.

four junctions $(\xi_{i-1}, \theta_{j-1})$, (ξ_{i-1}, θ_j), (ξ_i, θ_{j-1}), and (ξ_i, θ_j) within the rectangular cells are known. Interpolation was carried out using the formula

$$f(\xi, \theta) = (1-p)(1-q)f_{00} + q(1-p)f_{10} + p(1-q)f_{01} + pqf_{11} \qquad (5.114)$$

where

$$p = \frac{\theta - \theta_{j-1}}{\Delta\theta} \qquad \Delta\theta = \theta_j\theta_{j-1} = \pi/6; \quad j = 1, \ldots, n;$$

$$\theta_0 = 0; \quad \theta_n = 2\pi \qquad (5.115)$$

$$q = \frac{\xi - \xi_{i-1}}{\Delta\xi} \qquad \Delta\xi = \xi_i - \xi_{i-1}; \qquad i = 1, \ldots, m;$$

$$\xi_0 = \xi_{min}; \qquad \xi_m = \xi_{max} \tag{5.116}$$

ξ_{min} and ξ_{max} are the minimum and maximum values of the variable ξ. f_{00}, f_{01}, f_{10}, and f_{11} are values of the function in the junctions $(\xi_{i-1}, \theta_{j-1})$, (ξ_{i-1}, θ_j), (ξ_i, θ_{j-1}) and (ξ_i, θ_j), respectively.

Equation (5.114) allows us to interpolate the function $f(\xi, \theta)$ with an error proportional to $O(h^2)$ at constant steps of $\Delta\xi = \Delta\theta = h$. Because the measured value is complex, Eq. (5.114) is used twice. It is used first when the real part of the complex function $g(\xi, \theta)$ is interpolated, and second, is used to interpolate the imaging part of the function. In the program for a scan, it is supposed that an object rotates at angle $\Delta\theta$ $(0 \le \theta \le 2\pi)$ and then shifts on $\Delta\xi$ along the ξ-axis in the ξ, η coordinate system connected with an oscillator and receiver of radiation from an initial position characterized by a $\xi_{min} < 0$ coordinate where the distance between the η-axis of the ξ, η coordinate system and a line through the sample center parallel to the η-axis to the final position is characterized by the coordinate $\xi_{max} > 0$. The scanning results are established after obtaining angle θ in accordance with Eqs. (5.109) and the (5.111) and angles θ_{j-1} and θ_j (Fig. 5.90). The values of θ are substituted in Eq. (5.114) and the values of function $g(\xi, \theta)$ in 11 points of $\xi_\gamma(\gamma = 1, \ldots, 11)$ in the interval between ξ_{i-1} and ξ_i for all values of $i(i = 1 \ldots, m)$ where measurements of U and U_I are carried out. Values of $\xi_{\gamma=1}, \xi_{\gamma=11}$ are equal to ξ_{i-1} and ξ_i, correspondingly. To find the function $\hat{g}(K_\xi, \theta)$ using Eqs. (5.100) and (5.103), the function $g(\xi, \theta)$ is integrated by ξ at a given K_ξ (where θ is fixed).

After obtaining a Fourier image $\vec{Q}(\vec{K}_c)$ of the function $Q(\vec{r})$ in each of nine points of integration in the regions I_l $(l = 1, \ldots, 28)$ by formulas (5.100) and (5.102), for each region of I_l formula (5.113) is used. To obtain the resulting function $Q(\vec{r})$ and, consequently, the image (cross section) of the object under investigation, the integrals obtained are summarized.

Conclusions   The image is a reconstructed distribution function in the region of the cross section of the electrodynamic parameters of the object being investigated. The images are obtained using the methods of first-order diffraction tomography for scattering objects that have a simple structure. The reconstruction of images has been shown in the millimeter-wave band, for which the high-frequency approximation in the basic equation of diffraction tomography can be used. In this case, integration in the frequency space region is obtained in finite terms near the origin of the coordinates (e.g., $|K| \le 0.2$ mm^{-1} at $k \simeq 0.7 - 0.8$ mm^{-1}). Expansion of the region of integration in frequency space leads to a substantial increase in the noise level in object space.

Investigations have shown that there are difficulties in the reconstruction of object images of complicated structure (several short-range inhomogenities located in the object). In this case, a multiscattering case can be observed and the methods of first-order diffraction tomography do not allow the reconstructed function to be sufficiently accurately.

5.4.2   Experimental Study of the Multiview Tomographic Method in a Millimeter-Wave Band

The reliability of first-order diffraction tomography in the millimeter-wave band for object sounding (quasi-optical tomography) has been not studied thoronghly. There are some difficulties in the measurement of the amplitude and phase of a scattered field in the millimeter-wave band [0]. On the other hand, the application of millimeter waves in diffraction tomography can increase its reliability considerably. For example, increasing the sounding radiation frequency makeszt possible to obtain a Fourier image of an object at higher frequencies and therefore to improve the quality (resolution) of the tomographic image. New applications of tomography appear, for example, to permit visualization of surface or underground objects. Wave fields in may be represented a quasi-optical approach as narrow beams of plane wave diverging weakly when they propagate in a medium or in free space. Apertures of millimeter-wave antennas forming such beams are comparatively moderate in size ($A \approx 10\lambda$). For sounding volumetric objects, it is very suitable to use shaped quasi-optical beams. Besides, there are waveguides for millimeter waves such as dielectric waveguides in which different modes propagate. Such modes may be represented as a superposition of two plane waves traveling along a waveguide and be considered as a ray when reconstructing tomographic images of a surface object or imaging of an object located under the surface. In this case, a rotating waveguide line with a traveling mode may be located above the surface being investigated at a certain distance from it. In this section, results of the experimental reconstruction of volumetric body cross sections and images of objects by the theoretical methods considered are given in the millimeter-waveregion.

Experimental Setup and Examples of Tomographic Reconstruction   The millimeter-wave tomographic system shown schematically in Fig. 5.91 was used in this study. This system is operated at 30 to 50 GHz and the instruments are controlled by a personal computer that has a laboratory card to create analog/digital, digital/analog, and digital/digital outputs. Two stepper motors are used: to rotate the object and to move the object along the v-axis. Both stepper motors can be controlled through the digital/digital outputs of the computer. The oscillator [60] used in this setup generates signalsat 30 to 50 GHz. The transmitter antenna illuminates the target. Scattered waves from the target are received by the receiver antenna. Signals generated and received by the receiver antenna are detected using square detectors. A scalar network analyzer measures the transmission coefficient value using the outputs of square detectors. A network analyzer has an analog output for amplitude measurements. These analog signals are converted to digital signals using an analog-to-digital converter, and the resulting data are recorded by a personal computer. This procedure is repeated

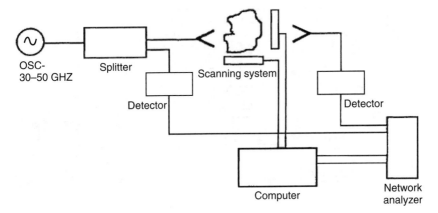

FIGURE 5.91   Microwave tomographic system.

until the object scanning is complete. Then the data measured are processed to get a two-dimensional image.

Figure 5.92 shows reconstructed tomography image of homogeneous dielectric cylinder with radius $a = 17$ mm using the tomographic reconstruction method considered in Section 5.4.1. The function $T(v)$ (5.87) is measured, and it is supposed that the scattering function is $k''(x,y)$. To obtain experimental data for images in Fig. 5.92, two rectangular horns were used as the transmitter and receiver antennas. The millimeter-wave tomographic setup allows us to measure the phase and

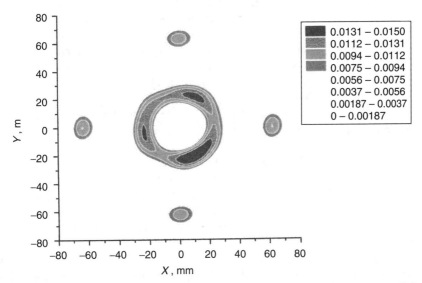

FIGURE 5.92   Experimental millimeter-wave tomographic image of a homogeneous dielectric cylinder cross section.

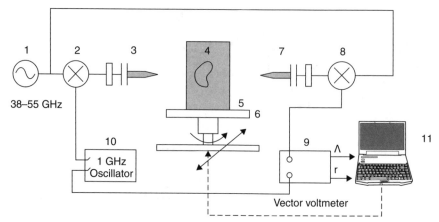

FIGURE 5.93 Tomographic system with dielectric inserted horn antennas that make it possible to measure the phase and amplitude of the total U and incident U_I field at a straight line $\eta = $ const.

amplitude of the U and U_I fields at the straight line $\eta = $ const in the ξ, η coordinate system shown in Fig. 5.93. The difference between this setup and the setup shown in Fig. 5.91 is the use of dielectric antennas to focus the beam. A signal from the oscillator (1) [60] with an operating frequency in the Δv band $\cong 38$ to 55 GHz is divided into two signals by a directional branch. One of these signals goes to the up-convertor (2) and the other to the down-convertor (8). A signal from the reference generator (10) with frequency $v \simeq 1$ GHz is also supplied to the up-convertor (2). As a result, the signal radiated by one antenna (3), which is a dielectric waveguide inserted in a rectangular waveguide [electric vector $\vec{E}$ of the wave is directed along a straight line (6)], is shifted up in frequency 1 GHz relative to the operating frequency of the oscillator (1). The sample (5) can be rotated around its axis and moved along the straight line (6). Step motors are used to move and rotate. The signal received by a second antenna (7) in the absence of sample is sent to the down-convertor (8).

The signal from the operating oscillator (1) reaches the down-converter, and the signal at the output of device (8) is shifted down and its frequency $v_{out} \cong 1$ GHz. This signal feeds the vector voltmeter (9). A signal of frequency $v \simeq 1$ GHz from the generator (10) is also supplied to the voltmeter. Amplitude A and phase Φ measured by the vector voltmeter (9) are equal to the amplitude and phase of the incident field U_I at the straight line (6). Analog signals corresponding to the amplitude and phase measured for field U_I are received by the system (11) to process data received to obtain an object's image. The system also controls the stepper motors.

When an object scatters an incident wave weakly and the Rytov or Born approximation is valid for the scattering field, the signal measured along the straight line (6) will correspond to the total field U received by antenna (7). The U_I measured along the line (6) is also substituted in Eq. (5.100) or (5.102) instead of plane–wave

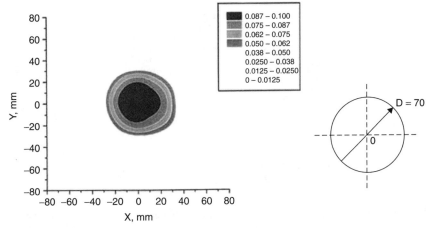

FIGURE 5.94 Reconstructed cross-sectional image of a circular foam plastic cylinder and its physical cross-sectional diameter.

field $e^{i\eta k}$ in the straight line $\eta = $ const. The function $g(\xi, \vec{\theta})$ at $\eta = $ const where $0 \leq \theta \leq 2\pi$, $\Delta\theta = \pi/6$ is obtained from Eq. (5.94) or (5.98). All measurements were carried out at $\eta \simeq 4$ to 5λ.

Figures 5.94 to 5.97(a) show images of the $|Q(\vec{r})|$ cross sections of objects under experimental investigation. The images represent four cylinders, each of height $h \simeq 7\lambda$ ($\lambda \simeq 8$ mm) and diameter $D \simeq 9\lambda$, which have circular cross sections made of foam plastic. Two of these samples also have axial circular holes of diameter

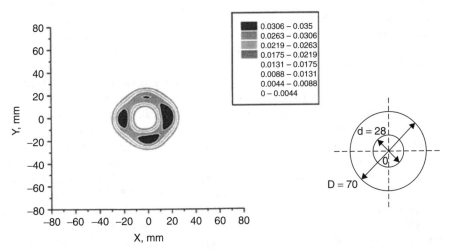

FIGURE 5.95 Reconstructed cross-sectional image of a circular foam plastic cylinder with a circular hole at the center and its physical cross-sectional diameters.

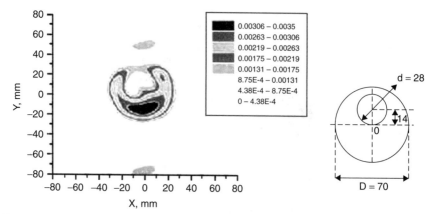

FIGURE 5.96 Reconstructed cross-sectional image of a circular foam plastic cylinder with a circular hole shifted from center and its physical cross-sectional diameters.

$d \simeq 3.5\lambda$ cut out. One is a foam plastic trapezoidal prism of height $h \simeq 7\lambda$ and cross section shown in Fig. 5.97(a).

Figure 5.94 gives the cross-sectional image of the circular cylinder obtained experimentally, which has a diameter $D = 70$ mm. The values of the function $|Q(\vec{r})|$ are shown in the inset. Figure 5.95 to 5.97(a) show similar cross sections of the foam plastic objects investigated. In Figs. 5.97b and 5.98, two analogous images are shown. Figure 5.97b shows the cross-sectional image obtained experimentally of a circular cylinder with a circular hole at the center with a circular slot cut out along the cylinder axis.

The reconstructed cross-sectional images in Figs. 5.94 to 5.98 were obtained using I_l for $l = 1, \ldots, 16$. A zero level of the function is taken equal to 30 to 50% from the maximum. To use all I_l integrals $(l = 1, \ldots, 28)$ improves the image quality and the resolution increases. However, this process also increases the noise level. Figure 5.99 shows cross sections of circular cylinders made of foam plastic and Teflon by choosing all I_l integrals $(l = 1, \ldots, 28)$ for the samples given in Figs. 5.95 to 5.98.

We note that image reconstruction is possible by approximate formulas obtained from Eqs. (5.100) and (5.102) for the condition $(K_\xi/k)^2 \ll 1$. This approximation is analogous to the high-frequency case in the framework of Rytov approximation [67]. In this approximation the function $Q(\vec{K}_c)$ is defined not at the circles, but on straight lines. The linear integrals of the function to be found are described by the scattered field out of the scattering object. When the parameter η is increased gradually from 0 to 5λ and higher, reconstruction of the image according to the data obtained experimentally using Eqs. (5.100) and (5.102) shows that the object image also deteriorated. At the same time, approximate formulas (5.100) and (5.102), obtained at $(K_\xi/k)^2 \ll 1$, do not depend on the parameter η and the image reconstruction is near the actual image. In Fig. 5.100a and b, one can see the reconstructed cross-sectional images of the circular cylinders made of foam plastic shown in Figs. 5.95, and 5.97b before high-frequency approximation. The

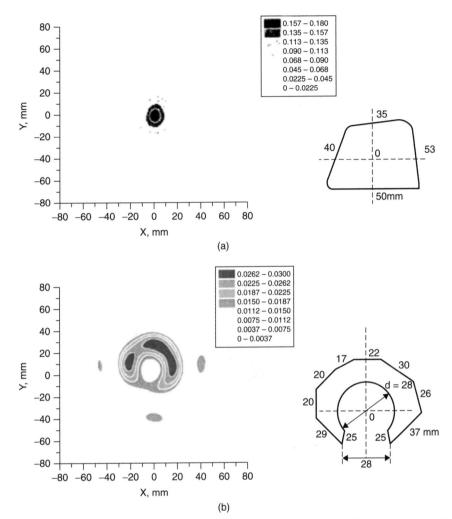

FIGURE 5.97   Reconstructed cross-sectional images of (a) a trapezoidal prism made of foam plastic and (b) a circular cylinder with a circular slot at the center and their physical cross-sectional dimensions.

approximation in Eqs. (5.100) to (5.102) and integrals I_l at $l = 1, \ldots, 16$ were used. Figure 5.101 gives the same images, but using I_l at $l = 1, \ldots, 28$.

Tomographic Imaging of Objects Using Open Waveguides   A method for solving the two-dimensional inverse scattering problem of a cylindrical scattering object has been described [69]. The object can be characterized by its electrodynamical parameters $\varepsilon(x, y)$, $\mu(x, y)$, and $\sigma(x, y)$, where ε and μ are the dielectric permittivity and magnetic permeability, respectively, and σ is the conductivity. All the parameters depend on the coordinates x and y. This object is under the surface

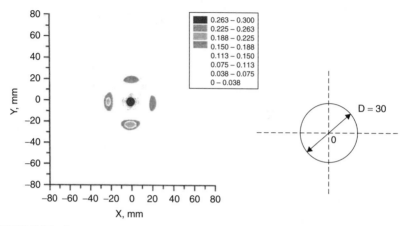

FIGURE 5.98 Reconstructed cross-sectional image of a circular cylinder made of Teflon and its physical cross-sectional diameter.

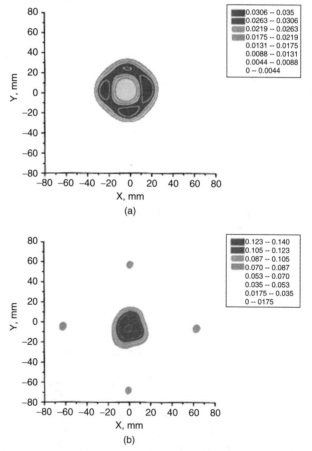

FIGURE 5.99 Reconstructed cross-sectional images of circular cylinders of foam plastic using I_l at $l = 1$ to 28.

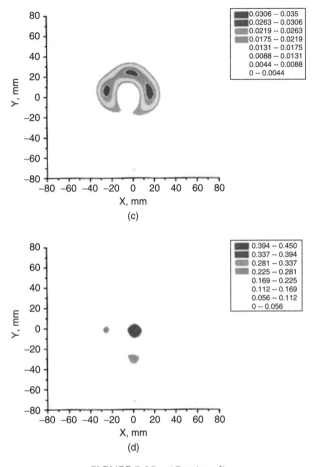

FIGURE 5.99 (*Continued*)

of homogeneous medium 2, characterized by ε_2, μ_2, and σ_2. A radiation source irradiates the object and a receiver receives the scattered field, which are located in homogeneous semi-infinite medium 1, characterized by ε_1, and μ_1 over the surface of medium 2. Media 1 and 2 have a common border. In this configuration the origin of the x,y coordinate system is at the boundary between media 1 and 2 at $x = 0$. The x-axis is directed from the surface into medium 2. The z-axis is directed along the cylindrical element and is perpendicular to the x,y plane. Electric vector $E_0\vec{z}$ of an incident wave is parallel to the z-axis. It is supposed that the field $U_I = E_0$ of the incident wave is known when the object is absent. If the scattering object is inserted into medium 2, a scattered field $E_S\vec{z}$, which occurs due to scattering of the incident wave by the object under investigation, is observed over the surface of the medium 1/medium 2 boundary. This field is assumed to be known at a straight line L and is a function of coordinate y at a distance $|x_0| = $ const. from the surface.

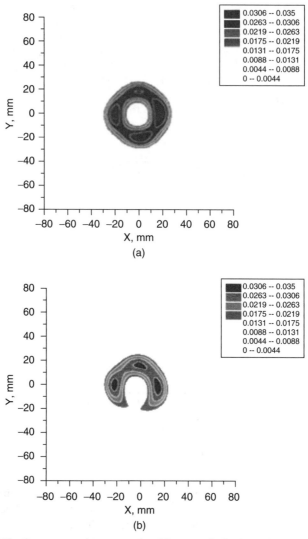

FIGURE 5.100 Reconstructed cross-sectional images of circular cylinders of foam plastic using I_l for $l = 1$ to 16 and the approximation in Eqs. (5.100) and (5.102).

The solution of the inverse problem described above is of practical importance, due to its connection with a Fourier-based diffraction tomographic (FBDT) scheme [69]. Obtaining data about the incident field and the field on the straight line L by changing the radiation frequency or incident angle θ, the field U_I at the surface can be calculated using an FBDT scheme for the object function

$$\chi(x, y) = [\sigma(x, y) - \sigma_2] - i\omega[\varepsilon(x, y) - \varepsilon_2] \qquad (5.117)$$

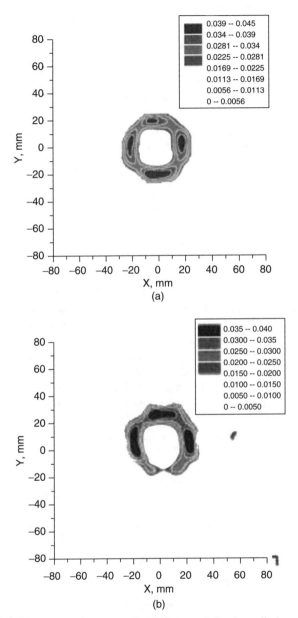

FIGURE 5.101 Reconstructed cross-sectional images of circular cylinders of foam plastic using I_l for $l = 1$ to 28 and the approximation in Eqs. (5.100) and (5.102).

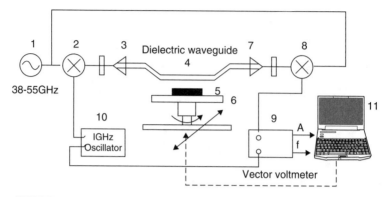

FIGURE 5.102 Tomographic setup using a dielectric waveguide structure.

In this way, an image of the dielectric properties of the object in the x,y -plane can be obtained.

There is a possibility of obtaining a similar image of an object located under the surface of medium 2 if the electrodynamical properties of this object do not change much. Block diagrams for obtaining tomographic images of such objects are shown in Figs. 5.102 and 5.103. These systems operate like the tomographic setup shown in Fig. 5.93. But instead of dielectric antennas (3) and (7), two waveguide pieces, (3) and (7), for matching line (4), which is a rectangular dielectric waveguide for the setup shown in Fig. 5.102 and a copper wire (Sommerfeld line) for the setup shown in Fig. 5.103, are employed. The waveguide lines (dielectric and copper wire) transmit the wave in a straight line. It is known [78] that for a lossy transmission-line, the amplitudes of fields $\vec{E}$ and $\vec{H}$ decrease exponentially along the wave propagation. For a distance interval Δz, the decrease in the transmitting power is

$$\Delta P = -2\alpha P \Delta z \tag{5.118}$$

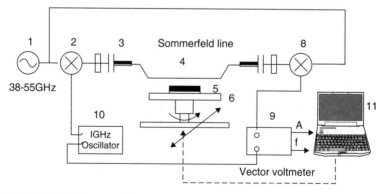

FIGURE 5.103 Tomographic setup using a Sommerfeld line waveguide structure.

where α is the attenuation coefficient of the wave, which is constant for a given transmission line. If $\alpha = \alpha(z)$ is a function of the coordinate z, Eq. (5.118) becomes a form similar to Eq. (5.87):

$$P = P_0 \exp\left[-2 \int_{z_0}^{z_1} \alpha(z)\, dz\right] \tag{5.119}$$

where P_0 is the transmitting power at the starting point z_0 of the line. The distance interval Δz is defined by the waveguide line with length $l = |z_1 - z_0|$ and P is the transmitting power at the endpoint z_1 of the line.

Let us consider an object under investigation which is in disk form and is characterized by the complex electrodynamical parameters $\varepsilon = $ const and $\mu = $ const. The object is located at a certain distance t below the waveguide line. In a scan, a shift in the waveguide line or the object is performed; the transmitting power P will be a function of this shift along the straight line perpendicular to the propagation wave direction. Thus, after measuring the ratio P/P_0, the projection of the function $a(x,y)$ onto the v-axis at a given angle θ can be found. Angle θ is an angle between the positive direction of the u and x axes of Cartesian coordinate systems x,y and u,z when the u,z coordinate system is rotated relative to the motionless coordinate system x,y connected with this disk. The function $\alpha(x, y)$ can be obtained by employing the Fourier method of tomographic reconstruction. In a more general case, for the reconstruction function

$$k_z(x, y) = \beta(x, y) + i\alpha(x, y) \tag{5.120}$$

where $\beta(x, y)$ is the phase constant. Information about amplitude and phase changes of guided wave after its interaction with the object located near the waveguide can be used.

Experimental images [the reconstructed function $\alpha(x, y)$, using the high-frequency approximation in Eqs. (5.100) to (5.102)] of ferrite disks placed under a dielectric waveguide are shown in Fig. 5.104. The distance between the top surface of the disk and the lower surface of the waveguide is fixed ($t \simeq 2$ mm). In Fig. 5.105, the experimental images of two ferrite disks placed under a Sommerfeld line (wire) at the frequency $\nu \simeq 38$ GHz are shown.

Conclusions The experimental results obtained by the tomographic systems described above show that the cross-sectional tomographic images of volumetric dielectric objects in millimeter wavelengths are possible. The images are obtained using methods of first-order diffraction tomography for weakly scattering objects that have simple structures. But even the first approximation for the scattered field and information obtained by measurements of its phase and amplitude allow us to study comparatively small objects [an object diameter of $D \simeq 9\lambda$; an inhomogeneity of dimension $d \simeq 4\lambda$ ($\lambda \simeq 8$ mm)] at a refractive index of the material of $n \simeq 1.03$ to 1.1. Employment of the fundamental equation of diffraction tomography (without approximation) and expansion of the integration region in frequency space do not essentially improve the results of reconstruction of the function.

5.4.3 Microwave Imaging of Immersed Dielectric Bodies: An Experimental Survey

Electromagnetic radiation such as x-rays is a well-known technique for CT tomography imaging and is widely used for diagnosis. However, x-rays are ionizing radiation and they can cause side effects to the human body. Microwaves can be used for imaging instead of x-rays, but the resolution is low. Although imaging with microwaves has low resolution, microwaves have special importance for subsurface detection and imaging, such as mine detection. The main reason for using microwaves in subsurface imaging is good penetration of electromagnetic waves into soil at microwave frequencies. Thus, imaging with microwaves is useful [5].

In the literature, the diffraction tomographic imaging technique was applied to biomedical diagnosis and geophysical diffraction tomography using electromagnetic and ultrasonic waves. In Devaney's paper [79], the foundations of geophysical

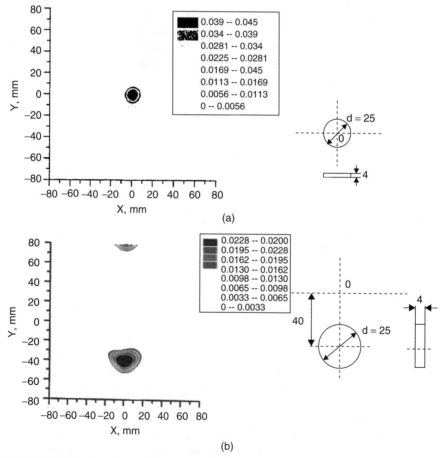

FIGURE 5.104 Experimental images of ferrite disks placed under a dielectric waveguide and their cross sections.

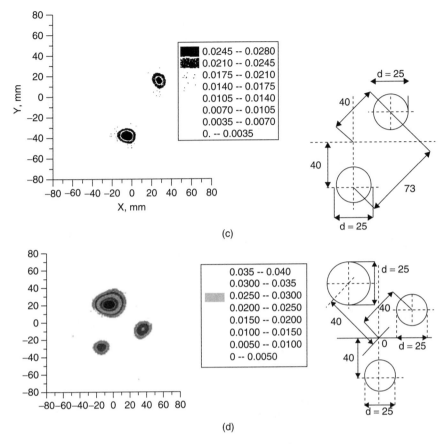

FIGURE 5.104 (*Continued*)

diffraction tomography for offset vertical seismic profiling and well-to-well tomography were presented for weakly scattering inhomogeneous formations on which Born or Rytov approximation can be employed. Low-level microwaves have also been used for the same purpose during the last two decades. There is a fundamental difference between tomographic imaging with x-rays, ultrasound, and microwaves. Being nondiffracting, x-rays travel in a straight line, but in the ultrasound and microwave methods the energy often does not propagate along straight lines [80]. This raises certain difficulties in diffraction tomography in a microwave region, due to reflection and refraction at the boundaries of a body [81]. Additionally, if the dimensions of an inhomogenity are comparable to the wavelength of the radiation, diffraction effects occur. For the last 20 years, the greatest interest in microwave tomography has been for its use in biomedical imaging [94]. Jacobi and Larsen's study [82] can be considered among the first studies of tomography in the microwave region. They described a method for producing microwave images that are improved through the reduction of multipath propagation effects.

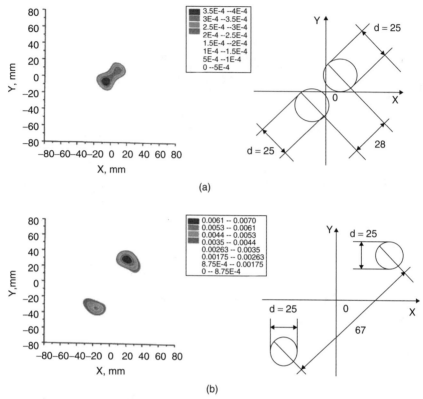

FIGURE 5.105 Experimental images of ferrite structures placed under a Sommerfeld line (wire) and their cross sections.

This method employs water-immersed antennas and microwave time-delay spectroscopy (TDS) at the S-band. After their study, Bolomey and Pichot's group demonstrated the possibilities of microwave diffraction tomography for biomedical applications [70]. An image reconstruction method from a biological object immersed in a homogeneous medium was presented by Baribaud et al. [83]. Later, water-immersed microwave antennas for microwave biomedical imaging were created by two groups [84,85]. A numerical method and experimental technique for microwave imaging of inhomogeneous bodies was presented in a widely cited paper by other members of Pichot and Bolomey's group [38]. Several biomedical prototypes were also created for biomedical applications [71,86,87]. Vertiy and Gavrilov extended diffraction tomography to millimeter-wave bands [58,81,88]. In these studies, millimeter-wave band tomographic images of some objects were obtained. There have also some recent investigations of microwave imaging. An inverse scattering method for reconstructing the contours of PEC cylinders of arbitrary cross section based on contour deformations using a level set formalism were presented by Ferraye et al. [89]. Some recent methods to improve image quality have also been announced. The first, an improved Rytov approximation and nonlinear optimization technique, was developed by Kechribaris et al. [90]. They applied their method numerically to both direct and indirect inverse problems, and it showed significant improvement over conventional diffraction tomographic methods. Another method to improve image quality was proposed by Tanaka et al. [91]. They use numerically produced lowpass filtered time-domain data and applied a forward–backward time-stepping (FBTS) algorithm. Both papers are numerical investigations and have not yet been applied to experiments with real data.

As mentioned earlier, diffraction tomography provides a method of reconstructing the cross sectional images of dielectric objects. The tomographic technique uses two main approximations, the Born and Rytov approximations. These approximations are first order and are used to reconstruct images of weakly scattered objects. Two restrictions due to the Born and Rytov approximations were formulated and simulated computationally in work by Slaney et al. [74] and Kak and Slaney [92]. It is shown that while first-order Born and Rytov approximations can produce excellent reconstructions for small objects with a low refractive index, they both break down quickly when their assumptions are violated. The simulations of Slancy et al. showed the first-order Born approximation to be valid for objects where the product of the relative refractive index and the diameter of the cylinder is less than 0.5λ, and the first-order Rytov approximation to be valid with essentially no constraint on the size of the object; however, the relative refractive index must be less than a few percent. These assumptions were extended for lossy dielectrics in Paoloni's paper [93]. The results of his study suggested that the reconstruction procedure is most accurate when the loss tangent of the outside medium is chosen to approximate that expected within the object to be imaged.

In this work, our aim is to investigate microwave images of some dielectric objects immersed in a medium using the first-order diffraction tomographic algorithms and to show the limitations of the Born and Rytov approximations experimentally. Although only free space was used as a medium in former studies, we also used sand as a second immersion medium in this study. The dielectric objects and surrounding medium are assumed to be lossless here. Experiments were done for cylindrical objects of different shapes and dielectric constants in two different immersion media (air and sand). We used wood, rolled soft paper, rolled paper, and poly (vinyl chloride) (PVC) as materials to be imaged.

Imaging Immersed Object Cross Sections by First-Order Diffraction Tomography   The basic equation of diffraction tomography can be obtained as a result of an inverse problem solution of scattering by a plane electromagnetic wave propagating to the object under investigation [67,73,92]. We consider a two-dimensional scattering object composed of a region filled by an immersion medium in which an investigating region (object) is located, characterized by the refractive index $n(\vec{r}') = 1 + n_\Delta(\vec{r}') + n_\delta(\vec{r}')$, where the function $n_\Delta(\vec{r}')$ is equal to zero outside the region filled with an immersion medium and the function $n_\delta(\vec{r}')$ is the difference between refractive indexes of the immersion and object mediums. The function $n_\delta(\vec{r}')$ is equal to zero outside the object region. An incident harmonic plane wave, $U_I(\vec{r}) = \exp[ik(\vec{\theta}\vec{r})] \cdot \exp(-i\omega t)$, illuminates the object, $\vec{\theta}$ is a unit vector pointing in the direction of the wave propagation, and $k = \omega/c$ (ω is the radiation frequency and c is the velocity of light).

In the case of direct scattering, the total field $U = U_I + U_S$ (where $U_S(\vec{r})$ represents the scattered field) satisfies the given wave equation

$$\nabla^2 U + k^2(1 + f)^2 U = 0 \tag{5.121}$$

and the boundary condition in infinity. $f = n_\Delta(\vec{r}') + n_\delta(\vec{r}')$ and ∇^2 is the Laplacian in the equation. The scattered field U_S outside the object can be found using Eq. (5.121) and is written in integral form:

$$U_S(\vec{r}) = \int G(\vec{r} - \vec{r}')Q(\vec{r}')U(\vec{r}')\,d\vec{r}' \tag{5.122}$$

where $G(\vec{r} - \vec{r}') = (i/4)H_0^{(1)}(k|\vec{r} - \vec{r}'|)$ is the Green's function. It is a solution of the differential equation

$$\nabla^2 G + k^2 G = -\delta(\vec{r} - \vec{r}') \tag{5.123}$$

where $\delta(\vec{r} - \vec{r}')$ is Dirac's δ-function of the argument $\vec{r} - \vec{r}'$. $H_0^{(1)}$ is a Hankel function of the first type of zero order, $Q(\vec{r}') = k^2[2n_\delta(\vec{r}') + n_\delta^2(\vec{r}') + 2(1 + n_\delta(\vec{r}'))n_\Delta(\vec{r}') + n_\Delta^2(\vec{r}')]$, and $U(\vec{r}')$ is the field within the object. As the Green's function is a radiation field of a two-dimensional point source according to Eq. (5.123), the scattered wave field expression in Eq. (5.122) can be considered as a superposition of fields of all two-dimensional point sources located in the region where $f \neq 0$. If the difference $n_\delta(\vec{r}')$ is small, the function $Q(\vec{r}')$ can be written in the form

$$Q(\vec{r}') = Q_0(\vec{r}') + Q_1(\vec{r}') \tag{5.124}$$

where

$$Q_0(\vec{r}') = k^2[2n_\Delta(\vec{r}')] \quad \text{and} \quad Q_1(\vec{r}') = k^2[2n_\delta(\vec{r}')] \tag{5.125}$$

Equation (5.122) becomes in this case

$$U_S(\vec{r}) = \int G(\vec{r} - \vec{r}')[Q_0(\vec{r}') + Q_1(\vec{r}')]U(\vec{r}')\,d\vec{r}' = U_{S0}(\vec{r}) + U_{S1}(\vec{r}) \tag{5.126}$$

where

$$U_{S0}(\vec{r}) = \int G(\vec{r} - \vec{r}')Q_0(\vec{r}')U_0(\vec{r}')\,d\vec{r}' \tag{5.127}$$

and

$$U_{S1}(\vec{r}) = \int G(\vec{r} - \vec{r}')Q_0(\vec{r}')U_0(\vec{r}') \, d\vec{r}' \tag{5.128}$$

In Eqs. (5.127) and (5.128), the function $U_0(\vec{r}')$ is the total field in the immersion region. It is supposed that $U_0(\vec{r}')$ is also a plane wave and that the scattered field within the object U_{St} satisfies the condition

$$U_{St} \ll U_0 \tag{5.129}$$

The field $U_{S1}(\vec{r})$ can be considered the Born approximation for a scattered field outside the object region produced by the inhomogeneity $n_\delta(\vec{r}')$. The function $U_{S1}(\vec{r})$ can be written in the form

$$U_{S1}(\vec{r}) \equiv U_{SR}(\vec{r}) = U_0(\vec{r})[\exp(ik\Phi_R(\vec{r})) - 1] \tag{5.130}$$

where $U_{SR}(\vec{r})$ represents the scattered field outside the object in the Rytov approximation [73]; $\Phi_R(\vec{r})$ is a required function and it is supposed that $|\nabla\Phi_R|^2 \ll 1$, where ∇ is a gradient.

If the scattering produced by the source function $Q_0(\vec{r}')$ is eliminated from the scattered field of an object, it is possible to reconstruct the function $Q_1(\vec{r}')$ by using the first-order diffraction tomographic method solution of the scattering inverse problem [67]. The elimination process can be done on a measurement stage as we describe next. Let us consider a setup consisting of a rectangular container (object) located between two dielectric rod antennas (one is irradiating and the other is receiving). The harmonic probing field is a beam formed by an irradiating antenna. This field illuminates a point x in front of the container at a normal angle and is characterized by the complex amplitude U_I. According to Eq. (5.126), the field measured at the same point x at the opposite side of the container can be written

$$U_S(x) = U_I(x)\tau + U_{S1}(x) \tag{5.131}$$

where τ is the complex transmission coefficient of the system. It is possible to present Eq. (5.131) in the form

$$\alpha U_S(x) = \alpha U_I(x)\tau + \alpha U_{S1}(x) = \alpha U_0(x) + \alpha U_{S1}(x) \tag{5.132}$$

where α is the correction coefficient obtained from the calibration condition so that $\alpha U_I(x)\tau = 1$ (when the object is absent). The following result is obtained from Eq. (5.132):

$$\alpha U_S(x) = 1 + \alpha U_0(x)[\exp(ik\Phi_R(\vec{r})) - 1] \tag{5.133}$$

One can see that the function $\Phi_R(x)$ is determined from the equation

$$\alpha U_S(x) = \exp[ik\Phi_R(x)] \tag{5.134}$$

where $\alpha U_S(x)$ is the corrected measured field.

In an inverse scattering problem, the function $Q_1(\vec{r}')$ should be found using the function $ik\Phi_R(x)$. Solution of such a problem using Eq. (5.128) allows us to obtain the main equation of diffraction tomography [38,71,80,81]. After that, we may find functions $Q_1(\vec{r}')$ and $n_\delta(\vec{r}')$. In this work, the tomographic algorithm used for imaging of object cross sections was constructed on the base of the Born and Rytov approximation for a scattered field, Eq. (5.130), and main equation of diffraction tomography was used [80,81]. The image function is $\varepsilon(x', y') = [1 + n_\delta(x', y')]^2 \simeq 1 + 2n_\delta(x', y')$ calculated after obtaining the function $1/k^2 Q_1(\vec{r}')$. The calculation method of the function $Q_1(\vec{r}')$ is identical to the method described by Gavrilov and Vertiy [81]. The following lines show how to calculate the dielectric permittivity distribution of the object $\varepsilon(x', y')$.

Let us consider this problem for a model of the experiment conducted. In Fig. 5.82, the setup structure is shown in the central cross-sectional plane. There are two rectangular systems of coordinates x', y' and x, y in this plane. The centers of the coordinate systems are superposed and located on the antenna axis at the center O between them. The probing wave propagates along the y-axis. It is assumed that this wave is a plane wave and that field data can be measured by a receiver at a point on the probing line that is parallel to the x-axis at $y = y_0$ ($y_0 > 0$). It is necessary to note that after correction of the measured field, Eq. (5.93) can be imagined as being a field formed by an incident wave propagating in free space with complex amplitude $U_I(x) \equiv 1$ on the probing line. In this case the object is also located in free space and its refractive index in the central cross-sectional plane is characterized by the function $1 + n_\delta(\vec{r}')$ in the x', y'-system of coordinates. This system can rotate on the angles ϕ_i relative to the fixed x, y-system of coordinates, in which measurements are conducted.

The function $Q_1'(\vec{r}') \equiv 1/k^2 Q_1(\vec{r}') \equiv 1/k^2 Q_1(x', y')$ is defined on all planes so that it is not equal to zero in the region of the object but does equal zero without the object. The Fourier transform of this function is written

$$\tilde{Q}_1'(K_{x'}, K_{y'}) = \int_{-\infty}^{+\infty} \int_{-\infty}^{+\infty} Q_1'(x', y') \exp[-i(K_{x'}x' + K_{y'}y')] \, dx' \, dy' \quad (5.135)$$

where $K_{x'}$ and $K_{y'}$ are space frequencies. The function $\tilde{Q}_1'(K_{x'}, K_{y'})$ is defined on a plane in Fourier space. The rectangle system of coordinates $K_{x'}$ and $K_{y'}$ with the center at point O_F is given in this plane. If the integral in Eq. (5.135) exists, the function $Q_1'(x', y')$ can be presented by using the inverse Fourier transform as

$$Q_1'(x', y') = \frac{1}{(2\pi)^2} \int_{-\infty}^{+\infty} \int_{-\infty}^{+\infty} \tilde{Q}'_1(K_{x'}, K_{y'}) \exp[i(K_{x'}x' + K_{y'}y')] \, dK_{x'} \, dK_{y'}$$

$$(5.136)$$

The function $Q_1'(x', y')$ can be also written in the x, y-system of coordinates as

$$Q_1'(x', y') = Q_1'[x'(x, y), y'(x, y)] = Q_2'(x, y) \quad (5.137)$$

where the functions $x'(x, y)$ and $y'(x, y)$ are defined by the coordinate transformation at the rotation of the x', y'-coordinate system relative to the fixed x, y-coordinate frame. The function $Q_1'(x', y') \equiv Q_2'(x, y)$ if $\phi_i = 0, 2\pi$, as the coordinate systems agree at these angles. The Fourier transform of the function $Q_2'(x, y)$ is $\tilde{Q}_2'(K_x, K_y)$, and it is defined by the integral in Eq. (5.136) in the rectangular system of coordinates K_x and K_y on the same plane in the Fourier space and the $K_{x'}$, $K_{y'}$-coordinate system. The coordinate systems in the Fourier plane have a common central point O_F. If the x', y'-system of coordinates is rotated about the fixed x, y-system of coordinates on the angle ϕ_i, the $K_{x'}$, $K_{y'}$-coordinate frame is rotated in the Fourier plane on the same angle about the fixed K_x, K_y-coordinate system as

$$\tilde{Q}_1'(K_{x'}, K_{y'}) = \tilde{Q}_1'[K_{x'}(K_x, K_y), K_{y'}(K_x, K_y)] = \tilde{Q}_2'(K_x, K_y) \qquad (5.138)$$

Here the functions $K_{x'}(K_x, K_y)$ and $K_{y'}(K_x, K_y)$ are defined by the same coordinate transformation with the functions $x'(x, y)$ and $y'(x, y)$(rotation of axes on the angles ϕ_i). The function $\tilde{Q}_1'(K_x', K_y')$ is identical to the function $\tilde{Q}_2'(K_x, K_y)$ at $\phi_i = 0; 2\pi$. There is the relation [74] between the function $\tilde{Q}_2'(K_x, K_y)$ and the Fourier transform $\tilde{g}(K_x)$ of the field $g(x) \equiv \ln[\alpha U_s(x)]_{\phi_i=0}$ at the angle $\phi_i = 0$:

$$\tilde{Q}_2'(K_x, K_y) = \frac{2i\sqrt{k^2 - K_x^2}}{k^2} \exp\left[-iy_0\sqrt{k^2 - K_x^2}\right] \tilde{g}(K_x) \qquad (5.139)$$

The field $g(x)$ can be considered in Eq. (5.139) as the normalized scattered field in the first-order Born approximation or as the function $ik\Phi_R(x)$ on use of the Rytov approximation for the normalized scattered field. In the K_x, K_y-system of coordinates, points K_c with coordinates K_x and K_y in Eq. (5.139) belong to the semicircle C defined by the equation

$$K_y = -k + \sqrt{k^2 - K_x^2} \qquad (5.140)$$

The semicircle center is located on the K_y-axis at $K_y = -k$. Hence, using Eq. (5.140) and the formula for the rotation transformation, it is possible to obtain from Eq. (5.139) that

$$K_x = K_{x'}\cos\phi - K_{y'}\sin\phi \qquad (5.141)$$

$$K_y = K_{x'}\sin\phi + K_{y'}\cos\phi \qquad (5.142)$$

at a position of the object in the x, y-system of coordinates to find the function $\tilde{Q}_1'(K_{x'}, K_{y'})$: namely, the Fourier transform of the object function desired in the $K_{x'}$, $K_{y'}$-system of coordinates at the points defined by Eqs. (5.140) to (5.142).

Let us consider next the situation when measurements are conducted at the angles $\phi_i = i \Delta\phi, i = 0, 1, \ldots, m$ with an angle step $\Delta\phi$ ($\phi_0 = 0$; $\phi_m = 2\pi$). The function $\tilde{g}(K_x)$ in Eq. (5.139) obtained at each angle ϕ_i is designated as $\tilde{g}_{\phi_i}(K_x)$.

As in the Fourier plane, the $K_{x'}$, $K_{y'}$-system of coordinates is rotated on the same angles. At each angle ϕ_i a set of points $K_{c'}$ with coordinates $K_{x'}$ and $K_{y'}$ will belong to the semicircle C defined in the K_x, K_y-system of coordinates. If we know the coordinates $K_{x'}$ and $K_{y'}$ of points $K_{c'}$, it is possible find the function $\tilde{Q}'_1(K_{x'}, K_{y'})$ from Eq. (5.139). It can be shown using Eqs. (5.140) to (5.142) that corresponding coordinates K_x and K_y and the angle ϕ_i are defined by the expressions

$$K_x = \pm\sqrt{K_{x'}^2 + K_{y'}^2}\sqrt{1 - \frac{K_{x'}^2 + K_{y'}^2}{4k^2}} \tag{5.143}$$

$$K_y = -\frac{K_{x'}^2 + K_{y'}^2}{2k} \tag{5.144}$$

where $\sqrt{K_{x'}^2 + K_{y'}^2} \leq k\sqrt{2}$ and

$$\tan\phi_i = \frac{K_{x'}K_y - K_{y'}K_x}{K_{x'}K_x + K_{y'}K_y} \tag{5.145}$$

It is sufficient to take a positive value for the root in Eq. (5.143), as the rotation angles were in the range $0 \leq \phi_i \leq 2\pi$. The object function $Q'_1(x', y')$ was obtained by the inverse Fourier transform of the function $\tilde{Q}'_1(K_{x'}, K_{y'})$ defined in Eq. (5.136). In this case, integration area was given as 0.4×0.4 mm^{-2} (for the operating frequency $f = 8$ GHz) in the $K_{x'}$, $K_{y'}$-system of coordinates. The area center agrees with the center O_F. The area sides are parallel to the coordinate axes. Near the center O_F the area was divided into $S_1 = 16$ square cells (0.05×0.05 mm^{-2}). The rest of the area was divided into $S_1 = 12$ square cells of 0.1×0.1 mm^{-2}. Nine points (four in corners, four in the center of the sides, and one in the center) were selected in each cell. The function $\tilde{Q}'_1(K_{x'}, K_{y'})$ was obtained at all points of the integration area using the technique described. An interpolation procedure has been developed to obtain the function $\tilde{g}_\phi(K_x)$ at the angles $\phi \neq \phi_i$, calculated using Eq. (5.145). In the inverse Fourier transform, the Simpson formula was applied to each cell of the area, and the 28 integrals obtained were summarized. The distribution $\varepsilon(x', y')$ has been calculated as

$$\varepsilon(x', y') = 1 + Q'_1(x', y') \tag{5.146}$$

As mentioned earlier, it is necessary to assume that the object is weakly scattered, so that either the Born or the Rytov approximation can be used [92]. This restriction can be formulated separately for the Born and Rytov approximations. The first-order Born approximation is valid for objects where the product of the deviation from the average of the refractive index n_δ and the diameter of the cylinder d is less than 0.5λ [74,92]; that is,

$$n_\delta d < \frac{\lambda}{2} \tag{5.147}$$

where λ is the wavelength of the radiation in the medium and n_δ is given by

$$n_\delta = \frac{|n_{\mathrm{ob}} - n_m|}{n_m} \qquad (5.148)$$

where n_{ob} and n_m are the refractive index of the object and medium, respectively, and d is the diameter of the object.

On the other hand, former simulations [74,92] show that the Rytov approximation is very sensitive to the relative refractive index and can easily deteriorate with a small change in the relative refractive index, but produces excellent reconstructions for objects as large as 100λ. The first-order Rytov approximation is valid with essentially no constraint on the size of the object; however, n_δ must be less than a few percent [74,92].

$$n_\delta \lesssim 0.03 \qquad (5.149)$$

Experimental Setup and Results   In this study we compare the images of four dielectric objects in air and in sand, which are used as the immersion media. The objects' cross sections are in the shape of a ring and a rectangle. The experimental setup used to reconstruct microwave images of dielectric objects located in the surrounding medium (i.e., sand) is shown schematically in Figs. 5.106 and 5.107.

The setup consists of a PC, a network analyzer, a step motor control unit, an antenna system, an imaging cylindrical object, a plastic box (3) filled with sand, a mechanical linear scanner (1) run by a step motor (2), another step motor (5) for

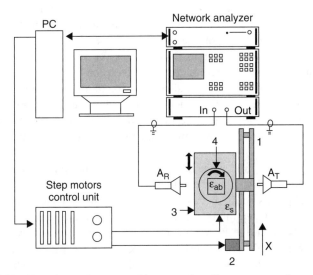

FIGURE 5.106   Experimental tomographic setup: 1, a linear scanner; 2, a step motor for linear scanning; 3, a plastic box filled with sand (the container); 4, a cylindrical PVC pipe and transmitter and receiver antennas after [5].

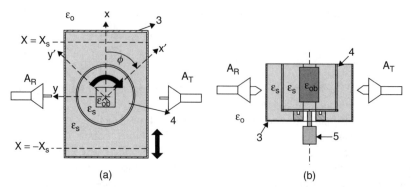

FIGURE 5.107   Details of a plastic box filled with sand (the container): (a) top and (b) side views. 3, a plastic box; 4, a cylindrical PVC pipe; 5, a step motor that turns the object and the PVC pipe. ε_{ob} and ε_s refer to the dielectric permittivity of object and sand, respectively after [5].

rotating the imaging object, and a PVC pipe (4). The symbols ε_{ob}, ε_s, and ε_o refer to the relative dielectric permittivities of the object, sand, and air, respectively. The network analyzer is composed of a sweep generator, a measurement set, and a frequency converter. This system operates in the frequency range 4 to 12 GHz. Only data at 8 GHz ($\lambda = 3.75$ cm) are used for image reconstruction. The step motor control unit controls both step motors. One of them scans linearly, and the other rotates the imaging object in the medium. The step motor for linear scanning is placed on the mechanical linear scanner, and the step motor for rotating is placed under the plastic box. The linear scanner moves the container with the object in the x-direction. Similarly, the other step motor rotates the object in the range 0 to 360°. The step motors are completely computer controlled by the step motor control unit. The antenna system consists of transmitting (A_T) and receiving (A_R) antennas, which have the same construction. Each antenna includes a metallic rectangular waveguide and partially inserted dielectric plate in the waveguide. Each dielectric plate operates as a longitudinal irradiator and longitudinal receiver, respectively, and these plates are set opposite each other so that their axes are coaxial. The antennas in the setup are fixed close to the wall of the box filled with sand. The transmitting antenna forms a narrow beam that propagates inside the sand and illuminates the object.

The top and side views of the container in which the imaged object is placed are shown in Fig. 5.107. In the container, there are two parts filled with sand. The imaged object is located inside the sand in the first part, and this part is rotated by the second step motor (5). The second part of the container is comprised solely of sand between the PVC pipe and the outer wall of the box, and does not turn. The pipe is used here to turn the object easily in the sand and to cancel the friction between the sand and the object due to the rotation. The container is moved by equal steps in the x-direction between two A_T and A_R antennas for each angle position of the object in the box. A special program is used to control the equipment drivers and collect experimental data during the measurement. Another program, based on the first-order diffraction tomographic algorithm using the Born and Rytov approximations for a scattered field is applied to process row data for image reconstruction of the object cross section.

Data collection is carried out as follows. The initial position of the angle ϕ is fixed at $0°$ (Fig. 5.107), and the center of the container is positioned at the point $x = x_s = 0.15$ m (the starting position of the linear scan). Then the complex amplitude of the signal passed through the container is measured and is calibrated by computer as $U_0 = 1.0 + i0.0$. It is supposed that the field scattered by the object is absent in this position. Then the container is scanned with the space step $\Delta x = 0.005$ m up to the final position $x = -x_s = -0.15$ m. On each step in scanning, the complex signal measured is multiplied by the adjusting multiplier received from the first scanning step. Then the container returns to the initial position, the object turns with an angle $\Delta \phi = 20°$, and the measurement repeats itself. Therefore, the number of angular slices is equal to 19. The function Φ_R in Eq. (5.129) is identical to $0.0 + i0.0$ in the absence of an object in a completely filled container, as the complex amplitude of the measured signal $\alpha U_S(x)$ defined in Eq. (5.134) is equal to $1.0 + i0.0$. Hence, the contribution $Q_0(\vec{r}')$ of the area filled by the immersion environment in the object function $Q(\vec{r}')$ in Eq. (5.124) is zero. Measured complex signal distinguished from $U_0 = 1.0 + i0.0$ in the presence of an object makes it possible to find the function Φ_R at all angular occupation of the object to find the function $Q_1(\vec{r}')$ after the restoration process.

For images of the object cross sections in air, a container filled with sand is not used to reconstruct images of the objects in air. It was used only for imaging an object cross section when an object was placed in sand. The gray scale shows the levels of the image function $\varepsilon(x, y) = [1 + 2n_\delta(x, y)]$ in the figures presented below. In Fig. 5.108 one can see a reconstructed microwave diffraction tomographic image of rolled soft paper ($\varepsilon_{ob} = 1.2$) in air only when a container is not used. A similar microwave image of rolled soft paper in air was also obtained in an earlier study [58]. Figure 5.108 is formed to show the capability and resolution of the tomographic setup for different cross sections of the object. The vertical (a) and horizontal (b) images show cross-sectional images constructed at a variety of starting positions. These positions are illustrated right sides of each image. Figure 5.108 proves that the tomographic setup gives us quite precise images for an object of this size ($d \simeq 3\lambda$), and it is possible to reconstruct images of weakly scattered objects that have different cross sections.

The structure of the object can easily be defined from the image. Because the resolution of the images is half a wavelength (i.e., 3.75 cm; the frequency is 8 GHz in our experiments), it is difficult to define details for smaller, more complex objects. All the other images shown in each figure were constructed both in air and in sand. One can see the microwave images of an object's cross sections in air and in sand: of cylindrical rolled soft paper (Fig. 5.109), rolled paper (Fig. 5.110), a rectangular wood block (Fig. 5.111), and hollow PVC (Fig. 5.112).

The restriction criteria for the Born and Rytov approximations were formulated in Eqs. (5.147) to (5.149). We plan to show the effect of these two restrictions on image quality. If a reconstructed image is not like the original cross section and

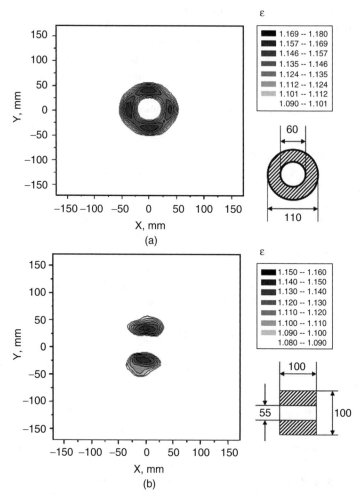

FIGURE 5.108 Reconstructed cross-sectional images of cylindrical rolled soft paper in air when it is (a) vertical and (b) horizontal. The real cross sections are shown on the right-hand side (in millimeters). ε presents the levels of an image function in gray scale after [5].

differs from the original object's dimensions, it is expected that one or two of the restrictions may be violated.

The deviation of the relative refractive index (n_δ) and its multiplication by the diameter of the object ($n_\delta d$) are calculated and tabulated separately in Table 5.2. The value of d is calculated as the sum of the radial thickness for cylindrical ring objects (rolled soft paper, rolled paper, and PVC) and the arithmetic average of two adjacent sides of a rectangular object (wood) in Table 5.2. Comparing each reconstructed and original cross section of the objects from Figs. 5.109 to 5.112, plus and minus signs are written for each image in Table 5.2. A plus sign represents a good resulting image, in that the reconstructed image is nearly identical to the

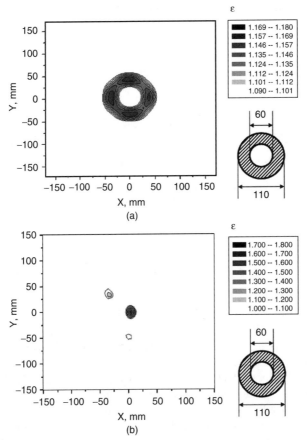

FIGURE 5.109 Reconstructed cross-sectional images of cylindrical rolled soft paper in (a) air and (b) sand. The real cross sections are shown on the right-hand side (in millimeters). The symbol ε presents the levels of an image function in gray scale after [5].

original cross section of the object, shown on the left side for each image. A minus sign represents a distorted image, with the reconstructed image not matching the original cross section. According to Table 5.2, we can test the relation against the success of the reconstructed image and the Born and Rytov restrictions. Comparing the n_δ and $n_\delta d$ values calculated against the conditions of Eqs. (5.147) and (5.149), it is found that the Born and Rytov approximations give us true predictions for objects made from wood, rolled soft paper, and rolled paper. These objects have a smaller deviation from the relative refractive index than that of PVC. Thus, we can expect that both the Born and Rytov approximations are valid for these types of objects but not for PVC. Although the Born approximation provides an accurate prediction, the Rytov approximation does not. This situation can be explained as follows. The PVC sample has a higher relative refractive index value than the others; thus, the Rytov approximation is completely violated. Additionally, the PVC sample has a different shape than the others. The thickness of the material walls

TABLE 5.2 Limitation Conditions of the Born and Rytov Approximations and Quality Image

Object	d(cm)	$n_{ob} = \sqrt{\varepsilon_{ob}}$	Medium	$n_m = \sqrt{\varepsilon_m}$	$n_\delta \gtrsim 0.03$ Rytov App.	$n_\delta d$ (>1.8750 for Air, >1.859 for Sand) Born App.	Success of Image[a]
Wood	6.75	1.5811	Air	1.0003	0.5806	3.8900	−
	(1.80λ)		Sand	1.5811	0.0000	0.0000	+
Rolled soft	5.00	1.0954	Air	1.0003	0.0951	0.1141	+
paper	(1.33λ)		Sand	1.5811	0.3072	1.5360	−
Rolled	5.50	1.5811	Air	1.0003	0.5806	3.1933	−
paper	(1.47λ)		Sand	1.5811	0.0000	0.0000	+
PVC	0.50	1.9235	Air	1.0003	0.9229	0.4615	+
	(0.13λ)		Sand	1.5811	0.2166	0.1083	+

[a]The plus sign represents an accurate reconstructed image: a minus sign, an inconsistent image.

is quite a bit smaller than that of rolled paper. It means that the dimensions of the PVC sample are smaller than those of the others. Because the Born approximation gives better results for smaller objects, it is obvious that the Born approximation may give us better results for a PVC sample.

On the basis of these results, it is fair to say that the Born and Rytov approximations give us true predictions if the limitations on microwave diffraction imaging of objects immersed in a medium are not violated.

There are also some experimental limitations on a diffraction tomographic experiment [92]: (1) limitations caused by ignoring evanescent waves, (2) limitations caused by the finite data sampling rate, (3) limitations caused by the finite receiver length, and (4) limited physical views of the object. The first can be overcome by choosing higher frequencies. The finite sampling rate limitation is caused by the finite size of the receiver. To prevent this situation, a point receiver must be used instead of a large antenna. The error due to the finite receiving length can be ignored by choosing a long enough receiving line length when collecting data. But we do not consider these limitations further here.

Conclusions Cross-sectional images of some dielectric objects of different cross-sectional shapes and dielectric properties located in air and sand were obtained. Data from an 8-GHz frequency ($\lambda = 3.75$ cm) were used in the experiment and the algorithms of microwave first-order diffraction tomography were used to reconstruct cross-sectional images. The limitations of the first-order Born and Rytov approximations were tested experimentally. If we ignore the experimental limitations, it is proved that the Born approximation is valid for small objects such as thin cylindrical rings and for objects that deviate widely in refractive index. The Rytov approximation gives a more accurate estimation for small deviations in the relative

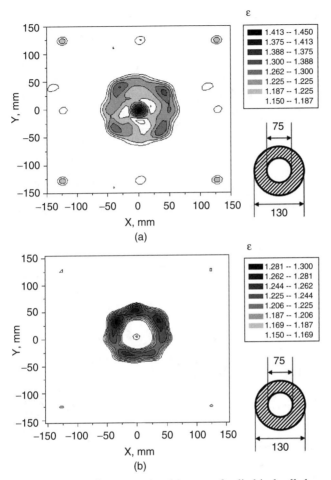

FIGURE 5.110 Reconstructed cross-sectional images of cylindrical rolled paper in (a) air and (b) sand. The real cross sections are shown on the right-hand side (in millimeters). The symbol ε presents the levels of an image function in gray scale after [5].

refractive index values and is also valid for large objects such as thick cylindrical hollow or filled objects if the first condition is satisfied.

These results are consistent with earlier theoretical investigations [74,92]. To get better images, the experiments show that we have to take into account the limitations of the theoretical simulations obtained by first-order diffraction tomography. In addition, imaging of objects located in a medium is possible using the first-order diffraction tomographic method and can be employed for any imaging purpose, such as military, biomedical, and NDT (nondestructive testing). The technique is especially promising to detect landmines embedded in different soil structures by the proper use of scanning and imaging software. It can also be utilized nondestructively to locate defects such as cracks and bubbles in a dielectric medium.

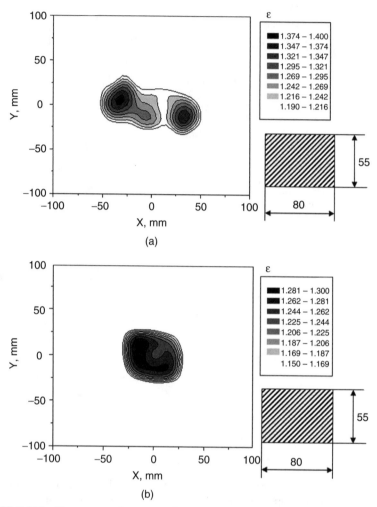

FIGURE 5.111   Reconstructed cross-sectional images of a rectangular wood block in (a) air and (b) sand. The real cross sections are shown on the right-hand side (in millimeters). The symbol ε presents the levels of an image function in gray scale after [5].

5.4.4   Subsurface Microwave Imaging Using the Angular Spectrum of an Electromagnetic Field

Microwave imaging is a technology that has enormous potential, and many studies have been published in recent years. The most popular topics in this field are biological imaging [82–87,94,], nondestructive testing [95–98], and geophysical applications [99]. Subsurface detection of buried objects using microwave imaging has also been developed by some researchers [43,49,58,99,100].

The microwave images of object cross sections located under the surface of the ground can be obtained using diffraction tomography (DT) if the complex

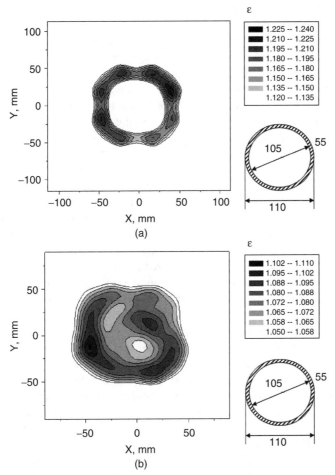

FIGURE 5.112 Reconstructed cross-sectional images of hollow PVC in (a) air and (b) sand. The real cross sections are shown on the right-hand side (in millimeters). The symbol ε presents the levels of an image function in gray scale after [5].

amplitude of a backscattered field on a direct line parallel to the ground surface is known (calculated or measured). [43,100]. However, in practice there are a series of problems. The first problem is the measurement of a scattered field, which may have a complex structure. The second problem is the creation of an antenna that ensures irradiation adequate to the computational algorithm. The third problem is the need to find the Fourier transformation of the backscattered field in the tomographic algorithm. If there is an error in the value measured for the complex amplitude of a scattered field and there are noises in the measuring system, the reconstructed image can be deformed.

It is also necessary to emphasize that inverse electromagnetic problems are ill-posed problems, which can be a primary reason for instability in the input data (i.e., the values measured may have large errors) or if the parameters chosen for these problems come from regions of instability.

A new approach to solving the third problem described above, reconstruction of microwave images of subsurface objects by tomography, is offered herein. The necessary measurement of the complex amplitude of a scattered field at points along a direct line is excluded in the method considered. Also excluded is calculation of direct Fourier transformation of the backscattered field. It is suggested that the plane-wave spectrum of the backscattered field be measured and used in the tomographic image reconstruction algorithm. Only the inverse Fourier transform for the plane-wave spectrum is necessary to reconstruct the image. It should be mentioned here that in this case the inverse Fourier transform of incomplete spectrum data obtained from the experiment may also cause the image to deteriorate. Using measurements of high spatial frequency to obtain information connected with evanescent fields, which decay exponentially on distance from an excited object, is a problem. All these questions require further investigation and are not considered further here.

A single point scatterer may be considered as a point source for the electromagnetic field described by the Green's function, and total radiation from a scattering inhomogenity is the field superposition created by all the point sources of a field, which are distributed in a region of inhomogeneity. We can try to employ the tomographic algorithms to reconstruct not only images of the scattering objects, but also images of the radiating objects under the surface. Generally, the method of image reconstruction described can be employed to obtain an image of any electromagnetic field source placed under the surface of the medium being studied.

A dipole source is used for imaging in our experiments. The problem of radiation of an arbitrary source placed into a medium also has some applications. Response of a point source in the presence of cylindrically stratified geometry was considered by Govel and Chew [102]. It can be used to study the effect of coating on the radar scattering cross section of buried objects. In geophysics, such a model can be used to predict the response of a source or antenna in the vicinity of a borehole drilled for exploration. A loop antenna in a borehole was used to determine the dielectric constant of rock formations [103]. Calculations of electromagnetic fields due to a loop antenna embedded in a dissipative medium (e.g., soil, rock, forest, seawater) have been the subject of a problem regarding the radiation of a source into a medium [104]. Jacobi et al. [84] and Guo et al. [85] considered water-immersed microwave antenna systems for medical imaging as examples of an active source placed into a medium.

Another example of an active source in a medium is a leaky buried coaxial cable. Some results of theoretical and experimental investigations of the radiation properties of buried leaky coaxial cable guarding a radar system have been presented by Blaunstein et al. [105] for various artificial and natural local inhomogeneous conditions along a cable system. These applications show the importance of investigating active sources in a medium.

Further, we concentrate on solving the imaging problem connected with an electromagnetic wave source embedded in a medium by measuring the propagating part of the angular spectrum of the electromagnetic field above the surface being investigated. We note that a scattering and a radiating source (an active dipole) have been placed underground in our experiments. But However, in the known DT algorithms, it is supposed that active sources are absent in the medium. Therefore, we should have constructed a new algorithm for our case. In general, our problem is very complex, so we used a standard DT algorithm (without change) for imaging subsurface object cross sections [43,100], to show that images can be obtained by using the measured components of the plane-wave spectrum of an electromagnetic field above the surface of the medium. In this case we assume that an active source is absent in the medium, but that there is an incident plane wave that travels in the direction of the normal to the medium surface and that the frequency of this wave coincides with the operating frequency of the active source.

Experiments were carried out at frequencies from 3 to 4 GHz [101]. The following steps were taken to obtain the images:

- Excitation of a dipole source placed in the medium
- Rotation (scanning along the circuit) of the receiving antenna with respect to the surface normal
- Measurement of the real and imaginary parts of the transmission coefficients (the spectrum of the plane waves) as dependent on the inclination angle of the receiving antenna at all frequency points
- Storage of experimental data in PC data files
- Calculation of the object image from the spectrum data

Data on the transmitted field at 32 frequencies in the operating band (with a constant frequency step) were used. A box 1.5×1.0 m^2 and 0.4 m deep containing objects was filled with sand. The cross-sectional images of the irradiating dipoles considered as the scattering object in the subsurface region were reconstructed from the measurement data. Experiments were carried out at frequencies from 3 to 4 GHz.

Theoretical Considerations Next we consider the image processing. For the image reconstruction, the plane-wave spectrum of a measured field is used. The field $\psi(x, y_1)$ measured at the line $y = y_1$ above the surface in the air (the one-dimensional case is shown in Fig. 5.113a) is represented in the form of the Fourier integral

$$\psi(x, y_1) = \int_{-\infty}^{\infty} \hat{\psi}(v, y_1) \exp(2\pi i x v) \, dv \qquad (5.150)$$

where $\hat{\psi}(v, y_1)$ is the Fourier transform of $\psi(x, y_1)$ and is defined as

$$\hat{\psi}(v, y_1) = \varphi(v) \exp(-i\gamma_1 y_1) \qquad (5.151)$$

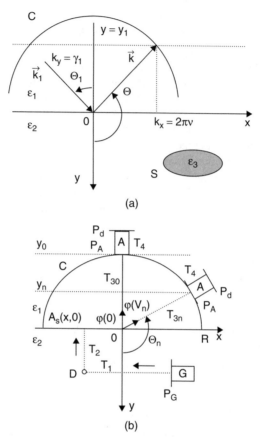

FIGURE 5.113 (a) Region under consideration in the xy-plane and projections of a plane-wave vector $\vec{k}$ from the plane-wave spectrum of a measured field on the axes of reference. (b) Measurement scheme.

Here $\gamma_1 = \sqrt{k^2 - (2\pi v)^2}$, where $k = \omega/c$ is the wavenumber of the plane wave in free space, ω is the cyclic frequency, and c is the velocity of light. The function $\varphi(v)$ is the angle spectrum of the measured field. The variable v is the space frequency in Eq. (5.150), and it also defines the direction of propagation of the plane wave in expansion of the measured field in terms of the plane waves. It is seen from Fig. 5.113a that the following relation connects value v and angle θ:

$$2\pi v = \frac{\omega}{c}\sin\theta \tag{5.152}$$

The function $\varphi(v)$ may be written as

$$\varphi(v) = c_1(v)c_2(v) \tag{5.153}$$

where

$$c_1(v) = \frac{ik_2^2 T}{\gamma_1 + \gamma_2} \tag{5.154}$$

T is the Fresnel transmission coefficient of the boundary between medium 1 and medium 2 with dielectric permittivities $\varepsilon_1 = \varepsilon_0$ (air) and $\varepsilon_2 = \varepsilon_{r2}\varepsilon_0$, respectively, ε_0 is the dielectric permittivity of a vacuum, ε_{r2} is the relative dielectric permittivity of medium 2. $\gamma_2 = \sqrt{k_2^2 - (2\pi v)^2}$, where $k_2 = \omega^2 \varepsilon_2 \mu_0 + i\omega\mu_0\sigma_2$, k_2, and σ_2 are the wavenumber and conductivity of medium 2, respectively, and μ_0 is the magnetic permeability of a vacuum.

The function $c_2(v)$ may be written in integral form as

$$c_2(v) = \iint_S K(x', y') \exp[-2\pi i(\alpha x' + \beta y')] \, dx' \, dy' \tag{5.155}$$

where $\alpha = \alpha(v, \omega, \theta_1)$ and $\beta = \beta(v, \omega, \theta_1, \varepsilon_{r2}, \sigma_2)$ are functions of the variable v. The quantities ω, θ_1, ε_{r2}, and σ_2 are parameters. $\varepsilon_{r2} = 3.0$, $\sigma_2 = 0$, $\theta_1 = 0$, and θ_1 is the angle of incidence of the conjectured plane wave at frequency ω. S denotes that integration is over the cross section of the object under investigation. ω assumes values of $\omega_i (i = 1, \ldots, J)$ in a working frequency band. The image function $K(x, y)$ represents the distribution of normalized polarization current, which is sought in medium 2 using Eqs. (5.153) to (5.155). Note that the results of Chommeloux et al. [100] were used to obtain the formulas in Eqs. (5.154) and (5.155):

$$\hat{\psi}[v(\theta), R \cos\theta] = \varphi[v(\theta)] \exp[-i\gamma_1(\theta)R \cos\theta] \tag{5.156}$$

Our aim is to obtain the function $\varphi[v(\theta)]$ from measurement of the complex amplitude of the plane wave (a component of the spectrum of the plane waves of a measured field) on a circle C of radius R (Fig. 5.113b, which shows the scheme of such a measurement).

Let us consider this scheme. We suppose that the x-axis divides the free space into two infinite half-spaces 1 and 2 filled by homogeneous dielectrics having dielectric constants ε_1 (air, $y < 0$) and ε_2 (sand, $y > 0$). The plane $y = 0$ is the surface of the sand. There is an irradiating center D(active dipole) under the surface. Generally, this center can be an inhomogenity excited by the field of an electromagnetic wave, whose source is placed in region 1. In our case the generator G(from the network analyzer) exits the center D by using coaxial line with a transmission coefficient T_1. The dipole D radiates an electromagnetic wave in the surface direction of half-space 2. We can characterize the propagation of this wave by the transmittance coefficient T_2 of the dielectric layer between the dipole and the surface of the half-space 2. It is assumed that this coefficient exists because in another case we could not measure the transmission coefficients in our experiment. Actually, if a transmittance coefficient T is defined as $T(x, y) = F_l(x, l)/F_{l_0}(x, l_0)$, where $y = l_0$ and $y = l$ are boundaries of the medium layer and $F_l(x, l)$ and $F_{l_0}(x, l_0)$

are complex functions, then, in general, $T(x)$ is also a complex function of the x-coordinate. Thus, the coefficient T_2 is the function of the x-coordinate and actualizes the space modulation of the electromagnetic wave produced by the source. The wave radiated by the dipole creates the distribution $A_s(x, 0)$ of an electromagnetic field on the surface of half-space 2. If we take the Fourier transform of this function, we obtain the angle spectrum $\varphi[\nu(\theta)]$ of the electromagnetic field in the half-space $y < 0$. Using the network analyzer, we can measure the complex transmission coefficient T between the reference planes P_G (generator) and P_d (detector) for each angle $\theta_n (-m \leq n \leq m$, where n and m are integer numbers and $2m + 1$ is the number of angles, $m = 0, 1, \ldots$) at the scanning site by antenna A along circle C of radius R. The transmission coefficient for $\theta_0(y_0 = -R)$ can be written in the form

$$T_0 = \frac{A_{y_0}}{A_G} = T_1 T_2 \tau_{30} T_{30} T_4 \tag{5.157}$$

where A_{y_0} is the complex amplitude of the electromagnetic field in the reference plane P_D when antenna A when the placed at $y = y_0$, A_G is the complex amplitude of the electromagnetic field in the reference plane P_G, τ_{30} is the coefficient of coupling of the field $A_s(x, 0)$ with the plane wave propagating in the direction of the angle θ_0, T_{30} is the transmission coefficient of the free-space layer between the plane

$$y = \tan(\pi - \theta_0)x \tag{5.158}$$

and the reference plane $P_A(y_0 = -R)$ of antenna A, and T_4 is the transmission coefficient of antenna A.

Analog equations can be written for any position y_n of antenna A on circle C:

$$T_n = \frac{A_{y_n}}{A_G} = T_1 T_2 \tau_{3n} T_{3n} T_4 \tag{5.159}$$

Here the transmission coefficient T_{3n} is defined for the free-space layer between the plane

$$y = \tan(\pi - \theta_n)x \tag{5.160}$$

and reference plane P_A of antenna A at position y_n; τ_{3n} is the coefficient of coupling of the field $A_s(x, 0)$ with the plane wave propagating in the direction of the angle θ_n. The transmission coefficient T_{3n} can be written in the form [46]

$$T_{3n} = \exp\left[-i\left(y_n\sqrt{k^2 - (2\pi\nu)^2} - 2\pi\nu x_n\right)\right] \tag{5.161}$$

The coefficients of coupling τ_{30} and τ_{3n} are defined by

$$\tau_{30} T_2 T_1 A_G = G_0 \tag{5.162}$$

$$\tau_{3n} T_2 T_1 A_G = G_n \tag{5.163}$$

where G_n and G_0 are complex amplitudes of the plane waves propagating in the directions θ_n and θ_0, respectively. They are the angle spectrum of the electromagnetic field in the space $y < 0$. Let us suppose that in the calibration process we can determine the complex coefficient $T_0 = 1$; then

$$T_1 T_2 = \frac{1}{\tau_{30} T_{30} T_4} \tag{5.164}$$

and

$$T_n = \frac{\tau_{3n} T_{3n} T_4}{\tau_{30} T_{30} T_4} = \frac{\tau_{3n}}{\tau_{30}} \tag{5.165}$$

as $T_{3n} = T_{30}$ for reference plane P_A tangential to line C. Using the definitions in Eqs. (5.162) and (5.163), Eq. (5.165) can be rewritten as

$$T_n = \frac{G_n}{G_0} \tag{5.166}$$

Using (5.166), one can see that if we measure the transmission coefficient T_n in each nth position of the antenna A, we can determine the angle spectrum G_n of electromagnetic field in half-space $y < 0$ up to the complex constant G_0:

$$G_n = T_n G_0 \tag{5.167}$$

After that it is possible, using the tomographic algorithm described in Section 5.2, to obtain the image of the center D.

It should be noted here that if we will shift source D in a distance x_0 along the x-axis, we can write the following equation for the Fourier transform of function $A_s(x, 0)$:

$$\hat{F}[A_s(x - x_0, 0)] = \exp[-i2\pi y(\theta)x_0]\hat{F}[A_s(x, 0)] \tag{5.168}$$

where the Fourier transform is denoted by $\hat{F}$. From (5.168) one can see that the change in dipole position in the x-direction brings to the phase changes in the spatial components of the field $A_s(x, 0)$. So if we know the function $A_s(x, 0)$ from the measurements for the dipole position at $x = 0$, we can calculate the spectrum of the dipole field $A_s(x - x_0, 0)$ for the position of the dipole at $x = x_0$.

It should also be mentioned that in our experiment we could not measure the total plane-wave spectrum of the field $A_s(x, 0)$ (only in the limits $120° \le \theta \le 240°$). Additionally, measurements of complex transmission coefficients are carried out with some errors. Therefore, the image of the displaced source from center ($x = 0$) can be reconstructed with the position error if x_0 is large enough.

Experimental Setup   The experimental setup shown in Fig. 5.114 has the parts:
(1) network analyzer, (2) microwave amplifier, (3) receiving antenna, (4) irradiating source (transmitter dipole antenna), (5) semicircular scanner, (6) step motor
control block, (7) foam plastic box filled with sand, (8) table for the fixing of the
scanner, and (9) absorber. A dielectric rod inserted into the WR-238 waveguide is
used as the receiving antenna, and the irradiating source is a dipole in our experiments. The setup operates as follows. Signal from the network analyzer operating
in the "transmission" regime exits the irradiating source. The signal transmitted is
received by the receiving antenna, which is fixed on the scanner and moves along
a semicircle of radius $R \simeq 0.4$ m with constant angle steps of $\Delta\theta = 1.7°$ in the
range $120° \leq \theta \leq 240°$. This radius was chosen because we measure the propagating part of the plane wave spectrum of the field $A_s(x, 0)$, and we need a distance
between the receiving antenna and the origin of the x, y coordinates of approximately several wavelengths. The angle step value $\Delta\theta$, the displacement values
$\Delta x = 0.01$ m, and the step $\Delta y = 0.05$ m on the reconstruction of the subsurface
region are selected so that the restored images do not depend on the values of these
steps. The number of frequencies, 32, is chosen so that the frequency dependencies
of amplitude and phase of the measured signal represent a good approximation
in the working frequency band $\delta f = 3.0$ to 4.0 GHz. Any increment of working
frequencies over 32 did not improve the image quality.

A decrease in frequency number leads to image deterioration because information about an object in the Fourier space is emasculated. Signals can be increased to
some level by the microwave amplifier, and after that the increased signal passes
to the network analyzer input. The plane-wave component from the plane-wave
spectrum of the field on interface propagating is received in the direction of the
receiving antenna. Signal from the antenna output acts on analyzer input.

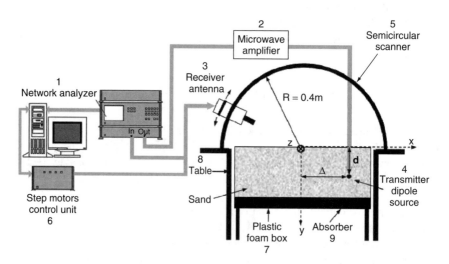

FIGURE 5.114   Experimental setup and position of a dipole inside sand.

The dipole was placed in different points on and under the surface of the sand. Before measurements are begun, the receiving antenna is placed in the middle of the semicircle and the dipole is placed under the receiving antenna on the surface of the sand and the system. It is calibrated such that on all 32 frequencies from the working band of $\delta f = 3.0$ to 4.0 GHz, taken with constant frequency step Δf and complex transmission coefficient $T_0(f_i) = 1$, $i = 1, 2, \ldots, 32$. It can be done by multiplying $T_0(f_i)$ by the complex coefficients U_i so that $T_0(f_i) = 1$ for all f_i. These coefficients are stored and used when the coefficients T_n are measured. Thus, after a calibration process the coefficients T_n are replaced by the coefficients $T_n U_i$ in Eq. (5.166). At the following step of the measurement, the antenna is reseted ($n = -m$) and data acquisition begins. The data are real and imaginary parts of complex transmission coefficients for each angle of the antenna θ_n at all 32 frequencies.

Results Several schemes of the dipole position were studied in this experiment. The images of the radiating centers were reconstructed using the measurement method and tomographic algorithm [100] using Eqs. (5.152) to (5.155). The image function is the modulus of the normalized polarization current $|K|$ distributed in the sand around the radiating center. In Fig. 5.115 one can see the images of the source placed on the surface of the sand. Small circles in the graphs symbolize the exact positions of the dipole. Figure 5.115a shows the source shifted a distance $\Delta \simeq -2\lambda_0 (\lambda_0 = 0.1 \text{ m})$ from the center of the semicircle. The situation $\Delta = 0$ is illustrated in Fig. 5.115b and $\Delta \simeq 2\lambda_0$ is shown in Fig. 5.115c. (See Fig. 5.114 for the position definitions of the dipole.)

Figure 5.116 illustrates the images of the source when the source is placed under the surface of the sand at the depth of $d \simeq 1.5\lambda_0$ for the distances $\Delta \simeq -2.5\lambda_0$, 0, and $2.5\lambda_0$. It is possible to see that the method described allows us to reconstruct an image of the irradiating source and to find the source position. The shape of the source is close to the tomographic image of the point scattering center [100]. The positions of the images of the source are nearly the same as their actual positions. However, for the shifted positions (in the case of $\Delta \simeq -2.5\lambda_0, 2.5\lambda_0$ $2.5\lambda_0$), there is an error in the vertical position of the source. When the source is shifted too much from the center, this small error comes out due to the plane-wave approximation violation, as we predicted in Eq. (5.168). We used a second dipole source to investigate the resolution of the method described. In that case we inserted two identical sources in the medium. Results of such an experiment are presented in Fig. 5.117, showing the images of two identical dipoles placed on the surface (a) and under the surface (b). In Fig. 5.117 dipoles were shifted in opposite directions along the x-axis at the distance $\Delta \simeq \pm\lambda_0$. The depth was $d \simeq 1.2\lambda_0$. A good resolution for the images can be seen in Fig 5.117. Experiments show us that the attainable resolution is near the appropriate resolutions in standard DT [100].

Additionally, experiments were carried out with a metal scattering rod instead of an active dipole. A rectangular marble block was used for the medium instead of soil. Two holes of diameter $\simeq 0.9\lambda_0$ were drilled inside the marble block (Fig. 5.118a). Then a rod, which is the same size in each hole, was placed in the first hole to a depth near the surface. Next, it is placed in the second hole at a depth $d \simeq 0.4\lambda_0$. Figure 5.118b and c show reconstructed images of the scattering object (i.e., the metal rod). In the first image the metal rod was located in hole 1, and in the second image the metal rod was located in hole 2.

A rectangular waveguide adapter was used to generate electromagnetic waves propagating inside the marble block. The adapter was inserted on the left side of the marble block with thickness of $h \simeq 0.7\lambda_0$. This construction is a form of dielectric waveguide in which inside the dielectric block the adapter excites a waveguide mode E, which propagates along the plate. This wave is scattered by the metal rod. The results of the reconstruction (Fig. 5.118b and c show that in the case considered, one can see the size (and shape) of the scatterer and its positions. So

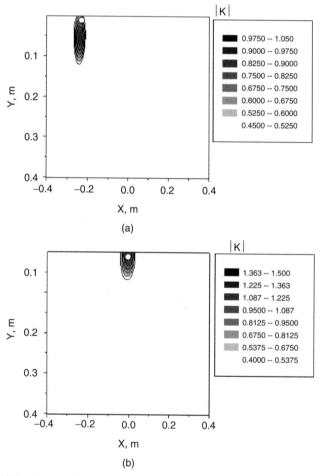

FIGURE 5.115 Images of a source placed just under the sand surface ($d = 0$) for the distances (a) $\Delta \cong -2\lambda_0$; (b) $\Delta = 0$; (c) $\Delta \cong 2\lambda_0$. Small circles show the exact positions of the dipole.

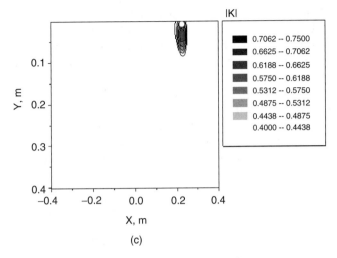

FIGURE 5.115 (*Continued*)

the image of the source is shown not only for an active source but also for the illuminated electromagnetic scattering center in the medium.

Conclusions A new approach to microwave image reconstruction of subsurface objects has been investigated. The necessity for complex amplitude measurement of an electromagnetic field is excluded in a point on a direct line above the surface in the method considered. Calculation of the direct Fourier transformation of the measured field is also excluded. In the experiment, data on a complex electromagnetic field (data from the transmission coefficients) at 32 frequencies in the operating band with a constant frequency step were used. Experiments were carried out at frequencies from 3 to 4 GHz. The images of the irradiating dipoles, considered as models of the scattering centers of an object, were reconstructed from the data measured.

The investigation proceeds as follows: There is the integral ratio between the object function $K(x, y)$ and the angular spectrum of the scattered field measured on the semicircle in a plane on the perpendicular surface of an inhomogeneous medium. Distribution of a normalized polarizing current under the surface of this medium (the image function) arises as an outcome of the effect of outside sources. Measurement of the angular spectrum of the scattered field at microwave frequencies under the scheme considered is executed using a network analyzer operating in the transmission mode. Thus, the angular spectrum is proportional to the transmission factor measured to within a complex constant. Using the measured part of the angle spectrum allows us to exclude measurement of the irradiated scattered field above the interface of the two mediums as well as calculation of the Fourier transform of the measured field and the unstable factor $\exp(i\gamma_1 y_1)$ in the inverse.

First, images of irradiating dipoles located under the surface of the sand are obtained using the multifrequency tomographic algorithm.

Using 32 frequencies in the range 3 to 4 GHz, the image of the scattering metal rod is located inside a dielectric layer with a thickness of 0.7 λ_0 (marble block), which is illuminated by the E-type wave propagating in this layer. The experimental investigation conducted in this frequency range showed that the setup and signal-processing methods described make it possible to determine the shape and to estimate the cross-sectional size of objects buried in sand. We can also evaluate the depth at which an object is located using the object cross-sectional image in the plane perpendicular to the medium surface.

The results obtained may be applied in practical microwave imaging systems for detection and observation of a variety of subsurface objects. It is also

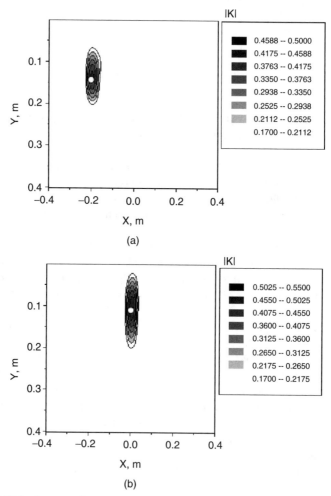

FIGURE 5.116 Images of a source placed under the sand surface ($d \cong 1.5\lambda_0$) for the distances (a) $\Delta \cong -2.5\lambda_0$; (b) $\Delta = 0$; (c) $\Delta \cong 2.5\lambda_0$. Small circles show the exact positions of the dipole.

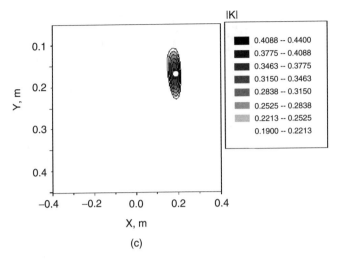

(c)

FIGURE 5.116   (*Continued*)

possible to use this technique for higher frequencies, such as the millimeter-wave region.

5.5 NONLINEAR INVERSION ALGORITHMS

The tomographic methods described in earlier sections represent a reliable way to tackle the inverse problem underlying subsurface diagnostics techniques exploiting electromagnetic waves. In particular, examples of their successful application in the framework of the GPR surveys reported in Chapter 12 show that these methods are presently ready for widespread application in the field. On the other hand, such proof of reliability is an input for new perspective advancements.

In particular, by considering the state-of-the-art standard processing techniques for GPR, as well as the limitations intrinsic to the methods discussed above, one can notice that such advancements should be directed toward the development of processing tools capable of achieving a more complete characterization of the targets. As a matter of fact, there is an obvious added value in an image that describes not only the target's morphology but also its electromagnetic characteristics. Of course, as mentioned at the beginning of the chapter, such a goal comes with a need to tackle the inverse scattering problem into a very difficult case: one in which a non-linear model is assumed (i.e., without introducing approximations on the scattering phenomenon such as those exploited within the linearized methods considered so far). In this section, we thus present an overview of the features of the class of nonlinear inversion algorithms, addressing some of the issues that are crucial in their development and the failures that may originate from a lack of care in their setting or adoption.

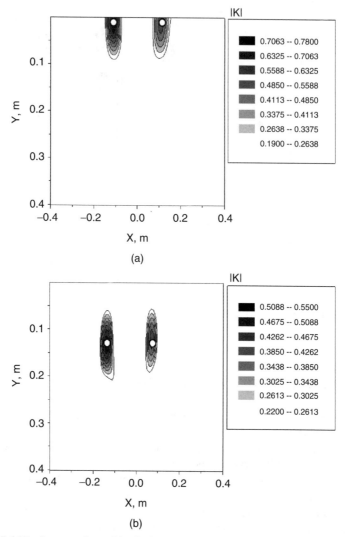

FIGURE 5.117 Images of two identical sources placed under the sand surface (a) for the depth $d = 0$ and the distances $\Delta \cong \pm\lambda_0$; (b) for the depth $d \cong 1.2\lambda_0$ and the distances $\Delta \cong \pm\lambda_0$. The small circles show the exact positions of the dipole.

5.5.1 Formulating the Inverse Problem as a Nonlinear Optimization

As discussed earlier, the common way to address inversion of the nonquadratic relationship between the scattered field data measured at the receivers and an unknown contrast function embedding the anomaly's features is to recast the inverse problem in terms of an optimization problem. To this end one must introduce a suitably defined cost functional, whose global minimum is attained in correspondence to the actual value of the contrast function.

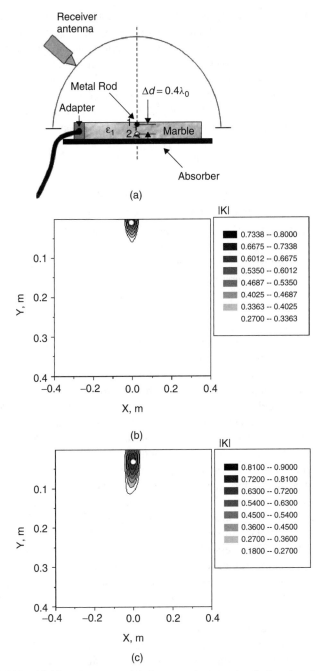

FIGURE 5.118 (a) Measurements with the scattering metal rod placed under the surface of a dielectric block. The reconstructed images of the rod at $\Delta = 0$ for the depths (b) $d \cong 0$ and (c) $d = 0.4\lambda_0$. Small circles show the exact positions of the dipole.

Without considering here the issue of regularization [106–110], which of course plays a role in this inverse problem (possibly even more critical than in the linearized case), two ways have been employed to define such a cost functional. The first (almost natural) way is to consider a cost functional that expresses the distance in the data space $S(L_\Gamma^2)$ between the measured data E_s and the "theoretical data," which arise from the electromagnetic scattering model adopted. According to the synthetic notation introduced in Section 5.2, these are given by $\mathcal{A}_e(\chi E)$. Within this framework, the inverse scattering problem is then cast as

$$\chi = \arg \min \| E_s - \mathcal{A}_e(\chi E) \|_{L_\Gamma^2}^2 \qquad (5.169)$$

wherein $\| \cdot \|_{L_\Gamma^2}$ denotes the L^2 norm in the data space $S(L_\Gamma^2)$, as weighted by the measured data energy $\| E_s \|_{L_\Gamma^2}^2$. This type of formulation is at the basis of several approaches, such as the Newton–Kantorovich method [111–114], the distorted Born method, and the distorted iterative Born method [115].

A straightforward way to address the optimization problem cast in (5.169) would be that of attempting an iterative least-squares minimization of the cost functional by means of a global optimization algorithm. Unfortunately, such a solution is not feasible, owing to the very large number of unknown parameters involved in the optimization. As a matter of fact, the computational cost of global methods grows exponentially fast with the number of unknowns, so that their computational burden rapidly becomes unaffordable [116]. As a consequence, the minimization task has to be tackled by means of local minimization approaches that rely on local derivatives. Newton's iterative methods [117,118], which approximate a nonlinear cost function with a quadratic model based on the function's first- and second-order derivatives, are the mostly widely adopted, with a variety of implementations, depending on how these derivatives are computed (i.e., Gauss–Newton, quasi-Newton etc.).

Regardless of the specific scheme applied to implement the optimization task, when facing the solution of a nonlinear inverse problem by means of the methods mentioned above, one cannot guarantee that a reliable outcome is achieved from the iterative process. As a matter of fact, local minimization by construction converges to the local minimum that is closer to the starting point of the procedure [119–121]. Therefore, owing to the nonlinearity of the functional relations between the data and the unknown, the effectiveness of these techniques in providing the true solution (or a reliable approximation of it) cannot be ensured, as it actually depends on the way in which the iterative process is initialized. In the inverse scattering literature, such a circumstance is usually referred to as the *local minima problem*, and several efforts have been made to overcome this or at least to reduce its influence on the solution methods developed [119,120].

To address issues arising from the nonlinearity of the problem, an alternative definition of the cost functional has been considered. The nonlinearity of the data-to-unknown relationship expressed through Eq. (5.7) essentially resides in the dependency of the total field E on the contrast, as expressed through Eq. (5.8). Such a circumstance suggests that it is possible to formulate the optimization problem with respect to the pair of unknowns given by the contrast and the total field. As observed from Eqs. (5.7) and (5.8), it is easy to notice that the integral operators are linear in the product within their two arguments (the contrast χ and the total field E). As such, these operators are bilinear [122] with respect to that pair, so that they are linear in one of the two arguments when the other one is fixed. Accordingly, optimization task is recast as

$$[\chi, E] = \arg \min \left\{ \| E_s - \mathcal{A}_e(\chi E) \|_{L_\Gamma^2}^2 + \| E - E_{inc} + \mathcal{A}_i(\chi E) \|_{L_D^2}^2 \right\} \quad (5.170)$$

wherein $\| \cdot \|_{L^2}$ denotes the L^2 norm in $T(L_D^2)$, as weighted by the incident field energy $\| E_{inc} \|_{L^2}^2$. This type of formulation is the one exploited in the modified gradient method [123], the contrast source inversion method [124], the bilinear approach [125], and the contrast source extended Born method [126].

Owing to the bilinear nature of the mapping involved, the minimization task underlying (5.170) is less sensitive than the previous formulation to the occurrence of local minima [120,125]. A physical insight on this alternative formulation is useful to better understand why it is so. As can be observed, the second term in (5.170) enforces, in the least-squares sense and throughout the iterative process, the physical consistence of the current solution with the "state" equation, Eq. (5.8). Hence, it can be interpreted as a constraining term which helps the minimization toward the achievement of a physically meaningful solution. It is also worth noting that while the type of nonlinear relationship of the functional in Eq. (5.169) cannot be clued, in this case, owing to the bilinear reformulation, the functional is of fourth order with respect to the unknown pair (χ,E).

As far as the minimization of the functional (5.170) is concerned, since the number of unknown parameters is further enlarged with respect to the previous case, the adoption of a global minimization algorithm is still (or even more) not viable, and local schemes are adopted. However, again by virtue of the involved bilinear mapping involved, this task can be performed efficiently by means of a conjugate gradient iterative scheme based on first-order derivatives, usually referred to as a *modified gradient* [123]. In the latter, as opposed to what happens in the previous case, the solution of a forward problem at each step of the iterative scheme (in order to update the optimization variable) is not required. In particular, given the polynomial nature (fourth order) of the data in relation to the unknowns, the optimal step along the minimization direction can be determined by means of a procedure based on closed-form expressions [125].

5.5.2 Regularizing Strategies for Quantitative Imaging

Nowadays, the modified gradient formulation of the nonlinear inverse scattering problem is the most popular way to address the issue of quantitative imaging. However, for the approach to be really effective, the inherent ill-conditioning of the inverse problem has to be tackled. To this end, a regularization aimed at opposing the loss of information intrinsic to the scattering phenomenon, and already discussed in the linear case, has to be enforced.

The usual Tichonov penalty term, which constrains the solution's energy [106], is, of course, one possible choice, as it represents a "general-purpose" tool for addressing the regularization issue. However, in the case of subsurface imaging, one must also face the reduction of the available information arising from the aspect-limited nature of the measurement configuration (i.e., only part of the field scattered by the targets can be gathered, and the probes cannot surround them completely). Therefore, different regularization strategies, which try to enforce the possibly available knowledge on the specific features of the problem or the peculiar nature of the targets, are worthwhile considering.

In this respect, the *regularization by projection* strategy provides a first useful tool [109]. Here, instead of assuming the straightforward discretization of the unknown contrast in terms of pulse basis functions (i.e., pixels), one takes advantage of a representation of the unknown function as the superposition of suitable basis functions, possibly matching the characteristics of the data-to-unknown mapping. By doing so, it is possible to consider a minimum number of unknown parameters (the expansions' coefficients rather then the local values of the function in the pixel), or at least a less redundant one. This is a crucial point, as it has been demonstrated that in nonlinear inverse problems ruled by fourth-order functionals like the one in Eq. (5.170), the occurrence of local minima in the iterative process can be reduced (or even avoided) if the ratio between the amount of independent data and the number of unknowns is sufficiently large [120].

Unfortunately, for a nonlinear mapping it is not possible to carry out an exact analysis of the most convenient basis functions, so that one often relies on an analysis of the linearized operator and checks its possible extension to "slight" nonlinearities, such as the quadratic case (wherein the Born series is truncated at the second term rather than at the first term as in the Born approximation) [16,17,34,127,128]. In particular, a similar type of analysis has shown that in the subsurface imaging case, being that the antennas are in close proximity to the scattering region, the shallower objects can be reconstructed with a higher spatial resolution level than that of the deeper ones [127]. Such an effect is clearly further enhanced by the presence of losses. From this observation it follows that a convenient way to expand the unknown contrast is to use a multiresolution expansion, so as to accommodate this type of inhomogeneous distribution of the information [127,129–132]. In practice, this can be achieved by expressing the contrast through the coefficients of a wavelet basis and by simply neglecting the finer details pertaining to the deeper region of the domain under test, where only the coarser details can be retrieved.

To illustrate the concepts described above, let us examine a numerical example related, for the sake of simplicity, to the same idealized two-dimensional geometry considered previously. Figure 5.119a shows a scenario in which a dielectric target mimics a plastic landmine is buried in soil. The field scattered by such a target has been computed by means of a forward solver based on solution of the method of moments [133]. When processing the simulated data concerning the scenario described above by assuming straightforward representation of the targets in terms of pixels, the reconstruction is dramatically different from

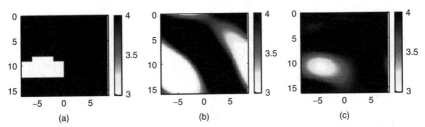

FIGURE 5.119 Numerical example showing the effect of the regularization by projection onto the inversion's outcome. (a) Dielectric permittivity map of a simulated scenario that mimics a plastic landmine. Result when considering a straightforward discretization in terms of pixels. (c) Result obtained using a multiresolution basis (Daubechies wavelets in the specific case) to expand the contrast. As can be seen only in the latter case, a good estimate of the target's position and permittivity is obtained.

the ground truth, as shown in Fig. 5.119b. This negative outcome is due to the large number of unknown parameters as compared to the amount of available independent data [119,120]. As a matter of fact, adopting a Daubechies wavelet representation [134], which makes it possible to improve the ratio among the data and the unknowns, the quality of the reconstruction improves remarkably (Fig. 5.119c).

In addition to the regularization by projection strategy described above, the inversion algorithm can also take advantage of the exploitation of some presumed knowledge regarding the targets, as this information allows us to reduce the unknown parameters to an even greater extent. For example, in many applications, the targets are or can reasonably be assumed to be electrically homogeneous (i.e., exhibiting a constant permittivity and/or conductivity). This is the case of many human-made structures such as buried walls or utilities, as well as plastic landmines or cavities and holes. In this case, a possible way to simply enforce in a very simple way this type of information within the iterative process is that of introducing an additional term that attains its minima at the origin and for a value of the contrast that corresponds to the target's permittivity. Accordingly, the optimization problem is recast as [135,136]

$$[\chi, E] = \arg\min \left\{ \|E_s - \mathcal{A}_e(\chi E)\|_{L_\Gamma^2}^2 + \|E - E_{\text{inc}} + \mathcal{A}_i(\chi E)\|_{L_D^2}^2 \right.$$
$$\left. + \|\chi\|_{L_D^2}^2 \|\chi - \chi_T\|_{L_D^2}^2 \right\} \tag{5.171}$$

wherein χ_T is the contrast that corresponds to the target's permittivity and itself can be assumed to be an unknown parameter throughout the optimization process.

To better illustrate the effect of the regularizing term introduced above, it is useful to look at the pictorial representation given in Fig. 5.120. In Fig. 5.120a the behavior of the bilinear approach functional (5.170) is sketched. For the sake of clarity, this multivariable functional is drawn as a single argument function: that is, by assuming that only one unknown parameter is changing while the others are fixed. Nevertheless, such a plot is useful to give an

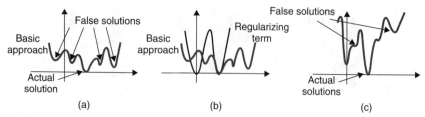

FIGURE 5.120 Effect of regularization enforcing the homogeneous nature of the targets. (a) Pictorial representation of the standard cost functional (drawn as a function of one variable for the sake of clarity). (b) Within the same conceptual framework, the basic cost functional, with the regularizing function superimposed. (c) Combination of the two terms from which one can appreciate the positive effect of the regularization that has been introduced on the "quality" of the local minima.

immediate representation of the issue faced in the solution of the nonlinear optimization problem underlying the inverse scattering, that is, the presence of several deep local minima that correspond to solutions remarkably different from the ground truth (this was noticeable, for example, in the central panel of Fig. 5.119b).

As explained earlier, owing to the fact that a deterministic optimization scheme converges to the closest extremum, the presence of these minima means that convergence to the *actual* global minimum is not ensured. Figure 5.120b shows the behavior of the regularizing term introduced in Eq. (5.171). This latter presents only two minima, which are located in $\chi = 0$ (where there is no target) and in $\chi = \chi_T$ (when the reconstructed contrast matches that of the target sought), respectively. Finally, the overall functional minimized in the optimization problem formulated in (5.171) is depicted in Figure 5.120. From this figure, one can clearly observe the effect of the regularization introduced, as now the cost functional exhibits local minima which are less pronounced than in the unregularized situation.

To illustrate the effect on the reconstruction of the regularization described above, let us consider the numerical example in Fig. 5.121, concerning the imaging of a plastic pipe filled with oil and buried in an inhomogeneous soil. A two-dimensional geometry has also been considered in this case. The scenario is depicted in Fig. 5.121a, and the reconstruction obtained by the regularization using projection strategy alone is shown in Fig. 5.121b. As can be seen, this regularization is not sufficient to tackle the inverse problem successfully. On the other hand, although the target is not exactly homogeneous, the small extent of the pipe's wall with respect to the diameter allows one to assume it to be as described (from the point of view of the inverse problem), as a homogeneous target having circular cross section and permittivity equal to that of oil. Therefore, the target can be imaged by making use of the regularization introduced in Eq. (5.171). As shown in Fig. 5.121c and d, the latter, indeed, provides good results, especially when the additional information on the value assumed by the unknown contrast sought is exploited.

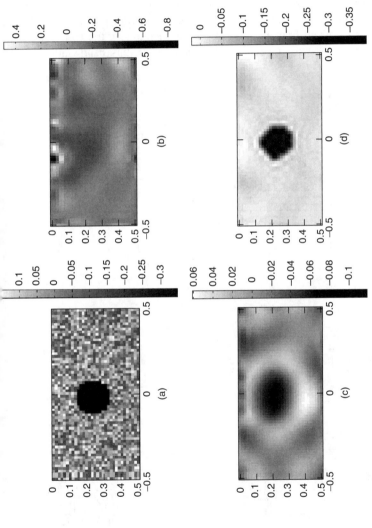

FIGURE 5.121 Effect of regularization enforcing the homogeneous nature of the targets onto the reconstruction results. (a) Simulated scenario consisting of a plastic pipe filled with oil and embedded in inhomogeneous soil. (b) Result achieved when adopting the inversion approach exploiting a multiresolution basis, but no information on the homogeneous nature of the target. Panel (c) Result achieved when adopting the regularization strategy in Eq. (5.171) when the value of the reference contrast χ_T is unknown. (d) Reconstruction obtained when the value of the reference contrast is known.

5.5.3   Example of a Laboratory-Controlled GPR Application

An example of the application of a nonlinear algorithm to the processing of GPR data is given here with respect to a controlled experiment concerned with masonry diagnostics (Fig. 5.122a). In this experiment [137], a wall 0.12 m thick, 1.49 m, high and 1 m wide made up of baked clay bricks and concrete is put in contact with a second brick wall of the same size that contains a cavity (0.50 m × 0.20 m × 0.12 m) 0.80 m from the floor. Behind the latter is a third wall, made of concrete.

The multiview, multistatic, multifrequency laboratory-controlled data were gathered by means of a stepped frequency radar system operating in the band 0.5 to 1.5 GHz, equipped with two shielded bowtie antennas. In this bandwidth, 201

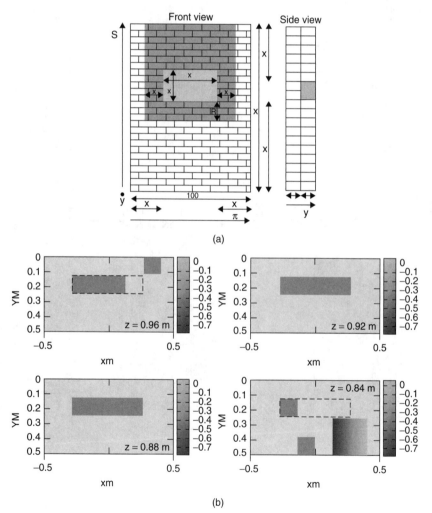

(a)

(b)

FIGURE 5.122   (a) Wall inclusion and measurement surface; (b) nonlinear reconstructions of the contrast function.

frequency steps are considered, and the area probed is 0.76 m × 0.89 m in size. Moreover, a multiple-receiver arrangement is considered wherein for each position of the transmitting antenna, a position of the receiving antenna is taken. In particular, by considering as a measurement region the rectangular domain of 0.89 × 0.76 m in the xz-plane corresponding to the shaded region in Fig. 5.122a, probes are moved along $N = 21$ lines parallel to the x-axis. For any fixed spot z, 14 source points along a line of 0.52 m with a spatial step of 0.04 m have been considered. For each source position, three measurements have been gathered at distances of 0.16, 0.2, and 0.24 m from the transmitter.

The result of the nonlinear algorithm [136] for four cross sections of the wall is reported in Fig. 5.122b. Each has been obtained by processing the data collected along the corresponding measurement line (from $z = 0.84$ m to $z = 0.96$ m). It is interesting to note that multifrequency data collected at only three frequencies (0.5, 0.75, and 1 GHz) have been processed. By exploiting the a priori information that a piecewise contrast is sought, a suitable choice to regularize the inverse problem is that of expanding the contrast into Haar wavelet coefficients [134]. In particular, those corresponding to the third-level coarse representation have been sought. Moreover, the a priori information that a void (having free-space permittivity) is searched for is exploited in order to further regularize the inverse problem. For each line, the real part of the reconstructed contrast function is shown in Fig. 5.122b. It is worth noting that despite a two-dimensional geometry being considered in the formulation of the inversion algorithm, a very accurate reconstruction is achieved. Of course, the result worsens at the cavity border.

5.5.4 Moving a Step Further: Stepwise Approaches to Quantitative Imaging

The previous examples show the advantages in terms of retrived information that can be achieved when the inversed problem is tackled without introducing approximation on the nonlinear data to unknown relationship. Nevertheless, owing to the difficulties noted above that arise from the nonlinear nature of the problem, the capability of an imaging procedure of providing reliable results actually depends on the availability of some knowledge of the targets and scenario as well as how such information is exploited. As a matter of fact, the introduction of suitable constraints on the space within which the solution is searched is the only dependable strategy to use to avoid or at least limit the possibility that the iterative process is trapped in a local minimum.

To address this issue, several studies have been carried out in the literature to understand which factors affect the degree of nonlinearity of the relationship between the data and the unknowns of such an inverse problem [119,138,139,159]. In particular, it turns out that the extent of the investigation domain, the average value of the unknown contrast, and the choice of the basis functions to adopt (e.g., pulse basis functions, Fourier harmonics, wavelet functions, splines) play important roles with respect to the actual possibility of achieving a good estimate of the ground truth.

In the framework of subsurface imaging, where the role of regularization is, as discussed, crucial, these results allow us to envision a different path to addressing the imaging problem. In particular, a possible strategy could be based on the

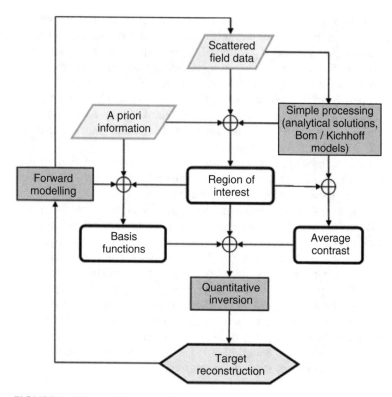

FIGURE 5.123   Architecture of a nonlinear imaging strategy. (From [140].)

architecture of Fig. 5.123, in which the inverse problem is faced in a stepwise fashion. Within such an architecture, measured data and a priori information possibly available on the targets are exploited to "build" a knowledge on those factors which allow to lower the nonlinearity of the problem and defeat its ill-posedness.

In particular, information about the extent of the investigation domain can be retrieved by processing the measured data thanks to linear inversion schemes [138,140] or, when a multistatic multiview measurement configuration is exploited, using more effective strategies [141,142,143,155–158]. Moreover, the average value of the unknown contrast can be achieved by taking advantage of some simple processing based on a comparison of the measured data with canonical solutions computed over region of interest the determined previously [119,138,139]. Finally, the choice of the basis functions is done by again merging the a priori available knowledge, information on the target's location, and analysis of the assumed forward model (e.g., by means of the SVD of the linearized operators or assuming a quadratic model).

Once these preprocessing steps are accomplished, partial information so achieved is exploited into the quantitative inversion stage. Note that at this stage, both a priori information and the knowledge gained through preprocessing steps can contribute to improve the reliability of the local optimization process, acting as additional regularizations (enforcing, for instance, the "homogeneous" nature of the targets) or as a convenient initialization for the iterations (using, for instance, the average estimated contrast). Also, it is worth remarking that the reconstructed contrast can be exploited to update the forward modeling in order to refine and repeat the overall process.

REFERENCES

1. Chommeloux, L., Pichot, C., Bolomey, J. 1986. Electromagnetic modeling for microwave imaging of cylindrical buried inhomogeneities. *IEEE Trans. Microwave Theory Tech.*, 34(10): 1064–1076.

2. Deming, R. W., Devaney, A. J. 1997. Diffraction tomography for multi-monostatic ground penetrating radar imaging. *Inverse Probl.*, 13(1): 29–45.

3. Fear, E. C., Hagness, S. C., Meaney, P. M., Okoniewski, M., Stuchly, M. A. 2002. Enhancing breast tumor detection with near-field imaging. *IEEE Microwave Mag.*, 3(1): 48–56.

4. Franklin, J. N. 1968. *Matrix Theory*. Prentice-Hall, Englewood Cliffs, NJ. Richmond, J. H. 1965. Scattering by a dielectric cylinder of arbitrary cross-section shape. *IEEE Trans. Antennas Propag.*, 13(3): 334–341.

5. Salman, A. O., Gavrilov S. P., Vertiy, A. A. 2005. Microwave imaging of immersed bodies: an experimental survey. *Electromagnetics*, 25(6): 567–585.

6. Tihonov, A. N., Ya, V. Arsenin. 1979. *Methods of Illconditioned Problems Solution* [in Russian] Science, Moscow.

7. Vertiy, A. A., Gavrilov, S. P., Aksoy, S., Voynovskyy, I. V., Kudelya, A. M., Stepanyuk, V. N. 2001. Reconstruction of microwave images of the subsurface objects by diffraction tomography and stepped-frequency radar methods. *Zarubejnaya Radioelektronika. Uspehi Sovremennoy Radioelektroniki*, 7: 17–52.

8. Vertiy, A. A., Gavrilov, S. P., Voynovskyy, I. V., Stepanyuk, V. N., Ozbek, S. 2002. The millimeter wave tomography application for the subsurface imaging. *Int. J. Infrared Millimeter Waves*, 23(10): 1413–1444.

9. Vertiy, A. A., Gavrilov, S. P. 2005. Subsurface tomography application for through-wall imaging. In *Proceedings of the 9th International Conference on Electromagnetics in Advanced Applications (ICEAA-05) and 11th European Electromagnetic Structures Conference (EESC-05)*, Torino, Italy, Sept. 12–16, pp. 223–226.

10. Wiskin, J. W., Borup, D. T., Johnson, S. A. 1997. Inverse scattering from arbitrary two-dimensional objects in stratified environments via a Green's operator. *J. Acoust. Soc. Am.*, 102(2): 853–864.

11. Persico, R., Alberti, G., Esposito, S., Leone, G., Soldovieri, F. 2000. On multifrequency strategies of use of G.P.R. systems. In *Proceedings of the Conference on Image Reconstruction from Incomplete Data*, SPIE's Annual Meeting, San Diego, CA, July–Aug., pp. 26–34.

12. Soldovieri, F., Persico, R., Leone, G. 2005. Effect of source and receiver radiation characteristics in subsurface prospecting within the distorted Born approximation. *Radio Sci.*, 40(3): RS3006–.

13. Soldovieri, F., Persico, R., Leone, G. 2005. Frequency diversity in a linear inversion algorithm for GPR prospecting. *Subsurf. Sens. Technol. Appl. J.*, Special Issue GPR 2004, 6(1): 25–42.

14. Soldovieri, F., Persico, R. 2004. Reconstruction of an embedded slab with Born approximation from multifrequency data. *IEEE Trans. Antennas Propag.*, 52(9).

15. Persico, R., Soldovieri, F. 2004. Reconstruction of a slab embedded in a three layered medium from multifrequency data under Born approximation. *J. Opt. Soc. Am. A*, 21(1): 35–45.

16. Leone, G., Soldovieri, F. 2003. Analysis of the distorted Born approximation for subsurface reconstruction: truncation and uncertainties effects. *IEEE Trans. Geosci. Remote Sens.*, 41(1): 66–74.

17. Pierri, R., Leone, G., Soldovieri, F., Persico, R. 2001. Electromagnetic inversion for subsurface applications under the distorted Born approximation. *Nuovo Cimento*, 24C(2): 245–261.

18. Daniels, D. J. 2004. *Ground Penetrating Radar*. IEE Radar, Sonar, Navigation and Avionics Series. IEE, London.

19. Chew, W. C. 1995. *Waves and Fields in Inhomogeneous Media*. Institute of Electrical and Electronics Engineers, Piscataway, NJ.

20. Soldovieri, F., Persico, R., Leone, G. 2006. A microwave tomographic imaging approach for multibistatic configuration: the choice of the frequency step. *IEEE Trans. Instrum. Meas.*, 55(6): 1926—1934.

21. Conyers, L. B., Goodman, D. 1997. *Ground Penetrating Radar: An Introduction for Archaelogists*. AltaMira Press, Lanham, MD.

22. Kim, Y. J., Jofre, L., De Flaviis, F., Feng, M. Q. 2003. Microwave reflection tomography array for damage detection in concrete structures. *IEEE Trans. Antennas Propag.*, 51: 3022–3032.

23. Colton, D., Kress, R. 1992. *Inverse Acoustic and Electromagnetic Scattering Theory*. Springer-Verlag, New York.

24. Persico, R., Soldovieri, F., Pierri, R. 2002. On the convergence properties of a quadratic approach to the inverse scattering problem. *J. Opt. Soc. Am. A*, 19(12): 2424–2428.

25. Tijhuis, A. G., Belkebir, A. C., Litman, A. C. S., de Hon, B. P. 2001. Theoretical and computational aspects of two dimensional inverse profiling. *IEEE Trans. Geosci. Remote Sens.*, 39(6): 1316–1330.

26. Bertero, M., Boccacci, P. 1998. *Introduction to Inverse Problems in Imaging*. Institute of Physics Publishing, Philadelphia.

27. Cui, T. J., Chew, W. C. 2000. Novel diffraction tomographic algorithm for imaging two-dimensional targets buried under a lossy Earth. *IEEE Trans. Geosci. Remote Sens.*, 38(4, Part 2): 2033–2041.

28. Cui, T. J., Chew, W. C., Aydiner, A. A., Wright, D. L., Smith, D. V., Abraham, D. J. 2000. Numerical modelling of an enhanced very early time electromagnetic (VETEM) prototype system. *IEEE Antennas Propag. Mag.*, 42(2): 17–27.

29. Meincke, P. 2007. Efficient calculation of Born scattering for fixed-offset ground-penetrating radar surveys. *IEEE Geosci. Remote Sens. Lett.*, 4: 88–92.

30. Witten, A. J., Molyneux, J. E., Nyquist, J. E. 1994. Ground penetrating radar tomography: algorithms and case studies. *IEEE Trans. Geosci. Remote Sens.*, 32(2): 461–467.

31. Pierri, R., Persico, R., Bernini, R. 1999. Information content of the Born field scattered by an embedded slab: multifrequency, multiview, and multifrequency-multiview cases. *J. Opt. Soc. Am. A*, 16(10): 2392–2399.

32. Idemen, M., Akduman, I. 1990. Two-dimensional inverse scattering problems with bodies buried in a slab. *Inverse Prob.*, 6(3): 749–766.

33. Hansen, T. B., Johansen, M. 2000. Inversion scheme for ground penetrating radar that takes into account the planar air–soil interface. *IEEE Trans. Geosci. Remote Sens.*, 38(1): 496–506.

34. Pierri, R., Leone, G., Soldovieri, F., Persico, R. 2001. Electromagnetic inversion for subsurface applications under the distorted Born approximation. *Nuovo Cimento*, 24C(2): 245–261.

35. Pommet, D., Marr, R. A., Lammers, U. H. W., McGahan, R. V., Morris, J. B., Fiddy, M. A. 1999. Imaging using limited-angle backscattered data. *IEEE Antennas and Propag. Mag.*, 39(2): 19–22.

36. Twomey, S. 1977. *Introduction to the Mathematics of Inversion in Remote Sensing and Indirect Measurements*. Dover Publications, New York.

37. Munson, D. C., O'brien, J. D., Jenkins, W. K. 1983. A tomographic formulation of spotlight-mode synthetic aperture radar. *Proc. IEEE*, 71(8): 917–925.

38. Pichot, C. H., Jofre, L., Peronnet, G., Bolomey, J.-Ch. 1985. Active microwave imaging of inhomogeneous bodies. *IEEE Trans. Antennas Propag.*, 33(4): 416–425.

39. Zhao, Z., Farhat, N. H. 1992. Tomographic microwave diversity image reconstruction employing unitary compression. *IEEE Trans. Microwave Theory Tech.*, 40(2): 315–321.

40. Mast, J. E., and Johansson, A. M. 1994. Three-dimensional ground penetrating radar imaging using multi-frequency diffraction tomography. In *Proceedings on Advanced Microwave and Millimeter Wave Detection. Proc. SPIE*, 2275, July.

41. Melamed, T., Ehrlich, Y., Hayman, E. 1996. Short-pulse inversion of inhomogeneous media: a time-domain diffraction tomography. *Inverse Probl.*, 12: 977–993.

42. Chen, F.-C., Chew, W. C. 1998. Development and testing of the time-domain microwave non-destructive evaluation system. In *Review of Progress in Quantitative Nondestructive Evaluation*, Vol. 17, D. O. Thompson, and D. E. Chimenti, Eds. Plenum Press, New York, pp. 713–718.

43. Vertiy, A. A., Gavrilov, S. P. 1998. Modelling of microwave images of buried cylindrical objects. *Int. J. Infrared Millimeter Waves*, 19(9): 1201–1220.

44. Noon, D. A. 1996. Stepped-Frequency Radar Design and Signal Processing Enhances Ground Penetrating Radar Performance. Thesis submitted for the degree of Doctor of Philosophy (Ph.D.), University of Queensland.

45. Cattin, V., Chaillout, J. J., Blanpain, R. 1998. Detection and localisation with a step frequency radar. In *Detection of Abandoned Land Mines*, Oct. 12–14. IEEE Conference Publication 458, pp. 86–90.

46. Goodmen, J. W. 1968. *Introduction to Fourier Optics*. McGraw-Hill, New York.

47. Davies, B. 1978. *Integral Transforms and Their Applications*. Applied Mathematical Sciences, Vol. 25. Springer-Verlag, New York.

48. Brigham, O. E. 1974. *The Fast Fourier Transform*. Prentice-Hall, Englewood Cliffs, NJ.

49. Vertiy, A. A., Gavrilov, S. P., Tansel, B., Voynovskyy, I. V. 1999. Experimental investigation of buried objects by microwave tomography method. In *SPIE Conference on Subsurface Sensors and Applications*, Denver, CO, July. *Proc. SPIE*, 3752: 195–205.

50. Bourgeois, J. M., Smith, G. S. 1998. A complete electromagnetic simulation of the separated-aperture sensor for detecting buried land mines. *IEEE Trans. Antennas Propag.*, 46(10): 1419–1426.

51. Mensa, D. L. 1991. *High Resolution Radar Cross-Section Imaging*. Artech House, Boston.

52. Daniels, D. J. 1998. Surface penetrating radar image quality. In *Detection of Abandoned Land Mines*, Oct. 12–14. IEEE Conference Publication 458, pp. 68–72.

53. Melamed, T., Heyman, E. 1997. Spectral analysis of time domain diffraction tomography. *Radio Sci.*, 32(2): 593–603.

54. Felsen, L. B. 1973. *Radiation and Scattering of Waves*. Prentice-Hall, Englewood Cliffs, NJ.

55. Pierri, R., Leone, G., Bernini, R., Persico, R. 1998. Tomographic inversion algorithms for permittivity reconstruction in subsurface prospection. In *Proceedings of the 7th International Conference on Ground Penetrating Radar*, University of Kansas, Lawrence, KS. May 27–30, Vol. 2, pp. 723–727.

56. Sean Lehman, K., Chambers, D. H., Candy, J. V. 1999. Spectral partitioning in diffraction tomography. In *SPIE Conference on Subsurface Sensors and Applications*, Denver, CO, July. *Proc. SPIE*, 3752: 314–325.

57. Daniels, D. J. 1996. *Surface-Penetrating Radar*. IEE, London.

58. Vertiy, A. A., Gavrilov, S. P., Voynovskyy, I. V., Aksoy, S., Kudelya, A. M., Salman, A. O. 2000. Diffraction tomography method applications in wide frequency range. *Int. J. Infrared Millimeter Waves*, 21(2): 321–339.

59. Reppert, P. M., Morgan, F. D., Tokzös, M. N. 1998. GPR velocity determination using brewster angles. In *8th International Conference on GPR*, Lawrence, KS, pp. 485–490.

60. Shestopalov, V. P., Vertiy, A. A., Ermak, G. P., Skryunik, B. K., Hlopov, G. I., Tsvyk, A. I. 1991. In *Generators of Diffraction Radiation*, V. P. Shestapalov, Ed. IRE NAS, Naukova Dumka, Kiev, Russia.

61. Daniels, D. J. 1999. System design of radar for mine detection. In *SPIE conference on Subsurface Sensors and Applications*, Denver, CO, July. *Proc. SPIE*, 3752: 390–401.

62. Joachimowicz, N., Pichot, C., Hugonin, J. 1991. Inverse scattering: an iterative numerical method for electromagnetic imaging. *IEEE Trans. Antennas Propag.*, 39(12): 1742–1752.

63. Detlefsen, J., Dallinger, A., Huber, S., Schelkshorn, S. 2005. Effective reconstruction approaches to millimeter-wave imaging of humans. In *Proceedings of the 28th URSI General Assembly*. New Delhi, India, Oct. 23–29.

64. Sheen, D. M., McMakin, D. L., Collins, H. D., Hall, T. E., Severtsen, R. H. 1996. Concealed explosive detection on personnel using a wideband holographic millimeter-wave imaging system. In *Proceedings of the Aerospace/Defense Sensing and Controls Conference*, Orlando, FL. *Proc. SPIE*, 2755: 503–513.

65. Jofre, L., Hawley, M. S., Broquetas, A., De Los Reyes, E., Ferrando, M., Elias-Fuste, A. R. Medical imaging with a microwave tomographic scanner. *IEEE Trans. Biomed. Eng.*, 37(3): 303–312.

66. Deans, S. R. 1983. *The Radon Transform and Some of Its Applications*. Wiley-Interscience, New York.

67. Natterer, F. 1990. *The Mathematics of Computerized Tomography* [in Russian]. Mir, Moscow.

68. Gamero, L., Franchois, A., Hugonin, J.-P., Pichot, C., Joachimowich, N. 1991. Microwave imaging: complex permittivity reconstruction by simulated annealing. *IEEE Trans. Microwave Theory Tech.*, 39(11): 1801–1807.

69. Lesselier, D., Duchene, B. Wave-field inversion of objects in stratified environments: from back-propagation schemes to full solutions. In *URSI: The Review of Radio Science, 1993–1996*, W. R. Stone, Ed. Oxford University Press, Oxford, UK, pp. 235–268.

70. Bolomey, J. Ch., Izadnegahdar, A., Jofre, L., Pichot, Ch., Peronnet, G., Solaimani, M. 1982. Microwave diffraction tomography for biomedical applications. *IEEE Trans. Microwave Theory Tech.*, 30(11): 1998–2000.

71. Jofre, L., Hawley, M. S., Broquetas, A., de Los Reyes, E., Ferrando, M., Elias-Fuste, A. R. 1990. Medical imaging with a microwave tomography scanner. *IEEE Trans. Biomed. Eng.*, 37(3): 303–311.

72. Murch, R. D., Chan, T. K. K. 1996. Improving microwave imaging by enhancing diffraction tomography. *IEEE Trans. Microwave Theory Tech.*, 44(3): 379–388.

73. Devaney, A. J. 1983. A computer simulation study of diffraction tomography. *IEEE Trans. Biomed. Eng.*, 30(7): 377–386.

74. Slaney, M., Kak, A. C., Larsen, L. E. 1984. Limitations of imaging with first-order diffraction tomography. *IEEE Trans. Microwave Theory Tech.*, 32(8): 860–874.

75. Chew, W. C., Wang, Y. M. 1990. Reconstruction of two-dimensional permittivity distribution using the distorted born iterative method. *IEEE Trans. Med. Imag.*, 9(2): 218–225.

76. Zhao, Z., Farhat, N. H. 1992. Tomography microwave diversity image reconstruction employing unitary compression. *IEEE Trans. Microwave Theory Tech.*, 40(2): 315–321.

77. Ney, M. M., Smith, A. M., Stuchly, S. S. 1984. A solution of electromagnetic imaging using pseudoinverse transformation. *IEEE Trans. Med. Imag.*, 3(4): 155–162.

78. Nikolskiy, V. V., Nikolskaya, T. I. 1989. *Electrodynamics and Propagation of Radio Waves*. [in Russian]. Nauka, Moscow.

79. Devaney, A. J. 1984. Geophysical diffraction tomography. *IEEE Trans. Geosci. Remote Sens.*, 22(1): 3–13.

80. Pan, S. X., Kak, A. C. 1983. A computational study of reconstruction algorithms for diffraction tomography: interpolation versus filtered backpropagation. *IEEE Trans. Acoust. Speech Signal Process.*, 31(5): 1262–1275.

81. Gavrilov, S. P., Vertiy, A. A. 1997. Application of tomography method in millimeter wavelengths band: I. Theoretical. *Int. J. Infrared Millimeter Waves*, 18(9): 1739–1760.

82. Jacobi, J. H., Larsen, L. E. 1980. Microwave time delay spectroscopic imagery of isolated canine kidney. *Med. Phys.*, 7(1): 1–7.

83. Baribaud, M., Dubois, F., Floyrac, R., Kom, M., Wang, S. 1982. Tomography image reconstruction of biological objects from coherent microwave diffraction data. *IEE Proc.*, 129(6): 356–359.

84. Jacobi, J. H., Larsen, L. E., Hast, C. T. 1979. Water-immersed microwave antennas and their application to microwave interrogation of biological targets. *IEEE Trans. Microwave Theory Tech.*, 27(1).

85. Guo, T. C., Guo, W. W., Larsen, L. E. 1984. A local field study of a water-immersed microwave antenna array for medical imagery and therapy. *IEEE Trans. Microwave Theory Techn.*, 32(8): 844–854.

86. Semenov, S. Y., et al. 1996. Microwave tomography: two-dimensional system for biological imaging. *IEEE Trans. Biomed. Eng.*, 43(9): 869–877.

87. Meaney, P. M., Paulsen, K. D., Chang, J. T. 1998. Near-field microwave imaging of biologically-based materials using a monopole transceiver system. *IEEE Trans. Microwave Theory Tech.*, 46(1): 31–45.

88. Vertiy, A. A., Gavrilov, S. P. 1997. Application of tomography method in millimeter wavelengths band: II. Experimental. *Int. J. Infrared Millimeter Waves*, 18(9): 1761–1781.

89. Ferraye, R., Dauvignac, J. Y., Pichot, C. 2003. Reconstructed of complex and multiple shape object contours using a level set method. *J. Electromagn. Waves Appl.*, 17(2): 153–181.

90. Kechribaris, C. N., Nikita, K. S., Uzunoglu, N. K. 2003. Reconstruction of two dimensional permittivity distribution using an improved Rytov approximation and nonlinear optimization. *J. Electromagn. Waves Appl.*, 17(2): 183–207.

91. Tanaka, T., Kuroki. N., Takenaka, T. 2003. Filtered forward-backward time-stepping method applied to reconstruction of dielectric cylinders. *J. Electromagn. Waves Appl.*, 17(2): 253–270.

92. Kak, A. C., Slaney, M. 1988. *Principles of Computerized Tomography Imaging*. IEEE Press, New York.

93. Paoloni, F. J. 1984. The effects of attenuation on the Born reconstruction procedure for microwave diffraction tomography. *IEEE Trans. Microwave Theory Tech.*, 34(3): 366–368.

94. Semenov, S. Y. et al. 1996. Microwave tomography: two-dimensional system for biological imaging *IEEE Trans. Biomed. Eng.*, (43)(9): 869–877.

95. Vertiy, A., Gavrilov, S. 2006. Imaging of buried objects by tomography method using multifrequency regularization process. In *Proceedings of the 11*[th] *International Conference on Mathematical Methods in Electromagnetic Theory (MMET-06)*, Kharkiv, Ukraine, June 26–29, pp. 152–157

96. Qaddoumi, N., Carriveau, G., Gachev, S., Zoughi, R. 1995. Microwave imaging of thick composite panels with defects. *Mater. Eval.*, 53(8): 926–929.

97. Quaddoumi, N., Zoughi, R. 1997. Preliminary study of the influences of effective dielectric constant and nonuniform probe aperture field distrubution on near field microwave images. *Mater. Eval.*, Oct, pp. 1169–1173.

98. Lockwood, S. J., Lee, H. 1997. Pulse-echo microwave imaging for NDE of civil structures: image reconstruction, enhancement, and object recognition. *J. Imag. Syst. Technol.*, 8: 407–412.

99. Park, S., Choi, H., Ra, J. 1998. Underground tomogram from cross-borehole measurements. *Microwave Opt. Technol. Lett.*, 18(6): 402–406.

100. Chommeloux, L., Pichot, C., Bolomey, J. C., Electromagnetic modelling for microwave imaging of cylindrical buried inhomogeneties. *IEEE Trans. Microwave Theory Techn.*, 34(10): 1064–1076.

101. Salman, A. O., Gavrilov, S., Vertiy, A. 2002. Subsurface microwave imaging by using angular spectrum of electromagnetic field. *J. Electromagn. Waves Appl.*, 16(11): 1511–1529.

102. Lovel, J. R., Chew, W. C. 1987. Response of a point source in a multicylindrically layered medium. *IEEE Trans. Geosci. Remote Sens.*, 25(6): 850–858.

103. Chew, W. C. 1984. Response of a current loop antenna in an invaded borehole. *Geophysics*, 49(1): 81–91.

104. Long, Y., Jiang, H., Lin, Y. 1987. Electromagnetic field due to a loop antenna in a borehole. *IEEE Trans. Geosci. Remote Sens.*, 34(1): 33–35.

105. Blaunstein, N., Dank, Z., Zilbershtein, M. 1999. Prediction of radiation pattern of a buried leaky coaxial cable. *Subsurf. Sens. Technol. Appl.*, 1(1): 79–99.

106. Tikhonov, N. Arsenin, V. Y. 1977. *Solutions of Ill-Posed Problems*. Wiley, New York.

107. Morozov, V. A. 1984. *Methods for Solving Incorrectly Posed Problems*. Springer-Verlag, New York.

108. Bertero, M. 1989. Linear inverse and ill-posed problems. In *Advances in Electronics and Electron Physics*, Vol. 75, P. W. Hawkes, Ed. Academic Press, New York, pp. 1–120.

109. Engl, H. W., Hanke, M., Neubauer, A. 1996. *Regularization of Inverse Problems*. Kluwer Academic, Dordrecht, The Netherlands.

110. Kirsch, A. *An Introduction to the Mathematical Theory of Inverse Problems*. Springer-Verlag, New York.

111. Joachimowicz, N., Pichot, C., Hugonin, J. P. 1991. Inverse scattering: an iterative numerical method for electromagnetic imaging. *IEEE Trans. Antennas Propag.*, 39(12): 1742–1753.

112. Pichot, Ch., Lobel, P., Blanc-Feraud, L., Barlaud, M., Belkebir, K., Elissalt, J. M., Geffrin, J. M. 1997. Gradient and Newton-Kantorovitch methods for microwave tomography. In *Inverse Problems in Medical Imaging and Nondestructive Testing*, H. Engl, A. Louis, W. Rundell, Eds. Springer-Verlag, New York.

113. Tijhuis, A. G. Franchois, A. 2001. A two-dimensional microwave imaging algorithm for a complex environment: preliminary results. In *Proceedings URSI International Electromagnetic Theory Symposium*, pp. 445–447.

114. Franchois, A., Tijhuis, A. G. 2003. A quasi-Newton reconstruction algorithm for a complex microwave imaging scanner environment. *Radio Sci.*, 38(2): 8011–.

115. Chew W. C., Wang, Y. M. 1990. Reconstruction of two-dimensional permittivity distribution using the distorted Born iterative method. *IEEE Trans. Med. Imag.*, 9(2): 218–225.

116. Nemirovsky, S., Yudin, D. B. 1983. *Problem Complexity and Method Efficiency in Optimization*. Wiley, New York.

117. Fletcher, R. 1990. *Practical Methods of Optimization*. Wiley, New York.

118. Nocedal, J., Wright, S. J. 2006. *Numerical Optimization*. Springer-Verlag, New York.

119. Bucci, O. M., Cardace, N., Crocco, L., Isernia, T. 2001. Degree of nonlinearity and a new solution procedure in scalar two-dimensional inverse scattering problems. *J. Opt. Soc. Am. A*, 18(8): 1832–1843.

120. Isernia, T., Pascazio, V., Pierri, R. 2001. On the local minima in a tomographic imaging technique. *IEEE Trans. Geosci. Remote Sens.*, 39(7): 1596–1607.

121. Tijhuis, A. G., Belkebir, K., Litman, A. C. S., de Hon, B. P. 2001. Theoretical and computational aspects of 2-D inverse profiling. *IEEE Trans. Geosci. Remote Sens.*, 39(6): 1316–1330.

122. Kolnogorov, A. N., Fomin, S. V. 1999. *Elements of the Theory of Functions and Functional Analysis*. Dover Publications, London.

123. Kleinman, R. E., van den Berg, P. M. 1992. A modified gradient method for two-dimensional problems in tomography. *J. Comput. Appl. Math.*, 42(1): 1735–.

124. van den Berg, P. M., Kleinman, R. E. 1997. A contrast source inversion method. *Inverse Prob.*, 13(6): 1607–1620.

125. Isernia, T., Pascazio, V., Pierri, R. 1997. A nonlinear estimation method in tomographic imaging. *IEEE Trans. Geosci. Remote Sens.*, 35(4): 910–923.

126. Isernia, T., Crocco, L., D'Urso, M. 2004. New tools and series for forward and inverse scattering problems in lossy media. *IEEE Geosci. Remote Sens. Lett.*, 1(4): 327–331.

127. Bucci, O. M., Crocco, L., Isernia, T., Pascazio, V. 2001. Subsurface inverse scattering problems: quantifying, qualifying, and achieving the available information. *IEEE Trans. Geosci. Remote Sens.*, 39(11): 2527–2538.

128. Bucci, O. M., Crocco, L., Isernia, T. 1999. Improving the reconstruction capabilities in inverse scattering problems by exploitation of close-proximity setups. *J. Opt. Soc. Am. A*, 16: 1788–1798.

129. Chiappinelli, P., Crocco, L., Isernia, T., Pascazio, V. 1999. Multiresolution techniques in microwave tomography and subsurface sensing. In *In Proceedings of the IEEE 1999 International Geoscience and Remote Sensing Symposium (IGARSS 1999)* Vol. 5, pp. 2516–2518.

130. Bucci, O. M., Crocco, L., Isernia, T., Pascazio, V. 2000. Wavelets in nonlinear inverse scattering. In *of the IEEE 2000 International Proceedings Geoscience and Remote Sensing Symposium (IGARSS 2000)*, Vol. 7, pp. 3130–3132.

131. Bucci, O. M., Crocco, L., Isernia, T., Pascazio, V. 2000. An adaptive wavelet-based approach for non-destructive evaluation applications. In *IEEE 2000 Antennas and Propagation Society International Symposium*, Vol. 3, pp. 1756–1759.

132. Baussard, E. L., Miller, E. L. Premel, D. 2004. Adaptative *b*-spline scheme for solving an inverse scattering problem. *Inverse Probl.*, 20: 347–365.

133. Harrington, R. F. 2001. *Time-Harmonic Electromagnetic Fields*. Wiley–IEEE Press, Piscataway, NJ.

134. Vetterli, M., Kovacevic, J. 1995. *Wavelets and Subband Coding*. Englewood Cliffs, NJ: Prentice-Hall.

135. Crocco, L., Isernia, T. 2004. Inverse scattering with real data: detecting and imaging homogeneous dielectric objects. *Inverse Prob.*, 17(6): 1573–1583.

136. Catapano, I., Crocco, L., Isernia, T. 2004. A simple two-dimensional inversion technique for imaging homogeneous targets in stratiffed media. *Radio Sci.*, 39: RS1012–.

137. Catapano, I., Crocco, L., Persico, R., Pieraccini, M., Soldovieri, F. 2006. Linear and nonlinear microwave tomography approaches for subsurface prospecting: validation on real data. *IEEE Antennas Wireless Propag. Lett.*, 5(1): 49–53.

138. Bucci, O. M., Cardace, N., Crocco, L., Isernia, T. 2000. 2-D inverse scattering: degree of non-linearity, solution strategies and polarization effects. In *Image Reconstruction from Incomplete Data*, M. Fiddy and R. Millane, Eds. *Proc. SPIE*, 4123.

139. Crocco, L., Soldovieri, F. 2008. From qualitative to quantitative inverse scattering methods for GPR imaging. In *Proceedings of the 12th International Conference on Ground Penetrating Radar*, Birmingham, UK.

140. Crocco, L., Soldovieri, F. 2003. Gpr prospecting in a layered medium via microwave tomography. *Ann. Geophys.*, 46(3): 559–572.

141. Catapano, I., Crocco, L., Isernia, T. 2008. Improved sampling methods for shape reconstruction of 3-D buried targets. *IEEE Trans. Geosci. Remote Sens.*, 46(10): 3265–3273.

142. Budko, N. V., van den Berg, P. M. 1999. Characterization of a two-dimensional subsurface object with an effective scattering model. *IEEE Trans. Geosci. Remote Sens.*, 37(5): 2585–2596.

143. Solimene, R., Soldovieri, F., Prisco, G., Pierri, R., Three-dimensional through-wall imaging under ambiguous wall parameters. *IEEE Trans. Geosci. Remote Sens.*, (Special Issue on Remote Sensing of the Building Interior), 47(5): 1310–1317.

144. Pierri, R., Soldovieri, F., Liseno, A., Solimene, R. 2001. In-depth resolution for a strip source in the Fresnel zone. *J. Opt. Soc. Am.*, A, 18(2): 352–359.

145. Persico, R., Bernini, R., Soldovieri, F. 2005. The role of the measurement configuration in inverse scattering from buried objects under the Born approximation. *IEEE Trans. Antennas Propag.*, 53(6): 1875–1887.

146. Meincke, P. 2001. Linear GPR inversion for lossy soil and a planar air–soil interface. *IEEE Trans. Geosci. Remote Sens.*, 39(12): 2713–2721.

147. Soldovieri, F., Hugenschmidt, J., Persico, R., Leone, G. 2007. A linear inverse scattering algorithm for realistic GPR applications. *Near Surf. Geophys.*, 5(1): 29–42.

148. Pierri, R. Soldovieri, F. 1998. On the information content of the radiated fields in the near zone over bounded domains. *Inverse Probl.*, 14(2): 321–337.

149. Bojarski, N. N. 1982. A survey of the physical optics inverse scattering identity. *IEEE Trans. Antennas Propag.*, 30: 980–989.

150. Pierri, R., Liseno, A., Solimene, R., Soldovieri, F. 2006. Beyond physical optics SVD shape reconstruction of metallic cylinders. *IEEE Trans. Antennas and Propag.*, 54(2): 655–665.

151. Soldovieri, F., Brancaccio, A., Prisco, G., Leone, G., Pierri, R. 2008. A Kirchhoff based shape reconstruction algorithm for the multimonostatic configuration: the realistic case of buried pipes. *IEEE Trans. Geosci. Remote Sens.* (IGARSS Special Issue), 46(10): 3031–3038.

152. Liseno, A., Soldovieri, F., Pierri, R., 2004. Improving a shape reconstruction algorithm with thresholds and multi-view data. *AEU Int. J. Electron. Commun.*, 58(2): 118–124.

153. Giannopoulos, A. 2006. *GprMax 2.0 (Electromagnetic Simulator for GPR)*.

154. Bucci, O. M., Crocco, L., Isernia, T., Pascazio, V. 2000. Inverse scattering problems with multifrequency data: reconstruction capabilities and solution strategies. *IEEE Trans. Geosci. Remote Sens.*, 38: 1749–1756.

155. Catapano, I., Crocco, L., Isernia, T. 2007. On simple methods for shape reconstruction of unknown scatterers. *IEEE Trans. Antennas Propag.*, 55(5): 1431–1436.

156. Catapano, I., Crocco, L., D'Urso, M., Isernia, T. 2007. On the effect of support estimation and of a new model in 2D inverse scattering problems. *IEEE Trans. Antennas Propag.*, 55(6): 1895–1899.

157. Cayoren, M., Akduman, I., Yapar, A., Crocco, L. 2007. A new algorithm for the shape reconstruction of perfectly conducting objects. *Inverse Probl.*, 23(3): 1087–1100.

158. Cayoren, M., Akduman, I., Yapar, A., Crocco, L. 2008. Shape reconstruction of perfectly conducting targets from single-frequency multiview data. *IEEE Geosci. Remote Sens. Lett.*, 5(3): 383–386.

159. D'Urso, M., Catapano, I., Crocco, L., Isernia, T. 2007. Effective solution of 3D scattering problems via series expansions: applicability and a new hybrid scheme *IEEE Trans. Geosci. Remote Sens.*, 45: 639–648.

CHAPTER SIX

Acoustic and Seismic Sensors

6.1 OVERVIEW

An elastic wave propagating in the Earth's crust is referred to as a seismic or earthquake wave, acoustic emission, a microseismic wave, or an earth tremor, depending on the discipline, the dominant frequency, and the propagation mode. The origin of these waves is either rupture of a fault or fracture (e.g., a natural or induced earthquake, microseismicity), an artificial explosion or vibration (e.g., a seismic wave in geophysical exploration), an oceanic wave or traffic (e.g., an earth tremor), or deformation of a rock mass (e.g., acoustic emission). An example of a subsurface elastic wave categorized by frequency is shown in Fig. 6.1.

High sensitivity, low noise level, and large dynamic range are necessary if seismic or acoustic sensors are to collect information on far-field earthquake and microseismicity from small-scale ruptures in faults. For example, a typical noise level (minimum detectability) expected for seismic and acoustic sensors for monitoring of natural earthquakes is around 1×10^{-6} g/$\sqrt{\text{Hz}}$, which is several orders smaller than that for conventional industrial velocity sensors. Information on the rupture mechanism at faults or fractures can be obtained by inversion of the signals collected [1]. Some of the acoustic characteristics of the Earth's crust, such as anisotropy and stress state, can be estimated by polarization analysis of the signals detected by multicomponent sensors. Hence, wideband nature, linear-phase characteristics, and low cross-sensitivity are required for earthquake or seismic sensors for the afore mentioned analysis. Harsh environment in deployment and operation, such as shock, temperature, pressure, and the presence of water, should also be

Subsurface Sensing, First Edition. Edited by Ahmet S. Turk, A. Koksal Hocaoglu, and Alexey A. Vertiy.
© 2011 John Wiley & Sons, Inc. Published 2011 by John Wiley & Sons, Inc.

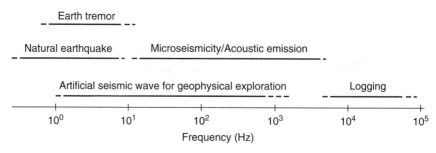

FIGURE 6.1   Frequency characteristics of the elastic waves that propagate Earth's crust.

considered in the design and manufacture of sensors. Monitoring of natural and induced seismic events in deep boreholes can be used effectively to collect data with a higher signal-to-noise ratio (SNR) and wideband power spectrum, because signals are not contaminated by acoustic noise, which propagates primarily as a surface wave, and signals have a wideband nature escaping from attenuation in the surface layer. Despite such advantages, the sensors and electronic circuits should sometimes work under temperatures exceeding 100°C, and therefore sensors for operation under high-temperature environments are of importance for downhole monitoring. In the case of stability monitoring of tunnels and cavities, long life and maintenance-free nature are strongly required. Cost for sensors, deployment, and recovery is also of practical concern, especially in exploration geophysics, where arrays of hundreds of sensors are used for three-dimensional seismic reflection surveys.

Considering the requirements and restrictions above for subsurface acoustic and seismic sensors, mechanical seismometers, electromagnetic sensors (geophone), piezoelectric sensors, strain sensors, capacitive sensors, and optical sensors have been developed and used in practice in the field. It should also be noted that characteristics of amplifiers and filters in the typical monitoring system (Fig. 6.2) should be understood as well as principles of sensors. In this chapter, principles behind mechanical seismometers, geophones, piezoelectric sensors, and capacitive sensors are described as well as fundamentals of sensor installation. Multicomponent seismic monitoring techniques, which take advantage of elastic waves in solid earth, are also covered in this chapter.

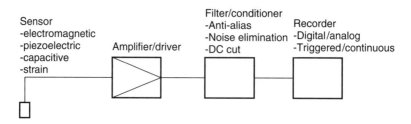

FIGURE 6.2   Seismic monitoring system.

6.2   OPERATING PRINCIPLES AND SENSOR PHYSICS

6.2.1   Mechanical Seismometers

A model of a simple pendulum is shown in Fig. 6.3, where a mass M is hung by a helical spring with a spring constant k. The mass can move freely along a vertical axis. The length of the spring without the mass is l_0 and with a mass is l_M. The gravity working on the mass and the contractile force from the spring are balanced. The contractile force from the spring can be written

$$Mg = k(l_M - l_0) \tag{6.1}$$

The equation of motion of the mass can be written using a displacement ξ from a balanced position as

$$M\ddot{\xi} = -k\xi \tag{6.2}$$

Hence, the natural period T_0 of the system is

$$T_0 = \frac{2\pi}{n} = 2\pi\sqrt{\frac{M}{k}} \tag{6.3}$$

In practice, the moving mass is hung by springs, comprising a physical pendulum. The natural period of a physical pendulum can be written

$$T_0 = 2\pi\sqrt{\frac{K}{MgH}} \tag{6.4}$$

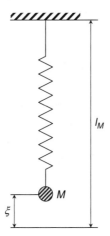

FIGURE 6.3   Simple equivalent model of a mechanical seismometer.

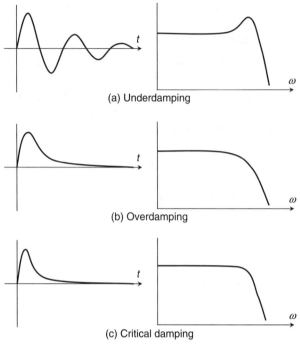

(a) Underdamping

(b) Overdamping

(c) Critical damping

FIGURE 6.4   Impulse responses and frequency characteristics of a mechanical seismometer for various dumping states.

where M is a total weight of the moving mass, K a moment of inertia around a axis of rotation, and H a distance between the center of gravity of the moving mass and the axis of rotation.

In most real sensors, a dumping force that is linear to the velocity works in a mass to avoid destruction of the sensor by resonance. The equation of motion of the dumped pendulum is in the style

$$\ddot{\theta} + 2hm\dot{\theta} + n^2\theta = -\frac{\ddot{x}}{l} \qquad (6.5)$$

where $n = \sqrt{g/l}$ (l is the length of the suspending string) and h determines the state of dumping (the dumping constant).

There are three types of behavior of the mass relating to the dumping constant h. Conceptual traces and spectra of the three types of motion are shown in Fig. 6.4.

Damped Oscillation   For $h < 1$, the solution of Eq. (6.5) can be written using integral constants α and γ as

$$\theta = \alpha e^{-hnt}\sin(\sqrt{1 - h^2}\,nt + \gamma) \qquad (6.6)$$

The motion of the mass is a damped oscillation (underdamping), where displacement of the mass decreases with a period $T_0' = T_0/\sqrt{1 - h^2} > T_0$. The amplitude ratio D of two succeeding peaks in the observed displacement is a constant (damping ratio):

$$D = \frac{a_1}{a_2} = \frac{a_n}{a_{n+1}} \tag{6.7}$$

and

$$D = \exp\left(\frac{\pi h}{\sqrt{1 - h^2}}\right) \tag{6.8}$$

Overdamping For $h > 1$, the solution of Eq. (6.5) can be written using integral constants A and B as

$$\theta = A \exp\left[-(h + \sqrt{h^2 - 1})nt\right] + B \exp\left[-(h - \sqrt{h^2 - 1})nt\right] \tag{6.9}$$

In this case, no oscillatory motion (overshoot) of the mass can be observed (see Fig. 6.4).

Critical Damping For $h = 1$, the solution of Eq. (6.5) can be written by using integral constants A and B as

$$\theta = (A + Bt)\exp(-nt) \tag{6.10}$$

This is a solution in boundary between over- and underdamping, and the settling time of the movement of the mass is the smallest among all possible solutions. In a case of detection of Earth movement, both the mass and the fulcrum of a pendulum (Fig. 6.5) can move. Assuming a fixed (global) coordinate system, the equation of movement of a pendulum for horizontal (x-axis) displacement of x can be written

$$\ddot{\theta} + 2hn\dot{\theta} + n^2\theta = -\frac{\ddot{x}}{l} \tag{6.11}$$

The frequency characteristic $H(\omega)$ of the displacement of the mass for $\theta \approx 0$ is

$$H(\omega) = \frac{\omega^2}{\omega^2 - n^2 - i2hn\omega} \tag{6.12}$$

The amplitude characteristic $|H(\omega)|$ and phase characteristic $\angle H(\omega)$ are

$$|H(\omega)| = \frac{1}{\sqrt{(1 - u^2)^2 + 4h^2u^2}} \tag{6.13a}$$

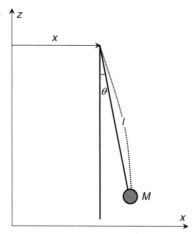

FIGURE 6.5 Simple equivalent model of a mechanical seismometer with a moving fulcrum.

and

$$\angle H(\omega) = \tan^{-1} \frac{2hu}{1 - u^2} \tag{6.13b}$$

where $u = n/\omega$. It is required that $|H(\omega)| = 1$ and $\angle H(\omega) = 0$ for detection of the displacement of the Earth (displacement seismograph). This condition can be satisfied if the frequency of the displacement is much smaller than the natural frequency of the pendulum.

The acceleration of the Earth is a second time differential of the displacement; hence, the conditions of $|H(\omega)| \propto \omega^2$ and $\angle H(\omega) = \pi$ are conditions to detect acceleration (an acceleration seismograph). This condition can be satisfied in approximation in much higher frequency than $2\pi/T_0$. The sensitivity of the acceleration seismograph is $1/n^2$. For a velocity seismograph, $|H(\omega)| \propto \omega$ and $\angle H(\omega) = \pi/2$. Around the conditions that $n \approx 1$ and $h \gg 1$, it is approximated that the seismograph has velocity sensitivity.

6.2.2 Electromagnetic Seismometers

The movement of a pendulum is transformed into voltage in electromagnetic seismometers. In the area of exploration geophysics, sensors with this principle are referred to as *geophones* or *pickup*. Moving-coil electromagnetic seismometers have typically been used for detection of the ground motion. An electromotive force at a cylindrical coil with a radius of a and a number of turns N in a homogeneous magnetic flux B is

$$E = 2\pi a N L B \dot{\theta} = G \dot{\theta} \tag{6.14}$$

where L is a distance between the axis of rotation of the pendulum and θ is an angle of rotation. A moment J working on the pendulum, which is induced by a current i in the coil, is

$$J = Gi \tag{6.15}$$

After connecting a resistance r to the coil, the equation of motion can be written

$$K\ddot{\theta} + D\dot{\theta} + C\theta = -Gi \tag{6.16}$$

where K is a moment of inertia.

Using the resistance R of the coil, current i in the closed circuit is defined as

$$i = \frac{G\dot{\theta}}{R+r} \tag{6.17}$$

Then Eq. (6.16) can be rewritten using damping constant h as

$$\ddot{\theta} + 2hn\dot{\theta} + n^2\theta = 0 \tag{6.18}$$

$$h = h_{01} + \frac{G^2}{2nK(R+r)} \tag{6.19}$$

where h_{01} is a damping constant without an external resistance r.

In the same manner as for the equation of movement for the mechanical seismometers with dumping [Eq. (6.5)], the equation of motion for an electromagnetic seismometer is

$$\ddot{\theta} + 2hn\dot{\theta} + n^2\theta = -\frac{\ddot{x}}{l} \tag{6.20}$$

The frequency characteristics of the electromagnetic seismometer are similar to those of the mechanical seismometer. However, a unique point regarding the electromagnetic seismometer is that the damping constant h is determined by both mechanical and electrical properties.

The equation of movement of an electromagnetic seismometer [Eq. (6.18)] can be modified for a feedback sensor, where a current linear to the angle of rotation θ is fed back to the coil:

$$\ddot{\theta} + 2hn\dot{\theta} + (n^2 + k^2)\theta = 0 \tag{6.21}$$

Such a feedback system can be used effectively to realize wideband acceleration and velocity seismometers, because the feedback can lower the natural frequency or enlarge the attenuation. The other advantage of the feedback system is that movement of the pendulum can be suppressed and the sensor can be used to detect strong ground motion.

6.2.3 Piezoelectric Sensors

Electrical potential (voltage) due to polarization charge appears between opposing faces of crystallized quartz, some of the ceramics and polymers when we apply stress, and differential voltage produces stress in the material in converse. Called the *piezoelectric effect*, this has been used for actuators, ultrasonic transmitters/transducers, ultrasonic motors, and acoustic filters.

The constitutive relations of the piezoelectricity, which describe the relationships among electrical and mechanical variables, are [2]

$$\mathbf{D} = \varepsilon_0 \mathbf{E} + \mathbf{P}$$
$$= \varepsilon^s \mathbf{E} + e\mathbf{S} \tag{6.22}$$

and

$$\mathbf{T} = c^E \mathbf{S} - e\mathbf{E} \tag{6.23}$$

where $\mathbf{D}$ and $\mathbf{E}$ are electric flux density and electric field. $\mathbf{T}$ and $\mathbf{S}$ are stress and strain, and ε, e, and c, are dielectric constant, piezoelectric stress constant, and elastic stiffness, respectively.

Mason's equivalent circuit of a piezoelectric transducer is shown in Fig. 6.6, where a piezoelectric element is represented as a three-port circuit with a transmission line (see the details in Kino [3] and Mason [4]). The relationship between

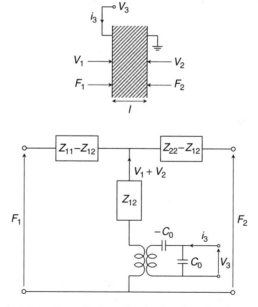

FIGURE 6.6 Maison's equivalent circuit of a piezoelectric accelerometer.

electrical and mechanical variables (F, force; V, voltage; i, current) is written using a matrix:

$$
\begin{bmatrix} F_1 \\ F_2 \\ V_3 \end{bmatrix} = \begin{bmatrix} Z_C \cot \beta l & Z_C \operatorname{cosec} \beta l & \dfrac{h}{\omega} \\[2ex] Z_C \operatorname{cosec} \beta l & Z_C \cot \beta l & \dfrac{h}{\omega} \\[2ex] \dfrac{h}{\omega} & \dfrac{h}{\omega} & \dfrac{1}{\omega C_0} \end{bmatrix} \begin{bmatrix} V_1 \\ V_2 \\ i_3 \end{bmatrix} \tag{6.24}
$$

where β is a propagation coefficient in the transmission line, and C_0 is a capacitance of the element. Z_c is a characteristic impedance of the transducer and can be represented by Z_{11}, Z_{12}, Z_{22}, and the other parameters in Fig. 6.6. As shown in Eq. (6.24), the piezoelectric transducers have resonant frequency determined by the size and acoustic and electric properties of the material. Most of the piezoelectric acoustic sensors for earthquake and seismic monitoring have resonant frequency on the order of 1 kHz to tens of kilohertz.

A simple model of a piezoelectric sensor detecting vertical motion of the ground is shown in Fig. 6.7. The equation of motion is

$$
m(\ddot{x}_1 + \ddot{x}_2) + r\ddot{x}_2 + kx_2 = 0 \tag{6.25}
$$

The solution of Eq. (6.25) can be written using the natural frequency of the accelerometer $\omega_0 = \sqrt{k/m}$ and the Q-factor $Q = \sqrt{mk}/r$ as

$$
x_2 = \frac{\left(\dfrac{\omega}{\omega_0}\right)^2}{1 - \left(\dfrac{\omega}{\omega_0}\right)^2 + j\left(\dfrac{1}{Q}\right)\left(\dfrac{\omega}{\omega_0}\right)} x_1 \tag{6.26}
$$

Equation (6.26) is approximated for a sinusoidal signal with an angular frequency $\omega(\omega \ll \omega_0)$:

$$
x_2 \approx -\frac{\ddot{x}_1}{\omega_0^2} \tag{6.27}
$$

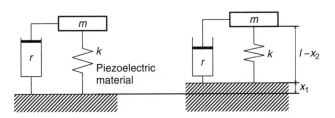

FIGURE 6.7 Simple equivalent model of piezoelectric sensor on a moving ground surface.

This result shows that a piezoelectric sensor works as an acceleration sensor in lower frequency than the resonant frequency. Piezoelectric accelerometers have advantages in flat amplitude and linear-phase characteristics in the frequency range of sensing. Lower cross-sensitivity and durability under a harsh environment without mechanical movement are the other advantages of piezoelectric sensors. Because the output impedance of the piezoelectric sensor is high, a charge amplifier or impedance converter is connected directly to the piezoelectric accelerometer to reduce noise during transmission.

Piezoelectric acoustic sensors are also used as ultrasonic sensors in geophysical logging in boreholes. A conceptual diagram of two major acoustic logging tools is shown in Fig. 6.8. One of the most commonly used acoustic logging tools is an ultrasonic tool. Several piezoelectric ultrasonic receives are mounted inside an ultrasonic logging tool, and they are used to detect several modes of acoustic waves which are radiated from a source inside the tool and propagate inside or near the field of a borehole. Travel time and attenuation are used to estimate formation velocity and attenuation. Full waveforms from the receivers can be used to identify location, orientation, and permeability of fractures crossing the borehole [5]. The other logging tool with piezoelectric ultrasonic sensors is the borehole televiewer (BHTV). Ultrasonic pulses are transmitted and received from a revolving piezoelectric device, and travel time or attenuation is used to obtain a visual acoustic image on a borehole wall. In the utilization of piezoelectric acoustic sensors for logging, signals are commonly used around the natural frequency of the sensor,

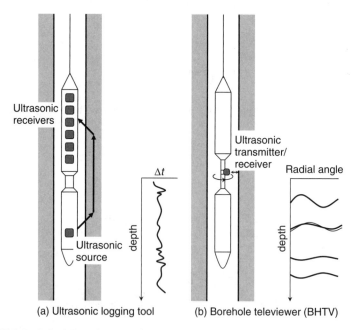

(a) Ultrasonic logging tool   (b) Borehole televiewer (BHTV)

FIGURE 6.8   Principles of (a) an ultrasonic logging tool and (b) a borehole televiewer.

because in the ultrasonic range attenuation on the formation or in the borehole fluid is not negligible, and higher sensitivity around the resonant frequency of the sensor can improve the SNR.

6.2.4 Capacitive Sensors

A capacitive sensor that detects the distance between a target and an electrode has been used as a noncontact displacement or velocity sensor. An electrostatic capacity C of a parallel-plate condenser with a gap d and a surface aperture S is

$$C = \frac{\varepsilon_0 S}{d} \tag{6.28}$$

where ε_0 is a dielectric constant in vacuum (Fig. 6.9). For a small change in the gap Δd, the associated change in the capacity ΔC is

$$\Delta C = -\frac{C}{d}\Delta d \tag{6.29}$$

This is because $\partial C/\partial d = -\varepsilon_0(S/d^2)$. The differential potential between the two electrode plates is

$$V = \frac{Q}{C} \tag{6.30}$$

where Q is an electrical charge.

Assuming constant charge on the electrode, the change in the potential associated with change ΔC in capacitance is

$$\Delta V = -\frac{V}{C}\Delta C$$
$$= \frac{V}{d}\Delta d \tag{6.31}$$

Thus, we can observe potential change linear to the displacement of the electrodes. Recent progress in the area of microelectromechanical systems (MEMS) enabled us to fabricate microsensors for detection of the ground motion. Several

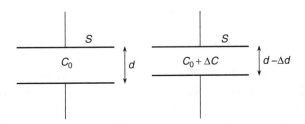

FIGURE 6.9 Capacitance with a variable gap length.

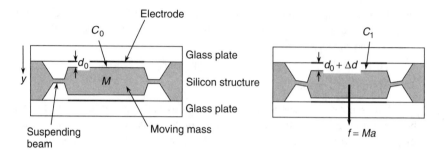

FIGURE 6.10 Principle of a MEMS capacitive accelerometer.

types of capacitive microaccelerometer have been developed for geophysical exploration and seismic monitoring within this decade [6,7]. A cut view of a MEMS accelerometer developed by Nishizawa et al. [6] is shown in Fig. 6.10. A moving mass suspended by beams is fabricated on a silicon plate and is bonded between two glass plates which have thin electrodes on the surface. This structure consists of two capacitors C_1 and C_2 at the top and bottom gaps. If both gap lengths are equal to d_0 without vibration, the initial capacitance is

$$C_{10} = C_{20} = \frac{\varepsilon_0 S}{d_0} \tag{6.32}$$

Displacement of a moving mass Δd by a small static acceleration a working along the y-direction induces change in capacitances:

$$C_1 = \frac{\varepsilon_0 S}{d_0 + \Delta d} \tag{6.33}$$

and

$$C_2 = \frac{\varepsilon_0 S}{d_0 - \Delta d} \tag{6.34}$$

The differential ratio of the two capacitances is linear to the acceleration as

$$\frac{C_2 - C_1}{C_2 + C_1} = \frac{\Delta d}{d_0} = \frac{M}{k d_0} a = \frac{1}{\omega_0^2 d} a \tag{6.35}$$

where k is a total spring constant of suspending beams.

It is clear that the sensitivity of this type of capacitive accelerometer is determined by the weight of the moving mass, the spring constant of the beams, and the length of the initial gap. However, reduction of the initial gap length is only a means to get higher sensitivity without change in the natural frequency. Hence, the length of the initial gap for sensitive accelerometer is typically on the order of micrometers. The air in such a narrow gap behaves as a dumper (squeezed film effect) and a noise source induced by Brownian motion. There is a risk of damage and sticking by hitting the glass. A feedback (servo) system is commonly used so an not to displace the moving mass and to control the frequency characteristics escaping from the squeezed film effect. Evacuation of the gap is also used to reduce the squeezed film effect.

6.2.5 Optical Sensors

Optical sensors using a laser and optical fiber can show advantages in size, cost, durability, and signal transmission to conventional mechanical and electrical sensors. Examples of optical acoustic sensors are shown in Fig. 6.11. Fiber Bragg grating (FBG), which is one variation of optical filters using artificially induced periodical change in the refractive index of the core, is sensitive to change in strain inside the fiber [8,9]. Therefore, FBG can be used as an acoustic sensor. The use of FBG sensors for the detection of a flow sensor inside a borehole, a distributed seismometer, and a low-frequency vibration sensor near the surface has been carried out experimentally [10]. Special manufacturing and signal-processing techniques are required for an the FBG sensor because it is sensitive to all the physical phenomena that bring a strain change to FBG.

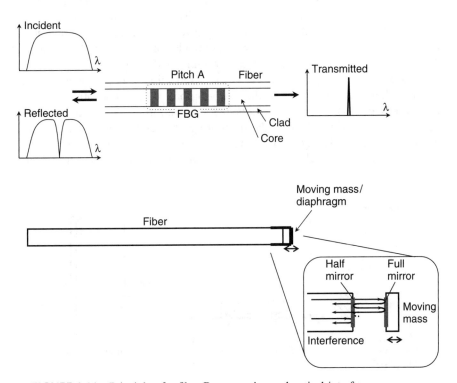

FIGURE 6.11   Principle of a fiber Bragg grating and optical interference sensors.

A laser signal is highly coherent, and interference can be observed where reflection occurs. Acoustic sensors for measurement of stress wave in water (hydrophone) using the principles of a Fabry–Perot interferometer has been fabricated using MEMS technologies [11]. A moving mass or diaphragm with a full mirror is bonded at the end of optical fiber and the intensity or spectrum of the laser is used to detect displacement of mass or diaphragm. This optical sensor can also be used to monitor the behavior of rock or concrete in laboratory tests, because the natural frequency can be designed up to several hundred kilohertz.

6.3   SENSOR INSTALLATION

One of the key factors in determining the bandwidth and SNR of collected signals is a method for the installation of sensors. It is necessary to detect the Earth's motion correctly at an observation point; however, coupling a sensor to the Earth sometimes strongly affects signal quality. Some sensor installation methods for earthquake or seismic monitoring are shown schematically in Fig. 6.12. When monitoring high-frequency AE (natural, induced, or operation related) in mines or tunnels, it is common for the sensors to be attached mechanically or chemically to the end of rock bolts inserted in the formation to increase the stability of subsurface cavities. Estimation of the source signature is often difficult for signals detected in such a way because resonant sensors are typically used to improve the sensitivity for small AE sensors, and the rock bolts behave as waveguides bringing strong frequency characteristics. Shallow boreholes are also often used for the installation of AE sensors to avoid noise from traffic, drilling, and blast.

Surface sensors are most commonly used in reflection or refraction geophysical and geotechnical seismic surveys where two- or three-dimensional arrays with tens to hundreds of seismometers or geophones are uniformly deployed to measure the wavenumber vector field. In the monitoring of microtremors, a circular array or random array is deployed to estimate velocity dispersion of Rayleigh waves, which is highly correlated to the velocity structure of S-waves. The seismometers

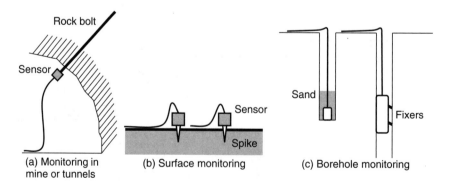

FIGURE 6.12   Installation of seismic–acoustic detectors.

are mounted on a spike that is inserted in the ground and covered by soft resin to protect them from water penetration and shocks in transportation. The surface layer is typically softer than the deeper formation, creating high attenuation, especially under high-frequency conditions and where there is poor coupling to the ground. It is also known that most acoustic noise propagates as a surface wave, where elastic energy is concentrated near the surface. Information on wavenumber vectors can be corrected by a surface array; however, elimination of surface noise sometimes becomes an analytical issue in the processing of data [12]. Surface monitoring also has a disadvantage in the detection of three-dimensional particle motion (a hodogram) because the sensor detects motion at a solid–air interface, not in bulk solid, and the coupling is no better than downhole measurement.

Signals that are wideband and low noise in nature can be collected by downhole monitoring, where seismometers, geophones, or accelerometers are mounted inside a pressure vessel (sonde) and installed inside a borehole using wireline. Improvement in coupling between the sonde and the borehole is one of the keys to collecting high-quality signals in downhole monitoring. One of the fixing methods is *sanding* of the sonde, where the sonde is hung reaching down to the bottom-hole and sand, cement, and drill mud are thrown into the borehole. Good coupling can be realized, and the structure of the sonde is simplified without a mechanism for fixing the sonde to the borehole in sanding. However, it may be difficult to retrieve the sonde because the coupling friction sometimes exceeds the strength of the wireline. The other practical method is to clamp the sonde to the borehole wall using mechanically or hydraulically driven fixers inside the sonde. The sonde can be installed or retrieved at any depth in the borehole using the fixing system, even though the sonde becomes larger, heavier, and expensive. Downhole measurement is suitable for multicomponent earthquake or seismic monitoring (see the next section) because of the installation of the sensor inside the solid earth. An electric compass or a mechanical or optical gyro is assembled inside the sonde for the orientation of each component. The number of monitoring stations is limited by practical cost considerations; hence, techniques to extract as much information from a "sparse network" is of importance in downhole monitoring. In the area of monitoring of microseismicity from oil and gas reservoirs or crosswell active seismic monitoring, a vertical array of sensors is often deployed inside a borehole.

Detected signals are transmitted to an amplifier or conditioner and then recorded to a medium such as magnetic tape, hard disks, or DVD ROMs. When monitoring earthquake and geophysical exploration, the length of the transmission line sometimes exceeds several kilometers, and noise suppression induced by the transmission might be required. Some patterns of signal transmission are shown in Fig. 6.13. Amplification of the signal in the sonde or at the wellhead works effectively to reduce the noise induced during transmission because of the low output impedance of the amplifier. However, an active circuit and power supply in the sonde or at the wellhead should be prepared for this transmission. Recent advances in integrated circuits have enabled us to digitize the signal at the sensor or the wellhead and transmit through a digital wireline, optical cable, or LAN. Problems in noise and attenuation do not appear in such a transmission system.

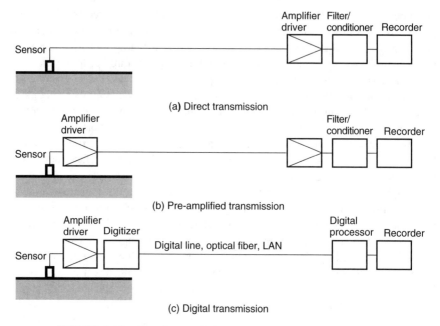

FIGURE 6.13 Signal transmissions in surface seismic monitoring.

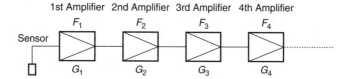

FIGURE 6.14 Cascade amplifiers.

Care must be taken when using a sequence of amplifiers (Fig. 6.14). The noise figure (NF) of the overall system can be written

$$\text{NF}_{\text{all}} = F_1 + \frac{1}{G_1}\left(\frac{F_2 - 1}{1} + \frac{F_2 - 1}{G_2} + \frac{F_3 - 1}{G_3} + \cdots\right) \qquad (6.36)$$

where F_i and G_i are the NF and gain of the ith amplifier. The NF is a measure of the reduction of the SNR using an amplifier or conditioner. We see that the NF_{all} for a sequence of amplifiers is determined primarily by the F_1 and G_1, features of the first amplifier, suggesting that the first amplifier, with a higher SNR (and smaller NF) and larger gain, improves the overall NF of the system.

The contamination of noise in signal transmission is also a practical problem in field monitoring. A *common-mode noise*, which is induced by differences in the ground potential level at the sensor, amplifier, and recording unit, is a typical noise in monitoring with a long transmission line. *Balanced signal transmission*, in which signal is transmitted through a shielded twisted-pair cable and a differential amplifier with a high common-mode rejection ratio (CMRR) is used at the receiving interface of the amplifier, is one solution to reducing common-mode noise. An isolator using an LED photocoupler or transformer can separate the transmission line and amplifiers, electrically and work effectively to eliminate the common-mode noise. A *normal-mode noise*, which is induced in a closed-loop circuit by electrostatic induction or electromagnetic induction, can be suppressed by the use of shielded cable and an appropriate ground.

Most earthquake and seismic sensors are sensitive to gravity, and signal is conditioned by a high-pass filter to eliminate the gravity component at the first stage of amplification or conditioning. Sampling theory does not allow the signal to be digitized to result in a frequency component higher than $f_2/2$ (f_s, sampling frequency) for correct recovery of the original signal from a digitized signal. Hence, the signal is filtered by an antialiasing low-pass filter at the interface of the digitizer. The gain at the amplifier should be chosen carefully to avoid saturation and bit-resolution error.

A continuous or triggered digital recording system is used most commonly to record signals. In a continuous system, all the signals, including noise, are recorded to a medium, typically to magnetic tape. On the other hand, fixed data length before or after a trigger signal is recorded in the trigger recording system. A trigger system is used in earthquake monitoring and exploration geophysics, because the data size can be reduced and tape change is not required. Triggered data are saved on a hard disk in binary format, which is standard in each trade. The sampling frequency should be determined by considering the bandwidth of the signals detected and by the sampling theory. A 16-bit digitizer can easily be found in the marketplace for a reasonable price. However, utilization of a 24-bit or higher digitizer can be expected in earthquake and seismic monitoring, because of the wide dynamic range of the sensor. Progress in semiconductor memory technologies has enabled us to record long-term earthquake and seismic data continuously with a higher sampling rate. A future data recording system would be a hybrid style of trigger or continuous recording system in which all the continuous data are stored as backup, and signals for analysis are extracted and stored in different media in the appropriate format.

6.4 MULTICOMPONENT TECHNIQUES

Solutions of the wave equation for elastic waves, which is derived from Hooke's law and Newton's law on motion, demonstrate that all the modes of elastic waves can propagate in solid media. Propagation of elastic waves in solid media is interpreted mathematically as a propagation of vector and scalar potentials [13] with different velocity and unique motion at a point (particle motion, hodogram) for each mode. An example of the complex behavior of three-dimensional particle motion associated with the arrival of a microseismic event is shown in Fig. 6.15. It is well known that P- and S-waves show perpendicular particle motion. More complex particle motion is observed at the onset of randomly scattered waves (coda), as shown in the figure.

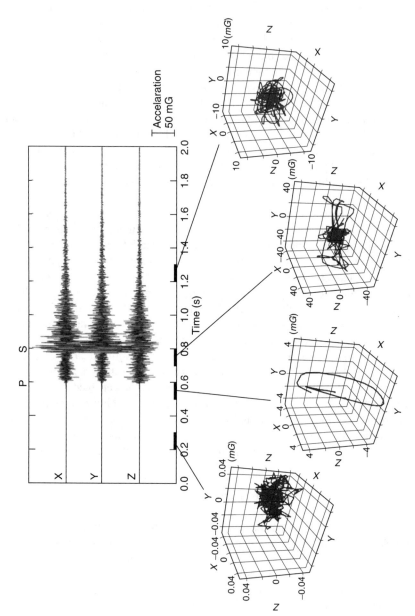

FIGURE 6.15 Typical three-dimensional motion of a microseismic event. (Adapted from [14].)

Multicomponent detection of earthquake and elastic waves is one of the most suitable means to collect information on such propagation of vector potentials. There is a wide flexibility in sensor configuration to detect and reconstruct three-dimensional particle motion correctly; however, the minimum configuration is a triaxial configuration in which three directional sensors that have equal sensitivity are mounted orthogonally (Fig. 6.16). The multicomponent sensor will be "overdetermined" if the number of sensors increases. We describe techniques for triaxial sensors in this book, although sensor evaluation and compensation techniques are applicable to an overdetermined multicomponent sensor [15].

Assuming identical flat sensitivity and zero-phase characteristics among the sensors, a time-variant vector $\mathbf{U}(n)$ of displacement, velocity, and acceleration can be written using the three components $s_x(n)$, $s_y(n)$, and $s_z(n)$:

$$\mathbf{U}(n) = [s_x(n), s_y(n), s_z(n)] \tag{6.37}$$

where

$$s_x(n) = |\mathbf{U}(n)| \cos \psi \cos \phi$$
$$s_y(n) = |\mathbf{U}(n)| \cos \psi \sin \phi$$
$$s_z(n) = |\mathbf{U}(n)| \sin \psi$$

n being a sampling sequence number. The angles are defined in Fig. 6.17.

The state of polarization can be evaluated by human observation of the trace of the three-dimensional particle motion (three-dimensional hodogram) using visualization software. However, mathematical analysis is indispensable for quantitative evaluation of polarization in the time and frequency domains. Principal component analysis (PCA) of a digitized 3C signal [16] is used effectively to evaluate the hodogram.

The variance and covariance matrix $\mathbf{C}$ of the digitized three-dimensional particle motion within an interval of $[n_1 T_s, n_2 T_s]$ (T_s, the sampling interval) can be written

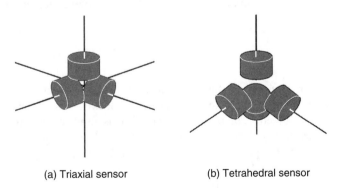

(a) Triaxial sensor   (b) Tetrahedral sensor

FIGURE 6.16 Seismic sensors for the detection of three-dimensional motion of the Earth.

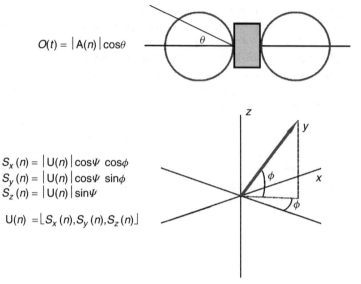

$$O(t) = |A(n)| \cos\theta$$

$$S_x(n) = |U(n)| \cos\psi \, \cos\phi$$
$$S_y(n) = |U(n)| \cos\psi \, \sin\phi$$
$$S_z(n) = |U(n)| \sin\psi$$

$$U(n) = \lfloor S_x(n), S_y(n), S_z(n) \rfloor$$

FIGURE 6.17 Conversion of three-dimensional motion of the Earth to three component signals.

$$\mathbf{C} = \begin{bmatrix} e_{xx} & e_{xy} & e_{xz} \\ e_{yx} & e_{yy} & e_{yz} \\ e_{zx} & e_{zy} & e_{zz} \end{bmatrix} \tag{6.38}$$

where

$$e_{ij} = \frac{1}{n_2 - n_1 + 1} \sum_{n=n_1}^{n_2} [s_i(n) - \bar{s}_i(n)][s_j(n) - \bar{s}_j(n)] \qquad i, j = x, y, z$$

Eigenvectors $v_l (l = 1, 2, 3)$ and eigenvalues $\lambda_1 (l = 1, 2, 3)$ of the variance and covariance matrix can be obtained by solving the following Eigen equations:

$$|\mathbf{C} - \lambda_l \mathbf{I}| = 0 \tag{6.39}$$

$$\mathbf{C} v_l = \lambda_l v_l \tag{6.40}$$

The eigenvector direction is that in which information regarding a digitized hodogram is most strongly projected, and eigenvalues shows power along each eigenvector. The direction of the first eigenvector is used as an estimate of the polarization direction of the hodogram. The state of polarization can also be estimated by the global polarization coefficient (GPC) [17],

$$\text{GPC} = \frac{\lambda_1}{\lambda_1 + \lambda_2 + \lambda_3} \tag{6.41}$$

The GPC takes values between 0.33 and 1. A higher GPC shows that the hodogram is linear, and a lower GPC means that the hodogram is spherical.

Short-time Fourier transform (STFT), which is Fourier transformation for a signal of short duration using a moving window, enables us to estimate the time–frequency representation of spectra [18]. The time–frequency representation $STFT_i(n_0, k)$ of the ith component of the signals within the interval $[n_1 T_s, n_2 T_s](n_0 = (n_2 - n_1)/2)$ can be written

$$STFT_i(n_0, k) = \sum_{n-n_1}^{n_2} s_i(n)e^{-j2\pi(n/N)k} \tag{6.42}$$

where $N = n_2 - n_1 + 1$. By using $STFT_i(n_0, k)$, the variance and covariance matrix can be rewritten in time–frequency representation style:

$$S(n_0, k) = \begin{bmatrix} s_{xx} & s_{xy} & s_{xz} \\ s_{yx} & s_{yy} & s_{yz} \\ s_{zx} & s_{zy} & s_{zz} \end{bmatrix} \tag{6.43}$$

where

$$s_{ij} = STFT_i(n_0, k) \, STFT_j^*(n_0, k)$$

This time–frequency matrix representation of the hodogram $S(n_0, k)$, referred to as a *spectral matrix* [19], enables us to visualize the behavior of a hodogram in the time and frequency domains. As we described in Section 6.3, the three-component detector may have the frequency-dependent detectability of a three-dimensional hodogram because of the resonance of the sonde and coupling to the borehole. The spectral matrix can be used to evaluate quantitatively the detectability of the hodogram in the time and frequency domains. Figure 6.18 shows two examples of the time–frequency representation of a hodogram at the onset of a P-wave for two types of three-component seismometers installed in a borehole. Although the bandwidth of a seismic signal is similar to that of the two detectors (-500 Hz), the hodogram from detector (a) showed highly linear and stable behavior up to 450 Hz, whereas a linear hodogram was observed up to 200 Hz for detector (b) and the hodogram is nonlinear and unstable at higher frequencies. The spectral matrix can be used for the in situ passive calibration of multicomponent detectors. Change in the state of polarization at the onset of P-waves is evaluated by a variation of the spectral matrix and is used for the precise detection of the time of arrival of seismic waves [20].

The concept of time–frequency representation of the three-dimensional hodogram has also been introduced in the passive (microseismic) reflection method, in three-dimensional time–frequency coherence analysis, and in the triaxial drill-bit reflection method [18,21,22], all of which have the ability to identify plane waves with small magnitude superimposed onto randomly scattered seismic waves.

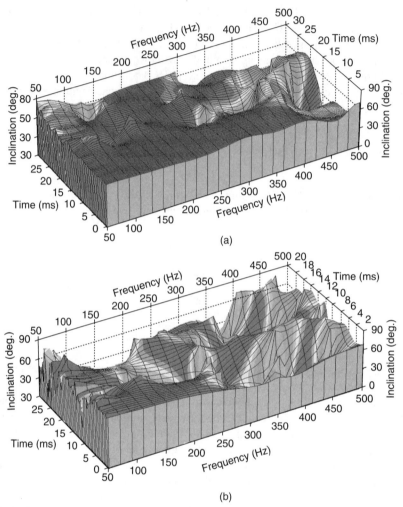

FIGURE 6.18 Time–frequency representation of three-dimensional motion detected by spectral matrix analysis.

6.5 LIMITATIONS

There are several limitations on the use of sensors for earthquake and seismic monitoring. Performance and cost are trade-offs as they are for other sensors. Cost considerations sometimes prevent the use of highly sensitive sensors, even in commercial exploration for oil and gas, which can result in considerable financial benefit when reservoirs are discovered. The cost of a highly sensitive earthquake monitoring system may exceed tens of thousands of dollars, and scientific researchers may find it difficult to deploy a wide-area high-density monitoring network even though the financing agency understands its importance.

Earthquake and seismic sensors for monitoring under high temperature are of importance for downhole monitoring of oil and gas, and of geothermal and natural earthquake events. As described earlier, the greatest advantage of downhole monitoring is that we can monitor near-field phenomena with high accuracy. Although it is highly dependent on the temperature gradient at a site, the operational temperature of a sensor may easily exceed 150°C at a depth of several kilometers. Few elements of an active electrical circuit can work under such a temperature, and the sensor itself may not fulfill its expected performance level and may suffer reduced durability. A heat insulation system such as a vacuum vessel or Peltier cooling system can improve durability under high-temperature conditions, but even such a system cannot be used for long-term monitoring and may create cost and size problems.

Sensor durability of 100 years or more is expected in the stability monitoring of nuclear disposal in tunnels or cavities in the earth, but such sensors do not exist currently in the marketplace. Long-life sensors are also in demand for use in intelligent structures where sensors buried in concrete are used to monitor stability. Long-life sensors are also appreciated for use in intelligent boreholes, where sensors are cemented to the outside of casing pipe and used for production and monitoring, reducing the cost of drilling and of monitoring and measurement. Sensors cannot be repaired or replaced in such built-in applications, so long life and reliability are expected.

Optical sensors have advantages in cost, size, and ability for multiplex signal transmission. However, most optical fibers for telecommunication may lose their low-attenuation nature or melt down under temperatures exceeding 100°C. The FBG also loses sensitivity under higher-temperature or high-water conditions. Some optical fibers for high-temperature are available, but any low cost advantage may be lost when for several hundred meters of such special fibers are required.

6.6 FUTURE PROSPECTS

The collection of information from inside the Earth within cost and time constraints is the main concern in practical earthquake and seismic monitoring, and effort are is directly mainly toward overcoming time and expense problems. Digital technologies are changing earthquake monitoring and seismic geophysical exploration drastically. Self-standing sensors with in a digital communication system accomplish monitoring at a reasonable expenditure of money and time through the use of widely deployed arrays. Digital amplifiers, filters, and processors can improve signal quality without distortion and make it possible to transmit signals from an array of thousands of sensors with a few digital lines. The availability of a high-bit-resolution analog-to-digital converter can produce a dynamic range exceeding 200 dB. The development of a sensitive seismometer for long-term monitoring in deep ocean boreholes is expected to clarify the physics of seismogenic data in the subduction zone.

Recent progress in the micro- and nanotechnologies has made it possible to fabricate various sensors and circuits on a small chip. Small and less expensive integrated micro-clinometers and micro-seismometers with digital radio telemetry could be used effectively as a disposal sensor to monitor active volcanos where people cannot install or retrieve sensors. Integrated sensors also have the ability to monitor the health of infrastructure such as tunnels, dams, bridges, and buildings. Acoustic sensors can provide information for modal analysis to estimate the strength of a structure using traffic or artificial vibration as a source. A nationwide and transocean optical fiber network can be used as a wide-area earthquake network for early warning of strong motion by installing optical micro-seismometers or micro-hydrophones every several kilometers. The characteristics of the coherent laser and the precise controllability of wavelengths may make it possible to realize ultrahigh-sensitivity seismometers.

Acknowledgment

The author wishes to acknowledge C. Watanabe for comments and suggestions regarding the writing of this chapter.

REFERENCES

1. Aki, K., Richards, P. G. 1980. *Quantitative Seismology*. W.H. Freeman, San Francisco.

2. David, J., Cheeke, N. 2002. *Fundamentals and Applications of Ultrasonic Waves*. CRC Press, Boca Raton, FL.

3. Kino, G. S. 1987. *Acoustic Waves*. Prentice-Hall, Englewood Cliffs, NJ.

4. Mason, W. P. 1950. *Piezoelectric Crystals and Their Application to Ultrasonics*. D. van Nostrand, Princeton, NJ.

5. Paillet, F. L., Cheng, C. H. 1991. *Acoustic Waves in Boreholes*. CRC Press, Boca Raton, FL.

6. Nishizawa, M., Niitsuma, H., Esashi, M. 2000. Miniaturized downhole seismic detector using micromachined silicon capacitive accelerometer (CD-ROM). In *SEG 2000 Expanded Abstracts*.

7. http://www.analog.com.

8. Hill, K. O., Fujii, Y., Johnson, D. C., Kawasaki, B. S. 1978. Photosensitivity in optical fiber waveguides: application to reflection filter fabrication. *Appl. Phys. Lett.*, 32:647–649.

9. Born, M., Wolf, E. 1974. *Principles of Optics, Electromagnetic Theory of Propagation, Interference and Diffraction of Light*. Pergamon Press, Elmsford, NY.

10. Takashima, S., Asanuma, H., Niitsuma, H. 2004. A water flowmeter using dual fiber Bragg grating sensors and cross-correlation technique. *Sens. Actuat. A*, 116:66–74.

11. Asanuma, H., Ohishi, H., Niitsuma, H. 2004. Development of an optical micro AE sensor with an automatic tuning system. *J. Acoust. Emission*, 23:64–71.

12. Kearey, P., Brooks, M. 1984. *An Introduction to Geophysical Exploration*. Blackwell Scientific, Oxford, UK.

13. Redwood, M. 1960. *Mechanical Waveguides*. Pergamon Press, Elmsford, NY.

14. Soma, N., Niitsuma, H., Baria, R. 2007. Reflection imaging deep reservoir structure based on 3D hodogram analysis of multi-component microseismic waveforms. *J. Geophys. Res.*, 112:B11303-1 to B11303-14.

15. Jones, R., Asanuma, H. 2004. The tetrahedral geophone configuration: geometry and properties. In *SEG 2004 Expanded Abstracts*, pp. 9–12.

16. Press, W., Teukolsky, S. A., Vetterling, W. T., Flannery, B. P. 2007. *Numerical Recipes: The Art of Scientific Computing*, 3rd ed. Cambridge University Press, New York.

17. Samson, J. C. 1977. Matrix and Stokes velocity representations of detectors for polarized waveforms: theory with some applications to teleseismic waves. *Geophys. J. R.*, 51:583–603.

18. Cohen, L. 1989. Time–frequency distributions: a review. *Proc. IEEE*, 77:941–981.

19. Moriya, H., Nagano, K., Niitsuma, H. 1994. Precise source location of AE doublets by spectral matrix analysis of triaxial hodogram. *Geophysics*, 59:36–45.

20. Moriya, H., Nagano, K., Niitsuma, H. 1996. Precise detection of a P-wave in low S/N signal by using time–frequency representations of a triaxial hodogram. *Geophysics*, 61:1453–1466.

21. Asanuma, H., Kizaki, T., Niitsuma, H. 2002. Analysis of seismic waves propagated through a pressurized fracture by the 3D-TFC method. In *SEG Expanded Abstracts*, pp. 2385–2388.

22. Asanuma, H., Liu, H., Niitsuma, H., Baria, R. 1999. Identification of structures inside basement at Soultz-sous-Foret (France) by the triaxial drill-bit VSP. *Geothermics*, 28:355–376.

CHAPTER SEVEN

Auxiliary Sensors

7.1 OVERVIEW

In this section, the auxiliary sensor types generally used for specific application areas or to improve fundamental detector performance are described briefly with their operating principles, current capabilities, limitations, and improvement potentials. Most of these sensor technologies, such as optical sensing, electrical impedance tomography, x-ray imaging, and explosive-material detection, are still at the stage of prototyping or technical research [1].

Optical sensors use infrared (IR) and hyperspectral (0.35 to 14 μm) bands specifically for the detection of shallow buried objects such as mines. The source of the electromagnetic signal radiated by the target may be either natural or artificial, corresponding to passive and active thermal detection concepts, respectively. The heating effect of solar energy causes thermal radiation detectable by a passive IR sensor. If a buried object absorbs heat very differently than the medium, it can easily be imaged with high resolution. Broadband IR sensors are commercially available in the world marketplace. However, day/night time zones and environmental conditions usually limit detection performance. Active thermal detection techniques benefit from externally controlled heat sources such as lamps, lasers [2], and high-power microwave (HPM) [3]. Here, the main trouble is long exposure times heat.

Electrical impedance tomography (EIT) is a basic low-cost measurement technique to obtain the conductivity distribution of ground surface. The detection principle is based on the discrimination of shallow buried objects whose electrical resistance differs from that of soil. Many low-current electrodes are used to probe

Subsurface Sensing, First Edition. Edited by Ahmet S. Turk, A. Koksal Hocaoglu, and Alexey A. Vertiy.
© 2011 John Wiley & Sons, Inc. Published 2011 by John Wiley & Sons, Inc.

the ground surface to image the surface resistivity with a required resolution [4]. The detection capability of EIT is limited to the probing depth, which is usually less than 20 cm. It is also not convenient for rapid (vehicle-mounted) wide-area scanning operations.

Nuclear quadrupole resonance (NQR) is an electromagnetic-based sensor technique based essentially on the magnetic resonance (MR) phenomenon [5]. The operational principle of an NQR is based on the electrostatic interaction of non-spherical nuclei with the gradient of the crystal electric field (CEF) produced by neighboring charges. Due to this interaction, the energy levels of a nucleus in the crystal electric field are split, and the electromagnetic field oscillating in the radio-frequency (RF) range 0.4 to 5.0 MHz can induce transitions between these levels. The resulting set of NQR transitions, called the *NQR spectrum*, it is a unique characteristic of the specific chemical compound. This sensor is employed specifically for the detection of explosive materials, including various solid-state crystalline substances. NQR can provide a highly specific and arguably unique frequency signature for the material of interest. Nevertheless, the NQR sensor is suffers greatly from environmental RF interference; for example, AM radio broadcast corresponds to TNT frequencies. Moreover, this technology is still not a rapid detector system. Thus, it is designed primarily for use as a confirmatory sensor for explosive object detection, in combination with other technologies, such as metal detector (MD) and ground-penetrating radar (GPR).

X-ray scattering is one of the highest-resolution buried-object detector technologies. A pencil beam of x-rays is emitted to the scanning area, and the backscattered rays are detected by a sensor [6]. The performance depends essentially on the x-ray absorbing characteristics of the target object. For example, organic materials, soil layers, and explosives have fewer absorbing and more scattering characteristics than those of metals. Solid objects are more easily distinguished than clutters. Since the higher x-ray energy sources are needed to penetrate deeper, such sensors are usually suitable for shallow buried targets.

In addition to these detector types, biological, electrochemical, and piezoelectric sensor systems are also utilizable for specific applications such as explosive material detection, which may be very important, especially for military applications (e.g., mine detection). Biological smell sensors can detect vapors or gases produced by the explosive components of objects. An electrochemical detector measures the changes in polymer electrical resistance of explosive vapors. A piezoelectric sensor uses the resonant frequency shifting of various materials when exposed to explosive vapors [1].

7.2 BIOLOGICAL AND CHEMICAL METHODS OF EXPLOSIVE DETECTION

7.2.1 Introduction

An increasing amount of effort is being spent on the development of new solutions for explosive detection, including chemical and biochemical sensor technology.

The importance of this type of sensor technology as a possible method for mine and explosive detection has recently been emphasized by Yinun [7]. Single sensors or sensor arrays for the detection of common explosives such as TNT or RDX are described here to summarize comprehensive work by several authors during the past decade. Substantial progress has been made, in this period, as indicated by the steeply increasing number of contributions to scientific journals and conferences. A description of sensitive chemical and biological materials, and a summary of the results achieved, are presented in this review on sensor technology for explosive detection. The relevant transducer technologies were described briefly in Section 2.7. In general usage the term *sensor* includes a wide variety of technologies, but we use it here only to refer to chemical sensors or biosensors.

Sensors do not detect a mine or buried unexploded ordinance (UXO) itself but, rather, the vapors emanating from the explosive contained in the device. Thus, gas-phase detection by gas sensors and liquid sensing in ground and surface waters are to be achieved. With the very few exceptions of special biosensor devices, for direct gas-phase detection only chemical sensors are utilized, whereas in ground- or seawater, the use of both chemical and biosensors is proposed. Other detection methods and instruments based on biological principles (e.g., assays using plants or algae) or physical principles (e.g., spectroscopy) are also available and may be called sensors, but the technologies do not comply with our definition of biosensors or chemosensors and so are included here only if the underlying method is suitable for use in chemical and biosensors or is related very closely to sensor technology. This applies, for example, to the various immunoassays techniques, as the antibodies can be used equally well in biosensors.

As of today, chemical and biochemical sensor technology is generally still judged to be an auxiliary method in the mine detection field, due to the limits of the current technology. In principle, sensor technology can be used alone for mine detection, but it is considered more suitable as a support technique: to classify objects detected by another detector (e.g., a metal detector) or to discriminate between mines or UXO and pieces of bare metal. However, in another section we deal with examples of sensor technology dedicated to mine detection. In other fields (e.g., security), sensor technology to detect explosives is considered much more advanced to meet the requirements of the application.

The main problem in gas-phase detection from a distance, especially when the explosive is buried and encapsulated, is the low vapor pressure of the most common explosives used in military applications. However, some explosives release degradation products or contain impurities (e.g., TNT releases DNT) that often display much higher vapor pressures. The same holds for liquid-phase detection, as the actual concentrations in water are extremely low, due to the low solubility and dilution effects. Thus, the sensors must be able to achieve very low detection thresholds. The second issue is the possible presence of interferants, such as hydrocarbons or other organic compounds contained in the ambient medium. Furthermore, the types of interferants and their concentrations may vary widely from location, to location, depending on the measurement conditions and other factors, such as soil type, vegetation, and the former use of the land. These two issues—low concentrations of the explosive and a high background level of interferants—need to be considered in sensor development. However, insufficient selectivity rather then sensitivity is the main obstacle in real applications. Highly selective sensors or sensor arrays are necessary to meet all requirements of the mine detection problems and continue to be the main challenge in development.

A variety of methods are employed to enhance detection limits and to counteract influences of interferants. The methods available are very often the same as those used in classical analytical methods. For example, detection can be improved by a two-step analyte enrichment: by sampling large amounts of air through a sorbent trap and by measurement of the desorbed analytes. Sample separation techniques such as chromatography or the use of filters may increase selectivity. However, the main objective of sensor development is to avoid sample enrichment and separation steps in the analysis procedure, as they increase the complexity and thus the total analysis time, and and eliminate some of the advantages of sensor technology over conventional analysis methods.

Following the foregoing procedure, to detect explosives under real conditions, sensors need highly specific interactions between the analyte molecules and the sensitive material. For this reason, the most common gas sensor type, a metal-oxide conductive sensor based on doped tin dioxide, tungsten dioxide, or other oxides, is not presented in the list of sensors suggested for explosive detection, because it offers only limited selectivity, despite its high sensitivity. Many different approaches have been employed to design sensitive materials in order to reach this objective. The first approach is to design the receptor molecules such that they are capable of specific chemical interactions with the target analyte molecules. Figure 7.1 shows examples of such chemical compounds selected for explosive detection with reported performance in quantification of the explosive vapor. A TNT molecule, for example, is an electron-deficient aromatic system and a hydrogen-bond acceptor. Consequently, electron donors and strong hydrogen donors have been proposed, and very low detection limits in the gas phase have been achieved using such receptors (Table 7.1). The second approach is to use highly specific properties of the analyte molecules, such as the quenching of fluorescence. Finally, the highly specific interaction pathways of biomolecules, such as antibodies and enzymes, favor the use of biosensors, as shown by the large number of publications on biosensors for explosive detection. However, whereas gas-phase detection is preferred to detect landmines, biosensors are usually proposed for explosive measurements in sea- and groundwater to assess pollution or to prepare and monitor remediation efforts.

To summarize, the challenge and the main focus of research in the field of explosive detection (for general cases, and specifically in mine detection) lies with the development of new sensitive materials, whereas the transducers utilized have usually been developed for other application fields. In the following section, the chemical sensors and biosensors proposed for the detection of explosives are presented categorized by transduction principles.

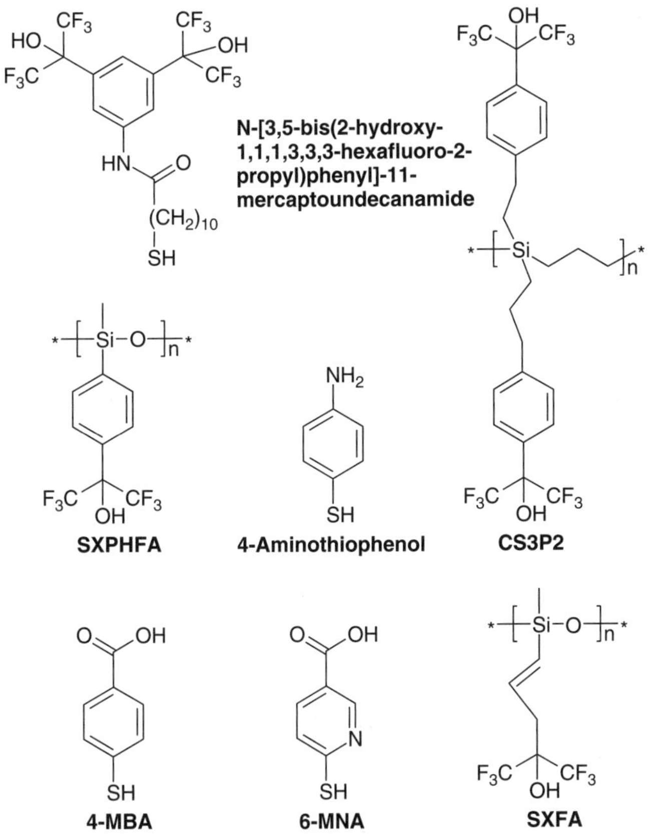

N-[3,5-bis(2-hydroxy-1,1,1,3,3,3-hexafluoro-2-propyl)phenyl]-11-mercaptoundecanamide

SXPHFA   4-Aminothiophenol   CS3P2

4-MBA   6-MNA   SXFA

FIGURE 7.1 Chemical structures of sensitive compounds capable of special chemical interaction with analyte molecules.

7.2.2 Electromechanical Transducers

The class of electromechanical transducers consists mainly of acoustic transducers such as the quartz crystal microbalance (QCM), the surface acoustic wave transducer (SAW), the cantilever, and the thin-film bulk acoustic resonator (FBAR). QCM and SAW devices have been used for many decades as chemical sensors for volatile organic compound (VOC) detection. The FBAR and cantilever are more recent developments made possible through advances in MEMS technology. They can easily be used to produce very compact microsensor arrays, and consequently, they are considered advantageous by many authors. Acoustic transducers are basically mass sensitive, whereas the cantilever can be used as a mass-sensitive device or as a device sensitive to surface stress produced by adsorbed analyte molecules.

A wide variety of sensitive materials suitable for use in these electromechanical transducers are available. The same sensitive material can often be used in all transducer types, but also in combination with others, such as optical and

TABLE 7.1 Compounds Used in Sorption-Based Chemical Sensors Capable of Special Chemical Interactions to Enhance Sensitivity and Selectivity Toward Explosives and Related Compounds

Sensitive Material	Strategy	Transducer	LOD/Lowest Concentration Tested	Analyte	Ref.
CS3P2	Strong H-bond donor via OH group	SAW (250 MHz)	92 ppt/31 ppb	2,4-DNT	11
SXPHFA	Strong H-bond donor via OH group	SAW (250 MHz)	235 ppt/400 ppb	2,4-DNT	13
N-[3,5-bis(2-hydroxy-1,1,1,3,3,3-hexafluoro-2-propyl)phenyl]−11-mercaptoundecanamide	Strong H-bond donor via OH group	QCM (9 MHz)	—/3 ppm	2,4-Dinitrotrifluoro-methoxybenzene	14
4-Aminothiophenol	Electron-rich aromatic system, π-electron donor	QCM (9 MHz)	—/3 ppm	2,4-Dinitrotrifluoro-methoxybenzene	15
β-Cyclodextrin functionalized with trimethylbenzyl groups	Molecular recognition via cage structure	SAW (250 MHz)	—/2 ppb —/—	o-Nitrotoluene TNT	17
Molecularly imprinted polymers based on acrylamide and methacrylic acid	Molecular recognition via target-specific arrangement of recognition sites in a cavity	QCM (10 MHz)	Very selective TNT uptake, no LOD given	TNT	18
4-Mercaptobenzoic acid	Hydrogen bonding through the carboxyl group	Cantilever	—/1.4 ppb	PETN	25
6-Mercaptonicotinic acid	Hydrogen bonding through the carboxyl group	Cantilever	—/290 ppt 20 ppt/0.1 ppb	RDX TNT	26, 27
SXFA	Strong H-bond donor via OH group	Cantilever	300 ppt/5 ppb	2,4-DNT	28

electrical (capacitive) transducers. Both bulk sorption into layers of polymer or macromolecules as well as adsorption at surfaces or thin layers down to the mono-layer level are utilized for explosive detection with electromechanical transducers. Direct liquid-and gas-phase sensing is possible with transducers in this category, generally depending only on the type of sensing material used. The QCM and SAW transducers are discussed together, as both are based on the same principle and yield comparable results when the same or similar sensitive layers are used.

A basic approach to the choice of sensitive material is to use thin layers of organic polymers without a special recognition element as the sorbent material on the transducer, similar to enrichment techniques for many analytical methods. This was reported by Tomita et al. three decades ago [8] as one of the first works on explosive detection with acoustic transducers. Different nitro aromatics such as DNT and DNB contained in TNT were the aim of a similar recent investigation using poly(ethylene gycol) and poly(dimethylsiloxane) (PDMS), both common column materials for gas chromatography columns and sorbent materials for analyte enrichment [9,10]. Detection of low levels of various nitro aromatic vapors exposed to the sensors was possible under laboratory conditions. However, selectivity is not expected to be as satisfactory in field applications, as both materials are known to absorb many other VOCs as easily.

A systematic and more directed approach to achieving higher sensitivity and selectivity for nitroaromatics in polymer sorbent design was developed by a group of authors specialized in this technique [11–13]. The layers selected were tested on SAW sensors. An extrapolated detection limit of 92 ppt for 2,4-DNT, which is one of the lowest detection limits reported for polymers on acoustic transducers, was achieved using hexafluoroisopropanol-functionalized sorbent materials. Those hexafluoroisopropanol functional groups are strongly hydrogen-bond acidic, due to the electron-withdrawing influence of the fluorine on the hydroxyl group, and thus they are able to interact very efficiently with nitroaromatics. This was proven by attenuated reflectance (ATR) infrared spectroscopy studies by clearly identifying the hydrogen-bond interactions. Self-assembled monolayers (SAMs) having similar functional groups able to undergo hydrogen bonding were also tested [14]. However, sensor response was quite low, as many fewer reactive sites are available in SAMs.

Similarly, different SAMs made of thiols of varying structures bearing amine, acid, or aromatic functional groups were created on QCM transducers, and their response to 2,4-dinitrotrifluoromethoxybenzene vapor was evaluated [15]. The compounds are able to interact with a nitroaromatic compound via hydrogen bonding between the CO_2H or NH_2 moieties and the nitro groups. Later, the selectivity of these sensors to exposures to solvents (e.g., ethanol, methyl ethyl ketone, toluene, and dichloromethane was evaluated. Compared with the VOC results, the SAM having an amino group showed the highest response to the nitro Compound. TNT and DNT are not among the analytes tested; nevertheless, they should give comparable results.

Instead of polymers or SAMs macromolecules, octaethylporphyrins (OEPs), and tetraphenylporphyrins (TPPs) with different metal centers—namely, (OEP)InCl,

(OEP)MnCl, (OEP)GaCl, (TPP)Pd, and (TPP)RhI—were used as sensing surface materials for nitroaromatics in a QCM sensor [16]. (OEP)MnCl was found to be the most sensitive and selective coating within this chemical class. 2,4-Dinitrotrifluoromethoxybenzene was used as a model compound for explosives containing nitro groups. Another type of macromolecule are the functionalized cyclodextrins, such as methylated or 2,4,6-trimethylbenzyl-substituted β- and γ-cyclodextrin [17]. SAW sensors coated with a macromolecule and the polymer poly(methylhydrosiloxane) showed a high sensitivity toward 2,4-DNT and o-nitrotoluene. In addition, sensor responses to possible common interferants in landmine detection were studied and the sensor sensitivities to them have been found to be much lower than those to DNT and explosives simulants. The response patterns for interferants and o-nitrotoluene of a sensor array containing the different sensors were constructed, and the device was able to detect and classify 2,4-DNT and TNT under ambient laboratory conditions. Unfortunately, here and as in many other cases, the effect of humidity on the sensor response was not studied but saved for further investigations. Changes in sensor performance of sorption-based sensors occur when operating in humid air, and they cannot be neglected in testing close to real conditions. General sensor stability and performance are often poor at high humidity levels. Besides, sorption of water changes the sorption characteristics, favoring polar compounds. Knowledge of humidity influence on sensor performance and the influence of interferant is crucial for the full characterization of a sensor system, but often is either not considered or is not reported.

A quite sophisticated approach to achieving molecular recognition with the help of well-defined interaction sites is the use of specially designed polymer materials. They have already shown enhanced sensor performance in terms of sensitivity and of selectivity in liquids. However, Bunte et al. used molecularly imprinted polymers (MIP), which mimic the principle of antigen–antibody interaction, as in immunosensors, for the gas-phase detection of explosives using QCM sensors [18]. Performance tests of TNT-imprinted polymer beads showed that acrylamide- and methacrylic acid–based MIPs possess enhanced selectivity for TNT. Absorption of 2,4-DNT by these MIPs was not detected. Using 2,4-DNT as a template, a methacrylamide-positive imprint effect for gaseous 2,4-DNT was achieved with no measurable cross-sensitivity for 2,4,6-TNT. Imprinted polymers showed a TNT uptake of about 150 pg/μg MIP per hour, which is substantially lower for nonimprinted polymers.

Molecular recognition using native antigen–antibody interactions was also studied for TNT and RDX detection in the gas phase using SAW sensors [19]. For this purpose, monoclonal anti-RDX and anti-TNT antibodies were immobilized on the transducer surface and exposed to TNT and RDX vapors. The final preparation of the sensor involved the deposition of a thin hydrogel layer to support the antibody layer and provide a semiaqueous environment on the sensor surface crucial for the antibodies to function. The detection of RDX was demonstrated in field tests, as the sensors were able to detect the low vapor concentration of the explosives. However, binding appears to be nonreversible, so the method does not seem suitable for screening purposes or continuous detection.

For sensing in the liquid phase, a competitive immunoassay using SAMs of oligo(ethylene glycol) (OEG) thiols terminated with a hydroxyl group and/or 2,4-DNT as a TNT analog on QCM and surface plasmon resonance (SPR) transducers was studied by Larsson et al. [20]. Three different SAMs were mixed in various proportions with hydroxyl-terminated OEG-thiols to obtain highly selective and sensitive biochips with a low nonspecific binding. Comprehensive instruction for the synthesis and preparation of all compounds was given. The detection limit for TNT in liquids was found to fall in the range 1 to 10 ppb. Similarly, for their dinitrophenyl–acetic acid receptor immobilized via OEG-thiols, a limit of detection of 80 ppt was reported by Mizuta et al. [21]. The LOD was reduced to 50 ppt by use of the secondary antibody. Because of their chemical stability and rather mild regeneration, the sensor surfaces are reported to be durable for more than 100 repeated uses without noticeable deterioration.

Another type of acoustic bulk resonator is the film bulk acoustic resonator (FBAR). Its working principle is very similar to that of a QCM, but it operates at a higher frequency. Lin et al. describe explosive trace detection in liquids through mass sensing using an FBAR coated with commercially available antibodies [22]. Further tests are considered necessary by the authors to fully explore the potential of their device. However, selective detection of TNT and RDX vapor traces are presented with detection limits of 8 ppb for TNT and 6 ppt for RDX.

Use of a cantilever for explosive detection has been studied intensively as well. A device using coated cantilevers has been used by Pinnaduwage et al. for the detection of plastic explosives [23,24]. In this device, a triangular microcantilever is coated with gold and 4-mercaptobenzoic acid on one side, making it capable of binding with PETN and RDX [25]. The same SAM was used later on QCMs [15]. When either of those substances binds, the microcantilever bends and the deflection is measured using a laser–photodiode system. The detection limit is 10 to 30 ppt for PETN and RDX. A similar approach has been pursued using 6-mercaptonicotinic acid [26], achieving a detection limit near 20 ppt for TNT. This material was also used in combination with hydrophobic heptadecafluorodecyltrimethoxysilane for nonspecific molecular-adsorption suppression [27]. A microcantilever coated with a polymer film of poly(1-(4-hydroxy-4-trifluoromethyl-5,5,5-trifluoro)pent-1-enyl)methylsiloxane achieved a detection limit of 300 ppt within a few seconds of exposure of the sensor to the vapor stream [28]. This polymer also has the hexafluoroisopropyl group, which is proved to be very efficient for TNT sensing on QCM sensors.

A prototype sensor array system using cantilevers has been developed by Rogers et al. for the detection of TNT and PETN [29–31]. However, not much detail on the sensitive material employed or experimental details has been given, probably because this device is intended for commercial development. Microcantilevers without a sensitive element have also been adapted to detect explosives via deflagration of adsorbed explosives. Apparently, they achieved some selectivity over nonexplosives, due to their much lower volatility than that of common VOCs [32]. Pinnaduwage et al. describe the detection of deflagration of TNT deposited on a piezoresistive microcantilever and point out its potential for explosive-vapor detection [33–35]. The deflagration of TNT causes the cantilever to bend, due to released heat, and its resonance frequency to shift, due to mass unloading. The minimum amount of TNT detected on the cantilever depends on the cantilever dimensions and was 50 pg for the cantilevers used.

A different thermal detection method using cantilevers is a variation of photothermal calorimetry, where an adsorbate-covered biomaterial cantilever is exposed to selected infrared waves to initiate thermal response due to the absorption of specific infrared waves by the adsorbate. A photothermal signature is obtained for the chemicals adsorbed by scanning a broadband wavelength region. For the wavelengths at which the adsorbed chemical absorbs photons, the temperature of those particular microstructures will rise in proportion to the number of molecules present on the surface. This method is employed for TNT [36–38], but also for RDX and PETN [39,40].

7.2.3 Optical Transducers

Optical sensors rely on the changes in frequency or intensity of electromagnetic radiation, normally in the ultraviolet, visible, or infrared range of the electromagnetic spectrum, to detect and identify the presence of chemical compounds. Fiber optics is often used as an optical probe to measure these changes, as they are suitable for many device configurations and allow special setups for remote sensing. Some authors term these sensors *optode* or *optrode*, equivalent to *electrode* for electrical measurements. A major class of optical transducers uses the evanescent field produced at the surface of an optical waveguide. This can be an optical fiber or a planar waveguide. On this scope, surface plasmon resonance (SPR) is the most commonly used method for explosive detection of such a transducer using an evanescent field. Most sensors described in the literature for explosives detection are based on optical transducers. Bioreceptors are often combined with optical transducers, but organic polymers or SAMs can be used equally well.

Optical chemical sensors reported for explosives detection rely primarily on luminescence phenomena. Fluorescence-based methods are especially sensitive, as fluorescence intensity can be measured very accurately at very low levels. Generally, two modes are feasible: quenching or amplification/modulation of luminescence. Many organic and inorganic compounds show luminescence effects and have been used as sensing material. Apart from this, few examples are based on colorimetry.

Quenching of fluorescence of a very strongly fluorescent material is quite effective in achieving very low detection limits for many analytes. This approach was developed by Swager and co-workers at MIT and was also applied to explosives detection [41–44]. The approach is described in detail by Thomas et al. [45]. The technology has also been exploited commercially for explosive detection [46] and is claimed to be superior to all other chemical sensor solutions. The sensoring device, called *Fido*, has also been suggested for use in landmine detection [47,48]. Pentiptycene-derived phenylene ethynylene polymers are used as sensitive materials. Strong interactions between the electron-rich polymer and the electron-deficient

TNT or DNT molecules are responsible for the effective quenching of fluorescence. Simonson et al. [49] use the same polymers in combination with UV–visible fluorescence lidar (light detection and ranging) technology for remote sensing. They demonstrated the detection of 1 ppm TNT in soil from 2-inch-diameter natural soil targets at a stand-off range of 0.5 km. A new approach focuses on induced lasing of the polymers instead of spontaneous light emission [50,51]. This method is claimed to show a 30-fold increase in sensitivity to a 1-s exposure of DNT compared to the old method.

However, those polymers are not readily accessible because the monomers require multistep syntheses. Thus, conjugated polymers such as different poly(phenylene vinylenes) and poly(diphenylacetylene) are also under investigation to detect TNT and several DNTs [52,53]. Pyrene can also be used for the detection of electron-deficient molecules [54–56]. Besides, polymers and copolymers containing tetraphenylsilole–or silafluorene–vinylene units [57] or the polymer poly[1-phenyl-2-(4-trimethylsilylphenyl)ethyne] [58] were tested against different explosive vapors.

Metallole-containing polymers show quenching behavior upon exposure to explosive analyte vapor. Quenching of luminescence is seen upon illumination with near-UV light. Detection limits were observed to be as low as 5 ng for TNT, 20 ng for DNT, and 5 ng for picric acid [59]. A three-step method uses quenching of green luminescence of the same polymetalloles by the explosive vapor, full elimination of the remaining luminescence by 2,3-diaminonaphthalene (DAN), and a reaction of the nitramines and/or nitrate esters with DAN, resulting in the formation of a blue luminescent traizole complex [60]. A similar three-step detection pathway uses blue-emitting silafluorene/fluorene-conjugated copolymers [61]. The three-step "turn-off/turn-on" method increases selectivity and makes it possible to cover a wide range of explosives, including nitramine (e.g., RDX) and nitrate ester–based (e.g., PETN) explosives and nitroaromatics (e.g., TNT) in the same assay.

In addition, the fluorescence of alkoxycarbonyl-substituted, carbazole-cornered, arylene–ethynylene tetracycle is found to be quenched effectively and reversibly by TNT [62]. A novel approach involves multiphoton excited fluorescence quenching of conjugated polymers containing iptycene units or dendrimers, which show good multiphoton absorption properties [63,64]. Inorganic materials can also exhibit luminescent effects. By monitoring the photoluminescence (PL) of a nanocrystalline porous silicon film on exposure to different explosive-related analytes, detection limits of 500, 2, and 1 ppb were observed for nitrobenzene, 2,4-DNT, and TNT, respectively [65,66]. DNT and TNT are very effective PL quenchers of porous silicon. Specificity to nitroaromatics can be increased by placing a platinum oxide or palladium oxide catalyst in the carrier gas line upstream of the porous silicon detector. PL quenching by NO_2 released in the catalytic oxidation of nitroaromatic compounds is less efficient than quenching of the intact nitroaromatic compound. Other organics possibly contained in the sample are not affected by the catalyst. This technique provides a means to discriminate nitro-containing molecules from other organic species and from NO_2. Silicon nanocrystals exhibit a similar quenching behavior upon exposure to nitroaromatics [67]. The principle of photolumines-cence quenching of a nanocrystalline porous silicon can easily be implemented in a microsensor. A device using photoluminescence quenching of porous silicon and polysilole nanowires was described by Sailor et al. ([68] and references therein). A detection limit of 1 ppb was demonstrated for TNT in air. Silicon modified with polydiacetylene membranes was used by Sabatani et al. [69].

Porphyrin-doped silica films with mesoporous structures [70–72] or as nanofibers produced in an electrospinning process [73] were employed to detect explosive compounds. All synthesized silica films showed high fluorescence quenching sensitivity toward the vapors of TNT, DNT, and NB. Detection of a trace vapor of TNT at a concentration of 10 ppb was achieved. Similarly, conju-gated polymers (i.e., poly [2-methoxy-5-(2-ethylhexyloxy)-p-phenylenevinylene] and a polyipticene containing polymer) entrapped in porous silicon microcavities have been studied as optical sensors for TNT [74]. The fluorescence spectra of entrapped polymers were modulated by the microcavity via a spectral "hole" that matches the resonance peak of the microcavity reflectance. Exposure of the porous silicon microcavity containing entrapped polymer to explosives vapor results in a red shift of the resonance peak and the spectral hole, accompanied by quenching of the fluorescence. This multiplexed response provides multiple monitoring parameters, enabling the development of an optical sensor array for the detection of target explosive vapor. The photoluminescence from single-walled carbon nanotubes is also found to be highly sensitive to the presence of nitroaromatic compounds such as nitrobenzene, 4-nitrotoluene, and 2,4-dinitrotoluene [75].

The fluorescence of ZnL [H2L = N,N'-phenylene-bis-(3,5-di-*tert*-butylsali-cylideneimine, (salophen)Zn] is quenched by nitroaromatics and 2,3-dimethyl-2,3-dinitrobutane [76]. Recently, a variety of common fluorophores were tested for their quenching efficiency in the detection of various nitrated compounds. Among the fluorophores investigated, purpurin, malachite green, and phenol red demonstrate the greatest sensitivity and selectivity for nitrated compounds [77] and are considered suitable for a variety of applications. Pyrene was found to be the most sensitive, but was not recommended for routine use due to health and waste disposal hazards. Other studies are based on dansyl chromophores immobilized on an epoxy-terminated self-assembled monolayer on glass slide surfaces [78]. However, in quenching experiments the sensors were shown to be more sensitive to nitrobenzene (NB) than to other nitroaromatics, including TNT, 2,4-DNT, p-chloronitrobenzene, m-dinitrobenzene, p-dinitrobenzene, and o-chloronitrobenzene.

Polymers and copolymers containing tetraphenylsilole or tetraphenylgermole with Si–Si, Ge–Ge, and Si–Ge backbones [79] and polymer such as poly(3,3,3-trifluoropropylmethylsilane) [80] have also been explored for the detection of nitroaromatic molecules such as NB, 2,4-DNT, TNT, and picric acid in liquid phase. Other systems under study were CdSe quantum dots covered with a ZnS core sensitive to TNT-containing samples dissolved in toluene [81], and quenching of the strong orange Mn2$^+$ photoluminescence of Mn2$^+$-doped ZnS nanocrys-tals upon exposure to TNT in water [82]. Aromatic polyoxadiazole derivatives containing 9,9'-dioctylfluorene show remarkable sensitivity for nitroaromatic compounds, exhibiting fluorescence quenching and UV–visible absorption changes in chloroform [83].

Chemical compounds that exhibit an increase in fluorescence upon interaction with analyte molecules have been used with similar success. Albert and Walt prepared optical microsensors for high-speed detection of low-level explosives and explosivelike vapors [84]. Changes in the fluorescence properties of Nile Red dye, which has been used as a fluorescent dye, during nitroaromatic compound vapor exposure was monitored. DNT and DNB are detected at low ppb levels. Each sensor within the array is a 3- or 5-μm porous silica bead impregnated with Nile Red. Due to the size of the individual sensors, very small sensor arrays can be realized. When Nile Red is sorbed to different porous silica microspheres with varying surface functionalities, the resulting excitation and emission spectra are shown to be dependent on the surface polarity. The entire array is monitored via a CCD camera.

Recently, the ability of dye-labeled solid-state DNA dried onto a surface to detect odors delivered in the vapor phase by changes in fluorescence was reported [85]. The sensors differ in the DNA sequence. Measurement results of selected sensors are reported for, among others, DNT. One sensor showed responses to DNT down to 6 ppb, whereas another sensor shows no response, demonstrating the discriminating power of the sensor array as an artificial odor detector.

In the liquid phase, a sensor based on fluorescence resonance energy transfer (FRET) using a dye-labeled anti-TNT antibody fragment that interacts with a co-functional surface-tethered DNA arm has been described [86–88]. The arm consists of a flexible biotinylated DNA–oligonucleotide base specifically modified with a dye and terminating in a TNB recognition element. Both of these elements are tethered to a neutravidin surface with the TNB recognition element bound in the antibody fragment binding site, bringing the two dyes into proximity and establishing a baseline level of FRET. Addition of TNT, or related explosive compounds, to the sensor environment alters FRET in a concentration-dependent manner.

Mimicking the enzymatic NADH-mediated reduction of RDX in a fluorescence-based sensor using the NADH analog 10-methyl-9,10-dihydroacridine (AcrH$_2$) was also evaluated for explosive sensing in the liquid phase [89]. This compound is able to form the N-methylacridinium fluorophore (AcrH$^+$) upon "H$^-$" abstraction. Both RDX and PETN generate the green-emitting AcrH$^+$ from the blue-emitting AcrH2 upon photolysis at 313 nm in deoxygenated acetonitrile solutions, whereas TNT is ineffective. Strong emission signals were observed with RDX and PETN concentrations as low as 7×10^{-5} and 1.3×10^{-4} M, respectively.

Colorimetric detection relies upon detecting the change in color arising, for example, in an amine-containing poly(vinyl chloride) (PVC) film when exposed to polynitroaromatics in the vapor phase [90]. This change is a result of the formation of visible light–absorbing complexes between primary or secondary amines and polynitroaromatics. Mesoporous organosilicas (PMOs) which incorporate a porphyrin into the material were used as an optical indicator of target binding [91]. The binding of p-nitrophenol, p-cresol, TNT, and RDX by porphyrin-embedded PMOs with selective adsorption of TNT over the other analytes is demonstrated.

The binding of each of the organics by the PMO results in unique changes in the spectrophotometric characteristics of the incorporated porphyrin. Explosives are dosed in micromolar concentration.

A specific color reaction between cyclopentadienyl manganese tricarbonyl (cymantrene) and trace levels of TNT has been reported [92]. After UV irradiation, a blue–green color is developed. The principle was first proposed using finger-prints containing DNT. Xie et al. [93,94] demonstrated a strong charge-transfer complexing interaction between the amino groups of 3-aminopropyltriethoxysilane and electron-deficient nitroaromatics. Both reactions can be followed with UV/visible spectroscopy.

In the class of *optical biosensors*, immunosensors are dominant. They utilize the quite specific interaction between an antibody (normally, the bioreceptor) and an antigen (the target analyte). They belong to the affinity-based sensor class of biosensors. Biosensors that use catalytically active species such as enzymes establish the other main class. Here, electrochemical transducers are generally employed. Examples of explosive detection are described in subsegment sections.

Closely related to immunosensors are immunoassays, such as enzyme immunoassays (EIAs) and the very common and long-established enzyme-linked immunosorbent assay (ELISA), a special type of EIA, as they are based on the same immunoreactions. The successful antigen–antibody binding is detected indirectly using an enzyme reaction of an enzyme linked to the antibody used for analyte detection. The product formation in the enzyme reaction provoked is monitored optically via changes in UV/visible adsorption characteristics or fluorescence. Immunoassays with labels other enzymes are also available. Most common of these are fluorescence labels, but detection is always achieved using a separate spectroscopic device. The same antibodies developed using immonoassays can be used in biosensors once the antibody or the antigen is immobilized on the transducer. The generally accepted definition of *biosensor* demands the close proximity of sensing element and transduction element, but this definition is not always followed strictly. For this reason, both systems using immunological methods, also often called biosensors when they are part of flow analysis systems allowing automated measurements, and immunoassays will be included in our overview of explosive detection methods.

Keuchel et al. [95] were the first to present an ELISA for the determination of TNT and other nitroaromatic compounds. It was possible using their competitive assay to detect TNT within the range 0.02 to 20 ppb. More recently, Altstein et al. [96] developed an assay for the detection of trace amounts of TNT on the basis of antibodies trapped in a sol–gel matrix instead of being absorbed by a polymer matrix like polystyrene. Detection limits were similar to those above. Bromberg and Mathies produced a homogeneous immunoassay on a microfabricated capillary electrophoresis chip for TNT and its analogs and found 1-ppb detection limits and a dynamic range of 1 to 300 ppb [97,98].

Goldman et al. [99] used selected recombinant anti-TNT antibodies isolated from the Griffin library of phage displayed scFvs. The best single-chain antibody showed a limit of detection of 1 ppb. Anderson et al. [100] evaluated four different commercial TNT monoclonal antibodies, two recombinant TNT antibodies, and a control antibody immobilized on beads. Use of a flow cytometer allowed simultaneous detection of TNT and other nitroaromatics. TNT could be detected at 0.1 ppb and quantified over the range 1 ppb to 10 ppm. In addition, the assay was shown to be effective in various matrixes, such as lake water, seawater, and acetone extracts of soil. Direct use of environmental samples was the focus of more application-oriented studies. TNT immunoassays were developed for the analysis of extracts from field soil samples [101], seawater [102–104], and river water [105]. Limits of detection were higher, due to the complex matrix, and could be as high as 25 ppb in seawater [104]. A biosensor for TNT in seawater achieved detection limits of 0.025 ppb [106,107]. Bromage et al. describe the development of a highly sensitive TNT immunosensor consisting of a highly specific monoclonal antibody [108,109]. They report sensitivity for TNT of 0.05 ppb in groundwater, which is higher than reported by other authors for real samples.

TNT is a focus in most work on immunological methods, but methods for RDX and other explosives, such as PETN, were developed starting with Bart et al. [110] in 1997. Detection limits for RDX of 15 ppb were achieved. A later method exhibited a detection limit of approximately 1 ppb in laboratory samples for RDX and TNT [111]. The system is based on previous work by the same authors [112]. Wilson et al. reported the first immunoassay for PETN [113]. The detection limits for TNT and PETN were 0.1 and 20 ppb, respectively. Narang et al. [114] built a dual-analyte detection system using capillary-based immunosensors. Two capillaries, one coated with antibodies specific for TNT and the other specific for RDX, were combined into a single device. The limits of detection for TNT and RDX in the multianalyte format are 0.1 ppb for TNT and 0.5 ppb for RDX. More immunosensors for the simultaneous detection of TNT and RDX are reported later [115,116]. Also, miniaturized field-portable immunosensors for the detection and quantification of TNT and RDX in groundwater were tested [117–119]. Van Bergen et al. [120] developed a semiautomated fiber optic biosensor for soil analysis of TNT and RDX. The limit of detection was determined to be 5 ppb for both TNT and RDX. A highly sensitive microcapillary immunosensor for the detection of RDX achieved detection limits for RDX of only 10 ppt, with a linear dynamic range from 0.1 to 1000 ppb [121].

A biosensor based on an evanescent field transducer was proposed by Walker et al. [122] for the detection of TNT. It uses a planar integrated optical waveguide for attenuated total reflection spectrometry. Sol–gels were molecularly imprinted for TNT using covalently bound template molecules. The sol–gel matrix was composed of bis(trimethoxysilylethyl)benzene (BTEB) and 2-(trimethoxysilylethyl)pyridine (TMSEPyr). Binding of TNT (and its subsequent conversion to the anion by abstraction of a methyl proton) results in the attenuation of light propagating through the waveguide, creating a spectrophotometric device. The limit of detection for gas-phase TNT is found to be 5 ppb in ambient air. Response to TNT is not reversible, which results in an integrating sensor device that can theoretically improve the ability to detect small amounts of the explosive if the exposure time is sufficient.

Surface plasmon resonance (SPR) is a generic transduction method based on evanescent fields for the readout of bioreceptors. Several examples describe the use of SPR for explosives detection. Bowen et al. [123] produced a biosensor from a monoclonal antibody for TNT covalently bound to an 11,11'-dithio-bis(succinimidoylundecanoate) self-assembled monolayer on a SPR chip. Matsomoto et al. used a polyclonal antibody against trinitrophenyl (TNP) derivatives, raised in rabbit, for the detection of TNT [124]. The immunoassay exhibited sensitivity for the detection of TNT in the concentration range 0.03 to 300 ppb. When an anti-rabbit IgG is used as secondary antibody, the detection limit is improved to 0.01 ppb.

Most promising work on SPR biosensors for explosives detection has been conducted by Shankaran and co-workers [125–130]. They describe the development of immunosensors for detection of TNT using different antigen conjugates for the production of antibodies. Polyclonal antibodies prepared from 2,4,6-trinitrophenol–bovine serum albumin conjugate (anti-TNP-BSA Ab, goat IgG) and another from 2,4,6-trinitrophenyl–keyhole limpet hemocyanine conjugate (anti-TNPh- KLH Ab, rabbit IgG) showed a high degree of affinity for TNT and varying degrees of cross-reactivity to the related nitroaromatic derivative compounds, such as 2,4-DNT, 1,3-DNB, 2-amino-4,6-dintitrotoluene (2A-4,6-DNT), 4-amino-2,6-dinitrotoluene (4A-2,6-DNT), and TNP. The anti-TNPh-KLH Ab showed a detection limit of 6 ppt. In another test, polyclonal anti-trinitrophenyl antibody (P-TNPh Ab) exhibited the highest sensitivity among several types tested, with a detection limit of 2 ppt TNT.

7.2.4 Electrochemical and Electrical Transducers

In addition to electromechanical and optical sensors, electrochemical sensors are also described for the detection of explosives in liquid media. Wang et al. [131] present a poly(guanine)-functionalized silica nanoparticle label-based electrochemical immunoassay for TNT. This immunoassay takes advantage of a magnetic bead–based platform for competitive displacement immunoreactions and separation, and uses electroactive nanoparticles as labels for signal amplification. The detection limit reported for this assay is 0.1 ppb TNT. Gwenin et al. [132] developed an amperometric biosensor based on genetically modified enzymes for in situ detection of trace vapors from a number of explosive compounds. The amperometric biosensor achieved a vapor detection limit of 6 ppt. The preparation and characterization of an amperometric TNT biosensor based on the surface immobilization of a maltose-binding protein (MBP) nitroreductase (NR) fusion (MBP-NR) onto an electrode modified with an electropolymerized film of N-(3-pyrrol-1-ylpropyl)−4, 4'-bipyridine (PPB) are described by Naal et al. [133]. Chaignon et al. investigated different enzymes to be used in a biosensor for detecting 3,5-dinitrotrifluoromethylbenzene [134]. *Bacillus* LMA, isolated from explosives-polluted effluents, was found to be the most efficient.

Different electrochemical methods are available for explosives detection as well. A recent summary on his comprehensive work is given by Wang [135].

Methods are reported by different authors based on voltammetry [136–142], amperometry [143], or impedance measurements [144–147] using electrode materials such as carbon or different metals. Some authors employ chemically modified electrodes to enhance sensitivity by enriching the analyte TNT at the electrode. Shi et al. use carbon electrodes modified with layers of mesoporous SiO_2 and poly(diallyldimethylammonium chloride) [139]. The detection limit was 10^{-9} mol/L for TNT. Hrapovic et al. propose metal nanoparticles (Pt, Au, or Cu) together with multiwalled and single-walled carbon nanotubes nanocomposites for TNT detection, among which the nanocomposite containing Cu and SWCNT in Nafion yielded the best results [141]. Riskin et al. show the use of gold electrodes modified with *p*-aminothiophenol or molecular imprinted oligoaniline-cross-linked gold nanoparticles [142]. Detection limits were 17 ppm and 46 ppt, respectively. Aminothiophenol is capable of strong π-donor–acceptor interactions with TNT. Also γ-cyclodextrin–modified electrodes were used for DNT enrichment [144–146].

Conducting polymer composites with carbon black are also employed for TNT detection experiments [148]. Single-molecule-detection TNT with a genetically engineered transmembrane protein pore, α-hemolysin (αHL), was proposed by Guan et al. [149]. Here the ionic current passing through a single pore is monitored at a fixed applied potential. Another transduction scheme uses a field-effect transistor (FET) [150] that has a gas-sensitive polymer transistor channel. The binding of nitroaromatic molecules to thin organic films increases the film conductivity and thereby the transistor electrical characteristics. Poly(3-hexythiophene) and dihexylquarterthiophene are used as transistor channel materials. The device has only reached prototype status, but FET-based sensors are interesting microsensors for mass-market applications. Patel and collaborators use a functionalized polymer-coated commercial micromachined capacitor device to measure the dielectric permittivity of selectively adsorbed analytes, including different nitroaromatics [151,152].

7.3 NUCLEAR QUADRUPOLE RESONANCE

7.3.1 Introduction

Detection of explosives and drugs is a problem of vital importance for our civilization. In particular, the detection and clearance of landmines after war conflicts is one of the most difficult global problems. It is an issue for Caucasia, the Middle East, Southeast Asia, the Balkan Peninsula, and for many other regions as well. This hinders the resumption of mined regions to normal activities long after periods of conflict. Despite all efforts, it has been estimated that clearing all existing mines could take 450 to 500 years at the current rate [153]. On the other hand, the problem of terrorism is an international issue for the modern world. Any country could be a subject of terrorist threats at any time and any place. For example, truck bombs were exploded at synagogues and the British Consulate in November 15, 2003, in Istanbul, Turkey. The explosions killed 57 people and injured more than

700. Aviation security is an outstanding problem, so the detection of explosives hidden on passengers and in luggage at airports is also a matter of great importance. Throughout the world, commercial aircraft have been lost due to suspected terrorist bombings, and hundreds of people have died in these tragedies.

Despite the availability of various detectors, there are a number of problems to be resolved, such as increasing the sensitivity, shortening the detection time, and decreasing the scanner cost. In aviation security, fast and reliable detection of explosives in passenger luggage is the requirement of greatest concern. For mine detection, there is the problem of a very low rate of detection and cleanup, with high false-alarm rates. Great emphasis has been placed on the development of new and improved techniques for explosives and mine detection. Recently, essential progress in the use of nuclear quadrupole resonance (NQR) for explosives detection has been observed. NQR is a sensitive tool for probing the electronic environment which depends on the molecular structure of a compound of a quadruple nucleus. It is also known that most explosive and narcotic substances contain nitrogen (^{14}N has nuclear spin $I = 1$) as the quadrupolar nuclei in their structure. Therefore, all explosive and narcotic substances in the solid state are best detected by the NQR method. Detection of RDX, PETN, tetryl, TNT and other ^{14}N-containing explosives are generic examples. Use of NQR on ^{14}N nuclei has a number of advantages in explosive and drug detection. In this section we focus on an explanation of the basic principles of NQR as well as on the use of this technique in explosives detection within the concept of subsurface sensing.

7.3.2   Principles of NQR

NQR is a phenomenon based on the electrostatic interaction of nonspherical nuclei with the gradient of the crystal electric field (CEF) produced by neighboring charges (i.e., by valence electrons). As a result of this interaction, the energy levels of a nucleus in the crystal electric field split and the electromagnetic field oscillating in the radio-frequency (RF) range (several kilohertz to hundreds of megahertz, depending on a nucleus and host compound) can induce transitions between these levels. Thus, there is a set of NQR transitions, called the NQR spectrum, and it is a unique characteristic of the specific chemical compound. Therefore, NQR provides the possibility of unambiguous detection of various solid-state crystalline substances.

The basic nucleus used to detect explosives and narcotics by the NQR method is ^{14}N. Fortunately, nearly all known explosives, narcotics, and medical products contain nitrogen in their structure. NQR signals of these substances are in the frequency range 400 kHz to 5.5 MHz. In the general case (i.e., for CEF of low symmetry, such as rhombic), there are three energy levels of ^{14}N nucleus with the spin $I = 1$, as shown in Fig. 7.2. Therefore, three NQR frequencies of the transitions between these levels are possible:

$$\omega_0 = \frac{eQq}{2\hbar}\eta, \qquad \omega_+ = \frac{eQq}{4\hbar}(3+\eta), \qquad \omega_- = \frac{eQq}{4\hbar}(3-\eta) \qquad (7.1)$$

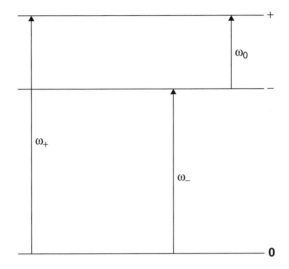

FIGURE 7.2 ^{14}N NQR energy levels.

where eQq_{zz} is the constant of NQR interaction, η the asymmetry parameter of the CEF gradient, and $\hbar$ is Planck's constant divided by 2π.

There are two main techniques of NQR registration: the continuous-wave (stationary) and pulse methods. Both techniques can be used in explosives detection. However, the choice of optimal method depends on the largest signal-to-noise ratio (SNR) available that determines the shortest detection time as well. One has to take into account the energy consumption, complexity, and cost of the construction of a detection device. For that reason, pulse modification of NQR is usually preferred over the stationary method. In this technique, the signal response is registered after RF pulse excitation with an output power of about 100 to 2000 W.

Comparing the scheme of a pulse NQR spectrometer, one sees many similarities with the usual transceiver–receiver structure of radio stations (Fig. 7.3). However, there is the difference that the transmitting–receiving coil (which is similar to a radio antenna and used to irradiate the sample and detect the NQR signal) is placed within a screened volume. The other principle blocks are as follows. A NQR console is an electronic unit that performs most of the operations of RF pulse formation and processing of signals detected. An amplifier amplifies RF pulses at the NQR console output up to power levels to 500 to 2000 W for a 50-Ω resistive load. The receiver is a low-noise preamplifier used to amplify very weak NQR signals before their transfer for processing in the NQR console. The sensor mentioned above consists of a transmitting–receiving coil antenna with a system for Q-factor damping and gating between the transmitting and receiving signals. For research purposes this detection unit has to be well screened in the NQR spectrometer. It can also be used as a luggage scanner, but for the distant detection of mines, it is not effective. Therefore, some special active and passive methods of signal filtering are used in mine detection (e.g., automatic subtraction of

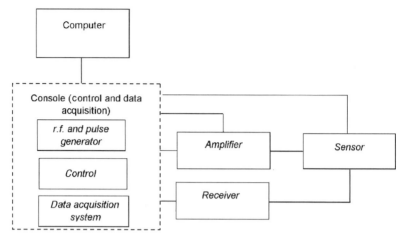

FIGURE 7.3 NQR system.

interference signals, digital processing). The final unit of a pulse NQR spectrometer is a computer equipped with special software to display the results of measurements, to perform additional postprocessing of signals, and for hardware diagnostics.

Let us consider an example of the simplest NQR/NMR experiment: one pulse sequence (Fig. 7.4). In this sequence the sample is irradiated by an excitation pulse with "reference" radio frequency (ω_0), which matches one of the NQR frequencies given in Eq. (7.1). Then a free induction decay (FID) NQR signal is acquired with a short time delay after an RF pulse (Fig. 7.4). The quadrature detection procedure is generally used to register NMR and NQR signals [154]. The NQR signal acquired is compared with the reference signal and any difference detected between signals

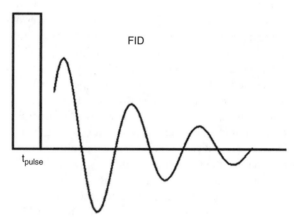

FIGURE 7.4 Excitation radio-frequency pulse with duration t_{pulse} and free induction decay NQR signal (in time domain) after the action of the pulse. The difference between the FID and excitation frequencies is usually detected.

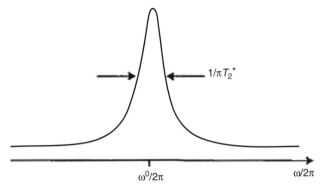

FIGURE 7.5 Frequency-domain NQR signal obtained after Fourier transformation of the time-domain signal in Fig. 7.4.

is recorded. The signal processing that follows includes Fourier transformation to the frequency domain to obtain the NQR spectrum (Fig. 7.5).

Any NQR signal can be characterized by its resonance frequency and related relaxation parameters. The resonance frequencies of various explosive compounds can be found in the extensive literature on NQR spectroscopic studies. The relaxation parameters are discussed below, due to their relevance with the SNR of the detection.

The NQR spectra have the following relaxation parameters [155]:

1. The spin–lattice relaxation time, T_1, is the time required to return the population difference of the two-level system to thermal (Boltzmann) equilibrium. It is known that to obtain the best SNR, the time period between the repetitions of pulse experiments has to be in the range 3 to $5T_1$.

2. The spin–lattice relaxation time in the rotating frame, $T_{1\rho}$, is the decay time of magnetization during a train of RF pulses in multipulse sequence.

3. The spin–spin relaxation time, T_2, is the parameter related to the lifetime of a coherent precession of an ensemble of nuclear magnetic moments of the sample and may be determined experimentally through the decay time of the "echo" signals in pulse experiments.

4. The time describing the width of a resonance line, T_2^*, is defined by a width at half-height, $\Delta v = 1/\pi T_2^*$, for the Lorentzian line shape (see Fig. 7.5).

Multipulse sequences are used to increase the small signal-to-noise ratio (SNR) of NQR signals. It is known that the signal detected is a mixture of noncoherent noise and coherent NQR signals. Therefore, signal accumulation through successive observation windows increases the ratio of coherent echo signals to the noise. During the accumulation process, SNR is increasing as $\sqrt{n}$, where n is the number of accumulations. There are the following commonly used pulse sequences for the detection of NQR signals:

1. Spin-locking spin-echo (SLSE), $t_i - (\tau - t_i^{90°} - \tau)^n$, where $90°$ is the phase shift of the next RF pulse with regard to the initial pulse

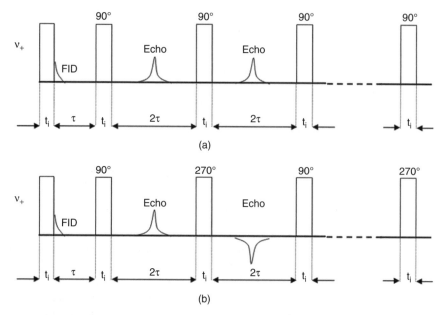

FIGURE 7.6   (a) Spin-lock spin echo (SLSE); (b) phase-alternated SLSE.

2. Phase-alternated SLSE, $t_i - \tau - (t_i^{90°} - 2\tau - t_i^{270°})^n$

3. Strong off-resonant comb (SORC) with alternated phase, $(t_i - \tau - t_i^{180°} - \tau)^n$, and without alternated phase, $(t_i - \tau)^n$

4. The composite SORC sequence, consisting of the combination of phase-alternated and non-phase-alternated pulse sequences (PAPS-NPAPS): $(t_i - \tau - t_i^{180°} - \tau)^{n/2} + (t_i - \tau)^n$

Examples of the SLSE and phase-alternated SLSE sequences are shown in Fig. 7.6. Some more complicated combinations of these sequences may be used to subtract spurious signals and for the better accumulation of NQR signal.

Factors Influencing SNR   The SNR of the NQR signal is also dependent on the probe design [156]:

$$\text{SNR} \propto \xi \gamma I (I + 1) \sqrt{\frac{\nu Q V T_2}{\beta T_1}} \tag{7.2}$$

where ξ is the filling factor of the coil with the sample, γ the gyromagnetic ratio of the nuclei, I the nuclear spin, ν the resonance frequency, Q the quality factor of the circuit, V the volume of the detection coil, T_1 and T_2 the relaxation parameters (defined above), and β the bandwidth of the receiver.

It is seen that SNR becomes smaller with a decrease in the resonance frequency. Therefore, a high Q value is needed to obtain the good sensitivity at lower frequencies. It is known that the bandwidth of the probe is proportional to the resonance frequency divided by Q, while the NQR line width for a specific compound does not change essentially for various lines with different transition frequencies. Thus, in the case of a probe with a high Q-factor used to detect a compound with low resonance frequencies, the large line width of the NQR line may become comparable to the probe bandwidth. This matter creates some difficulties in real applications. For example, the NQR line may appearout side the narrow frequency range of the tuned probe, due to the temperature drift of the resonance frequencies of the compound detected. There is also another complication. The high Q increases the "dead" time of the detector, which is the time for the detection coil to recover after an excitation pulse. It results mainly from ringing or from the oscillatory process of dissipation of energy stored in the detector circuit during transmission. To characterize the ringing process, a special parameter known as the probe ringing time, τ, is introduced. It is determined by the time required for the energy stored in the resonance detector circuit to decrease e^2 times (i.e., for voltage to decrease e^1 times). The time τ depends on the Q-factor of the detector as

$$\tau = \frac{2Q}{\omega_0} \tag{7.3}$$

Although the loaded Q-factor depends on the detector parameters and output impedance of the circuit, Eq. (7.3) is valid for any parameters, for any matching or loading of the detector circuit [157]. The quality factor of the tuned circuit Q is usually in the interval 100 to 500 (a typical value is 200). The typical NQR signal detected by the coils is as small as 20 nV, which corresponds to a power level of approximately-140 dBm for a 50-Ω load. For a 1-kW (60-dBm) RF excitation transceiver power, a power drop of about 200 dB is needed to recovery the sensitivity of the receiver preamplifier [155]. It is commonly accepted that the typical "ring-down" time for system recovery is about 20τ (the exact time for a 200-dB decrease is 23τ). Thus, one estimates that $\tau = 75$ μs for the detector with $Q = 200$ at a TNT frequency (842 kHz), and consequently, a 6-ms recovery time would be required [155]. In a short period, an increase of Q (i.e., lowering the internal losses in the coil) results in a longer ringing process, during which signal detection is impossible.

It is known that to obtain the most intense FID signal, one has to use an excitation pulse of optimal length, and the condition for such a pulse (called a 90° pulse) is written

$$\alpha \gamma B_{RF} t_P = \frac{\pi}{2} \tag{7.4}$$

where B_{RF} is the RF field, t_p the pulse duration, and α a parameter whose value depends on nuclear spin and the sample used (single crystal or powder). For NQR on ^{14}N nuclei with $I = 1$, the parameter $\alpha = 1$ and $\gamma B_{RF} t_P \approx 90°$, while for a powder sample, α is about 1.32 and $\gamma B_{RF} t_P \approx 119°$ [158]. On other hand, the bandwidth of the excitation pulse, $\Delta = \omega_i - \omega$, is related to the pulse duration as $t_P = \pi/2\Delta$. Therefore, the pulse length should be sufficiently short to cover a range of frequencies, where the resonance signal may be observed. The RF field acting on the sample is given as

$$B_{RF} \sim 3\sqrt{\frac{PQ}{\omega_0 V}} \qquad (7.5)$$

where P is the power of the RF field, V the volume of the detection coil, and ω_0 the excitation frequency [156].

There are some limitations for the power level of the excitation pulse and the quality factor of the detection coil. Equations (7.1) to (7.3) shows that to detect a signal with good SNR, one has to use the detection circuit with high Q as well as large a sample volume as possible. There is a set of contradictory requirements for the detector circuit, especially for detection of weak signals at low frequencies. For example, the detector coil should have a high quality factor Q for higher sensitivity; nevertheless, a moderate Q-value is necessary to obtain the optimal frequency bandwidth and the short "dead" time of the probe. Various methods have been proposed to match these conflicting objectives. A very simple solution to overcoming this issue is to use the spin echo rather than free induction decay to detect the NQR signal, since there is a time delay between the transmitter pulse and the echo signal, whereas the FID signal is observed just after excitation pulse [159]. Another approach is using a short phase-inverted pulse following the main transmitter pulse to suppress actively the ringing oscillation, which results in a decrease in the dead time of the probe coil [160]. However, the techniques manipulated by the Q-value are especially popular [157]. The Q-damping is the active way to change Q [161–164]. In this technique, Q is switched between a very low value for the period just after or even through the pulse transmission and a very high value for registration of the NQR signal. The low Q-factor essentially makes it possible to shorten the ringing time of the probe, while the high Q at the second detection stage increases the SNR. In passive methods of Q manipulation, which comprise *overcoupling* [165], the impedance of the probe circuitry is specifically mismatched to that of the preamplifier, so that the effective Q is reduced [157].

As mentioned above, the high Q-factor is very important in improving the probe sensitivity. According to Eq. (7.1), SNR improvement is proportional to the square root of the Q-value, while the probe bandwidth is proportional to ω_0/Q, where ω_0 is the resonance frequency of the tuned circuit. This case, called *super-Q detection*, is possible when the bandwidth of the probe is smaller than the line width of the signal detected. It has been shown [166] that in the absence of amplifier noise, SNR continues to increase with increasing Q even in the super-Q limit. In this case, the maximum obtainable SNR of a real device will be limited by the noise of the amplifier used in the system. One possible realization of the probe with very high Q is the superconducting coil. The probes with Q-values of up to 100,000 have been applied for NMR measurements (see, e.g., [167]).

Let us consider the probe circuits generally used to detect the NQR signals. These are the parallel and serial resonance circuits (Fig. 7.7). In both circuits the capacitance C_1 is used for impedance matching with a 50-Ω transmitter output

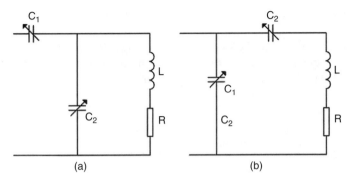

FIGURE 7.7 Circuit models for the probe of NQR detector: (a) parallel; (b) sequential.

cable, while C_2 is used to tune the circuit on the resonance frequency of an NQR line. A detailed analysis has been made of the use of parallel and serial circuits in NMR or NQR measurements [159]. Formulas given below describe the most important parameters of the probe. For a parallel circuit, the following expression is defined for the unloaded Q (the Q of the probe when it is not connected to an external load: e.g., to the preamplifier) [168]:

$$Q_0 = \frac{\left(1 - \sqrt{R/R_0}\right)\left(1 + C_2/C_1\right)^2}{\omega C_2 R_0} = \frac{1 + C_2/C_1}{\omega R_0 C_1} \tag{7.6}$$

where $R_0 = 50\ \Omega$ is the equivalent resistance of the circuit. The noise power density of the tuned detection circuit (neglecting the noise from amplifier) is given by [157,166]

$$P(\omega) = 4kT R_0 \frac{R_a^2}{(R_a + R_0)^2} \tag{7.7}$$

where R_a is the preamplifier resistance used in the NQR spectrometer and k is the Boltzmann constant.

In the case of $Q_0 \gg 1$ and $\omega \simeq \omega_0$ for the parallel and series tuned probes, the output voltage of the circuit is given by [157]

$$u_s(\omega) = j\omega\Phi(\omega)\sqrt{\frac{Q_0 R}{\omega_0 L}} \tag{7.8}$$

where $\Phi(\omega)$ is the Fourier transform of time-changing magnetic flux, $\Phi(t)$, and Q_0 is the unloaded Q.

As mentioned earlier, the pulse technique in NQR detection is usually preferred over the continuous-wave (CW) method, due to a better SNR than that obtained using the pulse version. For that we discuss briefly the basics of CW NQR spectroscopy. This technique is based on registration of the continuous absorption of RF power during irradiation of the sample. To calculate the absorbed power (i.e., the total energy absorbed per unit of time) one has to know the following parameters: the transition probability P_{ik} for NQR transitions between energy levels i and k, the populations differences between the same levels ΔN, the number of quadruple nuclei N, and the line shape function $L(\omega)$. In the high-temperature approximation ($\Delta E / kT \ll 1$) the population difference is [169]

$$\Delta N \approx \frac{1}{2I+1} \frac{\hbar \omega_Q}{kT} \tag{7.9}$$

where the NQR frequency $\omega_q = 2\pi \nu_q$, T the temperature, and k the Boltzmann constant. The probabilities of the transitions between the energy levels i and k is given as

$$P_{ik} = 2\pi \hbar (\gamma B_1)^2 \cos^2 \alpha_{ik} \tag{7.10}$$

where B_1 is the magnitude of the radio-frequency field and α is an angle between the direction of the RF field and the axes of the electric field gradient of the NQR crystal. The Lorentzian line shape is given as

$$L(\omega) = \frac{1}{\pi} \frac{T_2^*}{1 + \left[T_2^* (\omega - \omega_Q) \right]^2} \tag{7.11}$$

where the line width is $1/\pi T_2^*$.

Thus, one could obtain for the power of the absorbed energy as

$$P = \hbar \omega P_{ik} \Delta N L(\omega) = \frac{2\pi \hbar^3 (\gamma B_1 \cos \alpha_{ik})^2}{(2I+1)kT} \omega_Q \omega N L(\omega) \tag{7.12}$$

In this expression we suppose that the following condition is valid:

$$\gamma^2 B_1^2 T_1 T_2 \ll 1 \tag{7.13}$$

That is, the power applied to excite the NQR transition is small enough to prevent saturation of the NQR spin system. In this case one can use the linear approximation of the dynamic magnetization as a function of the RF field. It is known that the absorbed power is related to the dynamic susceptibility as

$$P = 2\omega \chi'' B_1^2 \tag{7.14}$$

The latter can be measured in the CW experiment through a change in the inductance of the probe coil. The imaginary part of the dynamic susceptibility measured in experiments is given as

$$\chi'' \propto \frac{\omega_Q N L(\omega)}{T} \tag{7.15}$$

It should be noted that the most important difference between the CW and pulsed techniques is that a much smaller power level of RF excitation must be used in the former. In this case, simpler and cheaper instrumentation can be employed. It is also much easier to search over large frequency ranges in the CW technique. However, an inherently low signal-to-noise ratio is a very severe constraint that currently prevents use of the CW method in real applications. There is probably a potential to improve the SNR in CW detection using narrowband detection and high-T_c superconductor (HTS) probes with very high Q-values.

Another way to improve the sensitivity of NQR is based on the use of double magnetic resonance methods. The double NMR/NQR system has the advantage of increasing SNR by the ratio $\gamma_H/\gamma_N \sim 15$, where γ_H and γ_N are the gyromagnetic ratios of ^{1}H and ^{14}N. Besides, Pusiol [170] took into account the fact that the decaying time of the cross-coherence of quadrupole nuclei is due primarily to the fluctuations in local fields produced by neighboring protons. In a weak magnetic field it is possible to use irradiation on the second NMR frequency in order to average to zero local fields induced by protons, which increases the decay time of magnetization of quadrupolar nuclei. As a result, the line width of quadrupolar spins suffers a narrowing and the SNR of the NQR signal is increased. The use of the cross-level double NMR-NQR method for NQR detection of explosives was proposed by Blinc's group [171–174].

The following double resonance approaches have been used to improve the SNR:

1. Detection of changes in NMR signals by level crossing in the double NQR-NMR system (Fig. 7.8)
2. Zeeman polarization enhanced NQR, or the method of increasing ^{14}N NQR signals by ^{1}H–^{14}N level crossing after application of a preparatory static (inhomogeneous) magnetic field (Fig. 7.9)
3. Triple resonance method or quadrupole–quadrupole enhanced NQR [172]

The method of increasing ^{14}N NQR signals by ^{1}H–^{14}N level crossing with static inhomogeneous magnetic fields, as low as 200 mT, was investigated by Luznik et al. [171]. Later, this method was developed by Thurber et al. [175], Rudakov et al. [176], and Mihaltsevich and Beliakov [177] using small magnetic fields (0 to 26 mT). Theoretical studies of the cross-relaxation processes in NQR-NMR double resonance [172,175] provide the possibility for further development of these methods. These cross-relaxation methods are based on a change of populations of proton and nitrogen subsystems, as predicted by the spin-temperature model. Recently, the rates of population change in double NMR-NQR resonance have been discussed [178]. It has been demonstrated that the contact between the NQR and NMR subsystems, realized via a double resonance conditions in the static magnetic field changes the relaxation rates for different transitions. It has been shown theoretically that relaxation T_1 under double-resonance conditions is characterized by three parameters, which are defined by cross-relaxation and the autocorrelation rates of nitrogen and proton. In experiments, the authors observed an essential change in the T_1 relaxation parameter for different nitrogen NQR transitions. That

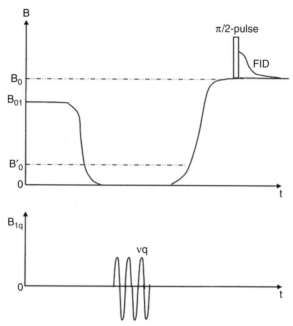

FIGURE 7.8   Detection of nuclear magnetic resonance (NMR) signal changes by the level crossing in the double NQR–NMR system. B_0 is the static magnetic field for the detection of NMR signal, B_{01} is the preparatory static magnetic field to polarize the proton subsystem, v_q is the NQR resonance frequency, B_{1q} is the radio-frequency field, B'_0 is the static magnetic field where the crossing of the energy levels of the proton and ^{14}N NQR subsystems happens (determined by the condition $2\pi v_q = \gamma B'_0$).

is, the NMR/NQR double system realized in small static magnetic fields makes it possible to increase the SNR for NQR signals of low-frequency explosives. Thus, it is obvious that double resonance can be used as another way to improve the SNR.

7.3.3   Detection of Explosives by NQR

The first research on the possibility of detecting explosives by pure NQR began in the United States in the early 1960s [179]. In recent years there has been growing interest in the use of the ^{14}N NQR technique for the detection of explosives and narcotics [180–183]. As mentioned above, all explosives and narcotics substances in the solid state, such as RDX, PETN, tetryl, TNT, and other ^{14}N-containing explosives, can generally be detected by NQR methods. Typical NQR frequencies of widespread explosives are shown in Fig. 7.10, and their relative resonance intensities are presented in Fig. 7.11. Each material has a set of resonance frequencies as a unique "fingerprint" which does not match any other materials in the field. Therefore, a specific feature of NQR technology compared with other techniques is the ability to detect and distinguish explosives from other items in the environment. In contrast to other bulk detection methods, NQR is specific to the material

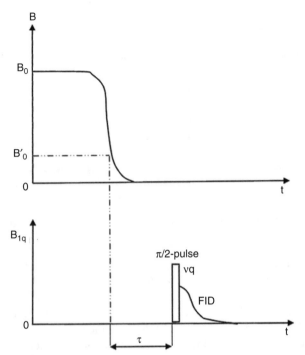

FIGURE 7.9 Zeeman polarization-enhanced NQR or the method of increasing the ^{14}N NQR signal by ^{1}H–^{14}N level crossing after application of a preparatory static magnetic field.

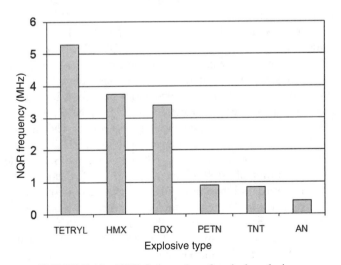

FIGURE 7.10 NQR frequencies of typical explosives.

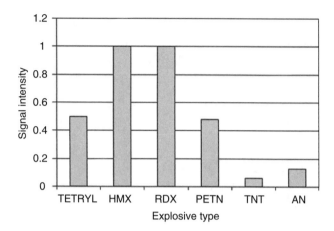

FIGURE 7.11 NQR signal intensities (in a.u.) for various explosives.

composition and provides a unique signature of the explosive or drug, regardless of geometry, packing, or location inside the item scanned [184].

In mine detection, NQR is immune to the presence of metal items, roots, vegetation, rocks, voids, or the nature of the terrain. One more important issue is that the method can be used in both wet and dry conditions. The NQR method is particularly suited for detecting mines with little or no metal content. Additionally, it is an inherently safe technique without any major risks of radioactivity or ionizing radiation. There is only a very low potential risk to the operator or the environment [185]. NQR is also capable of providing information on the amount of explosive present in the scanned volume because the signals detected are proportional to the mass of material. Furthermore, an NQR device is relatively inexpensive and may be made compact, depending on its application.

The advantage of NQR methods compared to NMR is that it is not necessary to use an externally applied magnetic field for NQR. However, the main issues of ^{14}N NQR detection of the explosives are the low signal-to-noise ratio (SNR) at low NQR frequencies and, for detection in the field, the influence of spurious signals from resonant acoustic ringing and radio-frequency interference signals. An especially difficult problem is detection of TNT by NQR, which is due to a number of factors. First, the ^{14}N resonance frequencies for TNT are in the range 714 to 870 kHz (i.e., very low in comparison with, e.g., RDX, which is at about 3.4 MHz) and the detection sensitivity is too low since the SNR scales approximately as the 3/2 power of NQR frequency (Fig. 7.12). Second, TNT has two distinct crystalline forms with two molecules per unit cell in both forms and three structurally nonequivalent ^{14}N sites per molecule. As a result, there are as many as 36 NQR resonances. Finally, the ratio of the spin–lattice to spin–spin relaxation times for TNT is large, which hampers coherent averaging. These factors result in the reduction in SNR of TNT, approximately 50 times relative to that of RDX.

It was noted above that multiple accumulation of NQR signal has to be employed to obtain better SNR. Let us consider a usual single-shot NQR experiment. In this

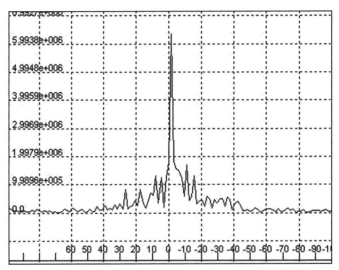

FIGURE 7.12   Fourier transform of NQR signal of RDX at a resonance frequency of 3410 kHz (the accumulation time is 1 s and the number of average NQR responses is 400).

case, the next single-pulse experiment can be repeated only three to five T_1 relaxation times later, because the spin system has to relax in the thermally equilibrium state before the next excitation shot applied. Therefore, the effective time of the signal registration using such single-shot detection is proportional to the ratio of the registration time of NQR signal (T_2^*) to the repetition time determined by T_1. This is not so critical for RDX detection because of the tolerable ratio $T_2^*/T_1 = 0.8$ ms/11 ms $= 0.07$, whereas for NQR signal of TNT this ratio is 0.4 ms/4 s $= 10^{-4}$. Thus, it is clear that multiple accumulations with a single shot (each exceeds 4 s!) cannot be used for the fast TNT detection. Although the simplest detection by accumulation of single-shot NQR signals could be employed for RDX, use of a multipulse sequence rather than a single pulse will further improve the signal registration. Very short T_1 and $T_2 \sim T_1/2$ times for RDX makes the use of a PAPS/NPAPS sequence for the detection of this compound very efficient [186]. An example of the frequency-domain NQR signal of RDX accumulated during 1 s is presented in Fig. 7.12.

TNT has very long T_1 and very short T_2 relaxation times. In this situation, application of an SLSE sequence to form multiple echo responses (each echo of length about T_2^*) will be more advantageous. For example, a series consisting of repeating 1-ms observation windows with a total sequence length of 50 ms will result in an SNR improvement of 2×50 ms/1ms $= 100$ times compared with a single-shot accumulation. In reality there is a smaller gain in the SNR, due to decay of the echo magnitudes during this sequence. A very long T_1 ($= 4$ s) parameter for TNT means that detection of TNT with the use of only one SLSE scan is practical in real applications. In fact, it is possible to use a total sequence of length larger than 50 ms, because the maximum length of a multiple-echo series is determined

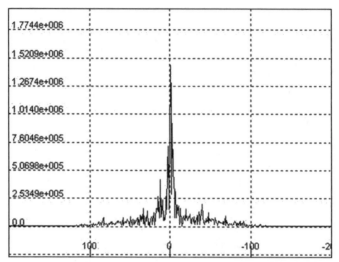

FIGURE 7.13   Fourier transformation of the NQR signal of TNT on the frequency of 842 kHz (the observation time is 5 min and the actual scans are 280).

by an effective T_{2e} parameter. This time is defined as the decay of multiple echoes in a sequence and depends on a specific sequence. T_{2e} is about 150 ms for the SLSE sequence applied to TNT at $T = 25°$ C. Therefore, the total length of an SLSE to detect TNT may be on the order of T_{2e} (i.e., 250 ms) (Fig. 7.13). Thus, we have seen how the relaxation parameters critically affect the sensitivity of explosive detection. As a result, the RDX detection is much easy than that of TNT.

Of course, several approaches to SNR improvement in TNT detection have been proposed. For example, the amplitude of the RF magnetic field can be increased to excite the NQR resonance [187]. Here, an RF amplifier with a 5.5-kW power output has been used to produce an 8-G RF field for TNT detection at 840 kHz, and another RF amplifier with a power output of 1.5 kW was run to produce a 3-G RF field for registration of RDX at 3.4 MHz. However, use of high output power is limited in real applications due primarily to high energy consumption, the technical restrictions (especially for portative and small weighting detectors), and the problem mentioned above resulting from elongation of the dead time of the receiver.

Some gain in the NQR spectrometer sensitivity is obtained by use of more advanced multiple-pulse techniques. It is known that appearance of the prototype devices for remote NQR detection is related to the discovery of the coherent steady states in multiple-pulse NQR [188–191], since in this case the transverse magnetization does not decay as long as is necessary for the multiple accumulations of NQR signals. An application of multiple-pulse sequences such as the SORC and SLSE sequences were discussed above. But even multiple-pulse sequences do not guarantee reliable remote detection of some compounds of interest (e.g., TNT, some narcotics). This stimulated further investigations of methods of sensitivity enhancement. It was proposed recently that the two-frequency excitation may improve the sensitivity of NQR technique [192–194]. However, in the authors' opinion, merely designing more complicated NQR sequences is not enough to make a breakthrough in the sensitivity of NQR detection.

Recently, a report on use of an RF atomic magnetometer for the detection of NQR has been published [195]. The detection of NQR by an atomic magnetometer has several advantages: The sample is placed outside the detector volume, and the detection does not use the resonance pickup coil as a probe. The present authors estimate the sensitivity limit of the magnetometer to be around 0.06 fT/Hz$^{1/2}$, whereas the sensitivity limit calculated for NQR using a conventional pickup coil is about 0.8 fT/Hz$^{1/2}$. In the experiment cited, the real sensitivity of 0.24 fT/Hz$^{1/2}$ was reached at 423 kHz of the resonance of NH_4NO_3. However, the number of accumulations was often as high as 65,536; that is, the experiment includes 32 repetitions of a 2048-echo sequence. As is well known, it is necessary to wait between successive sequences for a period of 3 to 5 T_1 for the best performance of a multipulse sequence. However, the T_1 parameter of ammonium nitrate is very long ($T_1 = 16.6$ s), which makes the NQR signal accumulation period very long. Therefore, tailoring the NQR relaxation parameters by making a contact between the proton and NQR subsystems is needed to increase the SNR of a NQR device based on both atomic magnetometer and the usual pickup probe coil. Consequently, it is obvious that further research on double resonance methods is necessary to change the relaxation parameters of an NQR system. These techniques could improve the conditions for the observation of NQR signals in any multipulse sequence as well as the final efficiency of any type of sensor (e.g., pickup coil, atomic or SQUID magnetometer, HTS-based resonator circuit).

It should be mentioned that there is also another type of very sensitive magnetometer the superconducting quantum interference device (SQUID), which has an excellent potential for measurement of very small magnetic fields (10^{-17} T) [196]. The spectrum of ^{14}N nuclei in cocaine hydrochloride and cocaine base has been detected by SQUID magnetometer [197]. It is remarkable that SQUID detection can be used in a double resonance system as well [198]. The cross-relaxation effects between nitrogen and protons have been used in the matching of transition frequencies of both subsystems under a static magnetic field. For that, a continuous RF magnetic field with a frequency swept through the NMR and NQR resonance has been used. Since there is dipolar coupling between the ^{14}N and ^1H nuclei, the change in the population of the matched levels is reflected in the proton magnetization, which, in turn, is detected by the SQUID. However, the SQUID used in this experiment operates at the very low temperature of liquid helium, 4.2 K.

There is a report on pulsed NQR experiments of ^{14}N performed on powdered ammonium perchlorate, NH_4ClO_4 at 1.5 K. It has been shown that a low-temperature SQUID detector enables one to excite and observe all transitions of a three-level system of ^{14}N nuclei. However, the low-temperature devices that mentioned above are not suitable for broad application in a system for the control and detection of explosives and narcotics. The main problem is the need in cooling with liquid helium to obtain very low temperatures. A possible solution of this problem is an application of high-temperature superconductors (HTSs), because cooling with liquid nitrogen ($T = 77$K) is much more suitable for commercial devices. Recently, an HTS radio-frequency superconducting quantum interference device (RF SQUID) was used for successful detection of NQR at about 887 kHz for ^{14}N in p-nitrotoluene [199]. Nevertheless, the HTS SQUID used in this article for NQR detection had an SNR similar to that of a low-noise preamplifier.

It is believed that the use of HTS for getting high-Q (the quality factor of the resonance circuit) resonators permits increasing the SNR [200]. Intensive development work on these devices continues for various NMR/NQR applications [201]. However, the principal problem in real-device applications of HTS resonators and HTS SQUIDs consists of the very long ringing time for high-Q probes (ca. 10,000 to 50,000). The use of the Q-switching (damping) methods in the pulse NQR is a challenging task for HTS devices, due to problems in matching the HTS and the usual electronic components. One of the possible ways to resolve these issues is the application of nonconventional NQR methods (such as stochastic NQR). Thus, there is optimism that further research could solve the sensitivity problem related to NQR detection, especially for the low-frequency TNT explosive.

7.3.4   Landmine Detection by NQR

Various prototypes of NQR landmine detection systems have been developed in the United States, the Russian Federation, the UK, and some other countries. For example, a high probability of detection and false alarm rejection at scan times of about 20 for TNT and a few seconds for RDX at 20°C have been demonstrated [202]. However, there are some issues to be resolved before the NQR technique will be realized in various applications for mine, explosive, or drug detection. The reliable remote detection at present is feasible only for a restricted number of compounds of interest (e.g., RDX). The method works well for RDX detection, where NQR frequencies of ^{14}N are in the range 3 to 5 MHz. However, the ^{14}N NQR signals are quite weak and require long accumulation times for TNT (with a frequency below 0.9 MHz) as well as for other explosives and drugs with low NQR frequencies [172]. Although essential progress in reducing the explosive detection time has been obtained, new methods of improving the SNR of the NQR signal have to be used.

The structure of a landmine detector is shown in Fig. 7.14. Unlike NQR detectors used in research or luggage scanning, the sensor in a landmine detector consists of an unshielded coil. The main difficulty in the NQR detection of landmines is the influence of outside radio-frequency interference signals (RFISs), since the frequency range of explosives coincides with the frequency range of broadcasting, navigation, amateur and communication radio stations, and other sources of RFIS. It is possible to use two methods to counter the RFIS issue: the active and passive techniques of RFIS attenuation. An example of passive method is a gradiometer developed by the Naval Research Laboratory [166]. The attenuation of the RFI signal of a gradiometer is 17 to 30 dB. The primary problems of gradiometers are the fast attenuation of the RF field emitted as a function of the distance from

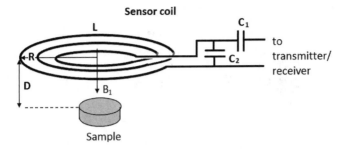

FIGURE 7.14 NQR setup for a landmine detector.

the plane of the coil and the decreasing magnitude of the NQR signal detected in the probe coil. In addition to the passive techniques of the RFIS cancellation, active attenuation of the RFIS can also be proposed. This method implies the use of at least two channels. The main channel is for the detection of an NQR signal together with an outside RFI signal. There is also an additional antenna connected with the second independent receiver channel and designed to receive the outside RFIS signal only. Thus, the construction makes it possible to process the NQR + RFIS and RFIS signals independently. Finally, the computer software subtracts the RFIS from the total (NQR + RFIS) signal of the first channel. The essential trouble in the active technique is fine-tuning of the second channel. The magnitude and phase of RFIS in channel 2 should coincide with that of channel 1. The adaptive filter and loopback are shown in Fig. 7.15 as a solution of this task. Experiments show that the attenuation of this system is at least 20 dB.

Figure 7.16 demonstrates the results of work of the two-channel system on a ^{14}N NQR frequency of 5190 kHz in $C_3H_6N_6O_6$ (RDX). Obviously, the system of

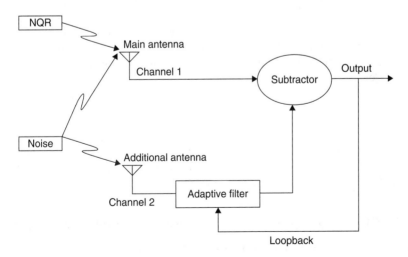

FIGURE 7.15 Arrangement of a two-channel NQR system for landmine detection.

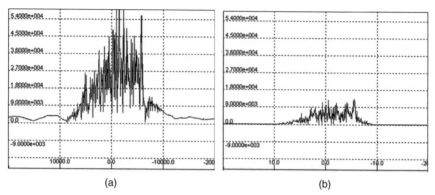

(a)   (b)

FIGURE 7.16 Signals after fast fourier transform in a two-channel system (Fig. 7.15): (a) one scan signal without the use of the second channel; (b) one scan signal after subtraction of noise detected by the second channel.

attenuation of a real device should include both approaches of the signal attenuation discussed above, and a number of antennas in the active method can be larger than two channels.

The current version of NQR landmine detectors, where the methods of RFIS cancellation are realized, can operate at a distance of about 15 cm for RDX and about 5 cm for TNT. However, the detection time (number of accumulations) is still a critical issue that strongly affects the operational efficiency of this detector. For that reason, most experts suggest using the NQR detectors as confirmation sensors in combination with other techniques.

Some commercial devices designed for the detection of illegal substances in vehicles have been described [203,204]. Ammonium nitrate (AN) is of special interest as the explosive used most often in car bombings. In the framework of the work cited, the special coil design, optimal sequences, and power amplifier were defined and created. It is believed that any car has good shielding against RF irradiation. Nevertheless, in reality, the holes and slots allow an RF field to enter a car. A slot of length a, the wavelength of the RF field used in NQR measurements, is much larger than $2a$ and the attenuation α of the magnetic mode penetrating such a slot is [203]

$$\alpha = \frac{27.3t}{a} \tag{7.16}$$

where t is the depth of the slot (i.e., the thickness of the metal), and the length of the slot is much larger than the slot's width. Thus, the attenuated RF field penetrates inside the car through the doors, windows, and slots.

7.3.5 Global NQR Applications for the Detection of Explosives in Luggage

Use of the NQR method for explosive detection in luggage have begun since 1994, when Quantum Magnetic Inc. announced commercial luggage scanners for explosives and narcotics on the basis of patents licensed by the Naval Research Laboratory. In the former Soviet Union, research on NQR detectors for luggage was conducted in the 1980s by cooperation between Kaliningrad State University and the Institute of Applied Physics (Novosibirsk) [206]. This collaborative work results in the development of the NQR scanner prototype. The first generation of NQR detectors (i.e., the scanners of Quantum Magnetic Inc. and of the Novosibirsk Institute of Applied Physics) were able to detect explosives in volumes of 120 to 160 L; however, the detection time was slow. Research in this direction has continued, and the next generations of NQR detectors with improved parameters have been developed in the United States, Russia, Australia, and China. For example, NQR-160, NQR-25, and NQR-15 model detectors have been put on the market by the Logys company (Fig. 7.17).

In 1998, an Australian company (QRSciences) began research on the creation of a landmine detector and luggage scanner on the basis of NQR technology. The first version of the device was based on the prototype developed in collaboration with Kaliningrad State University [205]. Afterward they produced new devices in collaboration with the British Technology Group using novel computer processing and highly sensitive sensors to achieve improved operational performance (Fig. 7.18).

Recently, a prototype of the NQR detector shown in Fig. 7.19 was made in the Qingdao Laboratory by collaboration with the China Research Institute of Radiowave Propagation and G. V. Mozzhukhin (one of the section authors). A hand luggage scanner developed by SpinLock Ltd. is one of the most advanced NQR detectors, based on use of the cross-relaxation method, enhanced multipulse

FIGURE 7.17 Parcel scanner NQR-15 of Logys Ltd., Russia (www.logsys.ru).

FIGURE 7.18 NQR scanner T3-03 of Qrsciences Ltd., Australia (www.qrsciences.com).

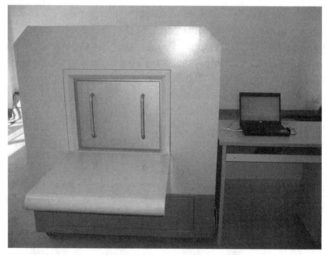

FIGURE 7.19 NQR detector for luggage at the China Research Institute of Radiowave Propagation.

sequences, and developed signal processing (Fig. 7.20). This scanner can employ a pulse external magnetic field to gain the SNR using cross-relaxation spectroscopy. The system yields a 10-fold increased sensitivity in RDX detection with localization of the NQR samples.

The most recent generation of NQR scanners have the ability to detect TNT, RDX, TETRYL, PETN, HMX, AN, AP, and mixtures of these substances. Nevertheless, the detection time of low-frequency explosives such as TNT and AN is still not suitable for the fast detection of these substances in luggage. It is obvious that performance improvement studies on NQR detectors are open for further research. Apparently, the application of the cross-relaxation methods, novel atomic and SQUID magnetometers, and HTS sensors will create breakthroughs in NQR explosive detection in luggage.

FIGURE 7.20   EDD CSC Q100 model NQR scanner of SpinLock Ltd., Argentina (www.spinlock.com.ar).

7.3.6   Conclusions

NQR is very prospective technology for explosive detection in both landmines and scanning luggage. However, up-to-date NQR applications for remote detection are far from in wide practical use. The main reason is the weak SNR of the method, especially for some important substances, such as TNT, AN, and narcotics. Therefore, the use of NQR detectors as confirmation sensors is currently much more feasible. For example, it has been proposed to use a NQR detector as a confirmation sensor for an x-ray scanning system or for a ground-penetration radar (GPR) detector [207]. Nevertheless, recent progress in tailoring of relaxation and cross-relaxation in quadrupole systems, use of HTS sensors, and use of atomic and SQUID magnetometers leads to expect an essential increase in SNR in NQR detection. It is thus highly probable that we will observe a wide application of NQR technology for explosives detection in a few years.

7.4   X-RAY, GAMMA-RAY, AND NEUTRON TECHNIQUES

Sensors based on radiation measurement are used widely in industry and medicine. Sensors of these types used in gauges are quite sensitive to particles and electromagnetic radiations, and they perform measurements nondestructively. In this section, basic principles of electromagnetic radiation and neutron measurement techniques are evaluated and some examples are presented.

7.4.1   X-ray Techniques and Gauges

X-rays consist of electromagnetic radiation produced following the ejection of an inner orbital electron and subsequent transition of atomic orbital electrons from states of high to low energy. When a monochromatic beam of x-ray photons falls onto a given specimen, three basic phenomena may result: absorption, scatter, or fluorescence. These lead to three important x-ray methods: the absorption technique, which is the basis of radiographic analysis; the scattering effect, which is the basis of x-ray diffraction; and the fluorescence effect, which is the basis of x-ray fluorescence analysis (XRF) spectrometry.

Interaction of the x-rays with matter can take place either by scattering or absorption. If the incoming x-ray intensity is I_0 and the absorbed intensity is I, then I can be expressed as

$$I = I_0^{-\mu x} \tag{7.17}$$

where x is the absorber thickness and μ is the mass attenuation coefficient of the absorber. It is clear that some x-ray photons are lost during the absorption process, due to photoelectric effect. Photoelectric absorption can occur at each energy level of an atom. The total photoelectric absorption is determined by the sum of individual absorptions within a specific shell. The value of the mass attenuation coefficient of an absorber depends on the incoming x-ray energy and the absorber density.

X-ray Sources   There are several different types of sources, which have been employed for the production of x-rays. Most common x-ray sources are either radioisotopes, which initiate the secondary fluorescent, or x-ray tubes. The primary x-ray source unit consists of a very stable high-voltage generator, which is capable of providing a potential of typically 40 to 100 kV. The current from the generator is fed to the filament of the x-ray tube, which is typically a coil of tungsten wire. The applied current causes the filament to glow, emitting electrons. A portion of this electron cloud is accelerated to the anode of the x-ray tube, which is typically a water-cooled block of copper with the required anode material plated or cemented to its surface. The impinging electrons produce x-radiation, a significant portion of which passes through a thin beryllium window to the specimen.

X-ray radiography is a widely used example and application of x-ray absorption. This is one of the basic nondestructive methods. An x-ray radiograph is a photographic record produced by the passage of the x-rays through an object onto a film, as shown in Fig. 7.21. When the film is exposed to x-rays, an invisible change called a *latent image* is produced in the film emulsion. The exposed areas become dark, and the degree of darkening depends on the amount of exposure (Fig. 7.21). If the thickness of the flaw is less than d, Eq. (7.17) becomes Eq. (7.18). Since the absorption of an x-ray is less, the exposure of that part will be greater and the image will be darker.

$$I_d = I_o e^{-\mu(x-d)} \tag{7.18}$$

Scattering of an x-ray occurs when the x-ray collides with one of the electrons of the absorbing element. Where this collision is elastic (i.e., when no energy is lost in the collision process), the scatter is said to be *coherent (Rayleigh) scattering*

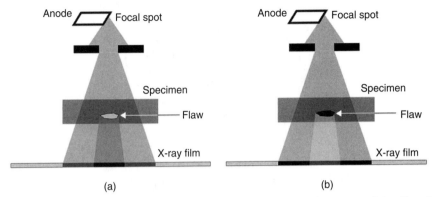

FIGURE 7.21 Absorption of x-ray radiations through matter. Exposure of the films is related directly to the attenuation of radiation.

[208]. Since no energy change is involved, the coherently scattered radiation will retain exactly the same wavelength as that of the incident beam. X-ray diffraction is a special case of coherent scatter where the scattered photons interfere with each other. It can also happen that the scattered photon gives up a small part of its energy during the collision, especially where the electron with which the photon collides is only loosely bound. In this instance the scatter is said to be *incoherent* (Compton scatter) and the wavelength of the incoherently scattered photon will be greater than 10 [208].

X-ray backscatter technology should be most applicable as a tool for identifying antipersonnel mines planted no more than 7.5 cm below the surface or that are lying on the surface but are hidden by camouflage or vegetation. Many groups have researched standard x-ray backscatter over the past years as a potential mine detection tool [209]. The intensity of x-rays scattered by a buried object is dependent on several factors: the intensity of x-rays emitted from the source, the attenuation of the x-rays before scattering, the probability of scattering in the back direction, the attenuation of the x-rays as they come back from the object and traverse the soil on the way to the detector, and the density and the thickness of the soil.

X-ray Fluorescence Logging Radioisotope XRF analysis in boreholes has been used since 1953 [210]. The construction of one XRF probe used as borehole logging (Fig. 7.22) illustrates schematically construction of an experimental XRF borehole probe. The image construction is adopted to allow a differential filter technique. Co–Ni filters are used to isolate CuK x-rays. These kinds of probes are widely used for the analysis of dry holes for elements with $Z < 50$.

X-ray gauges are noncontact measurement systems that are quite sensitive within the range 10 μm to 30 mm. These types of gauges are useful not only for measuring a strip with a sensitive surface, but also for any cold-rolling application. X-ray radiation gauges send ionizing radiation perpendicular through the strip. The moving strip absorbs a portion of radiation on its way from the source below the strip

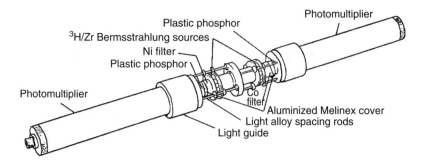

FIGURE 7.22 XRF borehole probe. (Adapted from [211].)

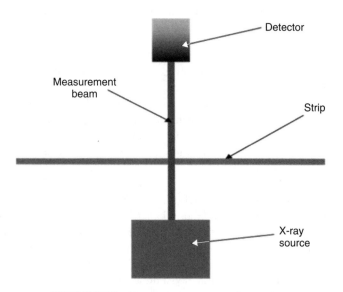

FIGURE 7.23 Noncontact x-ray gauging system.

to the detector. The thickness measurement is based on the reduction of radiation by the moving strip. The intensity of the measurement signal changes depending on the strip thickness according to the Eq. (7.17), where x is the thickness of the strip. A schematic diagram of a noncontact x-ray gauging system is shown in Fig. 7.23.

7.4.2 Gamma-ray Techniques and Gauges

Gamma rays are a type of electromagnetic radiation that results from a redistribution of electric charge within a nucleus. A γ-ray is a high-energy photon. The only thing that distinguishes a γ-ray from the visible photons emitted by a light bulb is its much shorter wavelength. For complex nuclei, there are many different possible ways in which the neutrons and protons can be arranged within the nucleus. Gamma rays can be emitted when a nucleus undergoes a transition from one such configuration

TABLE 7.2 Radioactive Isotopes Used in Industrial Radiography

Radioisotope	Half-Life	Energy Levels of γ-rays (MeV)	Practical Thickness Limits[a] (cm)
Thallium-170	127 days	0.084 and 0.54	1
Selenium-75	120 days	0.066 and 0.400	3.0–5.0
Iridium-192	70 days	0.137–0.651	3.50–6.25
Cesium-137	33 years	0.66	2.5–7.5
Cobalt-60	5.3 years	1.17 and 1.33	6.5–20.0

[a]For example, for steel applications.

to another. For example, this can occur when the shape of the nucleus undergoes a change. Neither the mass number nor the atomic number is changed when a nucleus emits a γ-ray in the reaction.

Gamma radiography is also widely used for the denser and thicker samples. The penetration capability of γ-rays is more than x-rays, because the wavelengths of γ-rays are shorter. They are distinguished from x-rays only by their source rather than by their nature. Gamma rays are emitted from the disintegrating nuclei of radioactive isotopes, and the quality (wavelength or penetration) and the intensity of the radiation cannot be controlled by the user. In industrial radiography, the artificial radioisotopes are used exclusively as sources of gamma radiation. Radioactive isotopes used in industrial radiography are shown in Table 7.2.

The basic operating principles of gamma radiography are also illustrated in Fig. 7.21. Application of the geometric principles of shadow formation to radiography leads to five general rules. Although these rules are stated in terms of radiography with γ-rays, they also apply to x-ray radiography. These rules are that the focal spot size of the radiation source should be as small as possible in order to have small amount of unsharpness, the distance between the radiation source and the material examined should always be as great as possible in practice; the film should be as close as possible to the object being radiographed; the central ray of the beam of radiation should be perpendicular to the film; and the plane of maximum interest of the object should be parallel to the plane of the film. A typical γ-radiograph of the radio taken by ^{192}Ir is shown in Fig 7.24.

Cobalt-60 Container Computer Tomographic System According to basic principles, the CT image is a slice of the object and the gray scale at different position is almost proportional to the density of the corresponding part of the object. Therefore, it is possible to determine the material of goods according to their density. The structure of a ^{60}Co container CT system is shown in Fig. 7.25. The detection subsystem (γ-ray source, detector, etc.) is installed on a big ring, which can be kept static or can rotate. The slice image of goods can be obtained according to the density data.

Gamma-ray gauging systems measure some subsurface properties of materials without contact. The penetrating nature of high-energy γ-ray radiation enables measurements to be made through the walls of sealed containers. Densities of materials

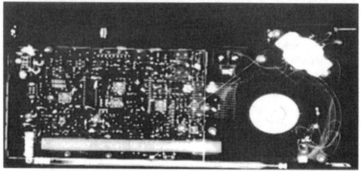

FIGURE 7.24 Radio radiograph obtained by γ-rays.

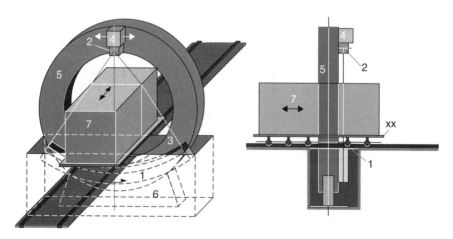

FIGURE 7.25 CT system structure from. (From [212].)

or levels of solids, liquids, or slurries in pipes and tanks can be determined using an externally mounted γ-ray source and detector. The working principles of these types of gauges are transmission or backscatter geometry, depending on the measurements. In the metal industry, level gauges are designed to measure the level of molten metal during the continuous casting of billets, blooms, or slabs. For example, a typical level gauge used in the steel industry is shown in Fig. 7.26. A radioactive source, usually ^{60}Co, is mounted externally with shielding that is located outside the water cooling jacket directly beside the mold. A scintillation detector is located on the opposite side outside the water jacket. The principle of operation of this system is based on a radioactive source that irradiates the mold. The molten steel rises and the radiation field is attenuated so that the scintillation detector, located on the opposite side of the mold, senses different field strength. Within the detector crystal, the radiation generates flashes of light that are converted to electrical pulses through a photomultiplier tube. Finally, the detector preamplifier creates and generates standard electronic pulses.

Gamma-ray density gauges are used extensively in the petroleum and coal industries. One of the most widespread applications is for interface detection in long pipelines carrying oil from oilfields to refineries or to seaports, where the oil is shipped abroad [213]. A schematic diagram of a method of measuring the amount of catalyst in an oil of varying density is shown in Fig. 7.27. The density of decanted oil is measured by the upper γ-ray gauge. The γ-ray gauge at the bottom of the settler gives the density of the catalyst or oil slurry. The amount of catalyst is determined from the difference between two signals [214].

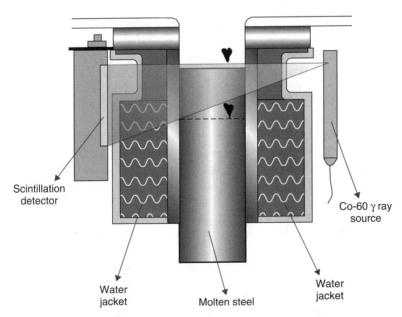

FIGURE 7.26 ^{60}Co radioactive-level gauges.

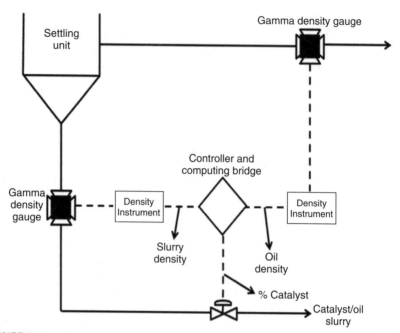

FIGURE 7.27 Method using two γ-ray density gauges to measure the amount of catalyst in an oil of varying density.

7.4.3 Neutron Techniques and Gauges

Along with protons, neutrons make up the nucleus, held together by the strong force. Various nuclides become more stable by expelling neutrons as a decay mode, known as *neutron emission*, and commonly occurs during spontaneous fission. Cosmic radiation interacting the Earth's atmosphere continuously generates neutrons that can be detected at the surface. Nuclear fission reactors naturally produce free neutrons. Their role is to sustain the energy-producing chain reaction. Neutrons exist in several energy states. Higher-energy neutrons have more energy than fission-energy neutrons. They are generated in accelerators or in the atmosphere from cosmic particles. They can have energies as high as tens of joules per neutron.

Weak interaction of neutrons with typical metals, such as Fe and Al, combined with a strong scattering of the neutrons by hydrogen, allows them to penetrate macroscopic metallic objects and to reveal the internal distribution of hydrogen-containing components, such as petrol, water, organic materials, and corroded parts. Depending on this property, several industrial applications are widely used in industry.

Neutron radiography is an imaging technique, which provides images similar to x- and γ-ray radiography. The difference between neutron and electromagnetic ray interaction is the mechanism that produces significantly different and often complementary information. While x- or γ-ray attenuation is directly dependent on atomic number, neutrons are efficiently attenuated by only a few specific elements.

For example, organic materials or water are clearly visible in neutron radiographs due to their high hydrogen content, while many structural materials, such as aluminum and steel, are nearly transparent. This is because of the high absorption cross section of hydrogen, carbon, and boron.

At the present time, neutron radiography, one of the main nondestructive testing techniques, is able to satisfy the quality-control requirements of explosive devices used in space programs. Most of the detonating devices of the space program have been submitted to systematic neutron radiography examination. The detection of 0.1-mm-thick cracks in the explosive charge is common, and the efficiency of the technique makes it easy to distinguish differences in the compression of the explosive even through different metallic containers, such as lead, aluminum, or steel [215]. The ability to detect compounds containing hydrogen atoms is also used to inspect oil levels and insulating organic materials. Neutron radiography also facilitates the checking of adhesive layers in composite materials and surface layers (polymers, varnishes, etc.). All types of O-rings and joints containing hydrogen can be observed even through a few centimeters of steel.

The basic layout of neutron radiography consists of a neutron source, a collimator functioning as a beam formatting assembly, a detector, and the object of study, which is placed between the exit of the collimator and the detector. Basic experimental neutron radiography assembly is shown in Fig 7.28. Neutron radiography detectors are generally plane-integrating position-sensitive imaging devices containing material with a high neutron cross section functioning as a neutron converter and recorder, which has the task of collecting the signal emitted by the converter during exposure. After the measurement, the detector signal is read out and may be digitized. In the context of neutron radiographic measurement, the output of the detector is called a *detector signal* or *radiographic image*; the terms *signal* and *image* are considered equivalent. Typical neutron radiography detectors are x-ray film/converter plate assembly, track-etch films, imaging plates, scintillation/camera and amorphous silicon, and flat-panel detectors.

Radiographs obtained by electromagnetic waves and neutrons show quite different images. Electromagnetic radiation interacts primarily with the electronic shell of the atom and therefore increases by approximately Z^n. Neutrons interact with

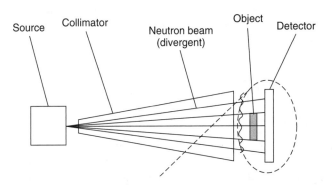

FIGURE 7.28 Neutron radiography assembly.

the atomic nucleus. The magnitude of the cross sections for the various neutron-induced processes depends on both the energy of the neutron and the structure of the target nucleus. Due to the high interaction probability of thermal neutrons with hydrogen-containing materials, neutrons deliver high-contrast images for thin layers of organic substances or samples consisting partially of hydrogen-rich components.

Since the attenuation of neutrons depends on the mass absorption coefficient of the elements, this gives the advantage of the development of neutron gauges. The principle of the nuclear method of moisture measurement is based on the principle that fast neutrons are slowed down by scattering hydrogen nuclei and so are hardly slowed down by scattering materials of higher atomic number. A beam of slow neutrons is created around a source of fast neutrons, and its concentration depends essentially only on the hydrogen or moisture content of the volume. If a fast neutron source is combined with a detector for slow neutrons, the result is a moisture probe that operates without contacting the measured medium temperature, pressure, pH value, or grain size, and the material's chemical bonds do not affect neutron deceleration. On the other hand, it should be noted that the chemically bound hydrogen content or that of the constituent water is detected, in addition to the hydrogen content of the free water or moisture. Thus, it is necessary to calibrate the neutron gauge according to these facts.

The fluctuations in constituent water content limit the accuracy attainable for moisture measurement. The same applies to fluctuations in the bound hydrogen content. However, it follows from the relationship of the molecular weight of hydrogen to water that a 1% fluctuation in hydrogen content simulates a 9% fluctuation in the moisture display. The display is approximately proportional to the moisture in percent by volume. The moisture in weight percent is given by the quotient of volume moisture and material density. The neutron display may be calibrated directly in "weight percent moisture" if the apparent density is sufficiently constant.

Configuration of one of the widely used neutron moisture gauge is shown in Fig. 7.29. It demonstrates that the bunker probe is installed in the outlet section of a bunker or a continuous-flow tank, so that the built-in Am−Be neutron source is sufficiently well surrounded by material to create an effective measurement volume. The probe, a BF_3 detector, is mounted in a dip pipe socket, closed at the bottom end, which passes through the tank or vessel wall at an angle of 30 to 45° to the horizontal or, alternatively, in a protective tube passing through both walls. If the temperature in the tank or vessel increases, air must be blown through the protective tube to cool the probe [216].

7.5   ELECTRIC IMPEDANCE TOMOGRAPHY

7.5.1   Operational Principle

Electrical impedance tomography (EIT) is a sensor technology developed to image the electrical conductivity distribution of a conductive medium. This technology is interesting due to its low-cost, easy-to-use measurement method, which yields direct information about the composition of the conductive medium. Buried objects,

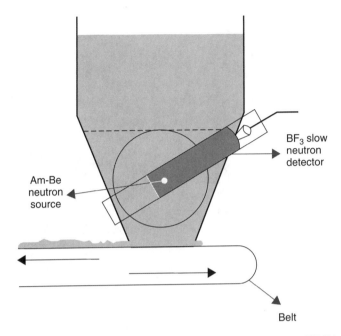

Am-Be
neutron
source

BF₃ slow
neutron
detector

Belt

FIGURE 7.29 Bunker probe neutron moisture meter. (From [216].)

whose conductivities differ from those of soil, cause a discontinuity in the soil conductivity that can be sensed by surface probing. Sensor probes are implemented in pairs and combined into array configuration to transform the discontinuities into surface conductivity maps. The detection of landmines buried at shallow depths is one of the particular application areas of EIT [217]. Characterization of the soil fields is another example in practice.

An EIT sensor probe pair consists of two electrodes that are inserted in the ground to a specified distance to measure the electrical resistance of the medium. The soil resistance is basically calculated by measurement of resulting electric voltage for the induced current. Typical stimulation current is on the order of milliampers, and the frequency of the stimulation is in kHz region. Electrodes are not like wide plates that can provide quasiuniform electric field and related current distribution in the measured piece of the medium. Therefore, soil conductivity (or resistivity) calculation is subject to some approximations, depending on the measurement scenarios, such as probe length under the ground and the distance between the probes. The Wenner four-pin method, which uses four metal electrodes driven into the ground along a straight line, is one of the measurement techniques often used. Soil resistivity is the function derived from the voltage drop between the center pair of pins, with current flowing between the two outside pins [218]. The calculation formula is given in Eq. (7.19) for the ground conductivity measurement illustrated in Fig. 7.30:

$$\rho = 2\pi a R \tag{7.19}$$

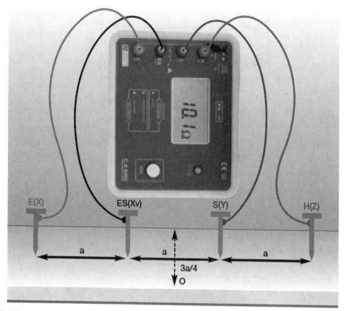

FIGURE 7.30 Ground conductivity measurement using the Wenner four-pin method. (From [219].)

where ρ is the soil resistivity ($\Omega \cdot$ cm), a the distance between probes (cm), and R the soil resistance read by the resistivity meter.

7.5.2 System Design

In the case of EIT array design for ground probing, a pair of electrodes are stimulated, and the electrical voltage is measured on the remaining pairs of the electrodes. After stimulation of independent combinations of interest, a data-processing algorithm is used to reconstruct an image of the electrical conductivity distribution under the surface. In this manner, both the metallic or dielectric objects buried in the ground can also be detected by conductivity anomalies. Further information about the buried object, such as size, shape, constitution, and depth, is also attainable by referring to postprocessing techniques. A typical EIT detector system, consisting of a 4 × 4 electrode array, a data acquisition unit, and a laptop for data storage and signal processing is shown in Fig. 7.31.

7.5.3 Capabilities and Limitations

The EIT technology has been implemented especially for medical diagnostic applications. The adaptation and development of EIT for the detection of buried objects are relatively recent in applications. Wexler was one of the first investigators to apply EIT to the detection of unexploded ordnance (UXO) in the ground [220].

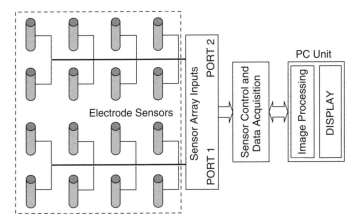

FIGURE 7.31 Typical EIT detector system.

The detection of UXO requires probing depths greater than the size of the surface electrode array. Such a limitation can degrade its operational usability for this particular application. Furthermore, the EIT technology is sensitive to noise, which prevents its use at depths that are large with respect to the electrode array size. Nevertheless, some promising performance results are reported in the detection of objects buried at shallow depths [221]. For example, it is reported [1] that pretty satisfactory detection performances were achieved by the EIT detector for the detection of antitank mines, which has been evaluated in soils of conductivity varying from 1 to 15 mS/m. The detector performed well for detecting antitank mines down to a depth of 15 to 20 cm, and reliable detections were obtained down to a range of 1.0 to 1.5 electrode spacing for objects with a size on the order of two-electrode spacing. The matched filter approach is also proposed to reduce the false alarms caused by objects of different sizes.

EIT technology has the following essential strengths:

- The hardware is relatively simple, easy to use, and low in cost.
- Soil fields can be surveyed; both metallic and nonmetallic objects are detectable.
- Wet grounds yield a better performance, due to better electrical contact.

EIT technology has the following major limitations:

- the EIT detector requires an electrically conductive environment. To work properly, In other words, it fails in environments containing dry sand or rock-covered surfaces.
- The requirement for an electrical contact between the electrode array and the soil is disadvantageous for high-speed vehicle-mounted scanning systems.
- The scan time of the system is also limited by data acquisition and processing time, which is determined by the number of independent configurations of

stimulating and recording pairs of electrodes. For example, there are thousands of independent configurations for large array detectors, and that could take minutes for each scan on the scanning direction.

Advanced EIT array systems can perform three-dimensional imaging of the soil field by introducing three-dimensional tomographic reconstruction algorithms. Figure 7.32 shows an example of EIT detector measurements carried out in a Montegrotto 64-electrode custom-built system used with 0.25- to 0.5-m grid spacing to obtain two- and three-dimensional resistivity images of the soil field. The depth slices show that different features are clearly detectable up to a depth of 0.60 m, due to strong resistivity contrasts between the archaeological structures and the soil. The detailed performance results are demonstrated in Section 9.6.

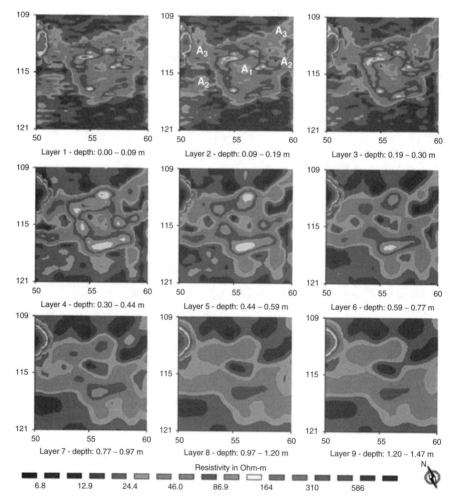

FIGURE 7.32 Electrical impedance tomographic images (three-dimensional depth slices).

7.6 INFRARED AND HYPERSPECTRAL SYSTEMS

7.6.1 Introduction

Hyperspectral imaging is used routinely in remote sensing for geological and geographical purposes. It can also be used in the detection of buried munitions; compact objects, including landmines; improvised explosive devices (IEDs); and unexploded ordnance (UXO). Hyperspectral imaging can also be used to detect surface-laid munitions. The focus in this section is on the detection of buried munitions, with landmines used as the main example. The principles are similar for the more general problem of buried munitions as well as for other small, compact buried objects.

Landmines are commonly used against land combat forces because they very effectively impede mobility and act as an excellent force multiplier. Because stationary troops are extremely vulnerable to modern weapons, the current military doctrine seeks a rapid method for detection of mined areas so that alternative routes that avoid them can quickly be planned and implemented. Detecting mined areas from a distance is also important in humanitarian demining operations to reduce the surface area that must be cleared painstakingly by hand. Thus, airborne mine detection is a high priority among a number of nations.

A brief description of the scenarios and constraints on the problem, as it pertains to buried mines, follows (see Morita and McKanney [222] for details). Landmines have a variety of shapes, with dimensions typically ranging from 10 to 30 cm. Outer case materials vary and include bare metal, plastic, paint, resin, and wood. Mined areas may consist of buried or surface-laid mines. Buried mines are put in place slowly, either mechanically or by hand, and may be detected by looking for patterns or random anomalies. The anomalies may be due to disturbed soil or vegetation immediately covering the mines or due to the presence of the mine itself. Disturbances due to mine-laying machinery may also cause anomalies. Surface artifacts associated with the mine-laying process may also be detected, such as vehicle or human tracks and paraphernalia such as containers, arming pins and rings, and parachutes. Mines may be detected along a slant direction from a land vehicle or airborne platform or along a direction closer to nadir from an airborne platform. Viewing near nadir minimizes vegetative obscuration and atmospheric attenuation while maximizing permissible sensor altitude for a given spatial resolution [222]. For this reason, down-looking sensors are generally preferred. Airborne sensors are preferred to land-based sensors for faster coverage, although land-based systems have better spatial resolution.

Because of the standoff distances involved, the only practical technologies are optical and radar based. Research on airborne optical detection of landmines has been conducted since the 1960s. Both active and passive approaches have been investigated, encompassing wavelengths from ultraviolet to thermal infrared. A number of studies have looked at possible technologies for remote detection of minefields. The Environmental Research Institute of Michigan (ERIM), under contract to the U.S. Army Mobility Equipment R&D Command (now the U.S. Night Vision & Electronic Sensors Directorate), reviewed a wide range of technologies for both buried and surface mines using such criteria as percentage of time usable, accuracy of location and delineation of minefield, rate of area coverage, vulnerability, and time delay to receive data [223]. Their initial screening concluded that only optical wavelength technologies would be at all feasible. A more detailed study then focused on panchromatic photogrammetric cameras, multispectral scanners, active and passive infrared scanners, image intensifiers, television cameras, and synthetic and real aperture radar. The study concluded that aerial photography and electrooptic scanners, particularly active ones at CO_2 wavelengths (ca. 10,600 nm), and millimeter-wave systems should be investigated further, although none stood out as clearly preferable [224]. The study also suggested that multispectral electrooptic scanners would be better than monospectral scanners for distinguishing mines and minimizing false alarms. The ERIM study covered detection of buried and surface-laid mines and mine-laying paraphernalia.

In 1984, DRDC Suffield initiated a set of studies to evaluate the feasibility of optical wavelength and microwave technologies for detecting surface-laid landmines from airborne platforms. Active and passive methods were evaluated at ultraviolet/visible [225], visible/infrared [226], millimeter- and microwave wavelengths [227]. As a result, research efforts concentrated on detection of surface-laid, and later buried, landmines by active and passive thermal infrared imaging, microwave radiometry, and visible-band passive hyperspectral imaging.

By the late 1980s, research groups in a number of countries were developing prototype hyperspectral imagers, and research was being directed toward using hyperspectral systems in place of multispectral systems for airborne mine detection. DRDC Suffield and Itres Research began collaborative investigations using the fledgling *casi* visible/near infrared (VNIR, 400 to 1000 nm wavelength) imager in 1989. Hyperspectral imaging can be *active*, employing artificial sources of illumination such as lasers or flashlamps, or *passive*, using natural sources such as the sun. Some efforts have been made to detect mines using active hyperspectral imaging, but detection was achievable only for surface-laid mines and signal-to-noise ratios were marginal, even at low altitudes and scanning speeds [228]. In the remainder of this section we focus on passive hyperspectral imaging, with emphasis on the results of the DRDC/Itres program.

7.6.2 Hyperspectral Imagers

Spectral imaging involves forming a set of images (collectively called a *data cube*) of the same scene, where each image corresponds to a different spectral wavelength range (*waveband* or *band*). Images may be categorized as multispectral, hyperspectral, and ultraspectral. There is some debate over how many bands and what bandwidth constitutes each category, but *multispectral* is usually taken to refer to noncontiguous bands (which may or may not be narrow), *hyperspectral* refers to many (typically, >100) contiguous narrow (<10 nm in VNIR range) bands, and *ultraspectral* refers to even finer spectral resolution (typically, >1000 contiguous narrow bands).

There are a few different image acquisition modes that are commonly used. *Whiskbroom imagers* image a point on the ground, after spectral dispersion, onto a linear array of sensor elements. The point [*instantaneous field of view* (IFOV)] is then scanned perpendicular to the direction of motion of the sensor (*across-track*) to form one spatial row of the spectral image. A complete two-dimensional image is then formed by collecting successive rows as the sensor platform moves *along-track. Pushbroom imagers* image a slit on the ground in the across-track direction. The slit image is focused and spectrally dispersed onto a two-dimensional array of sensor elements called a *focal plane array* or (FPA) (Fig. 7.33). The two-dimensional image is again formed by the along-track motion of the sensor. Whiskbroom imagers have inherently inferior spatial resolution compared to push-broom imagers because of limitations imposed by the mechanical scanning method. Staring sensors form a complete image without sensor motion.

All modes require some method of separation of spectral components. Whiskb-room and pushbroom imagers normally use an optical dispersive element such as a diffraction grating or prism. Some staring imagers use spectral filters in the fore optics or at the focal plane to switch between wavebands. These may be either multiple interference filters on a spinning wheel or tunable acoustoop-tic filters. Others use differential chromatic refractive focusing methods, where a lens is moved along the optical axis to obtain sequential images of a scene, each in focus for one particular wavelength at the focal plane. Computer processing then removes, at each wavelength, the effect of the defocused images from all other wavelengths, to produce the hyperspectral data cube. An example would be Pacific Advanced Technologies' IMSS line of handheld imagers covering the VNIR, shortwave infrared (SWIR; 1000 to 2500 nm wavelength), mid-wave

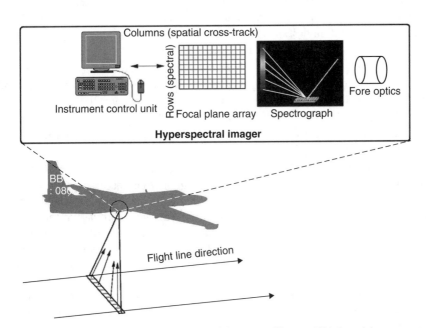

FIGURE 7.33 Image acquisition using an airborne pushbroom FPA-based hyperspectral imager.

infrared (MWIR; 3500 to 5000 nm), and thermal or long-wave infrared (TIR or LWIR; 8000 to 12,000 nm) bands. Some hyperspectral imagers use Fourier transform (FT) spectrometry methods, which generally have better spectral resolution than other dispersive methods but inferior spatial resolution for equivalent data acquisition times. Pushbroom imagers capture the spectra for all pixels simultaneously in an across-track line, whereas the spectral filter, differential chromatic refractive focusing, and imaging FT spectrometer methods do not. It is thus very difficult for the latter technologies to avoid mixing spectral signatures within a spatial pixel when measurements are changing temporally. In airborne applications, then, the rapidly moving platform will lead to poor spatial resolution and/or difficulties in spatially registering spectra.

To be suitable for airborne detection of landmines, previous research and analysis of the phenomenology of the scenarios has suggested that a hyperspectral imager should ideally have the following key requirements:

1. *High spatial ground resolution*. At least a few pixels must be placed on top of a mine to allow for false-alarm reduction. Thus, a resolution of 5 to 20 cm is needed. The required high spatial resolutions necessitate the use of pushbroom imagers. Further, this also generally requires sensor stabilization, geocorrection, or both.

2. *High spectral resolution for some spectral regions*. In the visible and SWIR bands, regions such as in the vicinity of the near-infrared chlorophyll edge may require resolution <10 nm. In the TIR band, the resolution required is a function of the local width of the relevant emissivity peaks. The narrowest spectral features of interest have full widths at half maximum of roughly 120 to 130 nm (see Fig. 7.39). A resolution of 125 nm (32 bands across 8000 to 12,000 nm) would yield roughly one or two points across the narrow spectral features.

3. *Spectral flexibility*. Although high resolution is required for certain spectral regions, lower resolution may be adequate in other regions, such as near blue and green. Further, the band set required can vary with environmental factors such as ground cover, mine type, sun angle, and illumination.

4. *Accurate spectral registration*. As noted earlier, this also favors the use of pushbroom scanners.

5. *Low noise*. In the TIR, the noise equivalent temperature difference (NETD) for each spectral band must be as low as possible. In previous studies [229], broadband TIR imagers could measure effective temperature anomalies from buried mines as low as the sensor NETDs (typically, 100 mK). This was also the typical value of background clutter variation, suggesting that there would be little value in having a broadband NETD much lower than 100 mK. Assuming that the signal/clutter ratio was approximately independent of wavelength, a 32-band imager would need a maximum NETD of 3 mK per band. Typical uncooled sensors have NETDs as low as 50 mK, and cooled ones can achieve 10 mK. Of course, if the sensor is limited by shot or thermal noise, increased integration time can ease this constraint somewhat.

6. *High dynamic range*.

7. *Near real-time analysis of imagery*. For military applications it is necessary to analyze the imagery within a few seconds to a few minutes of collection.

8. *Provision of true reflectivity for reflectance bands* (VNIR, SWIR). Scene radiance, which is normally measured, depends on incident illumination. The latter is a function of many scene and environmental factors, such as the time of day and season.

9. *Low weight, size, and power*. The imager must be capable of operation in a small fixed-wing aircraft or helicopter and, ideally, in an unmanned air vehicle (UAV).

Although not a strict technical requirement for airborne landmine detection, the design should consider nonmilitary airborne remote sensing applications. For example, design and construction to military specification (Mil-Spec) standards may not be essential to achieve mine detection goals, and the dramatically lowered costs of not obtaining such certification may make the instrument cost-effective for other commercial applications. This may seem irrelevant for airborne landmine detection, but the military instrument sales market is small compared to the nonmilitary market, and long-term product support will require sufficient market demand.

The requirements noted above point to limitations in the hardware and software that are common to many systems and must be overcome if they are to be suitable for mine detection. For dispersive systems these include FPA clock-out and readout electronics rates, which limit along-track spatial resolution, and fore-optics, which limit across-track resolution. For Fourier transform systems, spectral scan time and spatial resolution are limitations, while optical filter systems are limited by a lack of flexibility in changing spectral bands and filter switching times. Until recently, all systems have suffered from the inability to correct and analyze data in real time.

There are numerous air- and satellite-borne imagers. Since mines cannot be detected from satellites due to insufficient spatial resolution, we consider only airborne imagers. Most have been developed as general-purpose imagers, although a few, such as the Itres' *casi* [230], *sasi* [231], *tasi* [232] imager family, have been developed with airborne mine detection in mind. An exhaustive list and discussion of the pros and cons of each is beyond the scope of this section. Some examples include: visible and near infrared—*casi*, AISA Eagle (Specim); shortwave infrared—*sasi*, AISA Hawk (Specim), HYMAP (Integrated Spectronics), AVIRIS (JPL), Compass, and NVIS (NVESD); and thermal infrared—*tasi*, AHI (University of Hawaii) and SEBASS (Aerospace Corporation). Many are one of a kind used for dedicated research, although some of the controlling organizations will provide services to conduct customized flights and deliver data products. A few, such as the Itres imagers, are available commercially. Achal et al. [232] describe some of these imagers further, and Bolton [233] has compiled a good list of instruments available for airborne remote sensing, with emphasis on multi- and hyperspectral imagers.

7.6.3   Detection of Buried Landmines

Buried mines cannot be detected directly in the VNIR band, due to the high absorption of optical radiation by the soil and vegetation cover, but may be detected indirectly by observation of differences in reflectance spectra between compact regions above mines and regions of background materials due to:

1. Disturbed versus undisturbed vegetation and surface soil
2. Vegetative stress due to mine implantation even if the surface is not apparently disturbed
3. Vegetative stress due to explosive vapors or particulates originating either from leaks in the mine case or from residue contaminating the case

Stressed or senescent vegetation exhibits a measurable shift in the red edge of the reflectance spectrum relative to nonstressed or nonsenescent vegetation [234]. The shift can be estimated by least-squares fitting of the red edge (Fig. 7.34) using an inverted Gaussian model or by smoothing and differentiating with respect to wavelengths near the edge. It has long been known [235] that portions of the visible reflectance spectrum, in addition to the red edge region, can display marked differences between senescent and nonsenescent vegetation.

DRDC Suffield and HDI (Dartmouth, NS, Canada) conducted experiments on VNIR detection of surrogate mines and blocks of explosives, buried using standard mine-laying methods, in various areas of vegetative cover and bare soil [236,237]. A *casi* imager was placed in a personnel lift at a height of about 5 m above the

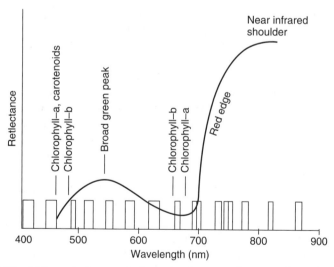

FIGURE 7.34   Major reflectance and absorption features of a vegetative reflectance spectrum. Features and associated pigments are labeled. Vertical axis is not to scale, but typical values for features may be obtained from Fig. 7.33. Bands used in *casi* vegetative stress detection experiments are shown as rectangles, whose widths signify respective bandwidths.

FIGURE 7.35 Image of hand-buried landmines under short natural grass. The image was obtained using a *casi* scanning horizontally at a height of 5 m above the ground approximately two months after burial. The 550.1-nm (green) band is displayed. Mine locations are not visible.

buried objects. Horizontal scanning was used to obtain images at various times over a $1\frac{1}{4}$-year period. Mines were detected by calculating for each pixel of each image, one of three spectral similarity metrics: the linear correlation coefficient, the fractional composition derived from orthogonal subspace projection (see Section 7.6.4), or the red edge shift. Compared to undisturbed background areas, reflectance differences were observed which were associated with vegetative stress due to soil disturbance (Figs. 7.35 and 7.36) and with disturbed bare soil. Within the limited data there was no evidence of a difference in vegetative stress due to explosives leaching into the soil, compared to stress induced by soil disturbance alone, even from the bare blocks of explosives.

Mines could be detected in short- to medium-length vegetation in an east coast Canadian summer–fall environment from $1\frac{2}{3}$ to $15\frac{1}{2}$ months after burial. Short vegetation gave better detection results than bare soil, which gave better results than medium-length vegetation. However, the probability of detection (P_d) varied from 55 to 94%, with a false-alarm rate (FAR) from 0.17 to 0.52 m^{-2}. Both were significantly poorer than those measured for surface-laid mines. Mines buried in bare soil were detectable for no more than a few days, depending on weathering. An exception occurred for a few mines, where the disturbed soil settled after the mines were buried. The resulting circular depressions trapped water and ice and were visible for several months.

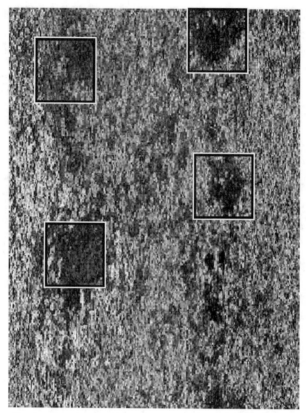

FIGURE 7.36 LCC image calculated from the hyperspectral data cube corresponding to Fig. 7.35. The four rectangles correspond to the location of the four buried mines. Average spectra obtained just prior to burial were used as the design vectors to detect nonsenescent vegetation; thus, the areas under the mines appear dark.

Heat flow anomalies caused by buried mines can be measured in the TIR band, and for many years numerous research groups have studied broadband thermal infrared imaging with this goal in mind. A broadband TIR imager can measure only the effective temperature of the soil surface, which is a function of its true temperature and its emissivity. The surface emissivity change is due to the surface soil disturbance caused by mine emplacement. The true surface temperature of an area of soil is a function of the thermal inertia below the area. The change in thermal inertia is due to the change in mass density and water content of the disturbed soil column above the landmine and the presence of the buried mine itself. Over time, the effects of the disturbed soil decrease due to natural weathering and compaction, until eventually only the effect of the landmine remains. If the mine is buried much deeper than the diffusion length of the soil (which is typically 10 to 15 cm), the mine cannot be detected. One of the problems with broadband TIR detection is the dramatic variation in P_d and FAR with time of day and environmental changes.

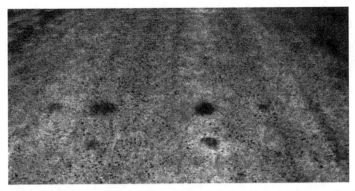

FIGURE 7.37   Nighttime (04h45 MST) broadband thermal infrared image of mines on a road.

FIGURE 7.38   Daytime (14h15 MST) broadband thermal infrared image of the same scene as that in Fig. 7.37.

For example, Figs. 7.37 and 7.38 are images of landmines buried in a dirt road. They were acquired at two different times of day in the diurnal cycle by a broadband thermal infrared camera. In the nighttime image (Fig. 7.37), two inverted L-shaped clusters of mines can be clearly seen, each consisting of two smaller antipersonnel mines and one antitank mine. Mines in the left-hand cluster were buried relatively recently compared to those in the right-hand cluster. The surface soil disturbance or scuffing, caused by the burial process, is visible in the daytime image well beyond the region immediately above each mine in the left cluster and is seen to obscure the thermal contrast due to individual buried mines. The contrast between the mined and nonmined areas reverses from night to day, due to the change in direction of subsurface heat flow. There is also much more clutter in the daytime image, due to variable solar heating and scattering of solar radiation.

Research has been conducted to develop methods to predict this performance by modeling the thermal transport physics or by measurement of subsurface heat flow [229]. Alternatively, spectral methods in the thermal infrared may be capable of reducing the variability by providing additional information about the wavelength dependence of the surface emissivity. Early research into such an approach looked at the relative responses in two infrared bands (3 to 5 μm and 8 to 14 μm), which allowed emissivity and temperature to be estimated when the emissivity was constant for the two bands [238]. In practice, this is usually not the case, and buried mine detection results were generally not dramatically better than those using simple broadband imagers. Multi- or hyperspectral methods in the TIR may be capable of reducing the variability by allowing better estimation of the wavelength-dependent emissivity and thus the true surface temperature.

Thermal infrared multispectral or hyperspectral imagers also have the capability to detect buried mines by measuring the recent disturbnce of certain types of soils. Early research, sponsored by the U. S. Defence Advanced Research Projects Agency (DARPA), showed some promise for quartz-bearing soils [239,240]. Quartz has significant restrahlen bands, whose reflectivities differ with particle size [241] (Fig. 7.39). Soil disturbance has been shown to measurably change the quartz reflectance spectrum (Fig. 7.40). This effect has been presumed to be due to a mixing of subsurface soil, which has one soil particle-size distribution, with surface soil, which has a different distribution [239]. (The original distributions of particle size are dictated by the type of soil and its environmental history, such as weathering, moisture, and state of compaction. Typically, well-aged soil will have a predominance of fine particles near the surface and an excess of coarse particles farther down.) Other materials, such as carbonates, exhibit similar TIR band spectral features, which may also be suitable for detection of soil disturbance. The reflectivity anomaly is not permanent. Over time, as natural weathering processes act on the soil, the particle-size distribution returns to its original state. The length of time to return to the original state is dependent on the environmental history and is generally not well known. Some anecdotal evidence suggests that the effect lasts up to a few months, which is significantly longer than visible band effects. Other questions that remain to be answered include what range of soil types and

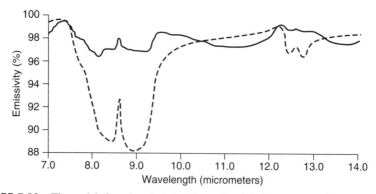

FIGURE 7.39 Thermal infrared emissivity (1 − reflectivity) spectra of fine (solid curve) and coarse (dashed curve) particles of quartz-bearing soil. (After [241].)

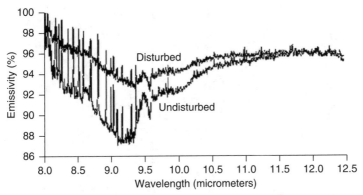

FIGURE 7.40 Thermal infrared emissivity spectra of disturbed and undisturbed quartz-bearing soil. (After [239].)

environmental conditions provide a measurable effect and what the P_d and FAR values are in practical situations for the detection of buried objects using this effect.

There is very limited information in the open literature regarding the performance of thermal infrared hyperspectral imagers for the detection of buried munitions. Some flights of TIR hyperspectral imagers, such as the University of Hawaii's AHI, have been done over minefields, but results have been inconclusive, due to a deficiency of acceptable ground truth or sensor motion stability. In the spring of 2008, DRDC and Itres conducted experiments on detecting buried and surface-laid landmines, improvised explosive devices (IEDs), and their components in a hot, arid, dusty, windy environment. Measurements were made around the clock for eight days and nights using state-of-the-art thermal and shortwave infrared hyperspectral imagers mounted on a personnel lift. While the detection results are very preliminary and much more analysis must be done, both TIR and SWIR bands appear to be useful and complementary, as expected. The TIR imager detected buried objects with a performance similar to broadband TIR at optimal times of the day when contrast across the TIR band was high. Some evidence also suggests that TIR hyperspectral imagery will provide improvement for low-contrast images (weaker detections). As an example, Fig. 7.41 is a grey scale rendition of a false color image, consisting of three bands extracted from the full 32-band data cube of a *tasi* TIR hyperspectral imager. The image includes a buried IED with its buried command wire. The dark blob is due to the IED, and the dark trail is due to the earth that was disturbed when the command wire was emplaced. In the original false color image, the background undisturbed soil appears reddish compared to the gray/black color of the IED and wire. The color suggests that the undisturbed soil has nonuniform emissivity across the wavelengths of the thermal infrared band, whereas the disturbed soil has a roughly constant emissivity across the band, both in qualitative agreement with Fig. 7.40. The average intensity across the band of the radiance of a scene element is governed by the apparent temperature of the soil surface and thus depends on the average surface emissivity and the presence of a shallow buried object or a disturbed subsurface soil column. The variation with

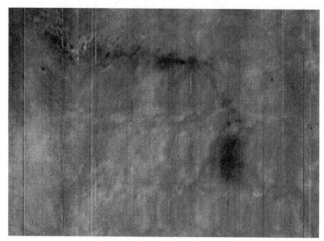

FIGURE 7.41 Three-band image of a buried IED and command wire, extracted from the data cube of a *tasi* TIR hyperspectral imager.

wavelength of the intensity across the band is governed solely by the disturbance of the soil surface. In principle, analysis of the TIR hyperspectral image can distinguish areas of disturbed surface soil with no subsurface object or disturbance, such as some of the scuffed soil surrounding the left-hand cluster of mines in Fig. 7.38, from areas with subsurface objects or disturbance. This could allow the images of recently buried objects to be sharpened to aid detection. Also, such analysis would allow objects buried a long time and having no remaining surface disturbance to be distinguished from recently buried objects.

In the same experiment, the SWIR imager detected all surface objects, even those with significant dust coating. It detected materials of interest on the surface at low area densities, distinguished artificial rocks from real ones, but had no easily recognized capability to detect buried objects. There are numerous minerals with strong spectral structural features in the SWIR band which might also be altered by soil disturbance, but this has not been seen in these or earlier experiments [231].

7.6.4 Detection of Associated Surface Artifacts

As mentioned previously, areas containing buried mines can be located by detecting surface disturbances and artifacts associated with the process of placing the mines. Surface disturbances were addressed in Section 7.6.3 and will not be discussed further. Although it seems that such surface paraphernalia should be easily visible, they are often very difficult to see, even when an observer is very close (an example of surface-laid mines is shown in Fig. 7.42). Some paraphernalia are camouflaged to be invisible to the naked eye at visible wavelengths. But while camouflage matches the object's coating reflectivity to that of the background in an average sense across the visible and near-infrared spectrum, exact matches occur at only a few points. Closely matching a coating to even a single background type across a large portion

FIGURE 7.42 Portion of a surface-laid minefield at DRDC Suffield, viewed at close range. Mines are invisible, even though no attempt has been made to hide them.

of the spectrum is very difficult [242]. Large differences in reflectivity between military camouflage and background occur in bands which may be quite narrow and have varying center positions and widths for different camouflage coatings and paints compared to different types of backgrounds. Such subtle mismatches between camouflaged material and background spectra can be discerned if the spectral range is finely and/or selectively divided (Fig. 7.43).

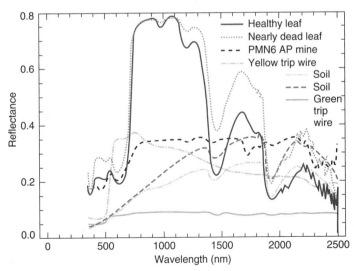

FIGURE 7.43 VNIR and SWIR reflectance spectra of mines, tripwires, and background materials.

Most previous research applicable to detecting camouflaged mine-laying paraphernalia has been directed at detecting surface-laid (camouflaged) landmines. Attempts in the visible wavelengths mostly used film–filter combinations to obtain a few (ca. eight per film type) relatively broad (ca. 50 nm minimum) visible bands [243]. Detection results were poor, in large part because the broad bands tended to average out spectral differences between mine and background [244]. Commercial multispectral scanners offered no substantial improvement. Although they typically offer better dynamic range than film, they generally have fewer than 10 spectral bands, which are >50 nm wide and are fixed in width and position. The fine spectral resolution available to hyperspectral imagers can provide that the necessary discrimination capability, provided that spatial resolution is sufficient to place at least one, and preferably several, pixels on the object of interest [245].

A major issue with using passive hyperspectral imaging to classify materials is that the radiance of a material in the field varies with a number of factors that are beyond the control of the person making measurements. To examine these factors and determine how to build an invariant target reference library, DRDC Suffield and Itres Research conducted a systematic investigation of the radiance of surface-laid replica mine images as a function of mine type and orientation, sun angle, diurnal and seasonal variations, background terrain, and other factors using a VNIR hyperspectral imager [246,247]. Calibrated radiance images were converted to "pseudoreflectance" images using a *down-welling radiation sensor* (also called *an incident light sensor*) or calibrated reflectance panels. The pseudoreflectance is not a true reflectivity but, rather, is proportional to the reflectivity, with zero offset, over a limited waveband. This is because the detected radiance spectra vary from mine to mine of the same type and from scene to scene for the same mine, due to a number of factors: spectral contamination in the near-IR from light scattered by nearby vegetation; near-IR absorption by mines when wet; offsets due to nonuniform lighting geometry between the mines and the reflectance panels; weathered or dirty versus pristine mines. Because of this variation, several reflectance spectra for each mine type were averaged over a central wavelength band where proportionality holds true to get a representative spectrum for the reference database used for target classification. Unconstrained least-squares fitting to a straight line yielded a "best fit" slope and an offset, which was generally nonzero.

Spectral reflectance values of the images were compared to reference spectra by estimating the linear correlation coefficient (LCC) for each pixel [247]. The LCC is a measure of spectral similarity, which is a generalization of the popular spectral angle mapper (SAM). LCCs have been shown [246] to be insensitive to the spectral scale and offset factors exhibited by the pseudoreflectance spectra typical of these experiments. End-member analysis and SAM do not have this property. The dynamic range of LCCs is typically very compressed, so a power-law stretch can be applied to the image to improve contrast [247]. Thresholding then provides a binary classification image for the reference target material. Although LCC is superior for images having several pixels across a target, orthogonal subspace projection (OSP) [248], a type of spectral unmixing pattern classification intended for sparse numbers of targets per image, has proven better for images with few or single pixels per

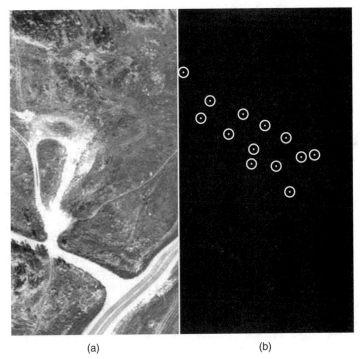

(a)   (b)

FIGURE 7.44 Airborne *casi* images of a field of surface-laid mine surrogates. Left image is a grey scale rendition of 3 color composite image of the area, derived from the 16-band raw data. Mines, which are green, are a few pixels wide and are invisible on both the color image and its grey scale rendition. Right hand image of the same scene is the threshold pixel-by-pixel product of the LCC and OSP classifiers. White dots signify pixels classified as "minelike." Each circle is the ground-truth location of a mine and in the full-sized image can be seen to contain a white dot. Also in the full-sized image, some false alarms can be seen.

target. The product of the two classifier images turns out to provide as good a P_d value as that of either classifier, but has a lower FAR (Fig. 7.44).

By processing the reflectance images using the LCC, OSP, and LCC × OSP product algorithms, individual mines could be distinguished from backgrounds in all seasons and times of day between sunrise and sunset, even when they were partially obscured by vegetation. High-spatial-resolution airborne imagery of real and surrogate minefields has also been obtained from fixed-wing (1.6 m × 0.5 m IFOV at 300 m altitude, 110 knots) and helicopter (0.1 m × 0.1 m at 75 m altitude to 0.2 m × 0.2 m at 150 m altitude, 35 knots) platforms [230]. Non-real-time analysis of images has yielded P_d values of nearly 100% with FARs of 0.0003/m^2 and 0.004/m^2.

The SWIR spectral region may offer further improvement in the detection of human-made surface materials. Most organic and complex inorganic compounds possess significant and unique absorption features in the SWIR. Some of these materials and pigments may be incorporated into the paints and plastics of containers and other articles associated with landmines. Earlier studies have shown that reflectivities for many mine coatings and background materials are significantly lower in the VNIR than the SWIR (Fig. 7.43). The contrast between background materials, particularly vegetation, and mine-related materials is much larger in the SWIR, and the dynamic range of reflectivities is also larger. The paints and plastics used in landmines and associated military articles possess general absorption features that may allow simple classifiers to distinguish human-made objects from natural features such as rocks, vegetation, and soil. For example, paints and plastics tend to be hydrophobic, while rocks and soils (due to porosity) and vegetation tend to be hydrophilic. Thus natural objects typically absorb more radiation than do human-made objects in the 1400- and 1900-nm water vapor absorption bands [249].

DRDC Suffield and Itres personnel have obtained field SWIR hyperspectral images of numerous surface-laid mines, human-made items, background materials, and people using a *sasi* imager mounted in a horizontally scanning personnel lift at an altitude of roughly 5 m [231]. Preliminary indications are that a simple generic classification decision boundary should be able to distinguish surface-laid landmines from many human-made artifacts and natural materials (Fig. 7.45).

Our previous studies have suggested that the ability of classifiers to separate camouflage coatings from backgrounds improves when the VNIR and SWIR spectra

FIGURE 7.45 Single-band image from a *sasi* data cube of landmines, surrogates, UXO, and people on grass. The waveband chosen has enhanced reflectivity contrast for certain hard plastics. Note the difference between the plastic lenses of the eyeglasses worn by the person on the left and the glass lenses of the person on the right.

are combined compared to the use of either band by itself. One approach is to design a suitable imager to provide the full band, while an alternative is to collect coregistered visible and shortwave images simultaneously with two imagers. For example, the spectral ranges of the *sasi* and *casi* overlap. If the two imagers are flown simultaneously, the spectral overlap range provides a common wavelength region that can be exploited to provide seamless images from 400 to 2450 nm. This has been demonstrated, and more information can be found in work by McFee et al. [250].

A more unorthodox example of detection of surface artifacts associated with buried mines is the detection of insect trails. During hyperspectral imaging trials to detect landmines and IEDs, ants of the genus *Messor* were observed to find and remove RDX and TNT explosives from surface piles under dry, hot, dusty, windy conditions [251]. Although this seems to be the first published reference to such behavior, there is anecdotal evidence of similar behavior having been observed previously. The most likely explanation for the collection and removal to their colonies of the grains of explosives is that the ants were mistaking the grains for seeds and were storing them for food. The detection and dispersal of explosives by ants or other such foraging insects may have applications in buried landmine or IED detection, provided that the ants are able to detect the tiny quantities of explosives on the device casing or soil surface, burrow to the device, and penetrate the casing. This may not be so farfetched. Humanitarian demining experts have recounted numerous incidents of ants cleaning out the solid TNT from inside mines. Of course, success will depend on the penetrability of the outer case of the mine. In any event, the slow nature of the process (on the order of days for detection) would probably restrict the method to use in humanitarian demining.

In practice, one could search for ant trails leading from buried mines or IEDs to established colonies. If no appropriate colonies existed, a collection of ants could be introduced to the area. Hyperspectral imaging could be used to detect the ant trails by spectral recognition of the ants' outer shell material, the explosives, or both. In the experiment above, spectral analysis of images from a *sasi* SWIR hyperspectral imager clearly revealed the trail of explosives to the anthill, despite being present in quantities not visible to the unaided eye (Fig. 7.46). It might also be possible to enhance the detection by modifying the ants genetically to increase their contrast in the visible or infrared bands, such as by giving them fluorescent shells.

7.6.5 Real-Time Detection

Most commercial pushbroom or whiskbroom HSI instruments can collect and store image data at aircraft speeds, but the imagery is analyzed off-line after returning to base. Delivery of processed data within a few days to weeks is acceptable for the remote-sensing community, the main market for these instruments. Although useful for humanitarian demining in the role of area reduction or level I clearance, it is unacceptable for military countermine operations, where a decision must be made within a few minutes. DRDC and Itres are developing a hardware and software system with algorithms that can process the raw hyperspectral data in real time

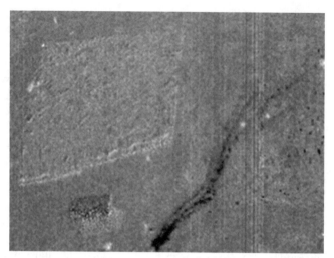

FIGURE 7.46   Trail of ants and explosives (black) detected with a *sasi* SWIR hyperspectral imager.

to detect mines. Radiometric correction is performed on the raw data stream, then custom mine classification algorithms such as those discussed in Section 7.6.4, are applied to the corrected data. A spectral signature library provides the reference spectral vectors. The classification results are stored and displayed in real time, that is, within a few frame times of data acquisition. Such real-time mine detection was demonstrated for the first time from a slow-moving (ca. 1 to 2 km/h) land vehicle in March 2000 [252]. An improved second-generation system has been completed which can achieve real-time detection of mines from an airborne platform, with its commensurately higher data rates. It is capable of capturing, correcting, and classifying imagery at a rate of 1.25 Mpixels/s. Assuming spectral summing to 16 bands and a 550-pixel across-track swath, this corresponds to an along-track platform speed of roughly 25 km/h for 5-cm pixels or 100 km/h for 20-cm pixels. The system is present in use at with the *casi*-550 and *tabi*−320, a thermal infrared broadband pushbroom imager. With minor modifications, it can also be compatible with the *sasi* and *tasi* imagers. Experiments to detect mines in real time from a slow airborne helicopter platform were conducted at DRDC Suffield in October 2006. Five to 10 reference mines were detected simultaneously in real time at 3.6 km/h and an altitude of 30 m (5-cm pixels) above ground (Fig. 7.47) [253]. This equates to 72 km/h for 20-cm pixels (maximum pixel size allowed for reliable classification) and one reference target.

7.6.6   Conclusions and Future Directions

The thermal infrared band has the most potential to aid in the detection of buried explosive objects by hyperspectral imaging. Limited data suggest that it performs as well as broadband thermal infrared imaging for high object/background contrasts

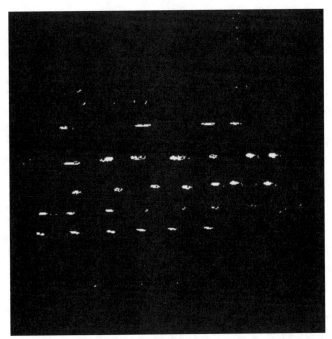

FIGURE 7.47 Portion of a binary output image of mines obtained from the second-generation real-time hyperspectral processing system mounted on a helicopter. White pixels have been classified as minelike, and indeed, correspond to mine surfaces.

and may improve detection for low contrasts. It may also allow a reduction in the diurnal variability in detection perfomance. Many more data are needed, and with the recent advent of suitable commercially available hyperspectral imagers such as the Itres *tasi*, it will be possible to acquire extensive data sets.

The detection of surface artifacts associated with buried explosive objects is a more mature field. Airborne detection of surface-laid mines, a problem that is technically similar to or even more challenging than detection of surface artifacts, has been demonstrated using visible/near infrared and shortwave infrared hyperspectral imaging. However, more research is needed specifically on detection of nonmine features found in minefields, such as tracks, parachutes, and packaging.

The key to reliable detection of buried objects such as landmines is having higher spatial resolution than the extent of the surface anomaly. Instruments specifically designed for such detection roles will be necessary and will be restricted to low, slow-flying aerial platforms to obtain useful resolution. Hyperspectral imagers may also be effective on ground vehicles, although it may be necessary to mount them on tall booms, to minimize obscuration by vegetation and to stabilize the imager.

Future development could rely on hyperspectral imagers with large numbers of narrow flexibly positioned bands for the collection of reference spectra and the design of advanced detection algorithms. Once optimum bands have been selected using the former instruments, multispectral imagers having higher speed and spatial resolution but a smaller number of fixed bands might be chosen for operational use. For military applications, real-time detection is necessary. Down-welling radiation sensors will be necessary for VNIR and SWIR band imagers to allow detection of surface artifacts independent of natural illumination, since calibrated panels are not practical in the field, particularly in real-time applications. Thus, real-time airborne detection of mines with real-time conversion of radiance to reflectance is an ultimate goal. For successful demarcation of the boundaries of areas contaminated by buried explosive objects using airborne imagery, geocorrection and/or platform stabilization are essential, although they may not be required for individual object detection. Thus, in-flight geocorrection is desirable, although real-time geocorrection is not necessary.

Most of the features noted above have been demonstrated alone or in limited combinations but have not been integrated into a single unified system. This will be a focus of future research. Limited research has shown that the wider the overall spectrum available, the less likely it is to have natural or intended (due to spoofing) false alarms and natural or intended (due to camouflage) missed detections. Future R&D will exploit this fact by examining the fusion of images collected simultaneously from instruments having different wavebands.

Acknowledgments

The authors of Section 7.3 acknowledge support under NATO SfP grant 982836. This work was also supported by grant 106T321 of TÜBITAK (the Scientific and Technological Research Council of Turkey) and in part by grant 00062.STZ.2007-1 of the Ministry of Industry and Trade of Turkey. The authors also very grateful for the support of the Gebze Branch of the Ministry of Internal Affairs: in particular the head of the branch, Ahmet Can.

REFERENCES

1. RAND. 2003. *Alternatives for Landmine Detection*. RAND Science and Technology Policy. RAND Corporation, Santa Monica, CA.

2. Poulain, D. E., Schuab, S. A., Alexander, D. R., Krause, J. K. 1998. Detection and location of buried objects using active thermal sensing. *Proc. SPIE*, 3392:861–866.

3. DiMarzio, C. A., Rappaport, C. M., Wen, L. 1998. Microwave-enhanced infrared thermography. *Proc. SPIE*, 3392:1103–1110.

4. Wexler, A. 1988. Electrical impedance imaging in two and three dimensions. *Clin. Phys. Physiol. Meas.* (Suppl. A); pp. 29–33.

5. Lee, Y. K. 2002. Spin-1 nuclear quadrupole resonance theory with comparisons to nuclear magnetic resonance. *Concepts Magn. Reson.*, 14:155–171.

6. Towe, B. Jacobs, A. 1981. X-ray backscatter imaging. *IEEE Trans. Biomed. Eng.*, 28:646–654.

7. Yinun, J. 2003. Detection of explosive by electronic noses. *Anal. Chem.*, 75(5):98A–105A.

8. Tomita, Y., Ho, M. H., Guilbault, G. G. 1979. Detection of explosives with a coated piezoelectric quartz crystal. *Anal. Chem.*, 51(9):1475.

9. Kannan, G. K., Kapoor, J. C. 2005. Adsorption studies of carbowax and polydimethyl siloxane to use as chemical array for nitro aromatic vapour sensing. *Sens. Actuat. B*, 110:312–320.

10. Kannan, G. K., Nimal, A. T., Mittal, U., Yadava, R. D. S., Kapoor, J. C. 2004. Adsorption studies of carbowax coated surface acoustic wave (SAW) sensor for 2,4-dinitrotoluene (DNT) vapour detection. *Sens. Actuat. B*, 101:328–334.

11. Houser, E. J., McGill, R. A., Nguyen, V. K., Chung, R., Weir, D. W. 2000. Recent developments in sorbent coatings and chemical detectors at the Naval Research Laboratory for Explosives and Chemical Agents. In *Detection and Remediation Technologies for Mines and Minelike Targets V*, A. C. Dubey, J. F. Harvey, J. T. Broach, R. E. Dugan, Eds. *Proc. SPIE*, 4038:504.

12. Houser, E. J., Mlsna, T. E., Nguyen, V. K., Chung, R., Mowery, R. L., McGill, R. A. 2001. Rational materials design of sorbent coatings for explosives: applications with chemical sensors. *Talanta*, 54:469–485.

13. McGill, R. A., Mlsna, T. E., Chung, R., Nguyen, V. K., Stepnowski, J. 2000. The design of functionalized silicone polymers for chemical sensor detection of nitroaromatic compounds. *Sens. Actuat. B*, 65:5.

14. Pasquinet, E., Bouvier, C., Thery-Merland, F., Hairault, L., Lebret, B., Methivier, C., Pradier, C. M. 2004. Synthesis and adsorption on gold surfaces of a functionalized thiol: elaboration and test of a new nitroaromatic gas sensor. *J. Colloid Interface Sci.*, 272:21–27.

15. Thery-Merland, F., Methivier, C., Pasquinet, E., Hairault, L., Pradier, C. 2006. Adsorption of functionalised thiols on gold surfaces: How to build a sensitive and selective sensor for a nitroaromatic compound? *Sens. Actuat. B*, 114:223–228.

16. Montmeat, P., Madonia, S., Pasquinet, E., Hairault, L., Gros, C. P., Barbe, J.-M., Guilard, R. 2005. Metalloporphyrins as sensing material for quartz-crystal microbalance nitroaromatics sensors. *IEEE Sens. J.*, 5(4):610–615.

17. Yang, X., Du, X.-X., Shi, J., Swanson, B. 2001. Molecular recognition and self-assembled polymer films for vapor phase detection of explosives. *Talanta*, 54:439–445.

18. Bunte, G., Hurttlen, J., Pontius, H., Hartlieb, K., Krause, H. 2007. Gas phase detection of explosives such as 2,4,6-trinitrotoluene by molecularly imprinted polymers. *Anal. Chim. Acta*, 591:49–56.

19. Lee, S. H., Stubbs, D. D., Hunt, W. D., Edmonson, P. J. 2005. Vapor phase detection of plastic explosives using a SAW resonator immunosensor array. *Proceedings of IEEE Sensors 2004*, pp. 468–471.

20. Larsson, A., Angbrant, J., Ekeroth, J., Mansson, P., Liedberg, B. 2006. A novel biochip technology for detection of explosives—TNT: synthesis, characterisation and application. *Sens. Actuat. B*, 113:730–748.

21. Mizuta, Y., Onodera, T., Singh, P., Matsumoto, K., Miura, N., Toko, K. 20008. Development of an oligo(ethyleneglycol)-based SPR immunosensor for TNT detection. *Biosens. Bioelectron.*, 24:191–197.

22. Lin, A., Yu, H., Waters, M. S., Kim, E. S., Goodman, S. D. 2008. Explosive trace detection with FBAR-based sensor. In *IEEE 21st International Conference on Micro Electro Mechanical Systems*, pp. 208–211.

23. Pinnaduwage, L. A., Ji, H.-F., Thundat, T. 2005. Moore's law in homeland defense: an integrated sensor platform based on silicon microcantilevers. *IEEE Sens. J.*, 5(4):774.

24. Pinnaduwage, L. A., Boiadjiev, V., Hawk, J. E., Thundat, T. 2003. Sensitive detection of plastic explosives with self-assembled monolayer-coated microcantilevers. *Appl. Phys. Lett.*, 83(7):1471.

25. Pinnaduwage, L. A., Hedden, D. L., Gehl, A., Boiadjiev, V. I., Hawk, J. E., Farahi, R. H., Thundat, T., Houser, E. J., Stepnowski, S., McGill, R. A., Deel, L., Lareau, R. T. 2004. A sensitive, handheld vapor sensor based on microcantilevers. *Rev. Sci. Instrum.*, 75(11):4554.

26. Li, P., Li, X.-X., Zuo, G.-M., Liu, J., Wang, Y.-L., Liu, M., Jin, D.-Z. 2006. Silicon dioxide microcantilever with piezoresistive element integrated for portable ultraresoluble gaseous detection. *Appl. Phys. Lett.*, 89:074104.

27. Zuo, G., Li, X., Zhang, Z., Yang, T., Wang, Y., Cheng, Z., Feng, S. 2007. Dual-SAM functionalization on integrated cantilevers for specific trace-explosive sensing and non-specific adsorption suppression. *Nanotechnology*, 18:255501.

28. Pinnaduwage, L. A., Thundat, T., Hawk, J. E., Hedden, D. L., Britt, P. F., Houser, E. J., Stepnowski, S., McGill, R. A, Bubb, D. 2004. Detection of 2,4-dinitrotoluene using microcantilever sensors. *Sens. Actuat. B.*, 99:223–229.

29. Rogers, B., Whitten, R., Adams, J. D. 2006. Multifunctional, self-sensing microcantilever arrays for unattended detection of chemicals, explosives, and biological agents. In *Unmanned/Unattended Sensors and Sensor Networks III*, Ed. M. Carapezza, Ed. *Proc. SPIE*, 6394:639409.

30. Rogers, B., Whitten, R., Adams, J. D. 2006. Self-sensing array (SSA) technology for homeland security applications. In *Sensors, and Command, Control, Communications, and Intelligence (C3I) Technologies for Homeland Security and Homeland Defense V*, Ed. M. Carapezza, Ed. *Proc. SPIE*, 6201.

31. Rogers, B., Whitten, R., Adams, J. D. 2005. The microcantilever array: a low power, compact and sensitive unattended sensor technology. In *Unmanned/Unattended Sensors and Sensor Networks II*, Ed. M. Carapezza, Ed. *Proc. SPIE*, 5986:598600.

32. Pinnaduwage, L. A., Thundat, T., Gehl, A., Wilson, S. D., Hedden, D. L., Lareau, R. T. 2004. Desorption characteristics of uncoated siliconmicrocantilever surfaces for explosive and common nonexplosive vapors. *Ultramicroscopy*, 100:211–216.

33. Muralidharan, G., Wig, A., Pinnaduwage, L. A., Hedden, D., Thundat, T., Lareau, R. T. 2003. Adsorption–desorption characteristics of explosive vapors investigated with microcantilevers. *Ultramicroscopy*, 97:433–439.

34. Pinnaduwage, L. A., Gehl, A., Hedden, D. L., Muralidharan, G., Thundat, T., Lareau, R. T., Sulchek, T., Manning, L., Rogers, B., Jones, M., Adams, J. D. 2003. Explosives: a microsensor for trinitrotoluene vapour. *Nature* 425:474.

35. Pinnaduwage, L. A., Wig, A., Hedden, D. L., Gehl, A., Yi, D., Thundat, T., Lareau, R. T. 2004. Detection of trinitrotoluene via deflagration on a microcantilever. *J. Appl. Phys.*, 95(10):5871.

36. Datskos, P. G., Sepaniak, M. J., Tipple, C. A., Lavrik, N. 2001: Photomechanical chemical microsensors. *Sens. Actuat. B.*, 76(1):393–402.

37. Datskos, P. G., Rajic, S., Sepaniak, M. J., Lavrik, N., Tipple, C. A., Senesac, L. R., Datskou, I. 2001. Chemical detection based on adsorption-induced and photoinduced stresses in microelectromechanical systems devices. *J. Vac. Sci. Technol. B*, 19(4):1173–1179.

38. Senesac, L. R., Yi, D., Thundat, T. 2007. Receptor-free nanomechanical sensors. In *Reliability, Packaging, Testing, and Characterization of MEMS/MOEMS VI*, A. L. Hartzell, R. Ramesham, Eds. *Proc. SPIE*, 6463:646302.

39. Krause, A. R., Van Neste, C., Senesac, L., Thundat, T., Finot, E. J. 2008. Trace explosive detection using photothermal deflection spectroscopy. *Appl. Phys.*, 103:094906-6.

40. Van Neste, C. W., Senesac, L. R., Yi, D., Thundat, T. 2008. Standoff detection of explosive residues using photothermal microcantilevers. *Appl. Phys. Lett.*, 92:134102.

41. Thomas, S. W., III, Amara, J. P., Bjork, R. E., Swager, T. M. 2005. Amplifying fluorescent polymer sensors for the explosives taggant 2,3-dimethyl-2,3-dinitrobutane (DMNB). *Chem. Commun.*, 4572–4574.

42. Williams, V. E., Yang, J. S., Lugmair, C. G., Miao, Y. J., Swager T. M. 1999. Design of novel iptycene-containing fluorescent polymers for the detection of TNT. *Detection and Remediation Technologies for Mines and Minelike Targets IV. Proc. SPIE*, 3710:402.

43. Yang, J.-S., Swager, T. M. 1998. Fluorescent porous polymer films as TNT chemosensors: electronic and structural effects. *J. Am. Chem. Soc.*, 120:11864–11873.

44. Yang, J. S., Swager, T. M. 1998. Porous shape persistent fluorescent polymer films: an approach to TNT sensory materials. *J. Am. Chem. Soc.*, 120:5321–5322.

45. Thomas, S. W., III, Joly, G., Swager, T. 2007. Chemical sensors based on amplifying fluorescent conjugated polymers. *Chem. Rev.*, 107:1339–1386.

46. Fisher, M., Sikes, J., Prather, M. 2004. Explosive detection using high-volume vapor sampling and analysis by trained canines and ultra-trace detection equipment. In *Sensors, and Command, Control, Communications, and Intelligence (C3I) Technologies for Homeland Security and Homeland Defense III*, E. M. Carapezza, Ed. *Proc. SPIE*, 5403:409.

47. la Grone, M., Cumming, C., Fisher, M., Reust, D., Taylor, R. 1999. Landmine detection by chemical signature: detection of vapors of nitroaromatic compounds by fluorescence quenching of novel polymer materials. In *Detection and Remediation Technologies for Mines and Minelike Targets IV*, Orlando, FL. *Proc. SPIE*, 3710:409.

48. Fisher, M., la Grone, M., Sikes, J. 2003. Implementation of serial amplifying fluorescent polymer arrays for enhanced chemical vapor sensing of landmines. In *Detection and Remediation Technologies for Mines and Minelike Targets VIII*, R. S. Harmon, J. H. Holloway, Jr., J. T. Broach, Eds. *Proc. SPIE*, 5089:991.

49. Simonson, R. J., Hance, B. G., Schmitt, R. L., Johnson, M. S., Hargis, P. J., Jr. 2001. Remote detection of nitroaromatic explosives in soil using distributed sensor particles. In *Detection and Remediation Technologies for Mines and Minelike Targets VI*, Ab. C. Dubey, J. F. Harvey, J. T. Broach, V. George, Eds. *Proc. SPIE*, 4394:879.

50. Rose, A., Zhu, Z., Madigan, C. F., Swager, T. M., Bulovic, V. 2005. Sensitivity gains in chemosensing by lasing action in organic polymers. *Nature*, 434:876–879.

51. Rose, A., Zhu, Z., Madigan, C. F., Swager, T. M., Bulović, V. 2006. Chemosensory lasing action for detection of tnt and other analytes. In *Organic Light Emitting Materials and Devices X*, Z. H. Kafafi, F. So, Eds. *Proc. SPIE*, 6333:63330Y.

52. Chang, C.-P., Chao, C.-Y., Huang, J. H., Li, A.-K., Hsu, C.-S., Lin, M.-S., Hsieh, B. R., Su, A.-C. 2004. Fluorescent conjugated polymer films as TNT chemosensors. *Synt. Met.*, 144:297–301.

53. Chen, L., McBranch, D., Wang, R., Whitten, D. 2000. Surfactant-induced modification of quenching of conjugated polymer fluorescence by electron acceptors: applications for chemical sensing. *Chem. Phys. Lett.*, 330(1–2):27–33.

54. Focsaneanu, K.-S., Scaiano, J. C., 2005. Potential analytical applications of differential fluorescence quenching: pyrene monomer and excimer emissions as sensors for electron deficient molecules. *Photochem. Photobiol. Sci.*, 4:817–821.

55. Goodpaster, J. V., McGuffin, V. L. 2001. Fluorescence quenching as an indirect detection method for nitrated explosives. *Anal. Chem.*, 73:2004–2011.

56. Zhang, S., Lu, F., Gao, L., Ding, L., Fang, Y. 2007. Fluorescent sensors for nitroaromatic compounds based on monolayer assembly of polycyclic aromatics. *Langmuir*, 23(3):1584–1590.

57. Sanchez, J. C., DiPasquale, A. G., Rheingold, A. L., Trogler, W. C. 2007. Synthesis, luminescence properties, and explosives sensing with 1,1-tetraphenylsilole- and 1,1-silafluorene-vinylene. *Polym. Chem. Mater.*, 19:6459–6470.

58. Liu, Y., Mills, R. C., Boncella, J. M., Schanze, K. S. 2001. Fluorescent polyacetylene thin film sensor for nitroaromatics. *Langmuir*, 17(24):7452–7455.

59. Toal, S. J., Sanchez, J. C., Dugan, R. E., Trogler, W. C. 2007. Visual detection of trace nitroaromatic explosive residue using photoluminescent metallole-containing polymers. *J. Forensic Sci.*, 52(1):79–83.

60. Sanchez, J. C., Toal, S. J., Wang, Z., Dugan, R. E., Trogler, W. C. 2007. Selective detection of trace nitroaromatic, nitramine, and nitrate ester explosive residues using a three-step fluorimetric sensing process: a tandem turn-off, turn-on sensor. *J Forensic Sci.*, 52(6):1308.

61. Sanchez, J. C., Trogler, W. C. 2008. Efficient blue-emitting silafluorene–fluorene-conjugated copolymers: selective turn-off/turn-on detection of explosives. *J. Mater. Chem.*, 18:3143–3156.

62. Naddo, T., Che, Y., Zhang, W., Balakrishnan, K., Yang, X., Yen, M., Zhao, J., Moore, J. S., Zang, L. 2007. Detection of explosives with a fluorescent nanofibril film. *J. Am. Chem. Soc.*, 129:6978–6979.

63. Narayanan, A., Varnavski, O., Mongin, O., Majoral, J.-P., Blanchard-Desce, M., Goodson, T., III. 2008. Detection of TNT using a sensitive two-photon organic dendrimer for remote sensing. *Nanotechnology*, 19:115502.

64. Narayanan, A., Varnavski, O. P., Swager, T. M., Goodson, T., III. 2008. Multiphoton fluorescence quenching of conjugated polymers for TNT detection. *J. Phys. Chem. C*, 112:881–884.

65. Content, S., Trogler, W. C., Sailor, M. J. 2000. Detection of nitrobenzene, DNT, and TNT vapors by quenching of porous silicon photoluminescence. *Chem. Eur. J.*, 6:2205–2213.

66. Sailor, M. J., Trogler, W. C., Content, S., Létant, S., Sohn, H., Fainman, Y., Shames, P. 2000. Detection of DNT, TNT, HF and nerve agents using photoluminescence and interferometry from a porous silicon chip. In *Unattended Ground Sensor Technologies and Applications II*, E. M. Carapezza, T. M. Hintz, Eds. *Proc. SPIE*, 4040:95.

67. Germanenko, I. N., Li, S., El-Shall, M. S. 2001. Decay dynamics and quenching of photoluminescence from silicon nanocrystals by aromatic nitro compounds. *J. Phys. Chem. B*, 105:59–66.

68. Sailor, M. J., Trogler, W. C., Létant, S. E., Sohn, H., Content, S., Schmedake, T. A., Gao, J., Zmolek, P. B., Link, J. R. 2001. Low-power microsensors for explosives and nerve warfare agents using silicon nanodots and nano wires. In *Unattended Ground Sensor Technologies and Applications III*, E. M. Carapezza, Ed., *Proc. SPIE*, 4393:153.

69. Sabatani, E., Kalisky, Y., Berman, A., Golan, Y., Gutman, N., Urbach, B., Sa'ar, A. 2008. Photoluminescence of polydiacetylene membranes on porous silicon utilized for chemical sensors. *Opt. Mater.*, 30:1766–1774.

70. Tao, S., Li, G. 2007. Porphyrin-doped mesoporous silica films for rapid TNT detection. *Colloid Polymer Sci.*, 285:721–728.

71. Tao, S., Li, G., Zhu, H. 2006. Metalloporphyrins as sensing elements for the rapid detection of trace TNT vapour. *J. Mater. Chem.*, 16:4521–4528.

72. Tao, S., Shi, Z., Li, G., Li, P. 2006. Hierarchically structured nanocomposite films as highly sensitive chemosensory materials for TNT detection. *Chem. Phys. Chem.*, 7:1902–1905.

73. Tao, S., Li, G., Yin, J. 2007. Fluorescent nanofibrous membranes for trace detection of TNT vapour. *J. Mater. Chem.*, 17:2730–2736.

74. Levitsky, I. A., Euler, W. B., Tokranov N., Rose, A. 2007. Fluorescent polymer–porous silicon microcavity devices for explosive detection. *Appl. Phys. Lett.*, 90:041904.

75. Kose, M. E., Harruff, B. A., Lin, Y., Veca, L. M., Lu, F., Sun, Y. P. 2006. Efficient quenching of photoluminescence from functionalized single-walled carbon nanotubes by nitroaromatic molecules. *J. Phys. Chem. B*, 110(29):14032.

76. Germain, M. E., Vargo, T. R., Khalifah, P. G., Knapp, M. J. 2007. Fluorescent detection of nitroaromatics and 2,3-dimethyl- 2,3-dinitrobutane (DMNB) by a zinc complex: (salophen)Zn. *Inorg. Chem.*, 46:4422–4429.

77. Meaney, M. S., McGuffin, V. L. 2008. Investigation of common fluorophores for the detection of nitrated explosives by fluorescence quenching. *Anal. Chim. Acta*, 610:57–67.

78. Li, H., Kang, J., Ding, L., Lu, F., Fang, Y. 2008. A dansyl-based fluorescent film: preparation and sensitive detection of nitroaromatics in aqueous phase. *J. Photochem. Photobiol. A*, 197:226–231.

79. Sohn, H., Sailor, M. J., Magde, D., Trogler, W. C. 2003. Detection of nitroaromatic explosives based on photoluminescent polymers containing metalloles. *J. Am. Chem. Soc.*, 125:3821–3830.

80. Saxena, A., Fujiki, M., Rai, R., Kwak, G. 2005. Fluoroalkylated polysilane film as a chemosensor for explosive nitroaromatic compounds. *Chem. Mater.*, 17(8):2181–2185.

81. Nieto, S., Santana, A., Hernández, S. P., Lareau, R., Chamberlain, R. T., Castro, M. E. 2004. Quantum dots for detection of trace amount of non-volatile explosives: the effect of TNT in the fluorescence of CdSe quantum dots. In *Sensors, and Command, Control, Communications, and Intelligence (C3I) Technologies for Homeland Security and Homeland Defense III*, E. M. Carapezza, Ed. *Proc. SPIE*, 5403:256.

82. Tu, R., Liu, B., Wang, Z., Gao, D., Wang, F., Fang, Q., Zhang, Z. 2008. Amine-capped ZnS-Mn^{2+} nanocrystals for fluorescence detection of trace TNT explosive. *Anal. Chem.*, 80:3458–3465.

83. Kim, T. H., Kim, H. J., Kwak, C. G., Park, W. H., Lee, T. S. 2006. Aromatic oxadiazole-based conjugated polymers with excited-state intramolecular proton transfer: their synthesis and sensing ability for explosive nitroaromatic compounds. *J. Poly. Sci. A*, 44:2059–2068.

84. Albert, K. J., Walt, D. R. 2000. High-speed fluorescence detection of explosives-like vapors. *Anal. Chem.*, 72:1947–1955.

85. White, J., Truesdell, K., Williams, L. B., Atkisson, M. S., Kauer, J. S. 2008. Solid-state, dye-labeled DNA detects volatile compounds in the vapor phase. *PLoS Biol.*, 6(1):e9.

86. Goldman, E. R., Medintz, I. L., Whitley, J. L., Hayhurst, A., Capp, A. R., Uyeda, H. T., Deschamps, J. R., Lassman, M. E., Mattoussi, H. 2005. A hybrid quantum dot-antibody fragment fluorescence resonance energy transfer-based TNT sensor. *J. Am. Chem. Soc.*, 127:6744–6751.

87. Medintz I. L. Goldman, E. R. Lassman, M. E. Hayhurst, A., Kusterbeck, A. W., Deschamps, J. R. 2005. Self-assembled TNT biosensor based on modular multifunctional surface-tethered components. *Anal. Chem.*, 77:365–372.

88. Medintz, I. L., Goldman, E. R., Clapp, R. A., Uyeda, H. T., Lassman, M. E., Hayhurst, A., Mattoussi, H. 2005. A fluorescence resonance energy transfer quantum dot explosive nanosensor. In *Nanobiophotonics and Biomedical Applications II*, A. N. Cartwright, M. Osinski, Eds. *Proc. SPIE*, 5705:166.

89. Andrew, T. L., Swager, T. M. 2007. A fluorescence turn-on mechanism to detect high explosives RDX and PETN. *J. Am. Chem. Soc.*, 129:7254–7255.

90. Gheorghiu, L., Seitz, R., Arbuthnot, D., Elkind, J. L. 1999. Amine-containing poly(vinylchloride) membranes for detecting polynitroaromatic vapors above land mines. In *Environmental Monitoring and Remediation Technologies II*, Boston. *Proc. SPIE*, 3853:296.

91. Johnson-White, B., Zeinali, M., Shaffer, K. M., Patterson, C. H., Jr., Charles, P. T., Markowitz, M. A. 2007. Detection of organics using porphyrin embedded nanoporous organosilicas. *Biosens. Bioelectron.*, 22:1154–1162.

92. Dorozhkin, L. M., Nefedov, V. A., Sabelnikov, A. G., Sevastjanov V. G. 2004. Detection of trace amounts of explosives and/or explosive related compounds on various surfaces by a new sensing technique/material. *Sens. Actuat. B*, 99:568–570.

93. Xie, C., Liu, B., Wang, Z., Gao, D., Guan, G., Zhang, Z. 2008. Molecular imprinting at walls of silica nanotubes for TNT recognition. *Anal. Chem.*, 80:437–443.

94. Xie, C., Zhang, Z., Wang, D., Guan, G., Gao, D., Liu, J. 2006. Surface molecular self-assembly strategy for TNT imprinting of polymer nanowire/nanotube arrays. *Anal. Chem.*, 78:8339–8346.

95. Keuchel, C., Weil, L., Niessner, R. 1992. Enzyme-linked immunosorbent assay for the determination of 2,4,6-trinitrotoluene and related nitroaromatic compounds. *Anal. Sci.*, 8(1):9–12.

96. Altstein, M., Bronshtein, A., Glattstein, B., Zeichner, A., Tamiri, T., Almong, J. 2001. Immunochemical approaches for purification and detection of TNT traces by antibodies entrapped in sol–gel matrix. *Anal. Chem.*, 73:2461–2467.

97. Bromberg, A., Mathies, R. A. 2004. Multichannel homogeneous immunoassay for detection of 2,4,6-trinitrotoluene (TNT) using a microfabricated capillary array electrophoresis chip. *Electrophoresis*, 25:1895–1900.

98. Bromberg, A., Mathies, R. 2003. Homogeneous immunoassay for detection of TNT and its analogues on a microfabricated capillary electrophoresis chip. *Anal. Chem.*, 75:1188–1195.

99. Goldman, E. R., Hayhurst, A., Lingerfelt, B. M., Iverson, B. L., Georgiou, G., Anderson, G. P. 2003. 2,4,6-Trinitrotoluene detection using recombinant antibodies. *J. Environ. Monit.*, 5:380–383.

100. Anderson, G. P., Moreira, S. C., Charles, P. T., Medintz, I. L., Goldman, E. R. 2006. TNT detection using multiplexed liquid array displacement immunoassays. *Anal. Chem.*, 78:2279–2285.

101. Goldman, E., Cohill, T., Patterson, C. H., Jr., Anderson, G., Kusterbeck, A., Mauro, J. M. 2003. Detection of 2,4,6-trinitrotoluene in environmental samples using a homogeneous fluoroimmunoassay. *Environ. Sci. Technol.*, 37:4733.

102. Goldman, E. R., Pazirandeh, M. P., Charles, P. T., Balighian, E. D., Anderson, G. P. 2002. Selection of phage displayed peptides for the detection of 2,4,6-trinitrotoluene in seawater. *Anal. Chim. Acta*, 457:13–19.

103. Goldman, E. R., Anderson, G. P., Lebedev, N., Lingerfelt, B. M., Winter, P. T., Patterson, C. H., Jr., Mauro, J. M. 2003. Analysis of aqueous 2,4,6-trinitrotoluene (TNT) using a fluorescent displacement immunoassay. *Anal. Bioanal. Chem.*, 375:471.

104. Green, T. M., Charles, P. T., Anderson, G. P. 2002. Detection of 2,4,6-trinitrotoluene n seawater using a reversed-displacement immunosensor. *Anal. Biochem.*, 10:36.

105. Kramer, P. M., Kremmer, E., Weber, C. M., Ciumasu, I. M., Forster, S., Kettrup, A. A. 2005. Development of new rat monoclonal antibodies with different selectivities and sensitivities for 2,4,6-trinitrotoluene (TNT) and other nitroaromatic compounds. *Anal. Bioanal. Chem.*, 382:1919–1933.

106. Charles, P. T., Rangasammy, J. G., Anderson, G. P., Romanoski, T. C., Kusterbeck, A. W. 2004. Microcapillary reversed-displacement immunosensor for trace level detection of TNT in seawater. *Anal. Chim. Acta*, 525:199–204.

107. Charles, P. T., Shriver-Lake, L. C., Francesconi, S. C., Churilla, A. M., Rangasammy, J. G., Patterson, C. H., Jr., Deschamps, J. R., Kusterbeck, A. W. 2004. Characterization and performance evaluation of in vivo and in vitro produced monoclonal anti-TNT antibodies for the detection of TNT. *J. Immunol. Methods*, 284:15–26.

108. Bromage, E. S., Lackie, T., Unger, M. A., Ye, J., Kaattari, S. L. 2007. The development of a real-time biosensor for the detection of trace levels of trinitrotoluene (TNT) in aquatic environments. *Biosens. Bioelectron.*, 22:2532–2538.

109. Bromage, E. S., Vadas, G. G., Harvey, E., Unger, M. A., Kaattari, S. L. 2007. Validation of an antibody-based biosensor for rapid quantification of 2,4,6-trinitrotoluene (TNT) contamination in ground water and river water. *Environ. Sci. Technol.*, 41:7067–7072.

110. Bart, J. C., Judd, L. L., Kusterbeck, A. W., 1997. Environmental immunoassay for the explosive RDX using a fluorescent dye-labeled antigen and the continuous-flow immunosensor. *Sens. Actuat. B*, 39(1–3):411–418.

111. Rabbany, S. Y., Lane, W. J., Marganski, W. A., Kusterbeck, A. W., Ligler, F. S. 2000. Trace detection of explosives using a membrane-based displacement immunoassay, *J. Immunol. Methods*, 246:69.

112. Rabbany, S. Y., Marganski, W. A., Kusterbeck, A. W., Ligler, F. S. 1998. A membrane-based displacement flow immunosensor. *Biosens. Bioelectron.*, 13:939–944.

113. Wilson, R., Clavering, C., Hutchinson, A. 2003. Electrochemiluminescence enzyme immunoassays for TNT and pentaerythritol tetranitrate. *Anal. Chem.*, 75:4244–4249.

114. Narang, U., Gauger, P. R., Kusterbeck, A. W., Ligler, F. S. 1998. Multianalyte detection using a capillary-based flow immunosensor. *Anal. Biochem.*, 255:13–19.

115. Bakaltcheva, I. B., Ligler, F. S., Patterson, C. H., Shriver-Lake, L. C. 1999. Multi-analyte explosive detection using a fiber optic biosensor. *Anal. Chim. Acta*, 399:13–20.

116. Bakaltcheva, I. B., Shriver-Lake, L. C., Ligler, F. S. 1998. A fiber optic biosensor for multianalyte detection: importance of preventing fluorophore aggregation. *Sens. Actuat. B*, 51:46–51.

117. Charles, P. T., Gauger, P., Patterson, C. H., Jr., Kusterbeck, A. W. 2000. On-site immunoanalysis of nitrate and nitroaromatic compounds in groundwater. *Environ. Sci. Technol.*, 34:4641–4650.

118. Gauger, P. R., Holt, D. B., Patterson, C. H., Jr., Charles, P. T., Shriver-Lake, L., Kusterbeck, A. W. 2001. Explosives detection in soil using a field-portable continuous flow immunosensor. *J. Hazard. Mater.*, 83:51–63.

119. Shriver-Lake, L. C., Charles, P. T., Kusterbeck, A. W. 2003. Non-aerosol detection of explosives with a continuous flow immunosensor. *Anal. Bioanal. Chem.*, 377:550–555.

120. Van Bergen, S., Bakaltcheva, I. B., Lundgren, J. S., Shriver-Lake, L. C. 2000. On-site detection of explosives in groundwater with a fiber optic biosensor. *Environ. Sci. Technol.*, 34:704–708.

121. Charles, P. T., Kusterbeck, A. W. 1999. Trace level detection of hexahydro-1,3,5-trinitro-1,3,5-triazine (RDX) by microimmunosensor. *Biosens. Bioelectron.*, 4(4):387–396.

122. Walker, N. R., Linman, M. J., Timmers, M. M., Dean, S. L., Burkett, C. M., Lloyd, J. A., Keelor, J. D., Baughman, B. M., Edmiston, P. L. 2007. Selective detection of gas-phase TNT by integrated optical waveguide spectrometry using molecularly imprinted sol–gel sensing films. *Anal. Chim. Acta*, 593:82–91.

123. Bowen, J., Noe, L. J., Sullivan, B. P., Morris, K., Martin, V., Donnelly, G. 2003. Gas-phase detection of trinitrotoluene utilizing a solid-phase antibody immobilized on a gold film by means of surface plasmon resonance spectroscopy. *Appl. Spectrosc.*, 57(8):906–914.

124. Matsumoto, K., Torimaru, A., Ishitobi, S., Sakai, T., Ishikawa, H., Toko, K., Miura, N., Imato, T. 2005. Preparation and characterization of a polyclonal antibody from rabbit for detection of trinitrotoluene by a surface plasmon resonance biosensor. *Talanta*, 68:305–311.

125. Shankaran, D. R., Gobi, K. V., Sakai, T., Matsumoto, K., Toko, K., Miura, N. 2005. Surface plasmon resonance immunosensor for highly sensitive detection of 2,4,6-trinitrotoluene. *Biosens. Bioelectron.*, 20:1750–1756.

126. Shankaran, D. R., Gobi, K. V., Sakai, T., Matsumoto, K., Imato, T., Toko, K., Miura, N. 2005. A Novel surface plasmon resonance immunosensor for 2,4,6-trinitrotoluene (TNT) based on indirect competitive immunoreaction: a promising approach for on-site landmine detection. *IEEE Sens. J.*, 5(4):616.

127. Shankaran, D. R., Matsumoto, K., Toko, K., Miura, N. 2006. Development and comparison of two immunoassays for the detection of 2,4,6-trinitrotoluene (TNT) based on surface plasmon resonance. *Sens. Actuat. B*, 114:71–79.

128. Shankaran, D. R., Kawaguchi, T., Kim, S. J., Matsumoto, K., Toko, K., Miura, N. 2006. Evaluation of the molecular recognition of monoclonal and polyclonal antibodies for sensitive detection of 2,4,6-trinitrotoluene (TNT) by indirect competitive surface plasmon resonance immunoassay. *Anal. Bioanal. Chem.*, 386:1313–1320.

129. Shankaran, D. R., Kawaguchi, T., Kim, S. J., Matsumoto, K., Toko, K., Miura, N. 2007. Fabrication of novel molecular recognition membranes by physical adsorption

and self-assembly for surface plasmon resonance detection of TNT. *Int. J. Environ. Anal. Chem.*, 87(10–11):771–781.

130. Kawaguchi, T., Shankaran, D. R., Kim, S. J., Gobi, K. V., Matsumoto, K., Toko, K., Miura, N. 2007. Fabrication of a novel immunosensor using functionalized self-assembled monolayer for trace level detection of TNT by surface plasmon resonance. *Talanta*, 72:554–560.

131. Wang, J.,. Liu, G., Wu, H., Lin, Y. 2008. Sensitive electrochemical immunoassay for 2,4,6-trinitrotoluene based on functionalized silica nanoparticle labels. *Anal. Chim. Acta*, 610:112–118.

132. Gwenin, C. D., Kalaji, M., Kay, C. M., Williams, P. A., Tito, D. N. 2008. An in situ amperometric biosensor for the detection of vapours from explosive compounds. *Analyst*, 133:621–625.

133. Naal, Z., Park, J., Bernhard, S., Shapleigh, J., Batt, C., Abruna, H. 2002. Amperometric TNT biosensor based on the oriented immobilization of a nitroreductase maltose binding protein fusion. *Anal. Chem.*, 74(1):140–148.

134. Chaignon, P., Cortial, S., Ventura, A. P., Lopes, P., Halgand, F., Laprevote, O., Ouazzani, J. 2006. Purification and identification of a *Bacillus* nitroreductase: potential use in 3,5-DNBTF biosensoring system. *Enzyme Microb. Technol.*, 39:1499–1506.

135. Wang, J. 2007. Electrochemical sensing of explosives. *Electroanalysis* 19(4), 415–423

136. Agui, L., Vega-Montenegro, D., Yanez-Sedeno, P., Pingarron, J. M. 2005. Rapid voltammetric determination of nitroaromatic explosives at electrochemically activated carbon-fibre electrodes. *Anal. Bioanal. Chem.*, 382:381–387.

137. Fu, X., Benson, R. F., Wang, J., Fries, D. 2005. Remote underwater electrochemical sensing system for detecting explosive residues in the field. *Sens. Actuat. B*, 106:296–301.

138. Pon Saravanan, N., Venugopalan, S., Senthilkumar, N., Santhosh, P., Kavita, B., Gurumallesh Prabu, H. 2006. Voltammetric determination of nitroaromatic and nitramine explosives contamination in soil. *Talanta*, 69:656–662.

139. Shi, G., Qu, Y., Zhai, Y., Liu, Y., Sun, Z., Yang, J., Jin, L. 2007. {MSU/PDDA}$_n$ LBL assembled modified sensor for electrochemical detection of ultratrace explosive nitroaromatic compounds. *Electrochem. Commun.*, 9:1719–1724.

140. Zimmermann, Y., Broekaert, J. A. C. 2005. Determination of TNT and its metabolites in water samples by voltammetric techniques. *Anal. Bioanal. Chem.*, 383:998–1002.

141. Hrapovic, S., Majid, E., Liu, Y., Male, K., Luong, J. H. T. 2006. Metallic nanoparticle-carbon nanotube composites for electrochemical determination of explosive nitroaromatic compounds. *Anal. Chem.* 78(15):5504–5512.

142. Riskin, M., Tel-Vered, R., Bourenko, T., Granot, E., Willner, I. 2008. Imprinting of molecular recognition sites through electropolymerization of functionalized Au nanoparticles: development of an electrochemical TNT sensor based on π-donor–acceptor interactions. *J. Am. Chem. Soc.*, 130:9726–9733.

143. Chen, J.-C., Shih, J.-L., Liu, C.-H., Kuo, M.-Y., Zen, J.-M. 2006. Disposable electrochemical sensor for determination of nitroaromatic compounds by a single-run approach. *Anal. Chem.*, 78:3752–3757.

144. Takahara, N., Yang, D.-H., Ju, M.-J., Hayashi, K., Toko, K., Lee, S.-W., Kunitake, T. 2006. Anchoring of cyclodextrin units on TiO$_2$ thin layer for effective detection of nitro-aromatics: a novel electrochemical approach for landmine detection. *Chem. Lett.*, 35(12):1340.

145. Ju, M. J., Yang, D.-H., Lee, S.-W., Kunitake, T., Hayashi, K., Toko, K. 2007. Fabrication of TiO_2/γ-CD films for nitro aromatic compounds and its sensing application via cyclic surface-polarization impedance (cSPI) spectroscopy. *Sens. Actuat. B*, 123:359–367.

146. Ju, M.-J., Yang, D.-H., Takahara, N., Hayashi, K., Toko, K., Lee, S.-W., Kunitake, T. 2007. Landmine detection: improved binding of 2,4-dinitrotoluene in a γ-CD/metal oxide matrix and its sensitive detection via a cyclic surface polarization impedance (cSPI) method. *Chem. Commun.*, 2630–2632.

147. Masunaga, K., Hayama, K., Onodera, T., Hayashi, K., Miura, N., Matsumoto, K., Toko, K. 2005. Detection of aromatic nitro compounds with electrode polarization controlling sensor. *Sens. Actuat. B*, 108:427–434.

148. Briglin, S. M., Burl, M. C., Freund, M. S., Lewis, N. S., Matzger, A., Ortiz, D. N., Tokumaru, P. 2000. Progress in use of carbon black–polymer composite vapor detector arrays for land mine detection. In *Detection and Remediation Technologies for Mines and Minelike Targets V*, A. C. Dubey, J. F. Harvey, J. T. Broach, R. E. Dugan, Eds. *Proc. SPIE*, 4038:530.

149. Guan, X., Gu, L.-Q., Cheley, S., Braha, O., Bayley, H. 2005. Stochastic sensing of TNT with a genetically engineered pore. *Chem. Bio. Chem.*, 6:1875–1881.

150. Bentes, E., Gomes, H. L., Stallinga, P., Moura, L. 2004. Detection of explosive vapors using organic thin-film transistors. *Proceedings of IEEE Sensors 2004*, pp. 766–769.

151. Patel, S. V., Hobson, S. T., Cemalovic, S., Mlsna, T. E. 2005. Chemicapacitive microsensors for detection of explosives and TICs. In *Unmanned/Unattended Sensors and Sensor Networks II*, E. M. Carapezza, Ed. *Proc. SPIE*, 5986:59860M-1.

152. Patel, S. V., Mlsna, T. E., Fruhberger, B., Klaassen, E., Cemalovic, S., Baselt, D. R. 2003. Chemicapacitive microsensors for volatile organic compound detection. *Sens. Actuat. B*, 96:541–553.

153. MacDonald, J., Lockwood, J. R. 2003. *Alternatives for Landmine Detection*. RAND Corporation, Santa Monica, CA.

154. Levitt, M. 2001. *Spin Dynamics: Basics of Nuclear Magnetic Resonance*. Wiley, New York.

155. Garroway, A. N., Buess, M. L., Yesinowski, J. P., Miller, J. B., Suits, B. H., Hibbs, A. D., Barrall, G. A., Matthews, R., Burnett, L. J. 2001. Remote sensing by nuclear quadrupole resonance. *IEEE Trans. Geosci. Remote Sens.* 39(6):1108–1118.

156. Farrar, T. C., Becker, E. D. 1971. *Pulse and Fourier Transform NMR: Introduction to Theory and Methods*. Academic Press, New York, p. 67.

157. Miller, J. B., Suits, B. H., Garroway, A. N., Hepp, M. A. 2000. Interplay among recovery time, signal, and noise: Series- and parallel-tuned circuits are not always the same. *Concepts Magn. Reson.*, 12(3):125–136.

158. Rowe, M. D., Smith, J. A. S. 1996. Mine detection by nuclear quadrupole resonance. In *The Detection of Abandoned Landmines: A Humanitarian Imperative Seeking A Technical Solution*, EUREL International Conference (Conf. Publ. No. 431), Edinburgh, UK, pp. 62–66.

159. Fukushima, K. E., Roeder, S. B. W. 1981. *Experimental Pulse NMR: A Nuts and Bolts Approach*. Addison-Wesley, Reading MA, p. 252.

160. Hoult, D. I. 1979. *Rev. Sci. Instrum.*, 50:193.

161. Kisman, K. E., Armstrong, R. L. 1974. Coupling scheme and probe damper for pulsed nuclear magnetic resonance single coil probe. *Rev. Sci. Instrum.*, 45:1159–1163.

162. Conradi, M. S. 1977. FET Q switch for pulsed NMR. *Rev. Sci. Instrum.*, 48:359–361.

163. Sullivan, N. S., Deschamps, P., Néel, P., Vaissière, J. M. 1983. Efficient fast-recovery scheme for NMR pulse spectrometers. *Rev. Phys. Appl.*, 18:253–261.

164. Andrew, E. R., Jurga, K. 1987. NMR probe with short recovery time. *J. Magn. Reson.*, 73(2):263–276.

165. Chingas, G. C. 1983. Overcoupling NMR probes to improve transient response. *J. Magn. Reson.* 54(1):153–157.

166. Suits, B. H., Garroway, A. N., Miller, J. B. 1998. Noise-immune coil for unshielded magnetic resonance measurements. *J. Magn. Reson.*, 131(1):154–158.

167. Black, R. D., Early, T. A., Johnson, G. A. 1995. Performance of a high-temperature superconducting resonator for high-field imaging. *J. Magn. Reson. A*, 113(1):74–80.

168. Jiand, Y. J. 2000. A simple method for measuring the *Q* value of an NMR sample coil. *J. Magn. Reson.*, 142:386–388.

169. Abragam, A. 1961. *The Principles of Nuclear Magnetism*. Clarendon Press, Oxford, UK.

170. Pusiol, D. J. 2005. Method, sensor elements and arrangement for the detection and/or analyses of compounds simultaneously exhibiting nuclear quadrupole resonance and nuclear magnetic resonance, or double nuclear quadrupole resonance. Patent Application G01N 24/00, US 2005/0202570 A1, Sept. 15.

171. Luznik, J., Pirnat, J., Trontelj, Z. 2002. *Solid State Commun.* 121:653–656.

172. Blinc, R., Apih, T., Seliger, J. 2004. Nuclear quadrupole double resonance techniques for the detection of explosives and drugs. *Appl. Magn. Reson.*, 25:523–534.

173. Blinc, R., Seliger, J., Apih, T., Lahajnar, G. 2001. Patent SI 20995, G01R 33/44, F42D 6/02, July 3.

174. Seliger, J., Blinc, R., Lahajnar, G. 2001. Patent SI 20551A, F42D 5/00, F42D 5/02, Oct. 30.

175. Thurber, K. R., Sauer, K. L., Buess, M. L., Klug, C. A. 2005. Increasing 14N NQR signal by ^{1}H–^{14}N level crossing with small magnetic fields. *J. Magn. Reson.*, 177:118–128.

176. Rudakov, T. N., Hayes, P. A. 2006. *J. Magn. Reson.*, 183:96–101.

177. Mikhaltsevitch, V. T., Beliakov, A. V. 2006. *Solid State Commun.* 138:409–411.

178. Prescott, D. W., Olmedo, O., Soon, S., Sauer, K. L. 2007. Low-field approach to double resonance in nuclear quadrupole resonance of spin-1 nuclei. *J. Chem. Phys.*, 126(20):204504 (9 pages).

179. Shaw, J. 1994. Brief history of explosives detection using pure quadrupole resonance. *NQI Newsl.*, 1(3):26–29.

180. Miller, J. B., Barrall, G. A. 2005. Explosives detection with nuclear quadrupole resonance. *Am. Sci.*, 93:50–57.

181. Garroway, A. N., Buess, M. L., Yesinowski, J. P., Miller, J. B. 1994. *Proc. SPIE*, 2092:318.

182. Sauer, K. L., Suits, B. H., Garroway, A. N., Miller, J. B. 2001. Three-frequency quadrupole resonance of spin-1 nuclei. *Chem. Phys. Lett.* 342:362–368.

183. Mozzhukhin, G. V., Bodnya, A. V., Fedotov, V. V., Chen, C., Zhang, G., Guo, H. 2005. The detection of the trinitrotoluene by pure nuclear quadrupole resonance. *Appl. Magn. Reson.*, 29:293–298.

184. Robert, H., Prado, P. J. 2004. Threat localization of QR explosives detection systems. *Appl. Magn. Reson.*, 25:523–534.

185. Peirson, N. F., Rowe, M.D., Smith, J. A. S. 2001. State of the art in NQR mine detection—an overview. *The Demining Technology Information Forum Journal* (DTIF), Issue 1.

186. Buess, M. L., Garroway, A. N., Yesinowski, J. P. 1994. Removing the effects of acoustic ringing and reducing temperature effects in the detection of explosives by NQR. U.S. patent 5,365,171, Nov. 15.

187. Barrall, G. A., et al. 2005. Advances in the engineering of quadrupole resonance landmine detection systems. *Proc. SPIE*, 5794:774–785.

188. Marino, R. A., Klainer, S. M. 1977. *J. Chem. Phys.*, 67:3388–3389.

189. Klainer, S. M., Hirshfeld, T. B., Marino, R. A. 1982. In *Fourier, Hadamard, and Hilbert Transforms in Chemistry*, A. G. Marshall, Ed., Plenum Press, New York, pp. 147–182.

190. Osokin, D. Y. 1982. *Phys. Status Solidi (b)*, 109(1):K7–K10.

191. Osokin, D. Y. 1982. The method of search and detection of nuclear quadrupole resonance spectra. SU 958935 AI G0IN 24/00, 15.09.1982.

192. Sauer, K. L., Suits, B. H., Garroway, A. N., Miller, J. B. 2003. Three-frequency nuclear quadrupole resonance of spin-1 nuclei. *J. Chem. Phys.*, 118:5071–5081.

193. Mozjoukhine, G. V. 2002. Application of the spherical tensor method for two frequency pure NQR of spin $I = 1$. nuclei. *Appl. Magn. Reson.*, 22:31.

194. Osokin, D. Y., Husnutdinov, R. R., Shagalov, V. A. 2004. Two-frequency multiple-pulse sequences in nitrogen-14 NQR. *Appl. Magn. Reson.*, 25:513–521.

195. Lee, S. K., Sauer, K. L., Seltzer, S. J., Alem, O., Romalis, M. V. 2006. Subfemtotesla radio-frequency atomic magnetometer for detection of nuclear quadrupole resonance. *Appl. Phys. Lett.*, 89:214108.

196. Augustine, M. P., TonThat, D. M., Clarke, J. 1998. SQUID detected NMR and NQR. *Solid State Nucl. Magn. Reson.*, 11:139–156.

197. Yesinowski, J. P., Buess, M. L., et al. 1995. Detection of 14N and 35Cl in cocaine base and hydrochloride using NQR,NMR and SQUID techniques. *Anal. Chem.*, 67(13):2256–2263.

198. Clarke, J., Braginski, A. I., Eds. 2006. *The SQUID Handbook*, Vol. II, *Applications of SQUIDs and SQUID Systems*. Wiley-VCH, Weinheim, Germany.

199. He, D. F., Tachiki, M., Itozaki, H. 2007. 14N nuclear quadrupole resonance of *p*-nitrotoluene using a high-Tc RF SQUID. *Supercond. Sci. Technol.*, 20:232–234.

200. Withers, R. S., et al. 1993. Thin-film HTS probe coils for magnetic resonance imaging. *IEEE Trans. Supercond.*, 3(1):2450–2453.

201. Schiano, J. L., Wilker, C. 2006. Patent WO2006060706, 2006-06-08, G01R33/44. McCambridge, J. D. 2006. U.S. patent 2006/0082368 A1, Apr. 20. Laubacher, D. B., McCambridge, J. D., Wilker, C. 2006. Patent EP1711840, 2006-10-18.

202. Deas, R. M., Cervantes, C., Schaedel, S. F. 2004. *Landmine Detection by Nuclear Quadrupole Resonance (NQR)*. Pentagon Report A456134. Dec.

203. Barras, J., Gaskell, M. J., Hunt, N., Jenkinson, R. I., Mann, K. R., Pedder, D. A., Shilstone, G. N., Smith, J. A. S. 2004. Detection of ammonium nitrate by nuclear quadrupole resonance. *Appl. Magn. Reson.*, 25:411–437.

204. Barrall, G. A. 2005. Private Communication and demonstration.

205. Grechishkin, V. S., Sinyavskii, N. Y., Mozzhukhin, G. V. 1993. A unilateral NQR explosives detector. *Russ. Phys. J.*, 35(7):635–637.

206. Anferov, V. P., Mozjoukhine, G. V., Fisher, R. 2000. Pulsed spectrometer for nuclear quadrupole resonance for remote detection of nitrogen in explosives. *Rev. Sci. Instrum.*, 71(4):1656–1659.

207. Prado, P. 2008. NATO ARW, Magnetic resonance in spotting explosives. Presented at a satellite meeting of the EUROMAR 2008 Conference, St. Petersburg, Russia, July 7–9.

208. Jenkins, R. 2000. X-ray techniques: overview. In *Encyclopedia of Analytical Chemistry*, R. A. Meyers, Ed., Wiley, Chichester, UK, pp. 13269–13288.

209. MITRE. 1996. *Approaches to Humanitarian Demining*. JSR-96-115. MITRE, McLean, VA.

210. Martin, P. W., Pringle R. W. 1960. X-ray analysis of geological formations. U.S. patent 2,923,824.

211. Clayton, C. G., Packer, T. W. 1971. UKAEA, unpublished work.

212. An, J. G., et al. Digital radiography inspection apparatus for large object. UK Patent GB 2368764, 2005.

213. Bierwolf, H. 1965. Density measurements with radioisotopes in the petroleum industry and the petroleum processing industry. *A.E.G. Mitt.*, 53:165.

214. Freeh, E. J. 1959. *Pet. Eng.*, 31(4):C10.

215. Peugeot, S. A. 2005. *Neutronographie*. DRN/DRE/SIREN -CEA/Saclay.

216. Berthold Industrial Systems. 2008. *Moisture Meter LB350*.

217. Alternatives for Landmine detection, RAND Science and Technology Policy, RAND Corporation, Santa Monica, CA 2003.

218. Roberge, P. R. 2006. *Corrosion Basics: An Introduction*, 2nd ed., NACE Press, Houston, TX.

219. Chauvin Arnoux Group. Application Notes of Earth and Resistivity Testers. http://www.chauvin-arnoux.com.

220. Wexler, A., Fry, B., Neumann, M. R. 1985. Impedance-computed tomography algorithm and system. *Appl. Opt.*, 24:3985–3992.

221. Church, P., Wort, P., Gagnon, S., McFee, J. 2001. Performance assessment of an electrical impedance tomography detector for mine-like objects. In *Detection and Remediation Technologies for Mines and Minelike Targets VI*, A. C. Dubey, J. F. Harvey, J. T. Broach, V. George, Eds. International Society for Optical Engineering, Seattle, WA, pp. 120–131.

222. Morita, Y., McKenney, H. 1980. *An Assessment of Technical Factors Influencing the Potential Use of RPVs for Minefield Detection*. Report 138300-57-T. Environmental Research Institute of Michigan, Ann Arbor, MI, July.

223. McKenney, H. 1980. *Remote Detection of Minefields*. Report 138300-65-F1. Environmental Research Institute of Michigan, Ann Arbor, MI, Sept.

224. Morita, Y., Suits, G., McKenney, H., Sattinger, I. 1979. *Identification and Screening of Remote Mine Detection Techniques*. Report 138300-22-T. Environmental Research Institute of Michigan, Ann Arbor, MI, June.

225. Robbins, J. 1988. *Report on a Study Relating to Optical Detection of Scatterable Mines*. DRES Contract Report 2/89. Scintrex Ltd., Toronto, Ontario, Canada, Sept.

226. Barringer Research. 1985. *A Study of Visible and Infrared Remote Minefield Detection*. DRES Contract Report 16–91. Barringer Research, Toronto, Ontario, Canada, July.

227. Keskinen, K. T., Tam, S. Y. K., Foley, P., Foo, S. L., Sinclair, I. 1989. *Feasibility Study of Microwave Detection of Minefields*. DRES Contract Report 24/89. MPB Technologies, Montreal, Quebec, Canada.

228. Simard, J. R., Mathieu, P., Fournier, G. R., Larochelle, V. 2000. Range-gated intensified spectrographic imager: an instrument for active hyperspectral imaging. In *Laser Radar Technology and Applications V*, G. W. Kamerman, U. N. Singh, C. Werner, V. V. Molebny, Eds. *Proc. SPIE*, 4035:180–191.

229. Russell, K. L., McFee, J. E., Sirovyak, W. 1997. Remote performance prediction for infrared imaging of buried mines. In *Detection and Remediation Technologies for Mines and Mine-like Targets II*, A. C. Dubey, R. L. Barnard, Eds. *Proc. SPIE*, 3079:762–769.

230. Achal, S. B., Anger, C. D., McFee, J. E., Herring, R. W. 1999. Detection of surface-laid mine fields in VNIR hyperspectral high spatial resolution data. In *Detection and Remediation Technologies for Mines and Mine-like Targets IV*, A. C. Dubey, J. F. Harvey, J. T. Broach, R. E. Dugan, Eds. *Proc. SPIE*, 3710:808–818.

231. McFee, J. E., Achal, S. B., Ivanco, T., Anger, C. 2005. A shortwave infrared hyperspectral imager for landmine detection. In *Detection and Remediation Technologies for Mines and Mine-like Targets X*, R. S. Harmon, J. T. Broach, J. H. Holloway, Eds. *Proc. SPIE*, 5794:56–67.

232. Achal, S. B., McFee, J. E., Ivanco, T., Anger, C. 2007. A thermal infrared hyperspectral imager (tasi) for buried land mine detection. In *Detection and Remediation Technologies for Mines and Mine-like Targets XII*, R. S. Harmon, J. T. Broach, J. H. Holloway, Eds. *Proc. SPIE*, 6553.

233. Bolton, J. 2008. Listing of Instruments Available for Airborne Remote Sensing. http://fullspectralimaging.net/LIAARS.aspx. Accessed Apr. 30, 2009.

234. Banninger, C. 1990. Fluorescence line imager (FLI) measured red edge shifts in metal-stressed Norway spruce forest and their relationship to canopy biochemical and morphological changes. In *Conference on Imaging Spectroscopy of the Terrestrial Environment*, Orlando, FL. *Proc. SPIE*, 1298:234–243.

235. Knipling, E. 1969. Leaf reflectance and image formation on colour infrared film. In *Remote Sensing in Ecology*, P. L. Johnson, Ed., University of Georgia Press, Athens, GA, pp. 17–29.

236. McFee, J. E., Ripley, H., Buxton, R., Thriscutt, A. 1996. Preliminary study of detection of buried landmines using a programmable hyperspectral imager. In *Detection and Remediation Technologies for Mines and Mine-like Targets*, A. C. Dubey, R. L. Barnard, C. J. Lowe, J. E. McFee, Eds. *Proc. SPIE*, 2765:476–488.

237. McFee, J. E., Ripley, H. T. 1997. Detection of buried landmines using a casi hyperspectral imager. In *Detection and Remediation Technologies for Mines and Mine-like Targets II*, A. C. Dubey, R. L. Barnard, Eds. *Proc. SPIE*, 3079:738–749.

238. Del Grande, N. K., Durbin, P. F., Gorvad, M. R., Perkins, D. E., Clarke, G. A., Hernandez, J. E., Sherwood, R. J. 1993. Dual-band infrared capabilities for imaging buried object sites. In *Underground and Obscured Object Imaging and Detection*, N. Del Grande, I. Cindrich, P. Johnson, Eds. *Proc. SPIE*, 1942:166–177.

239. DePersia, A. T., Bowman, A. P., Lucey, P., Winter, E. M. 1995. Phenomenology considerations for hyperspectral mine detection. In *Detection Technologies for Mines*

and Mine-like Targets, A. C. Dubey, I. Cindrich, J. Ralston, K. Rigano, Eds. *Proc. SPIE*, 2496:159–167.

240. DePersia, A. T., Bowman, A. P., Giles, A. L., Winter, E. M., Badik, F., Schlangen, M., Lucey, P., Williams, T., Johnson, J., Hinrichs, J., Horton, K., Allen, G., Stocker, A., Oshagan, A., Schaff, B., Kendall, B. 1995. ARPA's hyperspectral mine detection program. Presented at the International Symposium on Spectral Sensing Research (ISSSR), Melbourne, Australia, Nov.

241. Salisbury, J., Walter, L., Vergo, N., D'Aria, D. 1991. *Infrared (2.1-25 μm) Spectra of Minerals*. Johns Hopkins University Press, Baltimore.

242. Moore, R. L., Montgomery, J. P., Logan, K. V., Wells, T. B., Dean, A. S., Kuster, E. J. 1985. *Development of Multispectral Camouflage System*, Vol. VI A. Report ADB0931111. Georgia Institute of Technology, Athens, GA, Apr.

243. Suits, G. 1982. *Photographic Enhancement Techniques for Mine Detection*. Report 158700-33-T. Environmental Research Institute of Michigan, Ann Arbor, MI, July.

244. Testa, A. M. 1981. *Remote Minefield Detection Using 16mm Aerial Photography*. Report MERADCOM 2338. U.S. Army Mobility Equipment Research and Development Command, Fort Belvoir, VA, Sept.

245. Soffer, R. J., McFee, J. E. 1991. Preliminary investigations of a high resolution imaging spectrograph (casi) for detection of surface scattered land mines. In *Proceedings of the 14th Canadian Symposium on Remote Sensing*, May.

246. McFee, J. E., Achal, S., Anger, C. 1994. Scatterable mine detection using a casi. In *Proceedings of the First International Airborne Remote Sensing Conference*, Sept., pp. I587–I598.

247. Achal, S., McFee, J. E., Anger, C. 1995. Identification of surface-laid mines by classification of compact airborne spectrographic imager (casi) reflectance spectra. In *Detection Technologies for Mines and Minelike Targets*, A. C. Dubey, I. Cindrich, J. Ralston, K. Rigano, Eds. *Proc. SPIE*, 2496:324–335.

248. Harsanyi, J. C., Chang, C. I. 1994. Hyperspectral image classification and dimensionality reduction: an orthogonal subspace projection approach. *IEEE Trans. Geosci. Remote Sens.*, 32:779–785.

249. Nischan, M. L., Joseph, R. M., Libby, J. C., Kerekes, J. P. 2003. Active spectral imaging. *Lincoln Lab. J.*, 14(1):131–144.

250. McFee, J. E., Anger, C., Achal, S., Ivanco, T. 2007. Landmine detection using passive hyperspectral imaging. In *Chemical and Biological Sensing VIII*. *Proc. SPIE*, 6554.

251. McFee, J. E., Achal, S., Faust, A. A., Puckrin, E., House, A., Reynolds, D., McDougall, W., Asquini, A. 2009. Detection and dispersal of explosives by ants. In *Detection and Sensing of Mines, Explosive Objects, and Obscured Targets XIV*, R. S. Harmon, J. T. Broach, J. H. Holloway, Eds. *Proc. SPIE*, 7303.

252. Ivanco, T., Achal, S. B., McFee, J. E., Anger, C. D. 2001. Casi real-time surface-laid mine detection. In *Detection and Remediation Technologies for Mines and Mine-like Targets VI*, A. C. Dubey, J. F. Harvey, J. T. Broach, V. George, Eds. *Proc. SPIE*, 4394:365–378.

253. Ivanco, T., Achal, S. B., McFee, J. E., Anger, C. D., Young, J. 2007. Real-time airborne hyperspectral imaging of land mines. In *Detection and Remediation Technologies for Mines and Mine-like Targets XII*, R. S. Harmon, J. T. Broach, J. H. Holloway, Eds. *Proc. SPIE*, 6553.

CHAPTER EIGHT

Multisensor Fusion

8.1 PREVIEW

Sensors measure a physical quantity and convert it into a signal. In dynamic environments, no single sensor is capable of obtaining reliably all the information required at all times. The availability of sophisticated sensors and the existence of sophisticated information-processing systems in the last two decades made it possible to employ multiple sensors simultaneously, thus extracting as much information as possible. Diverse physical properties measured by these sensor technologies provide significant advantages over single-source data and make multisensory data fusion attractive. Nowadays, multiple sensors are used widely in the areas of detection and the identification of buried objects.

Military applications are one of the most active areas in sensor fusion. To enhance the efficiency of data fusion programs, the U.S. Department of Defense established the Joint Directors of Laboratories (JDL) Data Fusion Group in 1985. The group proposed a layout of a generic data fusion system in 1992 [1]. The data fusion model maintained by the group is the most widely used model. This architecture is comprised of a preprocessing stage, four levels of fusion and data management functions. The group also established a common language to provide a platform for such multidisciplinary area. The architecture was later refined by Hall and Llinas [2]. According to the fusion model developed by JDL, level 1 processing, which is called *object refinement*, performs four key functions:

1. Transforms sensor data into a consistent set of units and coordinates
2. Refines, and extends in time, estimates of an object's position, kinematics, or attributes

Subsurface Sensing, First Edition. Edited by Ahmet S. Turk, A. Koksal Hocaoglu, and Alexey A. Vertiy.

3. Assigns data to objects to allow the application of statistical estimation techniques

4. Refines the estimation of an object's identity or classification

Levels 2 through 4 are situation refinement, threat refinement, and process refinement, respectively. They are explained briefly below.

- *Situation refinement.* Identified objects' relationships to each other are discovered in the second level. This may include object clustering and relational analysis, force structure and cross-force relations, and so on.
- *Threat refinement.* At this level of processing, one tries to infer details about the future, such as intent estimation, event prediction, consequence prediction, susceptibility, and vulnerability assessment.
- *Process refinement.* Adaptive data acquisition and processing, and whether the other levels are or can be optimized are the issues of the fourth level.

Levels 2 to 4 are beyond the scope of this chapter and are not discussed. Among much other architecture, National Bureau of Standards' architecture has been known for a relatively extended period. This architecture is designed to control robotic systems, and its scope in multisensory fusion applications is rather limited. For more information about this architecture, see Luo and Kay [3]. Although the choice of architecture is an important issue, we limit our discussion to JDL's level 1 processing.

Level 1 processing is a well-studied area that deals not only with preconditioning the data for fusion but also with the fusion of sensor measurements, features, and decisions. The preconditioning includes, but is not limited to, converting each sensor measurement, feature, or decision into a common coordinate system. This issue is discussed further in Section 8.2. Fusion of sensor measurements, features, and decisions can be performed at a variety of levels. In addition to fusion from measurement to measurement, feature to feature, and decision to decision, any combination of the three can is possible. The main focus of this chapter is on the techniques used for fusing sensor measurements, features, and decisions.

The fusion techniques combine information from multiple sensors to achieve better accuracies. The theory behind these techniques is quite old and well established. Bayesian principles and Dempster–Shafer theory are among the commonly used approaches. These are followed by neural networks and fuzzy logic. Bayes' theory, which has the strongest foundation, is discussed in Section 8.4.1. Dempster–Shafer theory, which deals with measures of "belief" as opposed to probability, is discussed in Section 8.4.2. Fuzzy integrals, discussed in Section 8.5, are based on nonadditive measures and thus capture both positive and negative synergy among multiple inputs to a system. They are therefore useful not only to fuse multiple information sources but also to identify redundancy and complementarity among the sources. Neural networks can detect patterns in information that they have been used successfully in many classification problems. The use of neural networks for fusion, discussed is Section 8.6, focuses on Choquet integral–based artificial neural network structures. In the final section the implementation issues are summarized.

FIGURE 8.1 Multisensor fusion concept depicting a single soldier platform with GPR, a metal detector, and infrared technologies.

8.2 DATA ASSOCIATION

Data association is an important step in multisensor fusion. The association process is carried out to make sure that the fused data arise from the same target or physical location. Consider, for example, the soldier platform designed to locate buried landmines shown in Fig. 8.1. An infrared (IR) thermal imaging system is installed on the helmet and both GPR and a metal detector are integrated onto the handheld device. For this scenario, GPR and a metal detector have common apertures but differ in swath width and footprint. The association is rather simple, as one can interpolate the data to have a common swath width and footprint. On the other hand, the absence of real-time position information for the IR data makes the association difficult. In this case, a scene tracking method, which continuously monitors the position of the handheld device, needs to be employed. Thus, independent observations are "transformed" into a common coordinate system. Once a common coordinate system is established for each observation, fusing the observations is rather simple.

8.3 FUSION ARCHITECTURES

The signals obtained by the sensors carry information relating to some physical phenomenon. All such signals in a multisensor system may be fused at a variety of levels to make an inference about the state of nature. Sensor fusion is applied mainly at three levels: the signal level (data), the evidence level (features), and the dynamics level (decision). These three basic fusion architectures are summarized below. In addition to these architectures, there are situations in which a hybrid processing architecture is preferred. For example, consider a case in which decision-level data reveal the correlations among different attributes of variables. For such situations, features may be combined with raw data with a similar purpose. In these cases, a hybrid processing architecture is preferred.

8.3.1 Data-Level Fusion

Raw sensor data can be combined directly when the sensors observe the same physical manifestation of an entity. For example, an IR image and a visual image provide closely comparable observational data, hence their data may be merged directly. A data association process is required prior to the merging, as explained earlier. This step is usually difficult to apply if the sensors are displaced geographically. The raw sensor data from each sensor are then fused using pattern recognition techniques, physical models, or estimation techniques. Depending on the complexity of these techniques, data-level fusion may be costly in terms of computational requirements. This may affect the choice of the architecture. Figure 8.2 illustrates a conceptual view of the sensor data fusion process.

8.3.2 Feature-Level Fusion

In this architecture, the sensor data are represented by a feature vector. This is achieved by extracting features from each sensor. The extracted features are then subject to a fusion process similar to the previous architecture. An association step may be required to make sure that the fused features arise from the same physical location. The feature vectors are then combined to declare object identity. Neural networks, clustering algorithms, Bayes' rule, heuristic algorithms, and support-vector machines are some examples of techniques that could be used at this stage. The architecture is illustrated in Fig. 8.3. Multisensor systems frequently use this approach because each sensor may contribute a unique set of features with varying degrees of correlation with the objects to be identified.

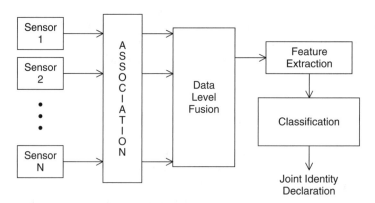

FIGURE 8.2 Direct fusion of sensor data.

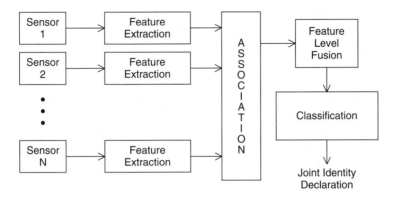

FIGURE 8.3   Fusion of feature vectors representing sensor data.

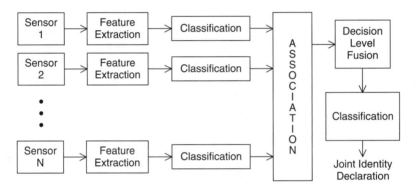

FIGURE 8.4   Fusing the sensor decisions.

8.3.3   Decision-Level Fusion

In this architecture, feature extraction and classification procedures are performed for each sensor independently. In other words, a feature vector is obtained by extracting features from each sensor's data and then pattern classification techniques are used to declare object identity. The individual decisions of each sensor are then associated and fused to come up with a final decision. Examples of decision-level fusion methods include Bayesian inference, the Dempster–Shafer method, and weighted decision methods. This approach is illustrated in Fig. 8.4.

8.4   PROBABILISTIC SENSOR FUSION

There are two probabilistic sensor fusion techniques that are considered to be standards in the level 1 fusion community: Bayesian taxonomy and the Dempster–Schafer method. We begin with an outline of Bayes' theory. We discuss well-known quantities such as priors, likelihood, and posteriors. We then discuss how to use the theory to fuse data from multiple sensors. We outline the ideas of Dempster–Shafer theory and present the basic Dempster–Shafer fusion equation.

8.4.1 Bayesian Theorem

In *An Essay Towards Solving a Problem in the Doctrine of Chances*, Bayes creates a methodology of mathematical inference. It can be seen as a way of understanding how the probability that a theory is true is affected by a new piece of evidence. It describes how the initial information E about a hypothesis H and the likelihood based on observations determine the posterior probability distribution. The law is a simple relationship of conditional probabilities:

$$P(H\,|E) = \frac{P(E\,|H)\cdot P(H)}{P(E)} \tag{8.1}$$

In this formula, H stands for a hypothesis that we are interested in testing, and E represents a new piece of evidence that seems to confirm or disconfirm the hypothesis. $P(H)$ represents our best of the probability of the hypothesis prior to consideration of a new piece of evidence. $P(H)$ is known as the prior probability of H. We call $P(E\,|H)$ the *likelihood* of H with respect to E. Observing the initial information E, Bayes' rule converts the prior probability $P(H)$ to the a posteriori probability $P(H\,|E)$.

Thus, the Bayesian posterior probability reflects our belief in the hypothesis [4]. It offers a direct means of combining prior information and the current observations. In the case of several events E_i that are distinguished from H in some way,

$$P(E) = \sum_i P(E\,|H_i)P(H_i) \tag{8.2}$$

By substituting this into Eq. (8.1), we obtain

$$P(H\,|E) = \frac{P(E\,|H)\cdot P(H)}{\sum_i P(E\,|H_i)P(H_i)} \tag{8.3}$$

Equations (8.1) and (8.3) are known as Bayes' rule. Bayes' formula can be expressed informally in English by saying that

$$\text{posterior} = \frac{\text{likelihood} \times \text{prior}}{\text{evidence}}$$

Notice that the denominator acts merely as a scale factor to guarantee that the posterior probabilities sum to 1. Hence, the posterior probability is determined mainly by the product of the likelihood and the prior probability.

8.4.2 Multisensor Fusion with Bayesian Inference

Bayes' theorem does not tell us how to assign the probabilities. It is a standard difficulty encountered when applying Bayes' theorem. Suppose that several sensors have supplied features to form a feature vector, data, to identify a buried object. The chance that the object is of type 1, or object$_1$, on the available evidence is

$$P(\text{object}_1 \,|\, \text{data}) = \frac{P(\text{data} \,|\, \text{object}_1) \cdot P(\text{object}_1)}{P(\text{data} \,|\, \text{object}_1) P(\text{object}_1) + P(\text{data} \,|\, \text{object}_2) P(\text{object}_2) + \cdots}$$

One is now confronted with the question: What are $P(\text{object}_1)$, $P(\text{object}_2)$, and so on? It is a basic issue with the use of Bayesian inference techniques. The choice of a priori probabilities and the likelihood values has a significant impact on performance. The priors are often assumed to be equal in the absence of any information, as in

$$P(\text{object}_i) = \frac{1}{N} \tag{8.4}$$

Using an iterative scheme when applying Bayes' rule, these priors are updated as described by

$$P_k(\text{object}_i) = P_{k-1}(\text{object}_i \,|\, \text{data}) \tag{8.5}$$

The a priori probability in iteration k is set equal to the value of the a posteriori probability from the previous iteration, $k - 1$. The update changes the priors unequally, acquiring more meaningful values in the process. Figure 8.5 illustrates this process.

The final decision about the surveyed area (e.g., deciding whether the buried object is of type 1, type 2 or type c, etc.) can be made by maximizing the posterior probability:

$$\text{object}_k = \arg\max P(\text{object}_i \,|\, \text{data}) \qquad i = 1, 2, \ldots, c \tag{8.6}$$

or by maximizing the likelihood, assuming equal *a priori* probability:

$$\text{object}_k = \arg\max P(\text{data} \,|\, \text{object}_i |) \qquad i = 1, 2, \ldots, c \tag{8.7}$$

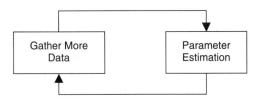

FIGURE 8.5 Applying Bayes' rule in an iterative scheme.

Given this, it is not surprising to see the use of Bayes' rule in sensor fusion problems. It is very fruitful in developing the ideas of sensor fusion. When combining information from multiple sources, one can use not only the features but also the likelihoods arising from experiments, the posterior probability distributions for each object from sensors, prior distributions, and so on. There are situations in which the decisions made by each sensor are different and a collective decision must be reached. To simplify our mathematical representation, let $data_i$ denote the information obtained from the ith sensor, where $i = 1, \ldots, N$. Then the posterior probability for the the ith sensor is $P_i(object_k \mid data_i)$ and the likelihood function is $c_i(object_k) = P_i(data_i \mid object_k)$, respectively. For example, suppose that we review the sensors' predictions in similar situations in the past and find that when the surveyed area contains $object_k$, the ith sensor's prediction (e.g., the confidence value, p), roughly follows the distribution $P(p \mid object_k)$. Using the notation c_i, for example, we can write the likelihood functions $c_i(object_k) = P_i(data_i \mid object_k)$. One can also consider $c_i(object_k)$ to be the sensors' posterior distributions for $object_k$. Having the information from different sensors, how should one combine them? Typical ways of combining different sources of information are the following:

1. *Linear combination.* The overall probability distribution for $object_k$ is obtained by assigning a weight to each sensor. Based on our confidence on each sensor, higher or lower weights, w_i, are assigned to each sensor:

$$c(object_k) = \sum_{i=1}^{N} w_i c_i(object_k)$$

where $w_i \geq 0$, $\sum_{i=1}^{N} w_i = 1$, and N is the number of sensors. For example, the posterior distributions from each sensor are combined as follows:

$$P(object_k \mid data) = P(object_k \mid data_1, data_2, \ldots, data_N)$$

$$= \sum_{i=1}^{N} w_i P_i(object_k \mid data_i) \tag{8.8}$$

A decision is then achieved by using Eq. (8.6) or (8.7). In the case of equal weights, the linear combination can give an erroneous result if one sensor is dissenting even if N is relatively large. This is because the linear combination gives undue credence to the opinion of the ith source.

2. *Geometric combination.* If the information sources seem to be independent, the overall probability distribution for $object_k$ can be obtained by multiplying c_i $(object_k)$'s. In the case, each sensor is given an equal chance to reinforce the opinion:

$$c(object_k) = \alpha \prod_{i=1}^{N} c_i(object_k)$$

where α is an appropriate normalizing constant.

Example 1   Suppose that two sensors are used to survey an area to find a buried object A. Let B denote the case that the surveyed area is blank or contains another object. Let the confidence levels for the first and second sensors be $c_1 = 0.6$ and $c_2 = 0.4$, respectively. Suppose that the sensors provide the probabilities given in the following table:

	Sensor 1	Sensor 2
A	0.6	0.4
B	0.4	0.6

Our confidence in the first sensor can be higher than that for the second based on some observations. Then one can assign the following weights:

	Sensor 1	Sensor 2
w	0.7	0.3

Using the linear combination and geometric combination, we obtain the following, respectively:

$$c(A) = 0.7 \cdot 0.6 + 0.3 \cdot 0.4 = 0.54$$
$$c(B) = 0.7 \cdot 0.4 + 0.3 \cdot 0.6 = 0.46$$

and

$$c(A) = \alpha(0.6 \cdot 0.4) = 0.5$$
$$c(B) = \alpha(0.4 \cdot 0.6) = 0.5$$

where $\alpha = 1/0.48$.

It can be observed immediately that when any of the sensors gives zero probability to object A, the linear combination will be in favor of the opinion of the other sensor. The choice between the two combinations therefore depends on the sensors' characteristics. Not only the parameters (e.g., w_i) used in the combination, but also the combination methods, are an optimization issue.

8.4.3   Dempster–Shafer Theory

Dempster–Shafer (DS) theory is often used as a method for sensor fusion. The theory owes its name to work by A. P. Dempster (1968) and Glenn Shafer (1976). It is also known as the *theory of belief functions*. It is a generalization of the Bayesian theory of subjective probability. This theory is suitable to reason with uncertainty. In a finite discrete space, Dempster–Shafer theory can be interpreted as a generalization of probability theory where probabilities are assigned to sets [5,6]. There are three important functions in Dempster–Shafer theory: the basic probability assignment function (m), the belief function (Bel), and the plausibility function (Pl).

As a primitive of evidence theory, the basic probability assignment (m) defines a mapping of the power set to the interval between 0 and 1. It verifies two axioms [7]:

$$\text{i. } m: 2^X \to [0, 1] \tag{8.9}$$

such that $\mathbf{m(\varnothing)} \, \mathcal{D} \, \mathbf{0}$:

$$\text{ii. } \sum_{A \in 2^X} m(A) = 1 \tag{8.10}$$

where X is the universal set and 2^X is the power set. An obvious aspect of the difference can be seen from the above: Probability distribution functions are defined on X, whereas basic probability assignments are defined on the power set 2^X. Observe also that when $A \subset B$, it is not required that $m(A) \leq m(B)$.

The value of m for a given set A, represented as $m(A)$, expresses available evidence that supports the claim that a particular element of X belongs to the set A. The upper and lower bounds of a confidence interval are determined by the belief confidence and plausibility confidence. Given a basic probability assignment m, a belief measure and a plausibility measure are calculated as follows:

$$\text{Bel}(A) = \sum_{B \subseteq A} m(B)$$

$$\text{Pl}(A) = \sum_{B \cap A \neq \varnothing} m(B)$$

for $A \in 2^X$. Evidence obtained from different sensors can be combined using Dempster's rule of combination. When only two sensors are available, the formula is

$$m_{1,2}(A) = m_1(A) \oplus m_2(A) = \frac{\sum_{B \cap C = A} m_1(B) \cdot m_2(C)}{1 - \sum_{B \cap C = \varnothing} m_1(B) \cdot m_2(C)} \tag{8.11}$$

where $A \neq \varnothing$ and $m_1(\cdot)$ and $m_2(\cdot)$ represent the degree of evidence in favor of the first and second sensors, respectively. It is straightforward to write the formula for more than two sensors [8]:

$$m_1(A) \oplus \cdots \oplus m_N(A) = \frac{\sum_{\cap A_i = A} \prod_{1 \leq j \leq N} m_j(A_i)}{1 - \sum_{\cap A_i = \varnothing} \prod_{1 \leq j \leq N} m_j(A_j)} \tag{8.12}$$

Uncertainty is the amount of uncommitted belief. When there is no uncommitted belief, Dempster's rule reduces to Bayes' rule. To see this, substitute $m_j(A_i) = 0$, where $A_i \neq A$, into Eq. (8.12). For this special case, Eq. (8.12) becomes

$$m_1(A) \oplus \cdots \oplus m_N(A) = \prod_{1 \leq j \leq N} m_j(A) \qquad (8.13)$$

Example 2 Suppose that two sensors are used to survey an area to find a buried object A. Let B denote the case that the surveyed area is blank or contains another object. Each sensor gives only one confidence level, indicating that the surveyed area contains the object A. Let the confidence levels for the first and second sensors be $c_1 = 0.6$ and $c_2 = 0.4$, respectively. For each sensor, this confidence level needs to be mapped onto the probability masses. A simple approach is the following: The unassigned mass is kept constant for each of sensor, and the probability mass assigned to object A is chosen proportional to the confidence level. The uncertainty levels $m_1(A \cup B)$ and can be used as a free parameter to optimize the detection performance. For the sake of simplicity, let $m_1(A \cup B) = 0.3$ and $m_2(A \cup B) = 0.1$. Due to Eq. (8.10), $m_i(A) + m_i(B) + m_i(A \cup B) = 1.0$ for $i = 1, 2$. Then the masses can be determined as follows:

$$m_i(A) = [1 - m_i(A \cup B)] \cdot c_i \qquad \text{for } i = 1, 2$$

In this case, the probability masses for the first sensor are $m_1(A) = 0.7 \cdot 0.6 = 0.42$ and $m_1(B) = 1 - m_1(A) - m_1(A \cup B) = 1 - 0.42 - 0.3 = 0.28$. Similarly, the figures for the second sensor are $m_2(A) = 0.36$ and $m_2(B) = 0.54$. These masses are now fused using Dempster's rule of combination:

$$m_{1,2}(A) = \frac{m_1(A) \cdot m_2(A) + m_1(A) \cdot m_2(A \cup B) + m_1(A \cup B) \cdot m_2(A)}{1 - [m_1(A) \cdot m_2(B) + m_1(B) \cdot m_2(A)]}$$

By substituting the values of the masses into the equation, we obtain the value of the fused mass for the set A:

$$m_{1,2}(A) = \frac{0.42 \cdot 0.36 + 0.42 \cdot 0.1 + 0.3 \cdot 0.36}{1 - (0.42 \cdot 0.54 + 0.28 \cdot 0.36)}$$

$$= 0.45$$

The other masses are found similarly.

$$m_{1,2}(B) = \frac{m_1(B) \cdot m_2(B) + m_1(B) \cdot m_2(A \cup B) + m_1(A \cup B) \cdot m_2(B)}{1 - [m_1(A) \cdot m_2(B) + m_1(B) \cdot m_2(A)]}$$

$$= \frac{0.28 \cdot 0.54 + 0.28 \cdot 0.1 + 0.3 \cdot 0.54}{1 - (0.42 \cdot 0.44 + 0.36 \cdot 0.28)}$$

$$= 0.51$$

$$m_{1,2}(A \cup B) = \frac{m_1(A \cup B) \cdot m_2(A \cup B)}{1 - [m_1(A) \cdot m_2(B) + m_1(B) + m_2(A)]}$$

$$= \frac{0.3 \cdot 0.1}{1 - (0.42 \cdot 0.44 + 0.36 \cdot 0.28)}$$

$$= 0.04$$

The results of the fusion of the two sensors are summarized in Table 8.1. The confidences for the first and second sensors were 0.4 and 0.6, indicating that the surveyed area does not contain any object. It is difficult to make a decision based on these confidence values, as they are equally far from 0.5. On the other hand, the fused masses are in the favor of labeling the surveyed area as a background. This result is due to the mass assigned to $m_1(A \cup B)$ and $m_2(A \cup B)$. The choices of these values are relevant with their ability to distinguish the target from the background. Since $m_2(A \cup B) < m_1(A \cup B)$, the second sensor is more reliable than the first. If they were equal, making the decision based on these masses would be equally difficult. For example, $m_1(A \cup B) = m_1(A \cup B) = 0.1$; then $m_{1,2}((A) = m_{1,2}(B)) = 0.49$.

The discussion above shows that the uncertainty levels play a key role in the decision. Therefore, they should be chosen to optimize the fusion performance on a training data set.

The foregoing approach does not take directly into account the problem of varying accuracy among different sensors. It is often true that the observation accuracies of different sensors are not the same. To approach this problem, a weighted version of Dempster's rule of combination is used. Assuming now that each sensor performs in similar situations, a weight w is determined to indicate its estimation correctness rate historically. This is related strongly to our trust in the sensor's current estimation from its current observation. The generalization of Eq. (8.12) using the weights is

$$m_1(A) \oplus \cdots \oplus m_N(A) = \frac{\sum_{\cap A_i = A} \prod_{1 \le j \le N} w_j \cdot m_j(A_i)}{1 - \sum_{\cap A_i = \varnothing} \prod_{1 \le j \le N} w_j \cdot m_j(A_i)}$$

where $w_i \ge 0$ and $\sum_{i=1}^{N} w_i = 1$.

TABLE 8.1 Mass Assignments for Example 2

Type	Sensor 1 (Mass m_1)	Sensor 2 (Mass m_2)	Fused Masses (Mass $m_{1,2}$)
Target (A)	0.42	0.36	0.45
Background (B)	0.28	0.54	0.51
Unknown (unassigned)	0.30	0.10	0.04
Total mass	1.00	1.00	1.00

8.5 FUZZY INTEGRALS FOR INFORMATION FUSION

Fuzzy integrals are one of the aggregation techniques commonly used for sensor data fusion and decision fusion. Fuzzy integrals integrate a real function with respect to a fuzzy measure. Both Choquet integral and Sugeno fuzzy models are popular and play an important role in multicriteria decision making. In the Sugeno fuzzy measure, identification of these fuzzy measures requires information regarding the fuzzy densities. Sugeno's argument when introducing the terms *fuzzy measures* and *integrals* in his thesis is this: Suppose that you pick an element ω in a set $\mathbf{X}$ but do not know which one. For a subset A of $\mathbf{X}$, you are asked whether $\omega \in A$. You may answer "yes" but are not quite sure. Human beings can express their degrees of trust or *grades of fuzziness* with values on [0,1]. Thus, to each A $\subseteq \mathbf{X}$, a value $\mu(A)$ is assigned, expressing a belief that $\omega \in A$ [7]. $\mu(A)$ is referred to as a grade of fuzziness. Sugeno's approach generalizes probability measures by dropping the additivity property and replacing a weaker one (i.e., monotonicity) [8–10]. Fuzzy measures include some important particular measures, such as plausibility and belief measures, in addition to the probability measure. Some well-known aggregation techniques, such as the weighted arithmetic mean and ordered weighted averaging aggregation operators, are special cases of the Choquet integral.

Next we define fuzzy measures and discuss their interpretations. We also define Sugeno and Choquet integrals for the discrete case.

8.5.1 Fuzzy Measures

Let $\mathbf{X} = \{x_1, \ldots, x_n\}$ be finite. A fuzzy measure, μ, is a real-valued function defined on the power set of $\mathbf{X}$, 2^X, with range [0,1], satisfying the following properties:

1. $\mu(\phi) = 0$ and $\mu(\mathbf{X}) = 1$
2. $\mu(A) \leq \mu(B)$ if $A \subseteq B$

A fuzzy measure can be interpreted as a representation of uncertainty, and it assigns a value to each crisp subset of the universal set signifying the degree of evidence or belief that a particular element belongs in the set. If A and B are disjoint subsets of $\mathbf{X}$, then $\mu(A \cup B) > \mu(A) + \mu(B)$ shows the existence of compatibility between A and B and $\mu(A \cup B) < \mu(A) + \mu(B)$ shows ineffective cooperation of members of $A \cup B$.

Fuzzy Density Function For a fuzzy measure μ, let $\mu^x = \mu(\{x\})$. The mapping $x \rightarrow \mu^x$ is called a *fuzzy density function*. We sometimes write μ^i instead of μ^{x_i} to make the formulas simpler. μ^i is called a *fuzzy density value* and is interpreted as the importance of x_i [11].

Sugeno Measure Let $\lambda \in (-1, \infty)$, $\lambda \neq 0$, be a real number. A fuzzy measure μ_λ is called a λ-additive, λ-fuzzy measure or a Sugeno measure if whenever

$$A \cap B = \varnothing \qquad \text{for all A, B} \subseteq \mathbf{X}$$

then [10]

$$\mu_\lambda(A \cup B) = \mu_\lambda(A) + \mu_\lambda(B) + \lambda\mu_\lambda(A) + \mu_\lambda(B)$$

For a finite set $\mathbf{X}$, the value of λ for any Sugeno fuzzy measure can be determined uniquely. Since $x = \bigcup_{i=1}^{n}\{x_i\}$ and $\mu_\lambda(\mathbf{x}) = 1$, the value of λ can be determined by solving the following equation [12]:

$$1 + \lambda = \prod_{i=1}^{n} 1 + \lambda\mu^i$$

Assigning densities appropriately is crucial for successful application of fuzzy integrals since the Sugeno measure is determined by the densities [11].

Shapley Value The value of μ_i alone does not indicate the importance of the element i. The value $\mu(A \cup x_i) - \mu(A)$ is the contribution of element i when it is added to set A. When there are N elements there are 2^n different coalitions, including the empty set. A particular element can be added to 2^{n-1} different coalitions. Therefore, to express the importance of an element, one must take into account all of these different coalitions [13]. Shapley [14] has proposed a definition of an importance index to define the importance of an element i. The Shapley index for every $x_i \in X$ is defined by

$$v_i := \sum_{k=0}^{n-1} \frac{(n-k-1)!k!}{n!} \sum_{A \subset X\backslash x_i, |A|=k} \mu(A \cup x_i) - \mu(A)$$

The Shapley value of μ is the vector $\mathbf{v}(\mu) := [v_1, \ldots, v_n]$. Note that indicates the contribution of the ith sensor, ith feature, and so on, the Shapley value may be used to reduce the inputs to the fusion system by eliminating those with small values.

Interaction Index The interaction index is an important concept in expressing interaction between elements. The value $\mu(A \cup \{x_i, x_j\}) - \mu(A \cup \{x_i\}) - \mu(A \cup \{x_j\}) + \mu(A)$ represents the interaction between elements x_i and x_j when they are joined to the set A. The interaction index of elements x_i and x_j is defined by taking into all coalitions [15]. It is defined as

$$I_{ij} := \sum_{k=0}^{n-1} \frac{(n-k-2)!k!}{(n-1)!}$$

$$\sum_{A \subset X\backslash\{x_i x_j\}, |A|=k} (\mu(A \cup \{x_i, x_j\}) - \mu(A \cup \{x_i\}) - \mu(A \cup \{x_j\}) + \mu(A))$$

The interaction index I_{ij} can be interpreted as the average added value given by putting x_i and x_j together, all coalitions being considered. If I_{ij} is positive, the interaction is said to be positive. In this case, it is concluded that using the ith and jth information sources together has a value (e.g., fusing them helps to achieve better accuracies).

8.5.2 Fuzzy Integrals

Integration of a real function with respect to a fuzzy measure is achieved via fuzzy integrals. The Sugeno fuzzy integral uses min and max operators for integration. Kruse [10] replaced min and max operators with product-like operators. Weber [16] replaced them by Archimedian t-conorms. These integrals, as well as the Sugeno integral, are not an extension of the Lebesque integral, whereas the fuzzy measure is an extension of the probability measure [17]. The Choquet integral, proposed by Murofushi and Sugeno [17] using the Choquet functional, reduces to the Lebesgue integral when the fuzzy measure is additive. Murofushi and Sugeno later introduced the fuzzy t-conorm integral. The class of fuzzy t-conorm integrals includes Choquet integrals and Sugeno integrals.

Sugeno Fuzzy Integral Let $\mathbf{X}$ be a finite set of size $\mathbf{n}$, $f: \mathbf{X} \to [0,1]$ a function, and $\mu: 2^X \to [0, 1]$ a fuzzy measure. Then the Sugeno fuzzy integral is defined by

$$S_\mu(f) = \overset{n}{\underset{i=1}{\max}}[\min(f(x_{(i)}), \mu(A_{(i)}))]$$

where $(\cdot)$ indicates a permutation on $\mathbf{n}$ such that $x_{(1)} \leq \cdots \leq x_{(n)}$ and $A_{(i)} = \{x_{(i)}, \ldots, x_{(n)}\}$.

Example 3 Let $\mathbf{X} = \{\text{sensor}_1, \text{sensor}_2, \text{sensor}_3\}$. To simplify the notation, let $x_i = \text{sensor}_i$, $i = 1, 2, 3$. Let $f(x_i)$ be the sensor measurement (e.g., confidence level) for sensor$_i$. Then, depending on the order of the sensor outputs, the Sugeno integral results in the fused data shown in Table 8.2. Similarly, x_i can be considered as the classifier$_i$ and $f(x_i)$ can be considered as the classifier output. The Sugeno integral output is then interpreted as the fused decision.

Choquet Integral Let f be a function on $\mathbf{X}$ and μ be a fuzzy measure defined on 2^X. The discrete Choquet integral of a function $f: \mathbf{X} \to \mathfrak{R}^+$ with respect to μ is defined by (Table 8.3)

$$Ch_\mu(f) = \sum_{i=1}^{n} \mu(A_{(i)})[f(x_{(i)}) - f(x_{(i-1)})] \tag{8.14}$$

An equivalent alternative representation of Choquet integral is the following:

$$Ch_\mu(f) = \sum_{i=1}^{n} f(x_{(i)})[\mu(A_{(i)}) - \mu(A_{(i+1)})]$$

TABLE 8.2 Results for Example 3

Possible Cases for Sensor Outputs	Sugeno Integral Output
$f(x_1) < f(x_2) < f(x_3)$	$\max\{\min\{f(x_1), \mu(\{x_1, x_2, x_3\})\}, \min\{f(x_2),$ $\mu(\{x_2, x_3\})\}, \min\{f(x_3), \mu(\{x_3\})\}\}$
$f(x_1) < f(x_3) < f(x_2)$	$\max\{\min\{f(x_1), \mu(\{x_1, x_2, x_3\})\}, \min\{f(x_3),$ $\mu(\{x_2, x_3\})\}, \min\{f(x_2), \mu(\{x_2\})\}\}$
$f(x_2) < f(x_1) < f(x_3)$	$\max\{\min\{f(x_2), \mu(\{x_1, x_2, x_3\})\}, \min\{f(x_1),$ $\mu(\{x_1, x_3\})\}, \min\{f(x_3), \mu(\{x_3\})\}\}$
$f(x_2) < f(x_3) < f(x_1)$	$\max\{\min\{f(x_2), \mu(\{x_1, x_2, x_3\})\}, \min\{f(x_3),$ $\mu(\{x_1, x_3\})\}, \min\{f(x_1), \mu(\{x_1\})\}\}$
$f(x_3) < f(x_1) < f(x_2)$	$\max\{\min\{f(x_3), \mu(\{x_1, x_2, x_3\})\}, \min\{f(x_1),$ $\mu(\{x_1, x_2\})\}, \min\{f(x_2), \mu(\{x_2\})\}\}$
$f(x_3) < f(x_2) < f(x_1)$	$\max\{\min\{f(x_3), \mu(\{x_1, x_2, x_3\})\}, \min\{f(x_2),$ $\mu(\{x_1, x_2\})\}, \min\{f(x_1), \mu(\{x_1\})\}\}$

TABLE 8.3 Choquet Integral Output

Possible Cases for Sensor Outputs	Choquet Integral Output
$f(x_1) < f(x_2) < f(x_3)$	$f(x_1)\mu(\{x_1\}) + f(x_2)\mu(\{x_1, x_2\}) + f(x_3)\mu(\{x_1, x_2, x_3\})$
$f(x_1) < f(x_3) < f(x_2)$	$f(x_1)\mu(\{x_1\}) + f(x_3)\mu(\{x_1, x_3\}) + f(x_2)\mu(\{x_1, x_2, x_3\})$
$f(x_2) < f(x_1) < f(x_3)$	$f(x_2)\mu(\{x_2\}) + f(x_1)\mu(\{x_1, x_2\}) + f(x_3)\mu(\{x_1, x_2, x_3\})$
$f(x_2) < f(x_3) < f(x_1)$	$f(x_2)\mu(\{x_2\}) + f(x_3)\mu(\{x_2, x_3\}) + f(x_1)\mu(\{x_1, x_2, x_3\})$
$f(x_3) < f(x_1) < f(x_2)$	$f(x_3)\mu(\{x_3\}) + f(x_1)\mu(\{x_1, x_3\}) + f(x_2)\mu(\{x_1, x_2, x_3\})$
$f(x_3) < f(x_2) < f(x_1)$	$f(x_3)\mu(\{x_3\}) + f(x_2)\mu(\{x_2, x_3\}) + f(x_1)\mu(\{x_1, x_2, x_3\})$

8.5.3 Representation of Some Traditional Aggregation Operations by the Choquet Integral

In this section we show that the Choquet integral generalizes some of the traditional aggregation operators.

Weighted Averaging   All attributes for the weighted averaging are noninteractive. Hence, their weighted effects are viewed as additive: for example, $\mu(A \cup B) = \mu(A) + \mu(B)$ whenever $A \cap B = \emptyset$. Then the Choquet integral defines a linear filter with weights $\mu(\{x_1\}), \mu(\{x_2\}), \ldots, \mu(\{x_n\})$ [18]:

$$\text{Ch}(f)(x) = \sum_{i=1}^{n} f(x_{(i)})\mu(A_{(i)}) - \mu(A_{(i+1)})$$

$$= \sum_{i=1}^{n} f(x_{(i)})\mu(x_{(i)})$$

$$= \sum_{i=1}^{n} f(x_i)\mu(x_i)$$

Note that $\sum_{i=1}^{n} \mu(x_i) = 1$ since $\mu(X) = \sum_{i=1}^{n} \mu(x_i) = 1$.

Order Statistics Filters   When the fuzzy measure μ is chosen to have the property

$$\mu(A) = \begin{cases} 0 & \text{if } |A| \leq n - k \\ 1 & \text{else} \end{cases}$$

the Choquet filter defines the kth-order statistics filter:

$$\mathrm{Ch}(f)(x) = \sum_{i=1}^{n} f(x_{(i)})[\mu(A_{(i)}) - \mu(A_{(i+1)})]$$

$$= x_{(k)}$$

Linear Combinations of Order Filters   The Choquet integral with fuzzy measure having the property that

$$\mu(A_{(i)}) - \mu(A_{(i+1)}) = w_t$$

for any f defines a linear combination of order statistics filters:

$$\mathrm{Ch}(f)(x) = \sum_{i=1}^{n} f(x_{(i)})[\mu(A_{(i)}) - \mu(A_{(i+1)})]$$

$$= \sum_{i=1}^{n} f(x_{(i)})w_i$$

Note that for this case the fuzzy measure depends only on the set cardinalities [e.g., $\mu(A) = \mu(B)$ whenever $|A| = |B|$].

8.6   ARTIFICIAL NEURAL NETWORKS

An artificial neural network (ANN), often just called a *neural network* (NN), is a computational model that is inspired by the way in which biological nervous systems such as the brain process information. Neural networks have remarkable ability to derive meaning from complicated or imprecise data. They are suitable for information fusion and feature extraction. The basic elements of artificial neural networks are perceptrons. A nonlinear activation function provides nonlinearity to the network. To make it more suitable for certain applications or to achieve better generalization, researchers modified the classical architecture. One of the attempts was to replace the conventional perceptron. Ritter and Beavers [19] introduced morphological perceptrons. Pessoa and Maragos [20] used a morphological–rank–linear (MRL) filter as the fundamental processing unit. Other attempts include using a network of linear correlational filters followed by a classical feedforward network. Won et al. [21] replaced correlational filters by morphological hit-and-miss transform. This architecture, called *morphological shared weight networks* (MSNNs), is discussed further later. It was later improved by introducing an architecture called *Choquet integral–based MSNN* (CMSNN) [22]. In this section we assume that the reader is familiar with basic knowledge of ANNs (a good reference is the text by Haykin [23]).

8.6.1 Choquet Integral–Based Morphological Shared-Weight Neural Networks

Shared-weight neural networks are feedforward networks with two layers. The first layer, the *feature extraction layer*, is used to learn features. The second layer is for classification based on the features extracted in the first layer. This architecture forms higher-order features from local features. In a standard shared-weight network, these local features are extracted using learned convolution kernels. On the other hand, in morphological shared-weight networks, these features are extracted using gray-scale morphological hit-and-miss transforms [24].

Gader et al. investigated the generalization aspect of shared-weight networks and concluded that a better generalization can be achieved using regularization [25]. Shared-weight networks were applied in many different areas, such as to discriminate textures and in handwritten character recognition problems. Shared-weight networks use linear correlational filters for feature extraction. Won et al. [21] replaced linear correlational filters in a standard shared-weight network with a morphological hit-and-miss transform. This architecture is called a *morphological shared-weight neural network* (MSNN). Since this transform uses min and max, it is not robust. Besides, the features that we can extract using a morphological hit-and-miss transform are limited when other morphological transforms, such as tophat and morphological gradient, are considered. Hocaoglu [22] replaced the linear correlational filters in shared-weight neural networks with differences between a pair of generalized Choquet integral filters. In this architecture the generalized Choquet integrals filters are not required to share the same domain. Thus, in addition to the morphological hit-and-miss transform, they have the ability to perform tophat and morphological gradient operations. To use a CMSNN to fuse features, one combines the features due to different sensors and/or algorithms to form a composite vector and use this vector as input to the CMSNN. To develop intuition about the use of a CMSNN in fusing information, we provide examples using its fundamental processing unit, a Choquet integral–based morphological operators neural node (CMONN). In relatively complex problems, CMSNNs are preferred over CMONNS.

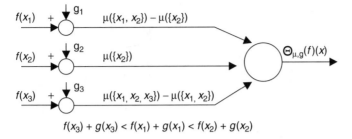

FIGURE 8.6 Choquet integral–based mathematical morphology.

The Generalized Choquet Integral The generalized Choquet integral is defined as follows:

$$\Theta_{\mu,g}(f) = \sum_{i=1}^{n} [\mu(A_{(i)}) - \mu(A_{(i+1)})] f_g(x_{\sigma(i)})$$

$$= \mathbf{e}^{\mathrm{T}} \mathbf{f}_g \qquad (8.15)$$

where $f_g(x_i) = f(x_i) + g(x_i)$:

$$f_g(x_{\sigma(1)}) \le f_g(x_{\sigma(n)}) \le \cdots \le f_g(x_{\sigma(n)})$$

$$A_{(i)} = \{x_{\sigma(i)}, \ldots, x_{\sigma(n)}\}$$

$$A_{(n+1)} = \varnothing$$

$$\mathbf{f}_g = \mathbf{f} + \mathbf{g} = \begin{bmatrix} f(x_1) + g(x_1) \\ f(x_2) + g(x_2) \\ \vdots \\ f(x_n) + g(x_n) \end{bmatrix}$$

$$\mathbf{e} = \begin{bmatrix} \mu(A_{\sigma^{-1}(1)}) - \mu(A_{(\sigma^{-1}(1)+1)}) \\ \mu(A_{\sigma^{-1}(2)}) - \mu(A_{(\sigma^{-1}(2)+1)}) \\ \vdots \\ \mu(A_{\sigma^{-1}(n)}) - \mu(A_{(\sigma^{-1}(n)+1)}) \end{bmatrix}$$

This operation is illustrated for three variables in Fig. 8.6.

Choquet Integral–Based Morphological Operators Neural Node Equation (8.15) can also be written as

$$\Theta_{\mu,g}(f) = f_g(x_{(1)}) + \sum_{i=2}^{n} [f_g(x_{\sigma(i)}) - f_g(x_{\sigma(i-1)})] \mu(A_{(i)})$$

$$= f_g(x_{(1)}) + \mathbf{v}^{\mathrm{T}} \mathbf{u} \qquad (8.16)$$

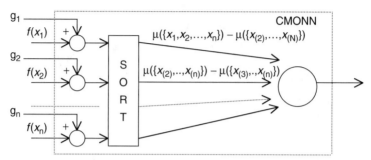

FIGURE 8.7 Neural node representation of Choquet integral–based morphological operators (CMONN).

where the vector $\mathbf{u}$ contains all the $2^n - 2$ values of the measure μ except $\mu(\mathbf{X})$ and $\mu(\varnothing)$; that is,

$$\mathbf{u} = [\mu(\{x_1\}) \quad \mu(\{x_2\}) \quad \cdots \quad \mu(\{x_n\}) \quad \mu(\{x_1, x_2\}) \quad \cdots \quad \mu(\{x_1, x_2, \ldots, x_n\})]^{\mathrm{T}}$$

and the vector $\mathbf{v}$ includes zeros and the differences of $f_g(x_i)$'s. This operation can be viewed as a neural node, as shown in Fig. 8.7. It is a Choquet Integral–based morphological operators neural node (CMONN).

CMSNN Architecture The MSNN architecture uses a hit-and-miss transform for feature extraction. The CMSNN architecture uses a pair of CMONNs in place of the hit-and-miss transform, as shown in Fig. 8.8. The purpose of this replacement is twofold. One is to obtain a robust feature extractor. The other is to obtain new features.

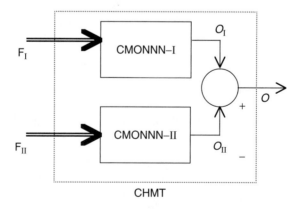

FIGURE 8.8 Choquet integral–based morphological hit-and-miss transform (CHMT).

8.6.2 Using CMONN for Sensor Fusion

CMONN is suitable for data and decision-level fusion. For data-level fusion, the inputs, $f(x_i)$, to the CMONN are simply the associated sensors' data; for decision-level fusion, the inputs are the decisions due to classifier outputs of each sensor. These inputs serve as a set of information sources. In the case of decision-level fusion, these sources are fused by CMONN to assign final a value indicating the confidence level at candidate locations. In the other case, it indicates the fused raw data of the sensors. As pointed out in Section 8.3.2, the information source to be fused can even be the likelihoods, the posterior probability distributions, and so on. Three fusion algorithms have been compared by Gader et al. [26] for the landmine detection problem. The first algorithm, CHS, uses Sugeno fuzzy measures, which were determined by trial and error with a variety of subsets of features. The second algorithm, COF, uses full measures (i.e., 32 measures for five features to be fused). The third algorithm, MOCOLOS, uses CMONN with linear order statistics (LOS) measures [i.e., $\mu(A) = \mu(B)$ if sets A and B have an equal number of elements]. In other words, the third algorithm is a special case of CMONN.

Parameters of standard Choquet fuzzy integrals are determined by a training algorithm developed by Grabisch and Nicolas [27], and also by a heuristic method using Sugeno measures [28]. CMONNs are trained by an alternating optimization method using a constrained set of parameters [22]. The comparison is made in terms of the probability of detection vs. the false-alarm rate. For similar probabilities of detection, the MCOLOS attains lower false-alarm rates than those attained by the other methods (Fig. 8.9).

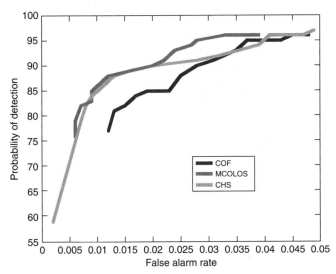

FIGURE 8.9 Plot of probability of detection vs. false-alarm rates for a Choquet integral with respect to optimal full measures (COF), morphological Choquet integral with respect to optimal LOS measures (MCOLOS), and Choquet integral with respect to a heuristically developed Sugeno measure (CHS).

8.6.3   Using CMONN for Reducing Redundant Information Sources

An important problem with using fuzzy integrals for fusion is the difficulty in determining the importance of each information source. It is even more difficult since the importance of each information source depends on the intrinsic properties and also on its rank for any given input. CMONN with LOS measures is an attractive tool for determine which information sources are less important (e.g., which sensors can be eliminated without reducing system performance). The structuring element, **g**, is the key to identifying important features.

When we are given a measure, it is not always easy to decide which information sources are more important than the others. An increasing number of information sources makes this problem more difficult harder since the number of parameters necessary to define a full measure increases exponentially. LOS measures are helpful in reducing the number of parameters. To see this, let us consider a fusion problem with three inputs, f_1, f_2, and f_3, as shown in Table 8.4. The last column shows the desired outputs, d. A simple solution is to threshold the first input by 0.5. This indicates that the other two features, f_2 and f_3, are redundant. Hence, the sensors providing these features may be eliminated.

A CMONN with LOS measures reduces Eq. (8.16) to

$$\Theta_{\mu,g}(f) = \sum_{i=1}^{n} f_g(x_{\sigma(i)}) w_i$$

where w is the LOS measures. The values of the parameters are optimized to increase system performance. For this small set of data, we would rather select the values for the sake of discussion. Using the weights $\mathbf{w} = (0, 0, 1)$, CMONN outputs the largest of $f_g(x_{\sigma(i)})$ for $i = 1, 2, 3$. If $\mathbf{g}$ is set to $(0.0, -1.0, -1.0)$, inputs f_2 and f_3 may not contribute to the output as $f_2 + g_2$, and $f_3 + g_3$ cannot be larger than $f_1 + g_1$. The output is wholly determined by the first input, f_1. The other two inputs are redundant. As this simple example shows, the structuring element may point out the sensors or features to be eliminated. This approach is also used to determine the domain of filters in CMSNN, to improve the performance, and to reduce the computation time [22].

TABLE 8.4 Classification Problem with Three Inputs, f_1, f_2, and f_3, and a Desired Output (d)

f_1	f_2	f_3	d
0.95	0.75	0.45	1
0.90	0.85	0.35	1
0.75	0.25	0.80	1
0.65	0.85	0.75	1
0.45	0.95	0.75	0
0.35	0.85	0.90	0
0.25	0.75	0.80	0
0.15	0.05	0.35	0

8.7 SUMMARY

Multisensor data fusion seeks to combine information from multiple sensors and sources to achieve inferences that are not feasible from a single sensor or source. Fusing sensors together can achieve several goals:

- *Improve confidence in the sensor readings.* One sensor may be fooled at a certain situation, but the others may not.
- *Improve performance.* Each sensor may provide diverse physical properties.
- *Improve reliability.* One sensor may fail, but the rest of the system will keep operating.

Fusion of sensor measurements, features, and decisions can be performed at a variety of levels: the signal level (data), evidence level (features), and dynamics level (decision). In addition to measure-to-measurement, feature-to-feature, and decision-to-decision fusion, a hybrid processing architecture may also be used to fuse data and features and for other purposes. Numerous techniques are available for the low-level data fusion process. Some conventional techniques, such as Bayesian inference and Dempster–Shafer evidential theory, were introduced. The use of fuzzy integrals and artificial neural networks, with special emphasis on the type of shared-weight networks, is discussed. These are relatively new techniques. Clustering and other pattern recognition techniques are not covered but are left to the reader to explore.

REFERENCES

1. Kessler, J., et al., 1992. *Functional Description of the Data Fusion Process*. Technical Report. Office of Naval Technology, Naval Air Development Center, Warminster, PA.

2. Hall, D. L., Llinas, J. 1997. An introduction to multisensor data fusion. *Proc. IEEE*, 85:6–23.

3. Luo, R. C., Kay, M. G. 1989. Multisensor integration and fusion in intelligent systems. *IEEE Trans. Syst. Man Cybern.*, 19(5):901–931.

4. Bolstad, W. M. 2007. *Introduction to Bayesian Statistics*. Wiley, Hoboken, NJ.

5. Dempster, A. 1967. Upper and lower probabilities induced by a multivalued mapping. *Ann. Math. Stat.*, 38(2):325–339.

6. Shafer, G. 1976. *A Mathematical Theory of Evidence*. Princeton University Press, Princeton, NJ.

7. Klir, G. J., Folger, T. A. 1988. *Fuzzy Sets, Uncertainty and Information*. Prentice Hall, Englewood Cliffs, NJ.

8. Barnett, J. A. 1985. Computational methods for a mathematical theory of evidence. In *Proceedings of IJCAI'81*, Vancouver, BC, Canada.

9. Grabisch, M., Nguyen, H. T., Walker, E. A. 1995. *Fundamentals of Uncertainty Calculi with Applications to Fuzzy Inference*. Kluwer Academic, Dordrecht, The Netherlands.

10. Kruse, R. 1983. Fuzzy integrals and conditional fuzzy measures. *Fuzzy Sets Syst.*, pp. 309–313.

11. Keller, J. M., Gader, P. D., Hocaoglu, A. K. 2000. Fuzzy integrals in image processing and recognition. In *Fuzzy Measures and Integrals*, M. Grabisch, T. Murofushi, M. Sugeno, Eds. Springer-Verlag, New York.

12. Keller, J. M., Osborn, J. 1996. Training fuzzy integral. *Int. J. Approx. Reason.*, pp. 1–23.

13. Grabisch, M. 1997. k-order additive discrete fuzzy measures and their representation. *Fuzzy Sets Syst.*, 92:167–189.

14. Shapley, L. S. 1953. A value of n-person games. In *Contributions to the Theory of Games*, Vol. II. Annals of Mathematics Studies, Vol. 28. Princeton University Press, Princeton, NJ, pp. 307–317.

15. Grabisch, M. 1996. The representation of importance and interaction of features by fuzzy measures. *Pattern Recognition Lett.*, 17(6):567–575.

16. Weber, S. 1984. Measures of fuzzy sets and measures of fuzziness. *Fuzzy Sets Syst.*, 13:247–271.

17. Murofushi, T., Sugeno, M. 1989. An interpretation of fuzzy measure and the Choquet integral as an integral with respect to a fuzzy measure. *Fuzzy Sets Syst.*, 29(5):201–227.

18. Grabisch, M. 1994. Fuzzy integrals as a generalized class of order filters. *Proc. SPIE*, 2315:128–136.

19. Ritter, G. X., Beavers, T. W. 1999. Morphological perceptrons. In *Proceedings of INNS and IEEE/NNC 1999 Joint Conference on Neural Network*, Washington, DC, July.

20. Pessoa, L. F. C., Maragos, P. 2000. Neural networks with hybrid morphological/rank/linear nodes: a unifying framework with applications to handwritten character recognition. *Pattern Recognition*, 33:945–960.

21. Won, Y., Gader, P. D., Coffield, P. C. 1997. Morphological shared weight networks with applications to automatic target recognition. *IEEE Trans. Neural Networks*, 8(5):1195–1997.

22. Hocaoglu, A. K. 2000. Choquet Integral Based–Morphological Operators with Applications to Object Detection and Information Fusion. Doctoral dissertation, University of Missouri–Columbia.

23. Haykin, S. 1998. *Neural Networks: A Comprehensive Foundation*. Prentice Hall, Upper Saddle River, NJ.

24. Won, Y. 1995. Non-linear Correlation Filter and Morphology Neural Networks for Image Pattern and Automatic Target Recognition. Doctoral dissertation, University of Missouri–Columbia.

25. Gader, P. D., Miramonti, J. R., Won, Y., Coffield, P. 1995. Segmentation free shared weight networks for automatic vehicle detection. *IEEE Neural Network*, 8(9):1457–1473.

26. Gader, P. D., Nelson, B. N., Hocaoglu, A. K. Auephanwiriyakul, S., Khabou, M. A. 2000. Neural versus heuristic development of Choquet fuzzy integral fusion algorithms for land mine detection. In *Neuro-fuzzy Pattern Recognition*, H. Bunke, A. Kandel, Eds. World Scientific, pp. 205–226, Hackensack, NJ.

27. Grabisch, M., Nicolas, J. M. 1994. Classification by fuzzy integral: performance and tests. *Fuzzy Sets Syst.*, 65:255–271.

28. Keller, J. M., Gader, P. D., Tahani, H., Chiang, J. H., Mohamed, M. 1994. Advances in fuzzy integration for pattern recognition. *Fuzzy Sets Syst.*, 65:273–283.

CHAPTER NINE

Geophysical Applications

9.1 INTRODUCTION

Every measurement assumes an underlying theoretical model. Without a theoretical model, a measurement result cannot be given meaning. The use of subsurface sensors and sensing technologies therefore requires understanding how the subsurface responds to the fields that we use to obtain subsurface properties in a nondestructive way. Although nonlinear geophysics exists, most applications use field strengths that cause small enough disturbances relative to an equilibrium state to which the subsurface responds linearly. Second, we assume that the subsurface does not change in time over the duration of a single measurement. Because of these two conditions, subsurface geophysical applications are investigations on a linear, time-invariant (LTI) system. Hence, all measurements can be understood from linear system theory and all operations that we apply are filters and can be understood from filter theory. As a direct consequence of the LTI system condition, the field strengths are linearly proportional to the applied source strengths, while medium property functions are independent of the applied fields and sources. They may depend on time relative to a reference time instant, usually chosen as the time at which the source is switched on or off. This allows for modeling all kinds of time-relaxation phenomena, which can be formulated mathematically by writing the medium property that shows time relaxation as a time-convolution operator. This notion goes back to Boltzmann [1]. A second direct consequence of the LTI system condition is that all time interactions are described by convolutions and the time axis can be transformed conveniently to the frequency domain. The advantage of such transformation is that time convolutions of two time-dependent functions transform to products of these frequency-dependent functions in the frequency

Subsurface Sensing, First Edition. Edited by Ahmet S. Turk, A. Koksal Hocaoglu, and Alexey A. Vertiy.
© 2011 John Wiley & Sons, Inc. Published 2011 by John Wiley & Sons, Inc.

domain, while the time-derivative operator is reduced to an algebraic factor. If a medium property function is a constant, the medium is said to be *homogeneous*, if it varies as a function of position in space, it is said to be *heterogeneous* or *inhomogeneous*. A homogeneous medium is shift-invariant, meaning that only the distance between two points is relevant; absolute positions are not. For such media a spatial Fourier transformation can be carried out with benefits similar to those for time-Fourier transformation. If a medium property function does not depend on orientation, the medium is said to be *isotropic*, whereas an *anisotropic* medium has a property function that is orientation dependent. For an anisotropic medium the medium property function is a tensor of rank 2 or higher, which depends on the geophysical method used. In this chapter we are interested primarily in electromagnetic soil properties, which are discussed in Section 9.2. The reason we restrict ourselves is that the remaining sections deal mainly with electric and electromagnetic methods for applications in hydrogeophysics in Section 9.3, contaminant remediation in Section 9.4, agricultural geophysics in Section 9.5, and archaeology and cultural heritage in Section 9.6.

9.2   ELECTROMAGNETIC PROPERTIES OF SOILS

For all geophysical applications measurements take place on a macroscopic scale. The measurements are carried out to obtain structural and/or property information about the subsurface. The scale at which this information can be obtained is called *resolution*. If the subsurface is thought to consist of blocks, the smallest box that can be used sets the resolution scale. The smallest box is represented by a constant medium parameter value and is called homogeneous.

For very high resolution surveys, this scale may relate to volumes of 1 cm^3 as the largest volume that must be considered homogeneous. For very low resolution surveys, this scale may relate to volumes up to 100 m^3. Generally, we try to work in resolution ranges where the soil grain is very small compared to the resolution scale. If the soil grain matches the size of the resolution scale, the dominant feature in the measurements is scattering. This results in data that are difficult to interpret.

The smallest volume that we must use as a homogeneous block to represent a piece of soil is then represented by some kind of average or effective value for that true volume of soil that exhibits heterogeneity at scales much smaller than the volume. How these small-scale heterogeneities result in an operational macroscopic parameter is an important area of research. Several theories have been proposed and are still under development. These theories must be validated using controlled laboratory measurements and then find applications in the field. Apart from this line of research, empirical relations exist that are usually based on a large number of measurements taken in the field. Empirical relations are often site specific, which means that these can be used after calibrating measurements in a field before carrying out geophysical surveys. All these results should tell us how we can understand our field measurements and how we can discretize our models for numerical simulations.

9.2.1 Effective Medium Models

If a macroscopic volume can be considered as homogeneous and if its constituents are isotropic, its effective medium property can be estimated in the static limit. Based on energy relations, Hashin and Shtrikman [2] found bounds for the elastic behavior of polycrystals and the magnetic permeability of multiphase materials. These bounds are known as the *Hashin–Shtrikman bounds*. For an N-component mixture with medium parameters given by $\chi_m, m = 1, \ldots, N$, sorted in value such that for $m = 1$ it attains the smallest value and for $m = N$ it attains the largest value, the minimum and maximum effective parameter values are given by

$$\chi_{min} = \chi_1 \left[1 + 3 \frac{\sum_{m=2}^{N} f_m/(\chi_m + 2\chi_1)}{1 - \sum_{m=2}^{N} f_m/(\chi_m + 2\chi_1)} \right] \tag{9.1}$$

$$\chi_{max} = \chi_N \left[1 + 3 \frac{\sum_{m=1}^{N-1} f_m/(\chi_m + 2\chi_N)}{1 - \sum_{m=1}^{N-1} f_m/(\chi_m + 2\chi_N)} \right] \tag{9.2}$$

These bounds can be used for χ being the complex electric permittivity $\hat{\varepsilon}$, the complex magnetic permeability $\hat{\mu}$, the complex elastic Lamé coefficients $\hat{\lambda}$ and $\hat{\upsilon}$, and other parameters that are used in linear time-invariant systems. A circumflex on a parameters indicate that it is a complex-valued function in the frequency domain. It is interesting to note that these bounds conform to the Clausius–Mossotti formula [3,4]. Both Mossotti and Clausius started a systematic investigation to correlate the macroscopic dielectric constant with the microscopic structure of materials, and they considered the dielectric to be composed of conducting spheres in a nonconductive medium. The formula also became known as the *Lorenz–Lorentz formula* [5], because the Dutch scientist Lorentz and the Danish scientist Lorenz derived independently the same formula for effective electric permittivity in the optical regime. Equations (9.1) and (9.2) can be regarded as the *Maxwell–Wagner mixing rule* when either medium 1 or medium N is regarded as the host and the other constituents as the inclusions, and it is also known as the *Rayleigh mixing formula*. In this sense these are dilute approximations for small volume fractions of the inclusions. This is equivalent to neglecting interactions between the inclusions.

Including Interaction Between Inclusions   Bruggeman attempted to include the effect of geometry in the computation of an effective parameter by considering the effective medium itself as the background medium [6]. Requiring that the medium is not changed when a spherical domain from the soil is replaced by a homogeneous sphere with the effective property value leads to

$$\sum_{m=1}^{N} f_m \frac{\chi_m - \chi_{eff}}{\chi_m + 2\chi_{eff}} = 0 \tag{9.3}$$

This formula is supposed to be valid for larger volume fractions of the constituents. In electromagnetic rock physics the formula is also known as the *Böttcher equation*

[7]. It is also known by number of other names: *effective medium approxima-tion* [8], *quasicrystalline approximation* or *coherent potential approach* [9], or the *Polder–van Santen model* [10], although the latter authors derived a different form that is equal to Eq. (9.3). A nonsymmetrical formula was also derived for two component mixtures by Bruggeman:

$$f_2 = \frac{\chi_1 - \chi_{\text{eff}}}{\chi_1 - \chi_2}\left(\frac{\chi_2}{\chi_{\text{eff}}}\right)^\alpha \tag{9.4}$$

known as the *BHS model* [11,12]. Bruggeman used $\alpha = 1/3$ for spherical inclu-sions. A complete analysis based on this method was given by Giardano [13], who developed directional-dependent depolarization factors α_j, $j = x, y, z$, for gener-ally oriented general ellipsoidal particles. Later, similar rules based on this rule were developed to accommodate more than two components in a mixture [14].

Power-Law Models A general model known as the *power-law model* can be written

$$\chi_{\text{eff}}^\alpha = \sum_{m=1}^{N} f_m \chi_m^\alpha \tag{9.5}$$

When applied to permittivity and assuming the magnetic permeability to be constant and equal to its free-space value, this formula represents a weighted average of the wave slowness of each constituent when $\alpha = 0.5$, known as the *complex refractive index method* (CRIM) [15] and a weighted average of velocities of each constituent when $\alpha = -0.5$. When $\alpha = 1/3$, it is known as the *Looyenga model* [16]. A special situation is obtained if we take the limit for $\alpha = 0$, known as *Lichtenecker's model*, although he introduced it as an empirical model:

$$\log(\chi_{\text{eff}}) = \sum_{m=1}^{N} f_m \log(\chi_m) \longrightarrow \chi_{\text{eff}} = \prod_{m=1}^{N} \chi_m^{f_m} \tag{9.6}$$

Two other choices for α yield models that represent anisotropy due to small-scale layering. If the field is perpendicular to the layering, the harmonic average is taken as

$$\chi_{\text{eff}}^{-1} = \sum_{m=1}^{N} f_m \chi_m^{-1} \tag{9.7}$$

and this is equivalent to replacing a lumped circuit of capacitors in series by its equivalent capacitor. Similarly, when the field is parallel to the fine layering, the arithmetic average is written

$$\chi_{\text{eff}} = \sum_{m=1}^{N} f_m \chi_m \tag{9.8}$$

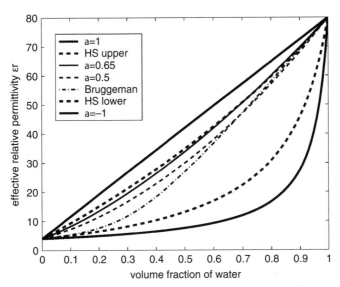

FIGURE 9.1   Example of a fully water-saturated porous medium with the possible porosity range taken to the extreme, between fully nonporous and pure water. The upper and lower bounds are the Hashin–Shtrikman bounds.

which represents the equivalent capacitor for a lumped circuit of parallel capacitors. This model is also generally used for mass density. In electromagnetic rock physics these models are known as *Wiener bounds* [17]. In elastodynamic rock physics they are known as *Reuss's* and *Voigt's bounds* [18,19]. The mechanical interpretation is that the Reuss bound gives the smallest value because it is the ratio of average stress to average strain, assuming that all constituents are under the same stress, while the Voigt bound gives the largest value because it is the ratio of average stress to average strain assuming that all constituents are under the same strain. The models discussed so far lead to the same values near the end members for two component mixtures. This is illustrated in Fig. 9.1, where the models are shown for a composite material involving water and an unconsolidated sand when the volume fraction of sand is $f_1 = 1 - \varphi$ and of water is $f_2 = \varphi(\varphi)$ is porosity. Especially in the interesting range of porosity values from 0.1 to 0.4, electric permittivity results depend strongly on the model used. Excellent review books exist on electromagnetic mixing models and effective medium theories [20,21], on seismic properties [22], and on general rock physics [23].

Behavior of Polar Liquids   In all these models it is important to know the medium parameters of the constituents. The electric parameters are taken as an example. Many reference sources exist of standard values (i.e., internationally accepted values obtained by measurements) [24–26]. For most solid earth materials in soils and soft rocks the electric permittivity is virtually constant from dc to the gigahertz range, which is the range where electromagnetic methods are used in geophysical applications. The relative electric permittivity values of these solids generally vary between 2 and 10. Gases are taken as free space and are given the electric permittivity value of free space, except for water vapor (steam) at high temperatures and pressures, where the static electric permittivity is between 1 and 3 [27]. Liquids that have electric dipole moments are known as *polar liquids*, usually have high relative permittivity values, between 10 and 100, which depend strongly on temperature [28] and on dissolved solids [29]. Water and ethanol are two well-known examples of polar liquids. Distilled water has a dc relative electric permittivity of $\varepsilon_s = 79.9$ at a temperature of $T = 20.6°C$ [30]. The water had a measured dc conductivity of $\sigma_s = 0.6 \pm 1$ μS/m. As a function of frequency, Debye [31] developed a model for electric behavior of distilled water given by

$$\hat{\varepsilon}_r = \varepsilon_\infty + \frac{\varepsilon_s - \varepsilon_\infty}{1 + jf\tau_r} - j\frac{\sigma_s}{2\pi f\varepsilon_0} \tag{9.9}$$

where $\hat{\varepsilon}_r$ denotes the frequency-dependent complex relative permittivity of water, j is the imaginary unit: $j^2 = -1$, f the frequency (hertz), $\tau_r = 1/f_r$ the characteristic relaxation time (f_r being the relaxation frequency) of distilled water (seconds), ε_∞ the relative permittivity in the limit $f \to \infty$, and ε_0 the free-space electric permittivity, defined as $\varepsilon_0 = 1/\mu c_0^2 (1/\Omega \cdot s/m)$, $\mu_0 = 4\pi \times 10^{-7} \Omega \cdot s/m$ being the free-space magnetic permeability and $c_0 = 299792458$ m/s the defined free-space wave velocity. Hasted found that $\varepsilon_\infty = 4.22$ and $f_r = 17$ GHz.

Generalized Models Based on Polar Liquids   This model can be generalized by a Cole–Cole model [32] as

$$\hat{\varepsilon}_r = \varepsilon_\infty + \frac{\varepsilon_s - \varepsilon_\infty}{1 + (jf\tau_r)^\alpha} - j\frac{\sigma_s}{2\pi f\varepsilon_0} \tag{9.10}$$

in which the effect of relaxation is modified by an exponential factor α, $0 < \alpha \le 1$. The deviation of the power of the frequency times the relaxation time is to allow different relations between the real and imaginary parts as a function of frequency. As an example, Fig. 9.2 shows the complex relative permittivity of water assuming zero dc conductivity and Hasted's values for the end values for relative permittivity and relaxation frequency for frequency values between 0 Hz and 100 GHz. At zero frequency all curves start at the dc value of 80, but their slopes deviate as a function of α. With decreasing α the imaginary part of permittivity also decreases, while the real value remains larger. These relations all satisfy Kramers–Kronig causality relations and are therefore valid model for physical response systems. The apex of each curve occurs at the same value of the real part of permittivity, given by $\varepsilon_{r,\text{apex}} = (\varepsilon_s + \varepsilon_\infty)/2$. The frequency where this occurs, called the *characteristic frequency*, is independent of α. In this example the characteristic frequency is $f_{\text{char}} = 2.8$ GHz.

A similar model for electric impedance was derived by Cole in 1928 [33], and a modified version for resistivity is given by

$$\hat{\rho} = \rho_\infty + \frac{\rho_0 \rho_1}{\rho_1 + \rho_0 (jf\tau_r)^\alpha} \tag{9.11}$$

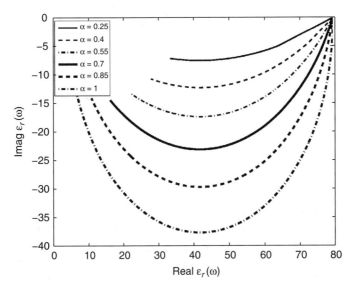

FIGURE 9.2   Argand or Wessel plot of the imaginary part of the relative permittivity as a function of its real part. Each curve represents this functional relationship as a function of frequency for a fixed value of α.

which is equal to the Cole model when $\rho_0 = \rho_1$, but the general model allows for a relaxation time that depends on the resistivities that is reported to be validated by experiments [34]. This is in accordance with relaxation theory, which interprets the time constant as the time necessary to discharge a lumped circuit of resistors and capacitors so that a large capacitance and resistance results in long time constants. The equivalent model of Eq. (9.11) for conductivity is given by

$$\hat{\sigma} = \sigma_s + \frac{\sigma_0 \sigma_1}{\sigma_1 + \sigma_0 (jf\tau_r)^{-\alpha}} \tag{9.12}$$

The relaxation time τ_r can be replaced by a different resitivity dependent relaxation time given by

$$\tau_{rm} = \tau_r \left(\frac{\rho_0}{\rho_1}\right)^{1/\alpha} \tag{9.13}$$

which is a necessary condition in time-lapse monitoring studies, where it is expected that the resistivity changes over time, because of which the relaxation time should also change. This then allows the apex of the curves in an Argand plot to move as a function changing the effective resistivity.

Maxwell–Wagner Effect   In the literature another polarization mechanism is explained. This mechanism occurs on a larger scale than the previous mechanisms. The polarization is contrary to the other mechanisms, a phenomenon that

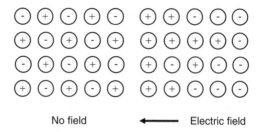

No field   ◄—————— Electric field

FIGURE 9.3   Principle of the Maxwell–Wagner effect.

takes place in heterogeneous materials. The medium is considered as nonconducting but contains mesoscale structures of isolated inclusions that can be conducting. When the material is exposed to an electric field, the inclusions act as polarizable macromolecules and the charges rearrange and create macroscopic polarization. This mechanism is known as *space-charge polarization* or the *Maxwell–Wagner effect*. This phenomenon occurs in liquids or in a porous medium saturated with liquids, as illustrated in Fig. 9.3.

Empirical Relationships   It is interesting to note that despite all the research into this subject, there is little agreement among researchers as to which is the best model. It seems to depend on many variables, which is a reason for many other scientists to find empirical relations that they can work with. For applications where water is the important parameter, the fact that water has a much higher electric permittivity value than that of other earth materials can be used. One such model that is in common use is the *Topp equation* [35]. Topp et al. tested four mineral soils with a range of texture from sandy loam to clay and presented an empirical relationship between the apparent dielectric constant and the volumetric water content, which was reported to be independent of soil type, soil density, soil temperature, and soluble salt content. This would imply that the formula could be used to determine volumetric water content, from air-dry to water-saturated samples. They reported obtainable precision of volumetric water content to within ± 0.01 from the measured electric permittivity by calibrating the relationship for the particular granular material of interest. An organic soil, vermiculite, and two sizes of glass beads were also tested successfully. Their empirical relationship is used a lot, but not always correctly. The Topp equation is actually a third-order polynomial fit carried out on many samples under various conditions, and many different values for the four coefficients that determine this third order polynomial relationship were presented [35]. The fact that a third-order polynomial relationship between volumetric water content and the real part of the electric permittivity works well for many soils is caused primarily by the high electric permittivity value of water. The equation can be written

$$\varepsilon_{r,\text{soil}} = A + B\theta_w + C\theta_w^2 + D\theta_w^3 \tag{9.14}$$

where the volumetric water content is given by θ_w and the coefficients are site specific.

A slightly different model that relates water content to relative permittivity is that of Ledieu et al. [36], which is given by

$$\theta_w = a\sqrt{\varepsilon_{r,\text{soil}}} + b \tag{9.15}$$

where the coefficients are again site specific and must be determined separately.

9.2.2 Simultaneous Determination of Macroscopic Electric and Magnetic Parameters

To measure the electric and magnetic parameters experimentally under controlled laboratory conditions is impossible. Experiments can be done with different types of devices, which measure voltage differences as a function of electrode distances and/or frequency. From these measurements the electric and magnetic parameters of the material under test can be computed. At low frequencies a four-electrode system is used, at higher frequencies a two-electrode system is used, and in the wave field regime either a waveguide or a transmission line can be used to measure electric parameters. Here the coaxial transmission line method is described.

Coaxial Transmission Line This type of tool has also been used for a long time, together with the coaxial-circular waveguide, which is not discussed here (see Taherian et al. [37]), and rectangular waveguides. Transmission lines are tools for the accurate determination of electromagnetic properties of samples using propagating waves. The full S-parameter matrix can be manipulated such that explicit expressions for the complex electric and magnetic permeability can be derived assuming that the sample in the transmission line is homogeneous. The properties so determined are related to electric and magnetic field lines that are perpendicular to the propagation direction of the TEM wave in the transmission line. It is important to design the tool such that over the desired frequency range, only the TEM mode propagates and no higher-order modes occur. The tool described here is designed to measure the electric and magnetic properties of partially saturated soils under controlled flow conditions in the frequency range 100 MHz to 3 GHz. The tool was designed such that it is possible to determine the full S-parameter matrix of the system (measuring both the transmission and reflection coefficients of a sample is crucial in the determination of permittivity). The geometry and size of the probe ensure that only the TEM mode propagates along the line, and it can therefore be described using transmission-line theory. However, by simple modifications the sample holder can be transformed into a circular waveguide similar to the one described by Taherian et al. [37], and it is possible to determine the permittivity of rigid cores with a full-wave description. The sample holder was chosen to be of a representative volume (10 cm long, 3 cm in outer diameter, and 0.9 cm in inner diameter) and to allow for fluid flow. It is gold plated to ensure low losses of energy in the line conductors, and its several parts can be characterized separately

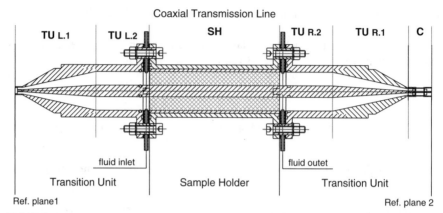

Coaxial Transmission Line

FIGURE 9.4 Coaxial transmission-line probe for permittivity and permeability determination from S-parameter matrix measurements on porous media. SH, sample holder; TUR, transition unit right; TUL, transmission unit left; C, connector.

for accurate modeling. The probe (Fig. 9.4) consists of three main sections: two transition units (TUs) and a sample holder (SH). Both transition units are again composed of three sections: a conical part, a cylindrical part, and a fluid distributor. They can be dismounted, enabling separate measurements of the two transition units together for high-accuracy calibration measurements. The transition units are Teflon filled and long enough to prevent any higher-order mode generated from reaching the measurement plane. The conical part of the transition unit eliminates the impedance jump between the cable connection and the sample holder such that the higher-order modes generated are negligible. When the line is completely filled with Teflon, the impedance throughout is very close to 50. The fluid distributors can be connected to four inlets and four outlets for fluids (Fig. 9.5). The fluid enters the sample through inlets located on the sides of the gold-plated outer ring and into the carved Teflon distributors; appropriate filters are placed on top. The carved fluid path ensures a reasonably homogeneous flow through the sample.

Transmission-Line Measurement of Porous Soils   The coaxial transmission-line introduced is a multisection apparatus that requires careful calibration to maximize the accuracy with which the electric and magnetic parameters can be obtained. This is realized through the design of the tool described above. Suppose that the network analyzer is calibrated to the endpoints of the cables to which the tool is connected. Then the tool can be described as a layered system, as depicted in Fig. 9.6, where each layer has a known thickness and is characterized by its complex electric permittivity and magnetic permeability, its complex impedance and propagation coefficient, or its scattering, or S, parameters.

Transmission-Line Equations   These representations are equivalent, and here a method is developed to obtain the complex electric permittivity and magnetic permeability from the transmission-line representation of Fig. 9.6b. Inside any layer

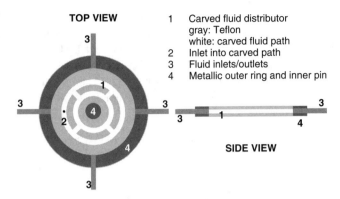

FIGURE 9.5   Fluid distributor unit.

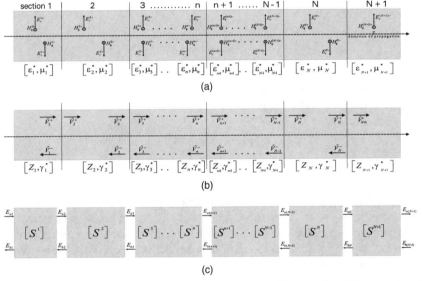

FIGURE 9.6   Equivalent representations of a sequence of layers with difference electric and magnetic parameters: (a) representation based on TEM waves propagating through the sequence; (b) representation based on transmission-line theory; (c) representation based on scattering (S) parameters of two-port networks.

the voltage and current wave satisfy the source free-wave equation, given by

$$\partial_x \partial_x F - \gamma^2 F = 0 \tag{9.16}$$

where F can be either the voltage V or the current I and γ is the propagation coefficient. For a transmission line it is given by

$$\gamma = \sqrt{(G + j\omega C)(R + j\omega L)} \tag{9.17}$$

where G (S) is the conductance and is related to the conductivity, C (F) is the capacitance and is related to the electric permittivity, R (W) is the resistance and is due to the nonperfectly conducting metallic parts, and L (H) is the inductance and is related to the magnetic permeability.

For a coaxial transmission line with inner radius b and outer radius a, these parameters are expressed in terms of the electromagnetic parameters and the geometrical factors of the line as

$$G = \frac{2\pi\hat{\sigma}}{\log(a/b)}$$

$$C = \frac{2\pi\hat{\varepsilon}}{\log(a/b)} \tag{9.18}$$

$$L = \frac{\hat{\mu}}{2\pi} \log\left(\frac{a}{b}\right)$$

$$R = \frac{1}{2\pi}\left(\frac{1}{a} + \frac{1}{b}\right)\sqrt{\frac{2\,\omega\mu}{\sigma_c}}$$

where R is determined by the properties of the metallic parts only and not by the sample material and $R=0$ when the lines are perfect conductors ($\sigma_c \to \infty$). The solution to Eq. (9.16) in any layer n can be written

$$V_n(x) = V_n^+ \exp(-\gamma_n x) + V_n^- \exp(+\gamma_n x) \tag{9.19}$$

$$I_n(x) = I_n^+ \exp(-\gamma_n x) + I_n^- \exp(+\gamma_n x) \tag{9.20}$$

where the superscript + refers to a wave propagating in the positive x-direction, the superscript − refers to a wave propagating in the negative x-direction, and the subscript n refers to the layer number. The characteristic impedance of the line in any layer is defined as the ratio between the voltage and the current waves propagating in the positive x-direction:

$$Z_n = \frac{V_n^+}{I_n^+} = -\frac{V_n^-}{I_n^-} = \sqrt{\frac{R_n + j\omega L_n}{G_n + j\omega C_n}} \tag{9.21}$$

Across any interface between the adjacent layers both the voltage and the current must be continuous, which leads to the following reflection and transmission coefficients:

$$r_n = \frac{V_n^-}{V_n^+} = \frac{Z_{n+1} - Z_n}{Z_{n+1} + Z_n} \tag{9.22}$$

$$t_n = \frac{V_{n+1}^+}{V_n^+} = \frac{2Z_{n+1}}{Z_{n+1} + Z_n} \tag{9.23}$$

In a two-port network analyzer the total reflection and total transmission coefficients are measured. The total reflection coefficient at the measuring plane is equal to the global reflection coefficient R_0, given generically for layer n by

$$R_n = \frac{r_n - R_{n+1}\exp(-2\gamma_{n+1}d_{n+1})}{1 + r_n R_{n+1}\exp(-2\gamma_{n+1}d_{n+1})} \qquad (9.24)$$

which is a recursive formula that is initialized for an N-layered medium by taking $R_{N+1} = 0$ because there are no waves propagating in the negative x-direction in the cable that is connected to the S_{12} port. To compute the total reflection response of the tool it is necessary to compute all the global reflection coefficients R_n, which allows for computing the total transmission response as

$$T_N = \prod_{n=0}^{N-1} \frac{(1 + R_n)\exp(-2\gamma_{n+1}d_{n+1})}{(1 + R_{n+1})\exp(-2\gamma_{n+1}d_{n+1})}(1 + r_N) \qquad (9.25)$$

Soil Characterization with Propagation Matrices   Rewriting the interface conditions in terms of $V^\pm$ using Eqs. (9.19) to (9.21) we obtain the matrix equation

$$\begin{pmatrix} \exp(-\gamma_n d_n) & 1 \\ \dfrac{\exp(-\gamma_n d_n)}{Z_n} & \dfrac{-1}{Z_n} \end{pmatrix} \begin{pmatrix} V_n^+ \\ V_n^- \end{pmatrix} = \begin{pmatrix} 1 & \exp(-\gamma_{n+1}d_{n+1}) \\ \dfrac{1}{Z_n} & -\dfrac{\exp(-\gamma_{n+1}d_{n+1})}{Z_n} \end{pmatrix} \begin{pmatrix} V_n^+ \\ V_n^- \end{pmatrix} \qquad (9.26)$$

which means that we can express voltage amplitudes inside a layer n in terms of the amplitudes inside an adjacent layer $n+1$ or $n-1$. We rewrite Eq. (9.26) as

$$\mathbf{L}_n\mathbf{v}_n = \mathbf{R}_{n+1}\mathbf{v}_{n+1} \longrightarrow \mathbf{v}_n = \mathbf{L}_n^{-1}\mathbf{R}_{n+1}\mathbf{v}_{n+1} \longleftrightarrow \mathbf{v}_{n+1} = \mathbf{R}_{n+1}^{-1}\mathbf{L}_n\mathbf{v}_n \qquad (9.27)$$

and we can concatenate several layers from the beginning of the tool to the end of the tool:

$$\mathbf{v}_0 = \mathbf{L}\mathbf{P}_s\mathbf{R}\mathbf{v}_N \qquad (9.28)$$

where the response matrices of the part to the left, $\mathbf{L}$, and to the right, $\mathbf{R}$, of the sample holder are given by

$$\mathbf{L} = \mathbf{L}_0^{-1}\prod_{n=1}^{s-1}\mathbf{P}_n \qquad (9.29)$$

$$\mathbf{R} = \left(\prod_{n=s+1}^{N-1}\mathbf{P}_n\right)\mathbf{R}_N \qquad (9.30)$$

where the matrix $\mathbf{P}_n$ is given by

$$\mathbf{P}_n = \begin{pmatrix} \cosh(-\gamma_n d_n) & Z_n \sinh(-\gamma_n d_n) \\ \dfrac{\sinh(-\gamma_n d_n)}{Z_n} & \cosh(-\gamma_n d_n) \end{pmatrix} \qquad (9.31)$$

which represents the **P**-matrix for the sample holder when $n = s$. Equation (9.28) can now be written

$$\mathbf{L}^{-1}\mathbf{v}_0 = \mathbf{P}_s\mathbf{R}\mathbf{v}_N \longrightarrow \mathbf{L}^{-1}\begin{pmatrix} 1 \\ R_0 \end{pmatrix} = \mathbf{P}_s\mathbf{R}\begin{pmatrix} T_N \\ 0 \end{pmatrix} \longrightarrow \begin{pmatrix} A \\ B \end{pmatrix} = \mathbf{P}_s\begin{pmatrix} C \\ D \end{pmatrix} \qquad (9.32)$$

which is easily solved for the impedance and propagation coefficient of the material under test as

$$Z_s = \sqrt{\dfrac{A^2 - C^2}{B^2 - D^2}} \qquad (9.33)$$

and

$$\gamma_s = \pm\dfrac{\text{acosh }\beta}{d_s} \qquad \beta = \dfrac{AB + CD}{AD + BC} \qquad (9.34)$$

where the sign for the acosh function is determined by the condition that $\Re(\gamma_s) \geq 0$. Considering a perfectly conducting line where the sample is located, the relative electric permittivity $\hat{\varepsilon}_r$ and the magnetic permeability $\hat{\mu}_r$ of the sample can be found as

$$\hat{\varepsilon}_r = \dfrac{1}{j\omega\varepsilon_0}\dfrac{\gamma_s}{Z_s} \qquad (9.35)$$

$$\hat{\mu}_r = \dfrac{Z_s\gamma_s}{j\omega\mu_0} \qquad (9.36)$$

To connect these results to all other known formulas, the reader is referred to Gorriti and Slob [38,39]. It turns out that it is very difficult to obtain accurate experimental results for the impedance using Eq. (9.33) because the denominator has a difference between two factors that can become almost equal and the result becomes meaningless. Since Eqs. (9.35) and (9.36) involve both impedance and the propagation coefficient, the experimental problems with the impedance makes it worthwhile to look for an expression for the electric permittivity assuming that the material under test is nonmagnetic. This is easily achieved from Eq. (9.34) as

$$\hat{\varepsilon}_r = \left(\dfrac{c_0\gamma_s}{j\omega}\right)^2 \qquad (9.37)$$

Stable and accurate results are obtained using this equation with a determined maximum error of 1% for frequencies where the internal wavelength is less than

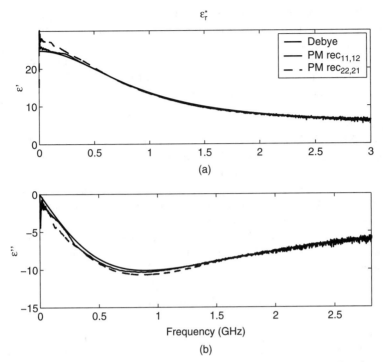

FIGURE 9.7 Reconstructed (a) real and (b) imaginary parts of the relative electric permittivity of ethanol measured with S_{11} and S_{12} (dotted lines) and the reverse situation S_{22} and S_{21} (dashed lines) together with a fit obtained from the Debye model (solid lines).

or equal to five times the length of the sample holder. As an example, the result for a 10-cm sample of ethanol is given in Fig. 9.7, where the real and imaginary parts of the relative permittivity are shown as a function of frequency for the experimental results from two sides of the sample. The fit obtained from the Debye model is shown for comparison. The fit is obtained through optimization of the two experimental results and optimizing for the static and infinite frequency permittivity values of ethanol, the relaxation frequency, and dc conductivity. The reconstructed permittivities and the Debye fit differ by less than 1%, with the measured curves between 300 MHz and 2 GHz. If any of the Debye parameters are changed by 1%, the error between the difference is increased two to three times. All parameters except the dc conductivity fall well within the error bounds of the values given by van Hemert [40].

9.2.3 Determination of Electric and Magnetic Properties

In contrast to the electrical properties of rocks, soil, and mixtures, only a limited number of measurements have been conducted on the magnetic properties of geomaterials in the frequency band of interest for GPR or other EM techniques. The interest for such properties started in the early 1970s, when small particles of iron or iron–nickel where found in lunar soil, and the dielectric and magnetic properties of lunar samples were measured [41,42]. However, recently, thanks again to planetary exploration, magnetic property measurements have gained new attention [43–53]. Two orbiting radar sounders, MARSIS [54], on-board the ESA satellite Mars Express, and SHARAD [55], on-board the NASA satellite Mars Reconnaissance Orbiter, were sent in 2003 and 2005 to explore the subsurface of Mars using electromagnetic waves in the frequency range 1.8 to 4 MHz (MARSIS) and 20 MHz (SHARAD). The main concern in these missions was, in fact, the presence of highly magnetic minerals in the Martian soil and rocks, which could cause strong attenuation, preventing the penetration of the signal at depth [42]. Despite the lack of data, several contributions to the magnetic behavior of single- or polycrystalline magnetite are available [57–59], as is information on the electromagnetic behavior of iron oxide/silica mixtures and iron oxide sand [60–63].

The magnetic properties of a material are characterized by the relative complex magnetic permeability μ^*, which relates the flux density B to the magnetic field H as follows:

$$B = \mu_0(H + M) = \mu_0\mu^* H \qquad (9.38)$$

with M being the magnetization of the tested material and $\mu_0 = 4\pi \times 10^{-7}$ H/m the permeability in the vacuum. The relative complex permeability can be defined as

$$\mu^* = \mu' - j\mu'' \qquad (9.39)$$

where the real part represents the magnetizability of the material, whereas the imaginary part takes all the magnetic losses into account.

Similar to the definition of dielectric losses, the power dissipation per unit volume associated with a time-varying magnetic field can be expressed as

$$P = \omega\mu_0\mu''\overline{H^2} \qquad (9.40)$$

where the bar denotes an average with respect to time. This term takes into account dissipation due to hysteresis, eddy current losses, and domain wall relaxations [61]. During the hysteresis loop, energy is dissipated as heat in a magnetic material, and the loss is controlled by sample properties such as porosity, grain size, impurities, and defects. For the eddy current loss the conductivity of the material is important: In metallic materials such loss is predominant but can still be important in ferromagnetic oxides such as magnetite, which has $\sigma \simeq 10^4$ S/m at room temperature. The entity of the eddy current loss also depends on the geometric dimensions of the thickness of a solid sample or the particle size of a granular sample. The effect on the relaxation of domain walls is quite complicated: In ferromagnetic materials, low-frequency relaxation (up to few tens of megahertz) arises as a result of domain wall bowing and displacement [61]. At higher frequencies (gigahertz) the domain

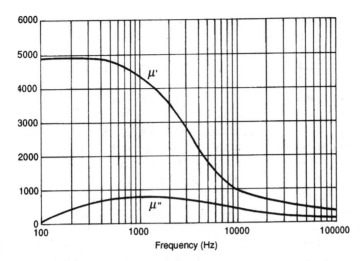

FIGURE 9.8   Real and imaginary parts of the permeability for a magnetite single crystal.

walls are unable to follow the field and the magnetization mechanism is due to spin rotation in the domains. Moreover, the relaxation frequency decreases as grain size increases; the domain wall resonance becomes less significant the smaller the grain and does not occur in the case of single domain grains. Goodenough has discussed losses in magnetic materials in detail [64].

The frequency dependence of magnetic properties can be described, in analogy with the electrical properties, using a Cole–Cole distribution [58]; see also Eq. (9.10):

$$\mu = \mu' - j\mu'' = \mu_\infty + \frac{\mu_s - \mu_\infty}{1 + (j\omega\tau)^\alpha} \tag{9.41}$$

where μ_∞ is the high-frequency limit of permeability, μ_s the low-frequency limit of permeability, ω the radiant frequency, τ the time constant, and α the distribution parameter. When $\alpha 1$, a single Debye-like relaxation mechanism is considered. As an example, Fig. 9.8 shows the frequency response of the real and imaginary permeabilities of a single-crystal magnetite [56]. Here the relaxation frequency is around 1 kHz; however, if magnetic measurements are made on polycrystalline magnetite, such a frequency tends to shift toward a lower value [58].

Finally, to calculate the permeability of multiphase mixtures, with magnetic and nonmagnetic components, mixing formulas has been developed [61,65]. Note that the results obtained for the effective permittivity (mentioned earlier) can be used directly to calculate the effective magnetic permeability as long as all ε parameters are changed into μ parameters in the mixing formulas [65].

Electromagnetic Measurement Through L–C–R Meters   Laboratory techniques for measuring the electromagnetic properties of granular materials at hertz-to-megahertz frequencies are usually based on the equivalent-circuit analysis performed through $L–C–R$ meters [66]. Conductivity and permittivity are obtained by measuring the magnitude and phase of the electrical impedance of a capacitive cell (i.e., parallel-plate capacitor) filled with the material being tested. Similarly, the magnetic permeability and relevant losses are obtained by measuring the impedance of a solenoid fully embedded in the material of interest.

Figure 9.9 qualitatively describes the method of the impedance measurements. The voltage generator V_g applies a constant voltage across the device under test (DUT) terminals and passes a current through the cell that contains the sample. The DUT may consist of either a capacitive cell (Fig. 9.9a) or a solenoid (Fig. 9.9b) for the dielectric and magnetic measurements, respectively. The low terminal of the DUT (indicated with L in the Fig. 9.9b) is connected to the virtual ground input of a current-to-voltage converter, which is represented schematically in the diagram by an operational amplifier whose feedback resistor (R_f) provides, at the output, a signal proportional to the DUT current. Note that when using capacitive cells the low DUT terminal is equipped with a ring guard electrode that is connected to ground, thus suppressing the possible perturbations caused by the stray field at the capacitor plate edges. The ratio of the generator voltage (V_g) to the DUT current (I) provides the electrical impedance of the sample. The electromagnetic properties of the material are therefore investigated by varying the frequency of the generator. In actual applications the current-to-voltage converter does not employ a simple operational amplifier, as indicated in Fig. 9.9, but rather, uses a more sophisticated system based on a combination of a null detector, phase-sensitive demodulator, and

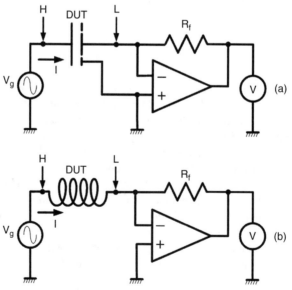

FIGURE 9.9 Auto balancing bridge system: (a) capacitive cell for dielectric tests; (b) solenoid circuit for magnetic tests.

integrator designed to assure high accuracy even at the higher frequencies, where the performance of operational amplifiers usually degrades significantly.

Dielectric Measurements Measurements of the dielectric properties of materials are performed through an analysis of the equivalent circuit. With reference to the lumped equivalent circuit shown in Fig. 9.10, the current in the cell can be expressed through its complex admittance Y^*:

$$I = V_g Y^* = V_g \left(\frac{1}{R_p} + j\omega C_p \right) \tag{9.42}$$

Using the geometrical features of the cell, the measured values of capacitance C_p and parallel resistance R_p can be related to the real and imaginary parts of the complex permittivity of the material enclosed between the electrodes. In particular, if A is the surface of the guarded electrode (the plate connected to the "low" terminal in Fig. 9.11) and d is the spacing between the two parallel plates, one

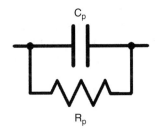

FIGURE 9.10 Equivalent electrical circuit of the capacitive cell.

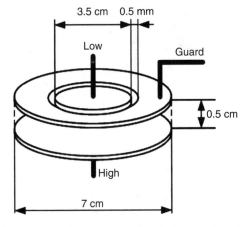

FIGURE 9.11 Capacitance cell to measure electrical properties of granular materials.

obtains

$$\varepsilon' = \frac{d}{A}\frac{C_p}{\varepsilon_0} = \frac{C_p}{C_0} \qquad \varepsilon'' = \frac{d}{A}\frac{1}{\omega\varepsilon_0 R_p} = \frac{1}{\omega C_0 R_p} \tag{9.43}$$

Note that a precise knowledge of the cell geometry is not strictly required to get accurate measurements, as both the real and imaginary terms of permittivity can be calculated using the previously measured value of the capacitance with the cell empty, as indicated in the right terms of Eq. (9.43). The effect of the fringing field at the edges of the capacitor plates is canceled by using a guarded electrode at the DUT low terminal.

Magnetic Measurements   The complex magnetic permeability is determined through a pair of electrical impedance measurements performed both with the solenoid empty and in the presence of the sample (i.e., with the solenoid embedded in the granular test material; Fig. 9.12). With reference to the lumped equivalent circuit shown in Fig. 9.13, the complex impedance of the empty solenoid is

$$Z_0^* = R_{Cu} + j\omega L_0 \tag{9.44}$$

where R_{Cu} represents the resistance of the solenoid (made in this example with copper wire) and L_0 its inductance. When filled with material, the inductance of the solenoid, being proportional to the complex permeability μ^*, becomes

$$Z_{mat}^* = R_{Cu} + j\omega\, L_{mat} = R_{Cu} + j\omega\, \mu^* L_0. \tag{9.45}$$

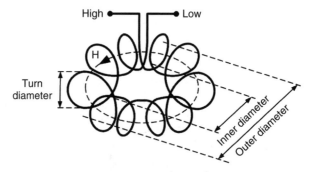

FIGURE 9.12   Inductor (toroid) to measure magnetic properties of granular materials.

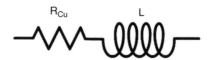

FIGURE 9.13   Equivalent electrical circuit of the solenoid.

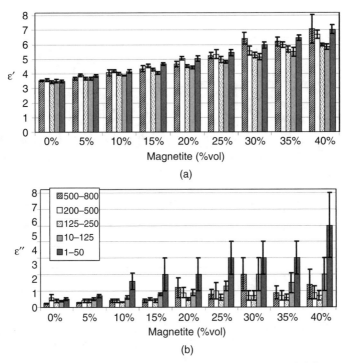

FIGURE 9.14 (a) Real and (b) imaginary parts of complex permittivity measured at $f = 1$ MHz for different grain sizes and magnetite volume fractions.

Using the equation of complex permittivity, the impedance, measured in the presence of material, can be rewritten as

$$Z_{mat}^* = (R_{Cu} + \omega L_0 \mu'') + j\omega\mu' L_0. \tag{9.46}$$

Similar to dielectric measurements, precise knowledge of the solenoid geometry is not strictly required, as both the real and imaginary parts of the permeability can be determined using the inductance L_0 and resistance R_{Cu} values measured with the toroid empty.

As an example, Fig. 9.14a and b show the real and imaginary parts of the permittivity measured with an $L-C-R$ meter, obtained for five different grain-size mixture samples (glass beads/magnetite) as a function of the volumetric content of magnetite measured at 1 MHz, whereas Fig. 9.15a and b show the real and imaginary parts of the magnetic permeability obtained for the same samples measured with an $L-C-R$ meter at 1 MHz (for details, see Pettinelli et al. [66]).

9.2.4 Electric Soil Characterization Under Flow Conditions

The coaxial transmission line described in Section 9.2.2 makes it possible to determine the complex electric permittivity of partially or fully saturated soil under

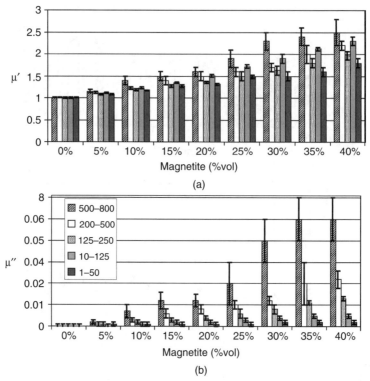

FIGURE 9.15 (a) Real and (b) imaginary parts of complex permeability measured at $f = 1$ MHz for different grain sizes and magnetite volume fractions.

flow conditions in the frequency range 300 MHz to 3 GHz. For frequencies below 3 MHz, a special design parallel-plate capacitor is described here.

Parallel-Plate Capacitor Parallel-plate capacitors have long been used and have certain attractive features, as described in Section 9.2.3, and a basic tool is shown in Fig. 9.11. One of the advantages is that the material parameters are measured in the direction of the field, so it is of interest principally for static and low-frequency properties where the modified Laplace equation is valid. For electromagnetic phenomena this applies to the equation of total divergence free electric current density where the electric field can be approximated properly as the gradient of an electric potential field. For field applications this is of interest in exploration, environmental, and engineering geophysics. Here we present a versatile tool that can be operated as a capacitive cell under controlled fluid-flow conditions at fixed elevated pressures and temperatures [67]. This allows us to study hysteretic response behavior as a function of saturation changes, due to reversing flow direction under quasistatic equilibrium conditions [68].

The experimental setup is based on the porous plate technique, combined with the micropore membrane technique [69,70]. We apply quasistatic conditions [71]

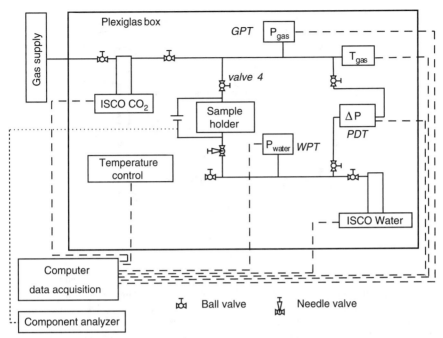

FIGURE 9.16   Experimental setup. PDT, pressure difference transducer; GPT, gas pressure transducer; WPT, water pressure transducer.

(i.e., small injection rates) such that viscous forces can be neglected. A schematic overview of the experimental setup is shown in Fig. 9.16. Two syringe pumps are used and can be set to a constant injection rate or a constant pressure. The gas and water phases are injected or produced at the top and bottom of the sample holder, respectively. The pressure difference between the gas and water phases is measured by a pressure difference transducer (PDT; accuracy ± 0.05 mbar), which is located at the same height as the middle of the sample. A temperature control system is used to maintain a constant temperature.

The sample holder (Figs. 9.17 and 9.18) consists of three parts: a PEEK [poly(ether ether ketone)] ring, which contains the sand sample, and two stainless steel end pieces. Two porous plates (SIPERM R, Cr–steel basis) with a permeability of 2×10^{-12} m^2 and a porosity of 0.32 support the sample and protect the hydrophilic membrane. Two stainless steel plates, both with 32 perforations ($D_p = 5$ mm), are used at the top directly above the sample in combination with a nylon filter. Concentric flow grooves in the end pieces redistribute the phases over the sample area to avoid preferential flow. The different parts of the sample holder are mounted together with four stainless steel bolts at both top and bottom.

To obtain the permittivity of the sample inside the PEEK ring, a precision component analyzer (Wayne-Kerr, 6640A) is connected to the sample holder (see Fig. 9.1). The electrodes are the two end pieces of the sample holder, including the support plates. The PEEK material is nonconductive, and hence the sample

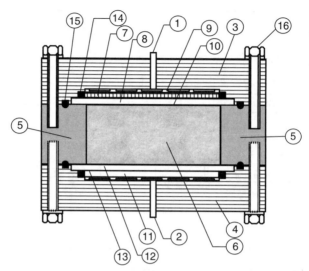

FIGURE 9.17 Sample holder: 1, gas inlet; 2, water inlet; 3, stainless-steel end piece 1; 4, stainless steel end piece 2; 5, PEEK ring; 6, porous medium (diameter is 84 mm, height is 27 mm); 7, perforated plate (diameter is 84 mm); 8, perforated plate (diameter is 90 mm); 9, concentric grooves; 10, nylon filter (pore size 210 μm); 11, SIPERM plate (diameter is 84 mm); 12, SIPERM plate (diameter is 90 mm); 13, water-wet filter (pore size 0.1 μm); 14, O-rings (2.1 mm); 15, O-rings (4 mm); 16, stainless-steel bolts.

holder acts as a parallel-plate capacitor. The impedance amplitude, $|Z|$ (ohms), and phase angle, θ (rad), are measured as a function of the frequency and is related directly to the effective complex permittivity ((ε_s^*)) of the sample, defined by $\varepsilon_s^* = \varepsilon_s' - i\varepsilon_s''$. Here ε_s' and ε_s'' represent, respectively, the real and imaginary parts of the permittivity. In this study we are interested in ε_s' and the results are presented as a function of S_w.

Determination of the Complex Electric Permittivity This setup is used in the frequency range from 1 kHz to 3 MHz. The macroscopic capillary pressure of a porous material is defined as the difference in gas and water bulk phase pressures. From the water volume produced, we obtain the water saturation. The complex capacitance, C^* (Farad) is inversely proportional to the complex impedance Z^* (Ohm) and frequency $\omega = 2\pi i f$ [f being frequency (1/s)], by $C^* = (i\omega Z^*)^{-1}$ and $Z^* = |Z|\exp(i\theta)$. The capacitor configuration results in a parallel circuit for which C^* is considered as the sum of the capacitance of the sample (C_s^*), the PEEK ring (C_{PEEK}^*), the cables (C_{cables}^*), the electrode design ($C_{\text{electrodes}}^*$), and the background noise (C_{noise}^*). Under the assumption that only C_s^* is changing for different samples, C^* can be written

$$C^* = C_s^* + C_{\text{residual}}^* = \frac{\varepsilon_0 \varepsilon_s A_s}{H} + C_{\text{residual}}^* \tag{9.47}$$

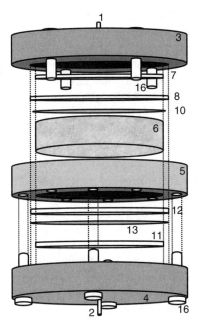

FIGURE 9.18 Three-dimensional representation of the sample holder (not to scale). The numbers correspond to the legend of Fig. 9.17. For visualization reasons, the rubber O-rings and the concentric grooves are not shown.

Here ε_0 is the permittivity of free space, ε_s the permittivity of the sample, $C^*_{\text{residual}}(s/\Omega)$ the capacitance of the residual contributions, and H (meters) and $A_s(\text{m}^2)$ are the height and cross-sectional area of the sample, respectively. Notice that this is a different formula from that given in Eq. (9.43), and the error analysis has shown that this leads to a more accurate determination of the electric permittivity. We use the complex refractive index (CRI) model [72] to evaluate the effective permittivity of the grain–water–gas mixture, given by [see Eq. (9.5)]

$$\varepsilon_s = \left[\phi S_w \sqrt{\varepsilon_w} + \phi(1 - S_w)\sqrt{\varepsilon_{\text{gas}}} + (1 - \phi)\sqrt{\varepsilon_{\text{grain}}} \right]^2 \tag{9.48}$$

where ϕ is the porosity and ε_w, ε_{gas}, and $\varepsilon_{\text{grain}}$ are, respectively, the water, gas, and porous medium permittivities. Furthermore, the Hashin–Shtrikman bounds [2] are used to investigate the validity of the results. The Hashin-Shtrikman bounds put limits on the effective electric permittivity of locally noninteracting and macroscopically homogeneous, isotropic mixtures.

Calibration and System Accuracy To obtain the most accurate data for the permittivity and the capillary pressure, different configurations of the sample holder are investigated. It appears that the type and combination of the support plates and the presence and the number of the stainless steel bolts do not influence the impedance measurements. The value for C^*_{residual} of Eq. (9.47) is obtained from air measurements using $\varepsilon_s = 1$. The capacitance C^*, of the air-filled sample holder is measured in the range $20 - 21 \pm 0.05$ pF for the frequency range 1 kHz to 3 MHz. A statistical analysis on 50 measurements [73] for air and six different calibration materials is performed to determine the maximum measurement accuracy. The mean and the relative error for both $|Z|$ and θ of 10 groups of two air measurements are on the order of 0.1%. The impedance tool is calibrated for $f = 3$ MHz using materials with known permittivity behavior within the range 2 to 25. High accuracy and good agreement are found for a wide range of permittivity values.

Sample Preparation and Experimental Procedure   The capillary pressure and complex permittivity behavior of an unconsolidated sand–water–gas (CO_2/N_2) system are measured simultaneously. For each experiment a new sand pack is used. The average grain-size fraction is determined and the porosity is obtained with helium at room temperature. After the porosity measurements the total system is evacuated for 1 hr. Subsequently, the sample holder is filled with distilled water (no salinity) at a pressure of approximately 8 bar to dissolve small air bubbles. Valve 4 (Fig. 9.16) is closed and the gas tubing and pump are filled with gas. The gas booster is used to bring up the gas pressure. We set a constant temperature and let the system equilibrate for two days. When the water and gas pressure are equal, a constant water refill rate is applied, the gas pump is set to a constant pressure, and valve 4 is opened. After the main drainage process, the main imbibition process starts when the water pump is set to a constant injection rate. During these flow cycles the electric response is measured over the entire frequency range and simultaneous capillary pressure measurements are taken as well. As an example, the result of such a measurement is shown for a sand pack of unconsolidated quartz sand samples with an average grain-size fraction of $360 < D_{50} < 410$ μm. The measurements are carried out under a temperature instability $\leq 0.5°$C, and the real part of the relative electric permittivity is shown in Fig. 9.19 as a function of water saturation for the first drainage and imbibition cycle carried out at 8 bar pressure. The hysteresis in electric permittivity as observed between the primary drainage (e.g., $S2a$ in Fig. 9.19) and secondary imbibition process (e.g., $S2b$ in Fig. 9.19) at 100 kHz can be explained by different phase geometries and surface water distributions (clusters).

The trajectory of the low-frequency permittivity as a function of water saturation is interpreted using percolation theory and depends strongly on the presence of the gas phase. During (primary) drainage the gas phase will first penetrate the larger pores, resulting in an increase in electric permittivity. This is the case during the first stage of the primary drainage curves, where for small amounts of gas, ε_s increases rapidly to a maximum. This may be explained by increasing water–gas interfaces. The sand sample is completely water-wet ($P_c > 0$), meaning that all the grains are coated with a thin water layer. These water layers can become extremely thin on the convex side of the grains, for which the nearby pores are filled with gas that protrudes into the water-filled pores. As the gas saturation increases and a large part of the water in the thin layers is "bound" to the grains

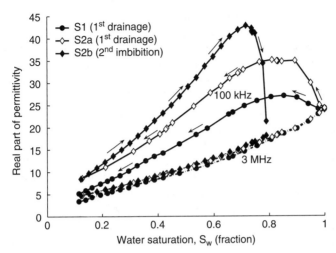

FIGURE 9.19 Real part of the complex permittivity, ε_s, as a function of water saturation for the N_2–sand–water system. S1a and S2a represent the primary drainage curves for the atmospheric and 8-bar pressure conditions, respectively. S2b is the secondary imbibition curve for the 8-bar case. The solid lines represent ε_s 100 kHz and the dashed lines 3 MHz. The permittivity at 3 MHz shows small hysteresis between the drainage (S2a) and the imbibition (S2b). The permittivity values obtained at 100 kHz show nonmonotonic behavior and hysteresis where the imbibition curve (S2b) is above the drainage curve (S2a).

(not free), rotation of the dipole orientations of the water molecules in these thin films becomes difficult under the presence of an electric field. We have found that the low-frequency permittivity behavior can be ascribed primarily to polarization of both the gas–water and water–solid interfaces, to different bulk phase geometries and distributions, and to the saturation history during drainage and imbibition. Our experimental data show that capillary pressure is a unique function of the permittivity and water saturation [68].

9.2.5 Time-Domain Reflectometry

Time-domain reflectometry (TDR) is a well-established electromagnetic technique that has been used since the 1930s to determine the spatial location of cable faults. At the end of the 1960s, Fellner-Feldegg [74] first used the TDR technique to measure the electrical properties of materials by examining alcohols held in coaxial cylinders. About a decade later, Topp et al. [35] extended this application to earth materials by determining the volumetric water content of soils in coaxial sample holders, and subsequently they developed a TDR system that was able to measure in situ soil electrical properties using a transmission line consisting of two parallel rods [75].

In the last 30 years, TDR has found its main application in agriculture and hydrogeology research, primarily for nondestructive measurements of soil water content as well as in environmental research for the detection of hazardous chemicals in soil; however, the technique is still suitable to evaluate the permittivity and conductivity of a solution or a granular material simply and rapidly [66,76]. For an overview of TDR applications, see O'Connor and Dowding [77]. The use of TDR for permittivity measurements is based on evaluating the velocity of a steplike signal which travels along a transmission line filled with the material under test or embedded in it (for details, see Topp and Ferré [78] and Robinson et al. [79]). The TDR signal has a broad band and the upper frequency in a sample can extend up to about 500 MHz [80].

A basic TDR system generally consists of a pulse generator, a sampler, an oscilloscope, a coaxial cable, and a probe, as depicted schematically in Fig. 9.20. In particular, Fig. 9.20a shows a TDR configuration that can easily be used in the field, where the probe can be directly inserted in the soil, and Fig. 9.20b shows a typical TDR configuration for laboratory measurements. The transmission line (i.e., the TDR probes in Fig. 9.20) is open at the far end; thus, a reflection coefficient equal to +1 produces the back reflection of the waves. This can be achieved in the laboratory but is, of course, an approximation in the field. When the probe is stuck into the ground, the end is not open but is matched in terms of the soil filling the sample. Of course, a reflection will occur because the metallic pins end, but the reflection coefficient is not unity. From the propagation time of the wavefront along the probe, the pulse velocity can be calculated according to the equation

$$v = \frac{2L}{t} \tag{9.49}$$

where t is the two-way travel time and L is the probe length. Assuming that the material is homogeneous within the sample, the electromagnetic wave velocity is

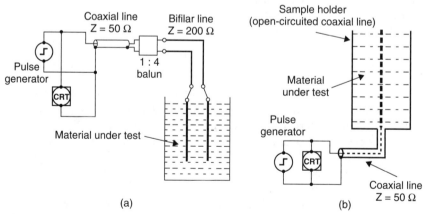

(a)   (b)

FIGURE 9.20   (a) Basic TDR system, consisting of a pulse generator, a sampler, an oscilloscope, a coaxial cable, a balun, a shielded two-wire transmission cable, and two parallel metal rods that are inserted in the material to be tested. (b) The same TDR configuration connected to a coaxial probe.

given by

$$v = \frac{c_0}{\sqrt{\varepsilon^* \mu^*}} = \frac{c_0 \sqrt{2}}{\sqrt{\mu' \varepsilon' - \mu'' \varepsilon'' + |\varepsilon^*||\mu^*|}} \tag{9.50}$$

where c_0 is the electromagnetic wave velocity in vacuum (ca. 3×10^8 m/s) and ε^* and μ^* denote the complex electric and magnetic parameters. Note that the complex permittivity and the complex magnetic permeability cannot be evaluated separately using the TDR technique in the time domain, since the measured velocity is a function of the product of these parameters. If magnetic materials are not present, TDR travel-time measurements can be used to estimate the *apparent relative dielectric permittivity*, ε_{ra}, through

$$\varepsilon_{ra} = \left(\frac{ct}{2L} \right)^2 \tag{9.51}$$

which takes into account both the real and imaginary parts of the permittivity. As an example, Fig. 9.21 shows a typical TDR waveform acquired in water. The apparent permittivity is calculated on the basis of Eq. (9.51) and the experimental evaluation of t_1 and t_2. Due to the nonideal nature of the system and to the dissipative and dispersive behavior of the materials, that evaluation of the two-way travel time is not straightforward and can be affected by a significant uncertainty [81].

Different approaches have been proposed to calibrate and/or to determine the value of the apparent permittivity from TDR waveforms. The first technique exploits the time-domain analysis and is based on the *tangent method* [35,80, 82–84]. Figure 9.21 shows an application of the tangent method, which provides

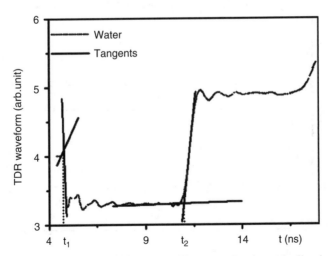

FIGURE 9.21 Tangent method applied to the TDR trace for the probe line in water. The times t_1 and t_2 are the limits of the two-way travel-time interval t.

some simple criteria for evaluating the time interval t and is largely used for its easy handling.

The second technique, developed in the time domain, is the derivative method [85,86]: the time derivative of the TDR response can be modeled by a series of Gaussian functions:

$$\dot{r}(t) = \sum_{j=1}^{n} A_j \exp\left(\frac{t - \tau_j}{\sigma_j}\right)^2 \tag{9.52}$$

where the index j runs over the successive reflections present in the TDR trace. Figure 9.22 shows an example of the fitting of Eq. (9.52).

A third method, developed in the frequency domain, focuses on the best fit of a theoretical transfer function to the experimental data [87–89]. As an example of this approach, Fig. 9.23 shows the real and imaginary parts of the theoretical scatter function fitted to the experimental data for a TDR probe filled with air.

Recently, calibration measurements in the time and frequency domains have shown that the tangent method can lead to retrieving the wrong probe parameters if the material under test is dispersive [86]. In fact, in the time domain, air and water provide different results for the probe length if the standard tangent line-fitting procedure is used. The application of the derivative method suggests that the anomalous result provided by the tangent method in water is determined by dispersion effects. To take the frequency dependence of the permittivity into account, a careful analysis in the frequency domain must be performed. Many

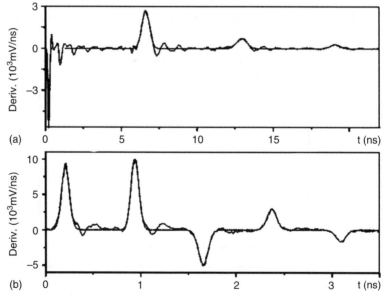

FIGURE 9.22 Experimental time derivatives, (dashed line) and the fit function of Eq. (9.52) (solid line) for demineralized water (a) and for air (b).

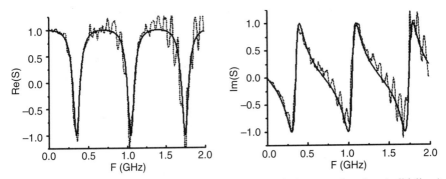

FIGURE 9.23 Real and imaginary parts of the theoretical scatter function (solid lines) fitted to the experimental ones (dashed lines) for the probe line filled with air.

dielectric materials, such as water, wet soils, and soils containing clay minerals, exhibit electric dispersion that causes TDR signal distortions. Consequently, time-domain studies return effective quantities whose values depend on the choice of the effective frequency [51].

Finally, analysis of the TDR waveform also allows one to estimate the dc conductivity using Eq. (9.53) [90]:

$$\sigma_{GT} = \sqrt{\frac{\varepsilon_0}{\mu_0}} \frac{1}{L} \frac{Z_0}{Z_c} \frac{2V_0 - V_f}{V_f} \tag{9.53}$$

where according to Fig. 9.24, L is the probe length, Z_0 the probe impedance in air, Z_c the cable impedance, V_0 the input voltage, and V_f the asymptotic voltage value. For a detailed description of the dc conductivity evaluation with TDR, see Robinson et al. [79].

Determination of Electromagnetic Parameters via a Scatter Function The TDR waveform carries information on the electromagnetic properties of the sample under test in a wide band of frequency. The frequency response of a TDR device can be studied conveniently by using its transfer function $S(\omega)$. This function allows one to estimate both the material electromagnetic properties and the transmission-line parameters, because it can be compared directly with the theoretical expressions.

The response $r(t)$ of a passive device to a step input function $x(t)$ is given by the convolution operation between $x(t)$ and the transfer function $s(t)$:

$$r(t) = x(t) * s(t) \tag{9.54}$$

In the case of TDR measurement, $s(t)$ is the transfer function of the line filled with the test material and r(t) is the response that contains all the information about the system features and the sample electromagnetic properties. In the frequency

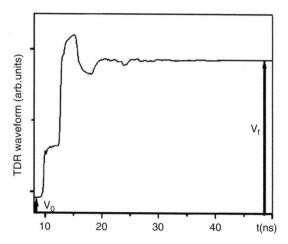

FIGURE 9.24   TDR waveform and relevant parameters used to estimate dc conductivity.

domain, by applying the convolution theorem, we have

$$R(\omega) = X(\omega)S(\omega) \tag{9.55}$$

so that the knowledge of r(t) and $x(t)$ allows us to calculate the transfer function $s(t)$. In practice, due to the $x(t)$ stepwise waveform, $S(\omega)$ is more easily determined by transforming the time derivatives of $x(t)$ and $r(t)$, using the property of the FT derivatives [i.e., $F'(\omega) = i\omega F(\omega)$, with i the imaginary unit]. From Eq. (9.55) we obtain

$$S(\omega) = \frac{R'(\omega)}{X'(\omega)} \tag{9.56}$$

where $R'(\omega)$ and $X'(\omega)$ are the FT of $r'(t)$ and $x'(t)$, respectively. Equation (9.56) allows derivation of $S(\omega)$ from the experimental data if the input function $x(t)$ is known. The theoretical scatter function $S_T(\omega)$ of an open-ended coaxial probe is given by [91,92]

$$S_T(\omega) = \frac{\rho^* + \exp(-2\gamma L)}{1 + \rho^* \exp(-2\gamma L)} \tag{9.57}$$

where ρ^* and γ are the reflection and propagation coefficients, respectively:

$$\rho^* = \frac{1 - z\sqrt{\varepsilon^*(\omega)/\mu^*(\omega)}}{1 + z\sqrt{\varepsilon^*(\omega)/\mu^*(\omega)}} \qquad \gamma = \frac{i2\pi f}{c_0}\sqrt{\varepsilon^*(\omega)\mu^*(\omega)} \tag{9.58}$$

The quantity z in Eq. (9.58) represents a normalized impedance given by $z = Z_C/Z_P$, where Z_C is the impedance of the cable connecting the TDR probe to the signal generator and Z_P is the impedance of the TDR probe in air. Frequency-domain analysis can be used to determine both the probe parameters (z and L) and the electromagnetic properties of the materials under test. To estimate the probe parameters [86,89], calibration measurements in some reference materials (e.g., water, air, ethanol) should always be performed. The fitting of $S_T(\omega)$ to the experimental scatter function, calculated by Eq. (9.56), allows the evaluation of z and L, since the electromagnetic properties of the reference materials are known from the literature. As an example of this approach, Fig. 9.25 shows the fitting results for the probe in water.

Once the probe parameters are known, TDR measurements can be performed in the probe filled with the material under test. The fitting of $S_T(\omega)$ to the experimental scatter function provides the electromagnetic constitutive parameters. In fact, for nonmagnetic materials, if the polarization process can be modeled with Debye-like relaxation, the relative complex permittivity is given by Eq. (9.57). $S_T(\omega)$ can be obtained applying Eqs. (9.57) and (9.58), whereas the minimization procedure allows evaluations of the model parameters and, consequently, the frequency dependence of test material EM properties. The same procedure can also be used assuming that the medium follows other relaxation models [65].

Another way to evaluate the electromagnetic parameters is the inversion analysis of TDR measurements [87,93,94]. In this case the model parameters are determined by matching the synthetic TDR signal with the signal measured in the time domain. If mixtures are analyzed, the use of dielectric mixing models in inversion analysis provides model parameters of each mixture component.

As an example of TDR application to estimate the electromagnetic parameters in the laboratory, Fig. 9.26 shows the experimental velocity values measured using the coaxial probe depicted in Fig. 9.20b on silica–magnetite mixtures with six different iron oxide volume fractions (0, 5, 15, and 25%) [50]. As expected, the wave velocity

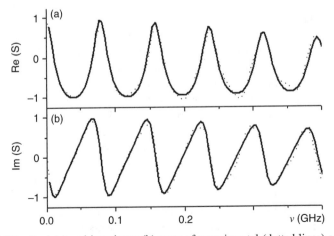

FIGURE 9.25 Real (a) and imaginary (b) parts of experimental (dotted lines) and theoretical (solid lines) scatter functions for a probe line filled with demineralized water.

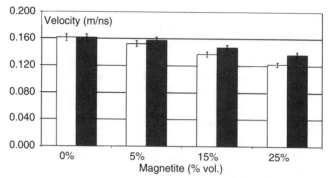

FIGURE 9.26 Velocity measured with TDR on two different mixture samples as a function of magnetite content. White Saro, natural grains; dark bers, crushed grains.

decreases with increasing magnetite volume fraction, since the refractive index n is higher for the magnetite with respect to the glass beads.

Conversely, Fig. 9.27 shows the results of a typical application of TDR technique in the field, where the spatial variability of both apparent permittivity and dc conductivity were studied, using a probe similar to the one depicted in Fig. 9.21a. Note that the strong lateral variability of the two parameters is due to a strong variation of the soil water content and the ionic concentration in the shallow subsurface water [95].

9.2.6 Ground-Penetrating Radar

Common GPR methods for soil characterization suffer from two major shortcomings. First, the forward model describing the radar data is subject to relatively strong simplifications with respect to electromagnetic wave propagation phenomena. Usually, wave propagation is assumed to be one-dimensional (plane wave) and the straight-ray approximation is used (see Section 9.4.3). In addition, wave propagation in the radar–cable–antenna system and the frequency-dependent antenna radiation pattern are usually not accounted for. This results in inherent errors and bias in soil properties' retrieval due to limited model adequacy. Furthermore, such simplifications do not permit exploitation of all the information contained in the radar data. In most applications, only the two-way travel time to a reflector is used and multiples or amplitude information is disregarded. To overcome such limitations, it is necessary to resort to full-waveform forward and inverse modeling approaches, involving three-dimensional solutions of Maxwell's equations [96–102]. Although such approaches still constitute a challenging task, they are progressively becoming a rational choice, owing to continuous progress in computing technologies [103].

A second problem arises from nonuniqueness and/or optimization issues in reconstructing subsurface images. The nonlinear inverse problem is usually constrained by providing additional information through smoothness conditions, via

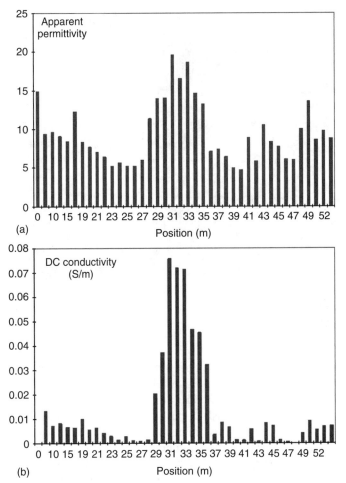

FIGURE 9.27   (a) Apparent permittivity and (b) dc conductivity measured along a profile (every meter) with a 30-cm-long vertical probe.

more advanced geostatistical models [104,105], or by including data from additional sensors [106,107]. A promising approach for time-lapse hydrological analyses is to resort to integrated, often referred to as joint, hydrogeophysical inversion techniques, where the geophysical and hydrodynamic models constrain each other [108–110]. In that way, the possible electromagnetic property distributions and correlated water content are not only limited to those respecting the physical laws of electromagnetics but also of hydrodynamics, thereby reducing significantly the solution space and complexity of the optimization problem. In the next two sections we describe new methods, where both the forward and inverse modeling issues outlined above are overcome. Applications of these methods, together with the use of classical methods as well, are presented in Section 9.4.

FIGURE 9.28   Handheld vector network analyzer with an off-ground horn antenna during a surface soil moisture measurement campaign in the irrigated areas in the region in Gabes, southern Tunisia. (From [111].)

Full-Waveform Modeling and Inversion of Off-Ground Monostatic GPR   We have recently developed a full-waveform electromagnetic model for the particular case of monostatic off-ground GPR; that is, a single antenna plays the role of emitter and receiver simultaneously and is situated at some distance above the soil [96]. The model includes internal antenna and antenna–soil interaction propagation effects and considers an exact solution of the three-dimensional Maxwell's equations for wave propagation in multilayered media. Both phase and amplitude information are used inherently for model inversion, thereby maximizing information retrieval capabilities from the measured radar data. The GPR system is set up with a vector network analyzer (VNA). The main advantage of VNA technology over traditional GPR systems is that the measured quantities constitute an international standard and are well defined physically with proper calibration of the system. Other advantages are the higher dynamic range and the possibility to avoid emitting in specific, narrow frequency bands, which may be sensitive to state regulations. The antenna consists of a double-ridged horn antenna, which is appropriate for off-ground operations, as it is coupled with air and highly directive. Figure 9.28 shows a handheld GPR system (VNA) used during a surface soil moisture measurement campaign in irrigated areas in southern Tunisia [111].

GPR Forward Modeling

Antenna Equation in the Frequency Domain   The radar signal as measured by the VNA consists of the frequency-dependent complex ratio $S_{11}(\omega)$ between the returned signal and the emitted signal, ω being the angular frequency. The antenna

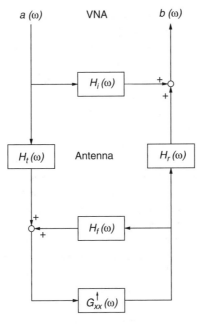

FIGURE 9.29   Radar-antenna-multilayered medium system modeled as linear systems in series and parallel. (From [96].)

is modeled using the block diagram depicted in Fig. 9.29. It relies on Maxwell's equation linearity and assumes that the spatial distribution of the backscattered electromagnetic field measured by the antenna does not depend on the air–soil (i.e., only the amplitude and phase change). This hypothesis corresponds to a local plane-wave approximation over the antenna aperture. This is expected to be a valid assumption if the antenna is not too close to the ground and assuming the soil to be described by a horizontally multilayered medium. The model consists of a linear system composed of elementary model components in series and parallel, all characterized by their own frequency response function accounting for specific electromagnetic phenomena.

The resulting transfer function relating $S_{11}(\omega)$ measured by the VNA (raw radar data) to the frequency response $G_{xx}^{\uparrow}(\omega)$ of the multilayered medium (see below) is expressed in the frequency domain by

$$S_{11}(\omega) = \frac{b(\omega)}{a(\omega)} = H_i(\omega) + \frac{H_t(\omega) G_{xx}^{\uparrow}(\omega) H_r(\omega)}{1 - H_f(\omega) G_{xx}^{\uparrow}(\omega)} \tag{9.59}$$

where $b(\omega)$ and $a(\omega)$ are, respectively, the received and emitted signals at the VNA reference calibration plane; $H_i(\omega)$, $H_t(\omega)$, $H_r(\omega)$, and $H_f(\omega)$ are, respectively, the return loss, transmitting, receiving, and feedback loss transfer functions of the antenna; and $G_{xx}^{\uparrow}(\omega)$ is the transfer function of the air–subsurface system modeled as a multilayered medium (referred to below as Green's function). All these functions are complex valued, thereby accounting for phase (related to propagation time) and amplitude effects. We can define $H(\omega) = H_r(\omega)H_t(\omega)$, so that the number of antenna transfer functions reduces to three. Due to inherent variations in the impedance between the antenna feed point, antenna aperture, and air, multiple wave reflections occur within the antenna. Under the assumption above, these reflections are accounted for exactly by the antenna transfer functions, which play the role of global reflectances and transmittances. In that way, the proposed model inherently takes into account the multiple wave reflections occurring between the antenna and the soil (through H_f). The characteristic antenna transfer functions can be determined by solving a system of equations such as (9.59) for these three unknown functions, by performing $G_{xx}^{\uparrow} = 0$ measurements for known model configurations (i.e., for which the Green's functions $S_{11}(\omega)$ can be computed) [96,112]. Generally, measurements are realized with the antenna at different heights above a perfect electric conductor, such as a copper sheet. A measurement can also be performed in free-space conditions, for which $G_{xx}^{\uparrow} = 0$. In that case, H_i is therefore directly equal to the measurement $S_{11}(\omega)$. Once the antenna transfer functions are known, antenna effects can be filtered out from the raw radar data, and the Green's function, containing only the response of the air–soil, can be derived as follows:

$$G_{xx}^{\uparrow}(\omega) = \frac{S_{11}(\omega) - H_i(\omega)}{S_{11}(\omega)H_f(\omega) - H_i(\omega)H_f(\omega) + H(\omega)} \tag{9.60}$$

Zero-Offset, Multilayered Media Green's Function   The solution of Maxwell's equations for electromagnetic waves propagating in multilayered media is well known. As an assumption to model the antenna, we define the Green's function as the backscattered (upward component denoted by the up arrow in $G_{xx}^{\uparrow}$, x-directed electric field (first subscript x in $G_{xx}^{\uparrow}$ at the antenna phase center for a unit-strength, x-directed electric source (second subscript x in $G_{xx}^{\uparrow}$ situated at the same position above a multilayered medium (Fig. 9.30).

The antenna is therefore modeled as a point source and receiver, and hence the radiation pattern is emulated by the pattern of a dipole assuming that a distributed source would permit as to account properly for the antenna radiation pattern. The antenna phase center represents the origin of the radiated field from which the far-field spherical divergence is initiated. Following Lambot et al. [96], the analytic expression for the zero-offset Green's function in the spectral domain (two-dimensional spatial Fourier domain) is found to be

$$\tilde{G}_{xx}^{\uparrow} = \frac{1}{8\pi} \left(\frac{\Gamma_n R_n^{TM}}{\eta_n} - \frac{\xi_n R_n^{TE}}{\Gamma_n} \right) \exp(-2\Gamma_n h_n) \tag{9.61}$$

where the subscript n equals 1 and here denotes the first interface and first layer (in practice, the air layer), R_n^{TM} and R_n^{TE} are, respectively, the transverse magnetic (TM) and transverse electric (TE) global reflection coefficients [113], accounting

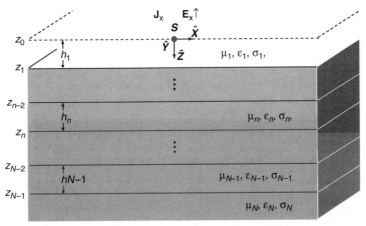

FIGURE 9.30 Three-dimensional layered medium with a point source and a receiver **S**. Each layer is characterized by the dielectric permittivity ε, electric conductivity σ, and thickness h.

for all reflections and multiples from surface and subsurface interfaces, Γ_n is the vertical wavenumber defined as $\Gamma_n = \sqrt{k_\rho^2 + \xi_n \eta_n}$, k_ρ is a spectral domain transform parameter, $\xi_n = j\omega\mu_n$, $\eta_n = \sigma_n + j\omega\varepsilon_n$ and $j = \sqrt{-1}$. The global TM- and TE-mode reflection coefficients at interface $n(n = 1, \ldots, N-1)$ are given by

$$R_n^{\mathrm{TM}} = \frac{r_n^{\mathrm{TM}} + R_{n+1}^{\mathrm{TM}} \exp(-2\Gamma_{n+1}h_{n+1})}{1 + r_n^{\mathrm{TM}} R_{n+1}^{\mathrm{TM}} \exp(-2\Gamma_{n+1}h_{n+1})} \tag{9.62}$$

$$r_n^{\mathrm{TM}} = \frac{\eta_{n+1}\Gamma_n - \eta_n\Gamma_{n+1}}{\eta_{n+1}\Gamma_n + \eta_n\Gamma_{n+1}} \tag{9.63}$$

$$R_n^{\mathrm{TE}} = \frac{r_n^{\mathrm{TE}} + R_{n+1}^{\mathrm{TE}} \exp(-2\Gamma_{n+1}h_{n+1})}{1 + r_n^{\mathrm{TE}} R_{n+1}^{\mathrm{TE}} \exp(-2\Gamma_{n+1}h_{n+1})} \tag{9.64}$$

$$r_n^{\mathrm{TE}} = \frac{\mu_{n+1}\Gamma_n - \mu_n\Gamma_{n+1}}{\mu_{n+1}\Gamma_n + \mu_n\Gamma_{n+1}} \tag{9.65}$$

where r_n^{TM} and r_n^{TE} denote the local plane-wave TM and TE mode reflection coefficients, respectively, at the interface n. These expressions are in a recursive form, similar to the one-dimensional version of Eq. (9.24). The recursion is initiated by the observation that there are no up-going waves from the lower half-space, such that $R_{N-1}^{\mathrm{TM}} = r_{N-1}^{\mathrm{TM}}$ and $R_{N-1}^{\mathrm{TE}} = r_{N-1}^{\mathrm{TE}}$. The local reflection coefficients determine that part of the electromagnetic wave that is reflected at a dielectric interface, while the other part is transmitted.

The transformation of Eq. (9.61) from the spectral domain to the spatial domain is carried out by employing the two-dimensional Fourier inverse transformation:

$$G_{xx}^{\uparrow} = \frac{1}{2\pi} \int_0^{+\infty} \tilde{G}_{xx}^{\uparrow} k_\rho \, dk_\rho \tag{9.66}$$

which has been reduced to a single integral in view of the invariance of the electromagnetic properties along the x and y coordinates. We developed an optimal procedure to properly evaluate that integral, which contains singularities [114]. As illustrated in Fig. 9.31, the integration path is deformed in the complex k_ρ plane by applying Cauchy's integral theorem. In addition to avoiding the singularities (branch points and poles), the path permits us to minimize the oscillations of the complex exponential part of the integrand, which makes the integration faster. Defining k_ρ as the complex number $(x + jy)$, the following relationship was found for the constant phase integration path:

$$y = \frac{x}{\sqrt{(xc_0/\omega)^2 + 1}} \tag{9.67}$$

where c_0 is the free-space electromagnetic wave velocity. This is illustrated in Fig. 9.31.

GPR Inverse Modeling   Inversion of the Green's function is formulated by a complex, least-squares problem as

$$\min \ \phi(\mathbf{b}) = |\mathbf{G}_{xx}^{\uparrow *} - \mathbf{G}_{xx}^{\uparrow}|^\mathrm{T} \mathbf{C}^{-1} |\mathbf{G}_{xx}^{\uparrow *} - \mathbf{G}_{xx}^{\uparrow}| \tag{9.68}$$

where $\mathbf{G}_{xx}^{\uparrow *} = G_{xx}^{\uparrow}(\omega)$ and $\mathbf{G}_{xx}^{\uparrow} = G_{xx}^{\uparrow}(\omega, \mathbf{b})$ are vectors containing, respectively, the observed and simulated radar measurements, from which major antenna effects have been filtered out using Eq. (9.60), is the error covariance matrix, and $\phi(\mathbf{b})$ is the parameter vector containing the soil electromagnetic parameters and layer thicknesses to be estimated. As function $\phi(\mathbf{b})$ usually has a complex topography, we use the global multilevel coordinate search algorithm [115] combined sequentially with the classical Nelder–Mead simplex algorithm for minimizing the function, thereby avoiding being trapped in local minima.

9.2.7 Combining Hydrological and GPR Modeling in a Closed-Loop Inversion Scheme

When the multilayered medium consists of a few layers only, the inverse problem is usually well-posed and only a limited number of unknowns have to be inverted for. In field conditions, however, the number of layers is typically unknown and soil moisture profiles vary piecewise continuously with depth, resulting inherently in an ill-posed electromagnetic inverse problem with a large number of unknowns. In that case, resorting to a joint electromagnetic–hydrodynamic inversion scheme appears to be necessary [100]. Such joint inversion strategies are attracting growing interest for solving hydrological problems, especially with the application of borehole GPR

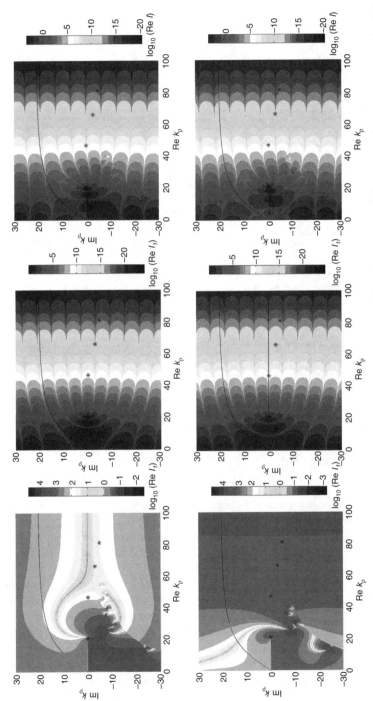

FIGURE 9.31 Optimal integration path in the complex plane for the real and imaginary parts of the two major components and of the integrand and for the full integrand. (From [114J.)

565

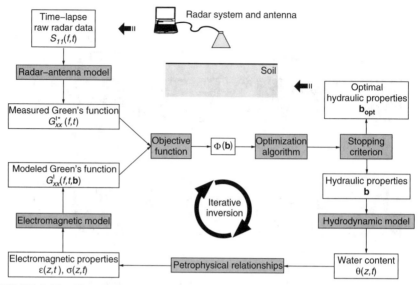

FIGURE 9.32 Flowchart representing the integrated electromagnetic and hydrodynamic inversion of time-lapse radar measurements for estimating unsaturated soil hydraulic properties and electromagnetic profiles (t is time, z is depth, and f is frequency). Shaded boxes denote operators, and white boxes denote variables. (From [119]).

tomography, where the classical geophysical inverse problem is typically ill-posed [104,109,116–118]. The proposed joint electromagnetic and hydrodynamic inversion procedure is depicted in Fig. 9.32 [110,119].

Inversion is formulated as a classical least-squares problem, where an objective function describing the discrepancies between the measured and modeled time-lapse radar data are minimized iteratively by means of an optimization algorithm:

$$\min \phi(\mathbf{b}) = \sum_t \sum_f |G_{xx}^{\uparrow *}(f, t) - G_{xx}^{\uparrow}(f, t, \mathbf{b})|^2 \qquad (9.69)$$

where t is the time for the hydrodynamic event. In this method, only time-lapse geophysical data are theoretically necessary and used effectively. When available, hydrological data may be included in the inversion flowchart [109]. It is worth mentioning that in the inversion flowchart, antenna effects are first filtered out from the raw radar data using the radar-antenna model [Eq. (9.60)]. With this inversion scheme, the traditional electromagnetic inverse problem for reconstructing electromagnetic and correlated water content profiles is constrained by a hydrodynamic model. In that case, the solution space is reduced significantly, as possible electromagnetic profiles cannot be arbitrary but should respect soil hydrodynamic laws. The parameters to be estimated reduce to the soil hydraulic properties, while the electromagnetic profiles, and correlated water content, are provided by the integrated model. Time-lapse radar data are therefore necessary as well as information regarding the hydrodynamic event, such as initial and boundary conditions. In addition, the soil-specific relationships relating the hydrological variables of interest to the soil electromagnetic properties should be known. Although the relationship between soil dielectric permittivity and water content is relatively constant for various soil types, the relation between electrical conductivity, water content, and salinity may be subject to larger uncertainties [35,120].

We showed that with such integrated inversion, elemental uniqueness and stability conditions of the inverse problem are theoretically satisfied for a range of different soils and hydrodynamic boundary conditions [121]. This means that enough information may be contained in the time-lapse, off-ground GPR data to estimate the soil hydraulic properties. Only in highly electrically conductive soils or in the absence of reflecting profiles are limitations expected due to the decreased information content of the radar data. The time-lapse GPR data can be acquired in a soil mapping context, which potentially provides a means to map the shallow (up to 30 to 50 cm) soil hydraulic properties with high spatial resolution at the field scale.

9.3 HYDROGEOPHYSICS

For laboratory measurements on soil samples, a parallel-plate capacitor, induction device, and transmission line (discussed in Section 9.2), can be used to determine the electric and magnetic properties of macroscopic volumes of soil under various conditions that are relevant to field studies. In this section we discuss briefly two ground-penetrating radar techniques that can be used for characterization of multicomponent three-phase porous media under field conditions. At the end of the section a hydrogeophysical application using geoelectric methods is described.

9.3.1 Role of GPR in Hydrogeophysics

Knowledge of the spatial distribution and dynamics of the surface and subsurface hydrological properties at various scales is essential in agricultural, hydrological, meteorological, and climatological research and applications. In particular, the *vadose zone*, defined as the unsaturated zone between the soil surface and the saturated aquifers, mediates many of the processes in the hydrological cycle that govern water resources and quality, such as the partition of precipitation into infiltration and runoff, groundwater recharge, contaminant transport, plant growth, evaporation, and sensible and latent energy exchanges between Earth's surface and its atmosphere. As an example, in catchment hydrology, the readiness of an area to generate surface runoff during storm rainfall is related to its surface storage capacity, a variable that can be computed accurately when surface water content is known. Disregarding the spatial variability of the surface water content may contribute significantly to errors on surface runoff estimation and flood forecasts, even at a limited spatial scale [122]. Given the predominant effects of soil moisture on the production of crops and soil salinization, agricultural and irrigation management practices also depend directly on a timely and accurate characterization of temporal and spatial soil moisture dynamics in the root zone [123] (see Section 9.6). Soil water dynamics also determine soil organic carbon sequestration and carbon-cycle feedbacks, which could affect climate change substantially.

Water flow and solute transport models in subsurface unsaturated porous media are now becoming readily available tools in the design and evaluation of management strategies for preserving the soil and water quality though sustainable exploitation and remediation plans. Nonetheless, the effectiveness with which these modeling tools in environmental management can be adopted relies heavily on the quality with which unsaturated flow parameters can be identified. The vadose zone's inherent spatial variability and inaccessibility precludes direct observation of the important subsurface processes. The development of appropriate methods for the determination and monitoring of the unsaturated flow properties of soils at scales that should be relevant to management practices remains a challenging task for the soil science community.

Commonly used techniques to characterize and monitor soil hydraulic properties are either suited to small areal scales (<0.1 m), such as soil sampling (e.g., the gravimetric method for soil water content determination), capacitive sensors, and time-domain reflectometry (TDR) [79], or to large areal scales (>10 to 100 m), such as airborne and spaceborne passive microwave radiometry and active radar systems [124–126]. In that respect, the development of field-scale techniques combined with hydrological models is needed not only to support field-scale management practices but also to upscale soil hydraulic properties and water flow processes and bridge the present scale gap between traditional ground probing and remote-sensing for improved remote sensing data products [127,128]. Present techniques still suffer from a lack of representative ground truths [127,129], as, for example, in the frame of new remote-sensing platforms such as SMOS [130] or SMAP (NASA, 2008; http://smap.jpl.nasa.gov). The characterization of spatial patterns and heterogeneities over a continuous range of scales is presently subject to intensive research for developing, calibrating, and testing distributed hydrological models, with, for example, the installation of field- to watershed-scale observatories [79,123,131,132].

In general, microwave remote-sensing methods have had the greatest success in producing surface soil moisture estimates suitable for assimilation in large-scale hydrological or land surface models [133]. Yet none of the available remote-sensing methods is sensitive to the vertical distribution of moisture in the soil, due to the narrowband and relatively high operating frequencies of the sensors (>1 GHz). Currently, vertical profiles of soil moisture can only be remotely estimated using soil hydrodynamic modeling, for which knowledge of the soil hydraulic properties is required. Field or watershed scale hydraulic parameters are often derived from soil texture information using pedotransfer functions [134], but the soil parameterization schemes often remain inadequate due to their inability to incorporate the natural heterogeneity of soils and the lack of detailed soil property maps. Recognizing these limitations, numerous studies have attempted to optimize large-scale hydraulic parameters using remotely derived surface soil moisture as an additional constraint [135–137]. Yet surface soil moisture time series are not sufficient to constrain the soil hydraulic properties properly, and little progress has been made toward the determination of more physically consistent parameter sets over large areal scales, including agricultural fields.

Among existing field-scale techniques such as electric resistivity or electromagnetic induction, which are quite widespread techniques nowadays to delineate management zones in quite a qualitative way (see Section 9.5), over the past decade GPR has become a very promising tool for mapping the soil hydrological properties at an intermediate scale (0.1 to 1000 m). GPR is adapted particularly to noninvasively determined soil water content [102,131,138]. As the dielectric properties of water overwhelm those of other soil components, water and its spatial distribution strongly control GPR wave propagation in the soil. Electric conductivity, which depends simultaneously on different soil properties, such as water content, ion concentration of the soil solution, and soil type [139], determines electromagnetic wave attenuation. In the areas of unsaturated zone hydrology and water resources, GPR has been used to identify soil stratigraphy [140], to locate water tables [141], to follow wetting-front movement [142], to estimate soil water content [138,143–146], to assess soil salinity, and to support the monitoring of contaminants. Time-lapse GPR measurements permit the monitoring of subsurface flow processes and inference of the soil hydraulic properties when coupled with soil hydrodynamic modeling. This has been applied particularly to cross-hole GPR, where one- or two-dimensional time-lapse images of the soil water content are obtained from travel-time tomography analysis and are subsequently inverted to retrieve key soil hydraulic parameters [108,116–118,147].

In this section we describe GPR techniques which are commonly used to identify surface and depth-dependent soil water content and retrieve the unsaturated soil hydraulic properties. Soil dielectric permittivity and correlated water content profiles are usually generated using ray-tracing-based algorithms, thereby implying the plane-wave approximation and straight or curvedrays [148,149]. Application examples are presented and the advantages and limitations are discussed.

9.3.2 GPR Techniques for Soil Moisture Retrieval

Surface Reflection Method for Soil Surface Water Content Estimation The common surface reflection method applies to air-launched GPR configurations, either monostatic or bistatic, and is based on the determination of the reflection coefficient of the soil surface interface. The following assumptions are considered in particular: (1) the antennas are located in free space (air) above a homogeneous half-space (soil) limited by a plane interface, (2) the reflection coefficient can be approximated by the plane-wave reflection coefficient, (3) antenna distortion effects are negligible, (4) the soil electric conductivity is assumed to be negligible, (5) the magnetic permeability is assumed to be equal to the free-space permeability, and (6) the dielectric permittivity is frequency independent. As a result, the reflection coefficient at the soil interface is a Dirac's delta function of time, and its amplitude is defined as the ratio between the backscattered (E_s) and incident (E_i) electric fields. For a normal incidence plane wave, the amplitude R of the reflection coefficient can thus be expressed as

$$R = \frac{1 - \sqrt{\varepsilon_r}}{1 + \sqrt{\varepsilon_r}} \tag{9.70}$$

where ε_r is the relative dielectric permittivity of the soil. The soil dielectric permittivity can therefore be derived as

$$\varepsilon_r = \left(\frac{1-R}{1+R}\right)^2 \tag{9.71}$$

The reflection coefficient R is usually determined from the measured amplitude of the soil surface reflection, A, relative to the amplitude measured for a perfect electric conductor (PEC) situated at the same distance as the soil: namely, A_{PEC}. The ratio between the reflection coefficient at the soil surface interface (R) and at a PEC interface (R_{PEC}) can be expressed as

$$\frac{R}{R_{PEC}} = \frac{E_s/E_i}{E_{s,PEC}/E_i} \tag{9.72}$$

Since $R_{PEC} = -1$, assuming E_i to be constant and assuming that the measured amplitude A is directly proportional to the backscattered electric field E_s (i.e., that there are no antenna distortion effects), Eq. (9.72) reduces to

$$R = -\frac{E_s}{E_{s,PEC}} = -\frac{A}{A_{PEC}} \tag{9.73}$$

Notwithstanding the practical appropriateness of this method for mapping applications, the method has been applied by a few authors only [143,150–152]. A major limitation of the method is related to the requirement of PEC measurements with the antenna at exactly the same height as for measurements above the soil. The concept is, however, commonly used in airborne and spaceborne radar remote sensing for the retrieval of soil surface water content [153].

Full-Waveform Inversion Adapted to Soil Surface Water Content Estimation
A more exact and more practical method for soil surface water content estimation is to resort to full-waveform inversion of off-ground GPR data (see Section 9.3). To identify the surface dielectric permittivity and correlated water content, we focus full-waveform inversion of the Green's function in the time domain on the surface reflection [154,155]. The measured and modeled frequency-domain Green's functions are first transformed in the time domain using the inverse Fourier transform:

$$g_{xx}^{\uparrow}(t) = F^{-1}(G_{xx}^{\uparrow}(\omega)) = \frac{1}{2\pi} \int_{-\infty}^{+\infty} e^{j\omega t} G_{xx}^{\uparrow}(\omega)\, d\omega \tag{9.74}$$

where ω is the angular frequency, t is time, and $j = \sqrt{-1}$. Then a time window is defined between t_{min} and t_{max}, focused on the surface reflection (Fig. 9.33). It is worth noting that since measurements are performed over a limited frequency range with a constant-amplitude sweep, the surface reflection in $g_{xx}^{\uparrow}(t)$ tends to be antisymmetric. The root t_i of the signal between the positive and negative peaks of

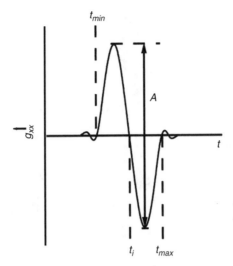

FIGURE 9.33 Radar signal in the time domain, $g_{xx}^{\uparrow}(t)$ from which antenna effects have been filtered out. Times t_{min} and t_{max} define the time window focused on the surface reflection in which inversion is performed; t_i is the time corresponding approximately to the surface interface in the space domain. A is the amplitude of the reflection considered when using the common surface reflection method, although the surface reflection has a different shape, as it contains some antenna effects.

the reflection corresponds exactly to the surface interface in the space domain only under hypotheses of the common surface reflection method. In practice, t_i does not correspond precisely to the surface interface because the reflection coefficient is time dependent because of (1) geometrical spreading in wave propagation (the reflection originates mainly from the Fresnel zone), (2) the delaying effect of electric conductivity, and (3) the distortion effects due to the frequency dependence of the soil electromagnetic properties.

The inverse problem is formulated in the least-squares sense and the objective function is accordingly defined as follows:

$$\Phi(b) - \left(\mathbf{g}_{\mathbf{xx}}^{\uparrow*} - \mathbf{g}_{\mathbf{xx}}^{\uparrow}\right)^{\mathrm{T}} \left(\mathbf{g}_{\mathbf{xx}}^{\uparrow*} - \mathbf{g}_{\mathbf{xx}}^{\uparrow}\right) \tag{9.75}$$

where

$$\mathbf{g}_{\mathbf{xx}}^{\uparrow*} = g_{xx}^{\uparrow*}(t)\big|_{t_{min}}^{t_{max}} \quad \text{and} \quad \mathbf{g}_{\mathbf{xx}}^{\uparrow} = g_{xx}^{\uparrow}(t)\big|_{t_{min}}^{t_{max}} \tag{9.76}$$

are the vectors containing, respectively, the observed and simulated time domain, windowed Green's functions, and $\mathbf{b} = [\varepsilon_{r,1}, d_0]$ is the parameter vector to be estimated, with $\varepsilon_{r,1}$ being the soil surface relative dielectric permittivity and d_0 being the distance between the antenna phase center and the soil surface interface. The subscripts 0 and 1 denote, respectively, the air and first soil layers. As for the common surface reflection method, the electric conductivity, magnetic

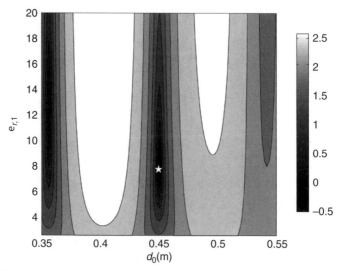

FIGURE 9.34   Objective function topography for the full-waveform inversion focused on the surface reflection, where $\varepsilon_{r,1}$ is the soil surface relative dielectric permittivity and d_0 is the antenna height above the soil surface. The white star represents the global minimum of the objective function. (From [155].)

permeability, and soil layering are assumed to have a negligible effect on the estimation of $\varepsilon_{r,1}$. The model configuration used for the inversion thus consists of a point source above a three-dimensional lower half-space.

The two-dimensional objective function in Eq. (9.75) is minimized using the local Levenberg–Marquardt algorithm. An initial guess for the antenna elevation d_0 is derived from the root t_i of the signal between the positive and negative peaks of the surface reflection (see Fig. 9.33). The initial guess for the relative dielectric permittivity is chosen arbitraryily in the range $2.5 \leq \varepsilon_r \leq 20$. We have verified that this procedure always leads to the global minimum of the objective function, which is characterized by a relatively simple topography, as shown in Fig. 9.34.

Surface Ground-Wave Methods   Surface soil dielectric permittivity and correlated water content can also be derived from the determination of the surface ground-wave velocity [95,156–162]. The *ground wave* is the signal that travels directly from source to receiving antenna though the soil surface. The depth of influence of GPR ground-wave data for soil moisture estimation has been estimated by various investigators to be between 10 and 50 cm, depending on acquisition parameters, mainly the operating radar frequency, and soil conditions. When the ground-wave has been identified using several measurements with increasing antenna separations, the ground-wave velocity can be determined from measurements with a fixed antenna separation [single trace analysis (STA)]. However, it was observed that STA leads to higher uncertainties in wave speed estimation compared to multiple measurements and suggested using GPR with multiple receivers for mapping applications [160]. The limitations of the technique for practical field applications are the required contact between the antennas and the soil, the identification of the ground wave, which may be ambiguous or even impossible in some conditions, and the presence of ambiguous guided waves when near-surface layering is present [101,163]. In addition, no information on the soil electric conductivity is provided. All of this may restrict the application of this method for agriculture and other environmental mapping applications.

A ground-coupled GPR system works with source and receiver antennas located at the soil–air interface; in this configuration, four fundamental waves can be distinguished and represented by ray paths and wavefronts sketched in Fig. 9.35. In particular, Fig. 9.35b shows the waves generated by an electric current point source which splits into two spherical waves propagating in the upper and lower media. The additional "conical wave" and "ground wave" guarantee continuity in the subsoil and in the air, respectively [164]. The conical wave is homogeneous

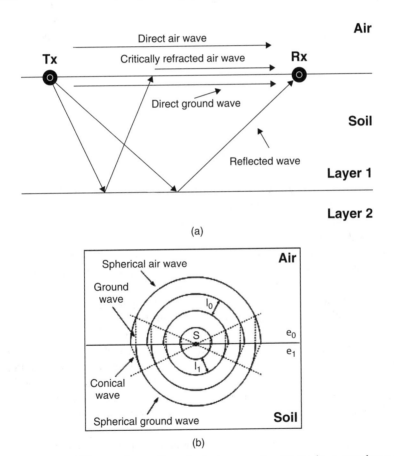

FIGURE 9.35 (a) Propagation paths of electromagnetic waves in a two-layer soil. (b) Wavefronts generated at the interface by an electric current source point (S) at the boundary interface between air and soil.

and propagates with the velocity of light, while the ground wave is inhomogeneous, propagates with the velocity of a wave in the soil and attenuates along the direction perpendicular to the wavefront. The ground wave is received early (after the air wave), but its evanescent character makes it difficult to detect with large antenna offsets. A more accurate discussion on this aspect is given by Annan [164], Brekhovskikh [165], and Berktold et al. [166].

The time delay Δt between the air and ground wave can be estimated using the equation

$$\Delta t = \frac{d}{c}(\sqrt{\varepsilon_r} - 1) \tag{9.77}$$

where d is the antenna separation, c the speed of light (3×10^8 m/s), and ε_{ra} the relative, apparent soil permittivity.

In for order to theoretically estimate the depth of the region intersected by the ground wave, it is possible to refer to the conclusions reached by van Overmeeren et al. [145], who, referring to the center of the antenna, have determined the depth z as

$$z = \sqrt{\lambda\, d} \tag{9.78}$$

which, in turn, is related to the permittivity by imposing

$$\lambda = \frac{c}{f}\sqrt{\varepsilon_r} \tag{9.79}$$

where λ is the wavelength and f is the central frequency of the GPR signal. Further estimations of ground-wave depth have been calculated by others authors, who found a sampling soil thickness ranging from $0.145\lambda^{1/2}$ to $\lambda/2$, with λ expressed in meters ([158] and references therein).

For soil water content estimation, two different approaches based on measurement of the ground-wave velocity—the multioffset and the single-offset configurations—have been tested and their accuracy has been evaluated [138]. The multioffset method requires the use of bistatic GPR equipment with detachable antennas (to achieve a suitable distance between the transmitter and the receiver) in order to unequivocally discriminate the air wave from the ground wave [160]. Once the ground wave is separated from other arrivals, its velocity can be determined by the slope of the linear relationship between antenna separation and ground-wave travel times [160]. The main disadvantages in estimating soil water content using multioffset GPR measurements are the time-consuming acquisition period and the fact that an average permittivity information over a large horizontal distance is obtained.

The ground-wave velocity can be determined from single-offset GPR measurements [167] when monostatic antenna or fixed (nonseparable) bistatic antenna configurations are used. This method allows for quicker data acquisition but is affected by some uncertainties since the arrival time of the ground-wave leading edge can be difficult to estimate [157], due to the superposition of the ground wave with the critically refracted and reflected waves. Moreover, in a single-offset configuration it might be difficult to choose an antenna separation for which the air and ground waves can always be separated when moving antennas across a field with varying soil water content.

Recently, a new approach to determining if the information on shallow subsurface soil dielectric permittivity (and therefore water content) carried by the early-time GPR signal's energy could be extracted properly has been explored [95]. The method is based on the assumption that the various antenna parameters, including the signal amplitude of the wavelet emitted by the transmitting antenna, are affected by the electromagnetic properties of the surrounding medium [168,169]. In this approach the average envelope amplitude of the first few nonoseconds of the GPR signal (acquired in a ground-coupled single-offset configuration) can be used conveniently for rapid mapping of the local variations of soil dielectric permittivity, without the requirements needed in reflection mode measurements and even if the air and ground waves are not decoupled. The *early-time method* has been tested on a site characterized by a very strong lateral gradient of the soil electrical properties, to achieve a large range of variation in the physical parameters, and the results indicate that this approach could be used conveniently for fast spatial variability mapping of the near-surface soil electrical properties [95].

Vertical Dielectric Profile Determination   A commonly used method to identify the depth-dependent dielectric permittivity that governs wave propagation speed is the common midpoint (CMP) method [170,171]. A different approach is to resort to borehole GPR. With this method, stacking velocity fields are extracted from multioffset radar soundings (i.e., measurements with different antenna separations), at a fixed central location. Wave propagation velocities in the ground can be obtained using the Pythagorean theorem or by tomographic inversion. The dielectric permittivity can then be computed directly and related to the water content. The CMP method has been used to estimate the groundwater level in an environment with multiple reflectors occurring at different depths [141]. The vertical dielectric permittivity distribution was estimated from the interval velocities obtained from the CMP. However, CMP-derived velocity estimates are generally characterized by low resolution and high uncertainty [172]. In addition, the most important practical disadvantage of CMP at present is that it cannot be used for real-time mapping, as it requires several measurements, with the antennas in contact with the soil, for a single profile characterization. The use of antenna arrays is a solution for fast surveys.

A different approach to reconstructing vertical dielectric profiles is to resort to borehole GPR, where the transmitting and receiving antennas are lowered at different depths in boreholes separated by some known distance. The soil dielectric permittivity is then determined from the determination of the wave propagation velocity between the boreholes from the travel times measured. Conducting multioffset tomographic measurements makes it possible in particular to reconstruct two-dimensional images of the soil dielectric permittivity and correlated water content between the boreholes [173–177]. The method is not appropriate for field-scale characterization, but permits high-resolution characterization and monitoring over distances of several meters.

9.3.3 High-Resolution Surface Soil Moisture Mapping in an Agricultural Field

This study was performed in the framework of the Hydrasens project (Belgian Science Policy Office, Stereo II Programme, project SR/00/100), where GPR-derived, high-resolution maps of soil surface water content are compared to estimations provided by spaceborne, synthetic aperture radar (SAR) remote sensing. The overall goals of the project are (1) to explore new strategies to integrate radar remote sensing, hydrologic, and hydraulic modeling for water management purposes through data assimilation, and (2) to demonstrate the applicability of advanced data assimilation schemes for a set of water management problems. The major outcomes should be fundamental knowledge about radar data assimilation in hydrology and a set of data and operational procedures with improved remote-sensing data products for supporting water management: in particular, implementation of the European Water Framework Directive and the forthcoming Directive on Floods.

The full-waveform method focused on how surface reflection (Section 9.3.2) was used to map surface soil moisture in an agricultural field. Figure 9.36 represents the GPR platforms designed for this application. The radar system is based on vector network analyzer (ZVRE, Rohde & Schwarz, Munich, Germany) technology, thereby setting up a stepped-frequency continuous-wave (SFCW) radar. The radar was controlled automatically by a computer and combined with a differential GPS for accurate, real-time positioning and mapping. The antenna system consisted of a linearly polarized, double-ridged broadband horn antenna (BBHA 9120 F, Schwarzbeck Mess-Elektronik, Schönau, Germany). The antenna was 95 cm long with a 68×96 cm^2 aperture area. The antenna nominal frequency range is

(a)   (b)

FIGURE 9.36 GPR system consisting of a vector network analyzer and a monostatic, off-ground horn antenna (200 to 2000 MHz) combined with a differential GPS for real-time mapping of surface soil moisture in agricultural fields. (a) Test site of the Forschungszentrum Jülich (FZJ) in Germany. (b) Test site of the Université Catholique de Louvain (UCL) in Belgium. The platform in (b) is also equipped with an EMI system (EM38, Geonics) for remote soil electric conductivity determination.

0.2 to 2 GHz, and its far-field isotropic gain ranges from 9 to 14 dBi. The high directivity of the antenna (45° 3-dB beam width in the E and H planes at 1 GHz) makes it suitable for use off the ground. Measurements were performed with the antenna aperture situated about 110 cm above the soil surface. This minimal height above the ground was necessary for this specific antenna to ensure adequacy of the electromagnetic forward model. The antenna was connected to the reflection port of the radar via a high-quality N-type 50-Ω coaxial cable. We calibrated the radar at the connection between the antenna feed point and the cable using an Open-Short-Match calibration kit. The frequency-dependent complex ratio S_{11} between the signal returned and the signal emitted was measured sequentially at 301 stepped operating frequencies over the range 200 to 2000 MHz with a frequency step of 6 MHz. Only lower frequencies, between 200 and 800 MHz, were considered for the inversions, to avoid roughness and crop effects [112]. The platform was also equipped with an electromagnetic induction (EMI) system (EM38, Geonics, Canada) situated about 25 cm above the soil surface to map soil electrical conductivity simultaneously. The EMI system was calibrated following standard Geonics' instructions, and the EM38 provided directly soil electrical conductivity values.

The agricultural field on which the GPR and EMI measurements were carried out in early April 2008 is situated in the loamy belt area in central Belgium (Walhain). The soil was a loamy soil over the entire field. Figure 9.37 shows elevation and slope maps of the 16-ha area. The slope ranges from 0% up to about 6%. A higher plateau is visible on the west side of the plot, while the elevation decreases progressively to the east. The soil was covered with winter wheat, as shown in Fig. 9.36b. The GPR and EMI measurements were taken following fertilization–cultivation tractor tracks, parallel to the SE–NW direction, to minimize crop damage. Spacing between the tracks was about 27 m. Along the tracks, measurements were acquired every 3 m, resulting in a total of about 3000 radar measurements over the entire field surface.

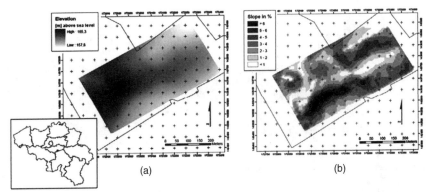

FIGURE 9.37   (a) Elevation map and (b) a corresponding slope map in an agricultural field in Walhain in central Belgium (Hydrasens project, Université Catholique de Louvain). The elevation model was obtained during the GPR and EMI measurements from the differential GPS. The maps were obtained using kriging, with about 3000 measurements over an area of 16 ha.

Figure 9.38 shows both GPR- and EMI-derived maps of soil surface water content and electrical conductivity, respectively. Compared to the topography, we observe some correlation, with in general higher water contents in the lower parts of the field and lower water contents in the upper parts, especially when the slope is high. It is worth mentioning that the surface water content is also influenced significantly by the local topography at a scale of a few meters, thereby explaining some complex patterns of soil moisture with respect to the field-scale topography. These patterns also originate from agricultural practices. The field picture (Fig. 9.38a) shows an area of ponding in a small basin, which is detected clearly in the corresponding area of the GPR image. In that basin we also observe a close correspondence between the soil water content distribution and the wheat characteristics (yellow color of the leaves). The wheat apparently suffers from a lack of oxygen in the root zone, due to water saturation. The SE–NW patterns observed are artifacts of the different resolutions used in mapping the field (perpendicular vs. parallel to the tracks).

The soil electrical conductivity map (Fig. 9.38b) is not fully correlated with the soil surface water content map. The origin of this difference is twofold. First, the measurement scales of GPR and EMI with respect to depth are significantly different. The GPR-derived soil moisture is expected to represent the top 10 cm of the soil profile, whereas the EMI system, used in vertical mode off the ground, is expected to be sensitive up to a depth of 50 to 100 cm. Second, it is worth noting that the electrical conductivity is multivariate and does not depend on soil moisture alone, but also on clay content, salinity, and density, which to some extent vary independently within a field. This makes its interpretation difficult and correlation with soil moisture relatively limited. Knowledge of soil electric conductivity is important in some conditions for surface water content estimation from the surface reflection, as above some threshold it influences the amplitude of the reflection significantly, resulting in overestimations of the surface dielectric permittivity if not

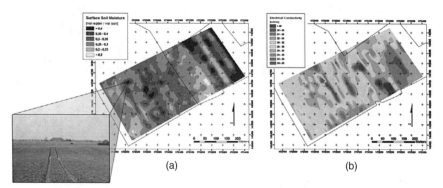

(a)   (b)

FIGURE 9.38 (a) Surface soil moisture and (b) bulk electrical conductivity in an agricultural field in Walhain in central Belgium. The maps were obtained from about 3000 GPR and EMI measurements using kriging. The field picture shows ponding at the soil surface, which corresponds well to a saturated area in the soil moisture map.

accounted for [155]. In another application of the full-waveform inversion [111], the technique was used in irrigated areas of southern Tunisia using a handheld radar system.

9.3.4 Early-Time GPR Signal Method in Field Measurements

A new approach based on the analysis of early-time GPR signals acquired in a single-offset configuration has been tested [178]. A site with strong lateral gradients in physical properties of the soil was selected on the basis of previous geophysical investigations performed on a gas vent, which have shown a strong variation in the electrical properties of the shallow subsurface. The gas vent is clearly visible on the surface as a 6-m-diameter subcircular area completely lacking vegetation (Fig. 9.39), surrounded by a series of quasi-symmetrical halos of progressively less stressed vegetation, referred in this paper as a sparsely vegetated zone (2 m wide), a moderately vegetated zone (3 m), and a yellow vegetated zone (6 m). Beyond this area the vegetation is a normal green color and does not appear to be affected by the gas released from depth.

A GPR survey was conducted over the gas vent using a Noggin^{PLUS} system (Sensors & Software, Inc., Canada) equipped with 250-MHz antennas and a SmartCart. Two long, preliminary profiles were first conducted across the center of the vent (one N–S, the other E–W), and then a grid of 61 parallel N–S profiles were performed at 1-m intervals across the structure to obtain a pseudo-three-dimensional image. All profiles were acquired with a fixed offset mode (FOM), with a 5-cm step size (measured with an onboard odometer), a time window of 60 ns, and a trace stacking of 16. To estimate soil relative permittivity and dc conductivity over a depth interval of 0 to 30 cm, TDR data along the central GPR line of the survey

FIGURE 9.39 Gas vent investigated with GPR and TDR techniques. Note the nonvegetated core of the vent.

grid (every 1 m) were collected using a Tektronix 1502C cable tester (Tektronix Inc., Beaverton, Oregon) connected through a 50-Ω coaxial line to a three-pronged probe. The probe, inserted vertically in the soil, consisted of three 30-cm-long stainless-steel parallel rods with a diameter of 0.4 cm and a distance of 4.5 cm between each rod.

GPR data have been processed using complex trace analysis [179]. This technique produces sections known as *instantaneous attribute displays* because the attributes are computed on a sample-by sample basis. In particular, amplitude information is encoded in the envelope amplitude attribute, also called *instantaneous amplitude*, which is a robust, smoothed, polarity-independent measure of the energy in the trace at a given time. Figure 9.40 is a time-slice map (0 to 4 ns) of the area investigated calculated using signal instantaneous amplitude. Note that the figure shows strong amplitude spatial variations, which correspond roughly to the three areas around the vent described above.

From the TDR measurements we calculate the apparent relative permittivity according to the equation

$$\varepsilon_{ra} = \left(\frac{ct}{2L}\right)^2 \tag{9.80}$$

where c is the speed of light in a vacuum, t the two-way travel time measured with the TDR, and L the length of the probe inserted in the soil. TDR dc conductivity

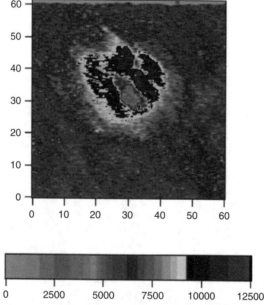

FIGURE 9.40 Time-slice map (0 to 4 ns) of the area around the gas vent calculated using an average envelope amplitude technique. The spatial distribution of the amplitude reflects the geometrical structure of the vent described above.

was computed according to the equation

$$\sigma_{dc} = \frac{\varepsilon_0^{1/2} Z_0}{\mu_0^{1/2} L Z_{probe}} \frac{2V_0 - V_F}{V_F} \tag{9.81}$$

where Z_0 is the impedance of the TDR feeder transmission line (50Ω), Z_{probe} the probe impedance in air, L the rod length, V_0 the TDR input step voltage, and V_F the final asymptotic voltage.

To compare GPR and TDR data, the central line of the GPR grid has been processed separately. The attribute representing the amplitude of the real trace envelope was computed on several time windows (0–4, 1–5, 2–6, 3–7, 4–8, 5–9, 6–10, and 8–12 ns) to measure the signal strength, which provides a value correlated to the wavelet energy content and its variation in time and space. Note that the time window of 4 ns was chosen on the basis of the nominal time length of the wavelet transmitted. Moreover, a radar trace was extracted every meter from the GPR section (i.e., in correspondence with the location of the TDR measurements) and used to compute a statistical correlation. Figure 9.41 is an example of the early-time GPR traces acquired in areas having different permittivities. It is clear from the figure that the waveforms differ in terms of amplitude and exhibit different "time stretching." Such variations appear to be related to the soil electrical parameters, with the higher amplitudes and shorter lengths associated with the lower permittivities.

A quantitative estimate of the degree of linear correlation between all the measured quantities is given by the coefficient r calculated using the pairs of measurements taken in the same location along the radar line. Table 9.1 summarizes the r coefficients obtained for the signal amplitudes and the soil parameters, computed for all time windows. The time windows shown in the table are measured starting from the air wave arrival time, which triggers the receiver data acquisition. The data presented in the table show that the amplitude/permittivity (A/ε_r) and

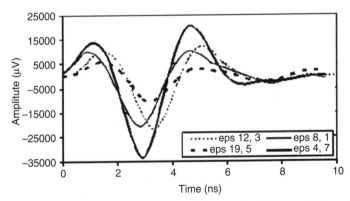

FIGURE 9.41 Early-time (0 to 10 ns) wavelets acquired for different relative soil permittivities from 4.7 to 19.5 ("eps" stands for $\varepsilon_{ra.}$).

TABLE 9.1 Correlation Coefficients(*r*) Between Amplitude (A), Relative Permittivity ($\varepsilon_{ra.}$), and Conductivity $\sigma_{dc.}$ for Various Time Windows

Time Window (ns)	$A/\varepsilon_{ra.}$	$A/\sigma_{dc.}$
0–4	0.80	0.71
1–5	0.81	0.73
2–6	0.82	0.75
3–7	0.83	0.78
4–8	0.83	0.77
5–9	0.81	0.71
6–10	0.77	0.58
8–12	0.49	0.12

amplitude/conductivity (A/σ_{dc}) correlations are high and essentially constant up to 9 ns. Such results suggest that the electromagnetic signals arriving at different times at the receiver antenna are affected equally by the soil dielectric properties through variations of the antenna parameters. In particular, the relatively high *r* value observed in the 0–4 ns window can be explained by a possible soil-antenna loading which affects the radiated power and consequently influences the overall field amplitudes, including the direct signal propagating in air [169].

In this study the hypothesis that the direct GPR signal amplitude is strongly correlated with the electromagnetic soil parameters was tested. The signal amplitude attribute evaluated on the early-time portion of the GPR trace was used to estimate the correlation between the soil electromagnetic properties measured with a TDR and the energy content of the GPR signals. The experimental method proposed was used on a site characterized by a very strong lateral gradient of the soil electrical properties, to achieve a large range of variation in the physical parameters and to ensure that the correlation coefficients estimated were statistically significant. The data have shown a relatively high degree of correlation and therefore indicate that the early-time method could be used conveniently for spatial variability mapping of the near-surface soil electrical properties (and water content) using GPR systems in single-offset mode (without employing detached-antenna configuration), which represents a potential for fast three-dimensional acquisition.

9.3.5 Water Table Depth Determination Using GPR and TDR Techniques

GPR and TDR techniques can be considered complementary methods in soil water content evaluation, due to their different horizontal and vertical spatial resolution. TDR can explore and give accurate information on a small volume of soil and is therefore sensitive to small-scale variations. From a practical point of view, however, TDR makes the assessment of spatial soil water content variation labor intensive, because probes need to be installed at each individual measurement location. On the other hand, GPR can be seen as the natural intermediate-scale counterpart of TDR for providing soil water content information with the accuracy required at the scale of interest of most common applications.

The scope of the study presented here is to better understand how the water table, the capillary fringe, and the unsaturated zone affect the results given by GPR and TDR. To this end, a specific test site in which the soil water content and the water table could be controlled was constructed [180]. The test site consists of a pit (area 4 m × 6 m, depth 1.2 m) excavated in the tuff and isolated with a poly (vinyl single chloride) (PVC) sheath. The pit was filled with about 1 m of dry sand overlaying 20 cm of gravel. The choice of sand allows minimizing the effect of soil surface roughness on GPR data. The system created to control the water level in the test site consisted of several vertical and horizontal PVC pipes (Fig. 9.42). Holes were drilled in the horizontal pipes, which were then covered with a wire mesh, which in turn was covered with a porous cloth. The pipes for water inflow and drainage were positioned in the gravel layer just below the interface between the gravel and the sand. The gravel, characterized by a very high porosity, was chosen to allow rapid water inflow and outflow and to produce fairly homogeneous soil wetting. A vertical pipe connected to the three horizontal buried pipes was fed by a 1000-L tank for water storage. Data were acquired for different water content conditions, and the water level was monitored using two piezometers inserted in the sand on opposite sides of the pit. Landmarks have been used to define the GPR acquisition grid (Fig. 9.42b). Finally, to investigate the water content of the first half-meter of the test site with a high vertical spatial resolution, close to one of the two piezometers, a multilevel TDR probe was inserted in the sand.

GPR measurements were carried out on the test site using a Noggin^{PLUS} system equipped with 250- and 500-MHz antennas and a SmartCard (Sensors & Software, Inc.) using the single-offset reflection method. Measurements were made on a grid (2.60 × 6.00 m) composed of 14 parallel lines, 20 cm apart, using an odometer, with a 5-cm step size for the 250-MHz antenna and 2.5-cm step size for the 500-MHz antenna. The time window selected was 60 ns and at each acquisition point a stacking of four traces was set. The same data set was acquired for both antenna frequencies in all water conditions.

TDR measurements were collected with a Tektronix 1502C cable tester (Tektronix Inc., Beaverton, Oregon) equipped with a RS232 interface for data recording into a laptop PC. The cable tester was connected (through a 50-Ω coaxial line) to a multilevel probe system. The multilevel probe consists of an array of 11 open-ended transmission-line segments deployed horizontally. Each transmission line consists of three parallel stainless-steel rods (length 18.8 cm, diameter 3 mm) with a 5-cm interelectrode distance. The vertical separation between each probe is 4 cm. The upper probe was put in the sand 8 cm below the surface, and thus the area investigated extended down to 52 cm.

The radargrams obtained with both antennas allowed good reconstruction of the geometry of the pit; all the events representing the sand–gravel interface and the bottom of the pit were clearly visible for all water-level conditions. The diffraction hyperbola generated by the buried pipes can also be recognized easily. Figure 9.43 shows a sequence of the same subsection (from 2.50 to 4.50 m) of the GPR profile

(a)

(b)

FIGURE 9.42 (a) Water system in the gravel; (b) test site completed with the grid for GPR and TDR acquisition.

acquired close to the TDR multilevel probe, collected by 500-MHz antennas, for different water table levels. Note that the reflection from the bottom of the pit becomes "deeper in time" when the water table increases, due to the higher water volume injected in the pit. Moreover, when the water level rises, the radar reflection from the water table cannot always be clearly detected because the water interface does not appear flat, due to the inhomogeneity of the capillary fringe and the unsaturated zone.

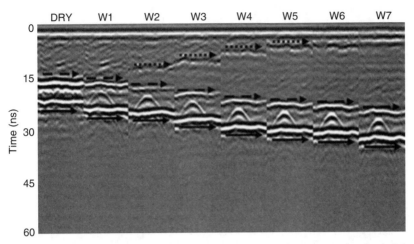

FIGURE 9.43 Sequence of the same GPR subsection for different water-level conditions. The bottom of the pit (solid arrows), the sand–gravel interface (dashed arrows), and the water table (dotted arrows) are emphasized.

The permittivity values obtained by analyzing the traces acquired with the TDR multilevel probe for several water levels are reported in Fig. 9.44. The same figure also shows the water table depth measured using the piezometer close to the TDR system. As expected, the dry vertical profile does not show any remarkable sand stratification; however, when the water is injected, some vertical permittivity variation is visible in the vertical profiles (W4, W5, 6, W7), due to local porosity inhomogeneity in the sand. By analyzing the different water condition profiles it is possible to estimate the extension of both the capillary fringe and the unsaturated zone, thus obtaining important hydrophysical properties of the soil examined. The thickness of the capillary fringe and the vadose zone are, respectively, 3 ± 2 cm and 12 ± 2 cm, where the uncertainty is associated to the vertical sensitivity of the TDR multilevel probe (e.g., the separation between each horizontal transmission-line segment).

To compare GPR and TDR data, the average propagation velocity in the sand layer was calculated from the two data sets. In particular, only data for the water level conditions where at least one TDR probe was lying under the water table (W4, W5, W6, W7) were used, and compared with GPR data coming from the profile closest to the TDR probe. The GPR velocity was compated using the diffraction hyperbola generated by the buried pipes, while the TDR velocity was computed by averaging the values calculated at each probe. Since TDR measurements investigate an area limited to 52 cm from the surface, the data were extrapolated down to the pipe position, to increase the accuracy of the wave velocity estimation, assuming the TDR velocities to be constant in the entire water-saturated region. The velocities obtained from the GPR hyperbola method and the TDR data were in good agreement, showing a discrepancy within 10% (probably due to the effect of the local porosity variation). Therefore, it was possible to use the velocity computed

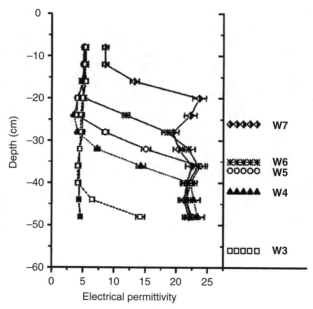

FIGURE 9.44 Electrical permittivity measured by TDR along the vertical. Values of different water-level conditions are shown (filled squares, dry, open squares, W3; filled triangles, W4; open circles; W5; asterisks, W6; rhombi, W7). On the right side of the figure the water table level measured by a piezometer is indicated.

from the TDR data to convert the time scale of GPR sections into depth. Once the time-depth scale is calibrated, the TDR data collected above the water table can be used to calculate the position of the water table precisely, as shown in Fig. 9.45, where the depth of the water table computed using the described method above is compared with the water level measured using a piezometer.

9.3.6 Subsurface Water Content Monitoring Using a Common-Offset Surface GPR Reflection Technique

In this section we describe the use of GPR travel times associated with subsurface reflectors to obtain information about moisture content variations in deeper soil layers. Following Eq. (9.70), a GPR reflection occurs where there is a dielectric contrast between two subsurface units, such as between a soil layer and bedrock, between unsaturated and saturated soils, or possibly between two different soil layers. The difference in dielectric constant between different soil textures alone typically is not large enough to give rise to a strong GPR reflection in the top few meters of the Earth's surface. However, different soil layers commonly retain different amounts of soil moisture, and these differences in soil moisture content may give rise to GPR reflections at the unit interfaces. As moisture increases, the dielectric constant increases, the velocity decreases, and the travel time of the GPR signal to the GPR reflectors increases. Because the depth to the interfaces does not change

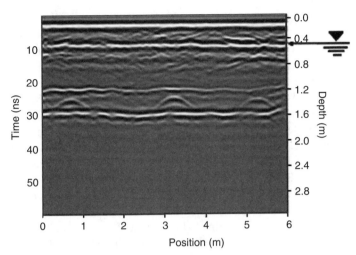

FIGURE 9.45 GPR section for a W4 water-level condition. The depth axis has been calculated using the TDR velocity value for the unsaturated zone (0.104 m/ns). On the right of the figure, the piezometric level of the water table is indicated.

over time, temporal changes in GPR reflection travel times to the reflectors can be used to estimate changes in shallow soil moisture content in a noninvasive manner.

Surface common-offset GPR reflection travel time data sets collected in a vineyard block several times during a grape-growing season at the Dehlinger Winery in Sonoma County, California were used to estimate the moisture content of a near-subsurface soil layer over a range of soil saturation conditions. The crops in this study area are 20-year-old Chardonnay vines, and no in-row tilling has been performed in the vineyard. At this site the water table is approximately 4 m below the ground surface, and the shallowest soils generally had a sandy loam texture. At a depth of 0.8 to 1.3 m below the ground surface, a thin (ca. 0.1 m), low-permeability clay layer was identified from borehole samples and logs. GPR data were collected during several data acquisition campaigns by pulling Sensors & Software's 100 Pulse Ekko MHz surface antennas along the ground surface between different vineyard rows. Several multioffset gathers were collected to help tie the reflectors to wellbore data. Comparison of GPR multioffset gathers, GPR common-offset reflections, and spatially distributed wellbore data suggested that the key reflector was associated with the thin clay layer. Analysis of the GPR common-offset reflection data suggested that this layer formed the base of a subsurface channel that cut diagonally through the vineyard block.

Field infiltration tests and neutron probe logs collected over the course of a year suggested that the thin clay layer inhibited vertical water flow and that the soils above the layer consistently had relatively high volumetric water content values. To test the concept of using time-lapse GPR reflection travel-time picks to estimate subsurface water content above a key reflector, the GPR reflection two-way travel time and the depth of the reflector at the borehole locations were used to calculate an average dielectric constant for the soil layer above the reflector. A site-specific relationship between the dielectric constant and the volumetric water content (obtained using laboratory TDR techniques) was then used to estimate the depth-averaged volumetric water content of the soils above the reflector. Compared to average water content measurements from calibrated neutron probe logs collected over the same depth interval, the estimates obtained from GPR reflections at the borehole locations had an error of 1.8%. To assess spatiotemporal variations in near-subsurface water content using the GPR reflection data sets, time-lapse GPR travel-time picks associated with all acquisition campaigns, wellbore information about the depths to the clay layer, and the site-specific petrophysical relationship were used within a Bayesian procedure (e.g., Hubbard et al. [183]) to estimate the depth-averaged volumetric water content of the soils above the clay layer reflector as a function of time.

Figure 9.46 illustrates the estimated volumetric water content for the zone located above the reflecting clay layer (i.e., at depths less than 1.5 m below the ground surface) at different times during the year. This figure indicates that there

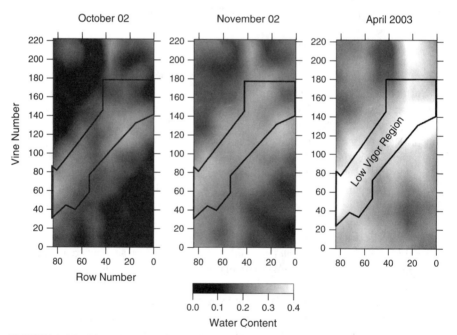

FIGURE 9.46 Plan view map of average volumetric water content of the top soil layer (<1.5 m below the ground surface), estimated using 100-MHz GPR reflection travel-time data at the Dehlinger Vineyards. The axes indicate distance in meters. The black line indicates the boundary of a low-vigor area identified by the vineyard manager, which is coincident with the consistently wetter area that is associated with a subsurface channel and that was identified using GPR reflection data. The Grayscale key at the down indicates relative volumetric water content in units of m^3/m^3, from dark (drier) to light (wetter) (Modified from © 2004 American Geological Institute and used witb their permision [182]).

is a consistent spatial pattern of water content variation over time at the Dehlinger site that might be difficult to detect using conventional measurement approaches. Figure 9.46 shows that a channel-shaped feature trends NE to SW through the study site, and that this feature influences water content distribution: Within this area, the soils are consistently wetter than the surrounding soils. The Dehlinger Winery vineyard manager has recognized that the area outlined by this channel-shaped feature consistently has lower grapevine vegetative growth (= lower vigor) than that of the surrounding area, based on annual measurements of fruit yield and pruning weight (Marty Hedlund, Dehlinger Winery, personal communication, 2005). The close correspondence of the soil moisture distribution and the vegetation variations at the Dehlinger Winery site indicate how soil variations may influence grapevine characteristics.

These results suggest that the two-way travel time to a GPR reflection associated with a geological surface can be used under some natural conditions to obtain estimates of average water content when GPR reflectors are present and detectable over time and when borehole control is available. However, if the subsurface horizon that causes the GPR reflection varies greatly in depth over short distances, it may be difficult to determine the depth of the reflecting horizon in all locations, and the absence of reasonably accurate depth information will reduce the accuracy of the GPR-obtained water content estimates. Additionally, if the soil has been disturbed (such as through farming practices) or if it is highly electrically conductive, the signal attenuation may be too large to permit the GPR signal to travel down to the reflecting interface and back up to the receiving antenna. However, where conditions are amenable to surface GPR reflection methods, the high-resolution characterization methods could be very helpful for guiding crop development. For example, such detailed soil information can be used to lay out crops according to natural geologic variations or to develop variable crop spacing in an attempt to encourage uniform development across a vineyard block. Details of the Dehlinger GPR reflection study are provided by Lunt et al. [184] and Hubbard et al. [182].

9.3.7   Joint Inversion of GPR Tomographic and Hydrological Wellbore Data Sets to Estimate Subsurface Permeability

The most typical approach to integrating hydrogeological and geophysical data involves a two-step hydrogeophysical procedure, whereby inversion of geophysical data is first performed to obtain estimates of geophysical attributes (e.g., radar velocity estimates are obtained through inversion of radar slowness measurements, and seismic attenuation estimates are obtained through inversion of seismic amplitude data). The geophysical attributes are considered hard data and are analyzed for correlation with direct borehole hydrogeological measurements, thereby providing a relationship between geophysical attributes and the hydrogeological parameters of interest. In the second step, estimates of hydrological parameters are obtained using the geophysical data, petrophysical relationship, and direct borehole data through direct mapping, geostatistical, or Bayesian methods. The majority of such hydrogeophysical studies have been performed using tomographic data, for which the measurement scales of geophysical and wellbore data are often comparable, and the geophysical data coverage is acceptable (e.g., Hubbard et al. [183]). Even under these favorable conditions, application of the two-step technique for mapping hydrological parameters can be limited by errors associated with geophysical data acquisition and inversion procedures, as well as inferred relationships of geophysical attributes with petrophysical properties. Recent research indicates that joint inversion of geophysical and hydrological data sets can reduce errors in both geophysical and hydrological parameter estimation. In this example we illustrate the joint inversion of geophysical and hydrological data, which effectively circumvents some of the obstacles commonly encountered during the two-step hydrogeophysical approach and takes advantage of the complementary nature of geophysical and hydrological data.

The use of GPR methods for mapping water distributions in the subsurface is now well established, as illustrated by the previous case studies; this mapping is made possible by the correlation between the soil water content and measured dielectric constant. However, in general, GPR measurements cannot be directly related to the soil hydraulic parameters needed to make hydrological predictions in the vadose zone (such as the absolute permeability and the parameters describing the relative permeability and capillary pressure function). On the other hand, time-lapse GPR data can contain information that can be indirectly related to the soil hydraulic properties, since soil hydraulic properties influence the time- and space-varying changes in water distribution, which in turn affect GPR data.

Kowalsky et al. [185] developed an approach for incorporating time-lapse GPR travel time and measurements of hydrological properties into a hydrological–geophysical joint inversion framework for estimating soil hydraulic parameter distributions. Coupling between the hydrological and GPR simulators is accomplished within the framework of iTOUGH2 [186]. Inversion is performed using a maximum a posteriori (MAP) method that utilizes concepts from the pilot point method. One of the benefits of this approach is that it uses GPR travel times directly without requiring the creation of velocity tomograms, thus alleviating difficulties inherent to tomographic inversion and allowing for sparser GPR data sets relative to those required for conventional tomography. This joint inversion method was later extended [109] to account for uncertainty in the petrophysical function (water content relationship to the dielectric properties) and to increase the flexibility of GPR data characteristics (to include multiple offset data acquisition in three dimensions), allowing increased resolution and accuracy of soil hydraulic parameter estimates.

This approach was applied to time-lapse neutron probe and tomographic radar travel-time data collected at the 200 East Area of the U.S. Department of Energy (DOE) Hanford, Washington site during an infiltration test. Unknown parameters estimated using the joint inversion approach included log-permeability values at pilot point locations (which are used to create three-dimensional permeability distributions), porosity, a parameter of the petrophysical function (the dielectric constant of the solid component of the soil), and the water injection rates, which were not measured precisely. Joint inversion resulted in estimates of hydrological parameter

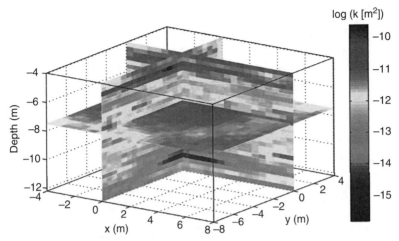

FIGURE 9.47   Realization of a log permeability model estimated through joint inversion of a neutron probe and GPR data sets. (Adapted from [109].)

distributions such as permeability (Fig. 9.47), which can be used to parameterize flow models for predicting fluid distributions at future times.

9.3.8 Subsurface Salinity Distribution Mapping: A Case History in The Netherlands

The importance of geophysical methods in groundwater studies [187] is under-pinned by the fact that (1) groundwater distribution is controlled by geological factors that can be mapped, (2) groundwater quality is controlled by geochemical factors, and (3) rock resistivity is related inherently to porosity, fluid content, and chemistry. The electrical conductivity of the subsurface is highly influenced by dissolved solids in groundwater, making electrical and electromagnetic (including GPR) methods useful for groundwater quality studies. Because of this, the use of electric and electromagnetic methods in hydrological research in coastal areas has long been well established [188–190]. Present-day computers and software [191] allow construction of two- and three-dimensional models within a reason-able amount of computation time compared with acquisition time. Depending on the size of the research area, use of these modern methods is feasible if used in an integrated way. This approach was investigated to map the subsurface salinity dis-tribution in de Nieuwe Keverdijkse Polder in The Netherlands. The main purpose in this study was to investigate the added value of this sequential approach in the detailed mapping, at a scale of several meters, of the subsurface fresh and brackish water distribution. Secondary objectives were (1) to compare the feasibility of each of these methods for mapping subsurface salinity variations, and (2) to delineate the distribution of fresh and brackish water in the NKP area.

9.3.9 Research Area

The Nieuwe Keverdijkse Polder (NKP) is located in the western part of The Netherlands near the towns of Weesp and Muiderberg (Fig. 9.48). The surface of NKP is approximately 1.2 m below mean sea level. The main aquifer is a sequence of alluvial sand layers up to 200 m thick, overlain by a 2- to 3-m-thick peat layer. Locally, sandy ridges protrude out of the peat layer, which are a remnant of the former Pleistocene morphology. This morphology was created by the Saalian glacial

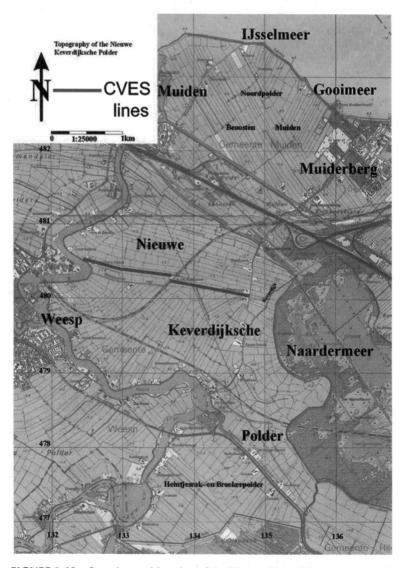

FIGURE 9.48 Overview and location of the Nieuwe Keverdijkse Polder (NKP).

stage processes and covered successively by younger sediments. This relative simple stratigraphy indicates that changes in the subsurface resistivity can be expected to relate primarily to changes in the salinity.

On a regional scale, the groundwater flows from relatively elevated areas, such as the river Vecht, the lake Naardermeer, and the ice-pushed ridge of Muiderberg toward two lowland polder areas like NKP. In the polders, groundwater level is controlled by a system of ditches and canals. Groundwater discharge is by vertical upward flow, and although the NKP aquifer is currently recharged by fresh water, the local discharging groundwater is mainly brackish. Measured chloride concentrations in shallow observation wells (<20 m deep) are as high as 125 mmol/L. The presence of this brackish water is generally attributed to former inundations during the Holocene [192].

Acquisition, Processing and Interpretation Strategies As a first coarse scale survey, EM-34 data were collected, with 400 Hz as the operating frequency and 40-m coil spacing and step size, along several lines in the NKP (Fig. 9.49). In the figure the dots indicate measurement lines, and the resulting apparent resistivity distribution is plotted on top of the area map. The total length of the measured lines is 20 km. The use of horizontal loops gave unstable results, so only vertical coplanar loops were used after the first day of measurements, and the plot in Fig. 9.49 is obtained from vertical loops only. In this mode, small misalignments of the loops have little effect on the measurement result. All measurements in the neighborhood of subsurface power lines and a gas distribution pipe have been discarded in the interpretation. From 517 measurements, 490 could be used to generate the plot of Fig. 9.49. The color scale of the plot is taken such that the last interval contains all values above 120 mS/m, as for the given frequency of operation and coil separation an average terrain conductivity exceeding 100 mS/m starts to deviate from the actual apparent resistivity, while from 120 mS/m it is really unreliable, due to the low induction approximation used by the EM-34 instrument. Changes in measured apparent resistivity values can be caused by changes in depth of the top of the brackish water or in the salinity of the brackish water.

Since apparent conductivity changes from EM-34 measurements alone are not conclusive with respect to the changes in lithology or changes in water quality in depth or laterally, a number of borehole water conductivity measurements were carried out. Samples have been taken since 1920, but the larger number of sampled data have been taken from 1972 to 2000, up to depths of 115 m. In principle, the conductivity of groundwater is related to the amount of total dissolved solids (TDS). At high amounts of TDS, with chloride as the dominant anion, a direct relation with electric conductivity (EC) and chloride concentration can be obtained, while at low amounts of TDS the HCO_3^- conc entration must be known to find a proper relationship between EC and chloride concentrations. The relation between the EC of the water samples and the chloride concentration is found as

$$C_{Cl} = 5.4 \times 10^{-3} \sigma_w^{1.456} \tag{9.82}$$

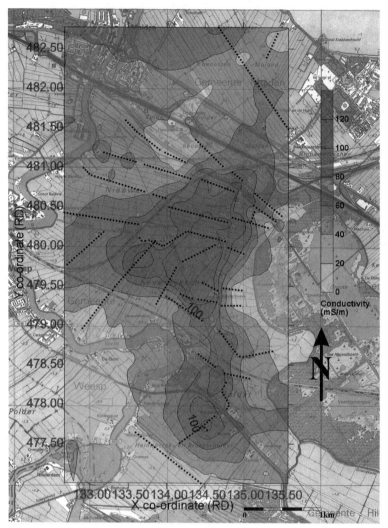

FIGURE 9.49 Apparent resistivity distribution from kriging contouring of EM-34 data acquired at locations indicated by the black dots.

where C_{Cl} is the chloride concentration in mg/L and σ_w is the water sample conductivity in μS/cm, or the EC value (Fig. 9.50). From the figure it is concluded that the EC values correlate extremely well with measured chloride concentrations. Via *Archie's law* [193], valid only for clay-free consolidated sediments, the bulk resistivity, ρ_b, is proportional to the water resistivity, $\rho_w = 1/\sigma_w$, where the proportionality factor is called the *intrinsic formation factor*, F. The law is given by

$$\rho_b = F\rho_w \qquad F = \lambda\phi^{-m} \tag{9.83}$$

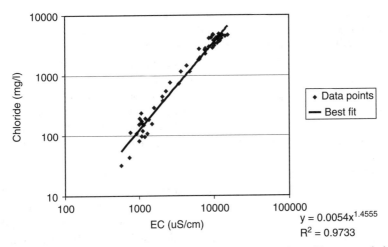

FIGURE 9.50 Correlation curve to calibrate chloride concentration with measured electric conductivity values of the water samples.

where λ is a geometrical factor depending on the medium, ϕ is porosity, and m is the cementation factor. This relation allows translating in situ water sample EC values to bulk resistivity values. The formation factor of the area has been obtained from samples taken in two boreholes. The formation factor was found to be $F = 3$ for the aquifer sands, indicating fine silt-rich sands. These relations are then used to estimate the chloride concentrations from the bulk resistivities as computed from the electric resistivity data [194]. From these results and from two boreholes that were made and logged with short and long normal resisitvity measurements, the following bulk resistivity ranges were obtained and used in the further planning and interpretation of surface resistivity data:

- Resistivity values range from 10 to 40 $\Omega \cdot$ m for peat containing fresh water
- Resistivity values range from 5 to 10 $\Omega \cdot$ m for peat containing brackish water
- Resistivity values range from 40 to 100 $\Omega \cdot$ m for sand containing fresh water
- Resistivity values range from 3.5 to 10 $\Omega \cdot$ m for sand containing brackish water

These ranges are obtained using the Dutch standard definition of brackish water: as having a chloride concentration between 150 and 1000 mg/L. At higher concentrations, the water is regarded as saline, while at lower concentrations it is regarded as fresh.

Based on the EM-34 results, indicating the highest conductivities are found in the north–south running area of the eastern part of the NKP area, standard vertical electric soundings (VES) have been performed at several points using Schlumberger arrays and were later compared with the EM-34 result. The first line is 415 m long, shown in Fig. 9.48 as the left line, while the second line is collected at the EM-34 line that starts at the end of line 1 and continues in a slightly different direction, parallel to the ditches bordering the cattle fields, and is 635 m long. The third line (rightmost red line) starts just east of the railway line, continues in the direction of line 2, and is 1 km long. The eight-channel SuperSting (Advanced Geosciences Inc.), with 72 electrodes available in one line, was used with 5-m electrode spacing. The dipole–dipole configuration was implemented for its high sensitivity for lateral variations in subsurface conductivity. Consequentially, the depth of investigation is limited. To test the reliability in depth of the dipole–dipole configuration, the reverse Schlumberger configuration with the same electrode spacing was also used in lines 1 and 2. The Schlumberger array lines are each 72 electrodes long of, so 355 m each. On the third line, two boreholes are located, borehole 1 at 735 m and borehole 2 at 420 m from the start of the line. In borehole 1, brackish water is found at the surface, while the top peat layer is 1.75 m thick. In borehole 2, the top peat layer is 1.5 m thick, while the fresh water zone goes down to depths of 5 m below the surface, which is the end of the borehole, and hence the transition to brackish water is not reached.

The inverse modeling (RES2DINV) results for the Schlumberger array data compare well with results obtained with the dipole–dipole configuration. The line is perpendicular to the regional trend, so we do not expect large influences from out-of-plane heterogeneities. Figure 9.51 shows the inverse modeling results for the dipole–dipole array for lines 1 to 3. The variable depth ranges are related to the total dipole length that could be used and that differed in the three lines. In line 3 the color bar mapping the resistivity values into colors has a wider range because of the sand-pushed ridge that is encountered in line 3 between 250 and 550 m from the start of the line, while in the rest of the lines peat was most common at the surface. The borehole results comply with the resistivity inversion results using either norm. The results indicate that the brackish water surfaces around 600 m from the start of the line.

Interpretation   The highest conductivities (lowest resistivities) are found in a north–south-running area in the eastern part of the NKP area. The transition from high to low resistivities (taken here as the 10-$\Omega \cdot$ m contour) is found at the westernmost end of the first line at a depth of about 25 m below the surface and then changes over a distance of some 300 m to approximately 15 m below the surface. This depth of 15 m is roughly maintained over the second line. The depth of the 10-$\Omega \cdot$ m contour in the third line is more or less constant at a depth of 10 m until it ascends quite abruptly toward the surface at a distance of about 600 m from the westernmost end of the line. Another interesting feature is the shallow (within 10 m below the surface) zone with high resistivities from a distance of 300 to 500 m. Although the inversion model simulates very high resistivities (locally, > 200 $\Omega \cdot$ m), the measurements from borehole 2 show that 130 $\Omega \cdot$ m is about the maximum restivity.

Using Archie's law, the formation factor (F) of the sandy aquifer sediment was estimated from the electrical conductivity of the natural groundwater in the borehole ($\approx$ 150 μS/cm) to be $F = 2$ for these surface sands ($F = 3$ was used for

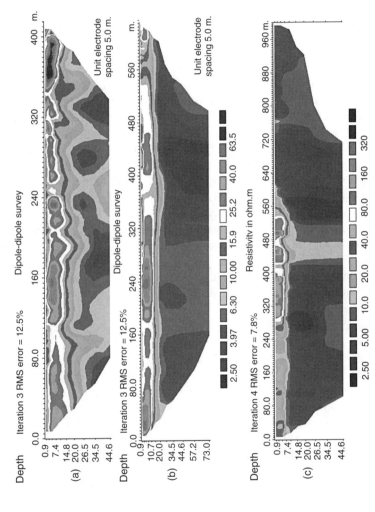

FIGURE 9.51 Resistivity distribution maps below the three lines of dipole–dipole measurements as obtained by inverse modeling for line 1 in (a), line 2 in (b), and line 3 in (c).

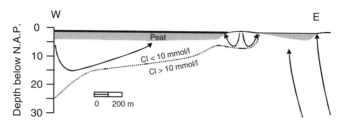

FIGURE 9.52 Salinity distribution in the NKP under the resistivity survey lines 1 to 3, where the transition from fresh to brackish water is taken as 10 mmol/L for the chloride concentration.

the aquifer sands), also indicating fine silt-rich sands. The thin peat layer is still visible at the top of all three plots, with resistivity values between 10 and 40 $\Omega \cdot$ m. It is clearly visible in the dipole–dipole results. Then just below the peat layer, values higher than 40 $\Omega \cdot$ m correspond to the aquifer filled with fresh water. The transition from the freshwater aquifer to the brackish water aquifer should not be interpreted erroneously as a freshwater peat layer. The vertical gradient is too large for the inversion model to reconstruct it properly. The values drop rapidly below 10 $\Omega \cdot$ m, and these correspond to brackish water. The total investigation revealed that the salinity of the shallow groundwater in the NKP has a complicated spatial distribution. It is believed that redistribution of the relic brackish water has taken place. This is influenced primarily by the variable thickness and heterogeneous composition of the confining peat layer and by the land surface elevation variations. This is illustrated in Fig. 9.52, where the transition from fresh to brackish water is put at the salinity level of 10 mmol/L for the chloride concentration.

Influence of the Confining Peat Layer   Figure 9.53 shows the lateral variations in the thickness of the confining peat deposits derived from the many shallow drillings that have been carried out (data provided by J. Dijkmans, TNO-Built Environment). The figure shows a striking resemblance between the peat thickness and the conductivity recorded by the EM-34 measurements: In areas where the peat is thinnest, the conductivity is highest. At the same time, local farmers report on high seepage rates in this area. An estimate of the vertical groundwater flow velocity based on (density-corrected) piezometric levels in an observation well yielded a rate of up to several meters per year. Clearly, the brackish water is transported toward the surface by upward groundwater flow, which tends to concentrate in areas with the lowest hydraulic resistance (i.e., where the peat layer is thinnest). Thus, by controlling groundwater flow patterns, the confining peat deposits determine the distribution of fresh and brackish groundwater.

Influence of the Land Surface Elevation   Differences in the land surface elevation in the NKP area are generally very small (<30 cm). Locally, however, the sandy ridges that protrude from the confining peat layer extend about 50 cm to 1 m above the surrounding surface. Their higher surface elevation compared to

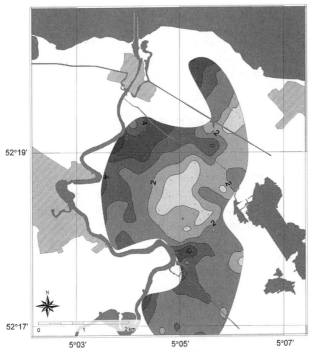

FIGURE 9.53   Estimated peat layer thickness distribution over the NKP area obtained from cores on an irregular grid of shallow drillings.

the surrounding terrain, together with a relatively high permeability, makes these areas a likely candidate for groundwater recharge. Indeed, the apparent resistivities measured with electric resistivity methods, as well as in the borehole, show higher value indicating that the brackish water is pushed down by infiltrating fresh water. The freshwater lens formed this way had a depth of at least 5 m (Fig. 9.51c) during the period of the survey (summer). It might well be that there is a seasonal trend in which the size of the freshwater lens varies depending on recharge conditions. It is striking that the same sandy ridges that form infiltration areas occur in the area where the strongest discharge takes place. There, they provide preferential pathways for the discharging groundwater due to their relatively high permeability. Apparently, the magnitude of the upward flow is so high that infiltration of meteoric water is prevented.

Conclusions   From the results shown here, the sequence of increasing data-acquisition complexity in the progress of the investigation has a positive effect on the choices made for the investigation. The time-consuming lines of multielectrode electric resistivity surveys are minimized and placed at the most interesting locations. The inverse modeling results, in combination with bulk resistivity values from borehole data, are decisive in the detailed reconstruction of the fresh and brackish water distributions.

The dipole–dipole array is efficient in measurement time for multichannel recording systems, with good lateral and vertical coverage provided that the data from large dipoles are used with care. It was found that major variations in salinity could occur over several meters distance, depending on local conditions, such as the heterogeneity of the confining layer and surface elevation.

9.4   CONTAMINANT REMEDIATION

9.4.1   Introduction

Contaminant flow and transport, natural attenuation, and contaminant remediation efficacy are all influenced by hydrogeological and biogeochemical properties. Hydrogeological properties, such as the spatial distribution of hydraulic conductivity, affect the migration and distribution of contaminant plumes and the delivery amendments that are introduced into the subsurface during in situ remediation processes. Biogeochemical properties of the contaminant and the aquifer material influence the evolution of a subsurface plume and its susceptibility to natural attenuation or engineered remediation treatments. Important biogeochemical properties for assessing plume evolution or remediation efficacy include the characteristics of the contaminant itself (mass, distribution, decay or degradation rates, sorption affinity), aquifer sediment properties (distribution of sediment geochemistry, sorption affinity, geochemical stability), and the groundwater chemistry as it affects the contaminant chemical speciation and sediment mineral stability and sorption (EPA, 2007) [181].

Developing a predictive understanding of plume fate and transport, natural attenuation capacity, or remediation success is complicated by the disparity of scales across which controlling hydrological, geochemical, and microbiological processes dominate. Gelhar [195] summarized the large hydraulic conductivity spatial variability that has been documented through many subsurface field studies. For example, the distributions of microfractures and geological layers both influence hydraulic conductivity, albeit over dramatically different spatial scales. Studies of shallow subsurface microbial communities suggest that microbiological properties appear to be spatially correlated with geologicol, hydrological, and/or geochemical properties [196] and heterogeneous deposits [197].

The complexity of subsurface hydrological and biogeochemical properties is magnified upon the introduction of plumes and remedial treatments into the subsurface, because these processes introduce strong perturbations to the geochemical equilibrium. Plume geochemical gradients naturally evolve with time as contaminant fluxes from the source zone changes and reaction of the plume with subsurface materials proceed. A series of redox gradients often become established adjacent to contaminant plumes; scales of redox zonation in natural environments can range from centimeters to kilometers [198]. In situ remediation treatments quickly induce dramatic perturbations to a subsurface system, typically through introduction of an injectate intended to cause biogeochemical reactions. Examples of treatments that cause large subsurface perturbations include in situ chemical oxidation, pH manipulation, and bioremediation. Potential alterations resulting from these in situ remediation treatments include the dissolution and precipitation of minerals, gas evolution, changes in soil water and oxygen levels, sorption, microbial attachment and detachment, oxidation and reduction, and biofilm generation. Some of these transformations can in turn alter the hydrological properties of a system. For example, the generation of gas bubbles or precipitates can clog pore spaces, thereby decreasing hydraulic conductivity and rendering it difficult to introduce subsequent treatment into the subsurface.

Although in situ remediation strategies are frequently used in practice, the spatiotemporal distribution of the biogeochemical transformations and the impact of transformational end products on hydrological characteristics on remediation efficacy are difficult to characterize. The level of effort that is required to characterize and monitor contaminant remediation treatments depends on many factors, including regulatory and risk drivers, the level of heterogeneity relative to the spatial extent of the plume, and the spatial and temporal scales of interest. In some cases, reconnaissance efforts that capture the major hydrobiogeochemical characteristics of the study site may be sufficient (such as for estimating the mean plume behavior in low-risk environments). Other investigations may require much more detailed initial characterization of hydrological and biogeochemical properties and ongoing detailed monitoring to ensure that the treatment is effective.

Conventional sampling techniques for characterizing or monitoring the shallow subsurface typically involve collecting soil or fluid samples using borehole techniques. For example, established hydrological characterization methods (such as pumping, slug, and flowmeter tests [199]) are commonly used to measure hydrological properties. The distribution of redox zonation is commonly inferred by observing patterns of electron acceptor measurements retrieved from wellbore aqueous samples [198]. When the size of the study site is large relative to the scale of the heterogeneity, when the heterogeneity is particularly complex, or when monitoring processes vary spatially, data obtained at point locations or within a wellbore may not capture sufficient information to describe the overall system characteristics or behavior. There is an inability to collect the necessary measurements using conventional characterization tools at a high enough spatial resolution, yet over a large enough volume for understanding and predicting flow and transport processes needed to guide environmental remediation.

Similar to how biomedical imaging procedures have reduced the need for exploratory surgery, integrating more spatially extensive geophysical with direct borehole measurements holds promise for improved and minimally invasive characterization and monitoring of the contaminated subsurface. Integration of geophysical data with direct hydrogeological or biogeochemical measurements can provide a minimally invasive approach for characterizing the subsurface at a variety of resolutions and over many spatial scales. In the last decade, many advances have been made that facilitate the use of geophysical data for hydrogeological characterization, as described by the books *Hydrogeophysics* [131] and *Applied Hydrogeophysics* [200]. The subdiscipline of biogeophysics, or the use of time-lapse geophysical methods for elucidating biogeochemical processes, is at even an earlier stage of evolution [201]. Whether for hydrogeophysical or biogeophysical applications, the main advantage of using geophysical data over conventional (wellbore based) measurements is that geophysical methods provide spatially extensive information about the subsurface in a minimally invasive manner. The greatest disadvantage is that the geophysical methods are indirect; they only provide proxy information about subsurface properties or processes relevant to contaminant remediation.

The use of geophysical data for aiding contaminant remediation investigation can generally be categorized into three broad categories: mapping of subsurface interfaces or features, hydrogeochemical property estimation, and dynamic process monitoring; all three categories are discussed in this section. Geophysical characterizations of engineered structures that contain or cap contaminants are not reviewed in this section. Readers are referred to Slater and Binley [202] and Majer [203] for geophysical characterization of contaminated barriers and to Meju [204] for a discussion of geophysical characterization of covered landfills.

9.4.2 Hydrogeophysical Mapping

Because geophysical attributes are often sensitive to variations in physical and geochemical properties, they are often useful for environmental remediation subsurface mapping. In this section we discuss three different examples of hydrogeophysical mapping, including the use of geophysical methods for mapping subsurface architecture and fracture zonation as well as for delineating plume geometries.

Subsurface mapping of units or interfaces is perhaps the best developed geophysical characterization approach, and it is most commonly performed using surface-based techniques. Examples of common objectives for environmental remediation include mapping the aquifer geometry, the top to bedrock or the water table, and the location of major preferential flowpaths. Common geophysical approaches for mapping these targets include surface GPR, electrical conductivity, and electromagnetic and seismic methods. Mapping relies on the sensitivity of geophysical methods to subsurface physical properties or contrasts of these properties. From these measurements, the nature and distribution of subsurface materials can often be deduced. For example, subsurface variations in elastic moduli and density associated with lithological heterogeneities can cause seismic waves to travel at different speeds. Information about these changes, and thus about the nature and distribution of the subsurface units, can be interpreted from analyzing the seismic arrival times. The contrasts in physical properties vary depending on which materials are juxtaposed, and the ability to detect these changes varies with the geophysical method employed. With methods such as surface seismic reflection or GPR, key interfaces are often manifested as laterally coherent wiggle traces and enable the production of a pseudo cross section of the subsurface. With methods such as electrical resistivity or seismic refraction, zonation of units that have distinct geophysical attributes (such as electrical conductivity or seismic P-wave velocity) is often used to infer lithological variations.

Comparison between direct measurements and geophysical attributes is necessary to interpret the geophysical data in a hydrogeologically meaningful manner.

The most common approach is one that capitalizes on expert skills and intuition in the comparison of wellbore information (such as depths to key lithological interfaces) with geophysical signatures. This methodology allows for incorporation of information that is often very difficult to quantify. For example, because the spatial distribution of hydrological properties is largely a function of ancient depositional and geological processes, interpretation of subsurface geometries between wellbores is often performed within the context of a conceptual geological model. This approach might be as simple as recognizing the lateral continuity of geological layers or the expected geometry of a particular depositional facies, or it might be as extensive as development of a hydrofacies model [205]. If a linkage between the direct wellbore and the indirect geophysical data set can be developed, the geophysical data set can then be used to extrapolate the information away from the wellbore. An example of such mapping is shown in Fig. 9.54a, where surface electromagnetic data were compared to wellbore data and used subsequently to delineate a buried channel [206]. In this case, the comparison suggested that the channel fill material had higher electrical conductivity than the surrounding geological materials. This figure shows a plan view map of the channel in the subsurface obtained by slicing through the geophysical data volume. At this site, the channel served as a significant subsurface flowpath, and delineation of this preferential path would have been difficult using wellbore data alone. Figure 9.54b illustrates the use of surface GPR to provide a pseudo cross section of the subsurface. Comparison of wellbore data at this site permitted interpretation of the architecture of key stratigraphic units within Atlantic Coastal Plain sediments that controlled groundwater flow in the shallow unconfined aquifer.

Because fracture zones often serve as fast paths for contaminants, they are also a common environmental remediation mapping objective. Delineation of subsurface

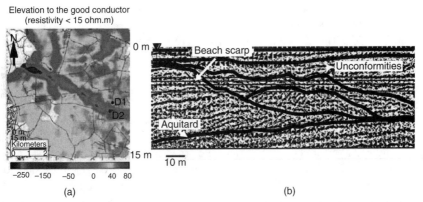

FIGURE 9.54 Example of the use of geophysical data for different mapping objectives. (a) Plan view map of electrical conductivity data interpreted in terms of elevation of a buried valley, obtained using surface electromagnetic methods (from [206]); (b) GPR cross section interpreted in terms of key stratigraphic features within a nearshore environment. (From [183].)

fractures is a challenging target due the typical disparity between the size of the fractures and the support scale of the geophysical measurements. However, studies have indicated that geophysical methods hold potential for delineating large individual fractures as well as fractured zones. Amplitudes in seismic transmission measurements are sensitive to the pressure state and aperture of isolated planar fractures [207]. Similarly, larger zones of fractured rock can be delineated using crosswell seismic surveys and may be visible as regions of decreased P-wave velocity [208], increased P-wave attenuation, or changes in shear wave polarization and/or splitting [209]. Due primarily to the differences in fluid dielectric constants in the fracture and the surrounding matrix, cross-well GPR measurements have also been used to both delineate fractures and to track the progress of conductive tracers within the fracture region [210]. Within a Bayesian framework, Chen et al. [211] inverted seismic slowness (the inverse of seismic velocity) and wellbore information to quantify fracture zonation within a uranium-contaminated aquifer at the Oak Ridge National Laboratory in Tennessee, where a biostimulation treatment was being performed. As seismic slowness is sensitive to the stiffness of a material, it was hypothesized that fracture zones would be less stiff than the surrounding competent rock and would thus have greater seismic slowness (lower seismic velocity). Figure 9.55 illustrates the geophysically obtained estimates, which indicate the "probability of being within a high hydraulic conductivity fracture zone" at the site. These estimates revealed that the target biostimulation zone has a varying thickness and dip and is sometimes laterally discontinuous. Indeed, comparison of tracer and uranium biostimulation experimental results at the study site with the fracture estimation suggests that the seismic method was useful for delineating zones that were hydraulically isolated from the amendment injection area.

Azimuthal electrical methods, which entail the acquisition of data along surface-based transects that extend radially from a common center point, have provided information about fracture existence and orientation. Azimuthal resistivity [212] and spontaneous potential [213] surface-based surveys have been used successfully to indicate hydraulic anisotropy along principal fracture strike orientations in crystalline bedrock. The azimuthal resistivity approach builds on the analogy between hydraulic conductivity and electrical conductivity in rock aquifer SP azimuthal results, whereas the azimuthal SP approach builds on electrokinetic theory that invokes a dependence of the streaming potential coupling coefficient on microgeometry in addition to hydraulic gradient.

In addition to delineating subsurface geometry and fracture zonation, geophysical methods have been used to delineate plume boundaries. Plume "mapping" has generally been performed using surface electrical methods, which capitalize on the commonly observed direct relationship between electrical conductivity and total dissolved solids (TDS). This is also a difficult mapping target, because electrical signatures respond to material properties as well as pore fluid ionic strength, and it is often difficult to deconvolve the different contributions to the geophysical signal. However, if the contrast between the concentration of the groundwater

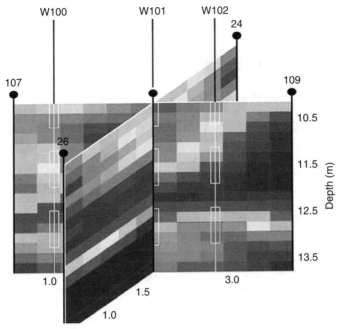

FIGURE 9.55 Estimates of natural fracture zonation obtained using seismic tomographic slowness measurements data within a stochastic estimation framework, where blue represents a high probability and red represents a low probability. (From [211].)

and the plume is great enough that the contribution of lithology on the electrical signature can be considered to be negligible, electrical methods can be used to delineate approximate plume boundaries. Figure 9.56a shows an example of the use of inverted surface electrical resistivity data to delineate a nitrate plume at the contaminated Hanford Reservation in Washington. Figure 9.56b shows a correlation between electrical resistivity and nitrate concentration obtained from colocated electrical and wellbore measurements above and within the interpreted plume region at the same site. Similarly, Watson et al. [215] used surface electrical methods to delineate the boundary of a high-concentration nitrate plume emanating from a settling basin at the contaminated Y-12 complex of the Oak Ridge National Laboratory in Tennessee.

9.4.3 Hydrogeophysical Property Estimation

In addition to providing information about plume boundaries and aquifer geometry, geophysical methods can potentially provide estimates of subsurface hydrogeological and geochemical properties that control flow and transport. The ability to

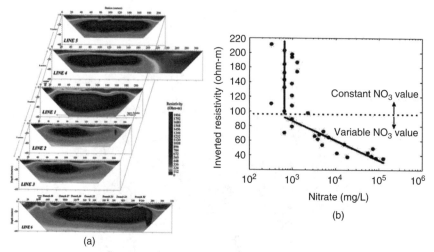

(a)

(b)

FIGURE 9.56 Inverted electrical resistivity profiles at the BC crib area of the contaminated Hanford, Washington reservation, where the low-electrical-resistivity (high electrical conductivity) regions were interpreted as the plume boundaries; from [214] (b) relationship observed between electrical conductivity and nitrate concentration at a single wellbore location within and above the plume interpreted. (Adpated from [195].)

estimate hydrogeological or geochemical properties using a particular geophysical approach is a function of the relation between the property under consideration and geophysical attributes, the availability of direct measurements for calibration, the magnitude of the property variations and contrasts, and the spatial scale of the property variations relative to the measurement support scale of the geophysical technique. In the simplest case, the property of interest can be inferred based on known geophysical responses to hydrological property variations. For example, electrical conductivity, which is typically greater in clay than in sandy sediments, and is greater in saturated sediments than in the same dry sediments, has been used to infer the distribution of sedimentary units or saturation state.

Quantitative estimates of subsurface hydrogeochemical properties are also potentially available using geophysical data sets. Such estimation requires a relationship or theory to link the geophysical measurements with the hydrogeological property of interest. Many of the relationships that link hydrogeological parameters (such as mean grain size, porosity, permeability, and type and amount of pore fluid) within typical near-surface environments to geophysical attributes (such as seismic velocity, electrical resistivity, and dielectric constant) are not well understood. Although a significant amount of research has been performed to investigate the relationships between geophysical and lithological or pore fluid conditions within consolidated formations common to petroleum reservoirs and mining sites, very little research has been performed to test these concepts within environments common to contaminated sites. As such, there are only a few relationships that are commonly used in near-surface geophysics, and even these are often used only to get a ballpark estimate. Examples include Topp's law (which relates dielectric constant values to moisture content [35]) and Archie's law (which relates bulk electrical conductivity to the electrical conductivity of water-saturated geologic materials [193]). One common approach in hydrogeophysical applications is to develop site-specific empirical relationships between the geophysical measurements and the parameters of interest, using colocated field data (such as at a wellbore) or by developing relationships using site material within the laboratory. Developing petrophysical relationships using colocated field data is often problematic, because of the different sampling scales, measurement directions, and errors associated with wellbore hydrological measurements and geophysical "measurements" located at or near a wellbore (e.g., [216]). Similarly, upscaling laboratory-derived relationships for use at the field scale also presents challenges (e.g., [217]).

Estimation of properties using geophysical methods also requires a method for integrating the sparse (yet direct) wellbore data with the more spatially extensive (yet indirect) geophysical data sets. Geostatistical methods are widely used for incorporating hydrological and geophysical data (e.g., [218]). These techniques rely on spatial correlation information to interpolate between measurements in a least-squares sense. Because natural geologic materials often exhibit strong spatial organization, these methods have been widely used within the hydrological sciences. Bayesian approaches have also proven to be useful for subsurface characterization, because they permit the incorporation of a priori information (such as that available from a conceptual model or from wellbore data sets), quantify uncertainty, and provide a framework for incorporating additional data sets as they become available.

An example of hydrogeophysical property estimation performed within a Bayesian framework is the use of cross-hole radar and seismic methods to provide multidimensional estimates of hydraulic conductivity at a DOE bacterial transport site located near Oyster, Virginia [183]. Tomographic data were used together with borehole flowmeter logs and a site-specific petrophysical relationship to provide estimates of hydraulic conductivity (and their uncertainties) at a very high spatial resolution (0.25 × 0.25 m; see Fig. 9.57). Comparison of geophysically obtained hydraulic conductivity estimates and tracer breakthrough data suggested that the tomographic estimates were extremely useful in helping to reduce the ambiguity associated with interpreting bacterial and chemical transport data collected during tracer tests [219]. Even though this site was fairly homogeneous (the range of hydraulic conductivity was approximately one order of magnitude) and had extensive borehole control (i.e., wellbores every few meters), it was difficult to capture with sufficient accuracy the variability of hydraulic conductivity using borehole data alone so as to ensure reliable transport predictions. By comparing numerical model predictions with tracer test measurements at the Oyster site [220] it was found that conditioning the models to the geophysical estimates of hydraulic conductivity significantly improved the accuracy and precision of the model predictions, relative to those obtained using borehole data alone. This study suggested that the geophysically based methods provided information at a reasonable scale and resolution for understanding field-scale processes. This is an

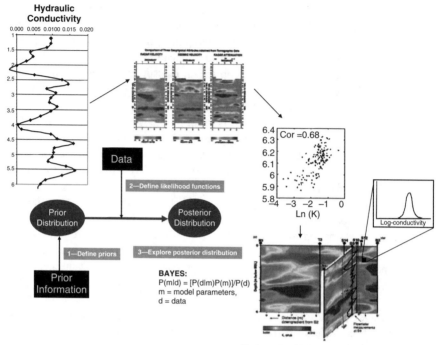

FIGURE 9.57 Example of the Bayesian approach for integrating disparate data sets for the estimation of hydraulic conductivity distributions, where the mean value of the hydraulic conductivity distributions estimated within each 0.25 × 0.25 cm pixel is shown on the bottom right. (Adapted from [183].)

important point, because it is often difficult to scale the information gained at the laboratory scale or even from discrete wellbore samples for use at the remediation field scale. Additional examples of hydrogeological property estimates using geophysical data are given in Section 9.3 (such as the estimation of permeability using time-lapse radar data sets within a coupled geophysical-flow model approach [131,185,200].

Estimation of sediment geochemistry is also extremely important for the prediction of contaminant evolution and transport, because sediment geochemistry controls sorption, bioavailability and bacterial attachment, and other critical processes. As such, sediment geochemistry greatly affects the mobility of contaminants and their susceptibility to remedial treatments. Traditional methods for characterizing geochemical heterogeneity typically involve collecting a subsurface sample and subsequently performing laboratory analysis (such as Fe(II) and Fe(III) extraction using the Ferrozine method [221]). Because these methods require core samples for analysis and are labor intensive, there are typically far fewer measurements of sediment geochemistry available for environmental remediation studies relative to hydrogeological measurements.

Geophysical methods also hold potential for providing information about sediment geochemistry using either qualitative inference or more quantitative estimation approaches. Because clays have higher electrical conductivity than sands, electrical methods have been used for decades to map the spatial distribution of clay-rich units. In some cases this is also useful for inferring information about the sediment geochemistry. For example, in shallow environments, clay-rich units are often accompanied by a higher level of organics and can be bioreduced naturally relative to neighboring units. Induced polarization (IP) methods have been used to detect the presence of a variety of reduced mineral phases at the effective support scale of the geophysical footprint. With the IP method, when voltage is applied in the subsurface in the presence of conductive minerals, ions can facilitate charge transfer reactions across fluid–mineral interfaces. This charge transfer can cause the voltage to lag the current, which creates an IP phase shift between the two parameters. Recent work has shown that the phase shift is enhanced in the presence of electroactive ions and conductive precipitates. For example, using laboratory geophysical and geochemical measurements of soil samples, Monsoor and Slater [222] found that IP responses varied as a function of Fe concentration in soils. The use of surface-based approaches was shown to allow for delineation of the location of IP phase anomalies within a shallow fluvial aquifer. Through comparing geophysical and geochemical measurements, it was found that the IP anomalies coincided with increased concentrations of reduced mineral phases (such as FeS) and extractable uranium within localized naturally bioreduced zones in the subsurface. These recent environmental studies build upon decades of research in the mining industry, where IP is a standard tool for locating buried ore bodies.

Because geophysical measurements often respond to physical property variations, it is also possible to exploit physiochemical relationships (if they exist) to estimate sediment geochemistry distribution quantitatively using geophysical methods. Recent research has illustrated the use of high-resolution tomographic GPR measurements and site-specific physiochemical relationships to estimate lithofacies and Fe(II) and Fe(III) within a Bayesian framework [223]. In this case the geophysical data did not sense the geochemical parameters directly. Instead, the radar amplitudes were sensitive to lithology (i.e., radar amplitudes were more attenuated in units with a larger clay fraction), and a relationship between the distribution of Fe(II) and Fe(III) and lithofacies existed for the site. The estimation approach developed exploited this mutual dependence to estimate lithology and sediment geochemical parameters along two-dimensional cross sections using tomographic radar amplitude data. Figure 9.58a illustrates the two-dimensional geophysical estimates of radar attenuation obtained from inversion of GPR amplitude data. Figure 9.58b and c illustrate the mean values of the estimated Fe(II) and Fe(III) distributions, respectively; variances associated with these estimates are available but are not shown in the figure. Cross-validation exercises revealed that the estimates obtained using the geophysical data were accurate and greatly improved the two-dimensional identification of the geological and geochemical properties. This study provided perhaps the highest-resolution field-scale characterization of geochemical properties performed to date. Variable lithofacies

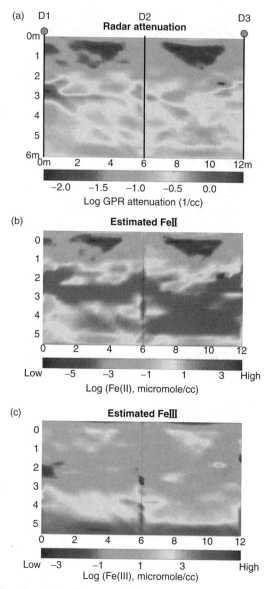

FIGURE 9.58 Two-dimensional images of the estimated (a) GPR attenuation, (b) Fe(II), and (c) Fe(III). (Adapted from [223].)

and sediment geochemistry obtained geophysically were used as parameterization for reactive transport modeling simulations [224]. The simulations illustrated the importance of understanding linked hydrological and sediment geochemical property distributions on the distribution of contaminants, remediation amendments, and subsequent biogeochemical reactions associated with remediation.

9.4.4 Biogeophysical Monitoring

As described in Section 9.4.1, in situ remediation approaches that strongly perturb subsurface systems typically lead to biogeochemical end products. Biostimulation, for example, involves the introduction of an electron donor and nutrients to the subsurface, typically via injection. Indigenous microbes can often gain energy though facilitating the transfer of electrons in a reduction process, and this process can lead to a less toxic or less soluble form of the original contaminant. Reduction typically occurs in a sequential fashion, with the most energetically favorable reactions occurring first followed by reactions that are less energetically favorable for the microorganism(s). Figure 9.59 illustrates the typical sequence of redox reactions that occur during bioremediation given the availability of electron donors and acceptors. As shown in this figure, the individual reactions can lead to biogeochemical end products [such as the production of N_2 from nitrate reduction or the production of Fe(II) from iron reduction]. Some of these products can in turn combine to form new remedial end products. As an example, the precipitate FeS can form from products associated with both iron and sulfate reduction. Because

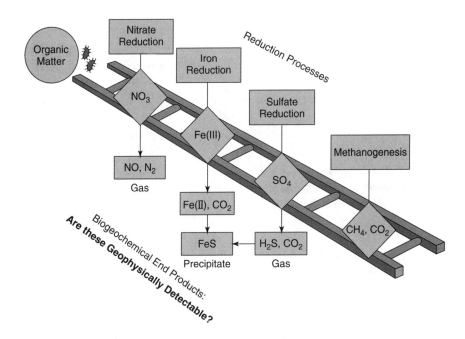

FIGURE 9.59 Bacteria decompose organic matter and produce molecular hydrogen and organic acids as metabolic by-products. These by-products are then utilized by respirative bacteria (such as nitrate reducers, iron reducers, sulfate reducer, and methanogens) in terminal electron-accepting redox processes. Important for geophysics is that these processes often yield end products (such as gases, precipitates, and biofilms) that may alter the geophysical signature. (From [198].)

these transformations vary significantly over space and time, it is challenging to monitor the dynamic complexity of the end products using only wellbore aqueous fluid samples.

The premise of biogeophysics is that some of the remediation-induced biogeochemical end products will be detectable using geophysical methods. A particularly powerful component in biogeophysics is the use of time-lapse data sets to monitor systems that have been perturbed, such as through in situ remediation processes. Geophysical data collected at the same location as a function of time can be used to monitor processes that are important for assessing attenuation capacity. Observing the data as time-lapse sections enhances the imaging of subtle geophysical attribute changes caused by natural or forced system perturbations and removes the dependence of the geophysical measurements on static contributions to the signal (such as lithology) and on data inversion artifacts. Although the biogeochemical transformations occur at grain–fluid boundaries, within pore spaces, and across pore throats, recent research has focused on exploring the signature of these changes over the effective support scale of the geophysical measurements. Some of the key environmental questions that are being addressed in the new field of biogeophysics include:

- Can geophysical methods be used to identify redox gradients associated with subsurface contaminant plumes or remediation processes?
- Which geophysical methods are most sensitive to which in situ remediation injectate and end products?
- Can we develop and test petrophysical models and associated parameters that link geophysical responses to end products?
- Can we distinguish between amendments and many possible biogeochemical end products using multiple geophysical methods?
- What is the macroscopic geophysical response to microscopic properties?
- Can geophysics be integrated with geochemical data to estimate end-product characteristics quantitatively?

Here, we briefly review some recent biogeophysical research that illustrates the potential of geophysical methods for providing information about biogeochemical changes associated with natural plume evolution or remediation treatments. Examples include the delineation of redox gradients at the field scale using spontaneous potential (SP) data sets and the use of time-lapse complex electrical, radar, and seismic methods to track transformations associated with remedial treatments at the lab and field scales, including the distribution of remediation amendments, changes in TDS, and the evolution of biominerals.

The distribution of redox processes is an important factor for assessing and monitoring biogeochemical gradients. Redox potential, or Eh, indicates the tendency for oxidation–reduction reactions to occur. Under equilibrium conditions, in situ measurements of redox potential can be obtained through wellbore measurements, although disturbance and contamination often corrupt these measurements.

The distribution of the redox zonation can also be deduced by observing patterns of electron acceptor consumption, final product production, and concentrations of dissolved hydrogen based on measurements retrieved from wellbores (e.g., [198]). However, many case studies suggest that delineation of redox processes using such approaches is not straightforward. Development of geophysical methods to detect redox zonation and biogeochemical end products remotely over various length scales could greatly increase our understanding of biogeochemical processes critical for environmental remediation.

SP methods have been used in recent studies to infer information about redox zonation. SP signals measure the potential difference between a fixed reference nonpolarizable electrode and a roving electrode using a high-input impedance voltmeter. In near-surface systems, the SP response depends on the groundwater flow (electrokinetic contribution) and redox conditions (electro-redox contribution). A recent study [225] assessed redox potentials associated with a landfill-contaminant plume using SP approaches. Variations in hydraulic head measurements were used in an aquifer study to estimate the electrokinetic contribution, which was subsequently "removed" from the effective SP signal [225]. They found that the residual SP signal correlated well with redox potential measurements collected in wellbores (Fig. 9.60). This study illustrates the value of an inexpensive and simple-to-operate

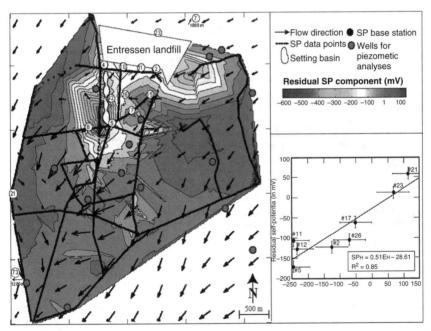

FIGURE 9.60   After effectively "removing" the streaming potential component of the self--potential signals associated with the groundwater flow the residual self-potential map was obtained that was proportional to the redox potentials measured in the aquifer in a set of piezometers. (Adapted from [225].)

geophysical technique to provide information about redox potential over relevant spatial scales and in a noninvasive manner. Such information can potentially be used to characterize or monitor redox zonation associated with natural attenuation or remedial treatments in a low-maintenance and minimally invasive manner.

Recent research has also explored the use of time-lapse geophysical methods for remote monitoring of biogeochemical transformations associated with changes in redox states, such as the evolution of gases and the development of precipitates. The majority of the studies have been performed at the laboratory scale. For example, Williams et al. [226] examined microbe-induced ZnS and FeS precipitation during a biostimulation experiment performed using *Desulfivibrio vulgaris* (a sulfate-reducing bacterium that couples incomplete oxidation of lactate to acetate with sulfate reduction). Geophysical, hydrological, and biogeochemical measurements were collected using an experimental suite of flow-through columns. In these columns, several pore volumes of lactate were initially flushed through the system, followed by introduction of bacteria into the middle of the column. Nutrients were introduced into the bottom of the upward-flowing column. Bioremediation-induced sulfate reduction was monitored over seven weeks; reduction was indicated by decreasing substrate and metals concentrations, increasing biomass, and visually discernible regions of metal sulfide accumulation. Figure 9.61a illustrates the sulfide precipitation front at one point in time of the biogeochemical measurement columns. The region of sulfide precipitation showed a shift toward the influent portion of the column over time as a result of microbial chemotaxis toward elevated substrate concentrations at the base of the column.

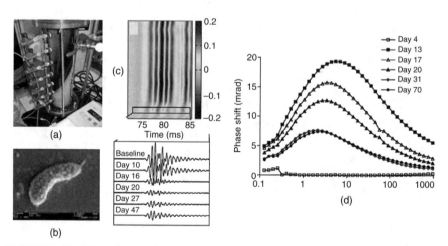

FIGURE 9.61 Example experimental column illustrating how measurements are collected down the length of the column and the presence of a developed sulfide precipitation front associated with sulfate reduction. (b) TEM image illustrating the mineralized encrustation on the experimental microbe due to the formation of sulfide precipitates (scale $\sim 1\mu$m), Changes in (c) seismic amplitude and (d) IP response associated with the initiation and aggregation of sulfide precipitates evolved at the laboratory scale. (Adapted from [226].)

Williams et al. [226] documented that the time-lapse geophysical signatures tracked the onset and evolution of the geochemical transformations over space and time. Regions of sulfide precipitation and accumulation resulted in substantial changes in seismic and complex electrical measurements (the latter conducted over the range 0.1 to 1000 Hz). The high-frequency seismic wave amplitudes were reduced by nearly 84%, and significant increases in complex electrical conductivity were observed with only minimal changes in the fluid conductivity (Fig. 9.61b and c, respectively). Changes in the electrical phase response and seismic attenuation were attributed to alterations in subsurface mineralogy arising from stimulated microbial activity within the pore space, including precipitation reactions, aggregation dynamics, and solid-state mineral transformations.

At the field scale, Hubbard et al. [227] illustrated the use of time-lapse geophysical methods for interpreting complex biogeochemical processes associated with a field scale, Cr(VI) bioremediation experiment performed at Hanford, Washington. They first performed a suite of laboratory experiments to understand the geophysical responses to (1) pore water replacement by the injected electron donor; (2) gas bubble formation; (3) variations in TDS; and (4) evolution, dissolution, or mineralogical alteration of solid-phase constituents. The laboratory experiments revealed that:

- Replacement of pore water by the remediation amendment (in this case, HRC, or a slow-release polylactate) dramatically increased seismic attenuation and electrical conductivity and slightly decreased dielectric constant.
- Gas generation associated with carbonate dissolution and denitrification processes decreased the dielectric constant and dramatically attenuated the seismic amplitudes.
- Changes in solute concentration or TDS (associated with processes such as sulfate reduction) directly correlated to changes in electrical conductivity.
- Evolution of disseminated minerals associated with reduction processes and calcite precipitation decreased the electrical conductivity; increased the seismic attenuation and velocity, and increased the dielectric constant.

Prior to geophysical monitoring at the field scale, they used wellbore flowmeter measurements and crosshole radar and seismic tomographic attributes within a discriminate analysis technique to estimate the zonation of hydraulic conductivity within the injection interval. Tomographic radar and seismic measurements were then collected during 13 acquisition campaigns that extended over a three-year experimental period (i.e., before, during, and after injection of the HRC amendment) Three different pumping campaigns were initiated during the experiment to restimulate activity in the injection zone. Figure 9.62 shows the changes in dielectric constant, electrical conductivity, and seismic attenuation obtained from inversion of the datasets and differencing from the baseline data set.

Using results from laboratory biogeophysical experiments and constraints provided by analysis of time-lapse wellbore fluid samples [Cr(VI) concentration, organic acids, $\delta^{13}C$ values of DIC, nitrate concentration, and sulfate concentration], Hubbard et al. [227] then interpreted time-lapse seismic and radar tomographic data

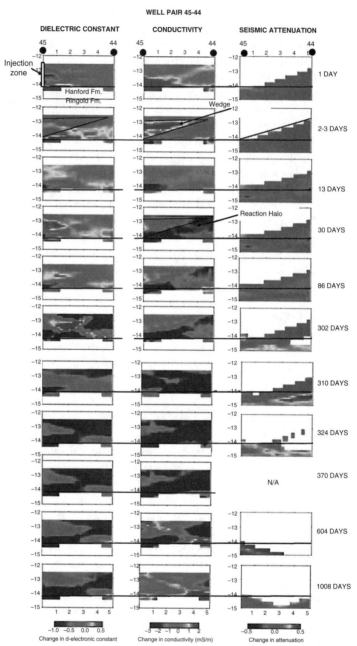

FIGURE 9.62 Time-lapse estimates of changes in a dielectric constant, electrical conductivity, and seismic attenuation (right) obtained using tomographic radar and seismic methods in association with a biostimulation experiment conducted to reduce and immobilize chromate at the Hanford, Washington Department of Energy Site. Day 0 indicates the date of amendment injection (i.e., initiation of biostimulation). Interpretation of the field geophysical monitoring data, constrained by laboratory biogeophysical experimental results and field aqueous geochemical data, permitted the spatiotemporal interpretation of amendment distribution, the evolution of gasses and precipitates, and the change in TDS associated with the biostimulation. (Adapted from [227].)

sets in terms of hydrological and biogeochemical transformations. The geophysical monitoring data sets were used to infer the spatial distribution of injected electron donor; the evolution of gas bubbles, variations in total dissolved solids (nitrate and sulfate) as a function of pumping activity, the formation of precipitates and dissolution of calcites, and concomitant changes in porosity. Although qualitative in nature, the integrated interpretation illustrates how geophysical techniques have the potential to provide a wealth of information about coupled hydrobiogeochemical responses to remedial treatments in high spatial resolution and in a minimally invasive manner. Particularly novel aspects of the study include the use of multiple lines of evidence to constrain the interpretation of a long-term field-scale geophysical monitoring data set and the interpretation of the transformations as a function of hydrological heterogeneity and pumping activity.

9.4.5 Challenges and Future Directions

The mapping, property estimation, and process monitoring examples provided in Sections 9.4.2 to 9.4.4 illustrate the potential that geophysical methods have for helping to improve our understanding of subsurface systems and how they respond to biogeochemical perturbations relevant to contaminant remediation. This information can, in turn, be used to gain a predictive understanding about plume transport, natural attenuation capacity of an aquifer, or efficacy of a remediation treatment. However, as described by Hubbard and Rubin [228], many obstacles still hinder the *routine* use of geophysics for quantitative hydrogeophysical characterization, such as issues of integration, petrophysical models, scale, nonuniqueness, and uncertainty. These obstacles are often exacerbated when using geophysical methods to monitor biogeochemical processes, in part because biogeophysics is at a very early stage of development, and our understanding of geophysical responses to biogeochemical end products is still developing. Scale matching issues can be even more significant with biogeophysics than with hydrogeophysics because many of the biogeochemical properties of interest occur at grain boundaries or in pore spaces that are much smaller than the support scale of the geophysical measurement. Nonuniqueness of the geophysical responses is problematic because geophysical signatures often respond to hydrogeological as well as biogeochemical heterogeneity. Because plume evolution or remediation treatments can lead to the occurrence of multiple geochemical processes that contribute simultaneously to the geophysical signature (i.e., the evolution of gases, biofilms, and precipitates), it can be challenging to deconvolve the influence of the individual contributions to the geophysical signature. Finally, biogeochemical end products can also alter the initial hydrological properties (such as through clogging or dissolution), which can further affect the geophysical signature.

Most of the case studies described previously were tested at the local scale (i.e., length scales of tens of meters or less), where the scale disparity between the wellbore and the geophysical measurements is not great. Although the importance of characterizing critical subsurface properties with sufficient confidence, yet over plume-relevant scales is recognized, there are few methods currently that can integrate key data sets that sample a variety of hydrobiogeochemical properties over a range of scales (from core samples to wellbores to surface data sets). A significant challenge is the development of organizing principles, strategies, and frameworks that can be used to guide the characterization and monitoring across scales, with the objective of developing approaches that can integrate multiple types of data collected at different support scale, that are tractable and cost-effective but that still capture the hydrobiogeochemical complexities of the system, and that govern the paths and rates of contaminant mobility and remediation success.

Finally, although improvements in shallow subsurface geophysical characterization approaches and reactive transport modeling (e.g., [229]) have been realized in recent years, there have been few attempts to test and document the synergies that come from integration of characterization, modeling, and monitoring data sets for improving environmental remediation. Recent research in the hydrogeophysical community has documented the benefit of jointly honoring the geophysical and hydrological data sets in forward and inverse flow modeling (e.g., [220,230]). Although it is intuitive and obvious that joint inversion of biogeophysical and biogeochemical data sets within reactive transport models also offer the potential for an improved understanding of in situ rates and mechanisms over field-relevant scales, this frontier area is relatively unexplored.

Understanding the full capacity of geophysical methods for environmental remediation characterization and monitoring is expected to improve through increased laboratory and field experimentation; development of petrophysical relationships and theory; improved understanding of geochemical to geophysical measurement scaling issues; through the development of estimation approaches for integrating multiscale hydrobiogeochemical–geophysical data sets; and through comparison and integration of field data and reactive transport model predictions. With the continuing hydrogeophysical and biogeophysical research that is currently under way within the community, we expect the use of geophysical methods for exploring complex and coupled hydrobiogeochemical subsurface processes associated with contaminant remediation to increase in frequency and become more quantitative in nature. Once developed, such geophysical approaches could conceivably be used in a semiautonomous and long-term monitoring mode. Although more research is needed before these methods can be considered as routine remediation monitoring tools, the examples provided in this section indicate that it is an avenue worthy of further pursuit.

9.5   AGRICULTURAL GEOPHYSICS

9.5.1   Important Agricultural Geophysics Considerations

The scale for agricultural geophysics applications is extremely small compared to the scale of geophysical investigation in the petroleum and mining industries. The geophysical investigation depth for agricultural applications is also typically much smaller than geophysical investigation depths needed for environmental or engineering site evaluations. Geophysical methods have been used in agriculture predominantly for some manner of soil investigation, although there have been even smaller-scale agricultural applications, one example being the geophysical imaging of interior portions of a tree trunk [231]. Consequently, agricultural geophysics tends to be largely focused on the soil profile within 2 m of the ground surface, which includes the entire crop root zone [232]. With respect to agricultural geophysics applications, this extremely shallow 2 m depth of interest is certainly an advantage, since most geophysical methods presently available have investigation depth capabilities that far exceed 2 m.

However, there are complexities associated with agriculture geophysics not typically encountered with the application of geophysical methods to other industries or disciplines. One such complexity involves the transient nature of certain soil conditions and properties that affect geophysical measurements. For example, apparent soil electrical conductivity, EC_a, measured using resistivity and electromagnetic induction methods, is influenced significantly by temperature and moisture conditions, and these conditions can change appreciably over a period of days or even hours, in turn significantly altering the EC_a measured over the same time frame. Moisture conditions also govern the soil dielectric content (or relative permittivity), κ, thereby affecting the ground-penetrating radar (GPR) results obtained within agricultural settings. Measured EC_a is also affected by soil nutrient levels and salinity, which sometimes exhibit little variation over long periods but will then change rapidly upon irrigation or fertilizer application. Other soil properties affecting EC_a, if they vary temporally at all, do so at a slower rate, and in this category are properties including pH, organic matter content, clay content, cation-exchange capacity, and specific surface.

Another complexity regarding agricultural geophysics is that the soil conditions and properties affecting geophysical measurements vary not only temporally, but also spatially, often exhibiting substantial variability over very short horizontal and vertical distances. Interestingly, it has been noted that although average EC_a values for an agricultural field may vary with changes in soil temperature and moisture, the EC_a spatial pattern itself within an agricultural field tends to remain relatively consistent over time, regardless of the transient temperature and shallow hydrologic conditions, thus indicating that EC_a spatial patterns are governed predominantly by the spatial variations in soil properties [233–235].

As an example, Fig. 9.63 shows maps of the soil electric conductivity measured on two different days (May 2 and May 22, 2001) using time-domain reflectometry (TDR) [79]. The corresponding semivariograms are also represented. Measurements were performed in a 120×120 m agricultural field from the Agricultural Research Center of Gembloux (Belgium). TDR measurements were performed with a spatial resolution of 5 m, resulting in 400 measurements, and correspond to the depth range 0 to 40 cm. The maps were obtained using kriging, based on the corresponding semivariograms. We can observe that the soil electric conductivity varies significantly at the field scale, and also during a short period of time, as a function of precipitation and evaporation events. The shape of the semivariograms indicates that the structure of the variability can change from one date to another, although we can recognize constant patterns, which were attributed to clay content. The correlation sill distance varies from about 25 m to 35 m. The changes observed

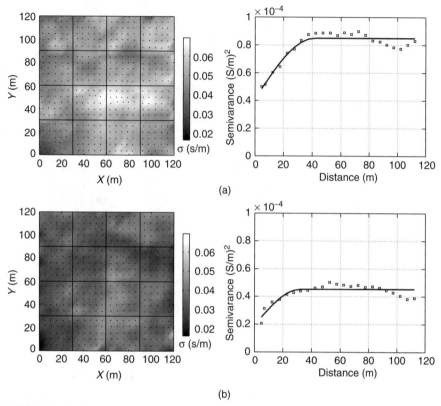

FIGURE 9.63   Spatial distribution of the soil electric conductivity (σ) within an agricultural field (Agricultural Research Center of Gembloux, Belgium) at two different dates, May 2, 2001 (a) and May 22, 2001 (b), and the corresponding semivariograms. Measurements were performed using time-domain reflectometry.

stem from the change in water content variability as a function of water content itself, thereby affecting the soil electric conductivity distribution [128]. The nugget effect is due mainly to the local variability (<5 m), and to a lesser extent, to TDR measurement errors.

A final complexity associated with agricultural geophysics is the need to detect objects of fairly small size. Examples include 5- to 10-cm-diameter drainage pipes [236–240] or 0.5-to 5-cm-diameter tree roots [241]. Resolving such small objects with geophysical methods can often be quite difficult.

9.5.2   History of Agricultural Geophysics

Some of the earliest agricultural geophysics research activity occurred in the 1930s and 1940s, focusing on soil moisture determination through EC_a measurement with resistivity methods [242–245]. Soil moisture determination using resistivity and electromagnetic induction, ground-penetrating radar (GPR), and time-domain reflectometry (TDR) methods can provide useful information for irrigation scheduling. The application of geophysical methods to agriculture did not gain substantial momentum until the 1960s, and to a greater extent the 1970s, with the use of resistivity methods for soil salinity assessment [139,246–248]. Through the use of resistivity or electromagnetic induction geophysical methods, EC_a measurements are employed successfully to monitor salinity levels in soil, so that field operations, such as soil profile water flushing, can be initiated well before salinity buildup causes crop damage.

Demographic growth is increasing the pressure on our environment through the development of agriculture, industry, and living and transport infrastructures to meet our vital needs and quality of life. Soil erosion, groundwater depletion, salinization, and pollution are major environmental threats to the sustainability of ecosystems and arable lands. Ongoing substitution of fossil fuels by biofuels for energy production further intensifies the exploitation of these resources. At the same time, climate change progressively alters the hydrological cycle with, for example, more severe droughts in some parts of the world and excessive rainfall in other parts. Consequently, developing sustainable and optimal management strategies of soil and water resources has become a societal priority [249].

In this context, there has been a rapid expansion in the use of geophysical methods for agriculture over the last 15 to 20 years, particularly in relation to the practice of precision agriculture [250,251]. Precision agriculture is a growing trend combining geospatial data sets, state-of-the-art farm equipment technology, geographic information systems (GISs), and global positioning system (GPS) receivers to support spatially variable field application of fertilizer, soil amendments, pesticides, and even tillage efforts [252,253]. The benefits of precision agriculture to farmers are maximized crop yields and/or reduced input costs. There is an important environmental benefit as well. Overapplication of agrochemicals and soil tillage is fairly common. Since precision agriculture operations result in optimal amounts of fertilizer, soil amendments, pesticides, and tillage applied on different parts of a field, there is potentially less agrochemicals and sediments released offsite via subsurface drainage and surface runoff. With reduced offsite discharge of agrochemicals and sediment, adverse environmental impacts on local waterways are diminished. In essence, precision agriculture techniques allow a farm field to be divided into different management zones for the overall purpose of optimizing economic benefits and environmental protection.

Horizontal spatial variations in EC_a have commonly been found to correlate relatively well with horizontal spatial variations in both crop yield [233,254] and soil properties [233,255,256]. Consequently, EC_a mapping with resistivity, electromagnetic induction, or TDR geophysical methods can often be used to delineate the horizontal spatial patterns in soil properties that strongly influence within-field variations in crop yield. These EC_a maps can then become an important guide for employing precision agriculture techniques to optimally manage different parts of a farm field for maximum economic and environmental benefit. It should be noted that in addition to the use of geophysical methods to aid precision agriculture practices, there were many additional agricultural geophysics advances over the past 15 to 20 years in areas such as buried infrastructure location [236], soil nutrient monitoring [257], and identification of subsurface flow pathways [258].

9.5.3   Geophysical Methods Commonly Employed for Agriculture

The four geophysical methods used most commonly for agricultural purposes are resistivity, electromagnetic induction, ground-penetrating radar (GPR), and time-domain reflectometry (TDR). Time-domain reflectometry has been used to measure apparent soil electrical conductivity (EC_a), although resistivity and electromagnetic induction are the methods typically employed to map lateral variations of EC_a within an agricultural field, which in turn often correlate with horizontal soil property spatial patterns. The resistivity method, employed in its most conventional form, uses an external power source to supply electrical current between two "current" electrodes inserted at the ground surface. The propagation of current in the subsurface is three-dimensional, as is the associated electric field. Information on the electric field is obtained by measuring the voltage between a second pair of "potential" electrodes, also inserted at the ground surface. The two current and two potential electrodes together comprise a single four-electrode array. The magnitude of the current applied and the measured voltage is then used in conjunction with data on electrode spacing and arrangement to determine an EC_a value.

Continuous measurement galvanic contact resistivity systems integrated with global positioning system (GPS) receivers have been developed specifically for agriculture. These resistivity systems can have more than one four-electrode array, providing investigations of shallow depths (0.3 to 2 m) with short time or distance intervals between the continuously collected discrete EC_a measurements. The location for each EC_a measurement is determined accurately by GPS. Steel coulters (disks) that cut through the soil surface are utilized as current or potential electrodes for these resistivity systems. Figure 9.64 shows two examples of continuous measurement galvanic contact resistivity systems that are currently employed for agricultural applications.

Electromagnetic induction equipment, field survey procedures, and data processing have been described in Chapter 4 and in other parts of Chapter 9. Some ground conductivity meters, which are a type of electromagnetic induction device used for measuring EC_a and magnetic susceptibility, have been developed specifically for agriculture. The ground conductivity meters typically employed for agriculture have intercoil spacings of around 1 m, and as a consequence, effective investigations depths of 1.5 m or less when positioned near the ground surface [259]. Vertical, horizontal, and perpendicular dipole orientations of the ground conductivity meter transmitter and receiver coils can provide different EC_a investigation depths within an agricultural setting. Although used primarily to map EC_a, ground conductivity meters can also be used to measure magnetic susceptibility, a property that has been demonstrated useful for delineating hydric soils [260]. Two examples of ground conductivity meters commonly used for agricultural applications are shown in Fig. 9.65.

FIGURE 9.64 Examples of continuous measurement galvanic contact resistivity systems. (a) Veris 3100 Soil EC mapping system (Veris Technologies, Salina, Kansas). (b) Close-up of steel coulters used for current and potential electrodes by the Veris 3100 Soil EC mapping system. (c) Automatic resistivity profiling (ARP)-03 device (Geocarta SA, Paris, France). (d) Close-up of steel coulters used for current and potential electrodes by the ARP-03 device.

(a)   (b)

FIGURE 9.65 Examples of ground conductivity meters used in agricultural settings: (a) DUALEM-1S (Dualem Inc., Milton, Ontario, Canada); (b) EM38-MK2 (Geonics Ltd., Mississauga, Ontario, Canada).

GPR equipment, field survey procedures, and data processing have been described in Chapters 3 and 12 and in other parts of Chapter 9. Applications of GPR in agriculture are widespread, ranging from drainage pipe detection [236], to tree root biomass determination [241], to soil moisture mapping [138]. The GPR systems utilized for agricultural purposes typically employ antennas with center frequencies in the range 200 MHz to 1.5 GHz. This antenna frequency range covers many scenarios where the goal is to image small, shallow buried features or objects within an agricultural setting. The anticipated depth and size of the subsurface feature or object of interest will provide guidance on the antenna frequency to use. For example, 250-MHz antennas are appropriate for locating a 20-cm-diameter subsurface drainage system pipe main at 1.5 m depth in a silt loam soil, while 1.5-GHz antennas might be a good choice for imaging 0.5-cm tree roots near the ground surface.

Finally, even though often omitted from standard geophysical texts, time-domain reflectometry (TDR) is a well-established electromagnetic geophysical technique that is often used in agricultural studies. The TDR method is nondestructive, rapid, and has a maximum depth of investigation of about 2 m (i.e., the entire root zone). Time-domain reflectometry can be used to measure not only soil dielectric content (or relative permittivity), and hence soil water content, but in addition, TDR can also determine EC_a [79,261]. Time-domain reflectometry systems usually employ sets of two or three thin metal rods comprising a waveguide probe that is inserted into the soil. Measurement of electromagnetic signal travel time, and hence velocity, along the waveguide probe allows determination of soil dielectric constant, which in turn can be used to calculate soil water content. Measured attenuation of the electromagnetic signal as it travels along the waveguide probe is employed to obtain EC_a values. For TDR, the electromagnetic signal frequency utilized to establish soil dielectic constant is around 1 GHz, while frequencies used to measure EC_a are in the low kilohertz range [262]. Consequently, TDR can have a wide rage of potential agricultural applications, from measurement of soil moisture conditions to mapping of EC_a patterns, that may reflect spatial variations in soil properties or soil salinity conditions.

Discussions have centered on resistivity, electromagnetic induction, GPR, and TDR, which, as stated previously, are the predominant geophysical methods used in agriculture today. However, there are other geophysical methods that have been used in only a limited manner, or not at all, that may find substantial application in the future. These other geophysical methods include electrical resistance tomography, magnetometry, nuclear magnetic resonance, seismic, and self-potential.

9.5.4   Overview of Agricultural Geophysics Applications

A review of past research indicates a wide range of potential applications for geophysical methods in agriculture. Table 9.2 serves to emphasize the variety of possible uses by listing just a few of the numerous ways that geophysical methods can provide valuable information for agriculture purposes. Table 9.2 emphasizes the expansion of agricultural geophysics applications that have occurred over just the last 15 to 20 years. With the many recent developments in near-surface geophysics equipment, field survey procedures, and data analysis techniques, it is expected that many innovative agricultural geophysics applications will be discovered in the not to distant future.

9.5.5   Agricultural Geophysics Case Histories

Location of Subsurface Drainage Pipes in Farmland and Golf Course Settings Using Ground-Penetrating Radar   A 1985 U.S. Department of Agriculture Economic Research Service survey showed that the states comprising the midwestern United States (Illinois, Indiana, Iowa, Ohio, Minnesota, Michigan, Missouri, and Wisconsin) had by that year approximately 12.5 million hectares that contained subsurface drainage systems [281]. Crop fields constituted by far the large majority of this land. The midwestern United States is itself one of the most productive agricultural regions in the world, due in large part to the widespread installation of subsurface drainage systems that remove excess water from soils. Increasing the efficiency of soil water removal on farmland that already contains a functioning subsurface drainage system often requires reducing the average spacing distance between drain lines. This is typically accomplished by installing new drain lines between the older ones. By keeping the older drain lines intact, less new drainage pipe is needed, thereby substantially reducing costs to farmers. However, before this approach can be attempted, the older drain lines need to be located in an effective and efficient manner.

By the year 2000, there were over 15,000 golf course facilities throughout the country [282]. Keeping these facilities in good condition requires continual maintenance and occasional remodeling of the various parts that comprise a golf course, especially the greens. In 1996 alone, $4.5 billion was spent for maintenance costs at U.S. golf courses [283]. One of the most important features of a golf course green

TABLE 9.2 Potential Agricultural Applications for Geophysical Methods

Geophysical Method	Agricultural Application	Refs.
Resistivity	Soil drainage class mapping	263
Electromagnetic induction	Determining clay-pan depth	264
	Estimation of herbicide partition coefficients in soil	265
	Mapping of flood deposited sand depths on farmland adjacent to the Missouri River	266
	Soil nutrient monitoring from manure applications	257
Ground-penetrating radar	Quality/efficiency improvement and updating of USDA/NRCS soil surveys	267,268
	Measurement of microvariability in soil profile horizon depths	269
	Bedrock depth determination in glaciated landscape with thin soil cover	270
	Plant root biomass surveying	241,271,272
	Identification of subsurface flow pathways	258,273
Ground-penetrating Radar magnetometry	Farm field and golf course drainage pipe detection	236,237,239,274,275
Resistivity and electromagnetic induction	Soil salinity assessment	120,246,247,275–278
	Delineation of spatial changes in soil properties	219,233,234,255,256
Resistivity and seismic	Tree trunk imaging	231
Resistivity, electromagnetic induction, ground-penetrating radar, and time-domain reflectometry	Soil water content determination	96,111,138,155,159 184,242,245,279,280

Source: Updated from [232].

is the subsurface drainage system. To repair or retrofit a subsurface drainage system within a green, golf course superintendents and architects need nondestructive tools for mapping the drainage pipes present.

Finding buried drainage pipe in a farm field or golf course green is not an easy task, especially when the drain lines were installed more than a generation ago. Often, the installation records have been lost, and the only outward appearance of the subsurface drainage system is a single pipe outlet extending into a water conveyance channel. From this, little can be deduced about the network pattern used in drainage pipe placement. Without records that show precise locations, finding a drain line with heavy trenching equipment causes pipe damage requiring costly repairs. The alternative of using a handheld tile probe rod is extremely tedious at best. Consequently, there is definitely a need to find better ways of locating buried drainage pipe, and GPR may provide the solution to this problem. The use of GPR to was shown to locate farmland drainage pipes in the Maritime Provinces of Canada [236], while employment of GPR could be shown to find drainage pipes beneath golf course greens [237].

Results   Aspects of GPR drainage pipe detection within farm field conditions typical of the midwestern United States have been documented [238,239]. The major findings for these investigations are listed as follows.

1. A GPR unit, (Noggin[PLUS], Sensors & Software, Inc.) with 250-Hz center frequency antennas seemed to work best for detecting buried drainage pipe in soils typical of the midwesterm United States.

2. Shallow hydrologic conditions with a wet soil surrounding a water-filled drainage pipe produce the poorest GPR drainage pipe response. Shallow hydrologic conditions with a wet soil surrounding an air-filled drainage pipe produce a much better GPR drainage pipe response. If frozen surface conditions exist, the GPR response for an air-filled pipe surrounded by wet soil improves further. The GPR drainage pipe response for a dry to moderately dry soil profile and air-filled pipes falls somewhere in between the best- and worst-case scenarios.

3. The type of drainage pipe present, whether clay tile or corrugated plastic tubing (CPT), does not seem to affect the GPR response.

4. The orientation of the GPR measurement transect with respect to the drain line, from perpendicular to directly along trend, governs the type of drainage pipe response found on a GPR profile, which ranges, respectively, from a tight, narrow reflection hyperbola, to a laterally stretched reflection hyperbola, to a linear banded feature.

5. GPR was successful in finding on average 74% of the total amount of pipe present at 14 test plots in central, southwestern, and northwestern Ohio. In seven test plots, 100% of the pipe was located, while in two test plots, none of the pipe was found. All in all, the GPR method worked quite well in finding clay tile and CPT drainage pipe down to depths of around 1 m within a range of different soil materials from sandy loam to clay. An example of the GPR results from one of the farm field test plots is provided in Fig. 9.66.

A GPR investigation of golf course greens using antennas with center frequencies of 250, 450, 500, 900, and 1000 MHz has been conducted [240]. The GPR results from this study indicate that a rather wide range of antenna center frequencies, from 250 to 1000 MHz, work reasonably well for mapping the buried drainage pipe network within a golf course green. This same range of antenna center frequencies also appears useful for delineating the areal extent of the constructed

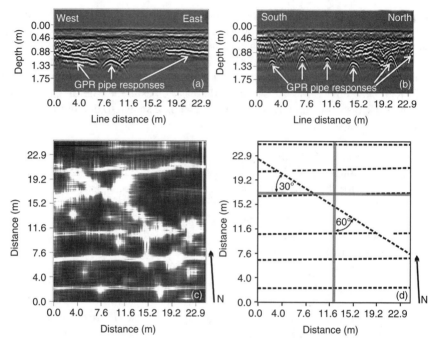

FIGURE 9.66 GPR results (250-MHz antennas) from a test plot in northwestern Ohio. (a) West-to-east GPR profile. Linear banded GPR responses, such as those shown at each end of this profile, occur when the GPR measurements transect directly overlies and is along the trend of a drainage pipe. The spreadout GPR reflection hyperbola response found near the center of the west-to-east GPR profile is due to a 30° difference between the orientation of the measurement transect and the orientation of the drain line. (b) South-to-north GPR profile. The drainage pipe reflection hyperbolas in this profile are, with one exception, representative of the GPR responses for pipes oriented about 90° to the measurement transect. The one exception is the fourth reflection hyperbola from the south, which is slightly elongated because the difference in orientation between the measurement transect and the drain line was only 60°. (c) A 25- to 40-ns GPR time-slice amplitude map with drainage pipe responses depicted by white or lightly shaded linear features. (d) Interpreted map of drainage pipe pattern [dashed black lines are drainage pipes, and solid gray lines show the position of the GPR measurement transects for profile (a) and profile (b)].

soil layers within golf course greens. However, higher antenna center frequencies (900 to 1000 MHz) seem to be the best alternative for determining depths and thicknesses of the constructed sand and gravel layers that are part of the greens. Figure 9.67 shows the GPR results from one of the golf course greens used in the investigation conducted [240].

Mapping Spatial Distribution of Tree Root Mass with Radar Tree root biomass studies provide valuable insight into belowground productivity in forest systems and are frequently used to test the effect of tree species, genetic selection, and

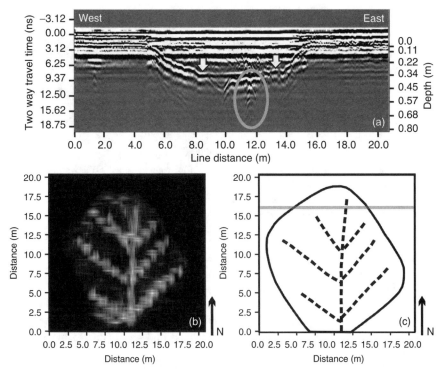

FIGURE 9.67 GPR results (1000-MHz antennas) from the Golf Club of Dublin (Dublin, Ohio) practice green. (a) West-to-east GPR profile. An oval-shaped gray line is used to highlight a typical golf course green GPR drainage pipe response. The tips of the downward-pointing arrows in this GPR profile indicate the position for the bottom of the sand layer within this green. (b) An 8- to 15-ns (0.34 to 0.61 m) GPR time-slice amplitude map with drainage pipe responses depicted by lightly shaded linear features. (c) Interpreted map showing the areal extent of the green (solid-line boundary), drainage pipe locations (dashed lines), and position of GPR measurement transect for profile (a). (From [240].)

subsequent management on carbon allocation. Quantifying belowground productivity of subsurface roots has always been difficult, but it is essential to accurately model carbon pools and predict fluxes. Up to half of the biomass in trees may be hidden belowground and needs to be accounted for when considering the potential for forest management to enhance carbon sequestration. Tree root systems are commonly evaluated via labor-intensive, destructive, time-consuming excavations. GPR has been shown to be a rapid means of detecting tree roots and measuring lateral root mass in well-drained electrically resistive soils [241,284–287]. The protocol for collecting root data with GPR can be separated into two procedures: (1) linking radargrams to destructively sampled soil cores to calibrate the mass of roots per area of soil [285–287], and (2) collecting radar data in a series of parallel transects around trees to add a spatial dimension to the GPR survey. Roots as small as 0.5 cm in diameter have been detected at depths of less than

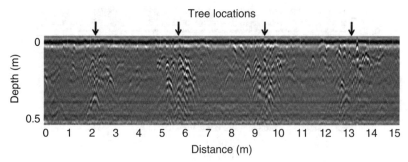

FIGURE 9.68 Radargram collected near a row of four trees with a 1.5-GHz antenna in the coastal plain region near Bainbridge, Georgia Roots produced strong reflection hyperbolas, which were clustered around the base of each tree. Most roots were in the upper 0.3 m and were sparsely distributed between trees.

50 cm with a 1.5-GHz antenna in well-drained, coarse-textured soils (Fig. 9.68). Greater depth penetration may be achieved with lower-frequency antennas, but the resolution of small roots suffers (i.e., 900 MHz $\sim$>2cm and 400 MHz $\sim$>4 cm). Being nondestructive, GPR allows repeated measurements that facilitate the study of root system development. It is important to note that surface-based radars used in reflection mode only measure lateral root mass and are unable to measure tap root mass.

A 900-MHz GPR system has been commercialized for root detection in suburban lawns and parks (Tree Radar Inc., Silver Spring, Maryland) to help arborists manage tree health and limit damage from construction by avoiding major rooting areas. Conditions found in most forests may be more challenging and the impact of the following needs to be considered: (1) uneven surfaces, (2) obstruction from understory plants, (3) nonroot reflectors (i.e., rocks, holes, surface and buried debris), (4) variable leaf litter depth, and (5) recent soil disturbance. That being stated, we have had a high degree of success in the Carolina Sand Hills in North Carolina, coastal plain in Georgia, Mississippi, and Florida, and on granular silt clay loams in Hawaii. High clay contents and rock volumes in the North Carlonia Piedmont and mountains have prevented wider application.

Lateral root surveys with GPR are accomplished with linear transects, When the goal is to map individual roots and create architecture maps or three-dimensional reconstructions, closely spaced grids are scanned and processed. This analysis is time consuming and not appropriate for stand-level analysis in forests. A rapid method was employed [287] for creating root mass distribution maps, which delineates the distribution and quantity of mass but does not attempt to detail the diameter or location of individual roots. This method was used to estimate lateral root mass in a replicated, intensive loblolly pine (*Pinus taeda* L.) management study (10 years after planting) near Bainbridge, Georgia: three blocks, four treatments: control, irrigation, fertigation, and fertigation plus pest control [287,288].

Field Procedure
1. Clear 2.5 × 3 m plots around 96 trees (24 rows of four trees), of leaf litter and woody debris.
2. Establish 10 transect lines 25 cm apart with the trees in the center.
3. Collect GPR data using a 1.5-GHz antenna connected to a survey wheel and a GSSI SIR-2000 GPR unit (data example Fig. 9.68).
4. Scan 84 core locations with GPR, collect and process cores, and correlate GPR data and actual root mass.

Data Processing
1. Apply GPR signal processing described by [284].
2. Use SigmaScan (SSPS, Chicago, Illinois) software to dissect 3-m GPR transects into 20 discrete 15-cm "virtual cores."
3. Use Surfer (Golden Software, Golden, Colorado) to project the root mass data on a grid for spatial analysis and to create contour maps.

Results
1. GPR data are readily calibrated to live root mass on this site, GPR data were highly correlated with mass from soil cores [root mass $(g \cdot m^2) = 12.44 + 0.00815 \times$ (GPR reflectance), $r^2 = 0.80$] and allowed the root estimates to be expressed on a mass per unit area basis.
2. Root mass contour maps showing areas of high and low root density relative to the location of each tree were created (Fig. 9.69).
3. No differences in lateral root mass were observed across the treatments (average 10.4 Mg/ha), but treatments that did not receive any fertilizer exploited a greater area of soil with low-density roots.

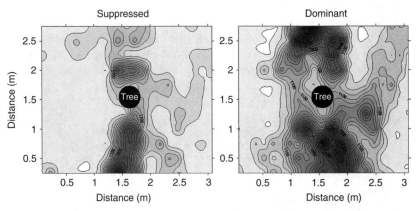

FIGURE 9.69 Root mass distribution maps $(7.5 \ m^2)$ created from 10 parallel GPR transects collected by a 1500-MHz antenna. Root mass values for each map are the sum total of all roots to a depth of 0.5 m and are equivalent to sampling (200) 15-cm-diameter soil core samples. Lateral root mass of the suppressed tree depicted here averaged 867 g/m^2 and the dominant tree was 27% higher at 1097 g/m^2.

4. Roots from the control trees occupied 72% of the available soil area compared to 55% in the fertigated treatment despite having 42% lower aboveground biomass (102 vs. 175 Mg/ha).

Conclusions On amenable sites such as this, GPR can be an excellent tool to measure lateral root mass and root distribution. The ability to sample large areas rapidly allows small treatment differences to be resolved nondestructively. This methodology will be useful for measuring belowground carbon accretion in aggrading forests over time.

Studying the Plant Root Zone with TDR In agriculture, TDR has been used to estimate several parameters in the plant root zone, such as plant water uptake [280,290–292], soilwater content and root density [293,294], evapotranspiration [295,296], and irrigation scheduling and plant water usage [297,298]. Several different strategies and TDR probe types can be used to estimate these parameters indirectly, based on measurement of the apparent permittivity and/or the electrical conductivity. One such approach, described below, involved the monitoring of water content variations in time using a multilevel vertical probe installed within the root zone of turfgrass planted under laboratory conditions.

Laboratory Procedure Two multilevel probes were used in this experiment, one with an electrode spacing of 2.5 cm (Fig. 9.70a) and the second with an electrode spacing of 5.0 cm (Fig. 9.70b). The data were acquired with a Tektronix 1502C cable tester (Tektronix Inc., Beaverton, Oregon) equipped with an RS232 interface. These probes were inserted in a 140-L cell containing a lower, 8.0-cm-thick gravel layer (for drainage purposes) and an upper, 37.0-cm-thick layer of essentially monogranular sandy soil ($\phi < 0.250$ mm $= 2\%$; ϕ 0.250 to 0.500 mm $= 15\%$; ϕ 0.500 to 1.0 mm $= 72\%$; ϕ 1.0 to 2.0 mm $= 11\%$) (Fig. 9.71).

The probes were positioned approximately 15 cm apart and then the sediment was poured around them without compaction (to minimize vertical stratification). The top electrode of both probes was placed 5.0 cm below the surface, resulting in the deepest electrodes being located approximately 2 cm above the gravel layer.

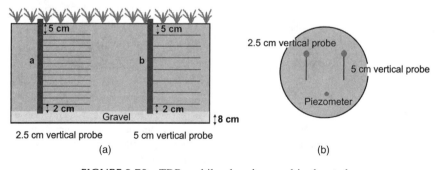

FIGURE 9.70 TDR multilevel probes used in the study.

(a)

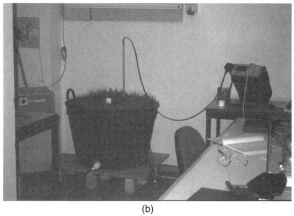

(b)

FIGURE 9.71 Cell during construction (a) and during the experiment (b).

The cell was saturated from the base by means of a flexible tube connected to a constant-level reservoir which could be positioned at different heights to control the location of the watertable. The cell was situated in a controlled environment and illuminated with fluorescent light for 12 hours a day with 150μ mol/s $\cdot$ m^2 of photosynthetic active radiation (PAR). The temperature was lowered to around $15°C$ during the period without light by means of an air conditioner.

Initial TDR measurements were taken in the dry sand, and then measurements were repeated after the cell was completely saturated. The cell was subsequently drained; then the water table was repositioned at a depth of 20 cm and the turfgrass (*Lolium perenne* L.) was planted. After sowing, daily measurements were taken for about a month.

Results Figure 9.72 shows the water-content vertical profiles for different experimental conditions: (1) dry sand, (2) saturated sand, (3) water table at 20 cm from the top with no grass, and (4) water table at 20 cm from the top with grass. As

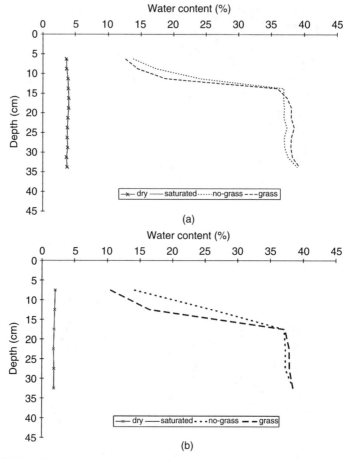

FIGURE 9.72 Water content profiles collected during various periods of the experiment with a 2.5-cm probe (a) and a 5.0-cm probe (b). Each of these plots shows the results collected during dry and saturated conditions and values when the water table was at a 20-cm depth with and without grass present.

can be seen, condition (1) shows a water content of about 4% at all depths, due to residual water in the sand. Condition (2) shows an average water content of about 37%, with some vertical variations due to sand stratification. Condition (3) shows the distribution of the water content in the unsaturated zone (0 to 15 cm), which is higher, for both probes, with respect to condition (4), when the grass is present (Fig. 9.72b). The difference between the two conditions (grass–no grass) has its highest value at a depth of 11.25 cm for the 2.5-cm probe and at 12.50 cm for the 5-cm probe. These results show that the depth of water-content reduction probably reaches a maximum where greatest uptake activity of the root system occurs.

Conclusions   The results of this case study indicate that TDR is a viable method for monitoring vertical soil-moisture distributions over time. In particular, the use of multilevel probes allows one to obtain stable and reproducible results, and therefore to perform an accurate estimation of water content variations along a vertical profile. These results also show how electrode spacing will affect how precisely one can locate the interval of maximum water uptake within the root system (i.e., the probe resolution).

Mapping and Monitoring Salinity Using Apparent Soil Electrical Conductivity

Although accurate worldwide data are not available, it is estimated that 7% of the world's land area (930 million hectares) is salt affected [299]. Even though salinity buildup on irrigated lands is responsible for a declining resource base for agriculture, the exact extent to which irrigated soils are salinized, the degree to which productivity is being reduced by salinity, the increasing or decreasing trend in soil salinity development, and the location of contributory sources of salt loading to ground and drainage waters are not known. The ability to map and monitor soil salinity at field scales and larger spatial extents is an invaluable tool to fill gaps in our knowledge of the extent of, and trend in, salinity development on agricultural land.

Geospatial measurements of apparent soil electrical conductivity (EC_a) have become one of the most frequently used measurements for characterizing soil spatial variability, particularly of salinity [278,300,301]. Corwin and colleagues have demonstrated the utility of geospatial EC_a measurements in a variety of agricultural applications, including (1) modeling salt loading to groundwater [302], (2) delineating site-specific management units in precision agriculture [278,303], and (3) salinity and soil quality assessment [304].

The practical technology and methodology for measuring, monitoring, mapping, and assessing soil salinity using geospatial EC_a measurements are presented for a case study on marginally productive saline-sodic soil located on the west side of California's San Joaquin Valley. The study characterizes spatiotemporal changes in soil salinity and other properties related to soil quality at a drainage water reuse site. This study is part of an ongoing, multidisciplinary collaboration investigating the sustainability of drainage water reuse on forage production as an alternative method for the disposal of drainage water in California's San Joaquin Valley [304,305]. For details of the methodology, see Corwin et al. [304,306].

Data Collection   The study site is a 32.4-ha saline–sodic field (fine, montmorillonitic, thermic, Typic Natrargid) located on Westlake Farm on the west side of the San Joaquin Valley. An initial EC_a survey was conducted on August 1999 12–16, using mobile electromagnetic induction (EM) equipment. The EC_a survey followed the protocols outlined by Corwin and Lesch [301]. The survey consisted of a grid of EC_a measurements arranged in a 4 (row) × 12 (position within row) pattern within each of 8 paddocks, for a total of 384 sites. The spacing between the 384 sites was approximately 20 m (N–S) and 30 m (E–W). All 384 sites were georeferenced to submeter precision with GPS. Measurements of EC_a were taken using a EM38 unit (Geonics Ltd., Mississaugua, Ontario, Canada) with the coil configuration oriented in the vertical (EM_v) and in the horizontal (EM_h) position. A follow-up EC_a survey was conducted in April 2002.

The EC_a measurements taken in 1999 were used to establish the location of 40 sites where soil core samples were taken. A model-based, EC_a-directed soil sampling approach, specifically a response surface sampling design, was used to locate the 40 sample sites. At each of the 40 sites, soil core samples were taken at 0.3-m increments to a depth of 1.2 m. To observe spatiotemporal changes, soil core samples taken August 19–23, 1999 were compared to samples taken 32 months later (i.e., April 15–17, 2002). Soil cores were analyzed for physical and chemical properties deemed important for soil quality assessment of an arid zone soil whose function was forage production for livestock. Soil chemical properties included electrical conductivity of the saturation extract (EC_e); pH_e; anions (HCO_3^-, Cl^-, NO_3^-, SO_4^{2-}) and cations (Na^+, K^+, Ca^{2+}, Mg^{2+}) in the saturation extract; trace elements (B, Se, As, Mo) in the saturation extract; $CaCO_3$; gypsum; cation-exchange capacity (CEC); exchangeable Na^+, K^+, Mg^{2+}, and Ca^{2+}; ESP; SAR; total C; and total N. Soil physical properties included saturation percentage (SP), volumetric water content (θ_v), bulk density (ρ_b), and clay content. To display and manipulate the spatial data, a commercial geographic information system (GIS) was used [98a].

Results Sustainability of drainage water reuse at the Westlake Farm site depends on spatiotemporal changes to soil properties that affect forage production or livestock health detrimentally. EC_e, SAR, Mo, and B were established as the most important properties for evaluating the study site's soil quality [304]. The 1999 and 2002 correlation coefficients between EC_a (both EM_h and EM_v) and soil properties over the depth 0 to 1.2 m showed that the properties of EC_e, SAR, Mo, and B correlated significantly with EC_a at the $p < 0.01$ level in both 1999 and 2002.

This indicated that the EC_a-directed sampling approach using the response surface sampling design accurately characterized the spatial distribution of EC_e, SAR, Mo, and B. Table 9.3 shows the shift in means from 1999 to 2002. In the case of EC_e, the mean levels were reduced ($p \leq 0.01$) in the first two depth increments. The salinity results suggested that leaching of salts had occurred in the near surface depth of 0 to 0.6 m, with negligible leaching below 0.6 m. Unlike EC_e, the mean SAR levels were reduced significantly ($p \leq 0.05$) across all four depth increments. For B, the mean level was reduced significantly ($p \leq 0.01$) in the depth increment 0 to 0.3 m, and elevated significantly ($p \leq 0.01$) in the depth increments 0.6 to 0.9 and 0.9 to 1.2 m. Mean Mo levels were reduced ($p \leq 0.01$) in all depth increments.

To visually evaluate spatiotemporal EC_e trends, 1999 baseline and spatiotemporal difference are shown in Fig. 9.73. Lightly shaded portions represent areas of decreases in salinity, and darkly shaded portions indicate areas of increases from 1999 to 2002. It was found that EC_e and SAR displayed similar spatial patterns and temporal changes. Overall, salinity was leached from the top 0.6 m, B was leached from the top 0.3 m and accumulated in the depth increment 0.6 to 1.2 m, and sodium and Mo were leached from the top 1.2 m.

Data and statistical analyses demonstrate the flexibility and utility of EC_a-directed soil sampling as a basis for assessing management-induced spatiotemporal changes in soil quality. Although only one type of management applied at one

TABLE 9.3 Mean, Range, Standard Deviation (SD), and Coefficient of Variation (CV) Statistics of EC_a, $N = 40$ SAR, B, and Mo, 1999 and 2002,

Soil Property	1999					2002				
	Mean	Min.	Max.	SD	CV	Mean	Min.	Max.	SD	CV
Depth: 0–0.3 m										
EC_a (dS/m)	13.0	5.6	35.7	7.5	57.8	11.43	4.83	30.60	6.06	53.05
SAR	28.2	8.3	70.2	16.5	58.7	23.46	5.62	59.50	14.4	61.39
B (mg/L)	17.0	1.1	42.5	8.2	48.2	14.21	2.64	33.23	7.35	51.75
Mo (μg/L)	862.3	442.0	3043	532.5	61.8	632.1	150.0	3291	592.1	93.66
Depth: 0.3–0.6 m										
EC_a (dS/m)	20.2	13.5	34.5	5.3	26.0	17.46	6.11	34.00	6.55	37.48
SAR	51.4	30.3	89.5	12.9	25.1	40.31	9.13	78.87	15.31	37.99
B (mg/L)	19.0	13.6	38.1	5.6	29.7	19.06	6.69	32.35	6.09	31.97
Mo (μg/L)	750.5	180.0	2488	430.2	57.3	576.5	220.0	1783	375.8	65.18
Depth: 0.6–0.9 m										
EC_a (dS/m)	22.5	9.7	43.2	6.5	28.7	22.49	7.94	37.90	6.96	30.96
SAR	59.0	24.0	107.6	16.6	28.1	53.35	16.26	91.90	16.00	30.00
B (mg/L)	17.5	9.4	31.3	4.8	27.2	21.49	11.17	34.19	5.84	27.17
Mo (μg/L)	780.5	183.0	1756	338.9	43.4	661.6	252.0	2372	451.5	68.24
Depth: 0.9–1.2 m										
EC_a (dS/m)	25.2	8.0	49.7	7.9	31.5	24.30	7.84	45.30	8.14	33.51
SAR	64.9	16.8	120.2	19.5	30.0	57.46	16.51	103.1	17.96	31.25
B (mg/L)	17.9	6.5	31.8	6.3	35.0	21.71	7.89	39.0	6.59	30.36
Mo (μg/L)	946.9	330.0	2856	450.7	47.6	720.7	240.0	2991	451.5	62.65

source: [289].

location was considered, the implication extends beyond the localized, though significant finding that EC_a can be used to monitor drainage water reuse in a saline–sodic soil system. More important, when EC_a is correlated with soil properties associated with soil quality (and/or productivity), EC_a-directed soil sampling is an effective tool to broadly evaluate the spatiotemporal impact of management on soil resources. Assessment and interpretation guidelines are currently available to document the effects of current and alternative soil and crop management strategies on soil resources [301,307].

Electromagnetic Induction to Manage Cattle Feedlot Waste Open-lot cattle feeding operations face challenges with respect to waste management. Relevant issues include: (1) collection of solid waste from feedlot surfaces to be utilized by crops, (2) control and utilization of nutrient-laden liquid runoff, and (3) feedlot surface management to reduce nutrient losses and gaseous emissions. Electromagnetic induction (EMI) tools have been valuable in assessing and managing nutrient resources from beef cattle feedlots. The work described here was all conducted at the U.S. Meat Animal Research Center (USMARC), Clay Center, Nebraska (40°32′ N, 98°09′ W, altitude of 609 m).

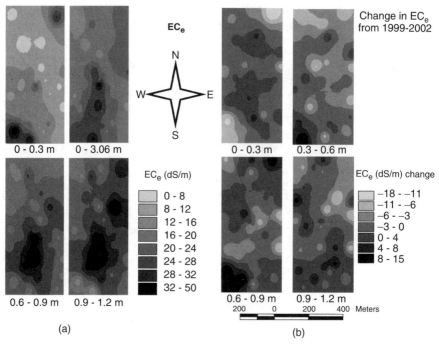

FIGURE 9.73   IDW interpolated maps of (a) EC_e for 1999 at depth increments of 0 to 0.3, 0.3 to 0.6, 0.6 to 0.9, and 0.9 to 1.2 m and (b) change in EC_e from 1999 to 2002 at depth increments of 0 to 0.3 and 0.3 to 0.6 m. (From [306].)

Cornfield Study   Electromagnetic induction methods were used to determine the agronomic effectiveness and environmental consequences of nitrogen fertilization for varying rates of compost, manure, and commercial fertilizer with and without a cover crop in a 5.9-ha corn silage research field. This field had the same treatments applied for 10 consecutive years. Sequential EMI surveys were examined as a tool in monitoring nitrogen-cycle dynamics for this field. A total of 114 EMI surveys were conducted on this field during the corn-growing seasons (1999–2003). A Geonics EM-38 was used to conduct soil surveys for the first two years, and a Dualem-2 was used in 2002 and 2003. The horizontal response was used from both instruments during the entire study. Figure 9.74 shows one such survey.

The four-year study concluded that apparent soil conductivity, EC_a, as measured by EMI clearly identified the effects of manure, compost, fertilizer, and cover crop on EC_a values. Compost and manure applied at the available nitrogen application rate of 200 + kg N/ha resulted in consistently higher conductivity and available nitrogen when compared with the commercial fertilizer 84 kg N/ha and compost at the phosphorus rate. Sequential measurements of profile-weighted EC_a effectively identified the dynamic changes in available soil nitrogen, as affected by animal manure and commercial anhydrous ammonia fertilizer treatments during the corn-growing season [308].

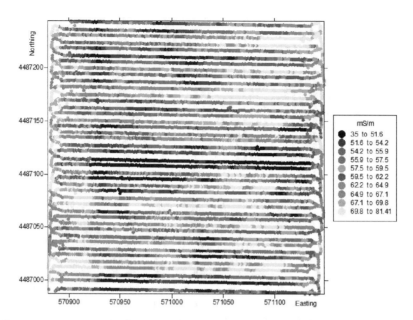

FIGURE 9.74   Survey of a research cornfield conducted June 17, 2004; light areas indicate higher conductivities.

Vegetative Treatment Area   Precipitation runoff from beef cattle feedlots can be managed by nontraditional methods; a vegetative treatment system is one such approach, with a design that includes a solid separation basin and a vegetative treatment area (VTA). The VTA is typically a hayfield designed to utilize nutrients and liquids discharged from the settling basin. Knowledge of liquid distribution is critical for proper management to ensure sustainability. The liquid discharge into the VTA contains high salt levels from the manure on the feedlots, making the use of EMI methods viable for tracking liquid flows in the VTA. A Dualem-1S (horizontal response) was used for this study. Figure 9.75 shows an EMI map of a VTA at USMARC which has been operating since 1996. The VTA conductivity map is located at the top of Fig. 9.75; light areas show regions of salt accumulation from the liquid discharge.

The EMI image of the VTA shows relatively uniform flow patterns of salt loading near the discharge areas from the berm. An earlier survey, conducted in August 2005 (not shown), clearly showed greater salt loading near the west end of the VTA [309]. An investigation revealed that the discharge tubes on that end had settled, allowing more flow into the VTA in that region. A modification was made to the inlets in the spring of 2006 that allowed a more even flow from the tubes. The 2008 image gives evidence of the success of that modification. Also, the image shows salts extending only about one-third the length of the VTA which demonstrate the conservative nature of the VTA design. Figure 9.75 is indicative of a sustainable system since much of the field does not show salt buildup; this

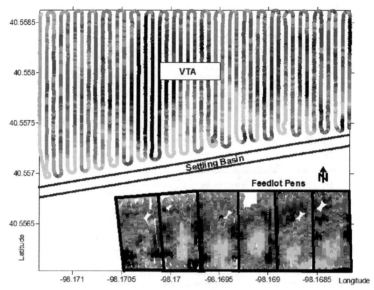

FIGURE 9.75 EMI map survey in 2008, showing feedlot pens that provide precipitation runoff to a VTA at USMARC. Light areas represent high EC_a values.

view has been supported by nutrient balances showing more nutrients leaving the hayfield in the hay crop than are deposited by the incoming effluent [310].

Cattle Feedlots   Figure 9.75 also shows EMI surveys of feedlot pens that drain into the VTA (shown in the lower portion of Fig. 9.75). The feedlot surface is the major source of nutrients to be managed by a feedlot. Understanding the distribution of nutrients on the surface allows the managers to make decisions that are environmentally sound. Electromagnetic induction surveys of feedlot surfaces using a Dualem-1S (horizontal response) have demonstrated that manure nutrient accumulations can be identified. Regression analysis shows that apparent soil conductivity, EC_a, is highly correlated with volatile solids, total nitrogen, total phosphorus ($R^2 = 0.92, 0.91, 0.93$, respectively [311]).

Maps illustrating zones of manure nutrient accumulation could be used to direct pen cleaning efforts. The concentrated scrapings from the feedlot would have added value as fertilizer for land application or composting. Also, zones of higher manure concentrations are more likely to have increased insect populations and pathogen buildups. The GPS coordinates associated with the mapping technique could be used for precision application of the pesticides or antimicrobial compounds. The same techniques could be applied to reduce malodorous emissions.

Developing Soil Similarity Zones in Agricultural Fields Using On-the-Go EC_a Mapping   Precision agriculture or site-specific management is a technology that uses advanced tools to map, and subsequently manage, the variability across the field. Among the many advanced sensors recently introduced, apparent soil electrical conductivity (EC_a) measuring devices provide the simplest and least expensive soil variability measurement. The value of EC_a mapping in site-specific management has been widely recognized as a surrogate spatial map for soil variability to identify within-field areas (or zones) of soil similarity. That provides the potential for EC_a mapping as a practical tool to delineate soil-based management zones for variable-rate application of agricultural inputs. In practice, the challenge is to identify the underlying differences in soil properties between the soil-based management zones.

In fields containing a high concentration of salts, EC_a measurements effectively portray both the nature and the main cause of EC_a variability (i.e., relative salinity). In contrast, EC_a in nonsaline fields depicts spatial variability without clearly identifying the dominant cause(s) of variability. Research shows that in nonsaline fields, soil water content (θ_w) and temperature, clay content and mineralogy, and organic matter content are among the dominating soil properties affecting EC_a (see Allred et al. [232] for a comprehensive review). Knowledge of the spatial distribution of clay content is of profound importance in site-specific management as it relates to soil texture. Soil texture is the most important factor affecting soil water movement and chemical transport and thus crop growth. The practical utility of EC_a to map texture (or clay content), however, is elusive and demands caution because of the complex interactions between EC_a and soil physical and chemical properties.

From theory (i.e., the dual-pathway EC_a model as formulated by Rhoades et al. [120] and applied by Corwin and Lesch [307] and Farahani et al. [312]), the relationship between EC_a and soil stable properties (such as clay content) is governed by soil transient properties of water content, ion concentration, and temperature at the time of the EC_a mapping. That is one of the main reasons that the strength of the literature-reported associations between EC_a and soil properties (e.g., clay content) varies widely with correlation coefficient values ranging from below 0.3 to above 0.8 [312]. Two other important considerations in EC_a mapping are the quality of the map and its temporal stability. These have been addressed in various studies, among which key findings from a more comprehensive study conducted in Colorado is presented below (for details, see Farahani et al. [312–316]).

Soil Properties   The objective in one Colorado study was to characterize the main soil properties that alter EC_a using multiyear measurements (1998 to 2003) in three center-pivot and non-saline-irrigated fields. Results show that EC_a correlated well with soil clay, water, and organic matter contents (with correlation coefficients between 0.66 and 0.96). Despite the strong correlations at each measurement date, there was no single unique relationship applicable across all measurement dates. The linear relationship between EC_a and the soil stable property of clay content changed when soil water content and solution concentration changed considerably. That finding was supported by theory (the dual-pathway EC_a model), with model predictions shown in Fig. 9.76. One immediate conclusion is that empirical EC_a vs. soil property relationships are inadequate for predictive purposes over time. However, this is not a problem when mapping stable soil properties (such as clay

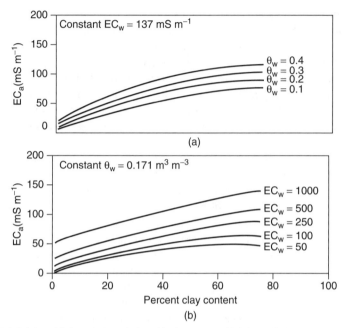

FIGURE 9.76 Sensitivity of the relationship between soil EC_a and clay content to changes in (a) volumetric soil water content (θ_w, m^3/m^{-3}) and (b) soil solution concentration (EC_w, mS/m^{-1}).

content) from EC_a maps since one would only conduct a single EC_a mapping and same-day soil sampling for calibration. It is unlikely that one would have a need to remap stable soil properties for many subsequent years.

Temporal Stability While the spatial variability of EC_a and its causative factors are of significant importance, understanding the temporal stability of the EC_a map is equally important. That is particularly true if delineated EC_a zones are to be used to manage agricultural inputs across the field for multiple years. In the three Colorado fields mentioned previously, multiyear measurements of near-surface (top 0.9 m of soil) EC_a produced maps that were highly stable over time when delineated into low, medium, and high zones (Fig. 9.77). This was mainly because the maps represented the stable soil properties, which remained unchanged over time. When salt concentration and buildup are low (as was the case herein), results suggest that single EC_a mapping should suffice to delineate soil-based management zones without a need for remapping. This is despite the fact that the absolute values of EC_a may change over time due to soil water and temperature variations.

Map Quality The effectiveness of site-specific management also depends on the quality of soil properties maps. The quality (or precision and accuracy) of EC_a delineation boundaries is dictated primarily by the interpolation techniques (such as kriging and inverse distance) used to map the point samples from the field and

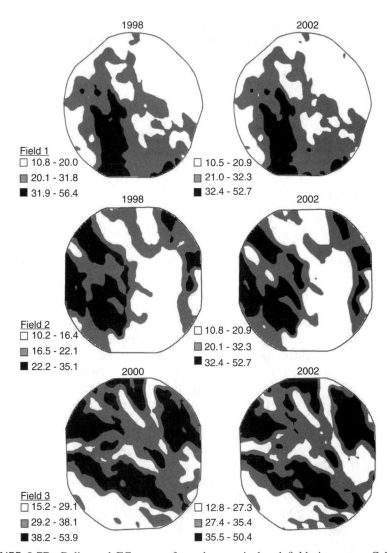

FIGURE 9.77 Delineated EC_a maps from three agricultural fields in eastern Colorado (measurement days were in 1998 and 2002 at fields 1 and 2 and in 2000 and 2001 at field 3).

the number of samples per unit area (i.e., the spatial sampling intensity). Literature suggests that the latter is more important than the interpolation technique [317,318]. Another Colorado study explored the quality of EC_a maps as affected by the spatial sampling intensity (i.e., varying swath width of the mobile EC_a sensing platform). Very dense EC_a measurements were collected from one field using the mobile EC_a sensor. This fine data layer, representing the base EC_a map with only 2.5 m between the centers of adjacent swaths, was then scaled up to produce coarser EC_a data layers with swath widths from 5 to 200 m. The patterns of low, medium, and high EC_a zones were produced for all data layers and compared. Results show less than 0.5% reduction in map prediction efficiency and map pattern agreement (as compared to the base map), respectively, per meter increased in swath width. Map quality deteriorated slightly as swath width increased to about 50 m, but at swath widths of 80 m and higher, the quality deteriorated severely and map delineations in the form of location-specific low, medium, and high EC_a zones became highly distorted and wrongly classified. Increasing the manufacturer's recommended swath width range of 12 to 18 m to a range of 30 to 50 m caused only a 10% reduction in map quality while reducing the mapping time by 65%.

9.6 ARCHAEOLOGY AND CULTURAL HERITAGE

Over the last 20 years, archaeologists have, increasingly, applied geophysical techniques to assess the presence of archaeological ruins in the subsurface. In fact, although excavation can expose buried structures for hands-on investigations, it is also irreversible, destructive, time consuming, and expensive. Furthermore, not all archaeological sites can be excavated, such as historic buildings, churches, mosques, the pyramids, parks, and areas which underlie modern urban development.

Among all geophysical methods, magnetic, ERT, and GPR are by far the most commonly used to investigate the presence, the dimension, the geometry, and the depth of an archaeological site. Of course, the suitability of one technique with respect to another depends on several factors, such as (1) the physical properties of the soil covering the site, (2) the physical properties of the material used for building the site (i.e., walls, roads, graves, etc.), (3) the depth and the hydrological conditions of the site (i.e., above or below the water table), and (4) the accessibility of the site and the presence of strong antropic noise. Moreover, a specific technique may be suitable to detect a certain archaeological target but ineffective for another, although this depends again on the environmental and geological conditions of the site. For example, in most countries magnetic techniques are not suitable to detect roads or graves, whereas in Italy, Roman roads (made of basalt blocks) and Italic graves (excavated in volcanic rocks) are easily recognizable using a magnetometer [319,320].

Interest in the use of geophysical techniques (especially GPR) in archaeology and cultural heritage is also related to their ability to assess the state of preservation of historical buildings. In big cities relevant damage to such buildings is usually a consequence of intense traffic, pollution and the specific destination of these structures. The restoration plans of such valuable buildings require the precise detection and location of the structural lesions, which can be done applying nondestructive techniques such as GPR, high-resolution ERT (with medical electrodes), or ultrasound.

9.6.1 Brief History of Geophysical Prospections in Archaeology

In early times, field archaeologists applied what can be considered as a precursor of the seismic survey to search for buried structures. By thumping the ground with a crowbar and listening to the sound variations, they were able to detect the presence of archaeological target buried in the ground. In fact, the first recorded application of a geophysical technique for archaeological prospecting can be attributed to Lieutenant-General Augustus Pitt-Rivers, who, in 1893, pounding on the ground with a bar, was able to detect a ditch from the variation of the sound [321]. Ironically, seismic surveys have not been used much in subsequent archaeological research. Actually, the resistance technique was the first geophysical method to be applied to archaeological prospection, whereas in time, magnetic survey became the most widely used method for this type of application, and GPR the most suitable to locate buried structure precisely and to achieve a three-dimensional reconstruction of the geometry of the subsurface ruins.

The first use of modern geophysical methodology was in 1938 over the site of a suspected buried vault adjacent to the Bruton Parish Church at Williamsburg, Virginia [322]. The resistance survey recorded the presence of an anomaly. However, subsequent excavation failed to locate any archaeological feature associated with it. According to Atkinson [323], the starting point for this discipline can be located in 1946, when a resistance survey at Dorchester on Thames (UK) was made.

Several geophysical surveys were also documented by Albert Hesse, a pioneer in the theory and application of geophysical techniques in archaeology, as well as the author of the first textbook on archaeological geophysics [324]. In particular, he reported on the work done at the Bleiche-Arbon site by various archaeologists, who used a metal detector to find several bronze daggers [325] and also discovered that tiles and bricks can cause a detectable electromagnetic signal [326].

The development of another successful technique, the magnetic survey, begun in 1958 by Martin Aitken, has introduced the use of a proton magnetometer into archaeology. Interest in this technique increased when it was discovered that not only fired kilns and ferrous objects would show magnetic anomalies, but also soil features such as ditches and pits. Shortly after, in 1963, Irwin Scollar introduced the differential sensor configuration for large-scale surveys [327] and in 1966 began automatic data recording on punched paper tape. Fluxgate magnetometers, which allowed continuous recording due to improved measurement speed, were also introduced in the 1960s by John Alldred and Frank Philpot, together with alkali vapor magnetometers for high-sensitivity measurements [328].

GPR application to archaeological prospections started in the middle of the 1970s, becoming more common in the late 1980s and early 1990s [329]. After this initial burst of development, there was an apparent decline of research interest in this discipline. Certainly, this period marked the unwelcome demise of the highly influential journal *Prospezioni Archeologiche*, published by the Lerici Foundation in Italy, reporting fundamental research in the field for two decades before it ceased publication in 1986.

At the beginning of the 1990s a new interest begun in the application of geophysical methods in archaeology; modern computer-based and real-time measurements techniques such as GPR became very common [329]. Moreover, the possibility to acquire a large volume of data with a fast, digital recording technique using different geophysical methods such as ERT, gradiometer, frequency-domain

TABLE 9.4 Potential Archaeological Targets for Principal Geophysical Methods

Archaeological Target	Archaeological Prospection			
	ERT	Magnetometry	Magnetic Susceptibility	EM/GPR
Roads	×			×
Building Structures		×		
Walls	×			×
Wall fabric				×
Occupation Soils			×	
Occupation Debris		×		
Cunicula				×
Hearthis		×		
Ditches	×			
Kilns		×	×	
Ovens			×	
Baked Clays		×	×	
Daub			×	
Graves	×			×
Complex stratigraphy				×

electromagnetic, and GPR opened new possibilities for archaeological and cultural heritage applications. At present, the most promising field of research is the *multi-techniques integrated approach*, which uses different geophysical methods, on the same site, to introduce some constrains in the geometrical and physical properties of the anomaly and to create more realistic inversion models of the subsurface. Excellent summaries of the history of geophysical prospections applied to archaeology have been published ([321] and [330]; more recently, [331–333]).

9.6.2   Archaeological Targets and Geophysical Techniques

As pointed out earlier, the suitability of a specific geophysical technique to detect an archaeological target successfully is based on several factors: nature of the target, contrast between the physical properties of the target and the background, target depth and dimension, soil conditions (e.g., water content), site accessibility, and antropic noise level. Table 9.4 shows the most common archaeological targets together with the most suitable geophysical techniques [331–333].

It should be emphasized, however, that there is no rigid rule in the choice of a method with respect to the target being sought, and the success of a survey is very site-specific. Nevertheless, if the aim of the survey is to find the general area where archaeological ruins are buried, the first step can be a reconnaissance prospection using a magnetometer or some fast frequency-domain electromagnetic techniques. Once a restricted region of interest has been delimited, a high-resolution method such as three-dimensional ERT or GPR can be applied. In particular, GPR is the only method that allows one to detect archaeological targets with both very high vertical and horizontal resolution [334–337], although soil conductivity can be an impenetrable barrier for this kind of technique, even when the targets are very shallow (within the first 30 cm of soil) [338].

GPR is also a very promising technique for nondestructive investigation of the state of preservation and the structural integrity of historical and archaeological buildings. Fractures, voids, water infiltration, and metal and concrete reinforcing structure can be detected in walls, pillars, and attics using high-frequency antennas [339,340]. In fact, building materials have, generally, good dielectric properties and allow for high penetration and good resolution. Moreover, the possibility to apply two- and three-dimensional acquisition often permit detailed reconstruction of the spatial distribution of the structural features, as shown in one of the case studies described below.

9.6.3 Geophysical Survey of a Roman Bath Complex

This case study report the results of a geophysical survey conducted in a thermal area in northern Italy [341]. The Euganean area, near the city of Padua, has long been known for its natural geothermal springs and has been documented since its early settlements (*Patavini fontes* and *fons Aponi* are cited frequently by classic authors). During the Roman Age, its thermal resources were exploited significantly and the Abano-Montegrotto neighborhoods became important religious, residential, and spa resorts. Rapid growth took place around the most important springs, and many complexes with pools and lodging facilities were created. Although thermal treatment was still in use during the Middle Ages, urbanization progressively hid the Roman heritage, and most of the buildings were demolished to recycle construction materials.

Early archaeological investigations in Montegrotto, dating to the eighteenth and nineteenth centuries, found no traces of the site: however, further excavations carried out in 1965 revealed the basement of a small theater and a building arranged around a central courtyard, probably a *nymphaeum*. In the late 1980s, during the restoration of the Hotel Terme Neroniane, new remains of the settlement were found near the railway station of Montegrotto.

Magnetometry   In the summer of 2005, a magnetic survey was carried out over an area of about 7000 m^2. Strong anomalies were expected, because in Roman buildings high-susceptibility minerals typical of the Euganean hills (trachyte containing variable concentrations of ilmenite and magnetite) were used. The aim of the survey was to investigate the southern part of the area, where GPR did not show any significant anomaly, whereas aerial photographs revealed the presence of several structures (Fig. 9.78). The survey was conducted with an Overhauser GSM-19 high-resolution magnetometer (0.1 nT), in gradiometric configuration, with the lower sensor at 0.30 m from the surface and a distance of 1 m between the two coils. Data were sampled on a regular grid of 1 m, oriented N–S (although the southwest corner of the area was not surveyed because of the presence of surface obstacles). Irregular spikes and very noisy data recorded along the eastern part of the area, close to the railway line, were removed, and the data were interpolated

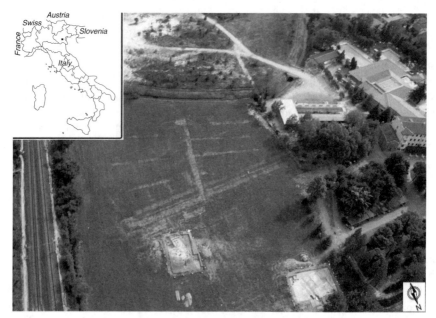

FIGURE 9.78   Low-altitude aerial photograph of the study area, showing the state up to 2003 and some soil marks from buried structures in the southern part (top).

with various gridding algorithms. The values of the magnetic field gradient varied between -50 and $+50$ nT/m. The resulting magnetic gradient gray-tone map (Fig. 9.79) is very rich in information, useful to identify features of various shapes.

In the figure, linear high-value dipole anomalies (-50 and $+50$ nTm) are visible, the shape of which may correspond to a large building complex, oriented along the cardinal axis. The southern edge of the map shows a slightly irregular square dipole anomaly due to spatial aliasing, corresponding to a square structure excavated in 1990 and detected also by GPR. At the southern boundary, two clear parallel anomalies outline the presence of a possible curvilinear arcade with lateral wings. Farther north, at the center of the complex, there is a weaker, circular anomaly (10 nT/m), with two correlated linear anomalies branching off from it toward the east and west.

The complex is closed to the north by two linear anomalies (20 nT/m) running E–W, one of which corresponds to the wall foundation detected by a GPR survey and belonging to the long walls excavated in 1992. Furthermore, a series of three strong orthogonal anomalies (50 nT/m), oriented N–S, may be associated with further subdivisions. In the southeastern corner a small area was selected for a more detailed magnetometric survey and also for an ERT prospection. In particular, a high-resolution survey (21 profiles on a $0.5 - m^2$ mesh) was carried out with a GSM-19 magnetometer. The data were interpolated and mapped (Fig. 9.80), putting in evidence the presence of an anomaly ($-40/+50$ nT/m) without aliasing, much clearer than the previous one, showing part of the exedra wall and the southern square room.

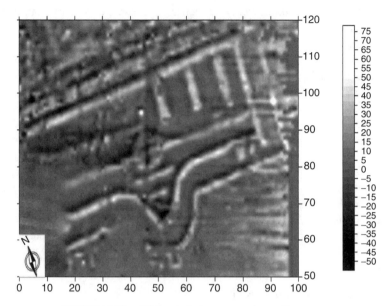

FIGURE 9.79 High-resolution magnetic gradient map.

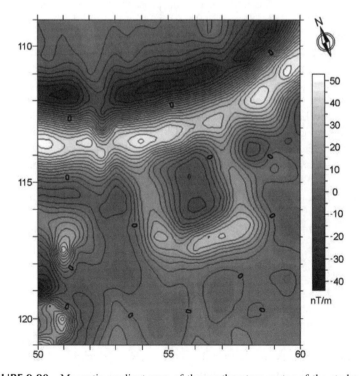

FIGURE 9.80 Magnetic gradient map of the southeastern sector of the study area.

Electrical Resistivity Tomography Electrical measurements carried out in Montegrotto, collected with a 64-electrode custom-built system, were aimed at obtaining two- and three-dimensional resistivity data. Several grid spacing values (from 0.25 to 0.5 m) for both two and three dimensions were employed in the field. Because of the limited number of channels available, roll-along geometry was used for data acquisition.

Pseudo-three-dimensional acquisition (area of 10×12 m^2, probe spacing of 0.25 m, profiles 0.50 m apart) was performed in the sector where a more detailed magnetic survey was acquired. The result of the electrical survey is illustrated in Fig. 9.81. The depth slices show that at least up to a depth of 0.50 to 0.60 m, three different features are clearly detectible, due to a strong resistivity contrast between the archaeological structures and the soil: the square central room (A1), the external wall of the nymphaeum (A3, tangent to the northern side of the room) and the median aisles (A2). The first two are consistent with the magnetic anomalies of Fig. 9.80, although the excavation made in the spring of 2006 confirmed the presence of all three features.

9.6.4 Cavity Detection with ERT

Another example of ERT use to detect archaeological ruins is presented in the following case study. The purpose of this prospection was to detect and locate precisely several buried cavities that were found during the building of some infrastructures close to the Prenestina Railway Station in Rome, Italy. The origin of the cavities can be ascribed to past extraction activities of pyroclastic deposits ("pozzolane"), which have been carried out in the area since Roman times.

Strong electromagnetic noise, due to presence of the railway (Fig. 9.82), prevented use of the magnetometer, and the high conductivity of the tuffs did not allow the use of GPR. ERT was the most suitable technique, because the resistive values of volcanic products are generally medium to low, whereas voids are highly resistive, so that a detectable contrast was expected. An RS-1 Syscal georesistivimeter by IRIS Instruments was used, with 48 electrodes and an electrode spacing of 2 m. Applying the "roll-along" technique, it was possible to extend the profiles up to a length of 142 m; moreover, Wenner configuration was used, so that the maximum exploration depth was about 16 m.

Two ERT profiles were performed along a direction parallel to the railway tunnel, with a partial overlap between the profiles, so that the origin of profile 1 coincides with the 80-m position in profile 2. Data inversion was carried out by Res2Dinv software, taking topographic variations into account. A reference elevation of 0 m was assigned to the bottom of a recent excavation made in the area. The root-mean-square error ranged from 6.8 to 7.7%, showing a satisfactory result for the inversion procedure.

Figure 9.83 illustrates the results obtained after data inversion. Two cavities were known from previous excavations in the surveyed area, and they were well detected with the technique (their entrances are indicated on the figure by the letter "E"). These highly resistive buried structures, having resistivity values up to 200

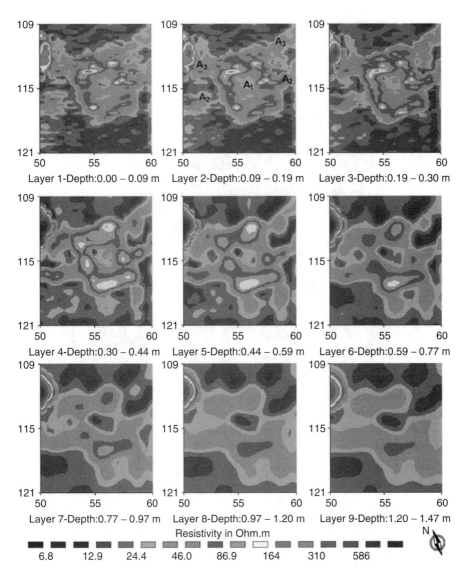

FIGURE 9.81 Electrical resistivity tomography pseudo-three-dimensional depth slices.

to 300 $\Omega \cdot$ m, are indicated by the letter "K." However, the profiles also show other resistive anomalies, with shape and resistivity values similar to the one indicated; therefore, they were interpreted as cavities. Generally, the buried cavities in this area have depths of a few meters and heights of about 3 to 4 m, so the bottoms of these structures are located 6 to 8 m from the ground surface. The existence of these cavities was subsequently confirmed with a borehole test performed on top of the main resistive anomaly.

FIGURE 9.82   Profile location. On the right is railway tunnel; on the left, below the slope, is the entrance of a known cavity.

9.6.5   Ground-Penetrating Radar Investigations into Construction Techniques

The setting of the Concordia Temple in Agrigento, Sicily is unique in the world; ratios are simple and proportions are perfect. Erected around 430 B.C., it is 19.75 m wide and 42.23 m long, slightly larger than a double square. It covers an area of 843.38 m² and is 13.48 m tall. The cell is preceded by a simple antechamber (*pronaos*, 5.11 × 7.65 m) with two columns and is followed by a back porch (*opistodomos*, 4.72 × 7.65 m), where the treasure, votive gifts, and archives of the temple were kept. In keeping with classical models, the elegant and airy colonnade has 6 × 13 columns which are 6.75 m high, constructed of four drums, and has 20 sharp-edged flutes.

Before the GPR survey, it was believed that the foundation of the temple consisted of a compact and homogeneous structure formed by an outside layer of bricks and various grouting materials, as depicted in Fig. 9.84. The scope of the radar investigation was to better define the foundation geometry and to detect the boundary between the bricks and the conglomerate [342].

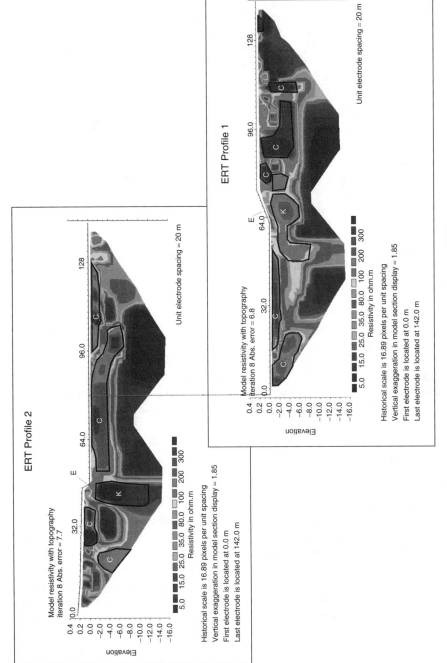

FIGURE 9.83 Cavity location on ERT.

653

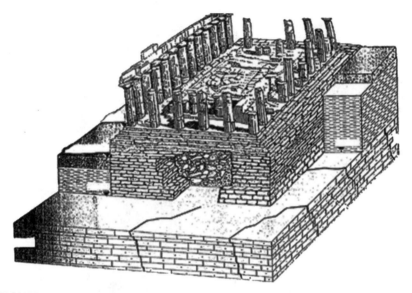

FIGURE 9.84 Hypothesized construction technique originally employed in the temple, consisting of a compact and homogenous structure consisting of an outside layer of bricks with various grouting materials.

The GPR profiles were acquired using a PulsEkko 1000 unit (Sensors & Software, Inc.) equipped with a 225-MHz bistatic antenna in order to obtain good vertical resolution and, at the same time, good signal penetration. The data were collected along parallel profiles, with an antenna spacing of 0.5 m, moving the antennas perpendicular to the direction of the profile (i.e., in perpendicular broadside mode. All data were acquired in step mode with a sampling interval of 0.1 m, a trace stacking of 16, and a time window of 100 ns. A total of 261 m of GPR data were collected within the external colonnade via four longitudinal profiles along the major axis of the temple and 10 transversal profiles (including two longer profiles along the minor axis of the temple and eight shorter along its northern side). A total of five longitudinal profiles parallel to the major axis of the temple and six transversal profiles were also acquired within the inner part of the temple with 1-m line spacing.

An analysis of the raw and processed sections collected outside the temple cell allows one to hypothesize that the top layer of the temple base is constituted of partially cemented orthogonal limestone and/or sandstone blocks that form the paving of the temple; the alignment of the blocks is probably not perfect, as demonstrated by the presence in all sections of several diffraction hyperbolas and corner reflectors (Fig. 9.85). In fact, the latter are typical GPR features generated by empty fractures or voids which are oriented approximately perpendicular to the profile.

In contrast, the bottom layer seems to consist of two different units: an artificial foundation made of structural kerbs, which correspond to the regular structure of alternating short reflectors and signal blanks detected by the radar (Fig. 9.86); and a natural foundation consisting of the limestone spur on which the temple rises.

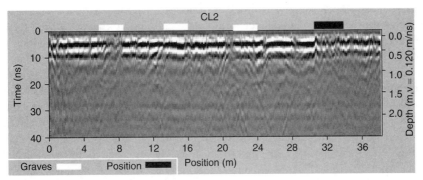

FIGURE 9.85 GPR-processed section collected along the major axis of the temple, showing clearly shallow scattering due to empty fractures or air gaps between the limestone blocks and to some archaeological remains.

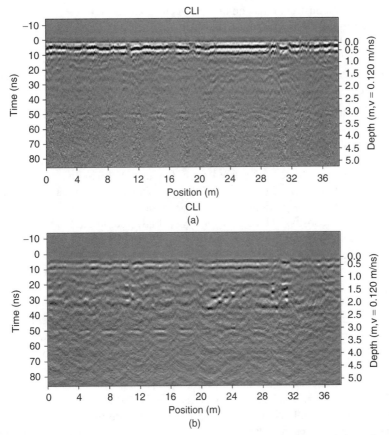

FIGURE 9.86 GPR-processed sections: (a) raw and (b) migrated, collected along the major axis of the temple, showing the events (at 50 ns) associated with the regular structure of short reflectors, alternating with areas of signal blanks.

The presence of shallow bedrock is supported by the continuous horizontal reflector visible in the data collected inside the cell and by the lack of the typical alternating structure in some parts of the longitudinal profiles collected along the southern side of the temple (Fig. 9.87). This technical solution was probably chosen because the flat part of the bedrock did not have the proper dimensions to contain the entire temple. As such, the natural surface was integrated with a structure consisting of parallel kerbs of differing lengths depending on the space required.

Due to the high scattering in the first 30 ns of the longitudinal sections, and therefore to the high attenuation of the GPR signal, it was not possible to detect the interface between the kerbs and the bedrock. It is expected, however, that the kerbs have different thicknesses, due to the irregular shape of the limestone spur, and that they may extend to considerable depths (as demonstrated by some temple photographs made at the beginning of twentieth century). In these images, in fact, a greater number of steps than those currently exposed are visible (a minimum of six steps in the northwestern part).

This type of construction technique has been adopted elsewhere, as demonstrated by similar foundations present in some temples in the archaeological valley near Agrigento (the Temple of Vulcan and the Temple of Demeter) (Fig. 9.88a and b) and the Temple E in Selinunte (Trapani). The use of GPR on the Concordia Temple made it possible to detect the location, depth, and size of the buried structures, and more important, the results obtained have drastically changed the original hypothesis about the building foundation.

9.6.6   GPR Survey in Pompeii

GPR investigations have been carried out in Pompeii, Italy, in an area of the Regio III not yet fully excavated (Fig. 9.89) [343]. In this area, as in many other parts of this Roman city, large portions of the archaeological features are still buried under thick volcanic deposits. The radar survey was conducted on the top of a long scarp parallel to Nola Street. This site was chosen because some remains are clearly visible at the front of the scarp, so they can be use to calibrate the radar sections. The survey was conducted using a Noggin^{PLUS} system (Sensors & Software, Inc.) equipped with 250-MHz antennas and a SmartCart. A time window of 100 ns, a 5-cm step size, and a trace stacking of 4 were used in all profiles. A first series of 30 lines 50 m long oriented E–W (i.e. parallel to the scarp illustrated in Fig. 9.90) were acquired. The distance between the lines was 0.5 m, to allow for a two-dimensional reconstruction of the area (30 × 50 m time-slice maps). Subsequently, 30 N–S oriented lines with the same interline spacing and a length of 10 m, were acquired perpendicular to the previous ones (30 × 10 m time-slice maps).

The data collected were generally of good quality and did not require any special processing. Figure 9.91 shows three sections acquired on the N–S grid. The most evident feature on each radar section is a series of undulated reflectors which correspond to the deposition of volcanic layers on top of the Roman buildings. Some of these reflectors overlay a hyperbola generated by the presence of a wall. The other hyperbolas visible in Fig. 9.91 (a, b, d, e, and f), are probably due

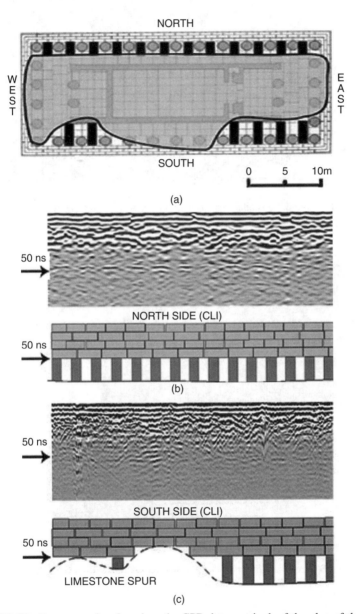

FIGURE 9.87 Reconstruction, based on the GPR data acquired, of the plan of the temple (a), the northern longitudinal section (b), and the southern longitudinal section (c). Note that the plan consists of two different units: the structural curbs (dark gray bars) and the limestone spur (tiles).

FIGURE 9.88   Foundations constructed of structural curbs in (a) the Temple of Vulcan and (b) the Temple of Demeter.

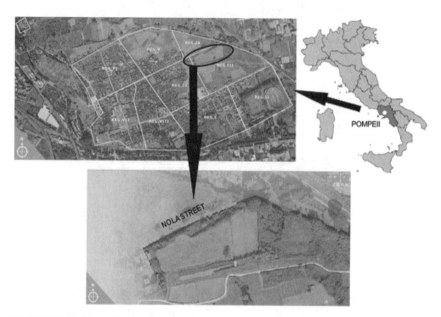

FIGURE 9.89   Localisation and partition of Pompeii delineating the area investigated.

to walls, part of a Roman *insula*, or human-made structures containing volcanic deposits. In particular, some walls can be followed through several sections and can be correlated with the walls that emerge from the excavated scarp parallel to Nola Street.

The presence of long structures such as walls can be better shown using time-slice representation. To this purpose, the data of the first grid have been processed

FIGURE 9.90 Reconstruction of GPR profiles, wooden fence, and the rest of the walls (a), emphasizing the scarp along Nola Street (b).

to produce 20 envelope average amplitude slices (one every 5 ns) which cover the entire time window. The time slices are shown in Fig. 9.92, where they have been numbered from most superficial to deepest. Because the most evident anomalies are present in the time interval between 10 and 40 ns, only the six slices included in this window are presented here. In slice 1, the large anomaly (inside the circle), which is also visible on slice 2, is due to the undulated reflectors mentioned before (feature c in the radargrams) and is therefore produced by the volcanic sediments overlaying the ruins. No anomalies correlated with human-made structures are visible at this depth interval. From slice 2 to slice 5 a series of elongated anomalies are visible, which correspond to the hyperbolas shown in the radargrams shown above (features a, b, e, and f). These anomalies have been interpreted as building structures located at a depth of about 1.20 m. Such anomalies are clearly correlated with the walls SM1 and SM2, as is clear from comparing the position of the anomalies with the bottom image of Fig. 9.92.

The GPR survey conducted in the Regio III of Pompei allowed for the reconstruction of the subsurface structure of the area and for the location of the main Roman ruins buried in the volcanic deposits. The interpretation has been carried out on single radargrams as well as on the time maps generated using the average envelope amplitude technique. The site chosen was particular suitable to correlating the structures exposed on the Nola Street scarp with the anomalies detected by radar. It is interesting to note that some walls are still partially intact inside the volcanic deposits, sometimes preserving the original structure of the buildings. In fact, in the time slices, several geometric structures are still clearly visible and have the same form as the Roman buildings inside a classical *insula*.

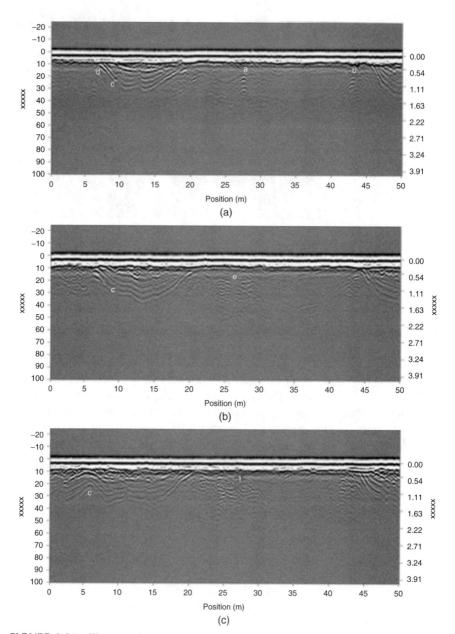

FIGURE 9.91   Three sections, belonging to the first grid, showing hyperbolas (a, b, d, e, f) due to archaeological remains; the anomalies (feature c) are correlated with volcanic deposits.

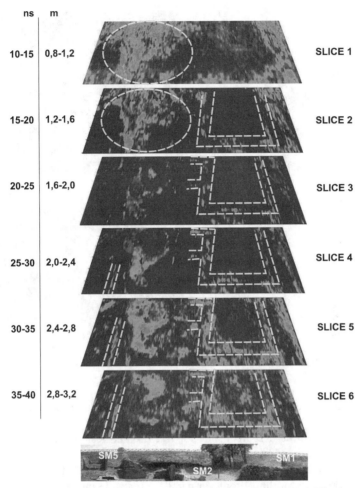

FIGURE 9.92 Time-slice maps with the main anomalies highlighted. The image at the bottom of the figure makes it possible to correlate the anomalies with Roman structures partially buried in the sediments.

9.6.7 Assessment of Fractures Using GPR

This case study shows the results obtained in GPR investigations carried out on the architrave of the Octavia Porch, located in downtown Rome, Italy, near the ancient Ghetto (Fig. 9.93). The goal of the research was the detection and location of fractures and internal structural lesions present in the marble blocks that form the architrave.

GPR data have been collected on the back part of the architrave of the *propy-laeum* of the Octavia Porch in the area shown in Fig. 9.93. The survey has been conducted using the Noggin[PLUS] system (Sensors & Software, Inc.), equipped with 500-MHz and 1-GHz antennas. The measures have been carried out in step mode,

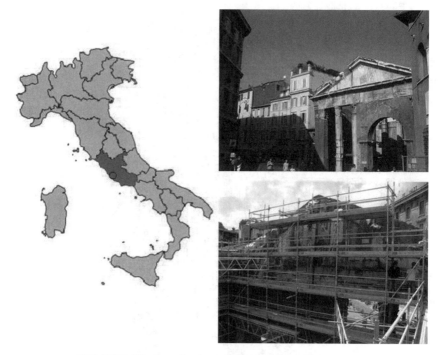

FIGURE 9.93 Localization of the Octavia Porch in Rome.

with a sampling interval of 0.05 m. For all the acquisitions, a time window of 30 ns and a stacking of 4 were used.

First, to choose the best frequency for fracture and lesion detection, two radar sections were acquired along the same horizontal profile (approximately 14 m long), using the perpendicular broadside mode for both the 500-MHz and 1-GHz antennas. Moreover, a third profile in parallel broadside mode was acquired with 1-GHz antennas on the same profile. Figure 9.94 shows the results obtained with the 500-MHz antenna. Note that the signal reflected from the marble–air interface (at about 18 ns) is particularly strong, indicating very low signal attenuation in the marble. In the figure, the main fractures are indicated by wide vertical arrows and correspond to the diffraction hyperbolas located at 2.7, 5.4, 7.3 (visible just on the external part of the architrave), 8.7, and 11.2 m. The fractures extend across the marble blocks and also produce strong hyperbolas at the marble–air interface. Moreover, the black arrows point out the hyperbolas generated by small internal fractures, microcavities, or detachments.

Figure 9.95 shows the results obtained on the same profile with a 1-GHz antenna used in parallel broadside mode. As it can be noted at high frequency the penetration is excellent and the marble–air interface is still clearly visible. As expected, the 1-GHz antennas allow for a more detailed image of the marble blocks, with fractures and lesions better resolved. Moreover, the parallel broadside mode put in evidence the difference in block compactness. In fact, between the position at 9 m and the

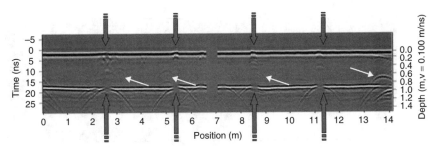

FIGURE 9.94 GPR raw data acquired using 500-MHz antennas. The inner lesions are indicated by black arrows, the filled fractures inside the architrave with wide vertical arrows.

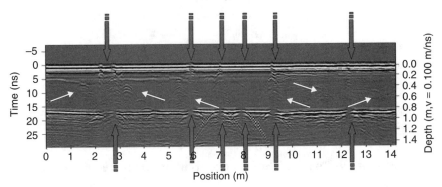

FIGURE 9.95 GPR section collected with a 1-GHz antenna. Again the filled fractures are indicated with (wide vertical arrows and the inner lesions with black arrow.

end of the section, the radargram seems less "transparent" to the electromagnetic pulses, probably due to a more intense alteration of the marble in that area.

The hyperbolas on the sections have been used to estimate an average pulse velocity of about 0.10 m/ns. This velocity has been used to convert the two-way travel time into depth, and to calculate a block thickness of about 0.80 to 0.90 m, which is in very good agreement with the measured thickness of the architrave. In the same area of the single profiles, a series of 10 parallel profiles with an interline spacing of 0.10 m were subsequently acquired. This multiprofile survey was conducted using 500-MHz antennas.

Time-slice maps have been constructed using envelope average amplitude and migration techniques to estimate the depth and dimensions of the main fractures. As an example, Fig. 9.96 shows the time slice at 17.70 to 19.20 ns, where a large fracture located at approximately 3.3 m along the profiles is clearly visible. This fracture (arrow in figure) has a complex geometry and crosses the entire thickness of the marble block.

Investigations on historical and archaeologically important buildings represent a very promising application of GPR, because they can provide both qualitative and quantitative data on structural damage that cannot be detected using any other

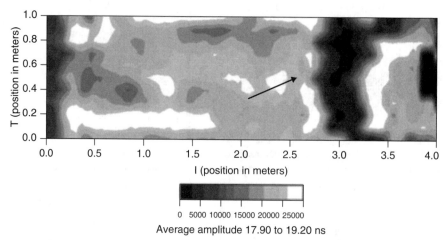

FIGURE 9.96   Average amplitude map of anomalies showing a large filled fracture (the black part of the map).

technique. This information regards not only the building damage but also the construction technique adopted, and can be of great value for restoration plans.

9.6.8   Microwave Tomography in the Search for *The Battle of Anghiari*

Microwave tomography has been applied to GPR data gathered on the painted walls of the Salone dei 500 in Palazzo Vecchio, Florence [344]. These investigations focused on the detection of possible remains of the legendary lost fresco *The Battle of Anghiari* by Leonardo da Vinci. In fact, a number of historians and scholars hypothesize that this valuable painting is buried in a wall of the Salone dei 500.

Scholars and technicians have spent decades searching for evidence that this Fresco was preserved by Vasari inside one of the present walls. In 2002, a new survey was promoted using a multitechniques approach, combining infrared camera, ultrasound transducer, and microwave surface penetrating radar (SPR) investigations. Pieraccini et al. [345] reported the results of the extensive measurement campaign carried out by the University of Florence in the Salone dei 500 through specially developed noncontact stepped-frequency surface-penetrating radar (SF-SPR) working at X-band. The results obtained encouraged further investigation; therefore, the University of Florence carried out a new measurement campaign in May 2003. To enhance penetration power maintaining the same high spatial resolution, a wide C-band system was used. The wavelength transmitted was increased from 30 mm to 75 mm, while the bandwidth was reduced only slightly with respect to the previous attempt.

Data have been processed independently by two different techniques based on different assumptions: conventional time-domain analysis [348] and inverse scattering. [346,347]. The results of these processing techniques are compared in this case study to show the evident similarities obtained using these techniques. The radar system employed was a continuous-wave step-frequency (CWSF) transceiver unit spanning a frequency band ranging from 2 to 6 GHz. CW-SF systems allow for higher mean radiated power than does pulsed radar, through the transmission of longer pulses, thus achieving good penetration and at the same time maintaining wide bandwidths for high-resolution applications.

The radar was used in a bistatic configuration, moving the receiving and transmitting antennas along a vertical rail fastened on a scaffolding about 6.5 m over the floor. This height was chosen on the basis of historical notes, the missing fresco is expected to cover an area of about 15 m^2. Many acquisitions were carried out, each one illuminating a slightly different volume while antennas were moved along the rail and the scaffolding was moved along the wall at a distance from it of about 2 m (Fig. 9.97). Table 9.5 summarizes the measurement parameters of the survey.

The two major side walls of Salone dei 500, the eastern and western walls, have been investigated. They are 50 m in length and are completely covered by

FIGURE 9.97 Measurement setup.

TABLE 9.5 Measurement Parameters for the Survey

Measurement Parameter	Parameter Value
Polarization	VV
Target distance	$\frac{1}{4}$ m
Transmitted power	20 dBm
Bandwidth	4 GHz
Central frequency	4 GHz
Number of frequencies transmitted	801
Vertical scan length	2.0 m
Vertical scan point number	111
Antenna separation distance	50 cm

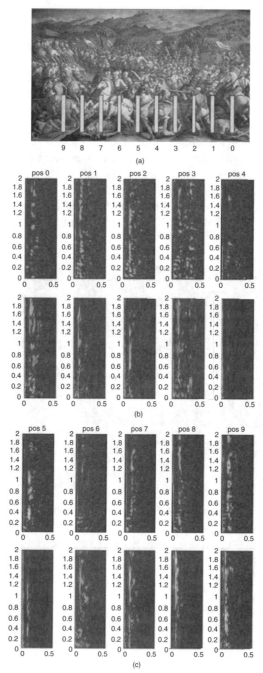

FIGURE 9.98 (a) South panel of the eastern wall with the antenna vertical scan positions indicated and enumerated. (b) With reference to part (a), the upper images are the corresponding results by the time-domain algorithm, and the lower images are those by the inversion algorithm. (c) With reference to part (a), the upper images are the corresponding results by the time-domain algorithm, and the lower images are those by the inversion algorithm.

great frescoes of Giorgio Vasari, representing victorious battles of the Florentine Republic. The scaffolding was moved in 28 steps along the eastern wall. From each of these positions, a complete vertical scan 2 m long was carried out. In particular, the south and north panels of the eastern wall were considered. According to the more reliable hypothesis, the south panel is likely to correspond to the position of Leonardo's fresco. On the western wall, six vertical scans were carried out, three scans for each panel.

Here, we compare the results of a classical signal-processing technique [3] with those achieved by microwave tomography [346–348]. In particular, we present results concerning the south panel of the eastern wall (see Fig. 9.98). The white bars drawn on the figure represent traces on the panel of the processed antenna vertical scans as enumerated sequentially in Fig. 9.98a. The upper and lower images in Fig. 9.98b and c are the results of time-domain analysis and an inversion algorithm, respectively, for each of the positions indicated on Fig. 9.98a. For all the images, the origin of the horizontal axis coincides with the air–wall interface. The direction is along the depth of the masonry and extends along the overall wall thickness, assumed to be 65 cm.

From both the upper (time-domain approach) and the lower images (microwave tomographic approach), it can be seen that behind the first interface, at a depth of about 15 cm, a clear vertical discontinuity all along the vertically scanned wall section can be identified. The discontinuity is more distinguishable on the last vertical scans of the panel considered. Evidence of the same deep discontinuity can be found in the images (not shown here) relative to the vertical scans over the north panel of the eastern wall, but it seems to vanish slightly [344]. The results on the south panel of the western wall (reported by Pieraccini et al. [344]) do not point out clear and geometrically defined discontinuities for both the time-domain analysis and the inversion algorithm.

REFERENCES

1. Boltzmann, L. 1876. Zur Theorie der elastischen Nachwirkung. *Ann. Phys. Chem.*, 7:624–654.

2. Hashin, Z., Shtrikman, S. 1962. A variational approach to the theory of the effective magnetic permeability of multiphase materials. *J. Appl. Phys.*, 33:3125–3131.

3. Mossotti, O. F. 1850. Discussióne analìtica sull'influènza che l'azióne di un mèzzo dielettrico ha sulla distribuzióne dell'elettricità alla superfìcie di piu corpi elettrici disseminati in èsso. *Mem. Mat. Fis. Soc. Ital. Sci.*, XXIV(Pt. 2), pp. 49–79.

4. Clausius, R. J. E. 1879. *Die mechanische Bahandlung der Electricität* Abschnitt III. F. Vieweg, Braunschweig, Germany.

5. Hippel, A. R. von 1954. *Dielectric and Waves*. Wiley, New York.

6. Bruggeman, D. A. G. 1935. Berechnung verschiedener physikalischer Konstanten von heterogenen Substanzen. *Ann. Phys.*, 24:636–664.

7. Böttcher, C. J. F. 1952. *Theory of Electric Polarisation*. Elsevier, Amsterdam.

8. Norris, A. N., Sheng, P., Callegari, A. J. 1985. Effective medium theories for two-phase dielectric media. *J. Appl. Phys.*, 57:1990–1996.

9. Kohler, W., Papanicolaou, G. 1981. Some applications of the coherent potential approximation. In *Multiple Scattering of Waves in Random Media*. North-Holland, Amsterdam, pp. 199–223.

10. Polder, D., van Santen, J. 1946. The effective permeability of mixtures and solids. *Physica XII*, 5:257–271.

11. Hanai, T. 1936. Dielectric theory on the interfacial polarization for two-phase mixtures. *Bull. Inst. Chem. Res.*, 39:341–368.

12. Sen, P. N., Scala, C., Cohen, M. H. 1981. A self-similar model for sedimentary rocks with application to the dielectric constant of fused glass beads. *Geophysics*, 46:781–795.

13. Giardano, S. 2003. Effective medium theory for dispersions of dielectric ellipsoids. *J. Electrostat.*, 58:59–76.

14. Johnson, R. H., Poeter, E. P. 2005. Iterative use of the Bruggeman–Hanai–Sen mixing model to determine water saturations in sand. *Geophysics*, 70:K33–K38.

15. Birchak, J. R., Gardner, C. G., Hipp, J. E., Victor, J. M. 1974. High dielectric constant microwave probes for sensing soil moisture. *Proc. IEEE*, 62(1):93–98.

16. Looyenga, H. 1965. Dielectric constants of mixtures. *Physica*, 31:401–406.

17. Wiener, O. 1910. Zur Theorie der Refraktionskonstanten. *Ber. Verh. Königlich-Sächsischen Ges. Wiss. Leipzig Math. Phys.*, 62:256–277.

18. Reuss, A. 1929. Berechnung der Fließgrenze von Mischkristallen auf Grund der Plastizitätsbedingung für Einkristalle. *Z. Angew. Math. Mech.*, 9:49–58.

19. Voigt, W. 1910. *Lehrbuch der Kristallphysik*. Leipzig, Germany.

20. Sihvola, A. 1999. *Electromagnetic Mixing Formulas and Applications*. Institution of Electrical Engineers, London.

21. Choy, T. C. 1999. *Effective Medium Theory: Principles and Applications*. Clarendon Press, Oxford, UK.

22. Mavko, G., Mukerji, T., Dvorkin, J. 1998. *Rock Physics Handbook*. Cambridge University Press, Cambridge, UK.

23. Guéguen, Y., Palciauskas, V. 1994. *Introduction to the Physics of Rocks*. Princeton University Press, Princeton, NJ.

24. von Hippel, A. R.. 1954. *Dielectrics and Waves*. Wiley, New York.

25. Landolt–Börnstein. 1996. New Series. Group 4. Physical Chemistry. Electronic version: http:\www.springermaterials.com/navigation/bookshelf.html

26. Weast, R. C., Astle, M. J. 1981. *CRC Handbook of Chemistry and Physics*, 62nd ed. CRC Press, Boca Raton, FL.

27. Uematsu, M., Franck, E. U. 1980. Static dielectric constant of water and steam. *J. Phys. Chem. Ref. Data*, 9:1291–1306.

28. Malmberg, C. G., Maryott, A. A. 1956. Dielectric constant of water from $0°$ to $100°C$. *J. Res. Nat. Bur. Stand.*, 56(1):RP2641.

29. Nörtemann, K., Hilland, J., Kaatze, U. 1997. Dielectric properties of aqueous NaCl solutions at microwave frequencies. *J. Phys. Chem. A*, 101:6864–6869.

30. Hasted, J. B. 1973. *Aqueous Dielectrics*. Chapman & Hall, London.

31. Debye, P. J. W. 1929. *Polar Molecules*. Dover Publications, New York.

32. Cole, K. S., Cole, R. H. 1941. Dispersion and absorption in dielectrics. *J. Chem. Phys.*, 9:341–351.

33. Cole, K. S. 1928. Electric impedance of suspensions of spheres. *J. Gen. Physiol.*, 12:29–36.

34. Grimnes, S., Martinsen, Ø. G. 2005. Cole electrical impedance model: a critique and an alternative. *IEEE Trans. Biomed. Eng.*, 52:132–135.

35. Topp, G. C., Davis, J., Annan, A. P. 1980. Electromagnetic determination of soil water content: measurements in coaxial transmission lines. *Water Resour. Res.*, 16:574–582.

36. Ledieu, J., Ridder, P. D., Clercq, P. D., Dautrebande, S. 1986. A method of measuring soil moisture by time domain reflectometry. *J. Hydrol.*, 88:319–328.

37. Taherian, M. R., Yuen, D. J., Habashy, T. L., Kong, J. A. 1991. A coaxial-circular waveguide for dielectric measurement. *IEEE Trans. Geosci. Remote Sens.*, 29:321–329.

38. Gorriti, A. G., Slob, E. C. 2005. Synthesis of all known analytical permittivity reconstruction techniques of non-magnetic materials from reflection and transmission measurements. *IEEE Geosci. Remote Sens. Lett.*, 2:433–436.

39. Gorriti, A. G., Slob, E. C. 2005. Comparison of the different reconstruction techniques of permittivity from S-parameters. *IEEE Trans. Geosci. Remote Sens.*, 43:2051–2057.

40. Van Hemert, M. J. C. 1972. Time Domain Reflectometry as a Method for Examination of Dielectric Relaxation of Polar Liquids. Ph.D. dissertation, Leiden University, The Netherlands.

41. Strangway, D. W. 1970. Possible electric and magnetic properties of near-surface lunar materials. In *Electromagnetic Exploration of the Moon*, W. Linlow, Ed. Mono Book Corp., Baltimore.

42. Olhoeft, G. 1998. Electrical, magnetic, and geometric properties that determine ground penetrating radar performance. In *Proceedings of the Seventh International Conference on Ground Penetrating Radar*, University of Kansas, Lawrence, KS, May, pp. 177–182.

43. Olhoeft, G. 1998. Ground penetrating radar on Mars In *Proceedings of the Seventh International Conference on Ground Penetrating Radar*, University of Kansas, Lawrence, KS, May, pp. 387–392.

44. Leuschen, C. 1999. Analysis of the complex permittivity and permeability of a Martian soil simulant from 10 MHz to 1 GHz. In *Proceedings of IGARSS'99*, Hamburg, Germany.

45. Heggy, E., Paillou, P., Ruffie, G., Malezieux, J. M., Costard, F., Grandjean, G. 2001. On water detection in the Martian subsurface using sounding radar. *Icarus*, 154:244–257.

46. Stillman, D. E., Olhoeft, G. R. 2004. GPR and magnetic minerals at Mars temperatures. In *Proceedings of the 10th International Conference on Ground Penetrating Radar*, Technical University, Delft, The Netherlands, June 21–24, pp. 735–738.

47. Pettinelli, E., Vannaroni, G., Pisani, A. R., Paolucci, F., Cereti, A., Riccioli, S., Del Vento, D., Dolfi D., Bella, F. 2005. Laboratory investigation into the electro-magnetic properties of iron oxide/silica mixtures as Martian soil analogues. *J. Geophys. Res. Planets*, 110:E04013.

48. Mattei, E., De Santis, A., Di Matteo, A., Pettinelli, E., Vannaroni, G. 2005. Time domain reflectometry of glass beads/magnetite mixtures: a time and frequency domain study. *Appl. Phys. Lett.*, 86:224102.

49. Stillman, D. E., Olhoeft, G. R. 2005. EM properties of magnetic minerals at radar frequencies. Presented at the Workshop on Radar Investigations of Planetary and Terrestrial Environments, Houston, TX, Feb. 7–10.

50. Pettinelli, E., Vannaroni, G., Mattei, E., Di Matteo, A., Paolucci, F., Pisani, A. R., Cereti, A., Del Vento, D., Burghignoli, P., Galli, A., De Santis, A., Bella, F. 2006. Electromagnetic propagation features of ground penetrating radars for the exploration of Martian subsurface. *Near Surf. Geophys.*, 4:5–11.

51. Mattei, E., De Santis, A., Di Matteo, A., Pettinelli, E., Vannaroni, G. 2007. Effective frequency and attenuation measurements of glass beads/magnetite mixtures by time-domain reflectometry. *Near Surf. Geophys.*, 5:77–82.

52. Cereti, A., Vannaroni, G., Del Vento, D., Pettinelli, E. 2007. Electromagnetic measurements on Martian soil analogs: implications for MARSIS and SHARAD radars in detecting subsoil water. *Planet. Space Sci.*, 55:193–202.

53. Mattei, E., De Santis, A., Di Matteo, A., Pettinelli, E., Vannaroni, G. 2008. Electromagnetic parameters of dielectric and magnetic mixtures evaluated by time domain reflectometry. *IEEE Geosci. Remote Sens. Neural.*, 5:730–734.

54. Picardi, G., Plaut, J. J., Biccari, D., Bombaci, O., Calabrese, D., Cartacci, M., et al. 2005. Radar sounding of the subsurface of Mars. *Science*, 310:1925–1928.

55. Seu, R., Phillips, R. J., Alberti, G., Biccari, D., Bonaventura, F., Bortone, M., et al. 2007. Accumulation and erosion of Mars south polar layered deposits from subsurface radar sounding. *Science*, 317:1715–1718.

56. Galt, J. K. 1952. Motion of a ferromagnetic domain wall in Fe_2O_4. *Phys. Rev.*, 85:664–670.

57. Parkhomenko, E. I. 1967. *Electrical Properties of Rocks*. Plenum Press, New York.

58. Olhoeft, G. R., Strangway, D. W. 1974. Magnetic relaxation and the electromagnetic response parameter. *Geophysics*, 39:302–311.

59. Olhoeft, G., Capron, D. E. 1994. Petrophysical causes of electromagnetic dispersion. In *Proceedings of the Fifth International Conference on Ground Penetrating Radar*, Kitchener, Ontario, Canada, June 12–16, pp. 145–152.

60. Robinson, D. A., Bell, J. P., Batchelor, C. H. 1994. Influence of iron minerals on determination of soil water content using dielectric techniques. *J. Hydrol.*, 161:169–180.

61. Klein, K., Santamarina, J. C. 2000. Ferromagnetic Inclusions in Geomaterials: Implications. *J. Geotech. Geoenviron. Eng.*, 126:167–179.

62. Van Dam, R. L., Schlager, W. 2000. Identifying causes of ground-penetrating radar reflections using time-domain reflectometry and sedimentological analyses. *Sedimentology*, 47:435–449.

63. Van Dam, R. L., Schlager, W., Dekkers, M. J., Huisman, J. A. 2002. Iron oxides as a cause of GPR reflections. *Geophysics*, 67:536–545.

64. Goodenough, J. B. 2002. Summary of losses in magnetic material. *IEEE Trans. Magn.*, 38:3398–3408.

65. Sihvola, A. 2000. Mixing rules with complex dielectric coefficients. *Subsurf. Sens. Technol. Appl.*, 1:393–415.

66. Pettinelli, E., Di Matteo, A., Paolucci, F., Bella, F., Mattei, E., Riccioni, S., De Santis, A., Vannroni, G., Cereti, A., Del Vento, D., Annan, A. P. 2005. Early-time GPR signal analysis: implications for water content measurement. In *Proceedings of the 3rd International Workshop on Advanced Ground Penetrating Radar*, Delft, The Netherlands, May 2–3.

67. Plug, W.-J., Moreno Tirado, L. M., Slob, E., Bruining, J. 2007. Simultaneously measured capillary pressure and electric permittivity hysteresis in multiphase flow through porous media. *Geophysics*, 72(3):A41–A45.

68. Plug, W.-J., Slob, E., van Turnhout, J., Bruining, J. 2007. Capillary pressure as a unique function of electric permittivity and water saturation. *Geophys. Res. Lett.*, 34:L13306.

69. Jennings, J. W., McGregor, D. S., Morse R. A. 1988. Simultaneous determination of capillary pressure and relative permeability by automatic history matching. *SPE Form. Eval.*, 3:322–328.

70. Christoffersen, K. R., Whitson, C. H. 1995. Gas/oil capillary pressure of chalk at elevated pressures. *SPE Form. Eval.*, 10:153–159.

71. Wildenschild, D., Hopmans, J., Simunek, J. 2001. Flow rate dependence of soil hydraulic characteristics. *Soil Sci. Soc. Am. J.*, 65:35–48.

72. Knight, R., Abad, A. 1995. Rock/water interaction in dielectric properties: experiments with hydrophobic sandstones. *Geophysics*, 60:431–436.

73. Gorriti, A. G., Slob, E. C. 2005. A new tool for accurate S-parameters measurements and permittivity reconstruction. *IEEE Trans. Geosci. Remote Sens.*, 43:1727–1735.

74. Fellner-Feldegg, H. 1969. The measurement of dielectrics in the time domain, *J. Phys. Chem.*, 73:616–623.

75. Topp, G. C., Davis, J. L., Annan, A. P. 1982. Electromagnetic determination of soil water content using TDR: II. Evaluation of installation and configuration of parallel transmission lines. *Soil Sci. Soc. Am. J.*, 46:678–684.

76. Pettinelli, E., Vannaroni, G., Cereti, A., Paolucci, F., Della Monica, G., Storini, M., Bella, F. 2003. Frequency and time domain measurements on solid CO_2 and solid CO_2–soil mixtures as Martian soil simulants. *J. Geophy. Res. Planets*, 108(E4):8029–8040.

77. O'Connor, K. M., Dowding, C. H. 1999. *Geomeasurements by Pulsing TDR Cables and Probes*. CRC Press, Boca Rator, FL.

78. Topp, G. C., Ferré, P. A. 2002. Water content. In *Methods of Soil Analysis*, Part 4, *Physical Methods*, J. H. Dane and G. C. Topp, Eds. Soil Science Society of America Book Series, Vol. 5. SSSA, Madison, WI, pp. 417–421.

79. Robinson, D. A., Jones, S. B., Wraith, J. M., Or, D., Friedman, S. P. 2003. A review of advances in dielectric and electrical conductivity measurement in soils using time domain reflectometry. *Vadose Zone J.*, 2:444–475.

80. Robinson D. A., Schaap, M., Jones, S. B., Friedman, S. P., Gardner, C. M. K. 2003. Considerations of improving the accuracy of permittivity measurements using time domain reflectometry: air–water calibration, effects of cable length. *Soil Sci. Soc. Am. J.*, 67:62–70.

81. Pettinelli, E., Cereti, A., Galli, A., Bella, F. 2002. Time domain reflectometry: calibration techniques for accurate measurement of the dielectric properties of various materials. *Rev. Sci. Instrum.*, 73:3553–3562.

82. Baker, J. M., Allmaras, R. R. 1990. System for automating and multiplexing soil moisture measurements by time domain reflectometry. *Soil Sci. Soc. Am. J.*, 54:1–6.

83. Heimovaara, T. J., Bouten, W. 1990. A computer-controlled 36-channel time domain reflectometry system for monitoring soil water contents. *Water Resour. Res.*, 26, 2311–2316.

84. Heimovaara, T. J. 1993. Design of triple-wire time domain reflectometry probes in practice and theory. *Soil Sci. Soc. Am. J.*, 57:1410–1417.

85. Robinson, D. A., Schaap, M., Or, D., Jones, S. B. 2005. On the effective measurements frequency of time domain reflectometry in dispersive and conductive dielectric materials. *Water Resour. Res.*, 41:W02007.

86. Mattei, E., Di Matteo, A., De Santis, A., Vannaroni, G., Pettinelli, E. 2006. Role of dispersive effects in determining probe and electromagnetic parameters by time domain reflectometry. *Water Resour. Res.*, 42.

87. Heimovaara, T. J. 1994. Frequency domain analysis of time domain reflectometry waveforms: 1. Measurements of the complex permittivity of soils. *Water Resour. Res.*, 30:189–199.

88. De Winter, E. J. G., van Loon, W. K. P., Esveld, E., Heimovaara, T. J. 1996. Dielectric spectroscopy by inverse modelling of time domain reflectometry wave forms. *J. Food Eng.*, 30:351–362.

89. Feng, W., Lin, C. P., Dechamps, R. J., Drnevich, V. P. 1999. Theoretical model of a multisection time domain reflectometry measurements system. *Water Resour. Res.*, 35:2321–2331.

90. Giese, K., Tiemann, R. 1975. Determination of the complex permittivity from a thin sample time-domain reflectometry, improved analysis of the step response waveform. *Adv. Mol. Relax. Process.*, 7:45–59.

91. Clarkson, T. S., Glasser, L., Tuxworth, R. W., Williams, G. 1977. An appreciation of experimental factors in time-domain spectroscopy. *Adv. Mol. Relax. Process.*, 10:173–202.

92. Ramo, S., Whinnery, J. R., Duzer, T. V. 1984. *Fields and Waves in Communication Electronics*. Wiley, New York.

93. Heimovaara, T. J. 1996. In *Second International Symposium and Workshop on Time Domain Reflectometry for Innovative Geotechnical Applications*, C. H. Dowding, Ed. Infrastructure Technical Institute, Northwestern University, Evanston, IL. http://www.iti.northwestern.edu/tdr/tdr2001/proceedings/final/TDR2001.pdf.

94. Huisman, J. A., Lambot, S., Vereecken, H. 2006. *Proc. TDR*, 2006, Purdue University, West Lafayette, USA, Sept., Paper ID 28. http://engineering.purdue.edu/TDR/papers.

95. Pettinelli, E., Vannaroni, G., Di Pasquo, B., Mattei, E., Di Matteo, A., De Santis, A., Annan, P. A. 2007. Correlation between near-surface electromagnetic soil parameters and early-time GPR signals: an experimental study. *Geophysics*, 72:A25–A28.

96. Lambot, S., Slob, E. C., van den Bosch, I., Stockbroeckx, B., Vanclooster, M. 2004. Modeling of ground-penetrating radar for accurate characterization of subsurface electric properties. *IEEE Trans. Geosci. Remote Sens.*, 42:2555–2568.

97. Ernst, J. R., Holliger, K., Maurer, H., Green, A. G. 2006. Realistic FDTD modelling of borehole georadar antenna radiation: methodolgy and application. *Near Surf. Geophys.*, 4:19–30.

98. Ernst, J. R., Maurer, H., Green, A. G., Holliger, K. 2007. Full-waveform inversion of crosshole radar data based on 2-D finite-difference time-domain solutions of Maxwell's equations. *IEEE Trans. Geosci. Remote Sens.*, 45:2807–2828. ESRI.

98a. ESRI, 1992. *ArcView 3.1*. Redlands, CA.

99. Gloaguen, E., Giroux, B., Marcotte, D., Dimitrakopoulos, R. 2007. Pseudo-full-waveform inversion of borehole GPR data using stochastic tomography. *Geophysics*, 72:J43–J51.

100. Soldovieri, F., Hugenschmidt, J., Persico, R., Leone, G. 2007. A linear inverse scattering algorithm for realistic GPR applications. *Near Surf. Geophys.*, 5:29–41.

101. Van der Kruk, J., Arcone, S. A. Liu, L. 2007. Fundamental and higher mode inversion of dispersed GPR waves propagating in an ice layer. *IEEE Trans. Geosci. Remote Sens.*, 45:2483–2491.

102. Lambot, S., Binley, A., Slob, E., Hubbard, S. 2008. Ground penetrating radar in hydrogeophysics. *Vadose Zone J.*, 7:137–139.

103. Sasaki, Y. 2001. Full 3-D inversion of electromagnetic data on PC. *J. Appl. Geophys.*, 46:45–54.

104. Giroux, B., Gloaguen, E., Chouteau, M. 2007. bhtomo: a Matlab borehole georadar 2D tomography package. *Comput. Geosci.*, 33:126–137.

105. Gloaguen, E., Marcotte, D., Giroux, B., Dubreuil-Boisclair, C., Chouteau, M., Aubertin, M. 2007. Stochastic borehole radar velocity and attenuation tomographies using cokriging and cosimulation. *J. Appl. Geophy.*, 62:141–157.

106. Ghose, R., Slob, E. C. 2006. Quantitative integration of seismic and GPR reflections to derive unique estimates for water saturation and porosity in subsoil. *Geophys. Res. Lett.*, 33:L05404.

107. Linde, N., Binley, A., Tryggvason, A., Pedersen, L. B. Révil, A. 2006. Improved hydrogeophysical characterization using joint inversion of cross-hole electrical resistance and ground-penetrating-radar travel time data. *Water Resour. Res.*, 42:W12404.

108. Rucker, D. F., Ferré, T. P. A. 2004. Parameter estimation for soil hydraulic properties using zero-offset borehole radar: analytical method. *Soil Sci. Soc. Am. J.*, 68:1560–1567.

109. Kowalsky, M. B., Finsterle, S., Peterson, J., Hubbard, S., Rubin, Y., Majer, E., Ward, A., Gee, G. 2005. Estimation of field-scale soil hydraulic and dielectric parameters through joint inversion of GPR and hydrological data. *Water Resour. Res.*, 41:W11425.

110. Lambot, S., Slob, E. C., Vanclooster, M. Vereecken, H. 2006. Closed loop GPR data inversion for soil hydraulic and electric property determination. *Geophys. Res. Lett.*, 33:L21405.

111. Lambot, S., Slob, E., Chavarro, D., Lubczynski, M., Vereecken, H. 2008. Measuring soil surface water content in irrigated areas of southern Tunisia using full-waveform inversion of proximal GPR data. *Near Surf. Geophys.*, 6:403–410.

112. Lambot, S., Antoine, M., Vanclooster M., Slob. E. C. 2006. Effect of soil roughness on the inversion of off-ground monostatic GPR signal for noninvasive quantification of soil properties. *Water Resour. Res.*, 42:W03403.

113. Slob, E. C., Fokkema, J. 2002. Coupling effects of two electric dipoles on an interface. *Radio Sci*, 37:1073.

114. Lambot, S., Slob, E., Vereecken, H. 2007. Fast evaluation of zero-offset Green's function for layered media with application to ground-penetrating radar. *Geophys. Res. Lett.*, 34:L21405.

115. Huyer, W., Neumaier, A. 1999. Global optimization by multilevel coordinate search. *J. Global Optim.*, 14:331–355.

116. Cassiani, G., Binley, A. 2005. Modeling unsaturated flow in a layered formation under quasi-steady state conditions using geophysical data constraints. *Adv. Water Resour.*, 28:467–477.

117. Looms, M. C., Binley, A., Jensen, K. H., Nielsen, L., Hansen, T. M. 2008. Identifying unsaturated hydraulic parameters using an integrated data fusion approach on cross-borehole geophysical data. *Vadose Zone J.*, 7:238–248.

118. Deiana, R., Cassiani, G., Villa, A., Bagliani, A., Bruno, V. 2008. Calibration of a vadose zone model using water injection monitored by GPR and electrical resistance tomography. *Vadose Zone J.*, 7:215–226.

119. Lambot, S., Slob, E., Rhebergen, J., Lopera, O., Jadoon, K. Z., Vereecken, H. 2009. Remote estimation of the hydraulic properties of a sandy soil using full-waveform integrated hydrogeophysical inversion of time-lapse, off-ground GPR data. *Vadose Zone J.* 8, 743–754.

120. Rhoades, J. D., Manteghi, N. A., Shouse, P. J., Alves, W. J. 1989. Soil electrical conductivity and soil salinity: new formulations and calibrations. *Soil Sci. Soc. Am. J.*, 53:433–439.

121. Jadoon, K., Slob, E., Vanclooster, M., Vereecken, H., Lambot, S., 2008. Uniqueness and stability analysis of hydrogeophysical inversion for time-lapse proximal ground penetrating radar. *Water Resour. Res.*, 44:W09421.

122. Merz, B., Bardossy, A. 1998. Effect of spatial variability on the rainfall runoff process in a small loess catchment. *J. Hydrol.*, 212:304–317.

123. Vereecken, H., Huisman, J. A., Bogena, H., Vanderborght, J., Vrugt, J. A., Hopmans, J. W. 2008. On the value of soil moisture measurements in vadose zone hydrology: a review. *Water Resour. Res.*, 44:W00D06.

124. Dobson, M. C., Ulaby, F. T. 1986. Active microwave soil moisture research. *IEEE Trans. Geosci. Remote Sens.*, 24:23–36.

125. Huisman, J. A., Snepvangers, J. J. J. C., Bouten, W., Heuvelink, G. B. M. 2002. Mapping spatial variation in surface soil water content: comparison of ground-penetrating radar and time domain reflectometry. *J. Hydrol.*, 269:194–207.

126. Jackson, T. J., Schmugge, J., Engman, E. T. 1996. Remote sensing applications to hydrology: soil moisture. *Hydrol. Sci.*, 41:517–530.

127. Famiglietti, J. S., Ryu, D., Berg, A. A., Rodell, M., Jackson, T. J. 2008. Field observations of soil moisture variability across scales. *Water Resourc. Res.*, 44:W01423.

128. Vereecken, H., Kasteel, R., Vanderborght, J., Harter, T. 2007. Upscaling hydraulic properties and soil water flow processes in heterogeneous soils: a review. *Vadose Zone J.*, 6:1–28.

129. Wagner, W., Bloschl, G., Pampaloni, P., Calvet, J. C., Bizzarri, B., Wigneron, J. P., Kerr, Y. 2007. Operational readiness of microwave remote sensing of soil moisture for hydrologic applications. *Nord. Hydrol.*, 38:1–20.

130. Kerr, Y. H., Waldteufel, P., Wigneron, J. P., Martinuzzi, J. M., Font, J., Berger, M. 2001. Soil moisture retrieval from space: the Soil Moisture and Ocean Salinity (SMOS) mission. *IEEE Trans. Geosci. Remote Sens.*, 39:1729–1735.

131. Rubin, Y., Hubbard, S. S. 2005. *Hydrogeophysics*. Water Science and Technology Library Series, Vol. 50. Springer, Dordrecht, The Netherlands.

132. Western, A. W., Grayson, R. B., Bloschl, G. 2002. Scaling of soil moisture: a hydrologic perspective. *Annul. Rev. Earth Planet. Sci.*, 30:149–180.

133. Thoma, D. P., Moran, M. S., Bryant, R., Rahman, M., Holifield-Collins, C. D., Skirvin, S., Sano, E. E., Slocum, K. 2006. Comparison of four models to determine surface soil moisture from C-band radar imagery in a sparsely vegetated semiarid landscape. *Water Resour. Res.*, 42:12.

134. Vereecken, H., Maes, J., Feyen, J., Darius, P. 1989. Estimating the soil-moisture retention characteristic from texture, bulk-density, and carbon content. *Soil Sci.*, 148:389–403.

135. Hogue, T. S., Bastidas, L., Gupta, H., Sorooshian, S., Mitchell, K., Emmerich, W. 2005. Evaluation and transferability of the Noah land surface model in semiarid environments. *J. Hydrometeorol.*, 6:68–84.

136. Peters-Lidard, C. D., Mocko, D. M., Garcia, M., Santanello, J. A., Tischler, M. A., Moran, M. S., Wu, Y. H. 2008. Role of precipitation uncertainty in the estimation of hydrologic soil properties using remotely sensed soil moisture in a semiarid environment. *Water Resour. Res.*, 44:22.

137. Santanello, J. A., Peters-Lidard, C. D., Garcia, M. E., Mocko, D. M., Tischler, M. A., Moran, M. S. Thoma, D. P. 2007. Using remotely-sensed estimates of soil moisture to infer soil texture and hydraulic properties across a semi-arid watershed. *Remote Sens. Environ.*, 110:79–97.

138. Huisman, J. A., Hubbard, S. S., Redman, J. D., Annan, A. P. 2003. Measuring soil water content with ground penetrating radar: a review. *Vadose Zone J.*, 2:476–490.

139. Rhoades, J. D., Raats, P. A. C., Prather, R. J. 1976. Effects of liquid-phase electrical conductivity, water content and surface conductivity on bulk soil electrical conductivity. *Soil Sci. Soc. Am. J.*, 40:651–655.

140. Davis, J. L., Annan, A. P. 1989. Ground penetrating radar for high resolution mapping of soil and rock stratigraphy. *Geophys. Prospect.*, 37:531–551.

141. Nakashima, Y., Zhou, H., Sato, M. 2001. Estimation of groundwater level by GPR in an area with multiple ambiguous reflections. *J. Appl. Geophys.*, 47:241–249.

142. Vellidis, G., Smith, M. C., Thomas, D. L., Asmussen, L. E. 1990. Detecting wetting front movement in a sandy soil with ground penetrating radar. *Trans. ASAE*, 33:1867–1874.

143. Chanzy, A., Tarussov, A., Judge, A., Bonn, F. 1996. Soil water content determination using digital ground penetrating radar. *Soil Sci. Soc. Am. J.*, 60:1318–1326.

144. Greaves, R. J., Lesmes, D. P., Lee, J. M., Toksov, M. N. 1996. Velocity variations and water content estimated from multi-offset, ground-penetrating radar. *Geophysics*, 61:683–695.

145. van Overmeeren, R. A., Sariowan, S. V., Gehrels, J. C. 1997. Ground penetrating radar for determining volumetric soil water content: results of comparative measurements at two test sites. *J. Hydrol.*, 197:316–338.

146. Weiler, K. W., Steenhuis, T. S., Boll, J., Kung, K. J. S. 1998. Comparison of ground penetrating radar and time domain reflectometry as soil water sensors. *Soil Sci. Soc. Am. J.*, 62:1237–1239.

147. Binley, A., Cassiani, G., Middleton, R., Winship, P. 2002. Vadose zone flow model parameterisation using cross-borehole radar and resistivity imaging. *J. Hydrolo.*, 267:147–159.

148. Cai, J., McMechan, G. A. 1995. Ray-based synthesis of bistatic ground penetrating radar profiles. *Geophysics*, 60:87–96.

149. Goodman, D. 1994. Ground penetrating radar simulation in engineering and archeology. *Geophysics*, 59:224–232.

150. Redman, J. D., Davis, J. L., Galagedara, L. W., Parkin, G. W. 2002. Field studies of GPR air launched surface reflectivity measurements of soil water content. In

S. K. Koppenjan and H. Lee (Editor), *Proceedings of the Ninth International Conference on Ground Penetrating Radar*, Santa Barbara, California, USA, *Proc. SPIE*, 4758:156–161.

151. Serbin, G., Or, D. 2004. Ground-penetrating radar measurement of soil water content dynamics using a suspended horn antenna. *IEEE Trans. Geosci. Remote Sens.*, 42:1695–1705.

152. Serbin, G., Or, D. 2005. Ground-penetrating radar measurement of crop and surface water content dynamics. *Remote Sens. Environ.*, 96:119–134.

153. Ulaby, F. T., Moore, M. K., Fung, A. K. 1986. *Microwave Remote Sensing: Active and Passive*, Vol. III, *From Theory to Applications*. Artech House, Norwood:MA.

154. Lambot, S., Weihermüller, L., van den Bosch, I., Vanclooster, M., Slob, E. C. 2005. Full-wave inversion of off-ground monostatic GPR signal focused on the surface reflection for identifying surface dielectric permittivity. In *Proceedings of the 3rd International Workshop on Advanced Ground Penetrating Radar*, S. L. and A. G. Gorriti, eds., Delft University of Technology, Delft, The Netherlands, pp. 113–118.

155. Lambot, S., Weihermüller, L., Huisman, J. A., Vereecken, H., Vanclooster, M., Slob, E. C. 2006. Analysis of air-launched ground-penetrating radar techniques to measure the soil surface water content. *Water Resour. Res.*, 42:W11403.

156. Du, S., Rummel, P. 1994. Reconnaissance studies of moisture in the subsurface with GPR. In *Proceedings of the Fifth International Conference on Ground Penetrating Radar*, Waterloo Center for Groundwater Research, University of Waterloo, Waterloo, Ontario, Canada, pp. 1241–1248.

157. Galagedara, L. W., Parkin, G. W., Redman, J. D. 2003. An analysis of the GPR direct ground wave method for soil water content measurement. *Hydrol. Process.*, 17:3615–3628.

158. Galagedara, L. W., Parkin, G. W., Redman, J. D., von Bertoldi, P., Endres, A. L. 2005. Field studies of the GPR ground wave method for estimating soil water content during irrigation and drainage. *J. Hydrol.*, 301:182–197.

159. Grote, K., Hubbard, S. S., Rubin, Y. 2003. Field-scale estimation of volumetric water content using GPR ground wave techniques. *Water Resour. Res.*, 39:1321–1333.

160. Huisman, J. A., Sperl, C., Bouten, W., Verstraten, J. M. 2001. Soil water content measurements at different scales: accuracy of time domain reflectometry and ground penetrating radar. *J. Hydrol.*, 245:48–58.

161. Huisman, J. A., Weerts, A. H., Heimovaara, T. J., Bouten, W. 2002. Comparison of travel time analysis and inverse modeling for soil water content determination with time domain reflectometry. *Water Resour. Res.*, 38:1224.

162. Weihermüller, L., Huisman, J. A., Lambot, S., Herbst, M., Vereecken, H. 2007. Mapping the spatial variation of soil water content at the field scale with different ground penetrating radar techniques. *J. Hydrol.*, 340:205–216.

163. Van der Kruk, J. 2006. Properties of surface waveguides derived from inversion of fundamental and higher mode dispersive GPR data. *IEEE Trans. Geosci. Remote Sens.*, 44:2908–2915.

164. Annan, A. P. 1973. Radio interferometry depth sounding: I. Theoretical discussion. *Geophysics*, 38(3):557–580.

165. Brekhovskikh, L. M. 1960. *Waves in Layered Media*. Academic Press, New York.

166. Berktold, A., Wollny, K. G., Alstetter, H. 1998. Subsurface moisture determination with the ground wave of GPR. In *7th International Conference on Ground-Penetrating Radar*, University of Kansas, Lawrence, KS, May, pp. 675–680.

167. Sperl, C. 1999. *Determination of Spatial and Temporal Variation of the Soil Water Content in an Agro-ecosystem with Ground-Penetrating Radar*. Technische Universitat Munchen, Munchen:Germany.

168. Turner, 1994. Subsurface radar propagation deconvolution: Geophysics, 59, 215–223.

169. Sbartaï, Z. M., Laurens, S., Balayssac, J. P., Arliguie, G., Ballivy, G. 2006. Ability of the direct wave of radar ground-coupled antenna for NDT of concrete structures. *Non-Destr. Test. Eval. Int.*, 39:400–407.

170. Garambois, S., Sénéchal, P., Perroud, H. 2002. On the use of combined geophysical methods to assess water content and water conductivity of near-surface formations. *J. Hydro.*, 259:32–48.

171. Mayne, W. H. 1962. Common-reflection-point horizontal data stacking techniques. *Geophysics*, 27:927–938.

172. Tillard, S., Dubois, J. C. 1995. Analysis of GPR data: wave propagation velocity determination. *J. Appl. Geophy.*, 33:77–91.

173. Binley, A., Winship, P., Middleton, R., Pokar, M., West, J. 2001. High-resolution characterization of vadose zone dynamics using cross-borehole radar. *Water Resour. Res.*, 37:2639–2652.

174. Cassiani, G., Fusi, N., Susanni, D., Deiana, R. 2008. Vertical radar profiling for the assessment of landfill capping effectiveness. *Near Surf. Geophys.*, 6:133–142.

175. Ernst, J. R., Maurer, H., Green, A. G., Holliger, K. 2007. Full-waveform inversion of crosshole radar data based on 2-D finite-difference time-domain solutions of Maxwell's equations. *IEEE Trans. Geosci. Remote Sens.*, 45:2807–2828.

176. Rucker, D. F., Ferré, T. P. A. 2004. Parameter estimation for soil hydraulic properties using zero-offset borehole radar: analytical method. *Soil Sci. Soc. Am. J.*, 68:1560–1567.

177. Zhou, C., Liu, L., Lane, J. W. 2001. Nonlinear inversion of borehole-radar tomography data to reconstruct velocity and attenuation distribution in earth materials. *J. Appl. Geophys.*, 47:271–284.

178. Pettinelli, E., Passeretta, A., Cereti, A., Menghini, A., Annunziatellis, A., Beaubien, S. E., Ciottoli, G., Lombardi, S. 2004. GPR and EM31 investigations on an active CO_2 gas vent. In *Proceedings of the Tenth International Conference on Ground-Penetrating Radar*, Delft, The Netherlands, pp. 563–566.

179. Taner, M. T., Khoeler F., Sheriff R. E., 1979, Complex trace analysis, Geophysics, 44, 1041–1063.

180. Di Pasquo, B., Pettinelli, E., Vannaroni, G., di Matteo, A., Mattei, E., De Santis, A., Annan, P. A., Redman, D. J. 2007. Design and construction of a large test site to characterize the GPR response in the vadose zone. In *Proceedings of the 4th International Workshop on Advanced Ground Penetrating Radar*, June 27–29, pp. 106–109.

181. EPA, 2007, Monitored Natural Attenuation of Inorganic Contaminants in Ground *water* EPA/600/R-07/139 and 140, http://www.epa.gov/oust/oswermna/mna epas.htm

182. Hubbard, S., Lunt, I. Grote, K., Rubin, Y., 2006. Vineyard soil water content: mapping small scale variability using ground penetrating radar. In *Fine Wine and Terroir: The Geoscience Perspective*, R. W. Macqueen, L. D. Meinert, Eds. Geoscience Canada Reprint Series 9. Geological Association of Canada, St. John's, Newfoundland.

183. Hubbard, S., Chen, J., Peterson, J., Majer, E., Williams, K., Swift, D., Mailliox, B., Rubin, Y. 2001. Hydrogeological characterization of the D.O.E. bacterial transport site in Oyster, Virginia using geophysical data. *Water Resour. Res.*, 37(10):2431–2456.

184. Lunt, I. A., Hubbard, S. S., Rubin, Y. 2005. Soil moisture content estimation using ground-penetrating radar reflection data. *J. Hydrol.*, 307:254–269.

185. Kowalsky, M. B., Finsterle, S., Rubin, Y. 2004. Estimating flow parameter distributions using ground-penetrating radar and hydrological measurements during transient flow in the vadose zone. *Adv Water Resour.*, 27:583–599.

186. Finsterle, S. 1999. *iTOUGH2 User's Guide*. Report LBNL-40400. Lawrence Berkeley National Laboratory, Berkeley, CA.

187. Meju, M. A. 2000. Environmental geophysics: the tasks ahead. *J. Appl. Geophys.*, 44:63–65.

188. Van Dam, J. C. 1976. Possibilities and limitations of the resistivity method of geo-electrical prospecting in the solution of geohydrological problems. *Geo-exploration*, 14:179–193.

189. Boekelman, R. H. 1991. Geo-electrical survey in the polder Groot-Mijdrecht. In *Hydro-geology of Salt Water Intrusion: A Selection of SWIM Papers*, (ed. W. de Breuck), Balkema, Rotterdam, The Netherlands, International Contributions to Hydrogeology, Vol. 11. pp. 363–378.

190. Nasir, A. S. S., Loke, M. H., Lee, C. Y., Nawawi, M. N. M. 2000. Salt-water intrusion mapping by geoelectrical imaging surveys. *Geophys. Prospect.* 48:647–661.

191. Loke, M. H., Dahlin, T. 2002. A comparison of the Gauss–Newton and quasi-Newton methods in resistivity imaging inversion. *J. Appl. Geophys.*, 49:149–162.

192. Appelo, C. A. J., Geirnaert, W. 1991. Processes accompanying the intrusion of salt water. In *Hydrogeology of Salt Water Intrusion: A Selection of SWIM Papers*, W. de Breuck, Ed. International Contributions to Hydrogeology, Vol. 11. Verlag Heinz Heise, Hanover, Germany, pp. 291–303.

193. Archie, G. E. 1942. The electrical resistivity log as an aid in determining some reservoir characteristics. *Trans. Am., Inst. Min. Metall. Pet. Eng.*, 146:54–64.

194. Post, V., Bloem, E., Ooteman, K., Slob, E. C., Groen, J., Groen, M. 2002. The use of CVES to map the subsurface salinity distribution: a case study in the Netherlands. Presented at the 17th Salt Water Intrusion Meeting, Delft, The Netherlands.

195. Gelhar, L. 1993. *Stochastic Subsurface Hydrology*. Prentice Hall, Englewood Cliffs, NJ.

196. Brockman, F. J., Murray, C. J. 1997. Subsurface microbiological heterogeneity. *FEMS Microbiol. Rev.*, 20:231–247.

197. Zhou, J., Beicheng, X., Huang, H., Palumbo, A. V., Tiedge, J. M., 2004. Micro-bial diversity and heterogeneity in sandy subsurface soils. *Appl. Environ. Microbiol.*, 70:1723–1734.

198. Chapelle, F. H. 2001. *Ground-water Microbiology and Geochemistry*. Wiley, New York.

199. Butler, J. 2005. Hydrogeological methods. In *Hydrogeophysics*, Y. Rubin, S. Hubbard, Eds. Springer, Dordrecht, The Netherlands, pp. 23–25.

200. Vereecken, H., Binley, A., Cassiani, G., Revil, A., Titov, K. 2006. *Applied Hydrogeo-physics*. Eds. H. Vereecken et al., Nato Science Series, Vol. 71. Springer, Dordrecht, The Netherlands, pp. 1–8.

201. Atekwana, E. A., Werkema D. D., Atekwana, E. A. 2006. Biogeophysics: the effects of microbial processes on geophysical properties of the shallow subsurface. In *Applied Hydrogeophysics*, H. Vereecken et al., Eds. NATO Science Series, Vol. 71. Springer, Dordrecht, The Netherlands, pp. 161–194.

202. Slater, L., Binley, A. 2006. Engineered barriers for pollutant containment and remediation. In *Applied Hydrogeophysics*, H. Vereecken et al., Eds. NATO Science Series Vol. 71. Springer, Dordrecht, The Netherlands, pp. 293–318.

203. Majer, E. W. 2006. Airborne and surface geophysical method verification. In *Barrier Systems for Environmental Contaminant Containment and Remediation*, L. G. Everett, Ed., CRC Press, Boca Raton, FL.

204. Meju, M. 2006, Geoelectrical characterization of covered landfill sites: a process oriented model and investigative approach. In *Applied Hydrogeophysics*, H. Vereecken et al., Eds. Springer, Dordrecht, The Netherlands, pp. 319–339.

205. Kolterman, C. E., Gorelick, S. M. 1996. Heterogeneity in sedimentary deposits: a review of structure-imitating, process-imitating, and descriptive approaches. *Water Resour. Res.*, 32:2617–2658.

206. Efferso, F., Auken, E., Sorensen, K. I. 1999. Inversion of band-limited TEM responses. *Geophys. Prospect.*, 47:551–564.

207. Majer, E. L., Peterson, J. E., Daley, T., Kaelin, B., Myer, L., D'Onfro, P., Rizer, W. 1997. Fracture detection using crosswell and single well surveys. *Geophysics*, 62:495–504.

208. Hayles, J. G. Serzu, M. H., Lodha, G. S., Everitt, R. A., Tomsons, D. K. 1996. Crosshole seismic and single-hole geophysical surveys to characterize an area of moderately fractured granite. *Can. J. Explor. Geophys.*, 32(1):6–23.

209. Meadows, M. A., Winterstein, D. F. 1994. Seismic detection of a hydraulic fracture from shear-wave VSP data at Lost Hills Field, California. *Geophysics*, 59(1):11–26.

210. Day-Lewis, F. D., Lane, J. W., Harris, J. M., Gorelick, S. M. 2003, Time-lapse imaging of saline-tracer transport in fractured rock using difference-attenuation radar tomography. *Water Resour. Res.*, 39:1290.

211. Chen, J., Hubbard, S., Peterson, J., Williams, K., Fienen, M., Jardine, P., Watson, D. 2006. Development of a joint hydrogeophysical inversion approach and application to a contaminated fractured aquifer. *Water Resour. Res.*, 42:W06425.

212. Lane, J. W. Jr., Heini, F. P., Watson, W. M. 1995. Use of a square-array direct current resistivity methods to detect fractures in crystalline bedrock in New Hampshire. *Ground Water*, 33(3):476–485.

213. Wishart, D. N., Slater, L. D., Gates, A. E. 2006. Self potential improves characterization of hydraulically-active fractures from azimuthal geoelectrical measurements. *Geophys. Res. Lett.*, 33:L17314.

214. Rucker, D. F., Fink, J. B. 2007. Inorganic plume delineation using surface high resolution electrical resistivity at the BC Cribs and Trenches Site, Hanford. *Vadose Zone J.*, 6:946–958.

215. Watson, D. B., Doll, W. E., Gamey, T. J., Sheehan, J. R., Jardine, P. M. 2005. Plume and lithological profiling with surface resistivity and seismic tomography. *Ground Water*, 43(2):169–177.

216. Day-Lewis, F., Lane, J. W., Gorelick, S. M. 2006. Combined interpretation of radar, hydraulic, and tracer data from a fractured-rock aquifer. *Hydrogeol. J.*, 14:1–14.

217. Moysey, S., Knight, R. J. 2004. Modeling the field-scale relationship between dielectric constant and water content in heterogeneous systems. *Water Resour. Res.*, 40:W03510.

218. Gomez-Hernandez, J. J. 2005. Geostatistics. In *Hydrogeophysics*, Rubin, Y., Hubbard, S. S., Eds. Springer, Dordrecht, The Netherlands, pp. 59–86.

219. Johnson, W. P., Zhang, P., Fuller, M. E., Scheibe, T. D., Mailloux, B. J., Onstott, T. C., DeFlaun, M. F., Hubbard, S. S., Radtke, J., Kovacik, W. P., Holbin, W. 2001. Ferrographic tracking of bacterial transport in the field at the narrow channel Focus Area, Oyster, VA. *Environ. Sci. Technol.*, 35:182–191.

220. Scheibe, T. D., Chien, Y. J. 2003. An evaluation of conditioning data for solute transport prediction. *Ground Water*, 41:128–141.

221. Roden, E. E., Lovely, D. R. 1993. Evaluation of Fe-55 as a tracer of Fe(III) reduction in aquatic sediments. *Geomicrobiol. J.*, 11:49–56.

222. Monsoor, N., Slater, L. 2007. On the relationship between iron concentration and induced polarization in march soils. *Geophysics*, 72:A1–A5.

223. Chen, J., Hubbard, S., Rubin, Y., Murray, C., Roden, E., Majer, E. 2004. Geochemical characterization using geophysical data and Markov chain Monte Carlo methods: a case study at the South Oyster bacterial transport site in Virginia. *Water Resour. Res.*, 40:W12412.

224. Scheibe, T., Fang, Y., Murray, C. J., Roden, E. E., Chen, J., Chien, Y., Brooks, S. C., Hubbard, S. S. 2006. Transport and biogeochemical reactions of metals in a and chemically heterogeneous aquifer. *Geosphere*, 2:220–235.

225. Naudet, V., Revil, A., Rizzo, E., Bottero, J.-Y., Bégassat, P. 2004. Groundwater redox conditions and conductivity in a contaminant plume from geoelectrical investigations. *Hydrol. Earth Sys. Sci.*, 8:8–22.

226. Williams, K. H., Ntarlagiannis, D., Slater, L. D., Dohnalkova, A., Hubbard, S. S., Banfield, J. F. 2005. Geophysical imaging of stimulated microbial biomineralization. *Environ. Sci. Technol.*, 39:7592–7600.

227. Hubbard, S. S., Williams, K., Conrad, M., Faybishenko, B., Peterson, J., Chen, J., Long, P. Hazen. T. 2008. Geophysical monitoring of hydrological and biogeochemical transformations associated with Cr(VI) biostimulation. *Environ. Sci. Technol.*, doi 10.1021/es071702s.

228. Hubbard, S., Rubin, Y. 2005. Hydrogeophysics. In *Hydrogeophysics*, Y. Rubin, S. Hubbard, Eds. Water and Science Technology Library, Vol 50. Springer, Dordrecht, The Netherlands, pp. 3–7.

229. Steefel, C. I, DePaolo, D. J., Lichtner, P. C. 2005. Reactive transport modeling: an essential tool and a new research approach for the Earth sciences. *Earth Planet. Sci. Lett.*, 240:539–558.

230. Revil, A., Linde, N., Cerepi, A., Jougnot, D., Matthäi, S., Finsterle, S. 2007. Electrokinetic coupling in unsaturated porous media. *J. Colloid Interface Sci.*, 313:315–327.

231. al Hagrey, S. A. 2007. Geophysical imaging of root-zone, trunk, and moisture heterogeneity. *J. Exp. Bot.*, 58:839–854.

232. Allred, B. J., Ehsani, M. R., Daniels. J. J. 2008. General considerations for geophysical methods applied to agriculture. In *Handbook of Agricultural Geophysics*, B. J. Allred, J. J. Daniels, M. R. Ehsani, Eds. CRC Press, Boca Raton, FL. pp. 3–16.

233. Lund, E. D., Colin, P. E., Christy, D., Drummond, P. E. 1999. Applying soil electrical conductivity technology to precision agriculture. In *Proceedings of the 4th International*

Conference on Precision Agriculture, P. C. Robert, R. H. Rust, W. E. Larson, Eds., St. Paul, MN, July 19–22, 1998. ASA, CSSA, and SSSA, Madison, WI, pp. 1089–1100.

234. Allred, B. J., Ehsani, M. R., Saraswat, D. 2006. Comparison of electromagnetic induction, capacitively coupled resistivity, and galvanic contact resistivity methods for soil electrical conductivity measurement. *Appl. Eng. Agric.*, 22:215–230.

235. Allred, B. J., Ehsani, M. R. Saraswat, D. 2005. The impact of temperature and shallow hydrologic conditions on the magnitude and spatial pattern consistency of electromagnetic induction measured soil electrical conductivity. *Trans. ASAE*, 48:2123–2135.

236. Chow, T. L., Rees, H. W. 1989. Identification of subsurface drain locations with ground-penetrating radar. *Can. J. Soil Sci.*, 69:223–234.

237. Boniak, R., Chong, S. K., Indorante, S. J., Doolittle, J. A. 2002. Mapping golf course green drainage systems and subsurface features using ground penetrating radar. In *Proceedings of the Ninth International Conference on Ground Penetrating Radar*, S. K. Koppenjan, H. Lee, Eds., Santa Barbara, CA. Apr. 29—May 2. *Proc. SPIE*, 4758:477–481.

238. Allred, B. J., Fausey, N. R., Peters, L., Jr., Chen, C., Daniels, J. J., Youn. H. 2004. Detection of buried agricultural drainage pipe with geophysical methods. *Appl. Eng. Agric.*, 20:307–318.

239. Allred, B. J., Daniels, J. J., Fausey, N. R., Chen, C., Peters, L., Jr., Youn, H. 2005. Important considerations for locating buried agricultural drainage pipe using ground penetrating radar. *Appl. Eng. Agric.*, 21:71–87.

240. Allred, B. J., Redman, D., McCoy, E. L., Taylor, R. S. 2005. Golf course applications of near-surface geophysical methods: a case study. *J. Environm. Eng. Geophys.*, 10:1–19.

241. Butnor, J. R., Doolittle, J. A., Kress, L., Cohen, S., Johnsen, K. H. 2001. Use of ground- penetrating radar to study tree roots in the southeastern United States. *Tree Physiol.*, 21:1269–1278.

242. McCorkle, W. H. 1931. *Determination of Soil Moisture by the Method of Multiple Electrodes*. Texas Agricultural Experiment Station Bulletin 426. Texas A&M University, College Station, TX.

243. Bouyoucos, G. J., Mick, A. H. 1939. A method for obtaining a continuous measurement of soil moisture under field conditions. *Science*, 89:252.

244. Edlefson, N. E., Anderson, A. B. C. 1941. The four-electrode resistance method for measuring soil-moisture content under field conditions. *Soil Sci.*, 51:367–376.

245. Kirkham, D., Taylor, G. S. 1949. Some tests of a four-electrode probe for soil moisture measurement. *Soil Sci. Soc. Am. Proc.*, 14:42–46.

246. Shea, P. F., Luthin, J. N. 1961. An investigation of the use of the four-electrode probe for measuring soil salinity *in situ*. *Soil Sci.*, 92:331–339.

247. Rhoades, J. D., Ingvalson, R. D. 1971. Determining salinity in field soils with soil resistance measurements. *Soil Sci. Soc. Am. Proc.*, 35:54–60.

248. Halvorson, A. D., Rhoades, J. D. 1974. Assessing soil salinity and identifying potential saline-seep areas with field soil resistance measurements. *Soil Sci. Soc. Am. Proc.*, 38:576–581.

249. McNeill, J. R., Winiwarter, A. 2004. Breaking the sod: humankind, history, and soil. *Science*, 304:1627–1629.

250. Stafford, J. V. 2000. Implementing precision agriculture in the 21st century. *J. Agric. Eng. Res.*, 76:267–275.

251. Zhang, N., Wang, M., Wang, N. 2002. Precision agriculture: a worldwide overview. *Comput. Electron. Agric.*, 36:113–132.

252. National Research Council. 1997. *Precision Agriculture in the 21st Century*. National Academies Press, Washington, DC.

253. Morgan, M., Ess, D. 1997. *The Precision-Farming Guide for Agriculturists*. John Deere Publishing, Moline, IL.

254. Jaynes, D. B., Colvin, T. S., Ambuel, J. 1995. Yield mapping by electromagnetic induction. In *Proceedings of Site-Specific Management for Agricultural Systems: Second International Conference*, P. C. Robert, R. H. Rust, W. E. Larson, Eds., St. Paul, MN. Mar. 27–30, 1994. ASA, CSSA, and SSSA, Madison, WI, pp. 383–394.

255. Banton, O., Seguin, M. K., Cimon, M. A. 1997. Mapping field-scale physical properties of soil with electrical resistivity. *Soil Sci. Soc. Am. J.*, 61:1010–1017.

256. Carroll, Z. L., Oliver, M. A. 2005. Exploring the spatial relations between soil physical properties and apparent electrical conductivity. *Geoderma*, 128:354–374.

257. Eigenberg, R. A., Nienaber, J. A. 1998. Electromagnetic survey of cornfield with repeated manure applications. *J. Environ. Qual.*, 27:1511–1515.

258. Gish, T. J., Dulaney, W. P., Kung, K.-J. S., Daughtry, C. S. T., Doolittle, J. A. Miller, P. T. 2002. Evaluating use of ground-penetrating radar for identifying subsurface flow Pathways. *Soil Sci. Soc. Am. J.*, 66:1620–1629.

259. McNeill, J. D. 1980. *Electromagnetic Terrain Conductivity Measurement at Low Induction Numbers*. Technical Note TN-6. Geonics Ltd., Mississauga, Ontario, Canada.

260. Grimley, D. A., Vepraskas, M. J. 2000. Magnetic susceptibility for use in delineating hydric soils. *Soil Sci. Soc. Am.*, 64:2174–2180.

261. Jones, S. B., Mace, R. W., Or, D. 2005. Time domain reflectometry coaxial cell for manipulation and monitoring of water content and electrical conductivity in variably saturated porous media. *Vadose Zone J.*, 4:977–982.

262. Jones, S. B., Wraith, J. M., Or, D. 2002. Time domain reflectometry measurement principles and applications. *Hydrol. Process.*, 16:141–153.

263. Kravchenko, A. N., Bollero, G. A., Omonode, R. A., Bullock, D. G. 2002. Quantitative mapping of soil drainage classes using topographical data and soil electrical conductivity. *Soil Sci. Soc. Am. J.*, 66:235–243.

264. Doolittle, J. A., Sudduth, K. A., Kitchen, N. R., Indorante, S. J. 1994. Estimating depths to clay pans using electromagnetic induction methods. *J. Soil Water Conserv.*, 49:572–575.

265. Jaynes, D. B., Novak, J. M., Moorman, T. B. Cambardella, C. A. 1995. Estimating herbicide partition coefficients from electromagnetic induction measurements. *J. Environ. Qual.*, 24:36–41.

266. Kitchen, N. R., Sudduth, K. A., Drummond, S. T. 1996. Mapping of sand deposition from 1993 midwest floods with electromagnetic induction measurements. *J. Soil Water Conserv.*, 51:336–340.

267. Doolittle, J. A. 1987. Using ground-penetrating radar to increase the quality and efficiency of soil surveys. In *Soil Survey Techniques*. W. U. Reybold, G. W. Petersen, Eds., SSSA Special Publication 20. Soil Science Society of America, Madison, WI, pp. 11–32.

268. Schellentrager, G. W., Doolittle, J. A., Calhoun, T. E., Wettstein, C. A. 1988. Using ground-penetrating radar to update soil survey information. *Soil Sci. Soc. Am. J.*, 52:746–752.

269. Collins, M. E. Doolittle, J. A. 1987. Using ground-penetrating radar to study soil microvariability. *Soil Sci. Soc. Am. J.*, 51:491–493.

270. Collins, M. E., Doolittle, J. A., Rourke, R. V. 1989. Mapping depth to bedrock on a glaciated landscape with ground-penetrating radar. *Soil Sci. Soc. Am. J.*, 53:1806–1812.

271. Wöckel, S., Konstantantinovic, M., Sachs, J., Schulze Lammers, P., Kmec, M. 2006. Application of ultra-wideband M-sequence-radar to detect sugar beets in agricultural soils. In *Proceeedings of the 11th International Conference on Ground Penetrating Radar*, Columbus, OH, June 19–22.

272. Konstantinovic, M., Wöckel, S., Schulze Lammers, P., Sachs, J., Martinov, M. 2007. Detection of root biomass using ultra wideband radar: an approach to potato nest positioning. *Agric. Eng. Int.*, 9:IT 06 003.

273. Freeland, R. S., Odhiambo, L. O., Tyner, J. S., Ammons, J. T., Wright, W. C. 2006. Nonintrusive mapping of near-surface preferential flow. *Appl. Eng. Agric.*, 22:315–319.

274. Rogers, M. B., Cassidy, J. R., Dragila, M. I. 2005. Ground-based magnetic surveys as a new technique to locate subsurface drainage pipes: a case study. *Appl. Eng. Agric.*, 21:421–426.

275. Rogers, M. B., Baham, J. E., Dragila, M. I. 2006, Soil iron content effects on the ability of magnetometer surveying to locate buried agricultural drainage pipes. *Appl. Eng. Agric.*, 22:701–704.

276. Hendrickx, J. M. H., Baerends, B., Rasa, Z. I., Sadig, M. Chaudhry, M. A. 1992. Soil salinity assessment by electromagnetic induction of irrigated land. *Soil Sci. Soc. Am. J.*, 56:1933–1941.

277. Doolittle, J., Petersen, M. Wheeler, T. 2001. Comparison of two electromagnetic induction tools in salinity appraisals. *J. Soil Water Conserv.*, 56:257–262.

278. Corwin, D. L., Lesch, S. M. 2005. Apparent soil electrical conductivity measurements in agriculture. *Comput. Electron. Agric.*, 46:11–43.

279. Sheets, K. R., Hendrickx, J. M. H. 1995. Noninvasive soil water content measurement using electromagnetic induction. *Water Resour. Res.*, 31:2401–2409.

280. Cereti, C. F., Pettinelli, E., Rossini, F. 1997. Water content measurements in fine-grained sediments using TDR and multilevel probes for turfgrass research. *Int. Turfgrass Soc. Res. J.*, 8:1252–1258.

281. USDA Economic Research Service. 1987. *Farm Drainage in the United States: History Status, and Prospects*. USDA Miscellaneous Publication 1455. U.S. Department of Agriculture, Washington, DC.

282. National Golf Foundation. 2001. *Golf Facilities in the U.S.* NGF, Jupiter, FL.

283. Beard, J. B. 2002. *Turf Management for Golf Courses*. Ann Arbor Press, Chelsea, MI.

284. Butnor, J. R., Doolittle, J. A., Johnson, K. H., Samuelson, L., Stokes, T., Kress, L. 2003. Utility of ground-penetrating radar as a root biomass survey tool in forest systems. *Soil Sci. Soc. Am. J.*, 67:1607–1615.

285. Cox, K. D., Scherm, H., Serman, N. 2005. Ground-penetrating radar to detect and quantify residual root fragments following peach orchard clearing. *HortTech*, 15:600–607.

286. Stover D. B., Day, F. P., Butnor, J. R., Drake, B. G. 2007. Application of ground penetrating-radar to quantify the effects of long-term CO_2 enrichment on coarse root biomass in a scrub-oak ecosystem at Kennedy Space Center, Florida USA. *Ecology*, 88:1328–1334.

287. Samuelson, L., Butnor, J., Maier, C., Stokes, T., Johnsen, K., Kane, M. 2008. Growth and physiology of loblolly pine in response to long-term resource management: defining growth potential in the southern United States. *Can. J. For. Res.*, 38:721–732.

288. Samuelson, L., Johnsen, K., Stokes, T. 2004. Production, allocation, and stemwood growth efficiency of *Pinus taeda* L. stands in response to 6 years of intensive management. *For. Ecol. Manag.*, 192:59–70.

289. Green, S., Clothier, B. 1999. The root zone dynamics of water uptake by a mature apple tree. *Plant Soil*, 206:61–77.

290. Gong, D., Kang, S., Zhang, L., Du, T., Yao, L. 2006. Two-dimensional model of root water uptake for single apple trees and its verification with sap flow and soil water content measurements. *Agric. Water Manag.*, 83:119–129.

291. Green, S. R., Kirkham, M. B., Clothier, B. E. 2006. Root uptake and transpiration: From measurements and models to sustainable irrigation. *Agric. Water Manag.*, 86:165–176.

292. Starr, G. C., Rowland, D., Griffin, T. S., Olanya, O. M. 2008. Soil water in relation to irrigation, water uptake and potato yield in a humid climate. *Agric. Water Manag.*, 95:292–300.

293. Starks, P. J., Heathman, G. C., Ahuja, L. R. Ma, L. 2003. Use of limited soil property data and modeling to estimate root zone soil water content. *J. Hydrol.*, 272:131–147.

294. Mojid, M. A., Cho, H. 2004. Evaluation of the time-domain reflectometry (TDR)-measured composite dielectric constant of root-mixed soils for estimating soil-water content and root density. *J. Hydrol.*, 295:263–275.

295. Ladekarl, U. L. 1998. Estimation of the components of soil water balance in a Danish oak stand from measurements of soil moisture using TDR. *For. Ecol. Manag.*, 104:227–238.

296. Hupet, F., and Vanclooster, M. 2004. Sampling strategies to estimate field areal evapotranspiration fluxes with soil water balance approach. *J. Hydrol.*, 292:262–280.

297. Coelho, E. F., Or, D. 1999. Root distribution and water uptake patterns of corn under surface and subsurface drip irrigation. *Plant Soil*, 206:123–136.

298. Camposeo, S., Rubino, P. 2003. Effect of irrigation frequency on root water uptake in sugar beet. *Plant Soil*, 253:301–309.

299. Szabolcs, I. 1994. Soils and salinisation. In *Handbook of Plant and Crop Stress*, M. Pessarakali, Ed. Marcel Dekker, New York, pp. 766.

300. Corwin, D. L. 2005. Geospatial measurements of apparent soil electrical conductivity for characterizing soil spatial variability. In *Soil–Water–Solute Process Characterization: An Integrated Approach*, J. Alvarez-Benedi, R. Munoz-Carpena, Eds. CRC Press, Boca Raton, FL, pp. 639–672.

301. Corwin, D. L., Lesch, S. M. 2005. Characterizing soil spatial variability with apparent soil electrical conductivity: I. Survey protocols. *Comput. Electron. Agric.*, 46:103–133.

302. Corwin, D. L., Carrillo, M. L. K., Vaughan, P. J., Rhoades, J. D., Cone, D. G. 1999. Evaluation of GIS-linked model of salt loading to groundwater. *J. Environ. Qual.*, 28:471–480.

303. Corwin, D. L., Lesch, S. M., Shouse, P. J., Soppe, R., Ayars, J. E. 2003. Identifying soil properties that influence cotton yield using soil sampling directed by apparent soil electrical conductivity. *Agron. J.*, 95:352–364.

304. Corwin, D. L., Kaffka, S. R., Hopmans, J. W., Mori, Y., Lesch, S. M., Oster, J. D. 2003. Assessment and field-scale mapping of soil quality properties of a saline-sodic soil. *Geoderma*, 114:231–259.

305. Kaffka, S. R., Corwin, D. L., Oster, J. D., Hopmans, J., Mori, Y., van Kessel, C., van Groenigen, J. W. 2002. *Using Forages and Livestock to Manage Drainage Water in the San Joaquin Valley: Initial Site Assessment*. Kearney Foundation Report. University of California, Berkeley, CA, pp. 88–110.

306. Corwin, D. L., Lesch, S. M., Oster, J. D., Kaffka, S. R. 2006. Monitoring management-induced spatio-temporal changes in soil quality through soil sampling directed by apparent electrical conductivity. *Geoderma*, 131:369–387.

307. Corwin, D. L., Lesch, S. M. 2003. Application of soil electrical conductivity to precision agriculture: theory, principles, and guidelines. *Agron. J.*, 95:455–471.

308. Eigenberg, R. A., Nienaber, J. A., Woodbury, B. L., Ferguson, R. 2006. Soil conductivity as a measure of soil and crop status: a four year summary. *Soil Sci. Soc. Am. J.*, 70:1600–1611.

309. Woodbury, B. L., Nienaber, J. A., Eigenberg, R. A. 2002. Effectiveness of a passive feedlot runoff control system using a vegetative treatment area for nitrogen control. *Appl. Eng. Agric.*, 21:581–588.

310. Eigenberg, R. A., Lesch, S. L., Woodbury, B. L., Nienaber, J. A. 2008. Geospatial methods for monitoring a vegetative treatment are receiving beef feedlot runoff. *J. Eniviron. Qual.*, 37:S68–S77.

311. Woodbury, B. L., Lesch, S. L., Eigenberg, R. A., Miller, D. N., Spiehs, M. J. 2009, EMI-sensor data to identify areas of manure accumulation on a feedlot surface. *Soil Sci. Soc. Am. J.*, 73:2068–2077.

312. Farahani, H. J., Buchleiter, G. W., Brodahl, M. K. 2005. Characterization of apparent soil electrical conductivity variability in irrigated sandy and non-saline fields in Colorado. *Trans. ASAE*, 48:155–168.

313. Farahani, H. J., Buchleiter, G. W. 2004. Temporal stability of soil electrical conductivity in irrigated sandy fields in Colorado. *Trans. ASAE*, 47:79–90.

314. McCutcheon, M. C., Farahani, H. J. Stednick, J. D., Buchleiter, G. W., Green, T. 2006. Effect of soil water on apparent soil electrical conductivity and texture relationships in a dryland field. *Biosyst. Eng.*, 94:19–32.

315. Farahani, H. J., Flynn, B. D. 2007. Map quality and zone delineation as affected by the width of parallel swaths of mobile agricultural sensors. *Biosyst. Eng.*, 96:151–159.

316. Shaner, D. L., Farahani, H. J., Buchleiter, G. W. 2008. Predicting and mapping herbicide–soil partition coefficients for EPTC, metribuzin, and metolachlor on three Colorado fields. *Weed Sci.*, 56:133–139.

317. Sawyer, J. E. 1994. Concepts of variable rate technology with considerations for fertilizer applications. *J. Prod. Agric.*, 7:195–201.

318. James, I. T., Waine, T. W., Bradley, R. I., Taylor, J. C., Godwin, R. J. 2003. Determination of soil type boundaries using electromagnetic induction scanning techniques. *Biosyst. Eng.*, 86:421–430.

319. Gaffney, C., Gater, J., Linford, P., Gaffney, V. L., White, R. 2000. Large-scale systematic fluxgate gradiometry at the Roman city of Wroxeter. *Archaeol. Prospect.*, 7:81–99.

320. Godio, A., Piro, S. 2005. Integrated data processing for archaeological magnetic surveys. *The Leading Edge*, 24:1138–1144.

321. Clark, A. 1990. *Seeing Beneath the Soil*. Routledge, London.

322. Bevan, B. W. 2000. An early geophysical survey at Williamsburg, USA. *Archaeol. Prospect.* 7:51–58.

323. Atkinson, R. J. C. 1953. *Field Archaeology*, 2nd ed. Methuen, London.

324. Hesse, A. 1966. Prospections geophysiques a fiable profondeur. In *Applications a l'archéologie*. Dunod, Paris.

325. von Bandi, H. G. 1945. Archäologische Erforschung des zukünftigen Stauseegebietes Rossens-Broc. *Schweiz. Gesell. Urgeschichte*, 30:100–106.

326. Laming, A., Ed. 1952. *La découverte du passe*. A. and J. Picard, Paris.

327. Scollar, I. 1965. Recent developments in magnetic prospecting in the Rheinland. *Prosp. Archeol.*, 1:43–50.

328. Ralph, E. K., Morrison, F., O'Brien, D. 1968. Archaeological surveying utilizing a high-sensitivity difference magnetometer. *Geoexploration*, 6:109–122.

329. Conyers, L., Goodman, D. 1997. *Ground Penetrating Radar: An Introduction for Archaeologists*. AltaMira Press, Londan, MD.

330. Scollar, I., Tabbagh, A., Hesse, A., Herzog, I. 1990. *Archaeological Prospecting and Remote Sensing*. Cambridge University Press, Cambridge, UK.

331. Schmidt, A. 2002 *Geophysical Data in Archaeology: A Guide to Good Practice*. Oxbow Books, Oxford, UK.

332. Gaffney, C., Gater, J. 2003. *Revealing the Buried Past*. Tempus Publishing, Stroud, UK.

333. Linford, N. 2006. The application of geophysical methods to archaeological prospection. *Rep. Prog. Phys.*, 69:2205–2257.

334. Imai, T., Sakayama, T., Kanemori, T. 1987. Use of ground-probing radar and resistivity surveys for archaelogical investigations. *Geophysics*, 52:137–150.

335. Basile, V., Carrozzo, M. T., Negri, S., Nuzzo, L., Quarta, T., Villani, A. V. 2000. A ground-penetrating radar survey for archaeological investigations in a urban area (Lecce, Italy). *J. Appl. Geophys.*, 44:15–32.

336. Neubauer, W., Eder-Hinterleitner, A., Seren, S., Melichar, P. 2002. Georadar in the Roman civil town Carnuntum, Austria: an approach for archaeological interpretation of GPR data. *Archaeol. Prospect.*, 9:135–156.

337. Quarto, R., Schiavone, D., Diaferia, I. 2007. Ground penetrating radar survey of a prehistoric site in southern Italy. *J. Archaeol. Sci.*, 34:2071–2080.

338. Doolittle, J. A., Minzenmayer, F. E., Waltman, S. W., Benham, E. C., Tuttle, J. W., Peaslee, S. D. 2007. Ground-penetrating radar soil suitability map of the conterminous United States. *Geoderma.* 141:416–421.

339. Ranalli, D., Scozzafava, M., Tallini, M. 2004. Ground penetrating radar investigations for the restoration of historic buildings: the case study of the Collemaggio Basilica (L'Aquila, Italy). *J. Cultural Heritage*, 5:91–99.

340. Cataldo, R., De Donno, A., De Nunzio, G., Leucci, G., Nuzzo, L., Siviero, S. 2005. Integrated methods for analysis of deterioration of cultural heritage: the crypt of "Cattedrale di Otranto". *J. Cultural Heritage*, 6:29–38.

341. Finzi, E., Praticelli, N., Vettore, L., Zaja, A. 2007. Multi-temporal geophysical survey of a Roman bath complex in Montegrotto Terme (Padova, northern Italy). *Archaeol. Prospect.*, 14:182–190.

342. Barone, P. M., Graziano, F., Pettinelli, E., Ginanni Corradini, R. 2007. Ground-penetrating radar investigations into the construction techniques of the Concordia Temple (Agrigento, Sicily, Italy). *Archaeol. Prospect.*, 14:47–59.

343. Barone, P. M., Bellomo, T., Pettinelli, E., Scarpati, C. 2007. Applications of GPR to archaeology and geology: the example of the regio III in Pompeii (Naples, Italy). In *Proceedings of the 4th International Workshop on Advanced Ground Penetrating Radar (IWAGPR 2007)*, Naples, Italy, pp. 64–68.

344. Pieraccini M., Noferini, L., Mecatti, D., Atzeni, C., Luzi, G., Persico, R., Soldovieri, F. 2006. Advanced processing techniques for step-frequency continuous-wave penetrating radar: the case study of "Palazzo Vecchio" walls (Firenze, Italy). *Res. Nondestr. Eval.*, 17:71–83.

345. Pieraccini, M., Mecatti, D., Luzi, G., Seracini, M., Pinelli, G., Atzeni, C. 2005. Non-contact intrawall penetrating radar for heritage survey: the search of the "Battle of Anghiari" by Leonardo da Vinci. *Non-Destr. Test. Eval. Int.*, 38:151–157.

346. Persico, R., Soldovieri, F. 2004. Reconstruction of a slab embedded in a three layered medium from multifrequency data under Born approximation. *J. Opt. Soc. Am. A*, 21:35–45.

347. Soldovieri, F., Persico, R. 2004. Reconstruction of an embedded slab from multi-frequency scattered field under Born approximation. *IEEE Trans. Antennas Propag.*, 52:2348–2356.

348. Pieraccini, M., Luzi, G., Noferini, L., Mecatti, D., Atzeni, C. 2004. Joint time-frequency analysis of layered masonry structures using penetrating radar. *IEEE Trans. Geosci. Remote Sens.*, 42:309–317.

CHAPTER TEN

Remote Sensing and Security

10.1 INTRODUCTION

In this chapter we address the most promising remote-sensing and security technologies related to applications of microwave and millimeter-wave imaging techniques. Active remote-sensing methods based primarily on through-wall tomography imaging and stepped-frequency radar are considered first. Then passive imaging methods in the millimeter-wave band which uses highly sensitive radiometric receivers are examined.

Active tomographic methods have been developed for image reconstruction of objects placed in a layer of a plane multilayered material half-space. This method is based on integral presentation of the scattered electric field for plane-wave excitation at different frequencies. Application fields are defined, mainly, by its advantages and disadvantages compared with other tomographic methods. The algorithm considered above can allow us to solve a very wide class of imaging problems, both for subsurface objects and for objects located in layers of planar multilayer structure. Medium losses can also be taken into account by this method. Results obtained by the algorithm described and subsurface tomographic method (STM) (see Section 5.3) for vertical slices in the case of buried objects are in good agreement. Although STM can be used for imaging of objects located in some layer structures, for example, as in the through-wall-imaging problem, adaptability of STM in the case of complex multilayer structures requires additional investigation. STM may have an advantage, due to brief time required for slice restoration. It allows obtaining images of buried or hidden objects in quasi-real time.

Subsurface Sensing, First Edition. Edited by Ahmet S. Turk, A. Koksal Hocaoglu, and Alexey A. Vertiy.
© 2011 John Wiley & Sons, Inc. Published 2011 by John Wiley & Sons, Inc.

Development of security and safety systems applicable to microwave radars and microwave tomographic methods for reconstruction images of horizontal and vertical slices of a probing area are quite useful technological steps. The investigation described in Section 10.2 is the first step toward the creation of multilayer structural tomography. Further, we intend to improve the algorithm described, which will lead to the development of procedures that will decrease the calculation time for slice reconstruction and development of tools that will allow us to take into account the characteristics of an antenna system in a real device. The experimental results show the effectiveness in the microwave region (2 to 4 GHz) of the tomographic method described. The application areas are wide ranging, including geophysics, medicine, nondestructive testing, and remote sensing.

This chapter consists basically of two parts. In Section 10.2 we describe the use of the microwave tomographic approach for through-the-wall remote sensing. Test and performance results are obtained from the framework of activity of the TUBITAK MRC MI International Laboratory for High Technology (ILHT). In Section 10.3 we introduce a millimeter-waveband passive remote-sensing method that uses array of mechanically and electronically scanning antennas. Systems used in practice and operating in real time are described in detail.

10.2 THROUGH-WALL IMAGING AND DETECTION

10.2.1 Tomographic Reconstruction in a Stratified Environment

Let us consider a multilayer plane structure formed by m loss homogeneous layers with a cylindrical inhomogeneity S embedded in its jth layer (Fig. 10.1) [5,11]. The medium with index 0 is air. The structure layers are characterized by material parameters $\varepsilon_1, \mu_1; \varepsilon_2, \mu_2; \ldots; \varepsilon_j, \mu_j; \ldots; \varepsilon_m, \mu_m$ for media 1, 2, ..., j, ..., m. Permeabilities $\mu_1, \mu_2, \ldots, \mu_j, \ldots, \mu_m$ are equal to a magnetic constant of vacuum μ_0. Permittivities $\varepsilon_1, \varepsilon_2, \ldots, \varepsilon_j, \ldots, \varepsilon_m$ may take complex values. The medium with index $m + 1$ is a substrate (a homogeneous half-space) characterized by complex permittivity ε_{m+1} and permeability $\mu_{m+1} = 1$. It is assumed that both the material and geometrical parameters of the multilayer structure are known. The object is a scattering inhomogeneity of arbitrary cross-section S located in the jth layer of the multilayer structure and characterized by complex permittivity $\varepsilon_s(x, y)$ and permeability $\mu_s = \mu_0$. The structure is irradiated by an incident plane wave at angle θ^i, and its electric field vector is perpendicular to the incidence plane and parallel to the cylinder axis (in the TM case). The problem is to define the object location and to obtain the image of the object cross section S from the diffracted field measurement at the probing (scanning) line at $y = y_s$. The following procedure may be employed for this purpose. For any point x, y inside or outside an object, the total electric field $E_z(x, y)$ may be expressed as a sum of the incident field E_z^I (defined as the field when the object is removed) and the diffracted field E_z^d [denoted further as $\psi(x, y)$]:

$$E_z(x, y) = E_z^I(x, y) + \psi(x, y) \tag{10.1}$$

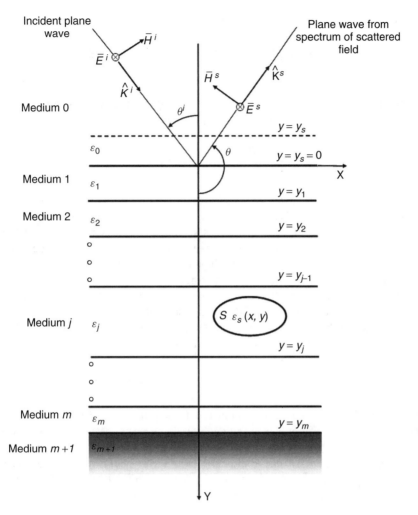

FIGURE 10.1 Wave scattering by the multilayer plane structure considered.

The exact representation for field $\psi(x, y)$ at point x, y (for time factor $e^{-i\omega t}$) may be written as [1,4]

$$\psi(x, y) = \iint_S [k_s^2(x', y') - k_j^2] E_z(x', y') G(x, y; x', y')\, dx'\, dy' \tag{10.2}$$

where $k_s^2(x, y) = \omega^2 \varepsilon_s(x, y)\mu_0$ and $k_j^2 = \omega^2 \varepsilon_j \mu_0 (j = 0, 1, \ldots, m + 1)$; $G(x, y; x', y')$ is the problem Green's function, which can be interpreted as an electric field created at point x, y by a line source of current located at point x', y'; and S is the region of the object cross section. The expression of the diffracted field at points at line $y = y_s$ may be presented in the form

$$\psi(x, y_s) = \iint\limits_S k_j^2 E_z^j(x', y') K(x', y') G_s(x, y_s; x', y') \, dx' \, dy' \tag{10.3}$$

where the function $E_z^j(x', y') \equiv E_z^i(x', y')$ in the jth layer of the structure, $K(x', y')$ is the normalized polarization current [1], defined as

$$K(x', y') = \left[\frac{k_s^2(x', y')}{k_j^2} - 1 \right] \left[1 + \frac{\psi(x', y')}{E_z^j(x', y')} \right] \tag{10.4}$$

and the function $G_s(x, y; x', y') \equiv G(x, y; x', y')$ at $y < 0$ is the problem Green's function. The function $K(x', y')$ equals zero outside the object and may be considered as the image of the object cross section. It follows from Eq. (10.3) that the object cross-section imaging is an inverse problem in which it is necessary to reconstruct the unknown characteristic function $K(x', y')$ from the known diffracted field in air at the probing line of finite length parallel to the multilayer structure interface. The solution of this inverse problem is reduced to the solution of the integral equation defined in Eq. (10.3), as described below.

Let us write the Fourier transform $\hat{\psi}(v, y_s)$ of the scattered field $\psi(x, y)$ with respect to x at probing line $y = y_s$ in the form

$$\hat{\psi}(v, y_s) = \int_{-\infty}^{\infty} \psi(x, y_s) \exp(-2\pi i v x) \, dx \tag{10.5}$$

where v is the space frequency and also defines a direction of propagation of the plane wave in expansion of the scattered field in terms of the plane waves because $2\pi v = (\omega/c) \sin \theta$ (see Fig. 10.1). Then, applying the operation in Eq. (10.5) to the integral equation in Eq. (10.3), it is possible to obtain an integral equation connecting the function $\hat{\psi}(v, y_s)$ with the image function $K(x', y')$. The modified integral equation takes the form

$$\hat{\psi}(v, y_s) = \iint\limits_S k_j^2 E_z^j(x', y') K(x', y') g_s(v, y_s; x', y') \, dx' \, dy' \tag{10.6}$$

where the function $g_s(v, y_s; x', y')$ is the Fourier transform of the Green's function $G_s(x, y_s; x', y')$ with respect to x.

Inversion of the integral equation of the type in Eq. (10.6) is considered first for the case of a cylindrical object placed in a homogeneous material half-space. Chommeloux et al. [1] showed that there is a relation between Fourier transform $\hat{\psi}(v, y_s)$ of the diffracted field $\psi(x, y_s)$ and two-dimensional Fourier transform $\hat{K}(\alpha, \beta)$ of the object function $K_1(x, y)$, defined as

$$K_1(x, y) \equiv \begin{cases} K(x', y'), & x, y \in S \\ 0, & x, y \notin S \end{cases} \tag{10.7}$$

This relation may be expressed as

$$\hat{K}(\alpha, \beta) = \eta(v)\hat{\psi}(v, y_s) \tag{10.8}$$

where the function $\eta(v)$ is a complex function of the real variable v.

Then the two-dimensional Fourier transform $\hat{K}(\alpha, \beta)$ takes the form

$$\hat{K}(\alpha, \beta) = \int_{-\infty}^{\infty} \int_{-\infty}^{\infty} K_1(x, y) \exp[-2\pi i(\alpha x + \beta y)] \, dx \, dy \tag{10.9}$$

where it is assumed that $\alpha = \alpha(v)$ and $\beta = \beta(v)$ are the real functions of the real variable v. This assumption simplifies the problem and reduces it to a two-dimensional inverse Fourier transform of the function $\hat{K}(\alpha, \beta)$ calculated in accordance with Eq. (10.8). However, in this case, a class of the solved problems is narrowed, as a medium that contains the object has to be lossless, and admissible values of v lie in the interval $[-k_0/2\pi, k_0/2\pi]$, where k_0 is the wavenumber of free space. The reconstruction for a more general and complicated case, when $\beta(v)$ takes complex values, was considered by Vertiy et al. [7,8].

When an object is in a layer of a multilayer structure, the expressions for the function $g_s(v, y_s; x', y')$ and field $E_z^j(x', y')$ in Eq. (10.6) are involved [10], and thus the integral equation in Eq. (10.6) is not reduced to a relation similar to Eq. (10.8). If so, a solution $K(x', y')$ of Eq. (10.6) can be obtained using the technique described next.

It can be shown that the function $g_s(v, y_s; x', y')$ in Eq. (10.6) takes the form

$$g_s(v, y_s; x', y') = B_0(v, x', y') \exp(-i\gamma_0 y_s) \tag{10.10}$$

where γ_0 can take complex values and is defined as $\gamma_0 = +\sqrt{k_0^2 - 4\pi^2 v^2}$ [i.e., $\text{Im}(\gamma_0) \geq 0$]. So the integral equation in Eq. (10.6) can be expressed as

$$f(v) \equiv \frac{\hat{\psi}(v, y_s) \exp(i\gamma_0 y_s)}{k_j^2} = \iint_S E_z^j(x', y') K(x', y') B_0(v, x', y') \, dx' \, dy' \tag{10.11}$$

Let us present a continuous variable over $[-v_0, v_0]$ to the complex-valued function $f(v)$ in terms of a linear combination of complex-valued functions:

$$f(v) = K_1\varphi_1(v) + K_2\varphi_2(v) + \cdots + K_N\varphi_N(v) \tag{10.12}$$

where $K_1, K_2, \ldots, K_N$ are complex constants.

Multiplying both parts of Eq. (10.12) in sequence by the functions $\varphi_1^*(v)$, $\varphi_2^*(v), \ldots, \varphi_N^*(v)$ (where the asterisk denotes complex conjugation) and integrating in the limits from $-v_0$ to v_0, it is possible to obtain a system of linear equations for finding constants $K_1, K_2, \ldots, K_N$. This system may be rewritten as

$$\sum_{i=1}^{N} K_i \int_{-v_0}^{v_0} \varphi_i(v) \varphi_k^*(v)\, dv = \int_{-v_0}^{v_0} f(v) \varphi_k^*(v)\, dv \qquad k = 1, 2, \ldots, N \qquad (10.13)$$

If functions $\varphi_1(v), \varphi_2(v), \ldots, \varphi_N(v)$ are orthogonal in the interval $[-v_0, v_0]$, namely,

$$\int_{-v_0}^{v_0} \varphi_i(v) \varphi_k^*(v)\, dv = 0 \qquad \text{when } i \neq k \qquad (10.14)$$

then the required constants $K_1, K_2, \ldots, K_N$ are determined by the formula

$$K_k = \frac{\int_{-v_0}^{v_0} f(v) \varphi_k^*(v)\, dv}{\int_{-v_0}^{v_0} |\varphi_k(v)|^2\, dv} \qquad k = 1, 2, \ldots, N \qquad (10.15)$$

On the other hand, the integral in Eq. (10.11) can be expressed as follows:

$$\iint_S E_z^j(x', y') K(x', y') B_0(v, x', y')\, dx'\, dy' = \iint_{S_j} K^j(x, y) \varphi^j(v, x, y)\, dx\, dy \qquad (10.16)$$

where the function $K^j(x, y)$ is defined in region S_j (image space) of the jth layer containing the object as

$$K^j(x, y) \equiv \begin{cases} K(x', y'), & x, y \in S \\ 0, & x, y \notin S \end{cases} \qquad (10.17)$$

and the function $\varphi^j(v, x, y) \equiv E_z^j(x, y) B_0(v, x, y)$ in the image space S_j.

The scheme of the jth layer containing the image space S_j with a grid is presented in Fig. 10.2. In this figure, the rectangular regions were denoted by $P_1, P_2, \ldots, P_i, \ldots, P_N$. These regions are pixels or image elements. Geometrical size of the image element is $\Delta x \times \Delta y$. An approximation of the integral in Eq. (10.16) can be obtained with the assumption that in the quadrate (rectangular) regions $P_1, P_2, \ldots, P_i, \ldots, P_N$, the normalized polarization current $K^j(x, y)$ takes constant values $K_1, K_2, \ldots, K_i, \ldots, K_N$, respectively.

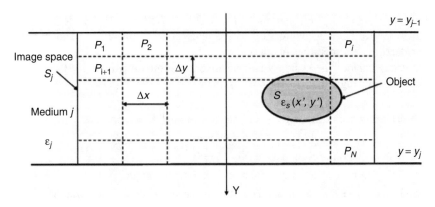

FIGURE 10.2 The jth layer, containing the object S and the image space S_j with a grid.

Moreover, we may introduce the functions $\varphi_1(\nu), \varphi_2(\nu), \ldots, \varphi_i(\nu), \ldots, \varphi_N(\nu)$ defined as follows:

$$\varphi_1(\nu) = \iint_{P_1} \varphi^j(\nu, x, y)\, dx\, dy$$

$$\varphi_2(\nu) = \iint_{P_2} \varphi^j(\nu, x, y)\, dx\, dy$$

$$\vdots$$

$$\varphi_i(\nu) = \iint_{P_i} \varphi^j(\nu, x, y)\, dx\, dy \qquad (10.18)$$

$$\vdots$$

$$\varphi_N(\nu) = \iint_{P_N} \varphi^j(\nu, x, y)\, dx\, dy$$

Then integral equation in Eq. (10.11) takes the form of Eq. (10.12), and the unknown constants $K_1, K_2, \ldots, K_N$, that is, normalized polarization current in regions $P_1, P_2, \ldots, P_i, \ldots, P_N$, can be found as a result of solving the linear equations system given in Eq. (10.13).

Let's write the system in Eq. (10.13) in matrix form:

$$A\vec{K} = \vec{b} \qquad (10.19)$$

where $\vec{K} = \begin{bmatrix} K_1 \\ K_2 \\ \vdots \\ K_N \end{bmatrix}$ is a vector of unknown coefficients K_i; $\vec{b} = \begin{bmatrix} b_1 \\ b_2 \\ \vdots \\ b_N \end{bmatrix}$ is a

vector of data $b_k = \int_{-\nu_0}^{\nu_0} f(\nu)\varphi_k^*(\nu)\, d\nu$; and A is a square $N \times N$ matrix with elements $a_{ki} = \int_{-v_0}^{v_0} \varphi_i(v)\varphi_k^*(v)\,dv, i = 1, 2, \ldots, N; k = 1, 2, \ldots, N.$ Equation (10.19) is an algebraic expression of the integral equation (10.11) and it may be considered as an operator equation.

Let us consider the system of linear algebraic equations in Eq. (10.19). If this system is nonsingular (i.e., det $A \neq 0$), then [4]

1. There is a unique inverse A^{-1} for which $A^{-1}A = I$ and $AA^{-1} = I$.
2. Every equation $A\vec{K} = \vec{b}$ has a solution.
3. The only solution of $A\vec{K} = 0$ is $\vec{K} = 0$.
4. The solution $\vec{K}$ to every equation $A\vec{K} = \vec{b}$ is unique.
5. Rank$(A) = N$; null space (i.e., the space of vectors $\vec{K}$ for which $A\vec{K} = 0$) has dimension 0.

If this system is singular, det $A = 0$, then:

1. There is no matrix A^{-1}, for which $A^{-1}A = I$ and $AA^{-1} = I$.
2. Some equations $A\vec{K} = \vec{b}$ have no solutions.
3. There are solutions of $A\vec{K} = 0$ with $\vec{K} \neq 0$.
4. The solution to $A\vec{K} = \vec{b}$ is never unique (if there is any solution $\vec{K}$, there are other solutions $\vec{K}^1 \neq \vec{K}$.
5. rank$(A) < N$; null space has dimension $N - \text{rank}(A) \geq 1$.

The singular system has matrix A with some vanishing eigenvalues. The system in Eq. (10.19) can also be ill-conditioned (large changes of the solution can be caused by small changes of the right part). Matrix A of such a system has eigenvalues close to zero. In practice, the order N of matrix A can be larger. If calculations are conducted with finite precision, then at sufficiently large values of N, accurate calculation of det A is a nonsolvable problem and the singular system can be indistinguishable from an ill-conditioned system in the frame of the given accuracy. Also, in adequate practical problems, the system in Eq. (10.19) with coefficients obtained from modeling and measurements can be considered as an approximation of some actual system $A_r\vec{K}_r = \vec{b}_r$, which can be obtained similar to Eq. (10.19) but with unknown coefficients. However, generally, solution $\vec{K}$ of the approximate system cannot be taken as an approximate solution of the system $A_r\vec{K}_r = \vec{b}_r$, because feasible solution $\vec{K}$ may not belong to the set of all solutions $\vec{K}_r$ of the actual system or it may be instable. Thus, to solve the practical inverse problems by the method considered, it is necessary to develop an algorithm that allows solving of the ill-posed problem in Eq. (10.19) and obtaining of stable solution $\vec{K}$ of the approximate system in Eq. (10.19), which would be close to solution $\vec{K}_r$ of the real system. See Tihonov and Arsenin [6] for more about the theory and variety of algorithms for the solution of ill-posed problems.

If there is a possibility of conducting the scattered field measurements using a set of frequencies $\omega_1, \omega_2, \ldots, \omega_q$, the following set of uncoupled linear systems of equations is obtained:

$$A_{\omega_1} \vec{K}_{\omega_1} = \vec{b}_{\omega_1}$$
$$A_{\omega_2} \vec{K}_{\omega_2} = \vec{b}_{\omega_2}$$
$$\vdots$$
$$A_{\omega_q} \vec{K}_{\omega_q} = \vec{b}_{\omega_q}$$

(10.20)

where the indexes $\omega_1, \omega_2, \ldots, \omega_q$ denote the matrixes and vectors for frequencies $\omega_1, \omega_2, \ldots, \omega_q$.

It is found from experiments that the solution (object function) of each individual system of linear equations (corresponding to certain frequency of excitation) in the set of Eq. (10.20) cannot be determined without a regularization process. The regularization process is necessary for stabilization of inversion against noise from measurement and numerical truncations. On the other hand, assuming that the object function is frequency independent, the systems of linear equations in Eq. (10.20) are related. A similar approach and a regularized pseudoinverse formula for the calculated object function were described by Deming and Devaney [2].

In the present procedure it is also assumed that the object function is a value without dispersion. This assumption is valid if the operating frequency band used is not very wide and if in the region of the dielectric object the scattered field is small in comparison with the field of the layer, as well as when a resonance response is absent. In practice, these conditions are often met approximately. The regularization procedure applied to the set of coupled equations in Eq. (10.20) was termed *multifrequency regularization*. As a result of multifrequency regularization, there was obtained a new nonsingular system of linear equations leading to approximate solution for the object function creating the backscattered field observed. To obtain this solution it is not necessary to employ methods that require matrix inversions or determinants calculation of high order.

Let us consider the multifrequency regularization process. If the systems of equations for different frequencies in the Eq. (10.20) set are coupled, it is possible to construct a new system of linear equations by the summation defined in

$$\tilde{A}\tilde{\vec{K}} = \tilde{\vec{b}}$$

(10.21)

where the matrix is

$$\tilde{A} = A_{\omega_1} + A_{\omega_2} + \cdots + A_{\omega_q} = \sum_{\tau=1}^{q} A_{\omega_\tau}$$

(10.22)

the vector of data is

$$\tilde{\vec{b}} = \vec{b}_{\omega_1} + \vec{b}_{\omega_2} + \cdots + \vec{b}_{\omega_q} = \sum_{\tau=1}^{q} \vec{b}_{\omega_\tau}$$

(10.23)

and the vector of unknown coefficients is $\tilde{\vec{K}} = \vec{K}_{\omega_1} = \vec{K}_{\omega_2} = \cdots = \vec{K}_{\omega_q}$.

As measurements can be conducted at any frequency in the interval of $\Delta\omega = \omega_q - \omega_1$, the summation can be replaced by integration. In this situation, Eqs. (10.22) and (10.23) take the following form:

$$\tilde{A} = \int_{\omega_1}^{\omega_q} A_\omega \, d\omega \tag{10.24}$$

$$\tilde{\vec{b}} = \int_{\omega_1}^{\omega_q} \vec{b}_\omega \, d\omega \tag{10.25}$$

where the elements of matrix $\tilde{A}$ and vector $\tilde{\vec{b}}$ are defined as follows:

$$\tilde{A} = (\tilde{a}_{ki}) = \int_{\omega_1}^{\omega_q} a_{ki}(\omega) \, d\omega \tag{10.26}$$

$$\tilde{\vec{b}} = (\tilde{b}_k) = \int_{\omega_1}^{\omega_q} b_k(\omega) \, d\omega \tag{10.27}$$

Although Eq. (10.21) is equivalent to any equation in the coupled set in Eq. (10.20), the structure of matrix $\tilde{A}$ essentially differs from the structure of matrixes $A_{\omega_1}, A_{\omega_2}, \ldots, A_{\omega_q}$. It was found that structure of matrix $\tilde{A}$ depends, essentially, on integration interval $\Delta\omega$. There is an optimal value of interval $\Delta\omega$ where diagonal and near-diagonal elements dominate in matrix $\tilde{A}$.

Matrix $\tilde{A}$ may be presented in expanded form as

$$\tilde{A} = A_1 + A_2 + A_3 + \cdots + A_N \tag{10.28}$$

where the matrixes $A_1, A_2, A_3, \ldots, A_N$ are formed from elements of matrix $\tilde{A}$ and they take the form

$A_1 = (\tilde{a}_{ki}), \tilde{a}_{ki} \neq 0$

    if $i = k(k = 1, 2, 3, \ldots, N)$ (10.29)

$A_2 = (\tilde{a}_{ki}), \tilde{a}_{ki} \neq 0$

    if $i = k + 1(N - 1 \geq k \geq 1)$ and   if $i = k - 1 \ (N \geq k \geq 2)$ (10.30)

$A_3 = (\tilde{a}_{ki}), \tilde{a}_{ki} \neq 0$

    if $i = k + 2(N - 2 \geq k \geq 1)$ and   if $i = k - 2 \ (N \geq k \geq 3)$ (10.31)

$$\vdots$$

$A_N = (\tilde{a}_{ki}), \tilde{a}_{ki} \neq 0$

    if $i = N(k = 1)$ and   if $i = 1(k = N)$ (10.32)

For optimal $\Delta\omega$, elements of matrix A_1 in absolute value are more than elements of matrixes $A_2, A_3, \ldots, A_N$. Matrix elements $A_2, A_3, \ldots, A_N$ in absolute value decrease with oscillations on increment of the matrix index. At reduction or widening of $\Delta\omega$ the matrix $\tilde{A}$ gradually loses these properties.

The regularized solution of Eq. 10.21 is taken in the form $\tilde{\tilde{K}} = R\tilde{\tilde{b}}$, where R is a regularizing matrix operator. Matrix R obtained from matrix $\tilde{A}$ at a value of $\Delta\omega$ may be expressed as

$$R = (A_1 + \alpha_1 A_2)^{-1} \tag{10.33}$$

where α_1 is a weighting coefficient. It is real value, which is chosen under additional information about the problem. If the weighting coefficient is equal to zero, the vector $\tilde{\tilde{K}} = A_1^{-1}\tilde{\tilde{b}}$ is taken as a solution of the problem. In this case, the coordinates of $\tilde{\tilde{K}}$ are determined by the formula

$$\tilde{K}_k = \frac{\int_{\omega_1}^{\omega_q} \int_{-v_0}^{v_0} f(v, \omega)\varphi_k^*(v, \omega)\,dv\,d\omega}{\int_{\omega_1}^{\omega_q} \int_{-v_0}^{v_0} \varphi_i(v, \omega)\varphi_k^*(v, \omega)\,dv\,d\omega} \qquad i = k = 1, 2, \ldots, N \tag{10.34}$$

where functions under an integral are defined at frequencies $\omega_1, \omega_2, \ldots, \omega_q$.

The investigation performed showed that the solution at $\alpha_1 = 0$ is close to the desired solution (good image contrast, real object shape, size, and position). If $\alpha_1 \neq 0$, the solution obtained at defined α_1 is better than the first one, but the difference between them is small.

10.2.2 Experimental Results

Experiments on imaging of different buried objects and a person behind a wall were carried out using the tomographic setup described by Vertiy et al. [7,9]. When using this method for imaging of buried objects, the subsurface area perpendicular to the surface planes at different values of z -coordinate (vertical slices) was divided along the y -axis (in depth) into 10 homogeneous layers of equal thickness and relative permittivities $\varepsilon_{r1} = \varepsilon_{r2} = \cdots = \varepsilon_{r10}$. The relative permittivity ε_{r11} of substrate is equal to the relative permittivity of the layers. Permittivities $\varepsilon_{r1}, \varepsilon_{r2}, \ldots, \varepsilon_{r11}$ are equal to relative permittivity ε_{rm} characterizing the medium in which the object is buried. In the general case, the thicknesses and permittivities of a layer system can be different. The area in the direction of the x -axis was also divided into 10 parts with size Δx, so 10×10 rectangular image elements were obtained. It is assumed that the structure is irradiated at an angle of incidence $\theta^i \approx 0°$ by a plane wave with an electric field amplitude of 1 V/m on the substrate surface.

A reconstructed vertical (horizontal) slice is an image map (height plot) illustrating polarization current absolute value $|\tilde{K}|$ in points of an image space. In this case, vertical or horizontal slices are obtained as a result of visualization of an interpolating function of two variables: x, y or x, z, respectively. For example, function the $|\tilde{K}(x, y)|$ is an interpolating function calculated at given points of an image space at $z = z_l$ ($l = 1, 2, \ldots, L$) using determined polarization current absolute values $|\tilde{K}(x_\xi, y_\eta)|$ ($\xi = 1, 2, \ldots, 10$; $\eta = 1, 2, \ldots, 10$) at centers of elements P_i for $i = 1, 2, \ldots, N = 100$. From the set of values $|\tilde{K}(x_\xi, y_\eta)|$ calculated at different planes $z = z_l$, it is possible to find polarization current absolute values $|\tilde{K}(x_\xi, z_l)|$ at points x_ξ, z_l of the image space at $y = y_\eta$ and then the interpolating function $|\tilde{K}(x, z)|$ calculated at given points of the image space at $y = y_\eta$. Obtained as a result of interpolation, the two-dimensional data set contained 81×81 elements. The interpolation process was also used for calculation of the function $\varphi^j(\nu, x, y) \equiv E_z^j(x, y) B_0(\nu, x, y)$ with fixed value of ν. Values of this function were determined at the corners of regions $P_1, P_2, \ldots, P_N$ using design methods of functions $E_z^j(x, y)$ [5,11] and $B_0(\nu, x, y)$ [1]. Values of function $\varphi^j(\nu, x, y)$ in regions $P_1, P_2, \ldots, P_N$ were found employing interpolation formula for four points.

Two experimental results obtained with buried objects are presented in Fig. 10.3. The first experiment is conducted with a VS-50 model plastic antipersonnel mine (APM) having the shape of a circular straight cylinder with diameter $d = 0.8\lambda_0$ and height $h = 0.35\lambda_0$ ($\lambda_0 = 0.1\ m$). This mine was buried at depth of $D = 1.2\lambda_0$ in a box with sand (relative permittivity of sand is $\varepsilon_{rm} \approx 3.0 + i0.0$). The length of the probing line is $2.9\lambda_0$. Data on the scattered field at the probing line are determined at 30 points with a constant step. The distance between the line of scanning and the medium surface is $0.2\lambda_0$. The measurements were conducted at 32 frequencies with a constant frequency step in the linear frequency range Δf from 2.5 to 4 GHz. The image space size is $2.9\lambda_0 \times 2.0\lambda_0$ (along the x, y -axes, respectively). Space frequencies ν in the range $2\nu_0 = 24\ m^{-1}$ with a step of $\Delta\nu = 0.12\ m^{-1}$ were used for reconstruction. The reconstructed vertical slice (image map) at the mine center is given in Fig. 10.3. The value $\alpha_1 = 0$ for vector $\vec{\tilde{K}}$ was used. The image size is $2.9\lambda_0 \times 2.0\lambda_0$. The black rectangle in the figure shows a contour of the real mine cross section on the vertical slice.

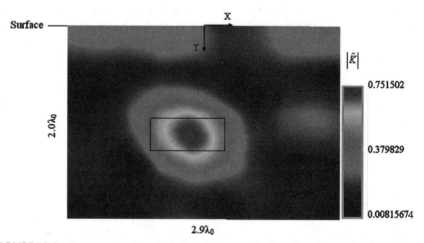

FIGURE 10.3   Reconstructed vertical slice (image map) of a subsurface region investigated using APM.

Another experiment is conducted with two buried objects: an APM and an antitank mine (ATM). The ATM had a circular straight cylindrical shape with diameter $d_2 = 3.0\lambda_0$ and height $h_2 = 1.0\lambda_0$. The mines were buried in different areas of a box with soil with relative permittivity of $\varepsilon_{rm} \approx 5.0 + i0.0$ at different depths D_1, D_2 ($D_1 \approx 0.6\lambda_0$ and $D_2 \approx 1.0\lambda_0$ for APM and ATM, respectively). The bases of the buried mines are parallel to the air–sand interface. Two-dimensional scanning was performed by shifting the probing line along the z-axis. The length of the probing line is $9.9\lambda_0$. Data on the scattered field at the probing line are determined at 100 points with a constant step at each position of the probing line. The distance between the probing line and the soil surface is approximately $0.2\lambda_0$.

The scanning length along the z-axis is $9.8\lambda_0$, and the number of the probing line positions was $L = 50$. Measurements were carried out at 32 frequencies with constant-frequency steps in the linear frequency range Δf from 2.6 to 3.7 GHz. The image space size is $9.9\lambda_0 \times 2.0\lambda_0$ (along the x and y axes, respectively). For the reconstruction space, frequencies v in the range $2v_0 = 24$ m^{-1} with a step of $\Delta v = 0.12$ m^{-1} were used. Horizontal slices of the investigated subsurface space at depths D_1 and D_2 were reconstructed at $\alpha_1 = 0.3$ for vector $\tilde{\vec{K}}$.

Figure 10.4a and b illustrate the reconstructed horizontal slices (image maps) at depths D_1 and D_2, respectively. The size of the images is $9.9\lambda_0 \times 9.8\lambda_0$ (along the x and z and axes, respectively). Black small and big circles show, respectively, contours of the real APM and ATM cross sections at the horizontal slices. In addition, for this experiment, Fig. 10.5a and b illustrate (as image maps) the structure (modules of complex elements $\tilde{a}_{ki}$) of the 100×100 matrix $\tilde{A}$ for narrow $\Delta f_1 = 3.59$ to 3.7 GHz and optimal $\Delta f_3 = 1.5$ to 3.7 GHz bands, respectively. Analogous to the image of the matrix $\tilde{A} = A_1 + A_2 + \cdots + A_N$ for the operating band $\Delta f_2 = 2.6$ to 3.7 GHz in which the measurements were conducted, one can see in Fig. 10.6 the structure of the matrix $A_1 + \alpha_1 A_2$ ($\alpha_1 = 0.3$) used in the reconstruction, results of which are presented in Fig. 10.4.

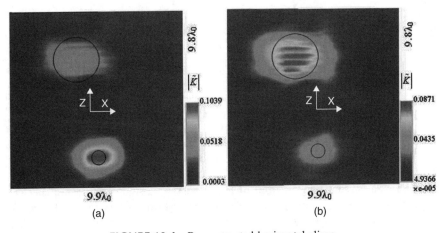

FIGURE 10.4   Reconstructed horizontal slices.

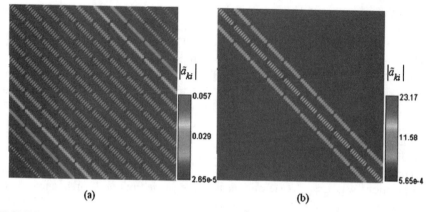

FIGURE 10.5   Image maps of a 100×100 matrix $\tilde{A}$ structure for different frequency bands (problem parameters as in experiments with APM and ATM): (a) narrowband, $\Delta f_1 = 3.59$ to 3.7 GHz; (b) optimal band, $\Delta f_3 = 1.5$ to 3.7 GHz.

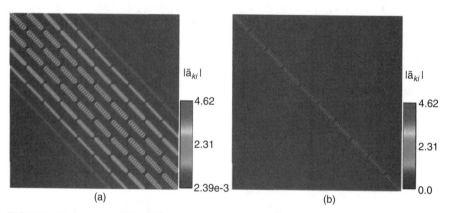

FIGURE 10.6   Image maps of a 100×100 matrix $\tilde{A}$ structure for the operating frequency band $\Delta f_2 = 2.6$ to 3.7 GHz (experiment with APM and ATM): (a) matrix, $\tilde{A} = A_1 + A_2 + \cdots + A_N$; (b) used in reconstruction, $\tilde{A} = A_1 + \alpha_1 A_2$ ($\alpha_1 = 0.3$).

It should be noted that after Fourier transformation of measurement data obtained in experiments with buried mines, the transformed data were filtered as described by Vertiy et al. [8]. Then they were used as values of function $\hat{\psi}(\nu, y_s)$ in the reconstruction process. In the process of filtering of the measurement data spectrum we eliminated spectral components near space frequency $\nu_0 = 0$ if $\theta^i = 0$. During the process, similar separation of the vertical slices on elements was used for imaging of a person lying in front of the tomographic setup behind a plane wall made from wood or foam plastic. The relative permittivities of the layers are $\varepsilon_{r4} = \varepsilon_{r5} = \cdots = \varepsilon_{r13} = 1.0$. The substrate relative permittivity is $\varepsilon_{r14} = 1.0$. The wall with thickness $t = 0.3\lambda_0$ has been divided into three plane layers with equal

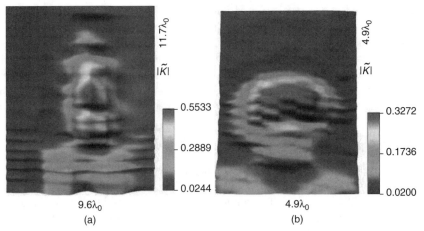

FIGURE 10.7 Reconstructed horizontal slices (height plots) of a person obtained using tomography: (a) behind a wooden wall (person's body); (b) behind a foam wall (person's head).

thicknesses of $0.01\lambda_0$. The relative permittivities of wall layers ε_{r1}, ε_{r2}, ε_{r3} are 3.0 for wood and 1.05 for foam plastic. It is assumed that the structure is irradiated at the angle of incidence $\theta^i \approx 0°$ by the plane wave with electric field amplitude of 1 V/m on the substrate surface.

Figure 10.7a presents a reconstructed horizontal slice (height plot) of the person lying behind a wooden wall. In this experiment the middle horizontal cross section of the person was at coordinate $y = 2.5\lambda_0$. The length of the probing line is $9.6\lambda_0$. Data on the scattered field at the probing line are determined at 33 points with a constant step at each position of the probing line. The distance between the probing line and the wall is $0.2\lambda_0$. The scanning length along the z-axis is $11.7\lambda_0$ and the number of probing line positions is $L = 40$. Measurements were carried out at 32 frequencies with a constant-frequency step in the linear frequency range Δf from 2.05 to 3.7 GHz. The image space size is $9.6\lambda_0 \times 4.0\lambda_0$ (along the x and y axes, respectively). For the reconstruction we employed space frequencies ν in the range $2\nu_0 = 20\ m^{-1}$ with a step of $\Delta\nu = 0.10\ m^{-1}$ and the value $\alpha_1 = 0.15$ for vector $\tilde{\vec{K}}$. The slice has been calculated at $y = 1.3\lambda_0$, and its size is $9.6\lambda_0 \times 11.7\lambda_0$ (along the x and z axes, respectively).

The person's head slice is given in Fig. 10.7b. In this case the foam plastic wall was used. The length of the probing line is $4.9\lambda_0$. Data on the scattered field at the probing line are determined at 40 points with a constant step function at each position of the probing line. Distance between the probing line and the foam wall is $0.2\lambda_0$. The scanning length along the z-axis is $4.9\lambda_0$ and the number of the probing line positions is $L = 40$. Measurements were conducted at 32 frequencies with a constant-frequency step in the linear frequency range Δf from 2.2 to 3.7 GHz. The image space size is $4.9\lambda_0 \times 3.5\lambda_0$ (along the x and y axes, respectively). For the reconstruction we employed space frequencies ν in the range of $2\nu_0 = 20$ m^{-1} with a step of $\Delta\nu = 0.10$ m^{-1} and the value $\alpha_1 = 0.15$ for vector $\tilde{\vec{K}}$. The slice has been calculated at $y = 1.875\lambda_0$ and its size is $4.9\lambda_0 \times 4.9\lambda_0$ (10×40 elements) along the x and z axes, respectively.

The slices obtained using the mathematical procedure and STM from the same measured data are similar. The STM slice corresponding to the slice in Fig. 10.7(b) is presented in Fig. 10.8a. This slice was constructed from a two-dimensional data array of 40×40 elements. It should be noted that the spatial resolution could be improved by submerging the body to be imaged in a medium with similar electrical properties [5].

A reconstructed three-dimensional distribution of the normalized polarization currents in the object can be used to construct an isosurface for visualization of the object shape or its internal structure. Figure 10.8b is a side view of an isosurface obtained from the person's head data using STM. In this experiment the scanning area size is $4.9\lambda_0 \times 4.9\lambda_0$. The number of scanning lines at constant values of z is 65. The number of samples measured on the scanning line is 65. The data volume calculated contained $65 \times 65 \times 65$ elements. The reconstructed image size is $4.9\lambda_0 \times 4.9\lambda_0 \times 4.9\lambda_0$. The NAG software (IRIS Explorer 5.0) was used for the isosurface construction. In this situation, one can easily see the structure of the person's head.

Tomographic method for image reconstruction of objects placed in a layer of a plane multilayered material half-space is considered. This method is based on integral presentation of the scattered electric field for a plane-wave excitation at different frequencies. Application fields are defined primarily by this method's advantages and disadvantages in comparison with another tomographic method: for example, with STM. The algorithm considered above makes it possible to solve a very wide class of imaging problems, for both subsurface objects and for objects located in layers of a planar multilayer structure. Medium losses can also be

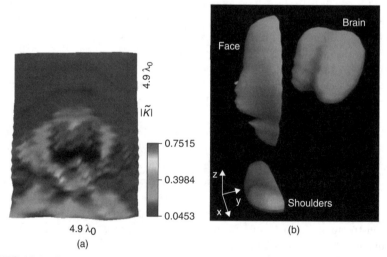

FIGURE 10.8 Reconstructed images of a person's head using STM: (a) horizontal slice (height plots); (b) isosurface.

taken into account using this method. Results obtained by the algorithm described and by STM for vertical slices in the case of buried objects are in good agreement. Although STM can be used for imaging of objects located in some layer structures, for example, as in the through-wall-imaging problem, the adaptability of STM in the case of complex multilayer structures requires additional investigation. STM may have the advantage, due to small time required for slice restoration. This allows us to obtain images of buried or hidden objects in quasi-real time.

The present investigation is the first step on the way to creating multilayer structure tomography and further, we wish to improve the algorithm described. It means development of procedures that will decrease the calculating time needed for slice reconstruction and the development of tools that will allow us to take into account the characteristics of an antenna system in a real device. The experimental results showed the effectiveness of the tomographic method in the microwave region (2 to 4 GHz). The results obtained may find applications in geophysics, medicine, and nondestructive testing for the development and creation of secure and safe systems using microwave radars and methods of microwave tomography for reconstruction images of horizontal and vertical slices of a probing area. For example, microwave imaging can be used as an alternative new technique for the early detection and imaging of breast cancer [3].

10.2.3 Subsurface Tomography Application for Through-Wall Imaging

In the present section a new approach to the through-wall imaging (TWI) problem using the microwave tomographic technique is considered. Cross-section restoration of the objects being studied is based on tomographic integral equation solution. Solution of this integral equation makes it possible to find an image function representing the normalized polarization current distribution in the probing cross section. It is shown that advanced through-wall images of a body can be obtained upon entering an effective dielectric constant instead of the wall and behind-wall media permittivities. This procedure substantially simplifies the through-wall reconstruction problem, as it is reduced to a subsurface reconstruction problem. The tomographic algorithm is employed for object cross-section imaging behind the wall.

Through-wall imaging is a new and very important direction in radar microwave imaging. Applications of this type of imaging include urban search-and-rescue scenarios and the search for objects located on the other side of a wall. The TWI development stimulated the design and creation of new types of radar imaging systems. To create such radar imaging systems, new research and development efforts in analysis and application software are needed. At the present time, the known through-wall radar structures are produced using analysis software, which is based on various techniques: short electromagnetic pulses; holographic, real aperture, projection, and diffraction tomographic reconstruction; and near-field methods. The microwave subsurface tomographic imaging technique (MSTIT) has been developed for identification of cylindrical inhomogeneities buried in a lossy half-space by Ferris and Currie [12], who presented the theoretical formulation upon which the imaging method is based. This technique was also considered by Benjamin et al. [13] and used for object imaging under the ground surface. The concept of scattered field expansion in terms of plane waves has been utilized for image reconstruction. It can also be applied in TWI, but this use requires further research. In the present work new results of the theoretical study and experimental investigation of TWI using the microwave tomography algorithm [12,13] are submitted. This tomographic algorithm is employed for the cross-section imaging of a person behind a wall. The results of theoretical study and experiment have demonstrated the successful use of MSTIT for imaging of objects located behind a wall.

Modeling and the experiments on the restoration of cross-section images of various objects embedded in material half-space with a plane boundary can be based on the tomographic integral equation [12–14]

$$\frac{\hat{\psi}(\nu, y_1)\exp(i\gamma_1 y_1)}{c_1(\nu)} = \iint\limits_{S} K(x', y')\exp[-2\pi i(\alpha x' + \beta y')]\,dx'\,dy' \quad (10.35)$$

where $\hat{\psi}(\nu, y_1)$ is the Fourier image of a complex scattered field $\psi(x, y_1)$, which should be measured above the surface at line $y = y_1$; the variable ν is the space frequency; $c_1[\gamma_1(\nu), \gamma_2(\nu)]$ is the complex function of ν; γ_1, γ_2, and β are complex functions of ν in the general case; α is the real function of ν; and the function $K(x', y')$ represents the normalized polarization current distribution in the region S (the cross section of an investigated object) which is sought for and which is the source of scattered field $\psi(x, y_1)$. The region investigated, S, is limited, so the integral in Eqs. (10.1) and (10.35) can be considered in infinite limits. It is supposed here that the Fourier transform of the scattered field exists, as this field is located in space.

The integral relation connecting functions $\hat{\psi}(\nu, y_1)$ and $K(x', y')$ in a similar case, when a cylindrical target characterized by permittivity $\varepsilon(x', y')$ is located in a homogeneous dielectric layer with ε_i of a plane multilayer structure placed on a substrate (bottom medium), can be written

$$\hat{\psi}(\nu, y_1) = \iint\limits_{S} k_i^2 K(x', y')E_z^t(x', y')\hat{G}_1(\nu, y_1 x', y')\,dx'\,dy' \quad (10.36)$$

where $K(x', y')$ is defined as in Eq. (10.35); k_i is the wavenumber of the medium for the layer in which the target is located; $E_z^t(x', y')$ is the incident field defined in the target region when the target is removed; and $\hat{G}_1(x, y_1 x', y')$ is the Fourier transform of the Green's function $G_1(x, y_1 x', y')$ and can be considered as the electric field created at points on the probing line by a line source of current situated at a point (x', y'). This function has been studied by Chommeloux et al. [1], who present its analytical expression. Then Eq. (10.1) can be obtained for the case of two adjoining homogeneous mediums with a plane boundary, by substituting explicit expressions for the fields $E_z^t(x', y')$ and $G_1(x, y_1 x', y')$ in Eq. (10.36). The wall structure can be complex in the TWI problem and contain several layers of different mediums. The solution of this problem—the reconstruction of $K(x', y')$ using Eq. (10.36)—is a difficult job, due to the complexity of expressions for electromagnetic fields in layers, and $\hat{G}_1(x, y_1 x', y')$, the inverse operator form, can differentiate from the inverse Fourier transform which is used in the solution of Eq. (10.35) and is well known. In addition, the inverse problem for Eq. (10.36) can be ill-posed.

Suppose that the multilayer structure and substrate in the TWI problem can be considered together as a material half-space characterized by an effective permittivity ε_{ef}. If such modeling is possible, Eq. (10.35) can be used instead of Eq. (10.36) to obtain the image function $K(x', y')$. The schemes of the TWI problem and its model, which have been investigated using numerical analysis, are shown in Fig. 10.9. In accordance with Fig. 10.9a, the TWI problem is to obtain the cross-section image of a dielectric cylinder (scatterer) characterized by permittivity ε_S which is located in material half-space (substrate) characterized by permittivity ε_3 behind a homogeneous wall layer with permittivity ε_2. The cross-section image is reconstructed from the scattered field data calculated in the points on the probing line at $y = y_1$ located above the wall layer in material half-space characterized by permittivity ε_1. All permittivities are relative and real. Values of ε_1 and ε_3 are equal to 1 and $\varepsilon_S = 5$. The value of ε_2 is varied from 2 to 10. It is assumed that the plane wave probes the structure under a normal angle of incidence. Frequencies f_j of the incident field are taken in the band $\Delta f = 0.68$ GHz from $f_1 = 3.32$ GHz to $f_{17} = 4.0$ GHz. The cross-section sizes of the cylindrical scatterer are small compared to the wavelength of the incident field ($\simeq 0.107\lambda_0 \times 0.107\lambda_0$, $\lambda_0 = 0.075$ m). The scatterer can be placed at two points (positions) 1 and 2, whose coordinates are $x_0 = 0.6\lambda_0$, $y_0 = 4.75\lambda_0$ for the position 1 and $x_0 = 2.61\lambda_0$, $y_0 = 7.95\lambda_0$ for position 2. The thickness of the wall layer is $3.2\lambda_0$. The coordinate is $y_1 = -0.014\lambda_0$. The probing line length is $11\lambda_0$.

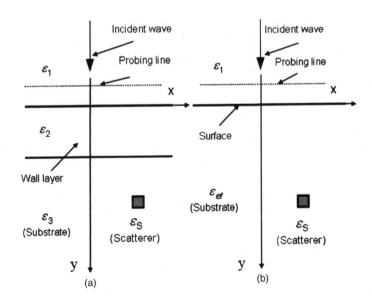

FIGURE 10.9   (a) TWI problem; (b) TWI problem model.

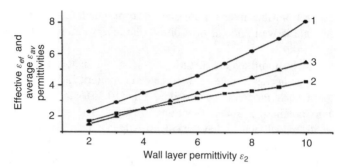

FIGURE 10.10 Dependencies of ε_{ef} on ε_2 at positions 1 and 2 of the target (curves 1 and 2, respectively) and ε_{av} on ε_2 (curve 3).

The Fourier transform of the scattered field at the probing line was calculated in accordance with Eq. (10.36) under the assumption that the scattered field is small compared with the incident field inside the scatterer. The space frequencies ν_k were taken in the band $\Delta\nu = 80$ m^{-1} from $\nu_1 = -40$ m^{-1} to $\nu_{401} = 40$ m^{-1}. The inverse of Eq. (10.36) yields the scattered field.

The scattered field data obtained are employed in the model (Fig. 10.9b). The parameters are same for both the TWI problem and its model. Nevertheless, in the model the wall layer is absent and two mediums with permittivities ε_2 and ε_3 are simulated by one medium with permittivity ε_{ef}, which is not known and is determined by the inverse of Eq. (10.35). In this case the permittivity of the substrate is varied in the model and taken as ε_{ef} if the sizes and position of the scatterer cross section and reconstructed cross-section image agree. In reconstruction space, frequencies ν_k were taken in the band $\Delta\nu = 10$ m^{-1} from $\nu_1 = -5$ m^{-1} to $\nu_{401} = 5$ m^{-1}. Modeling showed that the permittivity ε_{ef} can be found for the TWI problem considered. The permittivity ε_{ef} depends on the wall layer permittivity ε_2 and position (coordinate y_0) of the scatterer. The image coordinate x_0 is not changed with variations in ε_2. In Fig. 10.10 curves 1 and 2 illustrate obtained dependencies of ε_{ef} on ε_2 at positions 1 and 2 of the scatterer, respectively. Curve 3 shows the dependence of average permittivity $\varepsilon_{av} = (\varepsilon_1 + \varepsilon_2)/2$ on ε_2.

The results presented in Fig. 10.10 show that ε_{ef} is increased when the target is located near the wall layer. The effective permittivity is close to the permittivity of the walls if the target is located close to the wall and close to the permittivity of the surrounding medium if the target is far from the wall layer. The average permittivity can be used instead of the effective permittivity if the target is located behind the wall layer and the wall layer permittivity is between about 2 and 6.

10.2.4 Results of Experiments

In addition to simulation, experiments with the TWI radar were conducted. The through-wall version of the experimental system contains:

- A desktop or portable computer equipped with a data acquisition board

- Small lightweight vector measuring units (1 to 2 GHz and 2 to 4 GHz frequency ranges)
- A bowtie microwave antenna
- A two-dimensional scanner and control block

Positioned in front of wall, the transmitting antenna illuminates continuous waves with a frequency step. Frequencies f_j of the incident field were taken in the band $\Delta f = 2$ GHz, from $f_1 = 2.0$ GHz to $f_{32} = 4.0$ GHz. The waves pass through the wall material and are scattered by objects in the probing space. The return signal from the receiving antenna is fed to a measurement unit. This unit, together with transmitting and receiving antennas, forms the antenna block, which can be moved by the mechanical scanner in front of the target area along direct lines on a plane lying parallel to the wall surfaces. Signal-processing software is based on inverse problem solution (see Section 5.3) and allows image reconstruction of the scattering regions in the probing cross section, which is normal or parallel to the wall surface (vertical or horizontal slices). The spatial frequencies v_k were taken in the band $\Delta v = 20$ m^{-1} from $v_1 = -10$ m^{-1} to $v_{21} = 10$ m^{-1} in the image reconstruction. The effective permittivity has been selected as $\varepsilon_{effective} = 2$ in this experiment.

Measurements were conducted with a person located behind the wall. The experimental setup scheme, shown in Fig. 10.11, illustrates some components of

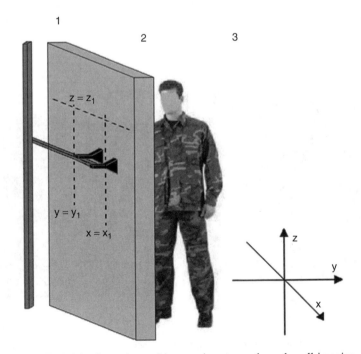

FIGURE 10.11 Scenario used in experiments on through-wall imaging.

the setup: a person; the setup outlook (XOZ); $y = y_1$, the probing plane behind the wall structure; the position of the antennas at each scanning step; and the scanning lines $z = z_1, x = x_1$ (dashed lines) in the plane $y = y_1$. This plane is located at coordinate $y \equiv y_1 = -0.2\lambda_0$. The scanning area is $17.3\lambda_0 \times 26.7\lambda_0$ along the x and z axes, respectively. The typical scanning step is approximately 0.14 to $0.4\lambda_0$. The wall structure layers are made from marble (layer 1, thickness of $0.4\lambda_0$) and wood (layer 2, thickness of $0.27\lambda_0$). The third person is located behind. Figure 10.12 represents a tomographic slice reconstructed using the method described above. A photograph of the multilayer wall containing wood, marble, and plaster layers is presented in Fig. 10.13. The results obtained show quite a good quality of reconstructed images if the target is located in the near-field zone of the scattered field.

Measurements were also conducted with the person is located behind a wooden wall (the marble layer is removed). These slices have been calculated at the coordinate $y = 2.4\lambda_0$. The coordinate y is approximately equal to $3.4\lambda_0$ in the experiment for the middle body cross section. As the wooden layer has permittivity $\varepsilon_2 = 3$, the permittivity ε_{ef} value is chosen to be 3, in accordance with results presented in Fig. 10.10 (curve 1), where one can see clearly the figure of the person behind the wall. It is easy to see that additional losses inserted by the marble wall lead to some distortions in the image in Fig. 10.14. One more reason for such distortions may be related to the filtering of the angular spectrum scattered by the object field. It is obvious that perturbation of images obtained will increase with an increase in the dielectric permittivity of the walls. We can also obtain three-dimensional images of a person behind a wall using the tomographic approach considered. Such images are shown in Fig. 10.15.

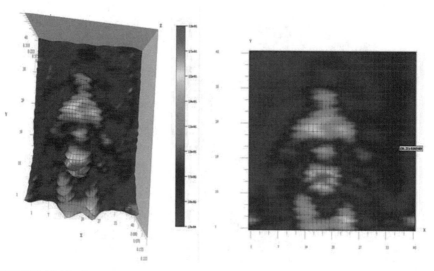

FIGURE 10.12 Tomographic image of a person placed behind a marble and wooden wall (the distance is 0.28 m).

FIGURE 10.13 Typical multilayer wall structures (i.e., marble and wood) used in tomographic measurements.

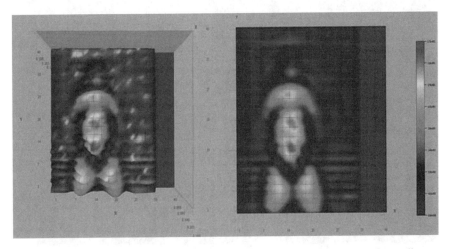

FIGURE 10.14 Tomographic images of people located behind a wooden wall.

A tomographic image and photograph of a person behind the wooden wall with an automatic gun are shown in Fig. 10.16. It is clear in this image that the bright region on the man's breast corresponds to strong reflection of electromagnetic waves from the metal gun, which is located on the person's chest. That result demonstrates the strong potential of the technology considered. Nevertheless, the method requires two-dimensional scanning and is not suitable for some applications. Further, we will show some tomographic results obtained using one-dimensional scanning and using no scanning but with a movable target.

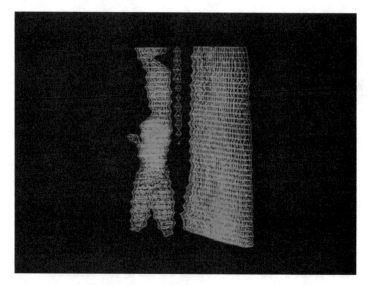

FIGURE 10.15 Three-dimensional image of a person behind a wall.

Automatic gun

FIGURE 10.16 Tomographic image and photograph of an armed person behind a wooden wall.

Next, let us consider other experiments for tomographic imaging. Figure 10.17 is a photograph of the setup (a) and a plastic antitank mine (b) placed behind a wall. Also shown are a one-dimensional scanner with a pair of horn antennas and a microwave vector measurement unit (Fig. 10.17) operating in the frequency range 2 to 5 GHz.

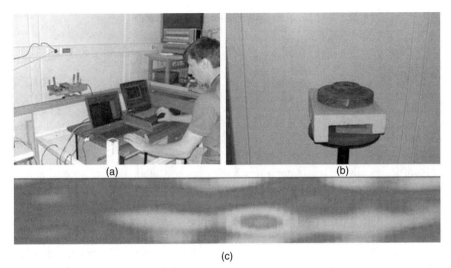

(c)

FIGURE 10.17 (a) 2- to 5-GHz through-wall tomographic set up; (b) plastic antitank mine; (c) horizontal slice image of the same mine.

Shown in Fig. 10.18 are two images obtained when a person has been moved along a wall at distances of 0.2 m (a) and 0.3 m (b). The antenna system of the tomographic setup is fixed in this experiment, and the scanner has not been used. The displacement of the person from the wall is clearly shown. Here the wall position is defined by the lower line. Thus, as one can see, the tomographic algorithm can be used successfully for imaging a cross section of a movable target. Of course, it is also possible that the target is moving in the near-field zone. In the far-field zone we could not reconstruct the image of a targets' cross section very successfully; however, we can define the distance between the target and wall and the target position. In the framework of a through-wall detection program, stepped-frequency radar has been designed with the specifications given in Table 10.1.

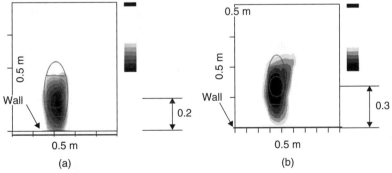

FIGURE 10.18 Tomographic images of a person who moved along the wall at different distances: (a) 0.2 m; (b) 0.3 m.

TABLE 10.1 Specifications of the Radar

B (MHz)	~100	~200	~500	~1000
ΔR (cm)	~150	~75	~32	~16

As a result of the well-known facts behind electromagnetic wave propagation phenomena, high-frequency radio waves give better resolution; however, the penetration depth decreases with increasing frequencies. Thus, there is a practically optimal operating frequency range lying in the approximate region 1 to 4 GHz (this also depends on the material properties of the walls). Furthermore, the radar range resolution (ΔR) is defined directly by the bandwidth of the radiated signal $\Delta R = vB/2$, where v is the velocity in a medium and $B = f_h - f_1$ is the signal bandwidth. The relations between the resolution and the bandwidth are presented in Table 10.1.

It is obvious that only ultrawide band systems (bandwidth/central frequency > 0.3) can provide suitable resolution in the frequency range 1 to 4 GHz. Most developers rely on a short-pulse generation technique to achieve ultrawideband operation (e.g., [12–14]) but here radar on alternative step-frequency technology is presented. The radar sweeps quickly through the frequency range, generating sequentially a set of equally distributed frequencies and collects the signal received on each frequency. Applying several data-processing sequences, including discrete inverse Fourier transform, input signal is mapped to the time domain, providing a reflector's profile. Repeating the sweep many times per second, the procedure is applicable for both static and relatively slow moving targets, such as a person.

Simplified illustration of the step-frequency operation principle is given in Fig. 10.19a. Vertiy et al. [7] and Noon [15] provide comprehensive information on step-frequency radar technology. Step-frequency radar has several important advantages over short pulse radars:

- Higher dynamic range and processed mean power, thus higher range and resolving power
- All radiated power contained within the effective antenna bandwidth
- Ability to apply "first reflection" suppression methods in hardware, thus increasing sensitivity further
- High reliability, stability, and relatively easy implementation
- Low power consumption, and thus long operation on an internal battery
- Ability to operate in continuous-wave (CW) mode, where the *maximum* detection sensitivity can be achieved by selecting the optimum frequency

10.2.5 System Description and Experimental Results

The radar presented here is a third generation of prototypes created in the course of through-obstacle detection and imaging systems development carried out by ILHT MRC TÜBITAK. The radar and the test area where the through-wall tracking mode

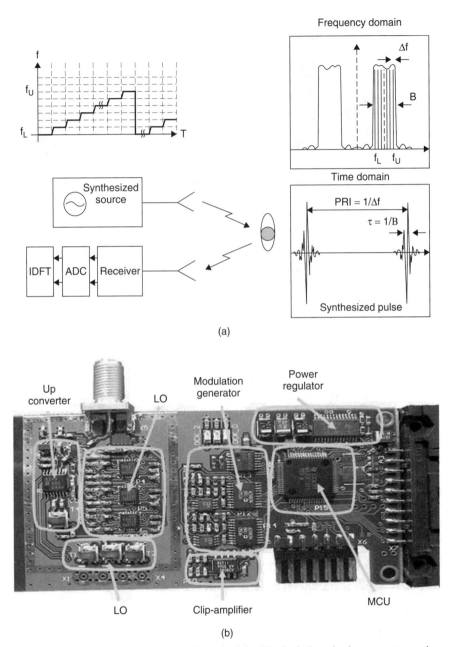

FIGURE 10.19 (a) Step-frequency radar principle; (b) single-board microwave transceiver for through-wall radar.

FIGURE 10.20 Radar prototype and test area used for through-wall mode development.

has been tested are shown in Fig. 10.20. The radar is based on the step-frequency principle explained briefly above. The radar consists of a measurement unit with integrated antennas, a mobile computer connected to the measurement unit through a wired or wireless interface, and system control and data-processing software. The lightweight battery-powered dust- and waterproof design of the measurement unit make it suitable for outdoor operation under harsh conditions.

The measurement unit consists of:

- A single-board complete microwave quadrature transceiver (see Fig. 10.19b)
- Small wideband SWR transmitter and receiver bowtie antennas (see Section 5.3.7)
- A lithium-ion battery

One of the most important features implemented in the radar is the ability to compensate for a constant input signal (i.e., reflection from first air–medium interface) in hardware and apply additional postamplification to the variable part of the input signal. Activation of this feature increases sensitivity to the signal reflected from the target, at least on 20 dB. The basic technical specifications of the radar are collected in Table 10.2.

The system is able to work in two modes, frequency-step and continuous wave. The CW mode can be used for detection of the target movement, even very small movements such as breathing and heartbeat:

- In caves and tunnels through the entrance
- In tunnels and communication lines through a layer of ground
- Inside the building through walls
- In free space

The frequency-step mode may be used not only for detection, but for also tracing the position of the target in one-dimensional (distance to the target) behind walls. Actually, the frequency-step mode can be used in all cases mentioned above, but

TABLE 10.2 Technical Specifications of the Radar

Parameter	Value	Unit	Comments
Common			
Detection range	a: 0–150 b: 0–20 c: up to 5	m	a. Straight tunnel b: Through walls c: Through ground layer (5 m for dry sand)
Position tracing range	a: 0–70 b: 0–20	m	a: Free space b: Three parallel brick walls (20 cm each) with a 5-m span between
Position resolution	0.5–1	m	
Position definition method	—	—	Readout from screen by operator
Position update rate	1–20	1/s	Update rate of covered area image on PC screen
Radar unit			
Frequency range	960–1800	MHz	
Frequency step	1–50	MHz	
Operation mode	—	—	CW or frequency step (32, 64, 128, 256 frequencies per sweep)
Dynamic range	≥ 100	dB	
Maximum output power	≤ 14	dBm	
PC interface	Serial	—	Wireless link or RS232
Weight	<5	kg	Without tripod
Power			
Voltage	9–24	V	Internal battery or external dc
Consumption	≤ 5	W	

sensitivity is higher in the CW mode, due to automatic selection of optimum frequency. The system has been tested in both artificially created and natural test environments. As an example a through-wall test area is shown in Fig. 10.20. A number of experiments have been conducted to optimize the system parameters and data-processing algorithms. Figure 10.21 presents a typical demonstration procedure along with the screenshot of the recorded trace of the target (human) moving between walls. The radar is located in front of the first wall. Along the y-axis the relative time is shown, and along the x-axis the distance between the radar aperture and the target is shown. At the upper part of the screenshot the target moves forward and backward perpendicular to the walls between the first and second walls. At the middle and bottom parts the same situation is repeated for movement between the second and third walls and behind the third wall. For better contrast, the steady object compensation mode is activated. In this way, the position of the target and the wall behind the target can be restored when the target appears between walls. As shown the target can easily be traced behind third wall

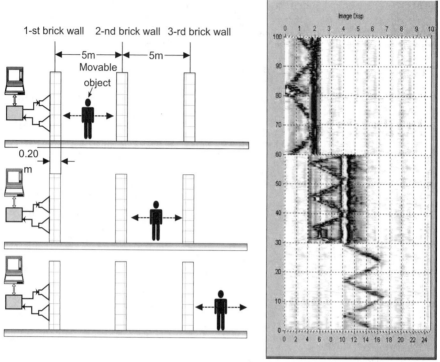

FIGURE 10.21   Record of a trace on a radar screen created by a person moving between walls.

as far as 18 m from the radar with an average resolution of the target position of about 0.5 m. The visibility and readability of the trace is pretty clear on the scan display. The radar in the CW mode has been tested in a number of scenarios, including the ruins of a building (Fig. 10.22).

The ability of a system to detect breathing and heartbeat is shown in Fig. 10.23. The recorded signal was obtained while a person was lying in a concrete tunnel at a depth of 2 m below the ground surface. To enable clear visibility of the heartbeat, breath was held twice during the record. Actually, there is no need to stop breathing because a clear breathing and heartbeat signature is visible when applying Fourier analysis to the signal.

A slightly modified scheme together with already developed tomographic software allowed us to convert this radar to a through-wall tomographic system. Radar in a through-wall tomographic application is shown in Fig. 10.24. In this experiment the radar is located in front of the first wall center and data are collected during movement of a target (a person) after an equal time interval. This procedure is equivalent to a scanning of the walls by the radar at a fixed target position. Some results on subsurface tomographic imaging of a person moving behind a brick wall of thickness 0.2 m are presented in Fig. 10.25. The movement occurs in parallel

FIGURE 10.22 Testing the initial prototype in ruins.

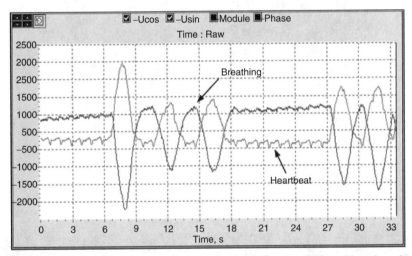

FIGURE 10.23 Record of the signal representing breathing capability and heartbeat detection (a person in a concrete tunnel 2 m below the ground).

to the wall at different distances from it. Figure 10.25a to c illustrate an image plane (a vertical slice) of 1.5×5 m. The reconstructed target position is denoted by number (1, 2, or 3) in these figures. In positions 1, 2, and 3 the target moved behind the wall at distances of 5, 3.5, and 2 m, respectively. Figure 10.25d to f illustrate an image plane of 1.5×6 m. In these figures the numbers also denote the reconstructed target positions. In position 4 the target moved behind the wall at a distance 0.2 m. In positions 5 and 6, the target is moved forward and backward, respectively, parallel to the wall at approximately same distance. Figure 10.25f illustrates a situation in which the target movements are in forward (positions 7 and 9) and backward (positions 8 and 10) directions at different distances from the wall.

FIGURE 10.24 Through-wall detection radar in a tomographic application.

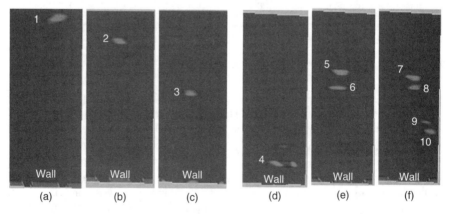

FIGURE 10.25 Tomographic images obtained with modified through-wall radar.

10.2.6   Conclusions

Two main directions in the field of through-obstacle detection and imaging, which are being developed by the TÜBITAK MRC MI ILHT, were described in this section. The first is based on the modification and redesign of existing radar, with the purpose of adding a second tracking coordinate and thus to create a real-time two-dimensional through-wall tracking system. Moreover, the second one is through-wall imaging based on tomographic processing of data obtained by multichannel step-frequency radar with a two-dimensional antenna array.

Preliminary results demonstrate real feasibility to create such a system. As an example, the tomographic image of a person placed 1 m behind the wall is shown in Fig. 10.7. The initial data were collected by frequency-step radar mounted on a two-dimensional mechanical scanner. The radar was scanned across the 1.0×1.5 m area from one side of the wall while the subject was placed on the other side of the wall. For the next step, it is planned to reach a real-time imaging system by substituting a mechanical scanner with a simultaneously operating array of transceivers and optimizing the computational algorithm.

10.3   MILLIMETER-WAVEBAND PASSIVE IMAGING

10.3.1   Introduction

Microwave radiometry is related primarily to measurements of natural thermally caused electromagnetic radiation of an object at a physical temperature above 0 K. In this section the microwave radiometry concept is discussed briefly with the essential technical details and real practical results. The reader is referred to a number of articles, reviews, and books that provide detailed explanations of the main principles of microwave radiometry [16–24,27].

As is well known, objects reflect and emit radiation in the millimeter-wave range just as they do in the infrared and visible ranges. The degree at which objects reflect or emit radiation is characterized by the emissivity (ε) of the object. A perfect radiator (absorber) has $\varepsilon = 1$ and is called a *blackbody* [22]. A perfect reflector (nonabsorber) has an emissivity of $\varepsilon = 0$. The Earth and the sky are basically blackbodies, and metal objects act as reflectors. Intermediate values of ε depend on the dielectric properties of the objects, angle of the observation, polarization parameters, surface roughness or coatings, wavelength, and other parameters. To measure such radiation for different objects it is more correct and understandable to use *radio brightness* (simply *brightness*) temperatures of the object, which can be expressed in temperature (T) units.

According to classic sources [23], a radiometer is a receiving device designed for measurements of the noise radio radiation level in a certain assigned band of Δf frequencies. The primary function of the receiver in a microwave radiometric sensor (radiometer) is to provide measurements of the input noise power expressed as an antenna temperature in equivalent blackbody temperature units. The sensitivity of the radiometer and the radiometric system (i.e., the minimum detectable signal) is determined by the amplitude of the fluctuations presented at the output indicator in the absence of the signal. Moreover, it can be expressed as a degree of the antenna temperature ΔT. (See the literature for more details in practical calculations and measurements of ΔT.)

The simplest explanation of radiometrical principles is shown in Fig. 10.26. As one can see, the radiometer is installed on an airplane and examines two measurements by radiometer. In the first spot of an antenna beam on the Earth's surface there are no metal objects, but an object (e.g., a car) exists in the second. It can be done by the simplest calculation, concerning the microwave power $P = kT\,\Delta f$, where k is Boltzmann's constant, T the brightness temperature, and Δf the bandwidth of the signal received, which is coming to the radiometer from these two antenna spots as the answer. The radiometer sensitivity should be sufficient to recognize the real contrast (difference) between these two spots. This task has real practical interest in special situations and in achieving high performance levels [25].

The difference between the brightness temperatures of the two spots in Fig. 10.26 (after two measurements) is defined as ΔT with the following simple approximation:

$$\left| T_{\text{back}} - \frac{T_{\text{back}}(S_{back} - S_t) + T_{\text{sky}}S_t}{S_{\text{back}}} \right| = |T_{\text{back}} - T_{\text{sky}}|\frac{t}{S_b} \tag{10.37}$$

where $S_{\text{back}} = \pi A^2$ is the area of the aperture antenna beam spot on the surface of Earth with the brightness temperature of T_{back}. $A = H1,22\lambda D$ is the radius of

FIGURE 10.26   Generic scheme of radiometer measurement.

the antenna beam spot on the surface of Earth, H the altitude, and S_t the area of the target with the brightness sky temperature T_{sky} (reflected from the sky). Therefore, when a radiometer has better sensitivity than the result according to Eq. (10.37), it is possible to say that the presence of metal in the second point of temperature contrast between the Earth (background) and the metal (sky) is high enough (Fig. 10.27) and depends on weather factors. Basically, a line-scanning radiometric system operates by moving a real aperture antenna beam across the desired field of view (FOV) and measuring the incoming noise power by the high-gain low-noise receiver shown in Fig. 10.28. Thus, the entire two-dimensional image is sampled pixel by pixel over time.

There are many practical applications of passive millimeter-wave imaging systems [27]. These systems have attracted increasing interest due to their ability to provide imaging through fog, clouds, drizzle, dry snow, smoke, and other obstacles, in situations where infrared and optic systems become inefficient. Minimum atmospheric absorption of millimeter-wave radiation occurs near 35 GHz and 94 GHz, and devices in these bands have been under very intensive development during the last 20 years. Hence, these frequency bands are used widely in modern millimeter- and sub-millimeter-wave imaging. The choice of operating frequency band depends on the particular application. It is known that better temperature contrast in a scene is obtained at 35 GHz than at 94 GHz, where higher spatial resolution is achieved. The main problem consists of obtaining an image in real time as in infrared thermal imagers. This can be fulfilled by simultaneous receipt of the radiation from various parts of a scene. To this end, the design of an array with a large number of receiving channels and a multibeam quasioptical antenna is required. Passive millimeter-wave imaging systems were developed from a single-channel scanning

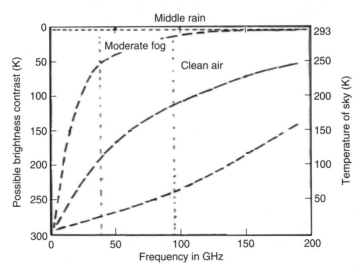

FIGURE 10.27 Brightness temperature difference of a metal at a point under $45°$ to the horizon and Earth.

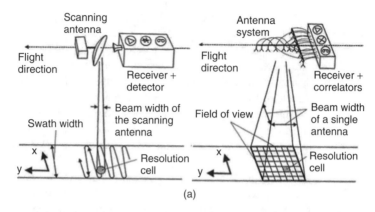

(b)

FIGURE 10.28 (a) Imaging principle of a line scanner (left) and a focal plane array (right). (b) Radiometric image from the fourteenth floor of the Paton Electric Welding Institute in Kiev.

imager to a full array which contains more than 1000 receivers [20,22,24,26,27]. Many collections [27] of passive millimeter-wave imaging systems are now in existence [27].

10.3.2 Development of a Practical Radiometric System

In this section we describe some passive millimeter-wave imaging systems designed at the State Research Center (SRC) Iceberg in Kiev in the last several years. Such imaging systems occupy an intermediate position between single-channel systems and full arrays. The main advantage of a scanning imaging system is the possibility of obtaining a wide FOV [28,29]. Described here is a 32-channel 8-mm system containing a linear array of radiometric sensors, a quasioptical multibeam antenna, a scanning mechanism with a control unit, a device for calibrating receiving channels, and a computer with an ADC/DAC card [30].

Laboratory Prototype of a Single-Channel 3-mm Imaging System   This system was first produced around 1987. At first it was mainly used primarily to carry out various experiments. At the same time, it was shown that the system is very convenient for research into and demonstration of the basic principles behind the detection of objects concealed under clothing. Passive detection of concealed weapons and other objects located under clothes is based on the fact that all bodies radiate, absorb, reflect, and pass electromagnetic energy. In addition, the quantity of the energy that is radiated, absorbed, and reflected depends on the material forming the objects, the shape of the objects, the brightness temperatures, and the surface properties, and also on the frequency of the radiation. Thus, clothing material is generally transparent to millimeter waves; the human body basically absorbs them, as they are equivalent to the body's own thermal radiation, and the material of which the weapon is made almost completely reflects the radiation. As a result, if the measurements are made indoors, a radiometric sensor can see the contrast between the radiation of a body with temperature around 309.6 K($= 36.6°$C) and and indoor room temperature of about 293 K (Fig. 10.29). Hence, passive formation of images in the 3-mm wave band represents an effective noninvasive remote method for detection and recognition of a weapon concealed under clothes, although there are certainly other problems in realizing a real system for security purposes. A simple demonstration system with a Cassegrain antenna and a scanning mechanism is shown in Fig. 10.30 (see [26]). The radiometric sensor represents a heterodyne receiver operating in the frequency band 90 to 94 GHz (Fig. 10.31). The fluctuation sensitivity in a compensation mode is not less than 0.05 K/ $Hz^{1/2}$.

Figure 10.32 shows the experimental results. The metal ring located under the clothes near the stomach and the plastic pistol above are seen through rather distinctly. Currently, there exists strong competition between leading microwave companies in the design and production of various passive microwave imaging systems: for example, for security control in airports [31]. Those who can produce really reliable, sensitive, and attractive systems will find a big market for their products, especially in currently, to deal with the prevention of terrorism and the design of various antiterroristic devices. The Technical specifications for such devices will be similar to those of the SRC Iceberg. After discussions with producers and customers, it is clear that such systems used for security [32] will probably not have such wide application areas, in that real working distances to the target objects will be limited to 3 m, which corresponds to the current technical level. Because such a restriction requires locating special enclosures at security control sites, the presence of there remote-sensing systems will be difficult to disguise.

Under the circumstances, it is proposed that a feasible system must be invisible and should have a working distance of 5 to 10 m. For a longer distance it will be necessary to use a receiving antenna with an aperture diameter matching that of a pixel in a microwave image, or a shorter wavelength should be used in the radiometers. At the same time, it will be prudent to remove all unnecessary microwave

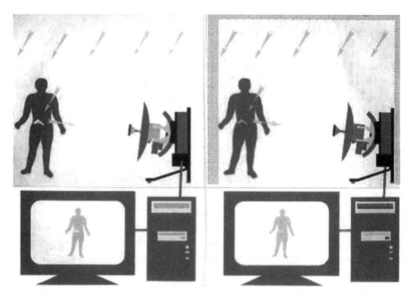

FIGURE 10.29   Outdoor and indoor difference in terms of the level of brightness contrasts.

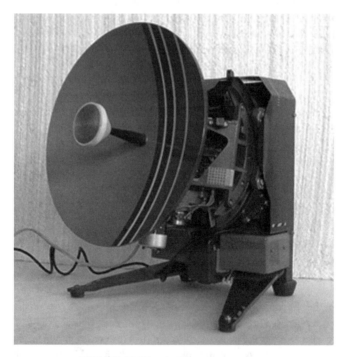

FIGURE 10.30   A 3-mm line scanner.

FIGURE 10.31 A 3-mm heterodyne radiometer.

FIGURE 10.32 Image of a human body with objects concealed under clothes.

FIGURE 10.33   A 3-mm radio image taken from a laboratory window.

losses before employing the radiometric sensors with the best sensitivity and to use a full-power-mode radiometric scheme [30]. Our line scanner (Fig. 10.30) was used successfully to obtain passive images at a 3-mm wavelength in various experiments related to the real radio vision. Figures 10.33 to 10.39 present highly promising results obtained using this system.

Passive 8-mm Microwave Imaging System with 32 Sensors   As mentioned earlier, the ability of millimeter waves to pass through fog, clouds, dry snow, smoke, and other obstacles makes millimeter-wave imaging systems the most efficient instruments available to solve a wide range of problems that cannot be managed using infrared and visible imaging systems. The main efforts are directed to achieving real-time images and to enhancing their quality. The time necessary to form an image depends on the number of sensors in the focal plane array and the scanning velocity. The better sensitivity of the sensors makes it possible to increase the scanning velocity due to the reduction in integration time required for good image quality. The image quality is also limited by the spatial resolution. The diffraction-limited spatial resolution of a passive millimeter-wave imaging system is inversely proportional to the diameter of the quasioptical antenna. Certain improvement in image quality can be achieved by digital signal processing of the images obtained, through the support of relevant mathematical methods [27,33–35].

The development of a practical millimeter-wave imaging system requires us to employ an increasing number of the highly sensitive receiving sensors (Fig. 10.40) in the focal plane array and a quasioptical antenna of sufficiently large diameter. The SRC Iceberg 16-channel radiometric imaging system [28] was capable of forming an image in 10 seconds. The 32-channel system with 8-mm wavelength imaging makes it possible to obtain an image three times faster and with better quality.

FIGURE 10.34   A 3-mm image of a scene near the Iceberg.

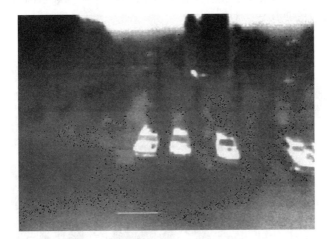

FIGURE 10.35   Image of a car parking lot.

FIGURE 10.36   A 3-mm image of two persons at the background of metal surface with metal list in hands.

FIGURE 10.37 A 3-mm radio image near parked cars Despite the nonoptical resolution, it is possible to identify car types.

FIGURE 10.38 Iceberg: polarization of a 3-mm radiometric scanning system.

Focal Plane Array The focal plane array is a main functional unit of the imaging system. It provides simultaneous receipt of signals arriving from different locations in the surveying scene. The array is built in modular form, beginning with a radiometric sensor, which is the basic component of the array. Characteristics of the sensors, such as fluctuation sensitivity, reliability, sizes, and cost, define the feasibility of the system. Creating an array is connected with solving some critical problems:

- The Good packing efficiency of the large number of receiving sensors in an array, preserving the high sensitivity of the sensors

FIGURE 10.39   Optical image (down) and two 100 images-GHz (up, vertical and horizontal polarizations) from near Kiev.

FIGURE 10.40   An 8-mm radiometric sensor.

- The sufficient identity of the receiving channels as related to the electrical parameters
- The low mutual influence of the receiving channels in the array
- The light weight, small size, and low cost of the array, which are especially important in commercial applications
- The problem of power dissipation, which is aggravated by an increase in the number of receiving channels

The circumstances that we have described were taken into account during the design of the radiometric sensor for use in an imaging system. Recent developments in GaAs HEMT technology make it possible to realize low-noise straight amplification receivers in the millimeter-wave band. Such receivers are more suitable for a radiometric array, due primarily to the absence of local oscillators. The array contains 32 sensors aligned in two vertical rows (Fig. 10.41) of 16 sensors each. The rows are shifted relative to each other to form an antenna with 32 beams which will draw 32 strips during scanning, with the antenna in the azimuth direction.

A large number of sensors comprising an array causes the problem of power dissipation. In our case, each sensor gives off about 0.5 W of heat, and hence the total power dissipation adds up to 16 W. To dissipate this heat, all sensors are fastened to an aluminum body with a radiator on the back side of the array. Supply sources with stabilized voltages of $+5$ V and ±12 V are common for all sensors. The output signals of the sensors are connected to an ADC card through a conductor bundle for digitization and subsequent processing.

The Sensitivity of the receivers in the total power mode is better than 10 mK at integration time $t = 1$ s. The microstrip construction of the receiving circuits make it possible to realize small sensors with dimensions of $14 \times 13 \times 80$ mm (see Fig. 10.40). The weight of each sensor does not exceed 40 g. The stretched configuration of the sensors helps to pack them efficiently into the receiving array. The waveguide input of each sensor is connected to the corresponding horn feed (Fig. 10.42). Dielectric rod antennas (with $\varepsilon_r = 2.1$) are used as feeders. This type of feeder has a smaller cross section than that of horn feeds, which makes it possible to arrange more compactly the sensors in a matrix design. The array construction make it suitable for integration in a larger two-dimensional receiving matrix. The structural block scheme of a matrix with 32 sensors is shown in Fig. 10.43.

Quasioptical Antenna, Calibration, and Scanning Mechanism   A multibeam quasioptical antenna consists of the main antenna and feeders located in the focal plane of the main antenna. The prime-focus-fed parabolic reflector is used as the main antenna. The reflector has a diameter $D = 900$ mm, providing a beam width of about $0.6°$ at the 3-dB level. The focal length of the antenna $F = 990$ mm, which satisfies the condition $F/D \geq 1$ for multibeam antennas. In the second experimental system (4×8 sensors) a dielectric lens with a diameter of 600 mm made of polyethylene was used as the antenna (see Fig. 10.44). Each feed is connected with its sensor and directed to the center of the main antenna. The reflector is seen by the feeds at an angle of $48°$ at the 7-dB level of its pattern. The reflector, with

FIGURE 10.41   Radiometric array with 32 sensors.

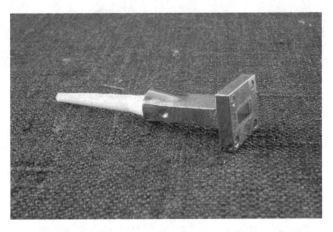

FIGURE 10.42   Dielectric rod at the front end of a radiometer.

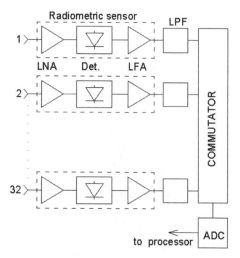

FIGURE 10.43   A 32-channel sensor array.

FIGURE 10.44   Passive 8-mm imaging system.

feeds in the focal plane, forms 32 beams. The angle between adjacent beams in the vertical plane is $0.48°$, and the total angle of view in the vertical plane is $16°$. The calibration system includes:

- Injection by turns of the calibrating signals to inputs of all the receiving channels
- Measurement of the levels of the output radiometric signals
- Calculation of the correction factors for each channel

The factors calculated are used to process the output radiometric signals obtained in the system's operation mode. The linear model of the conversion of the thermal radiation into radiometric receiver output signals is generally accepted. The output signal of each channel can be presented with high accuracy in the form of linear dependence:

$$C_i = a_i + b_i \quad \text{or} \quad T = \frac{C_i - b_i}{a_i} \tag{10.38}$$

where T_i is the effective temperature of radiation at the antenna output, C_i the output reading of the i th receiving channel that corresponds to T_a, and a_i and b_i are factors of linear dependence for the channel. To form an adequate image the certain output binary code C_a must correspond to each temperature T_a that is being measured. This code is proportional to a measured temperature of $C_a = AT_a$; hence, the corrected output signal for each channel C_{ai} can be calculated by the expression

$$C_{ai} = \frac{A(C_i - b_i)}{a_i} \tag{10.39}$$

where, a_i and b_i are correction factors for the i th channel and A is an arbitrary coefficient.

Correction factors a_i and b_i can be determined during active radiometric calibration of the channels by means of a connection to two reference radiators (called "cold" and "hot") of the inputs. These references have known temperatures of radiation T_1 and T_2. In this case, the a_i and b_i factors can be found according to.

$$a_i = \frac{C_{i2} - C_{i1}}{T_2 - T_1} \tag{10.40}$$

$$b_i = \frac{C_{i1} T_2 - C_{i2} T_1}{T_2 - T_1} \tag{10.41}$$

where C_{i1} and C_{i2} are the channel outputs.

An antenna with an array is attached to the scanning mechanism to provides antenna scanning in the horizontal plane at an angle of $\pm 60°$. The sensors are calibrated at the endpoints of each scan. The factors are calibrated using a digital processor. The additional circuits are designed to form a range of output voltages in the limits suitable for correct operation of the analog-to-digital (ADC) converters. In the imaging system the calibration process proposes placing noise reference sources before the antenna feeds. In this situation, the calibration determined for the receiving channels reduces considerably possible errors when determining the value of C_{ai}. Accuracy in this calculation also depends on the time stability of the receiving sensor parameters between the calibration cycles. Executing the calibration at a period of one cycle per scan makes it possible to eliminate the most intensive low-frequency components in the spectrum of gain fluctuations. Calibration is performed at the beginning of each scan. Due to the short duration of a scan (several seconds), correction factors between the next calibrations are considered invariable.

The calibration noise sources represent quasioptical wide-aperture reference radiators, which provide the calibration signals for all receiving channels. The radiowave-absorbing material is used to coat the radiators. The standing-wave ratio of the quasioptical reference source is less than 1.15. Metallic plates inside the radiators are cooled to $T_1 = 278$ K ($5°$C) by means of thermal-electric coolers or are heated at $T_2 = 323$ K ($50°$C) with an accuracy of ± 0.1 K by the electronic unit of temperature stabilization. Calibration reference sources are mounted on the fixed section of the system such that at the beginning of the each antenna scan, the "cold" and "hot" reference sources overlap all feeds in the calibration sections of the scanning sector.

During scanning, each beam draws a horizontal strip, and thus the entire observed scene is completely covered by the beams. The scanning mechanism includes the optical sensor of the angle, which provides information about the antenna position at any time during system operation. The signals are read through each $0.17°$ (1 pixel) in full angle of view in a horizontal plane of $120°$.

An image of the survey scene is formed during each scan (3 s). To enhance the quality of the image, a two-scan mode is provided in the system design. In this case, before reverse scanning, all beams are tilted at an angle of $0.24°$, with the help of the reflector. As a result, during two scannings the beams draw 64 strips, yielding better resolution in the vertical plane.

The scanning mechanism is designed on the base of a common asynchronous three-phase motor. The axis of the motor is mounted vertically. The reflector, with the array, is mounted on the motor's axis. Power is applied to the motor from a converter, which transforms a single-phase voltage (220 vac) into a three-phase pulse-duration voltage. The converter is controlled by the computer and provides switch-on and switch-off operation of the motor, a constant rotation velocity, and smooth stopping and reversing. To facilitate a load onto the motor when reversing, special springs are used. In practice, current does not go to the motor during reversing. An external view of this passive imaging system is shown in Fig. 10.44.

Performance Testing of the Imaging Systems   To form an image on the monitor screen, a complex software/hardware system has been developed. The hardware includes a personal computer (PC) with monitor and an ADC/DAC card (model L780). The card includes the following features:

- A 32-channel switchboard with an analog-to-digital converter used to read and digitize the signals from the sensors
- A digital-to-analog converter for computer control of the scanning mechanism
- Two digital inputs to transfer from the optical angle sensor to the computer information about the antenna's position.

The PC sets the operating modes for the scanning mechanism, processes the signals received, and forms the image on the monitor screen. The software specifies

(a)

(b)

FIGURE 10.45   (a) Buildings near Iceberg: 8-mm (120° on the horizon) and optical images. (b) Area in the Seoul city-8 mm image and optical images.

the sequence of system operation, to provide fast, flexible control of the imaging. The computer (with the monitor) is located at a convenient distance from the scanning receiver unit and is connected to the unit by a special cable.

Investigations using the system were performed under natural conditions, and the results shown in Fig. 10.45 to 10.47 were attained. In Fig. 10.45, an 8-mm radio images of the territories near Iceberg Kiev and Seoul obtained using the system developed are presented compared with the optical images. That the quality of the 8-mm waveband image is sufficiently good is clear.

Further we will give one more example of millimeter wave application for concealed weapon imaging [36,37]. Fig. 10.46 demonstrates 100 GHz radiometric imaging system designed at TUBITAK MRC, International Laboratory for High Technologies.

The millimeter waves are focused by a 0.5 m diameter dielectric lens. Image area is located 7 m away from object (staying person). To form 2D image linear scanner and flopping mirror have been used. Receiver module consists of horn antenna low noise amplifiers and detector. Radiometer receiver specifications are given in the frequency range of 98 GHz to 101 GHz, with 3 GHz bandwidth, 10

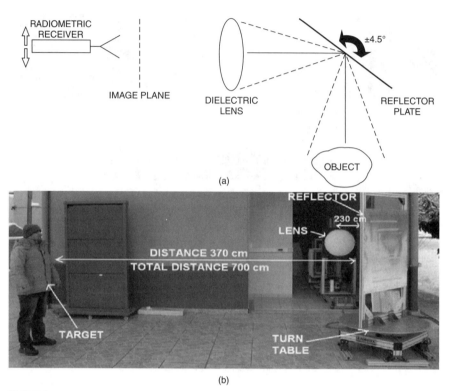

(a)

(b)

FIGURE 10.46 (a) Scheme of 100-GHz imaging system with linear scanner and flopping mirror and (b) radiometer for 100-GHz imaging with lens and flopping mirror.

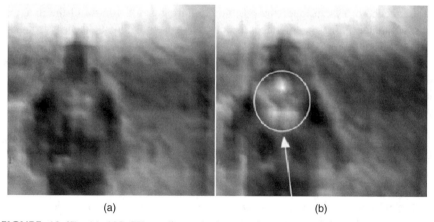

(a)   (b)

FIGURE 10.47 (a) 100-GHz radiometric image of a person without weapon and (b) 100-GHz radiometric image of a person with concealed weapon.

seconds integration time, 15 dB noise figure and 43 dB gain. Power Supply is $+12$ VDC and the radiometer sensitivity is $\Delta T \approx 0, 1K$

Figure 10.46 shows principle scheme of the radiometric imaging system and Fig. 10.47 exhibits results of the experiments for concealed weapon imaging using 3 mm wavelengths radiation. Obtained images demonstrate pretty satisfied sensitivity and resolution to distinguish the armed and non armed persons. As one can see in Fig. 10.47b, the image of the concealed weapon corresponds to white spots on the chest area of the staying person.

Figure 10.46(a) Scheme of 100 GHz imaging system with linear scanner and flopping mirror (upper), (b) Radiometer for 100 GHz imaging with lens and flopping mirror (lower).

Figure 10.47(a) 100 GHz radiometric image of person without weapon (left), (b) 100 GHz radiometric image of person with concealed weapon (right).

10.3.3   Conclusions

In this section we have demonstrated the fundamentals of microwave radiometry, considering its main problems in the field of real design of passive millimeter wave imaging system. The technical details, theoretical evaluation, experimental results, and promising ideas have been presented. The results demonstrated in this chapter have been obtained by two research groups in Ukraine (State Research Center "Iceberg") and in Turkey (TUBITAK MRC International Laboratory for High Technologies).

Acknowledgments

Some results in Section 10.2 were obtained with the assistance of I. V Voynovskyy and V. N. Stepanyuk.

REFERENCES

1. Chommeloux, L., Pichot, C., Bolomey, J. 1986. Electromagnetic modeling for microwave imaging of cylindrical buried inhomogeneities. *IEEE Trans. Microwave Theory Tech.*, 34(10):1064–1076.

2. Deming, R. W., Devaney, A. J. 1997. Diffraction tomography for multi-monostatic ground penetrating radar imaging. *Inverse Probl.*, 13(1):29–45.

3. Fear, E. C., Hagness, S. C., Meaney, P. M., Okoniewski, M., Stuchly, M. A. 2002. Enhancing breast tumor detection with near-field imaging. *IEEE Microwave Mag.*, 3(1):48–56.

4. Franklin, J. N. 1968. *Matrix Theory*. Prentice-Hall, Englewood Cliffs, NJ.   Richmond, J. H. 1965. Scattering by a dielectric cylinder of arbitrary cross-section shape. *IEEE Trans. Antennas Propag.*, 13(3):334–341.

5. Salman, A. O., Gavrilov, S. P., Vertiy, A. A., 2005. Microwave imaging of immersed bodies: an experimenal survey. Electromagnetics, 25(6):567–585.

6. Tihonov, A. N., Arsenin, V. Ya. 1979. *Methods of Illconditioned Problems Solution* [in Russian] Mir, Moscow.

7. Vertiy, A. A., Gavrilov, S. P., Aksoy, S., Voynovskyy, I. V., Kudelya, A. M., Stepanyuk, V. N. 2001. Reconstruction of microwave images of the subsurface objects by diffraction tomography and stepped-frequency radar methods. *Zarubejnaya Radioelektronika. Uspehi Sovremennoy Radioelektroniki* 7:17–52.

8. Vertiy, A. A., Gavrilov, S. P., Voynovskyy, I. V., Stepanyuk, V. N., Ozbek, S. 2002. The millimeter wave tomography application for the subsurface imaging. *Int. J. Infrared Millimeter Waves*, 23(10):1413–1444.

9. Vertiy, A. A., Gavrilov, S. P. 2005. Subsurface tomography application for through-wall imaging. In *Proceedings of the 9th International Conference on Electromagnetics in Advanced Applications (ICEAA-05) and 11th European Electromagnetic Structures Conference (EESC-05)*. Torino, Italy, Sept. 12–16.

10. Wiskin, J. W., Borup, D. T., Johnson, S. A. 1997. Inverse scattering from arbitrary two-dimensional objects in stratified environments via a Green's operator. *J. Acoust. Soci. Am.*, 102(2):853–864.

11. Gavrilov, S. P., Vertiy, A. A. 2007. Imaging of layer inhomogeneity in stratified environment via a tomographic reconstruction. *J. Electromagn.*, 256:473–494.

12. Ferris, D. D., Jr., Currie, N. C. 1998. A survey of current technologies for through-the-wall surveillance (TWS). *Proc. SPIE*, 3577:62–72.

13. Benjamin, R., et al. 2002. Through-wall imaging using real-aperture radar. In *Proceedings of the URSI General Assembly*, Aug.

14. Nag, S., Fluhler, H., Barnes, M. 2001. Preliminary interferometric images of moving targets obtained using a time-modulated ultra-wide band through-wall penetration radar. In *Proceedings of the 2001 IEEE Radar Conference*, pp. 64–69.

15. Noon, D. A. 1996. Stepped-Frequency Radar Design and Signal Processing Enhances Ground Penetrating Radar Performance. Ph.D. dissertation, University of Queensland.

16. Reinwater, J. H. 1978. Radiometers: electronic eyes that "see" noise. *Microwaves*, Sep., pp. 58–62.

17. Schuchardt, M., Newton, J. M., Morton, T. P., Galliano, J. A. 1981. The coming of mm-wave forward looking imaging radiometers. *Microwave J.*, June, pp. 45–62.

18. Paradish, F. J., Habbe, J. M. 1982. Millimeter wave radiometric imaging. *Proc. SPIE*, 337:170–181.

19. Appleby, R., Lettington, A. H. 1991. Passive millimeter wave imaging. *Electron. Commun. Eng. J.*, Feb., pp. 13–16.

20. Goldsmith, P. F., Hsieh, C. T. Huguenin, G. R., Kapitzky, J., Moore, E. L. 1993. Focal plane imaging systems for millimeter wavelength. *IEEE Trans. Microwave Theory Tech.*, 41(10).

21. Piechl, M., Suss, H., S. D., Greiner, M., Jirousek, M. 2004. Imaging technologies and applications in microwave radiometry. In *European Radar Conference*, Amsterdam, pp. 269–273.

22. Yujiri, L., Shoucri, M., Moffa, P. 2003. Passive millimeter-wave imaging. *IEEE Microwave Maga.*, Sept., pp. 39–50.

23. Esepkina, N., Pariisky, Y., Korol'kov, V. *Radioteleskopy i Radiometry* [Radiotelescopes and Radiometry]. Nauka Press, Moscow.

24. Skou, N. 1989. *Microwave Padiometer Systems: Design and Analysis*. Artech House, Norwood, MA.

25. Corrado, T. 1988. Smart munitions: Are they affordable? *MSN*, Dec., pp. 32–36.

26. Gorishnyak, V. P., Denisov, A. G., Kuzmin, S. E., Radzikhovsky, V. N., Shevchuk, B. M. 2002. Radiometer imaging system for the concealed weapon detection. In *Proceedings of CriMiCo'2002*, Sevastopol, Crimea, Ukraine, Sept. 9–13.

27. Passive millimeter wave imaging technology. *Proc. SPIE*, 3064, 3378, 3703, 4032, 4373, 5077, 5619, 5789, 6211, 6548.

28. Radzikhovsky, V. N., Gorishnyak, V. P., Kuzmin, S. E., Shevchuk, B. M. 2001. 16-channels millimeter-waves radiometric imaging system. In *MSMW'2001 Symposium Proceedings*, Kharkov, Ukraine, June 4–9, pp. 466–468.

29. Radzikhovsky, V. N., Gorishnyak, V. P. Kuzmin, S. E., Shevchuk, B. M. 2001. Passive millimeter-wave imaging system. In *Proceeding of CriMiCo'2001*, Sevastopol, Crimea, Ukraine, Sept. 10–14, pp. 263–264.

30. Gorishnyak, V., Denisov, A., Kuzmin, S., Radzikhovsky, V., Shevchuk, B. 2004. 8mm passive imaging system with 32 channels. Presented at EuroRAD 2004, Amsterdam, The Netherlands, Oct. 2–14.

31. http://www.brijot.com, http://www.sago.com, http://www.millivision.com, http://www.thruvision.com, http://www.xytrans.com.

32. Huguenin, G. R. 2006. The detection of hazards and screening for concealed weapons with passive millimeter wave imaging concealed threat detectors. Press release. http://www.millivision.com.

33. Reynolds, W. R., Hilgers, J. W., Schulz, T. J., Amphay, S. 1998. Super-resolved imaging sensors with field of view preservation. *Proc. SPIE*, 3378:134–147.

34. Lettington, A. H., Gleed, D. G. 1998, Image processing techniques for passive millimeter wave imaging. *Proc. SPIE*, 3378:161–175.

35. Terentiev, E., Pirogov, Y. A., Gladun, V. V., Ivanov, V. S., Terentiev, N. E. 2000, Additional enhancement of resolution in multi ray radio vision systems. *Proc. SPIE*, 4032:152–157.

36. Vertiy, A. A., Tekbas, M., Kizilhan, A., Panin, S., Ozbek, S., Sub-Terahertz Radiometric Imaging System for Concealed Weapon Detection. *PIERS 2010*, Cambridge, 5–8 July 2010, 348.

37. Cetinkaya, H., Tekbas, M., Kizilhan, A., Vertiy, A., 2010. Active microwave and millimeter-wave ISAR imagine and millimeter-wave passive radar receiver design. In *Proceedings of the Fourth World Congress-Aviation in the XXI-st Century Safety in Aviation and Space Technologies*, v.2 Kyiv, Ukraine, September 21–23, 22.97–22.100.

CHAPTER ELEVEN

Mine Detection

11.1 THE LANDMINE PROBLEM

Landmines have been used widely since World War II and have become an increasingly popular weapon. So far, more than 340 models of antipersonnel (AP) mines have been produced in 54 countries throughout the world [1,2]. Used initially as a defense weapon, landmines have been laid to limit the mobility of troops and military vehicles at or around battlefields. Later, however, mines became widely used in civil wars and ethnic conflicts as well as for border security. While military mine fields are laid in a certain order and are reasonably well documented, mines laid by militias are almost never marked or mapped, and when a conflict ends, the landmines are forgotten or deliberately left in fields, remaining active for decades. Landmines have been found to be an economically efficient weapon, as the cost of removing a single landmine is 100 to 300 times higher than the production and deployment cost.

The majority of antipersonnel mines are almost cylindrical in shape, with a diameter between 40 and 200 mm. The diameter of a typical modern AP mine varies between 55 and 100 mm, but roughly 50% of all mines laid have a diameter larger than 100 mm [3]. Typically, the mines are laid such that the cylinder's rotational axe is almost vertical. The height of the majority of antipersonnel mines varies between 30 and 80 mm. Typically having a cylindrical geometry, antitank mines have dimensions several times larger than those of antipersonnel mines. The operational burial depth of antipersonnel mines is less than 200 mm, whereas that of antitank mines is up to 150 cm. Regarding the explosive, it is estimated that from 80 to 90% of the total number of landmines now deployed contain TNT [4]. TNT is the most widely used explosive in all manufactured landmines [5].

Subsurface Sensing, First Edition. Edited by Ahmet S. Turk, A. Koksal Hocaoglu, and Alexey A. Vertiy.
© 2011 John Wiley & Sons, Inc. Published 2011 by John Wiley & Sons, Inc.

According to recent estimations, at least 60 million (some estimations say up to 200 million) undetected landmines are spread over countries on every continent. The countries most affected are Afghanistan, Iraq, Lebanon, Jordan, Cambodia, and Vietnam in Asia; Angola, Sudan, Mozambique, and Somalia in Africa; Bosnia and Croatia in Europe; El Salvador and Nicaragua in Central America; and Colombia in South America. The landmine problem is acute in 65 to 80 countries, mostly on the African continent. It is estimated that in Cambodia the ratio of population to the number of mines is 0.85 and that one of every 236 inhabitants has been maimed by such a device [6]. In Angola, there are 9 million landmines and one out of 470 inhabitants has suffered a mutilation, according to the International Committee of the Red Cross.

Widespread landmines strongly affect civil populations. Every year landmines kill and maim several hundred thousand civilians [7]. In 2007 the official number of victims was 5751 and the number of people injured was 473,000 [8]. In other terms, there is an injury every 70 seconds and a death every 90 minutes. However, these numbers, even if dramatically high, are underestimated, since many accidents are not reported. In addition to numerous casualties every year, the problem of nonusability of cultivation fields and infrastructure becomes major in countries with an agricultural-based economy. Typically, the discovery of two or three mines is a sufficient reason for the local population to abandon an agriculture field, or water resource or transportation infrastructure, causing an additional economical loss for the people who survived the conflict. Unremoved landmines also become a cause of displacement for affected populations and an obstacle to reconstruction.

After being recognized as a humanitarian problem, demining attracted notable attention from the United Nations (UN), national governments, and humanitarian funds. In 1980 the UN approved a convention on certain conventional weapons, annexed to the Geneva Conventions of 1949, which concerns the treatment of noncombatants and prisoners of war. It regulates the use in armed conflicts of certain conventional weapons that may be deemed to be excessively injurious or to have indiscriminate effects [9]. The convention has five protocols, one for each group of conventional weapons. Protocol II regulates the use of landmines. It prohibits the use of nondetectable AP landmines and their transfer; the use of non-self-destructing and non-self-deactivating mines outside fenced, monitored, and marked areas; the use of landmines that explode when detected, causing the injury or the death of the operator; the indiscriminate use of landmines; and calls for penal sanctions in case of violation.

Despite the good purposes, this convention failed to ban landmines, since every signatory country had the option to adopt a minimum of two of the five proto-cols, choosing those that better fit their political agenda. In 1997 the International Campaign to Ban Landmines [7], a coalition of nongovernmental organizations, launched a petition to ban the use, stockpiling, production, and transfer of antiper-sonnel landmines, and in 1999, the Convention on the Prohibition of the Use, Stock-piling, Production and Transfer or Anti-Personnel Mines and on Their Destruction (often referred as the Ottawa Treaty [10]), was signed by 135 counties and ratified by 84 states. All signatories who have ratified the treaty committed to report to the

Secretary-General of the UN on their progress in destroying their stockpiled mines by March, 1 2003 (Article 4) and destroying all mines in the territories under their jurisdiction by March, 1 2009 (Article 5). The provisions of the Ottawa Treaty, which has become international law, provide an impetus to the development and procurement of more efficient mine detection and disposal techniques. Civilian or humanitarian demining is aimed at restoring land to the population, and the goal is the complete removal of all landmines. It includes mine clearance (actual removal of landmines from the ground), as well as surveying, mapping, and marking of mine fields. Landmine removal thus becomes a topic of paramount importance for which large resources have to be applied under the umbrella of international cooperation. The main governments that fund humanitarian mine clearance agencies are the United States, the United Kingdom, Japan, and the Netherlands.

In the last 10 years, the states parties to the mine ban treaty has been augmented, thanks to the spread of public awareness of the landmine problem. The monitoring report of 2007 indicates the number of signatory countries to be 155, representing 80% of the world's nations [8]. Thirty-seven countries have not agreed to the ban, including China, India, Russia, and the United States. The Ottawa Convention has drastically reduced the number of new mines that are laid every year and the mines that are stockpiled by the signatory countries. However, there are still thousands of hectares of fields polluted by landmines that urgently need to be cleaned to be used for the economical revival of populations who suffered a war. With a current demining rate of approximately 100,000 mines per year, mine clearance is proving to be a much slower process than was thought in the beginning.

There are two principal types of mine clearance: military and humanitarian. Military demining aims at creating mine-free paths for troops that are leaving or moving into an area. This type of demining accepts certain casualties and uses fast methods. Humanitarian demining, on the other hand, aims at making land completely accessible and usable again for civilian activities. The UN statement of requirement defines the clearance criteria as follows [1]: "The area should be cleared of mines to a standard and depth.... The contractor must achieve at least 99.6% of the agreed standard of clearance. The target for all UN sponsored clearance programs is the removal of all mines to a depth of 130 mm." Demining time is not considered to be a major factor in humanitarian operations.

11.2 OVERVIEW OF DEMINING TECHNIQUES

Numerons techniques have been used to detect mines. At the moment, only four of them are used widely in humanitarian demining: ground-engaging machines, dogs, metal detectors, and prodders. For remediation of landmines spread over a large area, ground-engaging machines are used. These machines are intended to detonate mines via putting extra pressure on a ground surface and resisting mine explosions themselves. They are usually developed from machines intended to work in military minefield breaching, where complete clearance is not as important as speed. The main types of ground-engaging machines are flails, rollers, millers and tillers, sifters, and dozers and graders. Although time effective, these machines do not guarantee clearance standards as specified by the UN.

Dogs can be trained to smell explosive molecules in ground or in the air. The scent of explosive material permeates mine casings and rises up through the soil. The scent can stay either in the ground upper layer or on vegetation (bushes) around that area. Although dogs can locate newly laid mines, for mines laid long ago, dogs have difficulties with exact location. When a dog detects an explosive, the area is marked for detailed investigation by deminers. Regretfully, demining dogs can operate only a few hours per day, after which efficiency and reliability decline. Much research has been directed at replacing dogs with other animals (e.g., rats) [11] and at developing an electronic sensor to reproduce a dog's nose. In particular, the Apopo project in Mozambique trains rats to detect landmines. Rats are much faster than dogs to train, are active for much longer than dogs, and are lighter and thus can move freely without triggering the mines. However, the signals that rats produce during their work are vulnerable to misinterpretation; usually, only the personal trainer is able to interpret these signals.

Currently, the electromagnetic induction metal detector is the most popular device used by deminers to facilitate mine detection. Its origin as a mine detector dates to the time when mine cases were metal. Nowadays, the amount of metal in antipersonnel mines is typically quit low. Typically, an antipersonnel mine detonator contains several metal parts (the only known exceptions are the French Bakelite landmines placed in southern Lebanon in the late 1940s, which contained no metal at all). According to the GICHD [12], at present there are no known mines that contain any metal whatsoever. Minimum-metal mines, in which the only metal parts are related to the detonator, contain less than 5 g of metal (with a minimal weight of about 1 g) and are the most difficult to detect. Being the most sophisticated demining tool available until recently [12], the metal detector suffers from such problems as insufficient detection depth and a high false-alarm rate (false detection of subsurface inhomogeneities such as roots, rocks, water pockets, etc.) for antipersonnel mines with a low metal content. An example of metal-detector performance is given by the statistics of humanitarian demining in Cambodia between 1992 and 1998: only 0.3% of the 200 million items excavated by deminers were antipersonnel mines or unexploded ordinance (UXO) [13, p. xvi].

The most basic approach to detection of a single mine is prodding. Using prodders, rigid sticks of metal about 25 cm long, a deminer tests the soil at an angle of 30° (typically). Each time that he or she detects an unusual object, the deminer assesses the contour, which indicates whether the object is a mine. Although effective to a depth of 5 cm or less, this technique is slow and dangerous. The deminer might also encounter mines that have moved or have been placed so that they are triggered by prodding. Prodding remains a widely used procedure after initial detection of a suspicious object by a metal detector or a dog.

None of the methods described above satisfy the current needs of humanitarian demining. That is why considerable effort has been expended since the 1980s to develop high-tech sensors that can make demining faster, safer, and cheaper. The techniques investigated include:

- Ground-penetrating radar
- Electrooptical cameras
- Passive millimeter-wave sensors (radiometers)
- Nuclear quadruple resonance sensors
- Acoustic sensors
- X-ray backscatter
- Thermal and fast neutron technologies
- Chemical sensors
- Biosensors for explosive particle detection

All have demonstrated promising results, but many of them have a low maturity level for field application. Next we discuss briefly technological achievements to date using some of these sensors.

11.3 ADVANCED ELECTROMAGNETIC INDUCTION SENSOR

As discussed in Chapter 4, an electromagnetic induction (EMI) sensor uses low-frequency quasi-magnetostatic response characteristics to detect electrical inhomogeneities (first of all, metal) in the ground. When a primary magnetic field generated by a transmitter coil of a metal detector induces eddy currents in buried objects, they radiate a secondary magnetic field that is sensed by a receiver coil. If the receiver coil voltage induced exceeds some threshold, detection takes place. Range information for a detected object is not available. Due to the principle of operation, the sensitivity of a metal detector decreases sharply with an increase in range. Inductive metal detectors are used as landmine detectors due to the presence of metal in mines; EMI sensors are not capable of detecting explosives.

A soil compensation procedure is used widely in EMI sensors to support their operation under various soil and environmental conditions. However, the electromagnetic properties of certain soils may cause a serious loss of sensitivity and limit the performance of some metal detectors. In the majority of cases, a conventional metal detector has difficulty in discriminating between landmines and buried metallic clutter. Due to the high sensitivity required to detect minimum-metal-content landmines, the typical presence of a large number of small metal fragments, and the inability to discriminate between a landmine and metallic clutter, the false-alarm rate is high (as noted above, only 0.3% of the 200 million items excavated by deminers in Cambodia between 1992 and 1998 were antipersonnel mines or UXO [13]). To improve the discrimination of metal objects detected, some of EMI sensors used for landmine detection (e.g., pulse detectors) analyze the decay rate of the object's exponential response. The decay rate depends on the object's shape and constitutive parameters and thus can be used for landmine discrimination.

Another possibility for improving EMI's classification capabilities lies in results visualization. Some examples of continuous-wave EMI sensor data visualization are shown below. As discussed in Chapter 4, deconvolution of the sensor spatial response should be performed prior to visualization, and accurate sensor positioning is required. Detecting and recognizing a multi-ignition mine (e.g., a TMA-4 antitank mine with three detonators) can be done with high reliability, as shown in Fig. 11.1 (EMI data from the landmine test field in Benkovac, Croatia, uncooperative soil). The lateral distance between the three metallic objects with respect to the donators can be determined and matched with the original.

A blurred signature of captured two-dimensional metal detector data (e.g., in Fig. 11.1b, left) suggests a buried object in uncooperative soil. A "clear" signature, similar to a PSF, of a buried object indicates a single- or multitarget situation in cooperative soil; multitargets are observable in the raw two-dimensional data sets only if the lateral distance between the metallic objects is great enough in relation to the diameter of the searching coil. Any uncooperative soil generates blurred signatures, which is already observable in the raw two-dimensional data sets. With the next step, the deconvolution method for the reconstructed images $O(x, y)$, a single- or multitarget situation is detectable. For a single target, the pinpointing procedure extracts the optimal one-dimensional signature of a buried object and

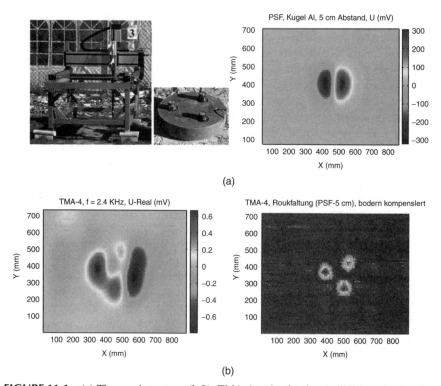

FIGURE 11.1 (a) Three-axis scanner (left), TMA-4 antitank mine (middle), and a two-frequency CW metal detector at the Benkovac test field in Croatia (right). (b) Raw data, real part, $f_1 = 2.4$ kHz (left) and the result of deconvolution using soil-compensated data; real part of the voltage, $f_1 = 2.4$ kHz (right).

a signal-processing method for object recognition (landmine or clutter), such as the use of level, phase-sensitive detectors or other pattern recognition methods, can be use.

For uncooperative soil, which is recognizable by the blurred signatures of the space-resolved two-dimensional raw data of the metal detector, the influences of the soil have to be compensated in the two-dimensional raw data sets with an adapted soil compensation procedure before the deconvolution method can be used. The deconvolution of soil-uncompensated two-dimensional data sets generates artifacts in the images which are not related to the original scene, as shown in Fig. 11.2. Furthermore, we cannot tell if the signature has been generated by a metallic object and the signature has been captured completely. Deconvolution of soil-compensated MD data sets provide better results (see Fig. 11.1b, right).

Limited discrimination capabilities and limited penetration depth are the most important weaknesses of EMI sensors for landmine detection. Furthermore, the performance of EMI sensors is seriously affected by the soil properties and elevation height of the sensor above the ground. Despite substantial progress in soil

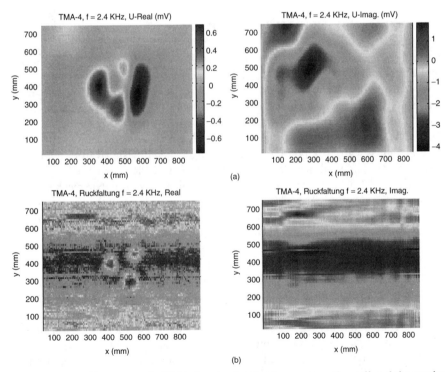

FIGURE 11.2 Signatures of a TMA-4 antitank mine in uncooperative soil and the results of deconvolution using two-dimensional raw data sets, no soil compensation of the data sets: (a) real and imaginary part of the voltage, $f_1 = 2.4$ kHz; (b) results of the deconvolution, real part (left) and imaginary part (right).

compensation techniques, substantial differences in the performance of different metal detectors at particular operational sites have been reported.

Due to its high degree of sensitivity to the presence of metal, EMI sensors were the first high-tech sensor to be deployed for landmine detection. Consequently, mine technology development went in the direction of metal content minimalization, resulting in the appearance of plastic mines. As a result, a stand-alone EMI sensor still is often used for close-in landmine detection—with a very high false-alarm rate, however. It is widely thought that substantial improvements of close-in detection could be achieved by combining an EMI sensor with a sensor capable of discriminating between metal debris (clutter) and landmines. Typically, a GPR sensor is considered to be such a secondary sensor, due to GPR sensor sensitivity to dielectric objects, deeper penetration capabilities, availability of depth information, and advanced classification capabilities.

EMI sensors are used in both handheld and vehicle-mounted mine-detection systems. In the latter, arrays of EMI sensors are used. Being moved together with a platform, such an array produces a two-dimensional map of detected metal fragments. Similar maps in handheld systems can be produced by adding either an ultrasonic or an optical reference system to an EMI sensor.

11.4   GROUND-PENETRATING RADAR

It has been demonstrated that ground-penetrating radar (GPR) is a useful sensor for landmine detection, especially for humanitarian demining [13]. Similar to an EMI sensor, GPR cannot detect an explosive; actually, GPR detects electrical (or magnetic, or both) inhomogeneities of the ground. These might be small metal parts in the trigger, explosive, water, and air within the mine case, and the case itself. The larger the volume of the inhomogeneity and the larger the electric (magnetic) contrast with the environment, the stronger the GPR signal from the object. Furthermore, field work has demonstrated that often, GPR does not detect a mine itself but a soil disturbance due to the mine, which may be either mechanical or hydrological in nature. The soil depth structure is usually disturbed mechanically by mine deployment, and a mine disturbs the vertical distribution of soil moisture. After a rain, soil moisture and, consequently, soil dielectric permittivity above the mine, are higher than those in the direct surroundings (the dielectric permittivity below the mine is also lower than that around the mine).

GPR has two important features, which makes it an important sensor for landmine detections. First, it is the only advanced sensor that provides a three-dimensional image of detected targets. Volumetric information is very important for discrimination between all types of clutter and landmines. Potentially, high-resolution images even allow for classification of landmines (Fig. 11.3). Furthermore, a three-dimensional image of the subsurface allows for accurate depth information of detected targets, which is especially essential for sensor fusion.

Second, as a GPR signal covers an ultrawide bandwidth (see Chapter 3), the scatter from a target signal can be used for target classification. The waveform of

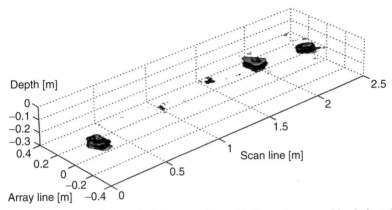

FIGURE 11.3 Focused image of a 2.5-m scan line with five antipersonnel landmines (from left to right: PMN2, M14, M14, PMN2, PMN2) buried at different depths under a rough sandy surface. Image is shown in a 15-dB dynamic range. (From [14].)

this signal depends on the waveform of the probing signal and the impulse response of a scatterer. The latter is determined by scatterer size, shape, and its contrast with the environment (Fig. 11.4). So the informational content of a GPR signal is much richer than that of an EMI sensor. However, extraction of target information from the GPR signal is not trivial and requires considerable computational effort.

For a particular object shape (i.e., cylindrical), a detailed analysis of the time-domain scattering transfer function and target response for different orientations of the cylinder has been carried out by Burganov et al. [15]. It has been shown that for an arbitrarily oriented cylinder, the time-domain scattering transfer function is described by four delta functions, whose mutual positions and amplitudes depends on geometry and orientation of the cylinder. Impulse responses for typical plastic antipersonnel landmines with low metal content buried in different orientations in dry sandy soil with a dielectric permittivity of 2.55 are given below.

Numerous field trials of different GPRs have proven that whereas for most ground types the radar can achieve a desirable detectability level, a decrease in the false-alarm rate remains the most important issue for practical applications. The classification of targets detected is a qualitatively new demand for conventional GPR, which has never been designed to measure accurately the waveform scattered from a target. Due to this demand a large number of dedicated GPR systems have been developed within recent decades. Yarovoy [16] provided an overview of basic approaches for landmine detection, major system designs, and signal processing.

The physical reason for the high false-alarm rate is GPR sensitivity to all (not only metal, as in the case of EMI) inhomogeneities of the ground [17]. The majority of them have a natural origin, which is different from the EMI situation. Furthermore, similar to all other electromagnetic sensors, mine response depends not only on the mine itself, but also on the environment. As a GPR signal originates from an electric (or magnetic, or both) contrast with an environment, the environment has an essential influence on mine response.

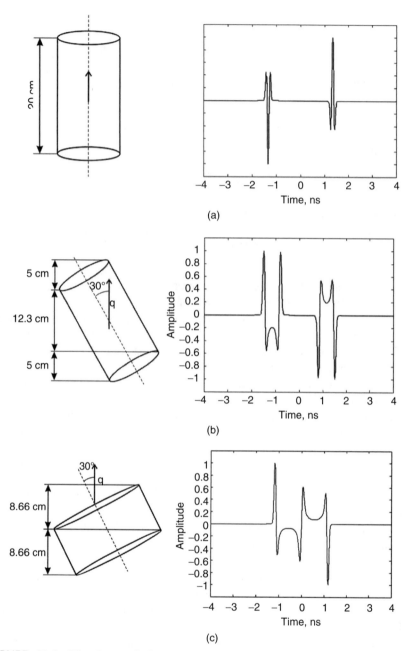

FIGURE 11.4 Waveforms of signals scattered from cylindrical objects with different orientations: (a) specular reflection from the bottom, $2L = 20$ cm, $R = 5$ cm, $\alpha = 0°$; (b) $2L = 20$ cm, $R = 5$ cm, $\alpha = 30°$; (c) $2L = 10$ cm, $R = 2L \cos \alpha = 8.66$ cm, $\alpha = 30°$; (d) $2L = 5$ cm, $R = 10$ cm, $\alpha = 30°$; (e) specular reflection from the element of a cylinder, $2L = 20$ cm, $R = 5$ cm, $\alpha = 90°$. (From [15].)

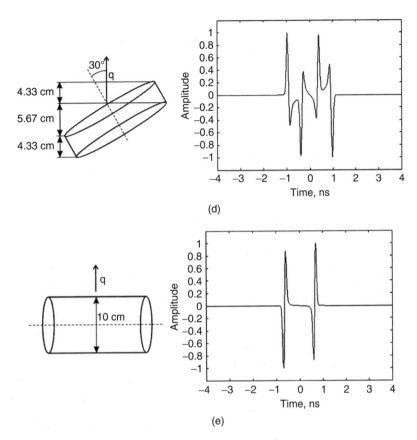

FIGURE 11.4   (*Continued*)

GPR sensors are used for both close-in and stand-off detection. In the former case, GPR is typically used together with EMI and electrooptical sensors, while in the latter case GPR often performs as a stand-alone sensor.

11.5   ELECTROOPTICAL SENSORS

Similar to EMI and GPR sensors, electrooptical (EO) sensors use an electromagnetic field to detect landmines. The similarity between these three sensors goes further: All three do not detect an explosive; they detect inhomogeneities of the ground. The principal possibility for detecting landmines using infrared radiation has been recognized since the 1950s [18]. There are three major physical phenomena that might play a role in the detection of surface-laid and buried mines using EO sensors: thermal contrast, contrast in (polarimetric) scattering properties, and a disturbance in vegetation/shallow subsurface properties (see Section 7.6, for a detailed explanation of EO sensors and their operations for subsurface imaging technologies).

The majority of antipersonnel and antitank mines have either a metal or a plastic case. Whereas the former material is a good thermal conductor (better than soil itself), the latter is a good thermal insulator. Thus, both types of mine cases disturb natural thermal energy transfer in the ground during a diurnal cycle: warming up the upper layer of soil in the morning hours and cooling down in the evening hours. Due to its thermal isolation properties, a buried plastic mine causes a local increase in soil temperature just above the mine in the morning and a decrease in the evening. This thermal contrast results in a variation of the radio-brightness temperature up to a few kelvin and can thus be detected by an EO sensor. As the temperature difference and its temporal behavior depend strongly on current environmental conditions (i.e., solar irradiance, soil moisture, air humidity, wind speed, soil properties, etc.) and their temporal development, the value of the thermal contrast cannot be used as a feature for landmine characterization or discrimination. Even more, the detectability of mines based on the thermal contrast at a given moment of time is not fully reproducible. Much better results can be achieved by diurnal observations and an analysis of the time sequence of EO images.

The smooth surfaces of human-made materials scatter impinging electromagnetic radiation (e.g., solar irradiance or artificial radiation) in a way different from those typical for soil surfaces and vegetation. While by unpolarized (e.g., sun or sky) illumination natural surface and vegetation scatter unpolarized electromagnetic waves, human-made (flat) surfaces produce polarized scattered waves if observed at low elevation angles. Similarly, the radio-emission from flat surfaces is polarized, and the polarimetric contrast reaches the maximal value near the Brewster angle. The total (apparent) radio-brightness temperature is affected both by its own radio-emission and by scattered waves, resulting in distinctly different polarimetric properties for natural (rough) and human-made (flat) surfaces. Polarimetric EO sensors can reliably detect (especially, surface-laid) mines based on these difference in polarimetric properties.

Finally, a mine disturbs the upward and downward flow of soil water. This causes an increase in soil moisture (and as a result, an increase in the soil dielectric permittivity and a decrease in the soil thermal emissivity) above the mine directly after a rain. The opposite effect takes place in the absence of rain. Differences in the thermal emissivity just above and farther away from a mine result in a spatial contrast of the radio-brightness temperature, which can be detected by an EO sensor. Furthermore, during a dry period, vegetation above the mine suffers from additional drought, due to lower soil moisture, and this can also be detected by EO sensor.

Mine contrasts as observed by EO sensors are highly dependent on environmental conditions (variations of sun irradiation, soil moisture, wind strength, etc. within a few previous hours and their values at the moment of observation). Being fundamental in nature, this dependence limits the application of EO sensors to some specific scenarios (surface-laid mines on roads or on bare soil, detection of triggering wires in vegetation, etc.) in which the detection and false-alarm rates are relatively stable. Furthermore, surface clutter (scattering of impinging electromagnetic radiation and Eigen thermal emission of soils on surface roughness) and thermal emission of foliage masks buried mines.

Due to their unreliable performance for detection of buried mines, EO sensors are thought to be used primarily for detection of surface-laid and flush-buried mines in bare soil conditions. As mine detection under such conditions can be done easily by a deminer through a visual inspection, EO sensors has found their place either in a multisensor vehicle-mounted systems or in airborne systems (mainly for the detection of mine fields). The large operational range [which varies typically from a few meters (vehicle-based systems) to a few kilometers (airborne surveys)] results in often selection of EO sensors for stand-off systems.

11.6   CHEMICAL SENSOR ARRAYS FOR MINE DETECTION

A number of studies (see, e.g., George et al. [19]) indicate that low (but significant) concentrations of TNT and other compounds found in the explosive charge of land-mines can slowly diffuse through most plastic landmine casings and through seams and seals in metal landmine casings. Because chemical signature compounds can escape from landmine casings, landmines represent a long-term source of vapors that, in theory, could be exploited as a means of locating landmines. Explosive vapor is used for landmine detection by dogs and other animals or insects and forms a basis for chemical sensors. In this section, work will be presented show-ing the application of chemical sensor-based detection technology to the detection of land mines. The principles and details of the underlying chemical sensors are described in Sections 2.7 and 7.2. Here, only work regarding testing on real sam-ples or the development of an application-specific device are included. Examples of sensor use on explosive detection in general were provided in Section 7.2.

Molecules of signature compounds can be transported through soil in the vapor phase, but the primary mode of transport is through the movement of soil water. As water moves through the soil, it very slowly disperses the explosive molecules through the soil. Water containing chemical signature molecules can evaporate readily when it reaches the ground surface. However, since the vapor pressure of the explosive is low, the rate of evaporation of these compounds is orders of magnitude less than that of water. As a result, the highest concentration of chemical signature compounds is found in the soil rather than in the air over the mine [20].

It has been observed that explosive charge is always accompanied by impuri-ties such as 2,4-dinitrotoluene (DNT), 2,6-DNT, 1,3-dinitrobenzene (DNB), and 2,4-DNB in military-grade TNT, and these impurities generate higher vapor con-centrations than does 2,4,6-TNT itself [21,22]. 2,4-DNT is an impurity in the range 9.32×10^{-5} to 7.43×10^{-4} g/g of military-grade TNT at 22°C. A headspace vapor constituent of 2,4-DNT was 6.9×10^{-11} to 1.9×10^{-9} g/g [21], which is signifi-cantly higher than 2,4,6-TNT. Similarly, 1,3-DNB is another impurity in the range 1.39×10^{-5} to 8.88×10^{-4} g/g of TNT at 22°C. A headspace vapor constituent of 1,3-DNB was 2.2×10^{-11} to 4.3×10^{-9} g/g. This is because the vapor pressure of 2,4-DNT and 1,3-DNB are approximately 40 to 1000 times greater than the vapor pressure of TNT. Recently, it was found that explosive detection techniques are very successful in recking out 2,4-DNT and 2,3-DNB vapors as a signature of TNT-based explosives [23].

An additional problem that is faced by chemical sensors is the localization of an explosive. The location at which explosive vapors are present at the highest concentration is often displaced from the mine location. Plume tracking to localize mines creates another burden in development. Consequently, fast screening of large areas, identification of objects found using other techniques, or verification of findings are seen as more suitable application scenarios for sensor systems.

Practical use of chemical sensors has begun, and a few successful applications are described next. Nomadics Inc. has developed a sensing technology, Fido, which was developed under the Defense Advanced Research Projects Agency's (DARPA) Dog's Nose Program, which detects the chemical signature of explosives emanating from buried landmines [24] using amplifying fluorescent polymer materials developed by the Massachusetts Institute of Technology. These materials enable detection of ultratrace concentrations of nitroaromatic compounds such as TNT, the most commonly utilized explosive in the production of landmines. When vapors of nitroaromatics are presented to the sensor, the fluorescent polymers emit light at a greatly reduced intensity, a property that enables rapid detection of trace quantities of explosives using relatively low-cost electronics and optics.

Cumming et al. [25] describe use of the same sensor system. The manually portable sensor prototype, similar in size and configuration to metal detectors currently in use for mine detection, has demonstrated performance comparable to that of canines during field tests monitored by DARPA at Fort Leonard Wood, Missouri. The same polymers were also used by White et al. [26], who have developed a portable artificial olfactory system, the electrooptical vapor interrogation device (EVID), for rapid detection and identification of volatile chemicals in the environment (>3 s). Brief air samples are drawn over an array of optically interrogated, cross-reactive chemical sensors. Biologically based pattern-matching algorithms automatically identify odors as one of several to which the device has been trained. In discrimination tests, after training to one concentration of six odors, the device gave 95% correct identification when tested at the original plus three different concentrations. Thus, as required in real-world applications, the device can identify odors at multiple concentrations without explicit training on each. In sensitivity tests, the device showed 100% detection and no false alarms for the landmine-related compound DNT at concentrations as low as 500 ppt (quantified by gas chromatography/mass spectrometry). To investigate landmine detection capabilities, field studies were conducted at Ft. Leonard Wood. In calibration tests, signals from buried PMA1A antipersonnel landmines were clearly discriminated from background. In a limited nine-site "blind" test, PMA1A detection was 100%, with false alarms of 40%. Although requiring further development, these data indicate that a device with appropriate sensors, exploiting olfactory principles, can detect and discriminate low-concentration vapor signatures, including those of buried landmines.

Despite substantial progress, at present the sensitivity of such detectors still lags some orders of magnitude behind desired low vapor concentration levels, down to 1×10^{-18}. New hopes for the future are associated with nanosensors with increased sensitivity and selectivity, and the ability to operate in a multimodal platform offers a potential paradigm for deploying a large number of sensors for detection. It has been reported that nanosensors have potential as highly sensitive and very selective signal transduction platforms for an integrated explosive sensor system [27].

11.7 SENSOR FUSION

As discussed above, at the moment there is no single sensor that can do the work. However, a combination of different sensors seems to provide reliable detection with a low false-alarm rate. The majority of multisensor systems developed so far combine a GPR sensor with a metal detector. In vehicle-based systems, an EO sensor is typically added [13]. These three sensors form the "detection" kernel of the system. In addition, some systems include "confirmation" sensors (e.g., a thermal neutron detector), which are used only for final classification of suspicious detected objects.

As of 2009, three multisensor handheld systems were operational: AN/PSS-14 (formerly HSTAMIDS) [28], Minehound [29], and ALIS [30]. All these systems include an EMI sensor as a primary detection sensor and a GPR sensor as a secondary sensor (i.e., for identification, classification, or confirmation). Numerous field trials have demonstrated a number of advantages of these systems over conventional metal detectors. In the next section, the operational and performance results of such dual-sensor systems are discussed.

In field trials of handheld mine detectors it has been demonstrated that a GPR sensor improves the overall performance of the entire detection system. Statistics of the operational use of AN/PSS-14 from April 2006 until March 2007, during which 22 mine fields with a total area of 238,365 m^2 were cleared, shows a 7- to 17-fold operational improvement compared with a metal detector [31]. Improved performance results in an increase in the area scanned per day from 25 m^2, which corresponds to the typical performance of a conventional detector, to 275 m^2. Operational testing of Minehound in 2005 has demonstrated a reduction in false alarms by a factor of better than $5:1$ in Cambodia and by a factor of better than $7:1$ in Bosnia and Angola [32]. A similar performance of the ALIS has been reported by Sato and Takahashi [33].

The majority of vehicle-based multisensor systems are aiming on scanning a 3- to 4-m-wide swath during mechanical movement of the carrier. As a result, these systems are equipped with an array of EMI and GPR sensors. In all systems known to the authors, GPR is a primary down-/forward-looking sensor. Some of these systems (e.g., GSTAMIDS, Minder) use their GPR essentially for detection of antitank, not antipersonnel, mines. Within the sensor suite (which includes EMI, GPR, and EO sensors), until recently GPR was the slowest sensor, typically limiting operational speed to a few kilometers per hour. Recently, application of digital beamforming together with SIMO antenna array topology has resulted in the removal of this limitation. An experimental GPR developed at the Delft University of Technology has demonstrated reliable detection of buried antipersonnel mines at a vehicle speed of 36 km/h (having a theoretical limit of about 150 km/h) [34]. An example of detectability maps produced by this system is shown in Fig. 11.5.

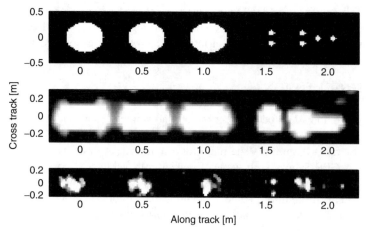

FIGURE 11.5 Scenario with buried antitank and antipersonnel mines (top); confidence map of a metal detector (middle) and radar confidence map (bottom). (From [35].)

To combine the output of different sensors in a joint detectability map, sensor fusion is employed. It is a process in which information from different sensors is used for a unified declaration of objects as detected by these sensors. Sensor fusion can be performed at three different levels: the data, feature, and decision levels [36]. In the majority of multisensor systems developed so far, data acquired by different sensors are fused at a decision level [13]. This is the simplest type of sensor fusion. In handheld systems such fusion is actually done by an operator. Fusion at lower levels (feature fusion or data fusion) requires considerably more computational power and an in-depth understanding of the performance of each sensor and access to the preprocessed data of these sensors. Despite evident difficulties with its realization, feature-level fusion can provide much better results in terms of detectability and false-alarm rate.

11.8 ALIS: A HANDHELD MULTISENSOR SYSTEM FOR LANDMINE DETECTION

The Advanced Landmine Imaging System (ALIS) has been under development by M. Sato at Tohoku University in Japan since 2002. In this section, the technical features and performance of ALIS are discussed to exemplify dual-sensor setups. Among other handheld dual sensors, ALIS is unique in its novel ability to track the sensor position, even though it is scanned by hand by deminers. Furthermore, provision of visual output from both sensors in ALIS helps us to understand the subsurface conditions much better than does a conventional audio signal.

11.8.1 ALIS System Design

Being a handheld dual-sensor system, ALIS includes a GPR sensor and an EMI sensor. Both sensors use the same sensor head and a sensor tracking system.

(a)   (b)

FIGURE 11.6   Different ALIS systems: (a) VNA-based ALIS, which is operated by two deminers; (b) VNA-based ALIS for a single operator (the deminer can observe the sensor scanned trace on the palmtop PC display in real time).

Information from both sensors and the tracking system is processed to produce two images (one from EMI, another from GPR), which are shown to a deminer. Based on these images, the deminer decides on detection and classification of objects detected. The original ALIS system configuration is shown in Fig. 11.6a. Two PCs are used: one palmtop PC is used for monitoring and one notebook PC is used for data acquiring and signal processing. According to radar designers, use of two PCs still has a great advantage for quality control of the operation. While the system can be operated by one deminer, the second operator can monitor the entire procedure of the operation via wireless LAN. All the instruments in this system can run on a rechargeable battery.

In 2007, a modification of ALIS to be used by a single deminer was developed (Fig. 11.6b). A VNA-based GPR unit, a metal detector controller, and a rechargeable battery unit are equipped in the backpack. The weight of the backpack is about 3 kg (the control unit of the pulse-based ALIS is lighter and smaller than that of the VNA-based ALIS). Both systems use the common sensor head and data acquisition and signal-processing software. Both systems of ALIS use only one palmtop PC, which is fixed to a flexible arm connected to the control unit. Data acquisition and processing can be done by this PC, and the operator can observe the trace of the metal detector response superimposed on the CCD-captured ground surface image on the PC display, which is the same as that used in the original ALIS.

ALIS uses two different radio-frequency technologies for GPR: stepped-frequency radar and impulse radar. The VNA (vector network analyzer) used to realize stepped-frequency radar was developed by Tohoku University with support from the Japanese Science and Technology Agency (JST). It is small, approximately $30 \times 20 \times 8$ cm, and light weight (less than 1.7 kg), but it has almost the same performance as the conventional commercial VNA, especially with regard to sweep speed and measurement accuracy. The VNA is a combination of a synthesizer and a synchronized receiver. It is controlled by a CPU and can store the measured data in its memory. The operational frequency of the GPR system can be adjusted depending on the soil condition by using a VNA, which is not easy for an impulse radar system. The calibration data can be stored in the memory of the VNA, and the output data can be calibrated by using these stored data. This calibration function is useful for better antenna impedance matching and can improve the radar data quality, because it suppresses the reflection from the antenna.

An alternative type of ALIS, the ALIS-PG, is operated by using an impulse GPR system. This impulse GPR system, also developed in the JST project, can generate a short pulse of approximately 200 ps, which covers the frequency ranging from dc to a few gigahertz. Compared with the VNA system, the impulse duration is fixed, and the operational frequency cannot be changed for dependence on the soil condition. The important advantages of using an impulse GPR system are its light weight and fast data acquisition. The approach is to operate the ALIS-PG under normal conditions, since the impulse GPR system is easier to operate; the ALIS–VNA is used under very wet soil conditions.

The MIL-D1 metal detector (CEIA, Italy) is used as an EMI sensor. The output data are digitized and used for two-dimensional imaging using positioning information calculated from a CCD picture. During the operation, the sensor operator can observe the metal detector response image together with a picture of the ground surface displayed on the palmtop PC in real time. Thus, the area, which shows a high metal detector response, can be scanned thoroughly.

The sensor head includes coils of the metal detector and GPR antennas. In prototypes of ALIS, both a Vivaldi antenna and a cavity-back spiral antenna have been used. In the latest ALIS version, cavity-back spiral antennas are molded with a metal detector sensor. The cavity spiral antenna is suitable for most normal operations of ALIS, but the Vivaldi antenna provides better performance, due to its wider frequency operation bandwidth. Therefore, the Vivaldi antenna is used in the vehicle-mounted ALIS. The weight of the sensor head is 2 kg.

The most unique feature of ALIS is its sensor tracking function, which is needed for imaging. ALIS uses a CCD camera fixed on the handle of the metal for sensor location tracking. The CCD camera captures several images of the ground surface per second; from this the relative movement on the ground surface is calculated, and the sensor position can be tracked. Figure 11.7 shows an example of the tracked sensor position acquired. The dots indicate the positions where ALIS acquired the data, including GPR, metal detector, and sensor position data. Figure 11.8 is an example of a metal detector signal image superimposed on the CCD-captured ground surface image, which the ALIS operator observes during hand scanning. This image is displayed on the PC screen, which the deminers hold in the hand, and the deminer can monitor in real time.

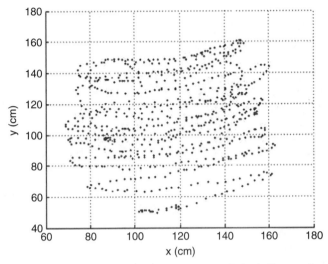

FIGURE 11.7   Trajectory of a sensor-tracked position. Each dot indicates a location of data acquisition. About a 1 × 1 m area is scanned in one data acquisition.

FIGURE 11.8   Visualized image of the signal response of a metal detector superimposed on a ground surface image.

This sensor tracking function has significant advantages:

1. The handheld scanning operation can be visualized, which improves the reliability of detection by a deminer.
2. A deminer can monitor the locus of scanning and can avoid the scanning blank area.

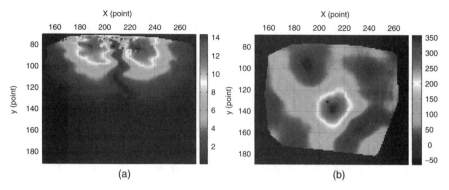

FIGURE 11.9 ALIS output image acquired at a CDS test site (Afghanistan, 2004).

3. The record of the locus of scanning by the deminer can be recorded, can be monitored in real time, and can be checked afterward. This record can be used for quality control of the demining. In addition, it can be used for training of deminers and can also be used to determine the cause of the mistake in the case of accident.

11.8.2 Data Processing and Display

The GPR data acquired with the sensor position information is processed after scanning the ALIS sensor over an area of about 1×1 m. At first, the entire data set acquired will be relocated on regular grid points. An Interpolation algorithm is used for this process. After relocation of the data sets, a metal detector signal can be displayed directly in a horizontal image, as shown in Fig. 11.9a.

A three-dimensional GPR image is reconstructed by a diffraction stacking algorithm (see Sections 3.5, 3.6, and 3.8 for details). The migrated GPR data gives a reconstructed three-dimensional subsurface image. However, in ALIS normally, only a horizontal slice of the three-dimensional image (Fig. 11.9b) is used for data interpretation. Experience from many trials shows that although a three-dimensional image has too high a clutter level, detection of buried landmine images in a horizontal slice remains reliable.

11.8.3 Evaluation Tests of ALIS

After laboratory tests, ALIS performance has been evaluated in several different locations. The first field trial test was carried out in Kabul, Afghanistan in December 2004. A field test was conducted at two locations. The first site (the CDS site) was a controlled flat test site, prepared for the evaluation of landmine sensors. The second site (Bibi Mahro Hill) is a small hill inside Kabul, which is a real landmine field where demining operations were being carried out. Afghanistan has relatively dry soil; even though we had frequent showers during the test, the soil moisture was about 10%. Under this soil condition, it has been found that ALIS can detect PMN-2 and Type 72 landmines buried at a 20-cm depth [37].

Then in April 2005, ALIS was demonstrated at JRC in Italy and at SWEDEC in Sweden with the support of ITEP. In May 2005, new tests were carried out in Egypt, where most of the landmines are buried in dry sand. We found that the condition in Egypt is suitable for the operation of GPR, but due to the extremely large area for landmine detection, we believed that unmanned vehicle-based ALIS should be used. In October 2005 ALIS was tested at the TNO test facility in the Hague, the Netherlands. This was a collaborative work between TNO and Tohoku University. Whereas, typically, ALIS acquires data within a 1 × 1 m square area by hand scanning, at the TNO test site an ALIS sensor was mounted on a mechanical antenna positioner to evaluate the characteristics of the sensor by without having to consider the operational skill of the operators. However, due to differences in the scanning method, signal-processing methods have also been modified.

Under the joint research leadership of JST and CROMAC, an evaluation of SARGPR in Croatia was carried out in February 2006. ITEP also supported this field trial test, and the detection results were evaluated by ITEP. Several test lanes having different soil properties were prepared, and three robotic machines equipped with a dual sensor and one handheld sensor were evaluated. Figure 11.10 shows one example of ALIS output in Croatia. The soil was wet; strong rain fell occasionally during the test.

Next, a field evaluation test was conducted in Cambodia from October to December 2006. This test was supported by the Ministry of Foreign Affairs of Japan as part of ODA to Cambodia. CMAC (the Cambodia Mine Action Center) conducted the test. A Tohoku University team has trained local deminers in ALIS operation. Afterward these deminers carried out a two-month blind test. No problems in the operation of ALIS have been reported, and it was confirmed that ALIS can be found acceptable by local deminers.

A systematic evaluation of ALIS was conducted in September–October 2007 in Croatia. This test was originally planned as an ITEP dual-sensor test, but due to the cancellation of other sensors, only ALIS was evaluated in this test. Therefore, it was not an ITEP test, but ITEP sent observers. The test was sponsored by JST (Japan Science and Technology Agency) and conducted by CROMAC (Croatian Mine Action Center)–CTDT, and the test lanes were designed by BAM. We used the

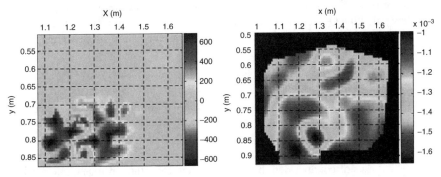

FIGURE 11.10 ALIS visualization output from a Cromac test site, Croatia. Lane 3, area 40.

ALIS-PG in this test. The ALIS team had trained Croatian deminers for two weeks in the operation of the ALIS-PG. Training included a tutorial on the fundamental principles of sensors, and on signal acquisition, processing, and interpretation. Then the ALIS team conducted training operations in calibration lanes. Two weeks of training was thought to be sufficient; however, longer experience in operating ALIS improves operator skills.

After the evaluation test carried out at the CROMAC-CTDT test site, we agreed to start evaluation tests with CROMAC-CTDT of ALIS-PG in mine fields in Croatia. In this test, ALIS-PG was tested in quality control (QC) operation. CROMAC is a demining organization within the Croatian government, but the actual demining activity is consigned to demining specialized companies. CROMAC's main operation is to assign demining areas to demining companies to execute the demining activities efficiently. More than 90% of demining work in Crotalaria is conducted by demining companies, and metal detectors have been used. During the demining operations, all detected metallic objects must, in principle, be removed. Demining companies must submit a daily log to CROMAC and CROMAC supervises the demining activity and conducts a final QC check after completion of the demining work. The supervision and QC check are conducted by a quality assurance officer. The final QC check is carried out directly by CROMAC-CTDT at the area where the demining work has been done and based on the location where mines or unexploded ordnance (UXO) had been detected and reported. There is a careful search for mines in selective areas where they believe there to be a high concentration of mines, therefore a great possibility of detecting buried mines that were missed during the demining work.

In the normal QC procedure in Croatia, QC is directed toward all demining methods: for example, machine demining, metal detectors, and demining dogs.

- The QC rules can be confirmed at the CTDT's Web site.
- QC is conducted at about 1 to 3% of the area where demining work had been completed.
- Currently, most detection is done by metal detectors, and mines are dug out one by one by hand.
- If two or more metallic objects larger than 3 cm^2 are found in 1 m^2, or if a piece or particle from a mine is found, it means that the demining company has failed the QC test and they must repeat the demining work in the assigned area at their own cost. Normally, the demining work in a single assigned area takes about a week.
- In the year under study, 10 cases failed QC testing. Parts of a mine or larger metal objects were found.
- CTDT's initial objective was to reduce the QC time by using ALIS.
- If ALIS can be used to estimate the size of a metallic object, it can eliminate the work involved in digging out the object.

CTDT does not conduct demining work. Therefore, their primary interest is in QC and not in adopting ALIS as primary demining sensor or equipment. However, they seem to be interested in instructing demining companies to evaluate the ALIS and to sell the ALIS to them. During December 2007 and April 2008, in more than 15 locations in Croatia, ALIS was used and evaluated in QC operation. We show some examples of these evaluation tests. The area was a mildly hilly terrain a few hundred meters away from a village with grass and trees. It is an area where sheep are pastured but is specified as an area suspected to be a dangerous zone. During 2 hours of operation on December 19, a number of metal objects (shown in Fig. 11.11) were detected. Each operator worked an area of about 5 × 10 m. The area where machine demining was conducted had no grass. Therefore, it was easier to use ALIS compared to manual search. The soil is plowed by the machine but that doesn't seem to affect the performance of ALIS.

As a QC procedure after machine demining, the soil is plowed and many small rocks appear on the surface. Figure 11.12 shows a buried object that was detected by ALIS in this site. It is a stone, and a piece of metal was located close to the stone. Figure 11.12a shows this composite object, and the ALIS images are shown in Fig. 11.12b (metal detector) and Fig. 11.12c (GPR). Therefore, the deminer has judged it to be a possible landmine.

On March 11, 2008, QC testing was carried out near Karlovac, which is located about 50 km south of Zagreb. The demining area is a former Yugoslavian army facility. The clearance area was very narrow near a fence; therefore, only manual demining was carried out. One AP mine case was found in this area, and many buried metal wires were detected.

In March 2009, the Cambodian Mine Action Center (CMAC) agreed to test ALIS as a primary sensor in real mine fields. In May 2009, the trial test was begun (Figs. 11.13 to 11.17). The minefield area had been a battleground between Vietnamese volunteer troops and Khmer Rouge troops during the period 1980–1986, then a

FIGURE 11.11   Metal fragments detected by ALIS.

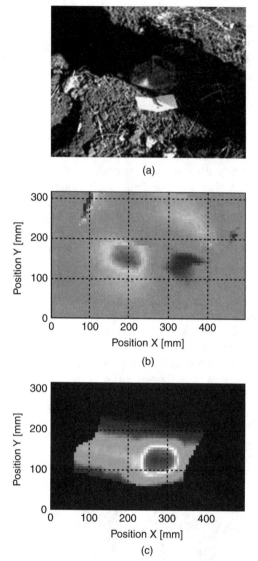

(a)

(b)

(c)

FIGURE 11.12 Stone with a piece of metal detected by ALIS: (a) photograph of the objects; (b) MD image; (c) GPR image.

battlefield between government troops and guerrilla during the period 1986–1990. Mines ware laid by Vietnamese volunteer troops and government troops to protect military camps and defensive lines. Two villagers had been injured; in 1981 one villager lost a leg by stepping on a landmine when walking across the minefield; and in 1984 another person was injured slightly. Many animals (e.g., cows) were killed by mines during 1980–1988. After the clearance, this minefield will be used by villagers as a rice field. During a 14-day operation in the minefield, the CMAC

FIGURE 11.13   CMAC ALIS team creating safe access between minefield boxes.

FIGURE 11.14   ALIS operation on a mine lane 25 × 0.5 m.

FIGURE 11.15   PMN2 mine under R.95 at a depth of about 15 cm (May 28 by PG-2).

FIGURE 11.16 PMN2 mine under R.13 at a depth of about 25 cm (May 28 by PG-1).

FIGURE 11.17 PMN2 mine laid under the root of a tree at a depth of about 15 cm, judged incorrectly to be metallic (May 20 by PG-2).

ALIS team found six mines of types PMN2 and POMZ2; five of these mines were identified correctly by ALIS operators. One UXO mortar armed 60-mm fuse was found by an ALIS operator and was judged to be a landmine.

11.8.4 Conclusions and Future Activities

Tohoku University has developed a handheld dual-sensor system, ALIS, which has high efficiency with better reliability for landmine detection by MD-GPR sensor fusion. The ALIS can visualize the signal, although it is a handheld sensor. Commercialization of the ALIS system is planned, and the ALIS was expected to be available commercially by the end of 2009. ALIS is now under evaluation by

CMAC in Cambodia and was scheduled to be tested in two international evaluation tests, Defuse and ITEP, in Germany in 2009.

11.9   CONCLUSIONS

Over the last two decades, considerable research and development effort has been undertaken to develop secure, reliable, and efficient high-tech means of landmine detection. A number of new technologies (e.g., chemical sensors) have been developed, and some known technologies (such as GPR and electrooptical sensors) have been improved considerably to become promising technologies for landmine detection. Under different environmental conditions, the EMI and GPR sensors demonstrated high detection rates for antitank and shallow-buried antipersonnel mines. Significant improvement in the classification of objects detected, and a drastic reduction in the false-alarm rate during demining operations have been reported using EMI and GPR sensors.

After more than a decade of intensive R&D work, three handheld systems with EMI and GPR sensors (AN/PSS-14, Minehound, and ALIS) came into production. This gives hope that operational deminers worldwide will finally receive tools that can make a real difference in the field. Further improvements in multisensor system performance are essentially related to the increased number of sensors available, the improvement in sensor fusion techniques, and adjustments in deminer training procedures.

REFERENCES

1. United Nations Information Center. http://www.cinu.org.mx/temas/asun_hum/minas. htm.
2. Vines, A., Thompson, H. 1999. *Beyond the Mine Ban: Eradicating a Lethal Legacy.* Research Institute for the Study of Conflict and Terrorism, London.
3. MineFacts. 1995. CD-ROM, v.1.2. U.S. Department of Defense, Washington, DC.
4. Patel, D. L. 1996. Mine explosive compounds. In *Proceedings of the Biological and Trace Gas Sensor Workshop*, Ft. Belvoir, VA.
5. King, C., King, J. 1999–2000. In *Jane's Mines and Mine Clearance*, 4th ed., C. King, Ed. Jane's Information Group, Coulsdon, UK.
6. The Cambodian Land Mine Museum. http://www.cambodialandminemuseum.org.
7. International Campaign to Ban Landmines. http://www.icbl.org.
8. Landmine Monitor Report. 2007. http://www.icbl.org/lm/2007/es/toc.html.
9. United Nations. The Convention on Certain Conventional Weapons. http://www. ccwtreaty.com.
10. http://www.un.org/Depts/mine/UNDocs/ban_trty.htm.
11. Habib, M. K. 2007. Controlled biological and biomimetic systems for landmine detection. *Biosens. Bioelectron.*, 23:1–18.
12. GICHD (Geneva International Centre for Humanitarian Demining) Research Report, June 2002, *Mine Action Equipment: Study of Global Operational Needs*, Genova.

13. McDonald, J., Lockwood, J. R., McFee, J., Altshuler, T., Broach, T., Carin, L., Harmon, R., Rappaport, C., Scott, W., Weaver, R. 2003. *Alternatives for Landmine Detection.* RAND Corporation, Santa Monica, CA.

14. Zhuge, X., Yarovoy, A. G., Savelyev, T. G., Aubry, P. J., Ligthart, L. P. 2008. Optimization of ultra-wideband linear array for subsurface imaging. In *Proceedings of the 12th International Conference on Ground Penetrating Radar*, June 1–5.

15. Burganov, B. T., Vinogradov, A. G., Sazonov, V. V., Yarovoy, A., Ligthart, L. 2008. Simulation of GPR returns from finite-length dielectric cylinders. In *Proceedings of the 5th European Radar Conference*, Amsterdam, The Netherlands, Oct. 30–31, pp. 292–295.

16. Yarovoy, A. 2009. Landmine and UXO detection and classification. In *Ground Penetrating Radar Theory and Applications*, H. M. Jol, Ed. Elsevier, New York, pp. 445–478.

17. Daniels, D. J., Ed. 2004. *Ground Penetrating Radar*, 2nd ed. Institute of Electrical Engineers, London.

18. Stewart, C. 1960. *Summary of Mine Detection Research*, Vol. I. Technical Report 1636-TR. U.S. Army Engineer Research and Development Laboratories (now NVESD), Corps of Engineers, Fort Belvoir, VA.

19. George, V., Jenkins, T. F., Leggett, D. C., Cragin, J. H., Phelan, J., Oxley, J., Pennington. J. 1999. Progress on determining the vapor signature of a buried landmine. In *Detection and Remediation Technologies for Mines and Minelike Targets IV. Proc. SPIE*, 3710 (P. 2):258.

20. la Grone, M., Cummings, C., Fisher, M., Fox, M., Jacob, S., Reust, D., Rockley, M., Towers, E. 2000. Detection of landmines by amplified fluorescence quenching of polymer films: a man-portable chemical sniffer form detection of ultra-trace concentrations of explosives emanating from landmines. In *Detection and Remediation Technologies for Mines and Minelike Targets V*, A. C. Dubey, J. F. Harvey, J. T. Broach, R. E. Dugan, Eds. *Proc. SPIE*, 4038:553.

21. George, V., Jenkins, T. F., Leggett, D. C., Cragin, J. H., Phelan, J., Oxley, J., Pennington, J. 1999. In *International Optics Engineering*, Part 1, *Detection and Remediation Technologies for Mines and Minelike Targets IV*, A. C. Dubey, J. F. Harvey, J. T. Broach, R. E. Dugan, Eds. *Proc. SPIE*, 3710:258–269.

22. Leggett, D. C., Jenkins, T. F., Murmann, R. P. 1977. *Composition of Vapors Evolved from Military TNT as Influenced by Temperature, Solid Composition, Age, and Source.* SR 77-16/AD A040632, U.S. Army Corps of Engineers, Cold Regions Research and Engineering Laboratory, Hanover, New Hampshire.

23. Kapoor, J. C., Kanan, G. K. 2007. Landmine detection technologies to trace explosive vapour detection techniques. *Def. Sci. J.*, 57(6):797–810.

24. Fisher, M., Sikes, J. 2003. Minefield edge detection using a novel chemical vapor sensing technique. In *Detection and Remediation Technologies for Mines and Minelike Targets VIII*, R. S. Harmon, J. H. Holloway, Jr., J. T. Broach, Eds. *Proc. SPIE*, 5089:1078.

25. Cumming, C. J., Aker, C., Fisher, M., Fox, M., la Grone, M. J., Reust, D., Rockley, M. G., Swager, T. M., Towers, E., Williams, V. 2001. Using novel fluorescent polymers as sensory materials for above-ground sensing of chemical signature compounds emanating from buried landmines. *IEEE Trans. Geosci. Remote Sens.*, 39(6):1119.

26. White, J., Waggoner, L. P., Kauer, J. S. 2004. Explosives and landmine detection using an artificial olfactory system. In *Detection and Remediation Technologies for Mines and*

Minelike Targets IX, R. S. Harmon, J. T. Broach, J. H. Holloway, Jr., Eds. *Proc. SPIE*, 5415:521.

27. Senesac, L., Thundat, T. G. 2008. Nanosensors for trace explosive detection. *Mater. Today*, 11(3):28.

28. C. Hatchard. 2003. A combined MD/GPR detector: the HSTAMIDS system. Presented at the International Conference on Requirements and Technologies for the Detection, Removal and Neutralization of Landmines and UXO (EUDEM2-SCOT 2003), Vrije Universiteit Brussel, Brussels, Belgium, Sept. 15–18.

29. Daniels, D. J., Curtis, P., Amin, R., Hunt, N. 2005. Minehound™ production development. In *Detection and Remediation Technologies for Mines and Minelike Targets X. Proc. SPIE*, 5794:488–494.

30. M. Sato, Fujiwara, J., Takahashi, K. 2007. The development of the hand held dual sensor ALIS. *Proc. SPIE*, 6553:6553C-1 to 6553C-10.

31. Doheny, R. C. 2007. Handheld Standoff Mine Detection System—HSATMIDS: program and operations update. Presented at the Mine Action National Directors and UN Advisors Meeting, Mar. 21, http://www.mineaction.org/. Doheny, R. C., Burke, S., Cresci, R., Ngan, P., Walls, R. 2005. Handheld Standoff Mine Detection System (HSTAMIDS) field evaluation in Thailand. In *Detection and Remediation Technologies for Mines and Minelike Targets X*, Orlando, FL. *Proc. SPIE*, 5794:889–900.

32. Daniels, D. J., MINEHOUND trials 2005–2006. Summary report, October 2006. http://www.itep.ws/pdf/MINEHOUND_ITEP_TrialsERAweb.pdf.

33. Sato, M., Takahashi, K. The evaluation test of hand-held sensor ALIS in Croatia and Cambodia. In *Detection and Remediation Technologies for Mines and Minelike Targets IV. Proc. SPIE*, 6553:1–9.

34. Yarovoy, A. G., Savelyev, T. G., Aubry, P. J., Lys, P. E., Ligthart, L. P. 2007. UWB array-based sensor for near-field imaging. *IEEE Trans. Microwave Theory Tech.*, 55(6):1288–1295.

35. Yarovoy, A., Savelyev, T., Zhuge, X., Aubry, P., Ligthart, L., Ligthart, L. P., Schavemaker, J., Tettelaar, P., den Breejen, E. 2008. Performance of UWB array-based radar sensor in a multi-sensor vehicle-based suit for landmine detection. In *Proceedings of the 5th European Radar Conference*, Amsterdam, The Netherlands, Oct. 30–31, pp. 288–291.

36. Waltz, E., Llinas, J. 1990. *Multisensor datafusion*. Artech House, Norwood, MA.

37. Sato, M. 2005. Dual sensor ALIS evaluation test in Afghanistan. *IEEE Geosci. Remote Sens. Soc. Newsletter*, Issue 136, 22–27 Sept. 2005.

CHAPTER TWELVE

Transportation and Civil Engineering

12.1 INTRODUCTION

The security and quality of the transportation are always an essential topic when assessing the development levels of states. State highway agencies spend millions of dollars every year for the maintenance, repair, rebuilding, quality control, and rehabilitation of roads and their connected units. Thus, there is a great need for usable nondestructive methods to provide fast and easy detection of critical deformities on roads, railways, highways, and bridge decks, and to develop reliable diagnostics for transportation materials and composites, such as asphalt and concrete construction blocks (Fig. 12.1).

The ever-expanding transport network as part of the economic and social development of Europe and other developed countries is extremely important: 79% of the transport of people and 44% of the transport of goods in Europe is over roads. Today's transport infrastructure relies on bridges, and even if in the next 15 to 35 years a doubling of road transport is expected, only a limited number of new bridges are planned. Bridges provide access to remote communities and a way to bypass obstacles and railway crossings, to cross rivers, and to improve traffic flow. Bridges are therefore essential components, so their serviceability and safety play a vital role day to day in keeping the transport network flowing smoothly and safely.

Bridges are very vulnerable and expensive elements in our transport network. Although some are much older, most bridges in the European Union (EU) national road network were built during the last 50 years. Most were designed for loads and traffic volume different from what is actually being experiencing. This obviously

Subsurface Sensing, First Edition. Edited by Ahmet S. Turk, A. Koksal Hocaoglu, and Alexey A. Vertiy.
© 2011 John Wiley & Sons, Inc. Published 2011 by John Wiley & Sons, Inc.

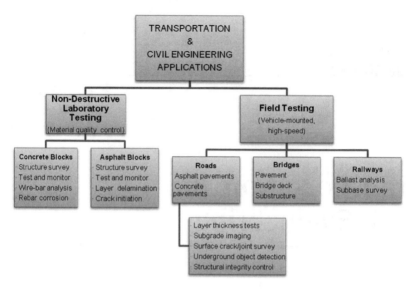

FIGURE 12.1 Subsurface sensing applications for transportation and civil engineering.

puts EU bridges in need of a higher frequency of repairs and possibly a reduced load-carrying capacity.

Each type of structure behaves differently, whether it is of reinforced or of pre-stressed concrete construction. Whatever the construction, each type of structure suffers from a different type of deterioration and has different maintenance needs. Defect identification is a fundamental aspect in defining the correct repair or maintenance solution. Visual inspection is thus the first step in the evaluation process. Generally, visual inspection reports of a bridge condition include an assessment of all structural elements. For each, deterioration levels are described by letter or numerical codes. Although the tables and codes may differ from country to country, the general idea is similar in most European states and in the United States.

Unfortunately, these descriptions do not cover all element or defect types, but they provide general guidance on the identification of severity states. Nevertheless, visual inspection remains mostly subjective. Furthermore, it should be noted that visual inspection ignores internal deterioration. Moreover, they can miss defects that occur and take place during different stages of the lifespan of a bridge. Some of these defects are not dangerous, as they remain stable. The others are under development during service loading, due to the variety of deterioration aspects of the structure. There are defects that begin to be active (or initiate activity) when the service conditions are changed or when adjacent materials or members change their properties because of deterioration. We also currently do not have a way to derive information about the interaction of defects and/or the severity.

Corrosion of reinforcement, concrete deterioration, and many other factors, such as service loading of a structure (which is very important for failure mechanic analysis), add to the difficulty of ensuring that bridges are maintained properly. The service loading depends not only on the weight of vehicles but is also influenced by the temperature and ground movement.

There is evidence that after a positive inspection report, where a visual inspection had been used to report on a structure's health, catastrophic failure may occur shortly thereafter. An example is the catastrophe of a viaduct in Canada, where the structure collapsed an hour after the inspection, when pieces of concrete began to peel away from the structure [1]. That terrible incident proves that inspection methods need to be carried out using more advanced techniques.

It is evident that evaluation of the health of road infrastructure components such as bridges, tunnels, and culverts is a sophisticated task. Large dimensions, many different structural elements with different materials, changeable service conditions, and environmental influences make evaluation difficult. In addition, local service conditions at the most critical locations, resulting from loading, thermal changes, freezing and defrosting processes, settlement of foundations, corrosion, and catastrophic accidents on the structure, complicate assessment of the deterioration processes. Structural defects such as shear and crack opening, material fatigue, loss of stability, and corrosion losses differ from structure to structure. Moreover, defect interactions make the deterioration processes unpredictable and impossible to evaluate analytically. Complicated structures, different types of cracks, deteriorated materials, together with impossible stress evaluation in the vicinity of the process zone make it impossible to use fracture mechanics methods for integrity assessment of road infrastructures.

Road infrastructure inspections in European countries are mainly visual and thus subjective. Such inspection reports cannot state with certainty details of the defects detected. We may not know if the defects are dangerous, or even understand their severity under service conditions. It is especially difficult to evaluate the health of pre- and posttensioned girders with high compressive stresses. Additionally, nondestructive testing (NDT) can significantly improve defect evaluation and data analysis. On the other hand, the choice of NDT cannot omit consideration of the structure type and its typically related defects.

12.1.1 Concrete Structures and Related Defects [2–4]

Reinforced Concrete Structures   Defect development in concrete structures can be caused by environmental factors, inferior materials, poor workmanship, inherent structural design defects, and inadequate maintenance. Environmental factors are generally the principal cause of concrete deterioration. Concrete absorbs moisture readily if not coated; this is particularly troublesome in regions of recurrent freeze–thaw cycles. Freezing water produces an expansive pressure in the cement paste and in nondurable aggregates. Carbon dioxide, another atmospheric component, can cause concrete to deteriorate by reacting with the cement paste at the surface.

Materials and workmanship in the construction of early concrete buildings are potential sources of problems. For example, aggregates used in early concrete, such as cinders from burned coal and certain crushed brick, absorb water and produce a weak and porous concrete. Alkali-aggregate reactions within the concrete can result in cracking and white surface staining. Aggregates were not always properly graded by size to ensure an even distribution of elements from small to large. The use of aggregates consisting of similarly sized particles normally produces a poorly consolidated and therefore weaker concrete.

Early builders sometimes compromised concrete inadvertently by using seawater or beach sand in the mix. A common practice, until recently, was to add salt to strengthen concrete or to lower the freezing point during cold-weather construction. These practices cause problems over the long term. In addition, early concrete was not vibrated when poured into forms as it is today. More often, it was tamped for consolidation; on floor slabs, it was often rolled with increasingly heavier rollers. These practices tended to leave voids (nonconcrete areas) at congested areas, such as at reinforcing bars (rebars) at column heads and other critical structural locations. Areas of connecting voids seen when concrete forms are removed, known as *honeycombs*, can reduce the protective cover over reinforcing bars.

Other problems caused by poor workmanship are not unknown in later concrete work. If the first layer of concrete is allowed to harden before the next one is poured next to or on top of it, joints can form at the interface of the layers. In some cases, these cold joints detract visibly from the architecture; in others they are harmless. Cold joints can permit water to infiltrate, and subsequent freeze–thaw action can cause joints to move. Dirt packed in joints allows weeds to grow; furthermore, they cause opening paths for water entrance. Inadequate curing can also lead to problems. If moisture leaves newly poured concrete too rapidly because of low humidity, excessive exposure to sun or wind, or use of a too-porous substrate, the concrete will develop shrinkage cracks and will not reach its full potential strength.

Structural design defects in concrete structures can be an important cause of deterioration. For example, the amount of protective concrete that surrounds reinforcing bars was often insufficient. Another design problem in early concrete buildings was related to the absence of standards for expansion and contraction joints to prevent stresses caused by thermal movements, which may result in cracking. Typical defects of reinforced concrete structures follow.

Cracks   Cracking occurs over time in virtually all concrete. Cracks vary in depth, width, direction, pattern, location, and cause. Cracks can be either active or dormant (inactive). Active cracks widen, deepen, or migrate through the concrete. Dormant cracks remain unchanged. Some dormant cracks, such as those caused by shrinkage during the curing process, pose no danger, but if left unrepaired, they can provide convenient channels for moisture penetration, which normally causes further damage.

Structural cracks can result from temporary or continued overloads, uneven foundation settling, or original design inadequacies. Structural cracks are active if the overload is continued or if settlement is ongoing; they are dormant if the temporary overloads have been removed or if differential settlement has stabilized. Thermally induced cracks result from stresses produced by temperature changes. They frequently occur at the ends or corners of older concrete structures built without expansion joints capable of relieving such stresses. Random surface cracks (also called *map cracks* because of their resemblance to the lines on a road map) that deepen over time and exude a white gel that hardens on the surface are caused by an adverse reaction between the alkali in cement and some aggregates. Since superficial repairs that do not eliminate underlying causes will only tend to aggravate problems, professional consultation is recommended in almost every instance where noticeable cracking occurs.

Spalling Spalling is the loss of surface material in patches of varying size. It occurs when reinforcing bars corrode, thus creating high stresses within the concrete. As a result, chunks of concrete pop off from the surface. Similar damage can occur when water is absorbed by porous aggregates and freezes. Vapor-proof paints or sealants, which trap moisture beneath the surface of the impermeable barrier, can also cause spalling. Spalling may also result from the improper consolidation of concrete during construction. In this case, water-rich cement paste rises to the surface (a condition known as *laitance*). The surface weakness encourages scaling, which is spalling in thin layers.

Deflection Deflection is the bending or sagging of concrete beams, columns, joists, or slabs, and can seriously affect both the strength and structural soundness of concrete. It can be produced by overloading, corrosion, inadequate construction techniques (i.e., use of low-strength concrete or undersized reinforcing bars), or concrete creep (long-term shrinkage). Corrosion may cause deflection by weakening and ultimately destroying the bond between the rebar and the concrete, and finally by destroying the reinforcing bars by themselves. Deflection of this type is preceded by significant cracking at the bottom of the beams or at column supports. Deflection in a structure (except for widespread cracking, sparing, and corrosion) is frequently caused by concrete creep.

Stains Stains can be produced by alkali–aggregate reaction, which forms a white gel exuding through cracks and hardening as a white stain on the surface. *Efflorescence* is a white, powdery stain produced by the leaching of lime from portland cement. Discoloration resulting from metals inserted into the concrete, or from corrosion products dripping onto the surface concrete, can be a most serious problem. Normally, embedded reinforcing bars are protected against corrosion by being buried within the mass of the concrete and by the high alkalinity of the concrete itself. This protection can, however, be destroyed in two ways. First, by carbonation; this occurs when carbon dioxide in the air reacts chemically with cement paste at the surface and reduces the alkalinity of the concrete. Second, chloride ions from salts combine with moisture to produce an electrolyte that effectively corrodes the reinforcing bars. Chlorides may come from seawater additives in the original mix or from prolonged contact with salt spray or deicing salts. Regardless of the cause, corrosion of reinforcing bars produces rust, which occupies significantly more space than did the original metal and causes expansive forces within the concrete. Cracking and spalling are frequent results. In addition, the load-carrying capacity of the structure can be diminished by the loss of concrete, by the loss of bond between reinforcing bars and concrete, and by the decrease in thickness of the reinforcing bars themselves. Rust stains on the surface of the concrete are an indication that internal corrosion is taking place.

Pre- and Posttensioned Structures   Deterioration in the condition of prestressing strands or tendons is not always reflected by distress visible on the concrete surface. Further, the effect of deterioration of prestressing steel is more disruptive than that of unstressed reinforcement. Strand, due to its high mechanical strength and metallurgical characteristics, is smaller in cross section than conventional reinforcing steel and is proportionally more impaired by loss of section. The material is also susceptible to less common and less predictable forms of deterioration, such as stress corrosion, hydrogen-assisted cracking, corrosion fatigue, and fretting corrosion.

Pretensioned Structures   The types of corrosion problems in prestressed structures are not significantly different from those in reinforced concrete structures. The absence of a posttensioning duct and anchorages in prestressed concrete makes it more similar to reinforced concrete in terms of the protection provided for the prestressing steel. The main influencing factors for corrosion in prestressed structures are the prestressing steel, concrete, and the severity of the environment. The effects of concrete and the environment are the same for prestressed and reinforced concrete structures. Although prestressing steel is more susceptible to corrosion, and the consequences of corrosion may be more severe than for unstressed reinforcement, corrosion of prestressing tendons in prestressed structures is rare for two principal reasons. Prestressed elements are always precast, generally resulting in improved overall quality control and good-quality concrete. Also, prestressed elements normally fit the classic definition of full prestressing; that is, concrete tensile stresses are limited to prevent flexural cracking of the concrete. Where corrosion has been discovered in prestressed structures, the cause is normally related to the structural form and details. Because prestressed elements are precast, the structure may contain a large number of joints or discontinuities. Poor design and/or maintenance of these joints may direct moisture and chlorides onto the prestressed elements of the structure in extremely localized areas.

Posttensioned Structures
Unbounded Single-Strand Tendons   Unbounded single-strand or monostrand tendons are greased and sheathed-type single-strand tendons commonly used in slabs. The majority of this steel was used in buildings (including parking structures) and in slabs-on-grade. Although monostrand applications would be very limited in bridge substructures, they have been used for transverse posttensioning in bridge decks and segmental box girders. Also, examination of corrosion problems in structures with monostrand tendons can provide insight into the overall picture of corrosion in posttensioned structures, including bridges.

Corrosion problems in monostrand tendons can be grouped into four areas: (1) damage to the sheathing, (2) poor anchorage protection, (3) system deficiencies, and (4) structural aspects. A very comprehensive discussion of corrosion of monostrand tendons has been provided by ACI/ASCE Committee 423 [5].

Inadequate anchorage protection can lead to multiple forms of corrosion problems in monostrand systems. Corrosion of the anchorage itself is a common problem. Failure of the anchorage in an unbounded system obviously leads to loss of the tendon. Corrosion of the anchorage typically occurs due to lack of a protective barrier or insufficient concrete cover. The concrete or mortar used to cover anchorage recesses after stressing is often of low quality, allowing moisture penetration to the anchorage. Many corrosion problems in monostrand systems have been related to the system itself. Most of these problems occurred in older monostrand systems; new systems have eliminated many of the problems found in older systems. Another common source of monostrand corrosion problems has been the discontinuity of sheathing and grease on the strand immediately behind the anchorage.

Also, some aspects of the design process may lead to further corrosion problems. Electrical contact between the monostrand tendon and other reinforcement may provide an opportunity for macrocell corrosion with a large cathode (reinforcement) and a small anode (monostrand tendon) in a nonisolated system. Large cathode-to-anode areas can lead to high corrosion rates and severe corrosion damage. An electrically isolated system should prevent this occurrence. Reinforcement congestion or reinforcement ties may lead to sheathing damage during posttensioning of monostrand.

External Multistrand Tendons The most common forms of external multistrand tendons occur in bridges. Cable stays may also be considered in this category. Corrosion protection for multistrand external tendons typically consists of a plastic or metal sheath normally filled with grout or corrosion-inhibiting grease. Corrosion-related failures or problems have resulted from a breakdown in the sheathing system or insufficient protection of the anchorages. These situations are worsened by poor or incomplete filling of the void space around the tendon with grout or with grease, which permits the movement of moisture along the tendon length after penetration.

Bonded Internal Posttensioned Tendons Many corrosion problems have resulted from various aspects of grouting. The effectiveness of the grout as corrosion protection is related both to its material properties and construction practices. The most common grout-related corrosion problems are attributed to incomplete grouting, that is, where the duct is not completely filled with grout. The extent of incomplete grouting may range from small voids to a complete lack of grouting. Common causes of incomplete grouting are construction difficulties, improper construction practices, blocked or damaged ducts, and improper placement or use of vents. The fresh properties of the grout may also affect the grouting process through insufficient or excessive fluidity and excessive bleed water, leading to entrapped air or the formation of bleed lenses. The severity of tendon corrosion is related to the extent of incomplete grouting and the availability of moisture, oxygen, and chlorides. In general, the most severe attack occurs when the tendon is intermittently exposed and embedded in the grout. In this situation, a concentration cell may occur due to variations in the chemical and physical environment along the length of the tendon. Concentration cells may result from differences in oxygen, moisture, and chloride concentration, and often lead to severe macrocell corrosion.

Tendon corrosion may also occur in situations where the entire length of the tendon is well grouted. The most common cause of corrosion in these situations has been sources of chlorides in the grout itself. Examples include seawater used as the mixing water or chloride containing admixtures. A combination of severe exposure conditions and low cover may lead to corrosion of the duct and subsequent penetration of moisture and chlorides from an external source.

The posttensioning duct is an important component of corrosion protection in posttensioned structures. Many forms of ducts exist, ranging from nonpermanent duct formers, to galvanized steel ducts, to plastic ducts, each providing an increasing level of protection. Incidents of corrosion have resulted from damaged ducts, improper splices between ducts, corroded ducts, and situations where nonpermanent duct formers have been used. Holes in the duct may allow concrete to enter the duct during casting. This may hamper placement and tensioning of the tendons and may cause difficulties during grouting. Damage or misalignment during construction or concrete placing may also lead to posttensioning and grouting difficulties.

12.1.2 Deck and Pavement Defects

Pavements and decks are peculiar types of "structures" (or better, "over-structures") characterized by an extremely high surface-to-volume ratio, and they are generally composed of different parallel layers. Due to these peculiarities, three main types of defect can be identified: surface defects (related primarily to the outer surface of the structure), transversal defects (mainly cracks parallel to the structure cross section), and subsurface defects (generally, internal cracks or delamination parallel to the surface). Two main types of pavement can be identified in relation to the material used for the outer layer: concrete and asphalt. Their applications are not related specifically to their typology, even if it is possible to identify some general uses.

Concrete Pavement Defects

Slab Cracking Full-depth cracking can occur in concrete slabs in a longitudinal, transverse or diagonal direction, dividing the slab into two or more parts. The causes of slab cracking are varied, the most common reasons being:

- Loads exceeding the slab bearing strength
- Loss of underlying slab support due to subgrade erosion
- Settlement or heaving
- Late sawing of transverse contraction joints during construction

Unless repaired and sealed, slab cracks will eventually break down, resulting in crack spalling and loose pieces of pavement material.

Slab Corner Cracking A full-depth crack can occur at the corner of a slab running from contraction to construction joint in a semicircular pattern around the slab corner. Corner cracks are usually caused by loads exceeding the slab bearing strength or by loss of foundation support at the slab corner. Fine hairline cracks can advance to more open cracks with eventual spalling and loss of material.

Scaling and Spalling Scaling and spalling of concrete slabs refer to the disintegration, ravelling, and breakup of a pavement surface with subsequent loss of material. *Scaling* is the peeling of a relatively thin layer from the concrete surface, usually fine material and aggregates from the concrete mix. *Spalling* is the loss of larger-sized material from the surface, usually to a depth equal to or greater than the maximum aggregate size. Causes of scaling and spalling are usually related to construction or material problems due to a weak layer of fine aggregate at the surface, as a result of overfinishing during construction, poor curing practices, poor-quality aggregates, action of water and freeze–thaw cycles, and the action of deicing chemicals.

Slab Joint Faulting (Stepping) Slab joint faulting (also called *stepping*) is the elevation or depression of a slab relative to an adjacent slab, resulting in a step when crossing the joint between slabs.

Asphalt Pavement Defects

Transverse and Longitudinal Cracking Surface cracking is considerably the most common defect found in pavements. Cracks can run perpendicular (transverse) or parallel (longitudinal) to the direction of traffic, or in a meandering random pattern across the surface. Transverse thermal cracks caused by cold temperature contraction of the surface layer will occur to some degree in all asphalt pavements.

Longitudinal cracks can be caused by opening of paving lane joints or by repeated overstressing of the vehicle wheelpath area by heavy traffic. Secondary cracking may eventually develop adjacent to the main crack, which can lead to pieces of pavement dislodging from the surface. Environmental factors also play a large role in the structural breakdown of pavement cracks. Water can enter the crack, washing out fines from underneath, and eroding support for the pavement.

Alligator Cracking Alligator cracking is a very serious load-associated defect normally found in vehicle wheelpaths. It is caused by severe structural overloading. In the initial stages, only fine hairline cracks may be evident. As development of the defect progresses, the cracks will widen and begin to interconnect to form a series of small polygons resembling the hide of an alligator. Spacing between cracks is usually 50 to 300 mm.

Map Cracking Map cracking is identified by large-scale patterned cracks intersecting at a spacing of 500 mm to 2 m. Map cracking occurs only in wheelpaths (load-associated); otherwise, the defect is named *block cracking* (see below). Map cracking may be accompanied by wheelpath rutting or settlement. Map cracking is caused by an excessive deflection of the pavement under load. It is similar to alligator cracking but the yielding layer is at a much deeper level in the pavement, most likely in the subgrade.

Block Cracking Block cracking is identified by intersecting transverse and longitudinal cracks forming rectangular blocks that may vary in size with a spacing of up to 3×3 m. Block cracking may appear to be similar to alligator cracking, but it is not limited to wheelpath locations (i.e., block cracking is not load-related) and will frequently occur throughout the entire pavement surface. Block cracking is normally found in older asphalt surfaces.

Ravelling Ravelling is the disintegration and subsequent loss of the asphalt surface. It is evidenced by the pop-out of surface aggregates and/or the loss of surface fines from the asphalt mix. Severe cases of ravelling are characterized by loose gravel and fine pavement material lying on the surface.

Wheelpath Rutting Rutting is a load-related depression found in the wheelpath area. Depending on the failure mode, rutting may be seen simply as the consolidation of pavement layers in the wheelpath or may be made worse by the upheaval of the pavement area adjacent to the wheelpath.

Asphalt Bleeding Bleeding refers to the accumulation of asphalt binder on the pavement surface normally in the wheelpath areas. Bleeding can be caused by excess asphalt cement and/or insufficient voids in the asphalt mix, with the excess asphalt being flushed to the pavement surface by wheel loads during hot weather.

12.1.3 Metal Structures and Related Defects

Steel alloys (e.g., carbon steel and stainless steel), and in some cases aluminum alloys, are frequently used as structural materials for many applications in the field of civil engineering and transportation, from buildings to tower, from bridges to steel lattice structures. In all these cases structures are obtained by joining together profiles, bars, squares, and angles of different section and geometry. In same cases, metal profiles are used in conjunction with concrete in mixed structures. Since metal elements are generally free from structurally significant defects, the main concern is about defects induced either by environmental factors (corrosion, fatigue) or by the joining process itself (e.g., cracks, voids). The most common defects encountered in practice follow.

Environmentally Induced Defects [9] All metallic materials exposed to the environment are affected by corrosion. Corrosion is indeed the electrochemical reaction of a material and its environment. All corrosion reactions are electrochemical in nature and depend on the operation of electrochemical cells at the metal surface. Corrosion attack can develop uniformly on the metal surface (uniform corrosion) or concentrate in specific locations (localized corrosion). Several types of localized corrosion can be defined; among them the most insidious are pitting, crevicing, and intergranular corrosion and stress corrosion cracking.

Uniform Corrosion Uniform corrosion, or general corrosion, is a corrosion process exhibiting uniform thinning that proceeds without appreciable localized attack. It is the most common form of corrosion and may appear initially as a single penetration, but with thorough examination of the cross section it becomes apparent that the base material has thinned uniformly. Uniform chemical attack of metals is the simplest form of corrosion, occurring in the atmosphere, in solutions, and in soil, frequently under normal service conditions. Excessive attack can occur when the environment has changed from that expected initially.

Pitting and Crevicing Pitting and crevice corrosion are forms of localized corrosion that are significant causes of failure in metal parts. Both forms attack passivated or otherwise protected materials (e.g., anodized aluminum, stainless steel, or carbon steel in concrete). Pitting, characterized by sharply defined holes, can cause failure by perforation while producing only a small weight loss on the metal. This perforation can be difficult to detect and its growth rapid, leading to unexpected loss of function of the component. Crevice corrosion is pitting that occurs in slots and in gaps at metal-to-metal and metal-to-nonmetal interfaces. It is a significant contributor to component failure because such gaps often occur at critical joining surfaces.

Intergranular Corrosion Intergranular corrosion is the preferential dissolution of the anodic component, the grain-boundary phases, or the zones immediately adjacent to them, usually with slight or negligible attack on the main body of the grains. Intergranular corrosion is usually (but not exclusively) a consequence of composition changes in the grain boundaries from elevated-temperature exposure (e.g., during welding). In general, grain boundaries can be susceptible to changes in composition, because grain boundaries are generally slightly more active chemically than the grains themselves.

Stress Corrosion Cracking Stress corrosion cracking (SCC) is a failure process that occurs because of the simultaneous presence of tensile stress, an environment, and a susceptible material. Removal of or changes in any one of these three factors will often eliminate or reduce susceptibility to SCC and therefore are obvious ways of controlling SCC in practice. SCC is a subcritical crack growth phenomenon involving crack initiation at selected sites, crack propagation, and overload final fracture of the remaining section. Failure by SCC is frequently encountered in seemingly mild chemical environments at tensile stresses well below the yield strength of the metal. Failures often take the form of fine cracks that penetrate deeply into the metal, with little or no evidence of corrosion on a nearby surface or distortion of the surrounding structure. Therefore, during casual inspection, no macroscopic evidence of impending failure is seen.

Fatigue Fatigue failure of engineering components and structures results from progressive fracture caused by cyclic or fluctuating loads (e.g., due to traffic or wind). The magnitude of each individual load event is too small to cause complete fracture of the undamaged component, but the cumulative action of numerous load cycles, often numbering in the hundreds of thousands or millions, results in initiation and gradual propagation of a crack or cracks. Complete fracture ensues when the crack reaches a critical size. Fatigue is an important potential cause of mechanical failure, as most engineering components or structures are or can be subjected to varying loads during their lifetime.

Joining-Induced Defects [10]

Welded Joints All welded structures contain discontinuities, to some level of examination. A discontinuity is not necessarily a defect until either it affects the fitness for service of a weld or is defined as such by an acceptance–rejection criterion related to an examination. Fitness for service is a concept of weld evaluation that seeks a balance among quality, reliability, and economy of welding procedure. Fitness for service is not a constant. It varies depending on the service requirements of a particular welded structure as well as on the properties of the material involved. The discontinuities found in welds vary considerably in their importance as failure origins. Their location (surface or subsurface) and geometry must be considered in evaluating the significance of weld discontinuities. The spatial location and shape of welding imperfections can be classified into three broad categories: planar, volumetric, or geometric. *Planar imperfections* are sharp crack-like features that can substantially reduce the fatigue strength of a welded joint or cause initiation of brittle fractures. Examples include hydrogen cracks, lamellar tears, lack of fusion, reheat cracks, solidification cracks, and weld-toe intrusions. *Volumetric imperfections* include porosity and slag inclusions. Because these types of imperfections tend to be nearly spherical in form, their notch effect is minor, and they usually have little or no influence on fatigue behavior. However, they do reduce the load-bearing area of the weld and hence reduce the static strength of the joint. *Geometric imperfections* include misalignment, overfill, stop/starts, undercut, and weld ripples. Geometric imperfections have the effect of locally elevating the stress over and above the nominal stress due to stress concentration from the joint geometry.

Weld discontinuities (or imperfections) can be caused by various design, processing, or metallurgical factors. Joint design, edge preparation, fit-up, cleanness of base metal and filler metal, shielding, and welding techniques all affect weld quality and must be carefully controlled to prevent porosity, cracks, fissures, undercuts, incomplete fusion, and other weld imperfections.

Fastener Joints The primary function of a fastener system is to transfer load. Many types of fasteners and fastening systems have been developed for specific requirements, such as higher strength, easier maintenance, better corrosion resistance, greater reliability at high or low temperatures, or lower material and manufacturing costs.

The selection and satisfactory use of a particular fastener are dictated by the design requirements and conditions under which the fastener will be used. For descriptive purposes, mechanical fasteners are grouped into threaded fasteners, rivets, blind fasteners, pin fasteners, and fasteners for composites. Rivets, pin fasteners,

and special-purpose fasteners are usually designed for permanent or semiperma-
nent installation. Threaded fasteners are considered to be any threaded part that
after assembly of the joint may be removed without damage to the fastener or
to the members being joined. Rivets are permanent one-piece fasteners that are
installed by upsetting one end mechanically. Blind fasteners are usually multiple-
piece devices that can be installed in a joint that is accessible from only one side.
When a blind fastener is being installed, a self-contained mechanism, an explosive,
or other device forms an upset on the inaccessible side. Pin fasteners are one-piece
fasteners, either solid or tubular, that are used in assemblies in which the load is
primarily shear. A malleable collar is sometimes swagged or formed on the pin
to secure the joint. In all these cases the most common defects may be related to
misalignment, improper fastening, and crevice corrosion, fatigue cracks originating
at the connecting holes, fretting, pack rust, and stress corrosion cracking.

Mechanical fasteners for composites are frequently used in combination with
adhesive bonding to improve the integrity of highly stressed joints. The usual
bolts, pins, rivets, and blind fasteners are used for composites; however, the many
problems encountered have stimulated the development and testing of numerous
special-purpose fasteners and systems. Some of these problems are drilling of and
installation damage to the composite, delamination of the composite material around
the hole and pullout of the fastener under load, differences in expansion coefficients
of the composite and the fastener, galvanic corrosion between the composite hole
wall and the fastener, moisture and fuel leaks around the fastener, and fretting.

12.2   PROPER SENSOR TYPES

Subsurface imaging of concrete or asphalt slabs, multilayer thickness analysis, surveying of the structural damages of roads, such as cracks, holes, bumps, bends,
voids, and water fillings, and railway ballast and bridge tendon characterizations
are main research subjects in transportation and civil engineering technology. Some
suitable methods and proposed sensor types are discussed briefly in this section,
together with their physical properties, operational adaptabilities, capabilities, reli-
abilities, costs, and multisensor approaches. Available sensor technologies and an
operational analysis chart are provided in Fig. 12.2 and Table 12.1, respectively.

12.2.1   General Description of Sensor Types Applicable to Civil Engineering Structures

The main concern in inspecting a structure is the location and identification of
defects, which in concrete structures are related primarily to steel reinforcement
corrosion. However, in consideration of the differences in construction typology it
is generally difficult to identify corrosion attacks directly. Since as reported above,
corrosion is generally related to construction defects, it is more convenient to try
to locate such types of defects as well as grouting defects in posttensioning ducts.
In other cases the presence of concrete degradation defects and corrosion-related

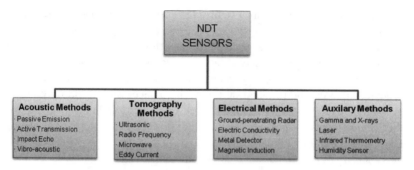

FIGURE 12.2   Nondestructive testing sensor technologies.

defects can be identified by the loss of performance of the structure itself (e.g., increasing deformity, crack opening, noncomplying mechanical behavior).

Several instrumental techniques have been developed with those aims during the past 20 years. The majority of these test methods fall within the classification of "nondestructive" on the basis of the stipulation that they should not influence the function of the structure or the element. According to the American Society of Nondestructive Testing (ASNT), NDT is defined as the examination of an object or material using technology that does not affect its future usefulness [7]. NDT can be used without destroying or damaging a product or material. There are many types of NDT methods for inspecting concrete structures: thermography, γ-ray, x-radiography, vibration analysis based on optical methods (i.e., holography), holography, neutron radiography, dye penetration, magnetic induction, electric potential mapping, radar, acoustic emission, ultrasonic time of flight, ultrasonic resonance spectroscopy, ultrasonic pulse echo, impact echo, ultrasonic guided wave, nonlinear vibroacoustics, and electromagnetic guided-wave testing. All of these methods have advantages and disadvantages. In recent years, several reports as well as national and international standards, have been published on this subject. However, concern about NDT method reliability and their cost/benefit ratio remain among owners and operators of large structures.

12.2.2   Global NDT: Acoustic Methods

For large structures where degradation phenomena are diffused or possible locations of faults are unknown, *local NDT methods* (even an optimization of testing mesh) are carried out by means of a very expensive statistical approach. In other cases, continuous monitoring of the structure is required, but the location of critical points is unknown. In all these cases, use of a *global NDT method* should be required. Among the various NDT methods, acoustic emission and vibroacoustic methods offer the best results.

Acoustic Emission   Acoustic emission (AE) occurs when energy is suddenly released from a material under stress, due to a change of state of the material,

TABLE 12.1 Operational Analysis Chart of NDT Sensors

Sensor Type	NDT	Operational Principle	Applications	Limitations
Ultrasonic	Acoustic emission	Detection of a mechanical wave generated by abrupt energy release in the material (e.g., cracks, phase transition).	Crack propagation monitoring	The method requires a means of loading the structure and complex electronic equipment and data analysis.
	Pulse velocity	Voltage pulses are generated and transformed into wave bursts of mechanical energy by a transmitting transducer. A receiving transducer is coupled to the specimen at a known distance to measure the interval between the transmission and the reception of a pulse.	Quality control, internal defect localization	Proper surface preparation is required. The work is very time consuming, as it takes only point measurements. Skill is required in the analysis of results, as moisture variations and the presence of metal reinforcement can affect results.
	Ultrasonic pulse echo	Pulsed compressional waves are induced in materials and when reflected back, are detected on the structure surface.	Quality control and integrity pile testing, internal defect localization	The technique cannot determine the net cross section of piles or their bearing capacity. Interpretation of results can be difficult, and calibrating standards are required.
	Resonance spectroscopy	Continuous sine-wave excitation is used to drive an ultrasonic transducer in contact with the structure. The response of the structure is detected by a receiver at a second location on the surface.	Flaws detection, structure integrity evaluation and monitoring	There is difficulty in interpreting the resulting data and an influence from the concrete aggregate size and range.

(continued overleaf)

TABLE 12.1 (*Continued*)

Sensor Type	NDT	Operational Principle	Applications	Limitations
	Impact echo	A stress pulse is introduced into a test object by mechanical impact on the surface.	Detection of partial or complete discontinuities, such as voids, abrupt changes in cross section, very weak concrete	The size of the test object plays an important role in the results obtained. Geometrical effects due to limited size are the cause of signals, which can be misleading. It is therefore necessary to perform an impact-echo test at several points on the surface to identify possible geometric effects.
Accelerometer	Vibroacoustic	Vibroacoustic signal analysis and evaluation of nonlinear behavior of a structure under dynamic excitation.	Structural integrity evaluation and monitoring, loading condition evaluation	It is difficult to interpret and quantify data; heavily damped masonry bridges yield little response.
Microwave antenna	Ground-penetrating radar	Radio-frequency waves (0.5 to 2 GHz) from a radar transmitter are directed into the material.	Detection of subsurface cracks, subpavement gaps, and water content in concrete and in soil	Any features screened by steel elements will not be recorded. With increasing depth, low-level signals from small targets are more difficult to detect, due to signal attenuation.
	Near-field microwave diffraction tomography	Microwaves (5 to 15 GHz) from a transmitter are directed into the material.	Strength and curing conditions of cement-based materials	A calibration line is necessary for every combination of materials used (i.e., concrete mix design).

Magnetic sensor	Cover meter	Magnetic induction in steel elements.	Used to determine the presence, location, depth, and dimension of rebars in concrete and masonry components	The results are affected by the presence of more than one reinforcing bar in the test area, by laps, by second layers, by metal tie wires, and by bar supports. The maximum range of the instrument for practical purposes is about 100 mm.
	Magnetic leakage flux	Measurement of magnetic flux leakage generated by discontinuity in steel elements.	Rupture of prestressing steel	The method is used for resolution in steel cross-section reduction. It is unsuitable for closely packed bar assemblies.
Electromagnetic circuit	Electromagnetic resonance	Measurement of the reflection caused by coupling of an electromagnetic wave of variable frequency into a reinforcement element embedded in concrete.	Rupture of prestressing steel	The method is used for radio-frequency signal damping in lossy materials. Fractures cannot be detected reliably if multiple or continuous contact exists between prestressing steel and other steel element.
X-ray-sensitive materials	High-energy radiography, gamma radiography	X-radiation attenuates when passing through a building component. The density and thickness of the materials of the building component will determine the degree of attenuation.	Can be used for locating internal cracks, voids and variations in density of materials, grouting of posttensioned construction as well as locating the position and condition of reinforcing steel in concrete	There are difficulties in placing the photographic films or sensible materials in a suitable position. Health and safety problems for both the operatives and for those in the vicinity, as it requires long radiation exposure time. Limited penetration depths. Areas must be isolated from the public.

(*continued overleaf*)

TABLE 12.1 (*Continued*)

Sensor Type	NDT	Operational Principle	Applications	Limitations
Radiation detectors	Gamma radiometry	Radiation attenuation in concrete and soil related to density.	Density determinations on soil, soil aggregates, and asphalt concrete	The equipment must be operated by trained and licensed personnel; requires radiation safety program.
Nuclear detectors	Neutron moisture gauge	It works on the principle that hydrogen retards the energy of neutrons. Water in the concrete contains hydrogen molecules, which scatter the neutrons.	Can be used to measure moisture content of concrete, soil and bituminous materials and to map moisture migration patterns in masonry walls	The equipment must be operated by trained and licensed personnel. A minimum thickness of surface layer is required for backscatter to be measured. It measures only the moisture content of the surface layer (50 mm). It emits radiation.
Infrared sensor	Thermography	An infrared scanning camera is used to detect variations in infrared radiation output of a surface.	Can be used to detect delamination, heat loss, and moisture movement through concrete elements, especially for flat surfaces	Reference standards are needed, and a heat source to produce thermal gradient in the test specimen may also be required. It is very sensitive to thermal interference from other heat sources. Moisture on the surfaces can also mask temperature differences.

such as crack growth or dislocations. This energy release causes transient elastic stress waves to propagate throughout a specimen. These stress waves are recorded by sensitive resonant piezoelectric sensors mounted on the surface. Traditionally, in AE testing a number of parameters are recorded from the signals, or sensor "hits"; from these parameters, the condition of the specimen is determined. Important parameters include hit arrival time, amplitude, and duration. It is important to take into account that wave propagation through a material depends on the material damping, geometry, and elastic properties. Concrete poses a problem for wave propagation because it is heterogeneous and contains microcracks. Its constituents, such as hydrated cement and aggregate, vary significantly in size and material properties. In addition, its properties can vary due to uneven consolidation, differential shrinkage, or bleed water. This nonhomogeneity causes considerable scatter in results, requiring proper data handling and management.

Vibroacoustics Vibroacoustic methods rely on the relationship between the parameters of the wave propagation as a function of stress occurring in a given object. This is manifested by changes to the modulation of the vibroacoustic signal parameters, which is caused by the disturbance of propagation of the sound wave in the material as a result of changes to the distribution of stress in a cross section of the prestressed structure. Defects appearing in the structure lead to a decrease in compressing stress, which results in a change in the distribution of stress in the cross section that is measurable to a degree, allowing one to detect the qualitative change of the effect of modulation of the vibroacoustic signal parameters.

12.2.3 Local NDT

When the location of the defective area is known and it is important to quantify the defects or the exact spatial location of the defects in a structure, the use of local NDT methods is more convenient. Defects could be related to concrete (e.g., emerging cracks, internal cracks, honeycombs, voids) or to steel elements (e.g., steel corrosion, prestressing steel cracks, cable or strand failure). There are many applicable NDT techniques. Each is based on a different theoretical principle, and as a result, yields different sets of information regarding the physical properties of the structure. The following five major factors need to be considered in the design of a nondestructive testing survey:

1. The required depth of penetration into the structure
2. The vertical and lateral resolution required for the anticipated targets
3. The contrast in physical properties between the target and its surroundings
4. The signal-to-noise ratio for the physical property measured at the structure under investigation
5. Historical information concerning the methods used in the construction of the structure

12.2.4   Identification of Concrete-Related Defects

Sonic/Ultrasonic Methods   Nondestructive sonic and ultrasonic testing methods are noninvasive and have been used in the assessment of civil engineering structures and materials for the past 30 years. The sonic method involves the transmission and reflection of mechanical stress waves through a medium at sonic and ultrasonic frequencies. Among several commonly used sonic techniques—sonic transmission, ultrasonic/sonic/seismic tomography, sonic/seismic/ultrasonic reflection, sonic resonance and impulse response methods—some emerging technologies seem very promising for application in the NDT of concrete structures. The most interesting are impact echo, advanced pulse echo imaging, and acoustic resonance spectroscopy.

Impact Echo (IE)   The fundamental principle of the impact-echo method is to introduce a low-frequency stress wave into the structure by hammer impact or steel spheres. In this way, certain resonance modes of the concrete structure under investigation are excited. In particular, thickness modes of vibration are used primarily to identify backwall or planar flaws. To that end, the time-domain signal of displacement or velocity is usually detected for a few centimeters beside the impact point. Then, performing a fast Fourier transform leads to significant peaks in the amplitude and phase spectra, which can be associated with the depth of backwall or flaws if the effective velocity of propagation of longitudinal waves in the structure is known.

An important problem in the interpretation of impact-echo data is concerned with the effects of lateral boundaries of the structure under investigation. These geometric effects, caused by reflections of wavefronts at the outer boundaries of the specimen, produce systematic errors in thickness and flaw depth determination. These uncertainties must be taken into account if the measurements are performed at specimens with lateral boundaries lying in the vicinity of the measuring point.

Advanced Pulse-Echo Imaging (APEI)   The operation of APEI equipment is based on the pulse-echo principle. In this manner, one transducer transmits the ultrasonic signal, while another receives it. The main advantage offered by the APEI equipment over other devices that use pulse-echo technology is the integration of multiple sensors on a mobile matrix that enables us to scan the area under study and to record the information and for image transformation in real time. This will greatly facilitate the interpretation of pulse-echo data by providing a good indication of the condition of the structure. Among other things, the APEI technique is able to detect the location of the opposite face of the structural element, the location and size of reinforcing bars and presetressing ducts, and the scale of corrosion of tendons and reinforcing bars.

Resonance Spectroscopy for In-Service Monitoring (RSIM)   RSIM equipment has been developed to meet the need to monitor the structural integrity of concrete structures over their life cycle. An RSIM network consists of a network that is made up of one or more RSIM sensors permanently attached to the structure being monitored, and computer-connected archive software written specifically for the project. The RSIM sensors are distributed over the structure, but primarily at those locations where damage or deterioration is anticipated. The sensors transmit and receive ultrasonic signals on a continuous basis at intervals predetermined by the engineer.

Signals are sent to the computer, either via a cable or by radio frequency, supporting the archive software. The software makes records of the spectrum of the signal transmitted, and the date on which the test was performed. Moreover, and most important, it can compare the information received with that in the historical file and can decide whether the new data are significantly different from the pattern for a sound structure.

Ground-Penetrating Radar (GPR)   GPR is a portable NDT technique that has a wide range of potential applications in the testing of concrete. It is gaining acceptance as a useful and rapid technique for nondestructive detection of delaminations and the types of defects that can occur in bare or overlaid reinforced concrete decks. It also shows potential for other applications, such as measurement of the thickness of concrete members, void detection, and rebar localization. The advantages and limitations of GPR in these applications, and performance results, are discussed in detail in Section 12.3.

GPR is the electromagnetic analog of sonic and ultrasonic pulse echo methods. It is based on the propagation of electromagnetic energy through materials of various dielectric constants. The greater difference between the dielectric constants at the interface of two materials causes a greater amount of electromagnetic energy to be reflected from the interface. Looked at another way, the smaller differences yield the smaller amount of wave reflections, and consequently much more electromagnetic energy continues to propagate through the second material. In this sense, a dielectric constant difference in the propagation of electromagnetic energy is analogous to an impedance difference in the propagation of sonic and ultrasonic energy. Therefore, GPR operation is based on soil characteristics, target depth, target size, required resolution, and scanning speed. Its performance is strongly connected with the best choice of operational parameters, such as frequency band, antenna model, and coverage height, depending on the test scenarios. Furthermore, the relevant signal-processing methods, such as filtering, averaging, background removing, and range gain, should also be applied to obtain better subsurface image visibility.

GPR is quite a common technique for mapping reinforcing. It is also rapid and can locate bars up to 500 mm from the surface, although normally, this is restricted to 200 to 300 mm. GPR has very high resolution and will identify bars that are closely spaced; for example, it will distinguish two bars with a separation of 30 mm at a depth of 100 mm. One advantage of this method is that it can identify several layers of reinforcement. It can also detect cable ducts that lie deeper (behind) the surface rebars.

Radiography   The intensity of a beam of x- or γ-rays suffers a loss of intensity while passing through a material. This phenomenon is due to the absorption or scattering of the x- or γ-rays by the object being exposed. The amount of radiation lost depends on the quality of radiation, the density of the material, and the thickness traversed. The beam of radiation that emerges from the material is generally used to expose a radiation-sensitive film so that different intensities of radiation are revealed as different densities on the film.

Unlike most metallic materials, concrete is a nonhomogeneous material, a composite with a low density matrix, a mixture of cement, sand, aggregate, and water, and high-density reinforcement made up of steel bars or tendons. Radiography can therefore be used to locate the position of reinforcement bar in reinforced concrete, and estimates can also be made of bar diameter and depth below the surface. It can reveal the presence of voids, cracks, and foreign materials, the presence or absence of grouting in posttensioned construction, and variations in the density of the concrete. The main limitation of the radiography, not considering those related to safety, is that high-energy radiation is often needed because of the thick sections to be radiographed. Moreover, x-ray equipment is usually very heavy, and thus can be difficult to set up rapidly in the field. In addition, possible requirement of long distance between focus and film can result, with longer exposure time and higher system cost.

Identification of Steel-Related Damage   At present there are some nondestructive techniques based on magnetic stray field measurements which were used successfully at real structures for the detection of prestressing steel fractures. These techniques use the ferromagnetic properties of prestressing steels. Magnetic stray fields are generated at fracture areas of magnetized prestressing steels. Therefore, ruptures can be detected from typical local anomalies of the magnetic leakage field using suitable magnetic sensors.

Magnetic Flux Leakage (MFL)   The magnetic flux leakage measurement method is capable of nondestructive and contactless detection of ruptures of prestressed wires or strands. This method has been used for some time on site. But there is still the problem of the capability of this method concerning the limit of detection of ruptures. In general, prestressed concrete members contain not only prestressed tendons but also additional mild reinforcements. This mild reinforcement also generates a magnetic stray field which disturbs typical fracture signals. Several methods for signal analysis have been developed to avoid this problem [6,8]. At best, ruptures or flaws generating signal amplitudes exceeding 15 to 20% of the amplitudes of mild reinforcement can be detected by measuring the residual magnetic field.

Use of this magnetic method for the investigations of pretensioned concrete members or single bar posttensioning tendons is not problematic. Due either to the larger distance between the pretensioning steel strands or the large cross section of single posttensioning bars, the magnetic interaction between broken steel and intact surrounding steel is not insignificant; such interaction could indeed cause magnetic shielding and reduction of sensitivity. For tendons containing bundles of wires, the situation becomes more complicated. It seems that only a loss of at least about 25% of the cross section of a bundle can be detected during a first measurement. However by repeating stray field measurements, the degradation of individual tendons can be observed with better resolution.

Electromagnetic Resonance Method (EMR) The basic idea of EMR is to consider the tendon itself as an unshielded resonator located in a material with electromagnetic loss (e.g., concrete). An electromagnetic wave of variable frequency is coupled into the end of the tendon. By systematic scanning of the reflection coefficient over a spectrum from low to high frequencies, resonance frequencies of the tendon are recorded. This method works pretty well for electrically isolated tendons [11]. Use on typical prestressed members becomes more difficult, due to the fact that all reinforcements can be connected with each other. Then interpretation of the signals is barely possible.

12.3 GROUND-PENETRATING RADAR FOR ROAD CHARACTERIZATION

GPR is one of the most promising nondestructive techniques, with a wide range of potential applications in the testing of road and pavement materials, such as of concrete slabs, asphalt roads, and bridge decks. It is gaining acceptance as a useful, accurate, and rapid sensor system for nondestructive probing of subsurface layers and as a diagnostic of material quality problems in civil engineering. GPR is a commonly used technique for mapping of reinforcing, as at bridge decks. It is also capable of measurement of the layer thickness of concrete members and the detection of voids or unwanted objects under the ground.

GPR operates using electromagnetic wave propagation and backscattering principles, similar to acoustic and ultrasonic pulse echo methods. Therefore, the greater difference between layers (or materials) in the sense of constitutional parameters, which correspond to dielectric permittivity, magnetic permeability and electrical conductivity constants, yields the greater amount of electromagnetic energy reflected from an interface, object, or void. In other words, the smaller difference decreases both the reflection at the interface and the attenuation loss in the material. This fact allows an electromagnetic wave to propagate much more easily into the next layer (Fig. 12.3). The detection depth can be up to meters with centimetric resolution. For example, two iron bars located in a concrete slab at a depth of 20 cm with 2-cm separation can easily be detected by proper arrangement of the GPR system. At this point, selection of an adequate frequency band is the major critical issue in GPR operation since attenuation of the electromagnetic waves propagated in a lossy medium (e.g., ground, concrete, asphalt) and reflectivity of subsurface objects (e.g., stone, void, pipe) rely heavily on the frequency. Low frequencies are needed for deeper analysis, and higher bands are required for small-object detection. Thus, ultrawide band GPR systems are preferred to obtain operational flexibility in applications [12].

Fundamental information about GPR sensor operation and the principal performance results are given in Chapter 3. In this section we focus primarily on the presentation of suitability and the use of GPR systems for road surveys, with the goal of detecting and identifying material damage problems.

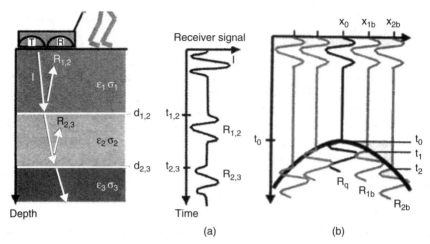

FIGURE 12.3   Operational presentation of GPR for a typical road survey: (a) A-scan image of impulse signal received from a multilayer reflection; (b) B-scan reconstruction of signals reflected from a buried object collected along the scanning direction.

12.3.1   Nondestructive Road Material Testing by GPR

The quality of roads is based primarily on the quality of the road materials. Multilayer thickness analysis and subsurface imaging of the asphalt, pavement, and concrete slabs are the essential issues in the concept. Accurate thickness measurements require knowledge of accurate electromagnetic properties of the related civil engineering materials, due to the strong correlation between physical (e.g., density, humidity, moisture content) and electromagnetic (e.g., permittivity, conductivity) parameters. Such a correlation is eventually a function of the frequency. On this scope, various studies have shown the possibility of the use of RF and microwave radar systems to extract the electromagnetic properties of the pavement subsurface. During the last decades, GPR has become one of the most useful nondestructive techniques for the survey of concrete and asphalt slabs [13].

Since the GPR method involves radiation and reception of short pulses of low-level electromagnetic energy, it is essential to determine the following issues for thickness evaluation and subsurface testing of the pavement:

1. Proper estimation of the signal velocity within the pavement, which depends on the constitutional parameters (ε, μ, σ) of the material, to convert the time delays of signals reflected from sublayers into the correct depth value. The pavement layer thickness (d) is calculated by recording the elapsed time (t) between transmission and reception of reflections according to the equation

$$d = \frac{ct}{2\sqrt{\varepsilon_r \mu_r}} \qquad (12.1)$$

where c is the speed of the light, ε_r the relative dielectric constant of the material, and t the time delay between reflections from the top and bottom of the surface. Note that $\mu_r = 1$ for most soil and pavement structures. Surface calibration should be used for conversion accuracy by placing a metal plate on the top surface.

2. Appropriate selection of the frequency band, the key factor in radar operation to meet both the required penetration depth and planar resolution for the material under test.

- Lower frequencies are needed for thick structure analysis, since surface reflectivity and soil attenuation losses increase sharply after 1 GHz, especially for compact concrete and wet pavement materials.
- Nevertheless, higher frequencies can yield better range and layer resolution to detect small objects, to image small iron bar misalignments, and to estimate small pavement layer thicknesses.
- Thus, handheld wideband GPR systems, which cover both lower and higher frequency bands, are usually proposed for such applications, due to the significant advantages mentioned above.

3. Arrangement of the radar pulse repetition frequency (PRF), depending on the data acquisition rate required. PRF is usually predefined as 100 kHz for most GPR systems. Nevertheless, it can be a very important parameter, especially for fast scanning, such as mobile road testing. For example, if GPR equipment is capable of a PRF rate of 500 kHz with 512 samples per scan, it is possible to collect around 600 impulse data per second, which may correspond to 1 cm of longitudinal resolution for a mobile GPR system with a 6-m/s scanning speed.

4. Use of surface and wheel-vibration calibration, signal preprocessing (e.g., filtering, background removal, time-varying gain), and image postprocessing (i.e., pattern recognition) techniques to obtain more accurate results and better visualization.

NDT performance results of handheld GPR from GSSI on reinforced concrete samples (Figs. 12.4a and 12.5) and on asphalt concrete road test site (Fig. 12.4b) are presented in Figs. 12.6 and 12.7, respectively.

12.3.2   Asphalt and Concrete Road Characterization by GPR

Motorway, highway, and railway road characterizations are evaluated primarily in the following classification:

1. Analysis of structural damages and pavement deformations
2. Determination of asphalt and concrete road thickness
3. Detection of joints, iron bars, dowels, ballasts, and gaps on the surface
4. Imaging of subgrades and objects buried deeply

Structural damage of road pavements (Fig. 12.8) is generally connected with the moisture level in deep layers or subgrade soils. Most such damage is not

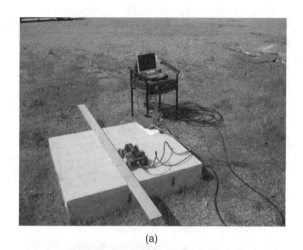

(a)

(b)

FIGURE 12.4   Laboratory NDT cases: (a) concrete slab test; (b) asphalt–concrete road test site.

FIGURE 12.5 Construction photograph of the shutter in Fig. 12.4a (side view).

visible until appearing as pavement cracks, curves, and cavities. Such pavement distortions can increase the cost of repair and maintenance in an inefficient way. Water infiltration and clayey soil pumping are two of the most important causes of the decrease in bearing capacity of the unbound layers, defined as the ability to carry a stated number of repetitions of a set load. Pavement engineers call this type of damage *pumping*. In this situation, some convenient techniques are available to

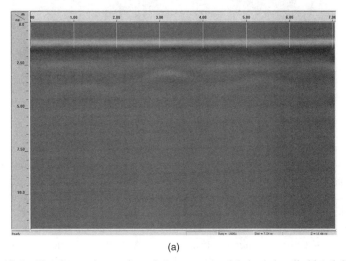

(a)

FIGURE 12.6 Nondestructive testing of the concrete slab by a handheld 1.6-GHz GPR. The GPR module scans along the middle line of the concrete slab in Fig. 12.5. (a) No data filtering (fullband: 250–3000 MHz), no range gain, and no background removal; (b) low-pass filtering (bandwidth: 200–1000 MHz) with range gain and background removal; (c) high-pass filtering (bandwidth: 1000–3000 MHz) with range gain and background removal.

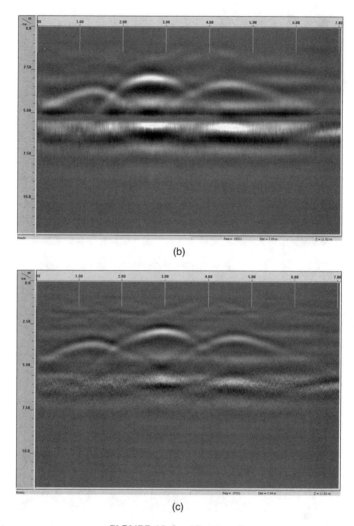

(b)

(c)

FIGURE 12.6   (*Continued*)

determine the stress and strain conditions of the roads due to the loading actions and temperature changes. Pavement engineers usually refer to the resilient modulus to characterize the elastic behavior of subasphalt unbound layers. Since this method only considers the ratio between the stress applied and returned, it is sometimes inadequate or useless for accurate assessments [14]. Layer thickness testing of a pavement structure is another important topic. A thickness deficiency of 12.5 mm on nominal 90-mm-thick pavement can lead to a 40% reduction in pavement life. Control quality is usually defined in contractions of pavement construction and demands one core for each 200 m of pavement. Obviously, this represents a huge number of cores and is greatly time consuming. It is assumed that in 95% of the cores the thickness should be equal to or greater than the design thickness, and that in the other 5%, tolerances are ±5 mm in the superficial layer, ±10 mm in the immediately subjacent layer, and 20 mm in the deeper layers [15].

GPR is used quite often in civil engineering and transportation to detect and measure subsurface cracks, subpavement gaps, and water content in soils. This method is very useful as an alternative to the traditional destructive techniques (i.e., core extraction) and achieves fairly satisfactory performances in many cases,

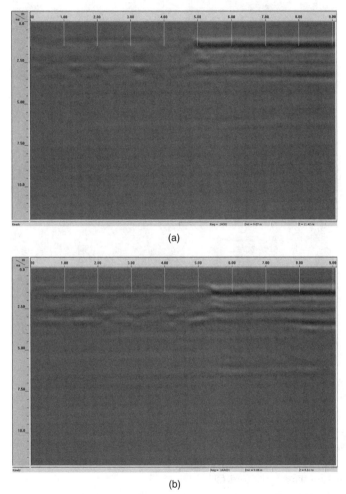

(a)

(b)

FIGURE 12.7   Subsurface survey of the road pavement by a handheld 1.6-GHz GPR (see Fig. 12.4b). Concrete block pavement-drainage asphalt; (a) no data filtering (full-band: 200–3000 MHz), no range gain, and no background removal; (b) high-pass filtering (bandwidth: 1000–3000 MHz), no range gain, but with background removal. Drainage asphalt-drainage asphalt (across wire bar joint): (c) no data filtering (fullband: 200–3000 MHz), no range gain and no background removal; (d) high-pass filtering (bandwidth: 1000–3000 MHz), no range gain, but with background removal.

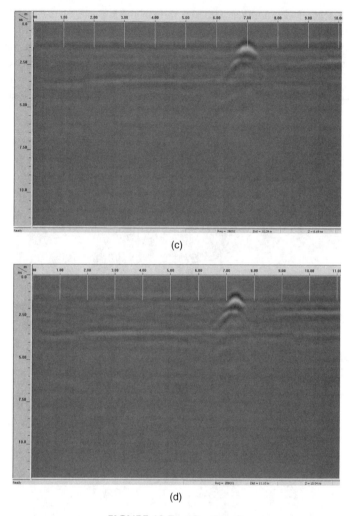

(c)

(d)

FIGURE 12.7   (*Continued*)

especially for air, asphalt/concrete pavement, and wet soil subgrade layers. As is well known from GPR operation theory, bigger differences in electric permittivity of the layers yield stronger electromagnetic wave reflections, which provide better contrast in subsurface imaging (see Chapter 3).

As mentioned previously, a typical GPR device consists principally of an RF transceiver block, a transmitter/receiver antenna head, and a system control/synchronization unit. This unit is an electronic device that generates radar trigger pulses to activate the RF impulse generator, and the transmitter antenna radiates the impulse signal into the medium being inspected. The receiver antenna senses the target reflections, and the control unit, simultaneously, stores the GPR data received, which can be processed on the platform by the user control interface.

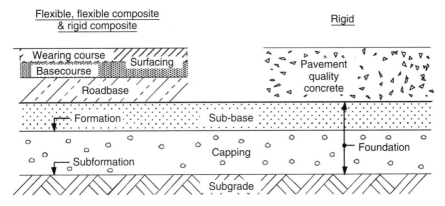

FIGURE 12.8   Multilayer view of typical road construction.

Vehicle-mounted GPR systems are generally utilized in road testing. Such GPR designs can mitigate the physical restrictions. Thus, high-gain three-dimensional antennas can be employed for wideband operations instead of ground-coupled planar antennas [12]. This type of antenna (i.e., horns) has the following major advantages over ground-coupled models:

1. High-speed road survey is available because of the air-launched antenna structure.
2. Surface calibration is also available, due to the clearly identified reflection signal from the pavement surface. It is done using a metal plate as a perfect reflector over the pavement and yields more accurate thickness measurements.
3. Since GPR can benefit from both low- and high-frequency bands, better detection and identification are obtainable both for large objects buried deeply and for very small targets at shallow depths.

Furthermore, the GPR equipment should be capable of collecting data at a high scan rate, in order to work at high speed with the desired resolution. GPR data are collected to provide real-time images that can be evaluated and preprocessed quickly in the field. The data can also be preprocessed in the office to obtain much more sophisticated assessments of structures, with great deal of detail and accuracy. Data correction is implemented by the calibration file obtained with the metal plate device, to take into account the antenna balance during a high-speed inspection survey. A vibration sensor can be mounted on wheels to compensate for the vehicle movement noise on a B-scan image. Longitudinal distance is also calibrated by rubber sheeting positioning using the data from the vehicle drive-shaft odometer. Use of GPS with proper precision can also be considered [15].

Three experimental studies have been implemented with a 1-GHz vehicle-mounted GPR system to investigate performance. The first test case is the pavement illustrated in Fig. 12.9, which consists of concrete pavement, drainage asphalt, cement concrete, and asphalt concrete blocks. The surface calibration

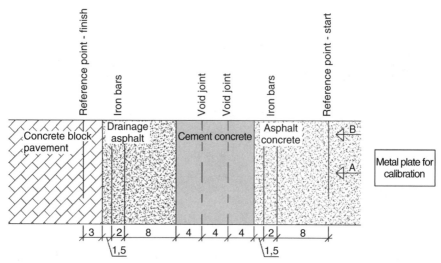

FIGURE 12.9 Pavement in test case 1. The GPR module is 32 cm above the ground; the units are meters.

process is first applied to the GPR system (Fig. 12.10). The performance results for three different preprocessing scenarios are shown in Fig. 12.11. The second test case was carried out on a recently constructed road, classified as a highway, which was selected to compare thickness values measured by cores and GPR equipment. Figure 12.12 is a photograph of the core taken from the asphalt road. Sublayer thickness plots by GPR are shown in Fig. 12.13 for various preprocessing scenarios. The EZ tracker function of Radan GSSI software allows

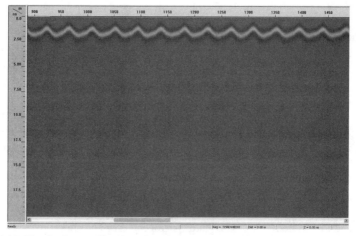

FIGURE 12.10 Surface calibration data collected by the bumper jump method for vehicle-mounted GPR operation.

us to define a continuous line representing interfaces of the pavement sublayers. The final test case is an old, corrupted pavement road with a subway bridge. The GPR measurement results shown in Fig. 12.14 are remarkably satisfactory for multilayer analysis and subsurface imaging of the road.

Remarks If we make a short interpretation here about the experimental results shown in figures for different test cases, it is obvious that adequate use of the GPR can provide great advantages to the inspection of pavement profiles at high speed with high accuracy. Many kilometers of roads a day can be surveyed in typical quality control tests, such as those for surface deformation, layer thickness, structure

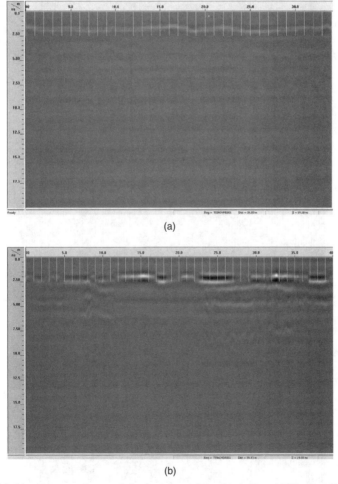

(a)

(b)

FIGURE 12.11   Test case 1: test of the pavement along the A-line. (a) No data filtering (full band: 250–3000 MHz), no range gain, but with back-ground removal; (b) no data filtering (full band: 250–3000 MHz) with range gain and background removal; (c) high-pass filtering (bandwidth: 1000–2500 MHz) with range gain and background removal.

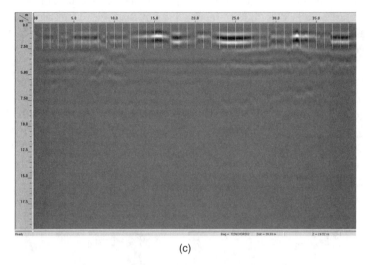

(c)

FIGURE 12.11 (*Continued*)

FIGURE 12.12 Core in test case 2. The first layer ACO 11S is 33 mm; the second layer ACP 22S is 76 mm.

damage, and subgrade object detection. Proper choice of the GPR frequency band has a majority to achieve better contrast for imaging of both near surface layers and objects buried in deep (Fig. 12.15).

12.3.3 Microwave Tomography Methods using GPR

Microwave tomographic approaches have been exploited widely in various applicative fields, such as civil engineering diagnostics, cultural heritage monitoring, archaeological prospecting, and security applications (e.g., through-wall imaging).

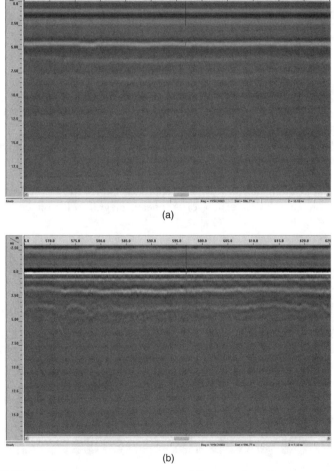

(a)

(b)

FIGURE 12.13   Test case 2: measurement around the core along the new asphalt road. (a) No data filtering (full band: 250–3000 MHz), no range gain and no background removal; (b) high-pass filtering (bandwidth: 1500–3000 MHz) with surface calibration, range gain but no background removal; (c) iterative interpretation of the interfaces of scenario b, using EZ tracker.

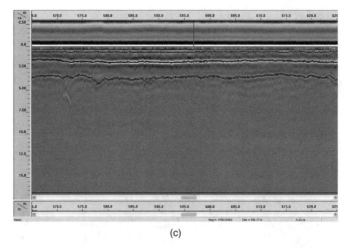

(c)

FIGURE 12.13   (*Continued*)

In this section we report some results of the use of microwave tomography–based approaches, leaving to the interested reader the opportunity of a more complete and detailed survey as presented elsewhere in this book and in the literature. In particular, the results of tomographic approach in the framework of civil engineering diagnostics and rebar location are presented here.

In this application, the tomographic approach exploited is based on the Born approximation and the inversion of a linear integral equation connecting the

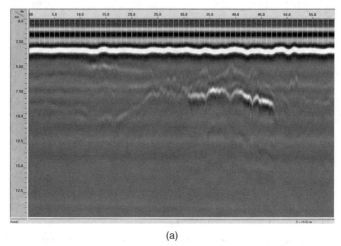

(a)

FIGURE 12.14   Test case 3: measurement along the corrupted (old) asphalt road: (a) across the railway bridge from left to right (filtered data without surface calibration); (b) EZ tracker interface plot of the data after surface calibration; (c) across the railway bridge from the opposite direction (EZ tracker interface after surface calibration).

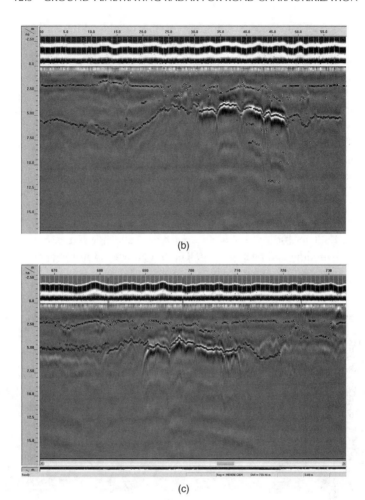

(b)

(c)

FIGURE 12.14   (*Continued*)

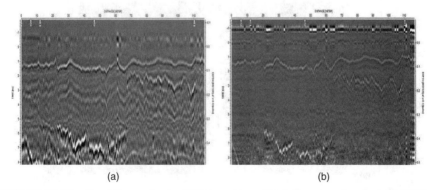

(a)   (b)

FIGURE 12.15   Road diagnostics by GPR for mid- and high-frequency bands: (a) 1.5 GHz;
(b) 4 GHz. Vertical axes: left, time (ns); right, depth (m); horizontal axis: distance (m).

unknown contrast function (accounting for the relative difference between the equivalent dielectric permittivity of the targets and that of the embedding medium) to the scattered field. The result of the overall diagnostic procedure is given in terms of the modulus of the contrast function. Regions characterized by significant values of the modulus of the contrast function especially provide information on the presence, location, and possibly, the geometry of the embedded targets [16–18].

Bridge Diagnostics The utilization of GPR as a fast and versatile tool to monitor the time behavior of transport infrastructures such as railways, tunnels, bridges, and highways is well known (Fig. 12.16). Here we are concerned with the cooperative use of high-frequency GPR measurements and linear inversion algorithms to perform the diagnostics on the Wyssenried road bridge in Switzerland [18]. The bridge was built in 1962 to cross a railway line, and it was demolished in 2004 after the railway line had been moved. The bridge was 25 m long and 5.6 m wide and was constructed of concrete with an asphalt pavement (Fig. 12.17a).

(a)   (b)

horn antenna 4 GHz   (c)   set of dipole antennas 1, 1.5 GHz

FIGURE 12.16 Mobile GPR setup.

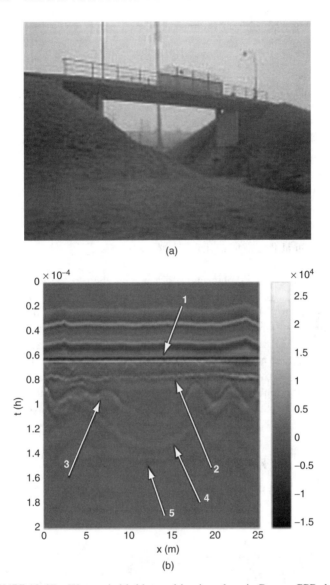

FIGURE 12.17 Wyssenried bridge and its time-domain B-scan GPR data.

GPR data were collected in December 2003 using EMPA's mobile acquisition unit, shown in Fig. 12.16a [19,20]. The system consists of a pair of GSSI model 4205 horn antennas with a nominal central frequency of 2.5 GHz, a SIR-20 radar unit, and a Trimble RTK GPS system for positioning. The acquisition parameters are summarized in Table 12.2.

A raw data set along the entire bridge length is shown in Fig. 12.17b, with a horizontal scale of 25 m and a vertical scale of 20 ns. Although the data are noisy, the following reflections can be identified:

TABLE 12.2 Data Acquisition Parameters of GPR

Parameter	Value
Acquisition speed	2.8 m/s
Horizontal sample rate	200 scans/m
Height of antenna casing above ground	0.25 m
Samples per scan	512

1. Reflection at the surface of the asphalt pavement
2. Reflection at the bottom of the pavement/the of the concrete
3. Reflection at the top layer of the reinforcing bars
4. Reflection at the tendon
5. Reflection at the bottom of the concrete slab

The microwave tomographic algorithm was applied to the two investigational zones shown in Fig. 12.17, each 2.0 m long (sections of the bridge from 4 to 6 m and from 6 to 8 m) with a depth of investigational range depth of 0.005 to 0.4 m. The "final" tomographic reconstruction result of the section from 4 to 6 m is shown in Fig. 12.18 (at a distance scale equivalent to that of the raw data). Such a figure depicts the modulus of the contrast function normalized with respect to its maximum in the overall vertical profile, and was achieved using the following strategy. First, we applied the tomographic reconstruction algorithm to subsections of 0.5 m extent, then the four reconstructions were joined to achieve the final image presented in Fig. 12.18. The parameters exploited in the inversion are reported in Table 12.3.

The value of the model relative dielectric permittivity was achieved by using the information about the depth of the top layer of the reinforcing bars and by measuring the two-way time delay. Figure 12.18 depicts an expanded scale reconstruction

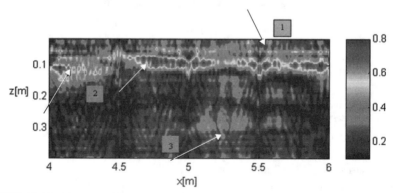

FIGURE 12.18 Tomographic reconstructions on an expanded spatial scale. The main elements of the bridge are indicated by arrows.

TABLE 12.3 Operational Parameters Used in the Inversion Algorithm

Parameter	Value
Model relative dielectric permittivity of the soil	16.4
Model conductivity of the soil	0.001 S/m
Spatial step of the measurements	0.01 m
Measurement domain	0.5 m (51 points spaced by 0.01 m)
Frequency band	300–2500 MHz
Frequency step	50 MHz (45 frequencies exploited in the inversion)
Investigation domain	0.5 m (horizontal) × (0.05–0.4 m) depth

along the depth. Different subsurface features are clearly evident and, in particular, we can recognize the following targets:

1. Bottom of the asphalt layer
2. Top layer of the rebars and tendon (in fact, these two structures cannot be separated in this section)
3. The bottom of the concrete slab

Finally, the reconstruction results relating to the bridge section between 6 and 8 m are also presented. Figure 12.19 is analogous to Fig. 12.18. In this instance it is possible to recognize the following features denoted by arrows:

1. Bottom of the asphalt layer
2. Top layer of the rebars
3. The tendon
4. The bottom of the concrete slab

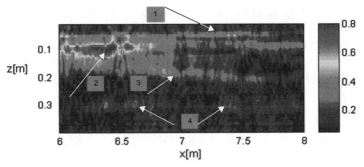

FIGURE 12.19 Tomographic reconstructions in an expanded spatial scale. The main elements of the bridge are indicated by arrows.

Rebar Detection and Localization   This test case is concerned with the detection and localization of the reinforcement bars in concrete ceiling slabs, thanks to data collected by commercial GPR [21]. This investigation was commissioned by Building Solutions, Islington (BSI), in the UK, and the local area provider of social housing. BSI were concerned that the modular blocks of flats on the Packington Estate should be reinforced adequately so that in the event of a gas explosion, the blocks would not add to the potential loss of life by collapsing. Some historical information had been obtained from archives, notably the series of patterns of reinforcement, but the record of the manner in which they had been assembled into the final blocks of flats was at best ambiguous. It was therefore necessary to establish the depth and position of reinforcement bars in the horizontal slabs of a significant sample of flats. The slabs were investigated through the ceilings rather than the floors, since there was a layer of reinforcement mesh above the rebars.

An initial investigation into the ceiling slabs of two flats, which had become vacant, revealed that these had been coated with Artex, which contains asbestos. This had significant time, health, and safety aspects, and as a result, cost implications, since extensive investigation would require safe removal of the asbestos. A further unexpected result of this trial was the discovery that in the context of a UK-wide shortage of steel in the 1960s when the flats had been constructed, there was considerable variation in the size and format of the rebars that had been used.

A trial investigation using GPR in the same flats showed good correlation in depth and position and confirmed that the radar was able to operate directly on the Artex surface (Fig. 12.20). On the basis of this trial, a more extensive GPR survey was commissioned to determine the position and depth of rebars and to attempt to measure rebar diameters, where possible. The radar used for the investigation was a Groundvue 5 GPR [21]. This radar uses twin horn antennas and has a central operating frequency of 4 GHz. The radar was mounted on a skid equipped with an optical encoder wheel for distance measurement. Measurements were taken along all four sides of the concrete slab to check for consistency of results both in the slab build and in the GPR data. The survey time sweep was 5 ns and the sampling interval 8 mm.

Here, we present the results of the tomographic approach used to process a data set of these time-domain GPR data [22]. In particular, we refer to the raw data depicted in Fig. 12.21, where a strong return at 0.5 ns is due to the direct transmission to receive antenna crosstalk. Moreover, the raw data show noticeable returns from shallower rebars and also a deeper reinforcement mesh, which are not clearly identified. Therefore, we have performed tomographic inversions to achieve clearer images of the structure under investigation.

In particular, attention has been focused on the subset in Fig. 12.22, where the contribution of the crosstalk and the air–wall interface has been erased by a temporal gating. In this radargram, the depth is reported in centimeters, making use of a propagation velocity of 0.1 m/ns. With this value, the height of the upper rebars above the ceiling is approximately 1.6 cm and the higher mesh is at a height of approximately 12 cm. In the inversion we have assumed the parameters reported in Table 12.4.

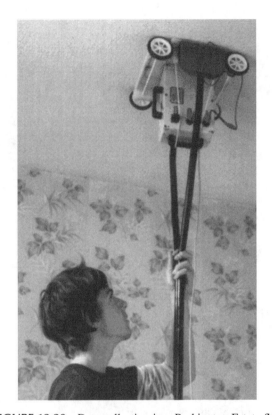

FIGURE 12.20   Data collection in a Packington Estate flat.

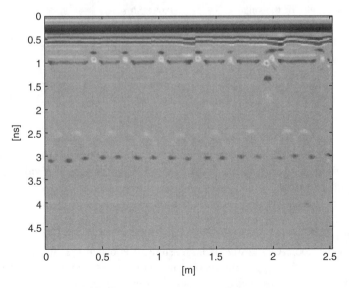

FIGURE 12.21   Raw data radargram.

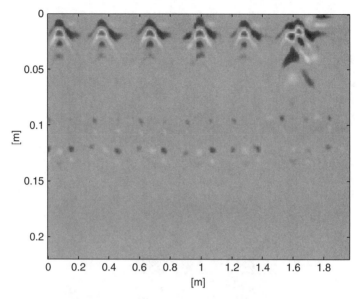

FIGURE 12.22 B-scan of the subsection investigated.

TABLE 12.4 Parameters Used in the Inversion

Parameter	Value
Model relative dielectric permittivity of the soil	9
Model conductivity of the soil	0.001 S/m
Spatial step of the measurements	0.008 m
Measurement domain	0.24 m (31 points spaced by 0.008 m)
Frequency band	1000–6000 MHz
Frequency step	200 MHz (26 frequencies exploited in the inversion)
Investigation domain	0.24 m (horizontal) × 0–0.2 m (depth)

To keep reasonable, the computational cost of the solution procedure the measurement domain of Fig. 12.22 (with extent 1.92 m) has been divided into eight subdomains of extent 24 cm. In particular, let us show where background removal has been carried out on the time-domain data. Figure 12.23 depicts the resulting radargram. From this figure it can be appreciated that the effect of the interface has been removed, but the signal from the deeper reinforcement mesh has been made weaker, too. As a result, it is possible to enhance the visualization of the rebars. Figure 12.24 depicts the reconstruction of the tomographic approach; now the rebars are clearly imaged by the adoption of a thresholding procedure on the amplitude level of the image.

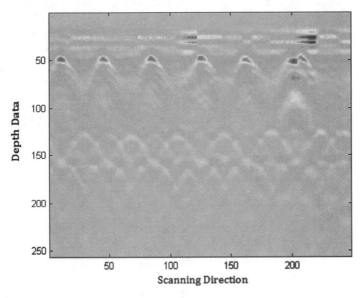

FIGURE 12.23 Raw data image after background removal.

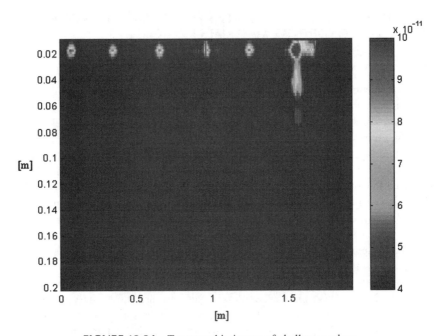

FIGURE 12.24 Tomographic image of shallower rebars.

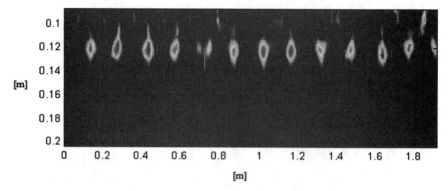

FIGURE 12.25 Tomographic image of reinforcement mesh.

The second result is concerned with the deeper reinforcement mesh. In this case we have processed time-domain data where we have removed the first 2.4 ns. Corresponding to this "cut" in the time domain, the reconstruction results have been achieved by investigating depths between 8 and 20 cm. The tomographic result, after the thresholding procedure, is presented in Fig. 12.25, and the mesh of rebars is well outlined.

12.4 EDDY CURRENT TOMOGRAPHY FOR THREE-DIMENSIONAL IMAGING IN CONDUCTIVE MATERIALS

12.4.1 Introduction

Electromagnetic methods of nondestructive control are frequently employed to restore unknown conductivity distribution in an area by measuring electrical signals detected by probes. Among them, eddy current (EC) technology is widely used in building, aircraft, chemical, power, and other industries for the detection of defects of multiple-layer metallic construction materials. The probe scans the space above a metal surface where the electrical currents (eddy currents) are excited. Methods of calculation and analysis of the behavior of ECs in metal structures, and methods of solution of inverse problems in EC testing, are considered in detail in the literature [23–26]. In this section we considered the results of a computer technology application on the basis of eddy current tomographic (ECT) imaging of defects around a system of rivets fastening two duralumin sheets.

Control of the metal product property is a complicated problem. The inner structure of the various metal machine parts may contain defects, such as cracks and flaws. Defects also arise when single parts are assembled into a module or fabric. These types of defects are often located under the surface of metal detail, or around a fastening element in an assemblage. In such cases, defects are invisible or poorly observable and ECT is a suitable technology because it makes possible the imaging of subsurface imperfections. For this purpose, the tomographic algorithm is employed to obtain a two-dimensional cross section and three-dimensional imaging of the defects. The cross-section restoration of objects is based on the tomographic integral equation solution. The integral equation is obtained by considering the inverse electromagnetic scattering problem. The experimental results obtained using standard eddy current probes operating in the frequency ranges 5 to 50 kHz and 100 to 1000 kHz are presented with a two-dimensional scanning arrangement and special software that supports collection, calculation, and visualization of the data being measured. We find that reasonably satisfactory images of the subsurface cracks in rivet heads can be obtained by using EC technology when the crack size (length) is larger than 2 mm.

An approach to imaging an anomaly can be made via almost straightforward application of a scalar, FFT-based diffraction tomography procedure [26]. An eddy current imaging method via a diffraction tomographic procedure starts with an equation that yields an anomalous electric field (E_A) on the probing line as a function of the eddy currents induced within the anomaly. Afterward, the EC problem can be solved if the inverse transform of this equation is found, and Bohbot et al. [26] proposed some feasible methods of reaching it. They considered the possibilities and limits of ECT using suitable modeling techniques. Moreover, the diffraction tomographic reconstruction methods developed so far are usually considered for far-field analysis, ignoring the excitation of near-field evanescent waves. These evanescent waves decay exponentially vs. the distance from the scattering object and contain high spatial frequency information. Therefore, they should be included into a reconstruction algorithm to achieve higher resolution.

The results of electromagnetic simulation on the imaging of subsurface objects using subsurface tomography are given by Vertiy et al. [29]. The objects investigated are dielectric cylinders embedded in the lossy dielectric interface. It is supposed that objects have weak contrast. A plane-wave spectrum of backscattered field is used for image reconstruction of the objects. It is determined that using an evanescent part of the scattered field spectrum yields images with higher resolution. The ECT method used for imaging of subsurface defects in metal samples has been considered by Vertiy et al. [27,28]. Furthermore, the experimental results obtained by EC near-field microscopy of metal structures using the tomographic imaging method are described in the following sections.

12.4.2   Description of EC Setup and Tomographic Method

The ECT measurement setup, which consists of a measuring unit with an eddy current probe, PC notebook, two-dimensional scanner, and scanner control block, is shown in Fig. 12.26. An eddy current detector is a probe that contains a coil (or set of coils) through which a time-varying electrical current is driven. The EC probe coil excites currents into the sample under test. Perturbation of this current by cracks or voids in a sample is detected by a special electronic scheme. The frequency used is sufficiently low, from a few hertz to a few megahertz; thus, the sample is placed at the near-field zone of the probes. The ECT system may be considered as a subsurface imaging device or as a scanning near-field microscope. But these

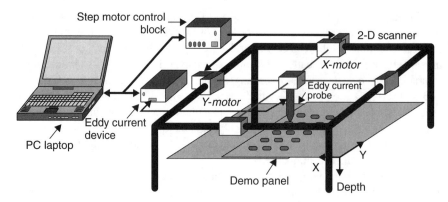

FIGURE 12.26 Eddy current tomographic setup.

types of two-dimensional eddy current images obtained without the tomographic reconstruction algorithm do not contain into depth information.

The eddy current tomographic method described by Vertiy et al. [27–29] is used for the cross-section image reconstruction of the subsurface region, which is orthogonal to the material's surface plane. It is supposed that a scattered field above the surface has only an E_y component of the electromagnetic field (the image reconstruction plane is the XOZ plane; see Figs. 5.23 and 5.39). The scattered field inside the object is formed by only one component from the plane-wave spectrum of the transmitted field, which propagates normally to the sample surface. As described in Section 5.3.1, only the real part of the complex variable of a propagation constant (β) is used at a reconstruction of the object function $K(x', y')$, which represents the normalized polarization current in the subsurface region S. The scattered field $\psi(x, y)$ measured above the sample surface is produced by that current. The tomographic image function is $|K(x', y')|$.

The evanescent part of the backscattered near field is used in the reconstruction algorithm [29]. The tomographic method described allows us to reconstruct images of cross sections of small inhomogeneity placed under the metal surface. The details of the method are described in Section 5.3.

12.4.3 ECT System Design

The ECT system shown in Fig. 12.27 has been realized as a universal device at the Tubitak Marmara Research Center (MRC), International Laboratory for High Technologies (ILHT). The ECT device consists of:

- A compact EC instrument designed and optimized for data collection sequences required for application of tomographic processing
- A mechanical three-dimensional scanner subsystem synchronized with the EC instrument to provide complicated scanning scenarios
- An Integrated software package

FIGURE 12.27 Eddy current tomographic system.

Basically, the EC instrument is implemented as a combination of a personal computer (PC), EC unit an external to the PC compact, and a low-power EC unit, which contains electronics required for operation as an EC tester. The external EC unit provides the following features:

- Operation as a conventional single- or multifrequency EC instrument working with any type of EC probe (absolute, differential, reflective)
- Operation in quick sweep-frequency mode required for application of EC tomographic processing
- Advanced multifrequency compensation circuitry to extend the dynamic range
- Embedded hardware to control a three-dimensional mechanical scanner and to synchronize its movement with data collection

The three-dimensional mechanical scanner provides positioning and scanning of samples. At present, the scanner has three linear orthogonal axes and is suitable for relatively small objects. The scanner provides a scanning area of about 150 × 150 × 150 mm with a a minimum step of 10 μm and a maximum speed 20 mm/s for each axis.

The integrated software package consists of firmware executed on a micro-controller inside the EC unit which includes a kernel and communication driver, hardware and scanner control library, main control and data collection program, and software executed on the PC (which includes the EC unit and scanner control library), data preprocessing and visualization packages, user interface, and tomographic processing program.

12.4.4 Performance Results

There are described experiments on the rivet cracks imaging by means of the ECT method used. The scheme of the sample of a duralumin panel with riveting is shown in Fig. 12.28. Here, black circles denote junctions. There are 20 rows and three columns (A, B, C) in the matrix of junctions. Thus, each rivet junction can be marked by two indexes. The size of a junction crack is defined by the number disposed close up. For example, the crack size of rivet junction A12 is 5 mm, and the scanned area is 10×15 mm. The reference coordinate system and the size of the eddy current probe head are also illustrated in Fig. 12.28.

In this experiment, 16 frequencies have been used in the bands 5 to 50 kHz and 100 to 1000 kHz with a constant frequency step. The portion of the plane-wave spectrum used is limited by values of v from -1000 to 1000 m^{-1}. Step Δv in the Fourier domain at the integration is 50 m^{-1}. The conductivity and permeability parameters of the material $\sigma_2 = 3.54 \times 10^7$ S/m, $\varepsilon_{r2} = 0.8$. The scanning has been executed both along and across the panel.

The results of experiments for the frequency bands 5 to 50 kHz and 100 to 1000 kHz are presented in Fig. 12.29a and 12.29b, respectively. Figure 12.30 illustrates horizontal and vertical (into the depth) cross sections (slices) of the A12 rivet junction on the panel shown in Fig. 12.31. The sequence of images illustrates the dynamics of three-dimensional tomographic image formation. A similar procedure has been carried out for the A12 rivet junction, which is covered by an aluminum layer of 0.5 mm thickness (see Fig. 12.31a). For step-by-step scanning, the depth slicing is presented in Fig. 12.31b.

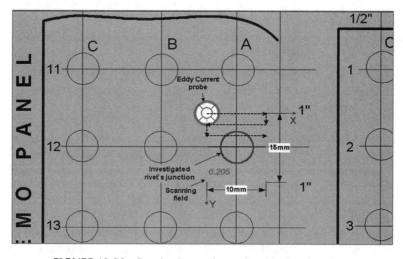

FIGURE 12.28   Duralumin panel sample with rivet junctions.

Comparing the images in Figs. 12.30 and 12.31, one can see the influence of an aluminum metal sheet on the image structure. The rivet image goes under the surface if the rivet head is covered by metal (Fig. 12.31). The technology considered here will be also useful for imaging of conductive composite material structure. The vertical slice image shown in Fig. 12.32a illustrates defects in a carbon–carbon composite block. The upper image is a photograph of the sample, and the bottom image corresponds to a tomographic slice image of the same sample.

Eddy current tomography can also provide important information related to the sample properties. Figure 12.32b illustrates the deformation field inside the metal plates connected by strong pressing. A hole has been drilled in the central part of plate. In the central part of the image (corresponding to the horizontal slice) is a horizontal slice where the deformation area is clearly seen.

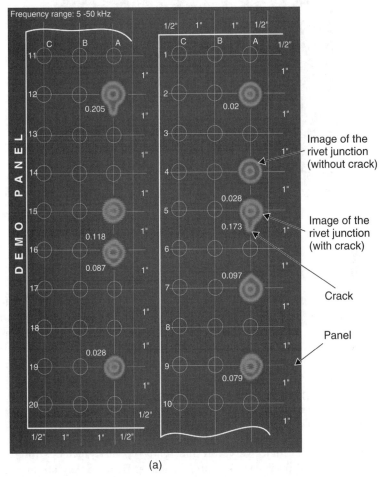

(a)

FIGURE 12.29 Demonstration panel with images of rivet junctions reconstructed by using (a) a frequency range of 5 to 50 KHz; (b) a frequency range of 100 to 1000 kHz.

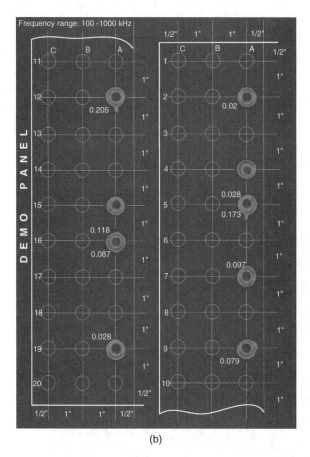

(b)

FIGURE 12.29   (*Continued*)

12.4.5   Conclusions

The performance results described in this section show the eddy current tomo-graphic technique to be a useful three-dimensional subsurface imaging technology in practice for nondestructive testing and near-field microscopy of materials in fab-rication, construction, electronics, biological, chemical, and medical applications.

12.5   ULTRASONIC METHODS FOR NONDESTRUCTIVE TESTING

12.5.1   Review of Conventional Ultrasonic Applications

Sound waves and ultrasonic waves are inherently sensitive methods for probing virtually all structural materials, including concrete. Because they are mechanical waves, sound waves and ultrasonic waves are intrinsically sensitive to the mechan-ical condition of any material; by way of contrast, electromagnetic waves are only sensitive to changes in the electrical properties of materials. Therefore, they are inherently suited to assessing mechanical properties. Moreover, sound waves and ultrasonic waves are also safe since there is no ionizing radiation, unlike γ-rays, x-rays, and microwaves.

Ultrasound waves are used extensively to test steel and other metals. They have also been used for many years in transmission testing to infer the strength of concrete through which the waves pass [30,31]. One popular embodiment of an instrument to perform this test is the PUNDIT, which has been around since the 1950s and has remained virtually unchanged, in terms of ultrasonic functionality, since then. In this application, two transducers are brought into contact with the surface of the concrete at some known distance apart. A high-voltage electrical spike is applied to a piezoelectric component inside the transmitter, which causes mechanical excitation and ultrasonic waves to be generated. An electronic counter is set to start regular timed clock pulses when the spike is applied to the transmitter. The time taken to reach the second transducer is recorded continuously, and this "time of flight" is converted into velocity using the flight length. Since the velocity of the signal is greater in denser materials, there is reasonable correlation with the strength of the concrete, and graphs such as that in Fig. 12.33 can be used to assess compressive strength indirectly.

However, the correlation is not perfect, not least because density is not the sole parameter affecting concrete strength. The type of cement, type of aggregate, and presence of additives will all influence strength but may not influence density and therefore signal velocity. Nevertheless, ultrasonic testing is a valuable tool,

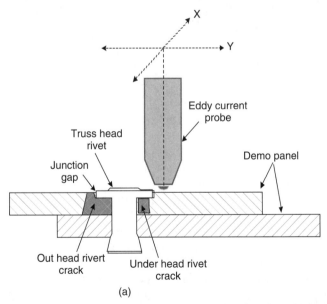

(a)

FIGURE 12.30 (a) Simplified schematic diagram of a rivet junction; (b) images of horizontal and vertical (depth) cross sections of the A12 rivet junction on the panel.

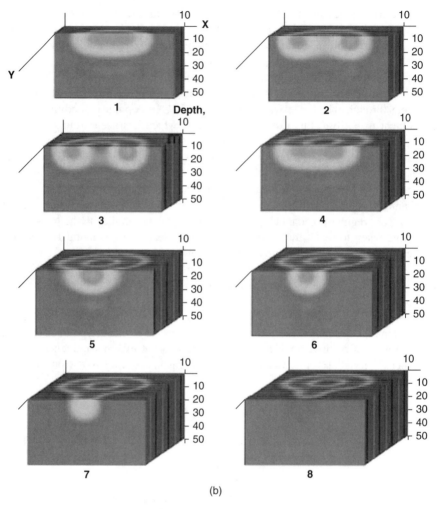

(b)

FIGURE 12.30 (*Continued*)

particularly for detecting a batch of concrete that varies from the norm within a project. The transducers can be located on opposite sides of a concrete section (through transmission) or on the same face, although existing equipment does not work well in back reflection.

In the past 20 years there have been considerable advances in the use of ultrasound in other industries and sectors. Ultrasonic pulseecho is now a widely used test technique in the medical field with high-quality safe images produced in real time. The main obstacle to similar advances in concrete is the heterogeneity of concrete with the aggregate particles, causing dispersion, multiple reflection, and attenuation of the signal. However since 1995, significant advances have been made using wideband transducer arrays and advanced signal-processing methods in order to detect reinforcement bars, tendon ducts, voids, and honeycombing in the range 0.5 to 1.8 m [32].

12.5.2 Pulse-Echo Testing

Transducer Developments Developments in transducers have led to transducers with the following new features: a broader band of frequencies, generation of known patterns of waves (and pattern detection in received signals), higher power signals, and a means of dry or no coupling [33]. High power is not an essential feature for probing concrete, because disruption of ultrasonic waves by random scattering from aggregates is the main difficulty when testing by pulse echo. High power is not essential because the absorption of ultrasound in concrete is low and because detection of know patterns of waves boosts the quality of signals substantially [34]. The essential feature is that the receiving transducer should be tolerant of the phase disruption caused by random scattering. An array of point receivers is the only successful option found so far, since large-aperture plane transducers are particularly intolerant of phase disruption and so are not suitable. It is interesting to note that large-aperture transducers have dominated concrete testing for many years, which largely explains why, traditionally, pulse-echo testing has failed to work on concrete. Medical ultrasound imaging by pulse echo is similar to concrete testing. Body tissue is a random scatterer of the wavelengths used and arrays of point receivers are used very effectively.

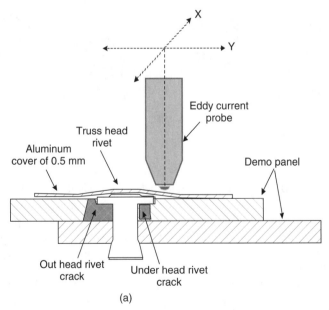

(a)

FIGURE 12.31 (a) Simplified schematic diagram of a metal sheet–covered rivet junction; (b) images of horizontal and vertical (depth) cross sections of the A12 rivet junction on the panel.

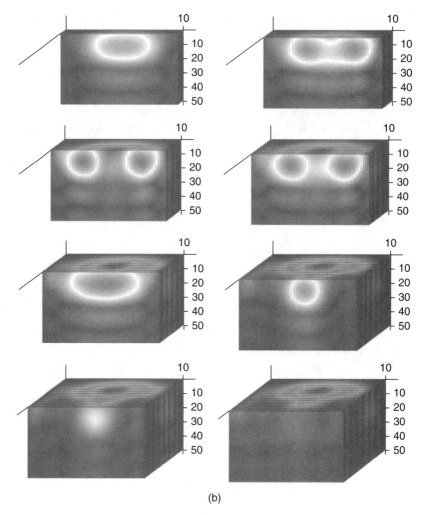

(b)

FIGURE 12.31   (*Continued*)

In the mid-1990s, a new type of transducer array was developed in Russia by Shevaldykin et al. [36]. These low-frequency broadband transducers, which have a nominal frequency of 50 kHz, emit transversal waves and use dry point contact. The transducers have a wear-resistant ceramic tip, which allows continuous use on rough and unprepared surfaces. Each transducer features an independent spring load that allows measurement on uneven surfaces.

In some applications it is an advantage if transducers can be attached permanently to concrete surfaces using coupling agents: for example, for long-term monitoring of structures. New materials matched acoustically to concrete have been developed, from which transducers can be made, that bond to cementitious mortars and consequently can be attached permanently or semipermanently to existing

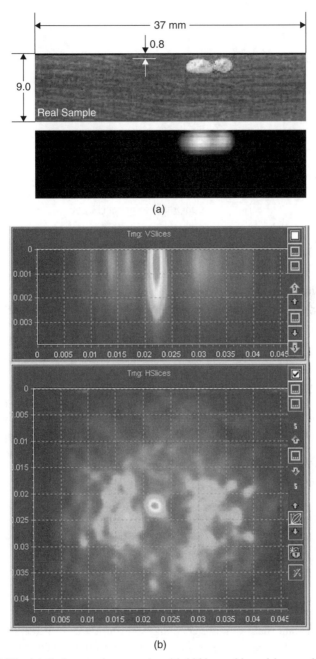

(a)

(b)

FIGURE 12.32 (a) Carbon–carbon sample with hidden voids and image obtained by an ECT system; (b) images of horizontal and vertical (depth) slices of the deformation field inside two metal plates.

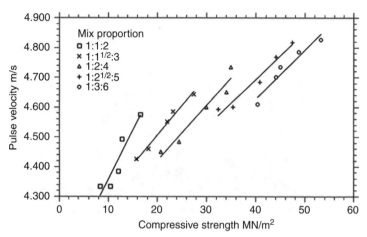

FIGURE 12.33   Relation between ultrasonic pulse velocity and compressive strength. (According to [35].)

concrete structures. This combination of properties of materials has created new possibilities for developing instruments using pulse-echo methods that assess the integrity of a structure over a long period and can be used to direct inspection to parts of a structure found to be changing, or in extreme situations, to give advance warning of a structural collapse.

Avoidance of the need for a coupling agent was the key to the development of an instrument for inspection imaging of concrete, since the process involves moving the transducers over large areas with measurements taken at relatively close centers. However, it is well known that the reliability of single-point results is seriously affected by scatter in the concrete, due to aggregates, pores, and so on. As a consequence, there may be many peaks in the A-scan, and it is difficult to differentiate that relating to a significant reflector, such as a reinforcement bar, a tendon, or a void (Fig. 12.34).

Spatial Compounding   As stated earlier, the main difficulty in pulse-echo testing is caused by the contribution of aggregate particles in concrete, causing random scattering of ultrasonic waves, particularly those close to the test surface and test location. Scattering by aggregates cannot be suppressed by averaging in time, but spatial averaging has been shown to have a positive benefit in reducing the interference effect of individual aggregate particles or reinforcement bars.

Consistently better results can be generated by combining (adding with and without phase cancellation) more echo signals from several different nearby locations, termed *spatial averaging* or *spatial compounding*. Peaks from aggregates tend to average out. The more signals used, the better the results, which lead to the use of circular arrays of multiple transducers. In these arrays of transducers, each is, in turn, a transmitter and the others are receivers. Thus, a 12-transducer array yields 132 (11 × 12) A-scans. These signals are combined and filtered (e.g., match

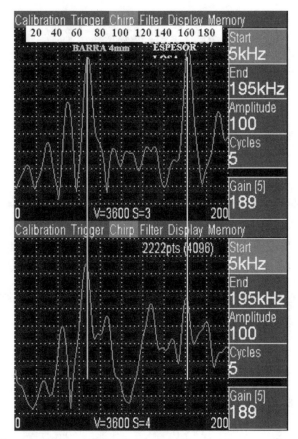

FIGURE 12.34   A-scan using APEI. Peaks marked by first line are echoes from the reinforcement bars, and peaks marked by second line are echoes from the back wall of the sample.

filtered with Hilbert transform) to detect the envelope of the signal. However, as the area covered by transducers gets larger, the lateral resolution of the system is degraded, limiting the usefulness of the system.

Synthetic Aperture-Focusing Technique (SAFT)   The SAFT provides a solution to this problem: It allows many more signals from many more test points, covering a large area, to be combined without degrading lateral resolution [37,38]. Indeed, the resolution from SAFT improves as more signals are used, so it is ideally suited to imaging in concrete.

The visualization of perpendicularly arranged tendons or bars is illustrated in Fig. 12.35a, which is a B-scan for measurements taken on the sidewall of the vertical web of a box girder bridge. This type of display gives more diffuse information and a less complete picture. One can have reasonable confidence in the location of the tendon ducts. Figure 12.35b shows the SAFT-B projection of the longitudinal

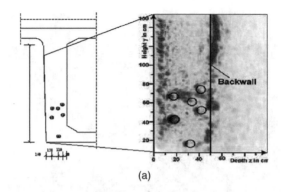

(a)

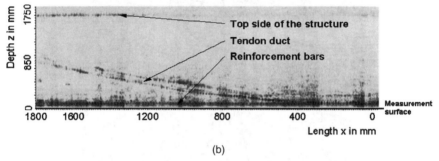

(b)

FIGURE 12.35 (a) Arrangement of tendon ducts in the cross section of a box girder web. Left: according to construction drawings. Right: located at a SAFT-B projection by ultrasonic echo. (b) SAFT-B projection of ultrasonic echo data, imaging the curvature of a tendon duct in the longitudinal section of a T-beam bridge. (From [39].)

curvature of a tendon in a bridge T-beam. Ultrasonic waves were reflected on the fully grouted tendon ducts with a low intensity compared to the reflection on the near-surface ordinary reinforcement. The backwall echo from the top surface of the beam is just discernible at 1750 mm.

However, for assessment of the grouting conditions of the tendon ducts, the arrangement of the tendons in the element and the type of tendon would need to be taken into account in the interpretation. Furthermore, the intensity of reflection depends on the arrangement of the cables in the tendon duct, the distance between the surface and the duct, and so on. Therefore, an increase in intensity is not universally associated with a void in the duct grouting.

Automation Automation of nondestructive scanning methods helps to overcome some of the disadvantages of manual point testing. Point measurements need to be repeated for statistical validation. This is time consuming and prone to manual fatigue and error, with the need to record the coordinates of the point for each test result. One approach to this, adopted within the SGIM project, is mounting the array of transducers on wheels. The array assembly is spring loaded, so it can be wheeled from one point to the next, and then by applying pressure, the

FIGURE 12.36 APEI transducer assembly prototype with transducers (internal) built by CU in 2003 for Sonatest.

transducers are brought into contact with the surface (Fig. 12.36). The rotation of the wheels is recorded, so the coordinates of the location can be generated and stored automatically. Four shear transducers (square) have not been inserted, but eight of them are present. There are also seven compression transducers (circular). The wheels and shaft encoders are visible; these are mounted on vertical sliding mechanisms to allow the array to engage the surface for a test and to allow the array to disengage and to be moved, with wheels rolling over the surface, to another position for reengaging and testing. The sheet on the floor is a silicone rubber gel to be used for rapid coupling. Figure 12.37 shown an alternative array of five

FIGURE 12.37 Array of five compression transducers mounted on an assembly.

compression transducers mounted on an assembly, which fits into predetermined positions in a frame attached temporarily to the concrete.

At the Federal Institute for Materials Research and Testing in Berlin (BAM), a number of scanners have been developed for both horizontal and vertical flat surfaces (Fig. 12.38). The large scanner covers an area of 10×4m; smaller versions exist for 1.8×1.8 m areas. Recent successful tests have supported the concept of vacuum mounting of a lightweight frame which does not require fixed mounting to the concrete surface. Mounting and testing times have been reduced significantly, with fully automated testing of a 10-m^2 area within one day being within reach [39].

Data Fusion   At BAM the on-site tests at bridges have shown that ultrasonic echo and radar methods complement each other very well. Data fusion can aid considerably in simplifying the complex data interpretation and assessment of the structures [40,41]. A useful combination of the data sets obtained with radar and ultrasonic echo in one data set is presented in Fig. 12.38 for a test area on a box girder web. Measurements were done from the inner side of the box girder. SAFT B-scans from the fused data set, with the multitude of reflections from the reinforcing bars near the surface and the reflection from the tendon duct on the left side of the B-scan, were measured mainly with radar. Radar measurement in

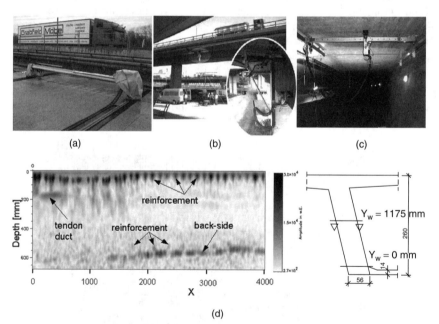

(d)

FIGURE 12.38   Automation systems developed at BAM. (From [39].) (a) Scanning system on top of a bridge deck, (b) Vertical application of a scanning system on a box girder web. (c) Application of a scanning system from the inside of a bridge deck. (d) B-scan through a box girder web at the level $y_w = 1175$ mm from a fused data set from reconstructed data from radar and ultrasonic echo.

this depth range is more suitable and useful than ultrasonic data. However, the reflection of the back side and signals from the reinforcing bar at this side, both at depths of between 45 and 60 cm, were measured exclusively with ultrasonic echos. These reflectors are located too deep to get a significant reflection with radar.

12.5.3 Structure Monitoring

Structure monitoring makes use of the high degree of repeatability of ultrasonic backscattered waves from a concrete structure [33,42]. It incorporates artificial neural networks (ANNs) to detect any significant changes in the backscattered signals caused by damage to the structure in the vicinity of each of one or many sensors. One form of structure monitoring, RSIM, uses a distributed network of up to 500 low-cost intelligent sensors all communicating with one personal computer (PC). The controlling program on the PC creates one ANN for each sensor. Output signals from each ANN are in the form of probabilities of significant structural change; each probability is calculated using Baysian methods. ANN probabilities are combined in a decision tree within the controlling PC program, which gives great flexibility to a structural engineer to incorporate particular risks pertinent to each structure. The decision tree gives a probability of significant structural change for major components of the structure and for the entire structure, providing an online risk analysis for the structure over periods of years, with no operator required. RSIM enables owners and operators of large concrete structures to assess the quality of their structures over long periods of time; operators will be able to assess the repair needs of the structures under their control more efficiently and to prioritize their repair budgets for optimum safety.

An assessment of the technique can be gained from a test in which sensors were installed on a beam at the positions shown in Fig. 12.39. The concrete beam was then subjected to bending to failure on a large compression machine. Table 12.5 shows the loads and cracks observed. Some RSIM data collected are presented in Fig. 12.40.

In the final version of RSIM software used in the tests reported here, the measured data taken from sound concrete (representing "good" data) can be migrated to become representative of "bad" data from concrete that has undergone a significant structural change. The process of migration is a form of signal processing and requires no experimentation or changes to the concrete sample. With these good and bad data, one can train an artificial neural network. After training, the ANN can then classify any RSIM data presented to it.

For the bending test, the spectra before loading were used as examples of good data, and these were migrated to synthesize bad data, then both sets were used for training. Experimental results covering all loads were then submitted to the trained ANN for classifying; the results of classification are shown in Fig. 12.41. The classification results from both sensors installed for the bending test (Fig. 12.39), are in good agreement with observations of the initiation of cracking (0.05 to 0.1 mm) after loading to 50 kN (see Table 12.5). Tests have also been implemented on samples that suffered from cracking due to corrosion of internal steel reinforcement

FIGURE 12.39 Experimental equipment used by SP to perform bending tests on a concrete beam under laboratory conditions (From [43].)

TABLE 12.5 Loads and Cracks Observed During the Bending Test

Time	Load (kN)	Crack 1	Crack 2	Crack 3	Crack 4	Crack 5	Other Remarks
8:28 to 10:28	0–40						
10:58	50	<0.05	<0.05				
11:28	60	0.1	0.1				
11:58	70	0.15	0.15				
12:28	80	0.15	0.15	0.05			
12:58	90	0.18	0.17	0.1	0.05	0.05	
13:28	100	0.22	0.22	0.1	0.05	0.05	
13:58	110	0.25	0.25	0.15	0.1	0.1	
14:28	120	0.25	0.3	0.2	0.15	0.1	
14:58	130	0.28	0.32	0.22	0.15	0.12	Shear cracks
15.28	140	0.39	0.35	0.23	0.16	0.14	Shear cracks
15:58	0			No measurement			

as opposed to external loading. The RSIM signals showed significant changes at all stages of the development of corrosion.

12.6 IMPACT ECHO

Among stress wave–based nondestructive methods, the impact echo (IE) is becoming a leading method for evaluating the quality of concrete structures [44,45]. The impact-echo method is based on the application of a short-duration stress pulse on the concrete surface by mechanical impact. The impactor can be a hammer or

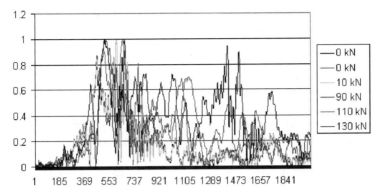

FIGURE 12.40   RSIM data for the varying loads. The horizontal axis is time, and the vertical axis is processed signal amplitude in units of volts. Gray scale in the key shows the various loads. (From [31].)

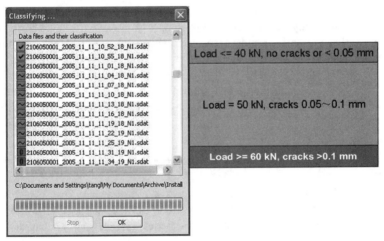

FIGURE 12.41   Classification results. The left box shows the results from the trained ANN classifying files of experimental RSIM data. Each file at the left box corresponds to a coded icon showing each classification: tick = no significant structural change (region 1: load <= 40 kN), zero = significant structural change detected (region 2: load > = 60 kN), and tilda = uncertain (region 3: load = 50 kN). Good correlation performance was achieved for ANN.

a steel ball with diameters ranging from 5 to 12.5 mm. Fast Fourier transform is generally used to analyze the impact-echo signals in the frequency domain. In recent years, the method has also been employed to detect and quantify cracking caused by alkali–silica reactivity or delayed ettringite formation [46,47]. Well-established limits and the potentiality of the impact-echo technique may be found in the literature [48]. Some laboratory results have demonstrated that the features of impact-echo waveforms provide a reliable measure of damage in concrete slabs [49]. Abraham and Côte used the impact-echo method to assess the quality of cable duct grouting in prestressed concrete structures [50]. The use of combined nondestructive methods has also been suggested for evaluating concrete model slabs containing voids and cable tunnels [51]. The impact-echo method has many advantages. It requires measurements and access to only one side of the object to be tested. Because of the easiness of performing impact-echo measurements, a larger area can be tested at greater resolution and lower cost than by using invasive methods. Impact echo is absolutely safe and does not require any special safety precautions. According to Ghorbanpoor and Benish, this method is one of the most appropriate techniques for evaluating fire damage on reinforced concrete [52].

12.6.1   Basic Principles

The IE method is an auscultation method which is nondestructive to the structures under study. It is applied mostly to concrete slabs with two parallel surfaces but is not limited to this application. It is based essentially on a frequency analysis of seismic response of the slab when subjected to a shock [48]. The basic principle was explained traditionally by the fact that the compression wave (P-wave) periodically reflects on the free surface and on surfaces with cavities (i.e., there are multiple echoes) or, more generally, on the interface of two layers of different mechanical impedances. If a slab of thickness T possessing a cavity of depth d is considered with a sensor placed near the impact point, the amounts of time necessary for the P-wave going g back and forth, τ_T and τ_d, respectively, are given by the following equations:

$$\tau_T = \frac{2T}{C_p} \qquad \tau_d = \frac{2d}{C_p} \tag{12.2}$$

where C_p is the velocity of the compression wave in the slab.

The periodicity of the many echoes, in theory, produces a discretization of the Fourier transformation (FT) at multiples of frequencies f_T and f_d that should be equal to

$$f_T = \frac{C_p}{2T} \qquad f_d = \frac{C_p}{2d} \tag{12.3}$$

It is well accepted that the thickness frequency f_T should be multiplied by shape factor $\beta = 0.96$ for the concrete [48]. In 2005, Gibson and Popovics [53] clearly explained the origin of the shape factor. They showed that it depends only on the Poisson ratio of the material and that the actual physical phenomenon for a slab is due to stationary Lamb waves. More precisely, the thickness resonance frequency of the impact echo method corresponds to the frequency at which the group velocity of the symmetric first mode of the Lamb wave (S1) is zero. When the Poisson ratio is 0.22, a classical value for concrete, shape factor β is 0.95. An open question is the pertinence of keeping the shape factor β above a void, especially a circular one, when dealing with tendon duct inspection. One can envision two possibilities:

1. The thickness is known at a particular place on the slab. Equation (12.3) gives us the value of βC_p. Knowing the value of f_d generates the value of d.
2. The velocity βC_p is not known. The FT gives information on the homogeneity of the slab and detects any "suspect" zones in which their thickness is concerned. The variation of the frequency content can also indicate the presence of a cavity.

The principal limit of the IE method when the frequency domain of the source is well chosen is the interpretation of the maxima in the Fourier domain, which becomes more difficult in the presence of noise, attenuation, and multiple waves [54]. The attenuation in reducing the observation time window will broaden the characteristic peak for thickness and will increase the amplitude of the secondary lobes. The amplitude of these secondary lobes adds to the noise measurement and is detrimental to the performance of the method. Thus, the IE method can only very rarely be transferred to the identification of soil behaviors, due to the strong internal damping of the surface materials.

For measurement above a plate that can be visualized as consisting of two layers of different acoustic properties (e.g., structures with reinforcing bars, posttensioned structures with a tendon duct fully grouted, concrete slabs with asphalt overlays), wave reflection and refraction at interfaces have to be considered. The response of a two-layer plate is, in fact, characterized by multiple reflection of the P-wave within the top layer and through the composite structures. The waveform is therefore more complex than the response of a simple plate. When the lower layer has higher acoustic impedance than the upper layer (e.g., tendon duct), the frequency corresponding to the reflections from the layer (steel bar) f_{steel} is

$$f_{steel} = \frac{C_p}{4d} \tag{12.4}$$

where d is the depth to the tendon.

If the duct is not grouted, the frequency corresponding to the slab thickness is shifted to a lower value f_T', and the frequency corresponding to the reflection from the empty duct, f_{void}, is

$$f_{void} = \frac{\beta C_p}{2d} \tag{12.5}$$

The easiest way to detect a void in a tendon duct or other density defect is to look for the thickness frequency shift from f_T to f_T' (Fig. 12.42). Figure 12.43 illustrates the frequency shift of the thickness frequency above different density defects in a concrete slab. The measurements have been taken along a 10×10 cm grid. When there is no defect, the thickness frequency is equal to 7.33 kHz. Above one massive honeycomb defect, the thickness frequency of the slab is not present; the frequency measured is the defect frequency.

Detection of the resonance frequency characteristic of a void is more a matter of conjecture. Because this resonance frequency f_{void} is very high and the surface

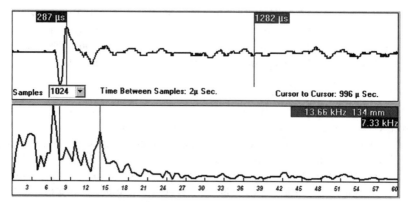

FIGURE 12.42   Impact echo result obtained over an empty PVC pipe embedded in a concrete slab: (a) time-domain signal; (b) FFT spectrum.The f_t frequency is 7.33 kHz, and the f_d frequency is 13.66 kHz, corresponding to a depth of 134 mm. The left shift of the first peak with respect to the theoretical values (the back vertical line) is noticeable.

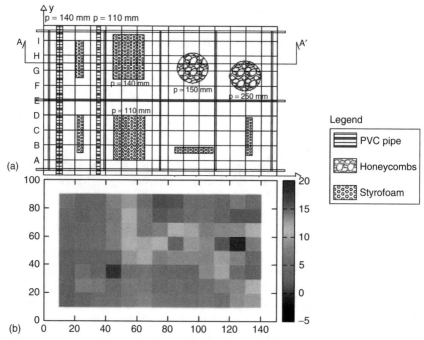

FIGURE 12.43   Application of IE techniques to detect density defects in concrete: (a) scheme of defect positioning in a concrete slab; (b) impact echo map (frequency shift in kHz).

of the reflection on the void side is small, it cannot be measured clearly with present measurement devices. Indeed, the intrinsic attenuation of concrete material and scattering of the wave due to the aggregates are so high that the resonance phenomenon under study is quite weak. Abraham has observed f_{void} experimentally above the empty duct in cases where the ratio of the void size to its depth was higher than what can be expected in current bridge beams [50]. In Fig. 12.44, the effect of increasing cover (along the arrow direction) on the thickness resonance frequency above an empty duct is evident.

12.6.2 Basic Signal Processing

Fast Fourier transform (FFT) provides information about the frequency content of a recorded impact echo signal. Unfortunately, the modulus of FFT sometimes proves to be difficult to interpret, especially when detection of a void is considered. For better extraction of frequencies, many signal-processing techniques have been tested. Consequently, it is often recommended to use a source with a frequency range close to the frequency to be determined. When the targeted frequency is raised, use of such a source always leads to dissipation phenomena, due to the diffraction and rapid attenuation caused by the aggregates and inhomogeneities of the concrete [54]. Another difficulty comes from surface waves, which provide information on the frequency content of the source: Its amplitude is often paramount in the Fourier domain and can mask useful peaks. Sansalone and Streett proposed the technique of "clipping" to reduce the influence of surface wave in clipping a signal when it exceeds a value chosen beforehand [48]. However, the nonrepeatability of the source and experimental conditions make this technique arbitrary and introduce artifact in the Fourier transform. In addition, the actual IE test signal often has a lot of noise and contains many frequencies. Thus, improvement in signal interpretation

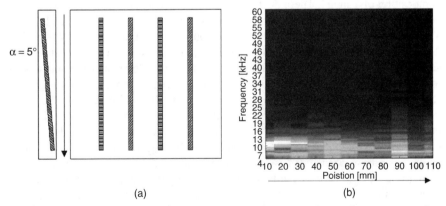

(a)   (b)

FIGURE 12.44 Example of the thickness resonance frequency change above an empty duct with increasing cover (arrow direction): (a) scheme of a concrete slab (300 mm thickness) embedding different ducts; (b) impact echogram (frequency amplitude versus position: arrow direction).

requires a sufficiently powerful processing tool. Among today's signal-processing techniques for improvement of the impact echo method, the idea of implementing time–frequency analysis on the IE signal is very reasonable, since it allows for consideration of a priori knowledge about the IE test.

Among other techniques for time–frequency analysis, the Hilbert–Huang transform (HHT) has been tested regarding its use for impact-echo data analysis. This is a rather new signal-processing method, which was developed in 1998 by Huang at the Goddard Space Flight Center of NASA [55]. The method consists basically of two parts. The first part is a decomposition of the signal into its intrinsic oscillation modes, and the second part is determining the instantaneous frequency function for each mode. The decomposition is based on an iterative algorithm called empirical mode decomposition (EMD). This approach makes it highly adaptive to the signal, as the basis for the decomposition is not a priori defined (as is the case with most other techniques, such as Fourier transforms or continuous wavelet transform) but is derived from the data itself. The output of the decomposition is in the form of intrinsic mode functions (IMFs). They are supposed to reveal the oscillation modes embedded in the signal. The instantaneous frequency obtained from the Hilbert transform makes it possible to obtain a relatively sharp representation of frequency vs. time and to reveal frequency fluctuations within the signal. Laboratory as well as on-site measurements have shown that HHT succeeds in identifying short transient signals which could not be identified using FFT. Furthermore, its application for signal filtering and its capability to analyze data from nonlinear systems could be demonstrated [56]. HHT is highly efficient due to its iterative nature. Nevertheless, it must be used carefully as a theoretical performance evaluation, which is only partially possible.

12.6.3 Continuous Wavelet Transform

Conventional Fourier analysis gives only mean coefficients representing spectral composition of a signal. These coefficients are independent of time, and thus FFT is not suitable for nonlinear and transient signals such as impact-echo signals [57,58]. The computational efficiency of the continuous wavelet transform (CWT) makes it a good tool for representing the transient features contained in nonstationary signals. The effect of CWT is similar to that of a multichannel bandpass filter that can separate the frequency components of a signal. A brief description of the wavelet theory and its application has been given by Reda Taha et al. [59].

The wavelet transform was introduced at the beginning of the 1980s by Goupillaud et al. [60] and Morlet et al. [61], who used it to evaluate seismic data. Since then, various types of wavelet transforms have been developed, and many other applications have been found [47,62]. Successful application of the CWT in analyzing impact-echo signals can be found in the literature [57,63]. The result of the application of CWT to a signal is a time–frequency domain diagram (scalogram). A scalogram is the squared magnitude of the wavelet transform of the signal and is an energy distribution of the signal in the time-scale plane (Parseval's theorem). Scalograms can be represented as images in which intensity is expressed by different shades of color. In the literature, real and complex wavelets can be found.

A wavelet is analytic if its Fourier transform is zero for negative frequencies [64]. IE signals can be processed successfully with analytic wavelets, allowing the separate evaluation of the signal phase as well as its analytic amplitude. An analytic wavelet is necessarily complex and is characterized entirely by its real part [64]. Thus, analytic wavelets can be chosen to visualize the distribution of the signal energy vs. time and frequency (or scaling). Resolution of the scalogram is limited by the Heisenberg uncertainty principle: The size of a time–frequency window cannot be made arbitrarily small, and a perfect time-frequency resolution cannot be achieved [65].

For IE data the complex Morlet wavelet can be used:

$$\Psi(x) = \sqrt{\pi f_b} \, e^{2i\pi f_c x} e^{-x^2/f_b} \tag{12.6}$$

where f_b and f_c are bandwidth parameter and wavelet center frequency, respectively.

The Morlet wavelet is similar to an impulse in geometric shape, especially when the scaling function is adjusted to be a small value. According to the matching mechanism of the wavelet transform, the better the wavelet function matches the signal in geometric shape, the more accurate the representation of the signal by wavelet coefficients [66].

Equation (12.6) does not satisfy the admissibility condition for a wavelet (strictly speaking, it does not have a zero mean). However, it is possible to enforce the admissibility condition by choosing proper parameters of f_b and f_c [65]. The condition of $2 f_c > 5$ is often sufficient to satisfy the admissibility condition [67] and leads to a negligible error in assessment of the signal frequency. In the following example the values of bandwidth and center frequency were set as $f_b = 1$ and $f_c = 4$ to process the signals acquired [68]. Setting $f_c = 1$ would have been enough to avoid the computational error explained above, but the choice of $f_c = 4$ allowed improvement in the frequency resolution of the scalogram, and according to the Heisenberg principle, the temporal resolution became lower. Analyses using scale values from 1 to 150–200 (low frequencies) in steps of 1 were performed, and the corresponding scalogram is shown in Fig. 12.45. The coloration is performed with all the coefficient values, and the wavelet coefficients at all scales are used to scale the coloration; the one-gray color map is selected. The CWT gives a good representation of the signal. The discontinuity of the signal, different frequency components, and their duration are clearly identified in the scalogram.

One of the most valuable features of wavelet transform is that it allows very precise analysis of the regularity properties of a signal. It is possible to locate very precisely those points where the signal displays abrupt changes. This is possible by analyzing the scaling behavior along special lines (i.e., local maxima) where the modulus of the CWT is concentrated. Connected lines of the local maxima are called maxima lines. Singularities in the signal can be identified by a jump in maxima modulus at specific time and frequency points. This method allows determination of the regularity of the signal at a given time point through the Hölder exponent of the signal at a given time point [69]. To calculate the Hölder exponent, the absolute value of the wavelet coefficients at a given time is needed.

Frequency 1: 4.88 kHz   Frequency 2: 11.72 kHz
Thickness 1: 300 mm   Depth target: 120 mm

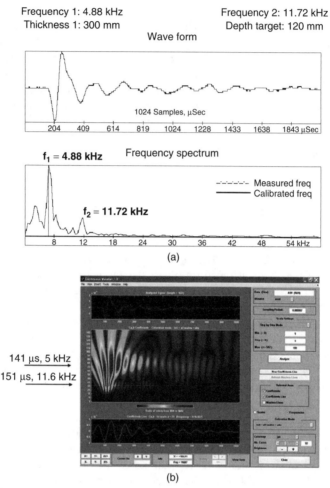

FIGURE 12.45 (a) IE signal and FFT for an empty duct embedded (120 mm depth) in a concrete slab (300 mm thickness); (b) CWT scalogram of the same signal.

12.7 DIAGNOSTIC METHODS FOR CONCRETE AND BRIDGES BY ACOUSTIC EMISSION

12.7.1 Introduction

The acoustic emission (AE) system can be used for long-term monitoring and short-term examination of structures. AE monitoring of bridges and their components requires application of mechanical or thermal stimulus. Using AE data acquisition systems during load testing of girders and following the analysis of measured data and interpretation of results are described in this section. At present, no standards have been set for field monitoring using acoustic emission techniques on structures such as concrete bridges.

12.7.2 Information Sources and Terminology

ARCHES (Assessment and Rehabilitation of Central European Highway Structures, http://arches.fehrl.org). ARCHES is an FP6 European project that was solved between 2006 and 2009. Work Package 2: Structural Assessment describes the experience using AE for soft, diagnostic, and proof-load testing of bridges.

RILEM (International Union of Laboratories and Experts in Constructions Materials, Systems and Structures, www.rilem.net). RILEM Technical Committee TC 212-ACD: Acoustic Emission and Related NDE Techniques for Crack Detection and Damage Evaluation in Concrete, began its activity in 2004. The TC prepares recommendations that concentrate on testing concrete structures using AE.

COST 534 (New Materials, Systems, Methods and Concepts for Durable Prestressed Concrete Structures, www.cost534.com). This COST action was completed in November 2007 and the final report is available. The possibilities of the acoustic emission method were studied as a part of Working Group 3: New Assessment Methods.

EWGAE (European Working Group on Acoustic Emission, www.ewgae.eu). The working group was established in 1972 to cover broad applications of the AE method in engineering practice. Every two years there is a conference where the progress in application of AE method is presented.

Important studies realized at various universities and research institutions in Europe, Japan, and the United States are described below. These studies, together with activities mentioned above, created the base for general analysis of current situation in nondestructive testing of reinforced concrete bridges and prestressed bridges by the acoustic emission method. The basic terms used in acoustic emission testing are described in the EN 1330–9 standards:

- *Acoustic emission* is a term used for elastic waves caused by sudden release of energy in material or during some process.
- *Acoustic emission testing* is a general term used to encompass all forms of examination and monitoring by AE.
- *Acoustic emission examination* is inspection or examination of an object during the application of defined loads using acoustic emission instrumentation to detect sources of AE, such as growing defects and discontinuities.
- *Acoustic emission monitoring* is continuous surveillance by AE of a component or structure in operational service or monitoring of a process.
- *Acoustic emission source* refers to the physical origin of one or more AE events.

Wave Propagation   Stress waves are generated by a sudden strain release within a solid body. The simplest case is that of an infinite medium. In air or water, only one type of wave exists. In this *compression* or *dilatation wave* (often called *P-wave*), the particles move in the direction of the wave. The propagation of stress waves in a solid such as concrete is more complex because solids can resist shear forces. In a wave mode that is independent of the P-wave, called a *distortion* or *shear wave* (*S-wave*), the particle motion is perpendicular to the direction of wave travel.

With the introduction of boundaries and interfaces such as free surfaces or cracks, a third type of wave exists, a *surface* or *Rayleigh wave* (*R-wave*). Surface waves are typically large in amplitude compared to P- and S-waves from the same source and therefore easier to detect, but their travel path can be complicated [70]. All three velocities depend on material properties E, v, and ρ, but not on the frequency t, which means that they are not dispersive.

Wave Damping   Concrete consists of aggregates in different sizes and types as well as cement and water. Conventionally (nonprestressed) reinforced concrete has its own very unique characteristics due to material heterogeneity and embedded steel reinforcement. To activate the steel reinforcement, reinforced concrete usually cracks at load levels well below capacity. Cracks dampen the progressing wave or, when wide enough, can become insurmountable barriers to wave transmission. All of them, plus the fact that concrete is a porous material that can contain water or air or both, influence the propagation of stress waves [70].

The following two basic mechanisms cause damping of a stress wave in an infinite medium.

1. Geometric attenuation is caused due to wave propagation from its origin over a larger distance. This attenuation is not frequency dependent.
2. Material attenuation is caused by absorption and scattering due to internal friction at aggregate boundaries. Material attenuation is frequency dependent and can only be determined experimentally.

Generally, higher frequencies attenuate at a higher rate than lower ones. Damping due to embedded reinforcing bars has an influence on the signal amplitude but is relatively unimportant for small-diameter rebars. Pencil lead breaks performed on a concrete surface with different lengths are generally recommended to evaluate the damping characteristics.

Data Acquisition Systems   There are two basic possibilities for recording acoustic emission signals: using a commercial system or assembling your own system. Many engineers prefer to use systems sold on the market. One example of an assembled system is the four-channel acquisition system used at the CDV Institute in the Czech Republic. It contains an efficient personal computer, a 12-bit sampling card, piezoelectric sensors (up to 1 MHz), two amplifiers, and a preamplifier. Recorded signals are processed by using software designed in an NI LabView environment and the results are presented with the help of NI Diadem software. A connection diagram for one channel of the AE data acquisition system is shown in Fig. 12.46. Employing that system in a car during in situ measurement is shown in Fig. 12.47.

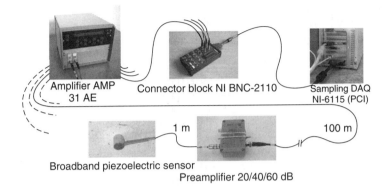

FIGURE 12.46   Connection diagram for one channel of the AE data acquisition system.

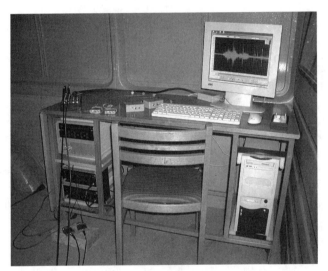

FIGURE 12.47   Placement of a four-channel AE data acquisition system in a car during a field test.

12.7.3   Basic Parameters of an AE Signal

Two types of emission can arise during application of stimulus on a structure: burst emission and continuous emission. In burst emission, acoustic emission events can be separated in time, as in bridge/girder loading. The basic (primary) parameters calculated from the time domain of the burst signal are presented in Fig. 12.48: peak amplitude, signal duration, rise time, and ring-down count (EN 1330–9). The other basic parameter is AE energy [71]. There exists a variety of definitions on AE energy, but it is generally defined as MARSE measured area under the rectified

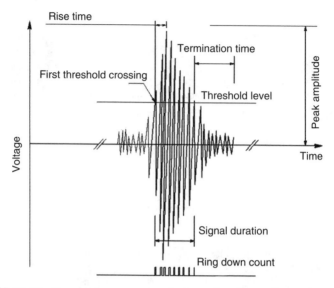

FIGURE 12.48   Four basic parameters of a burst signal according to EN 1330–9.

linear voltage time signal envelope or signal strength (known previously as relative energy). (Fig. 12.49). Signal strength is the area under the envelope of the linear voltage time signal, which normally includes the absolute area of both the positive and negative envelopes.

There are other parameters, which are calculated from the basic ones [72,73]:

- *Average frequency:* ring down count (count) divided by the signal duration
- *Initial frequency:* number of counts to the peak divided by the rise time
- *Reverberation frequency:* analogous to the initial frequency (from the peak to the end of the signal)
- *RA value:* rise time divided by peak amplitude

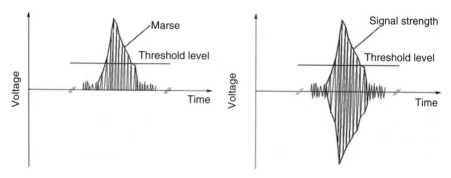

FIGURE 12.49   AE energy: MARSE and signal strength.

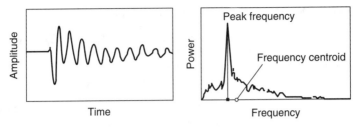

FIGURE 12.50 Time-domain and corresponding frequency spectrum of a burst signal.

Another option is to calculate a frequency spectrum from the time domain of the burst signal. For such calculations, a record of all the characteristics of the signal (waveform) is required. Fast Fourier transformation (FFT) is commonly used for this purpose (Fig. 12.50). The parameters used most often are [74]:

- *Peak frequency:* point in the power spectrum at which the peak magnitude is observed
- *Frequency centroid:* sum of magnitude times frequency divided by the total magnitude, equivalent to the first moment of inertia

More detailed analysis of each hit is possible in the form of time–frequency analysis. Wavelet transformation is often employed for this purpose.

12.7.4 Acoustic Emission Sources

The basic function of an AE system is to detect, locate, and classify emission sources. There are four basic categories of sources [70]:

1. Primary AE is generated when new damage or disintegration occurs (e.g., by an overload event where the maximum previous stress level is approached or exceeded).
2. Secondary AE is generated when the maximum stress level does not exceed the previous threshold.
3. Undesired noise is transmitted through the test setup (between the specimen/structure and the force application plates/vehicles).
4. Artificial sources are commonly used for calibration purposes or to study wave propagation.

Reinforced concrete is a composite material and has a larger number of source mechanisms than that for steel or plastic. An overview of possible AE sources in reinforced concrete is given in Table 12.6.

Classification of AE Sources AE sources are usually classified with respect to their acoustic activity and acoustic intensity [75]. Acoustic activity of a source is commonly measured by a number of AE events. Three different types of sources

TABLE 12.6 Possible AE Sources in Reinforced Concrete

Effect	Cause/Description	Category
Microcrack generation	Shrinkage, temperature, creep, low load effects, rebar corrosion	Primary, distributed
Macrocrack formation and propagation	Load effect due to shear, moment, tension forces, or rebar corrosion	Primary, from crack tip
Concrete crushing (plastic deformation)	Concrete in compression zone	Primary
Steel rebar yielding and fracture (plastic deformation)	Steel in tension, overload event, low-cycle fatigue	Primary
Rebar de-bonding (at crack planes, after crack formation)	Repeated differential loads (i.e., live loads)	Primary
Crack surface rubbing, interaction between steel rebars and concrete	Repeated differential loads (i.e., live loads)	Secondary
Artificially generated signals	Sensor pulse/pencil break	Calibration, surface
AE generated from outside the body	Experiment: slip/friction in test frame and bearings. Field: tire friction, uneven surface causes vehicle bouncing, studded tires	Undesired noise, surface
Artificial AE from within the electrical circuit/AE system	Power supplies, cables, cell phones	Undesired noise, electricity

Source: [70].

with respect to their AE activity are noted in Fig. 12.51. A source is considered to be active if its AE activity increases continuously with constant or stimulus. Such a source is considered to be critically active if the rate of change of its AE activity increases consistently with increasing stimulus or with time under a constant stimulus. The intensity of the AE source is often expressed with the help of average energy per event, average count per hit, average peak amplitude per hit, and other signal parameters. The AE source is considered to be intense when it is active and its intensity measure consistently exceeds the average intensity of active sources. The AE source is considered as critically intense when its intensity increases consistently with increasing stimulus or with time under constant stimulus. Different types of sources with respect to their AE intensity are illustrated in Fig. 12.52.

Location of AE Sources   When using source location algorithms, another characteristic of each AE source detected should be considered for source classification. The new characteristic is the position of the source located. The position of an

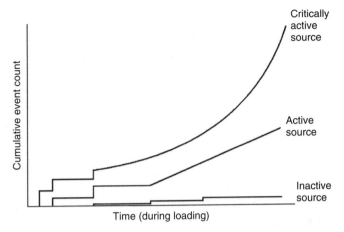

FIGURE 12.51   Different types of sources with respect to AE activity. [75].

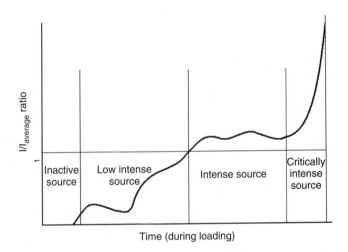

FIGURE 12.52   Different types of sources with respect to AE intensity. [75].

acoustic emission source is determined by analyzing the arrival times of an array of AE sensors. According to the number of sensors and their positions, four basic types of locations can be distinguished: zonal (near the AE sensor; one sensor is enough in this case), linear (at least two sensors; two P-waves need to be recorded for the same event), areal (three or more sensors are needed), and spatial (the minimum required number of sensors is six; at least six P-waves should be recorded for the same event).

The clustering of events located from a sharp discontinuity such as cracks is usually dense, whereas regions of plastic deformation associated with corrosion pits, for example, result in source areas that show more uncertainty in the definition of their size. In most cases, a growing crack is considered to be the more serious defect.

12.7.5 Acoustic Emission Analyses

Six basic factors should be considered when evaluating AE signals recorded during testing of a composite such as reinforced concrete girders:

1. The stress level at which the AE events occurred is important. The lower stress corresponds to the poorer structure.

2. Large increases in AE amplitudes normally indicate that a structure is near the failure level. The higher energy (amplitude) of an AE event yields the greater damage to the structure.

3. The total number of AE events is significant. The largest number of AE events indicates the greater damage to a structure.

4. The location of the AE sources within a structure is of key significance. Of much greater importance are AE events which originate at the same location. These AE events indicate a growing region of damage and a potential serious damage to the structure.

5. The rate of accumulation of AE events as a function of increasing stress (time) is quite important. When the slope of such a curve changes significantly, the rapid growth of damage indicates changes in source mechanism or flow growth becoming unstable as a precursor to total failure.

6. The value of the felicity ratio (described below) is a significant factor. The lower the value of the felicity ratio, the poorer the structure or specimen. The key experience is gained by monitoring good vs. bad specimens.

Different approaches have been developed to analyze AE signals recorded during loading of structures or specimens. The basic ones are mentioned below.

Kaiser Effect and Felicity Effect The *Kaiser effect* is defined as the absence of detectable acoustic emission until the previous maximum applied load level has been exceeded. The appearance of significant acoustic emission at a load level below the previous maximum applied level is called the *felicity effect* (Fig. 12.53). The felicity effect is observed in composite materials such as concretes: for example, when testing a reinforced concrete beam in a three- or four-beam bending mode.

At the beginning of loading, tensile cracks are nucleated at the bottom of a span, due to the bending moment. Diagonal shear cracks start to generate in the shear span before final failure. With respect to AE activity in a reinforced concrete beam, a relation between crack-mouth opening displacement (CMOD) and the presence of the Kaiser effect is known. The Kaiser effect disappears when the values of CMOD due to bending-mode failure become wider than 0.1 to 0.2 mm or if shear-mode failure is observed. It is noted that CMOD values above approximately 0.1 mm correspond to the serviceability limit of a reinforced concrete beam [76].

A common application of the Kaiser effect is in the determination of the maximum prior stress in the structure. In concretes, the Kaiser effect is merely temporary. After a long period of time the structure can heal itself so that it will

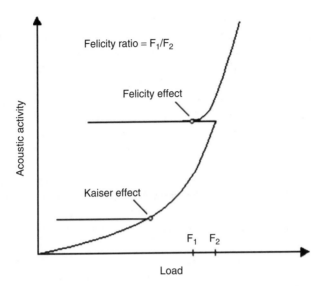

FIGURE 12.53   Kaiser effect and felicity effect.

produce acoustic emission on subsequent loading at levels lower than applied previously [71].

Felicity Ratio Analysis   The felicity ratio is defined as a term that gives a measure of the severity of damage a induced previously:

$$\text{felicity ratio} = \frac{\text{load at which significant emission restarts}}{\text{maximum load applied previously}} \qquad (12.7)$$

A decreasing felicity ratio corresponds to growing damage in the structure that is being monitored. A felicity ratio value greater than 1 implies that the Kaiser effect still holds for a structure. To identify the onset of significant emissions, one can use the force when a designated percentage of the total cumulative AE hits in the loading phase are reached (e.g., 5 to 10%) [70,77], or historic index criteria can be used [78]. The historic index is described below.

Parametric Analysis   For parametric AE data analysis, only a few key parameters, which are assumed to describe an AE signal waveform characteristically, are extracted and stored. Parameters can be processed and presented in real time and do not require large data storage space. All the parameters described in Section 12.7.3 can be used in this analysis. The most commonly used parameters are hits, counts, peak amplitude, energy, rise time, and duration. Deeper analysis of running processes in structure is possible with a combination of basic parameters, or by the use of other parameters, which can be calculated from the basic ones.

A test method for the classification of active cracks in concrete structures by the acoustic emission method, which was proposed by RILEM Technical Committee

212-ACD [76,79], is mentioned as an example. This is based on the moving average of more than 50 hits. Two parameters are calculated: the RA value (combination of rise time and peak amplitude) and the average frequency (combination of ringdown counts and duration time) [73]. The qualification of cracks (tensile and other more serious cracks) is shown in Fig. 12.54. The selection of AE sensors does not have much effect on the results. This method is suitable for in situ monitoring.

AE parameters are influenced by many factors, including specimen geometry, variability of material properties, characteristics of AE sensors, amplifiers, and data acquisition systems, as well as the selection of acquisition parameters (threshold, sampling rate, etc.). Therefore, it is recommended to set these factors as close in value as possible. Using parametric analysis requires access to the database of reference AE signals are produced by known deterioration processes. For reinforced bridge girders, this means carrying out a set of tests on the exact or at least a similar type of girder loaded with a different load level and with the occurrence of different defects.

Source locations and mechanisms are limited when monitoring large structures. Nevertheless, when examining linear elements such as girders, it is possible to make a zonal location in a group of sensors placed in a single line. A similar approach is possible for a zonal location of areal elements, such as a bridge deck.

NDIS Criterion The NDIS criterion was developed by the Japanese Society for Nondestructive Inspection. The technique basically recommends the use of load and calm ratios [74,79]. According to this standard, the loading effects characterized by the load ratio, which corresponds to the felicity ratio (sometimes called the concrete beam integrity ratio), and the unloading effects characterized by the calm ratio are combined into a damage classification chart that characterizes the current state of damage in a structure or specimen as on one of three levels: (1) minor damage, (2) intermediate damage, and (3) heavy damage.

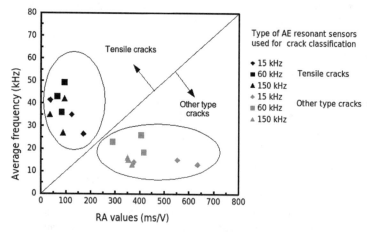

FIGURE 12.54 Qualification of damage using AE parameters. (From [80].)

The load ratio is the ratio of a load at the onset of AE activity in subsequent loading to the previous load. The *calm ratio* is the ratio of cumulative AE activity during the unloading process to the total AE activity during the last loading cycle. The load ratio is based on the Kaiser effect. A ratio greater than 1 is an indication of a structure in good condition; a value of less than 1 indicates the presence of damage. The generation of AE activity during unloading is also an indication of structural instability, as in a structure in good condition, no acoustic emission is generally recorded in the unloading phase. The amount of AE activity produced during unloading phase of the load cycle increases as shear cracks form and develop. The limit value in the calm ratio must be set individually for different objects of testing and distinct test procedures. The ratios obtained are plotted into a chart that is allocated into zones of damage as shown in Fig. 12.55.

Intensity Analysis This analysis method was developed by Fowler et al. in 1989 [81]. Intensity is a measure of the structural significance of an AE source. It consists of two parameters: the historic index and severity. The *historic index* compares the signal strength of the most recent hits with the signal strength of all hits up to that point. It is a method for determining changes in slope in the cumulative amplitude vs. number of hits curve, and is useful for identifying new damage as it occurs in the loading curve. The equation is

$$H(t) = \frac{N}{N - K} \frac{\sum_{t=K+1}^{N} S_{0t}}{\sum_{i=1}^{N} S_{0i}} \tag{12.8}$$

where N is the current number of hits up to time t, S_{0i} the signal strength (e.g., AE energy, peak amplitudes) of the ith hit, and K the empirically derived constant based on material and N. For concrete: $N < 50$, $K = 0$; for $51 \le N < 200$, $K = N - 30$; for $201 \le N < 500$, $K = 0.85N$; and for $501 \le N \le 2000$, $K = N - 35$

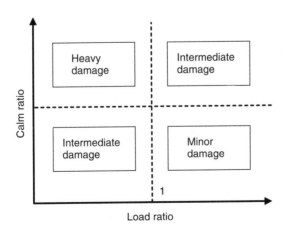

FIGURE 12.55 NDIS diagram.

[82]. For metals: $N < 15$, $K = 0$; for $16 \leq N < 75$, $K = N - 15$; for $76 \leq N < 1000$, $K = 0,8N$; and for $N > 1000$, $K = N - 200$ [83].

The *severity*, S_r computes the average of the largest signal strengths. A significant increase in the severity can indicate the onset of more serious structural damage as loading progresses. The equation is

$$S_r = \frac{1}{J} \sum_{m=1}^{J} S_{0m} \tag{12.9}$$

where S_{0m} is the signal strength (e.g., AE energy, peak amplitudes) of the mth hit and J is the number of hits over which the average should be computed. It is an empirically derived constant based on material and N. For concrete: $N < 50$, $J = 0$; for $N > 50$, $J = 50$ [82]. For metals: $N < 10$, $J = 0$; $N > 10$, $J = 10$ [83].

Both indexes are computed for each sensor independently. Zones of different stages of damage are well established from the pressure vessel industry. Sensors that plot toward the upper right of the chart (zone E: severe AE activity) indicate the greatest significance and sensors that plot either at the lower left or below a minimum severity are considered to be of lesser or no significance (zone A: insignificant AE activity) (Fig. 12.56). This method appears to work well when the loading is controllable and known. The method has recently been evaluated by several research groups for use of structural in-service testing on bridges [77,82]. The grading zones must be verified and empirical constants for J and K should be checked for different types of structures.

b-Value Analysis This analysis is based on statistical evaluation of peak amplitudes of AE hits recorded during the loading process. The basic relationship was established by Gutenberg and Richter in 1949 [84] and was used to characterize earthquake amplitude distributions. The modified magnitude–frequency distribution

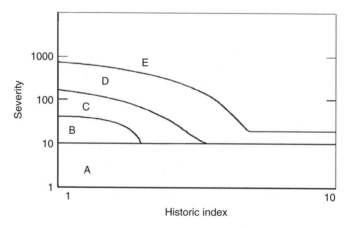

FIGURE 12.56 Typical intensity chart for a concrete structure. (Modified from [82].)

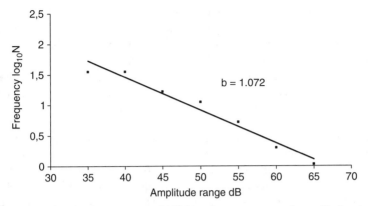

FIGURE 12.57   *b*-value from one set of 100 hits, frequency vs. peak amplitude, step 5 dB. (Adapted from [85].)

relationship for AE analysis is defined as follows [85]:

$$\log_{10} N = a - b' A_{\mathrm{dB}} \tag{12.10}$$

$$b = b' \times 20 \tag{12.11}$$

where A_{dB} is the maximal amplitude of AE hits in decibels, N the number of AE hits with magnitudes in the range $A_{\mathrm{dB}} \pm \Delta A$, a and b' are empirical constants, and b is the b-value (modified slope of the magnitude–frequency diagram). The number of consecutive AE hits used for calculation of the b-value is commonly set between 50 to 100, or can be calculated for an entire loading cycle in the case of cyclic loading (Fig. 12.57). The basic concept is that the b-value (the slope) drops significantly when stresses are redistributed and damage becomes more localized.

The b-value analysis appears especially well suited for implementation in a structural health monitoring system, since it is computationally inexpensive and theoretically, only one sensor is needed. Development of the b-value during a loading process is in good agreement with the damage observed. The onset and occurrence of damage are characterized by a sudden drop in the b-value. b-value time histories can differ from sensor to sensor. This is caused by the different sensor locations that affect the AE hit amplitude data, and the b-values are therefore a function of the sensor location with respect to the damage source [70]. The b-value must be taken into consideration during analysis of test results.

Relaxation Ratio: Energy Analysis   The relaxation ratio it is based on seismic application of energy release during fore- and aftershocks of earthquakes. The method involves looking at the AE energy that is released during loading and unloading [86]. The *relaxation ratio* is the ratio of average energy during the unloading to the average energy during the loading phase, where the average energy is calculated as the cumulative energy recorded for each phase of loading divided by the number of hits recorded. A relaxation ratio greater than 1 implies that the average energy recorded during the unloading cycle is higher than the average energy recorded during the corresponding loading cycle. Therefore, the relaxation (aftershock) is dominant; or vice versa, the loading (foreshock) is dominant.

Considering that AE activity during the unloading process is generally an indication of structural instability, a relaxation ratio (corresponding to the calm ratio in NDIS criterion) greater than 1 implies a defective structure. AE energy can be seen as a combination of two basic parameters: peak amplitude and signal duration.

Moment Tensor Analysis   This method allows localization of acoustic emission sources and distinction of different types of cracks generated during loading of concrete samples or structures. The location, type (such as shear, tensile, and mixed), and orientation of the cracks can be identified quantitatively by the SIGMA (simplified Green's functions for moment tensor analysis) procedure. Crack kinematics is represented theoretically by the moment tensor, which is defined by elastic constants and two vectors. The first vector is crack motion vector b, and the second is unit normal vector n to the crack surface. In the SiGMA procedure, two parameters are read from the signal waveform: the arrival time and the amplitude of the first motion. By applying the location procedure, the crack location can be determined from the arrival-time differences at more than five sensors. Classification of AE source into a particular crack type is performed by analysis of the moment tensor. It is assumed that dislocations of displacements at AE source consist of an opening motion (a tensile crack) and a lateral sliding motion (a shear crack).

The shear ratio (X) is a noticeable parameter for the classification of cracks. AE sources, of which the shear ratios X are less than 40%, are classified as tensile cracks. The sources of $X > 60\%$ are classified as shear cracks. Between 40% and 60% cracks are referred to as mixed mode [74].

One example of visualized generation of different types of cracks in a reinforced concrete sample during loading is shown in Fig. 12.58. This method is suitable primarily for laboratory testing of failure mechanisms of concrete structures during loading processes. The SIGMA software is built into some AE equipment.

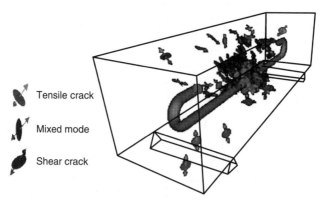

FIGURE 12.58   Three-dimensional results of the data recorded during loading by a Cosmo Player VRML viewer. (From [87].)

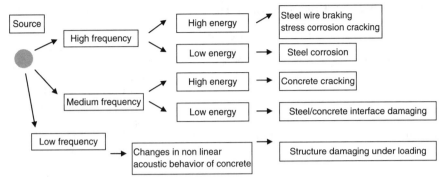

FIGURE 12.59 Connection between frequency and energy of an AE signal and relevant type of deterioration of reinforced concrete. (From [71].)

Frequency Analysis The frequency spectrum is calculated from the time domain of an AE signal. Such calculations are needed for a record of all the characteristics of the signal (waveform). The most commonly used parameters are peak frequency and frequency centroid (see Section 12.7.3). In the case of one short loading cycle, more complex analysis of the frequency spectra can be used. In many cases, the frequency spectrum of recorded signals provides detailed information about AE sources, which cannot be obtained by parametric analysis of the time domain of an AE signal. The connection between the frequency, the energy of the AE signal, and the relevant type of deterioration of the reinforced concrete can be settled as in Fig. 12.59.

The attenuation of the AE signal in concrete is frequency dependent. The high frequencies are attenuated much more quickly than low frequencies of the AE signal. It has a clear connection to the choice of AE sensor used, the distance of sensor placement on the surface of the structure, and the objective of the AE test. For example, a resonant AE sensor of 55 kHz is generally employed for the diagnosis of crack formation and propagation in concrete, whereas higher-band (such as 150 kHz) AE sensors are preferred for the analysis of more serious defects [82].

12.7.6 Laboratory and Field Test Results

Remarkable differences can be observed between the analysis of AE tests carried out on laboratory specimens (small and mid-sized samples, or full-scale samples such as girders) under controlled conditions and those carried out in the field on bridge structures (in situ). The basic differences are noted in Table 12.7.

Suitability of Different AE Analyses for Use During Laboratory or Field Testing It is difficult to say strictly which type of analysis is best for specific objective of the AE test. It can differ for different objectives that are sought. It is also based on the experience, equipment, and software tools used by the team that performs the

TABLE 12.7 Basic Differences Between Laboratory and Field AE Testing

Test	In Laboratory Conditions	In Situ
Preparation of the AE test	Usually, without serious problems	Special conditions: accessibility to examined parts of a structure, safety, traffic on a bridge, limited time, power supply, etc.
Control of conditions and environment	Full (governed)	Partial (attempt to take all conditions into consideration during evaluation); undesired noise from the suburbs
Loading	Often, cyclic loading with a given load increment and maximum loads; normally, there is a record from the loading process	Normally, only one loading cycle is carried out: static, pseudostatic (slow speed of lorries), or dynamic (traffic conditions); there is no exact record of the loading process
Used AE analysis	Often, a combination of different AE parameters; more complex calculations	Easier principles
AE source localization	Complex methodology using algorithms developed (e.g., moment tensor analysis)	Linear or zonal location is often used
AE evaluation	Even small differences are evident among a set of samples	Only definite differences are interpretable

test and analyzes the results. Table 12.8 presents the basic AE analyses described in this chapter, and the suitability of their use for laboratory or field testing based on the frequency of their use as found in the literature. The basic parameters used in the specific analysis are noted as well.

12.8 VIBROACOUSTIC MONITORING OF CONCRETE STRUCTURES

12.8.1 Introduction

Vibroacoustic monitoring and dynamic evaluation techniques can be employed successfully in structural condition assessment of prestressed concrete structure. The basic assumption is that increase and decrease of prestressing forces are accompanied by a change of stress distribution in the cross section of concrete structure, and that propagation path parameters will change due to this phenomenon.

Using a monitoring technique based on a change of dynamic characteristics to classify structures, it is necessary to determine a numerical relationship between the measurable dynamic characteristics and the degree of deterioration. The availability

TABLE 12.8 Frequency of Use of Basic AE analyses for Laboratory or Field Testing

Type of Analysis	AE and Other Parameters	Used for Laboratory Testing/Field Testing[a]
Parametric analysis	All basic AE parameters and their combination	*****/*****
Felicity ratio analysis	Hits loading/unloading	*****/**
NDIS criterion	Hits loading/unloading	*****/***
Intensity analysis	Historic index, severity	****/****
b-Value analysis	Amplitude	*****/****
Relaxation ratio	Energy ratio loading/unloading	***/*
Moment tensor analysis	Arrival times (waveform)	*****/—
Frequency analysis	Frequency (waveform)	**/*

[a]Range: *, rarely used; *****, used very often; —, not used.

of several physical models regarding the deterioration of a prestressing structure enables us to establish relationships between defects resulting from degradation of prestressed steel and stress distribution in the structure cross section. As the stress level affects the velocity of wave propagation, vibroacoustics is used to provide quantitative information on the general stress. This information is used to evaluate the overall condition with respect to the load-bearing capacity of structures using appropriate numerical modeling tools. Based on the development of stress distribution over time, a predictive evaluation of structure condition to establish the residual service life can be performed.

Research to date concerning components of machines (especially shafts) that are beam-type structures show that changes of stress in the outer layer of concrete can be detected by methods of vibroacoustic signal demodulation. Accordingly, the occurrence at the quantitative change of stress distribution leads to a change in the distribution of amplitudes around the relevant carrier frequency, usually the natural frequency of structure vibration. Thus, apart from the selection of the relevant model of the modulation phenomenon, the main issue is to define the method and the criteria of selection of modulated bands of diagnostically essential carrier frequencies.

For real-life structures, continuous acoustic monitoring using surface-applied sensors (called accelerometers) is a practical method of condition evaluation. The results obtained can indicate the occurrence and location of tendon failure. The method is capable of providing useful information, even in a noisy traffic environment, applying appropriate filtering techniques. The data can be used to support decisions regarding management and maintenance of structures.

12.8.2 Vibroacoustic Methods for Prestressed Concrete Structures

During the past few years many research tasks were linked directly or indirectly to the issue of defect detection while relying on analysis of the dynamic response of structures. Among others we point to the application of identification techniques

[88,89], methods of system transmittance analysis, FRF (frequency response function) [90], the intensively developed technique of random decrement of damping [91], as well as diagnostic techniques using information on changes in modal parameters [92]. All these methods are based on use of the relative change in frequency of proper vibration of a structure which occurs in connection with the appearance of cracks, notches, or other types of damage having an impact on the dynamic properties of structures [93,94].

Despite numerous research exercises, including analysis of the values characterizing changes in the shape of a structure's own function [95], and despite obtaining numerous interesting results (vibration-based damage detection), the method has so far failed to fully fulfill the expectations, especially when we do not have the model vibroacoustic signal for an undamaged system or under conditions of high uncertainty regarding the modeling, measurement, and analysis of the signal obtained. These difficulties increased attempts at defect identification in their early stages of development.

12.8.3   Amplitude-Modulation Phenomena in Prestressed Concrete Structure Assessment

The method of diagnosis examined relies on the relationship between the parameters of wave propagation as a function of stress occurring in a given object. This is manifested by changes in the modulation of the vibroacoustic signal parameters, which is caused by a disturbance of propagation of the sound wave in the material as a result of changes in the stress distribution in the prestressed structure cross section. Defects appearing in the structure lead to a decrease in the compressing stress, which results in a change in the stress distribution.

A useful way to evaluate the condition of a prestressed structure while underscoring the possibilities proposed by using the amplitude-modulation effects that occur in the vibroacoustic signal has been described by Radkowski and Szczurowski [96,97]. This method is based on the development of an algorithm of failure detection and identification. The authors carried out a series of tests aimed at investigating the dynamic response behavior of prestressed structure under changeable loading conditions (i.e., which is the structure's own frequency) and how it depends on its conditions and failure evolution.

A preliminary set of measurements tests were carried out on prestressed concrete beams made of B20 class concrete with dimensions of $1500 \times 100 \times 200 mm$ at the Kielce University of Technology. Specimens were placed in the bed of a strength-testing device, supported by two symmetrically placed supports. During measurements a predefined force was loaded by the machine's bending punch. For a detailed description of the tests, see the literature [98,99]. The experimental tests are performed with a load changing from 0.5 kN to 70 kN (beam failure occurring at 75 kN). The first small cracks usually emerged when a 45-kN load is applied. To observe the changes in terms of wave propagation through the beam being examined, vibration sensors were placed on the beam. Their exact locations, the locations of the impulse excitation force, and the point of loading are reported in Fig. 12.60a.

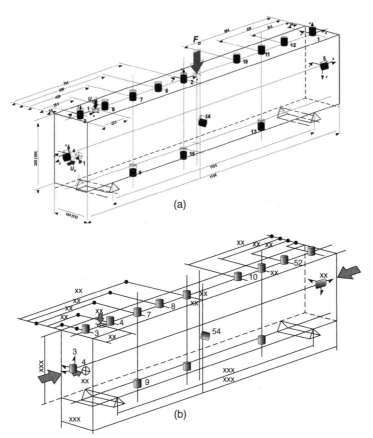

FIGURE 12.60 Location of vibration sensors and directions of vibration-causing forces in a prestressed concrete beam: (a) scheme of measurement made in Kielce and Warsaw (with perpendicular force); (b) scheme of measurement made in Poznań (without perpendicular force).

The results of measurements are to define the conditions of wave propagation caused by a pulse input. Distinct differences were observed in the values of response delay for the detector located at the opposite side of the beam (detector 5). Figure 12.61a presents an example of a signal spectrum obtained from one of the measurements. Figure 12.61b shows sidebands observed in the previous spectrum concentrated around 4800 Hz.

The change in the natural frequency as a function of crack expansion is evidenced in Fig. 12.62. It shows that load increase causes bending of a specimen, which leads to a change in the stress distribution in the beam and changes from compressive to tensile. This causes cracks to arise and expand further. It can be seen that before cracks arise there is no change in natural frequencies, but that after cracks arise, the change is pretty significant. Depending on the change of natural frequency, it is possible to infer quite a bit about crack appearance and evolution.

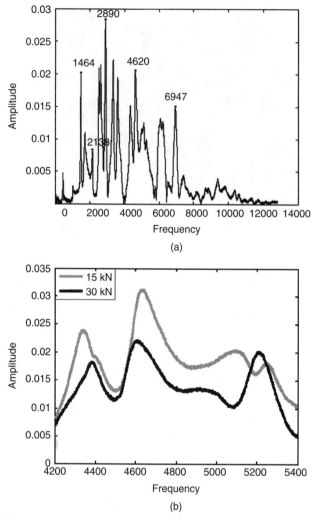

FIGURE 12.61   (a) Signal spectrum obtained from one measurement; (b) selected frequency band: changes in amplitude modulation resulting from a change of load.

This is related to the change in mechanical properties of the specimen and the stress distribution in a structure with cracks. However, this method gives no information about stress distribution at the stage before failure formation, so it is impossible to predict how close to failure the structure is.

It was also evidenced that due to the existence of dispersion phenomena in concrete, it is possible to observe not only phase velocity but also group velocity. Moreover, it is observed that modulation bands are present around the natural frequencies. These bands do not distend equivalently from the carrier frequency, and they are shifting due to the change in load.

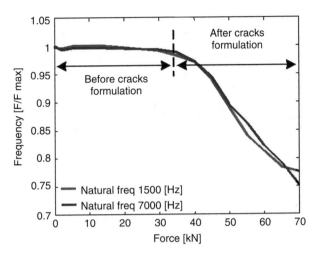

FIGURE 12.62 Normalized change of natural frequency as a function of crack propagation.

Furthermore, a set of 11 prestressed concrete beams was tested at Poznań University of Technology [100,101]. Each specimen differs in its degree of prestressing force. Beams had prestressing from 0 to 100 kN, with a step of 10 kN. Beam dimensions of 1300 × 140 × 110 mm were chosen. Specimens were produced in a factory using standard production procedures. During the investigations, the beams under test were supported once on steel prisms and then on expanders, allowing for free vibrations of a beam. This was done to verify which support gives more informative results with fewer disturbances. External forces were applied to a specimen during the measurement, to investigate the influence of prestressing force on the spectrum distribution of natural vibrations. Excitation of vibrations was made by a modal hammer, and the dynamic response of the structure was measured by a set of vibration sensors attached to the specimen surface. Here, data acquisition is provided using NI equipment with high sampling frequency.

A recorded response signal can be divided into two parts: stationary and non-stationary. The nonstationary part ends approximately 0.05 s after the signal start, and its spectrum contains many informative frequencies, such as higher natural frequencies, modulations frequencies, and additional peaks. A typical problem is the signal duration, which influences spectrum frequency resolution. The spectrum of the stationary part (from 0.05 to around 0.25 s) contains many peaks of lower frequencies related to free vibration modes (Fig. 12.63). Despite the increasing frequency resolution, the analysis of combined signals from nonstationary and stationary parts does not provide any improvement in the information contained in the signals, due to decreasing peak amplitudes.

An interesting result achieved by signal analysis is that changes in prestressing forces have no influence on wave phase velocity. Figure 12.64 shows the changes of phase velocity as a function of prestressing force, excitation waveform (below zero curves) and structure response. It is also seen that there are no unique and significant

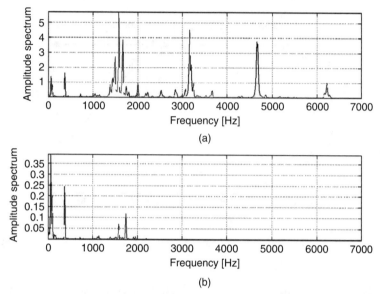

FIGURE 12.63 Spectra of (a) nonstationary and (b) stationary parts of a vibroacoustic signal.

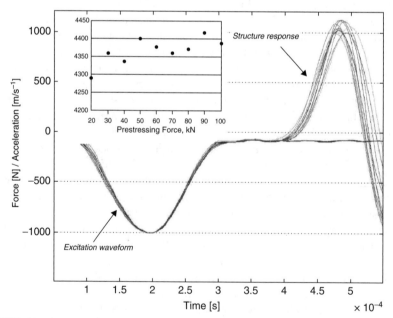

FIGURE 12.64 Changes in phase velocity as a function of prestressing force (inset), a waveform of excitation, and the structural response of the beam.

changes in the signal spectrum for different prestressing conditions (Fig. 12.65). Taking into consideration that the diagnostic experiment was performed without perpendicular forces, it is obvious that no dispersion phenomenon in prestressed structure is indicated in this way.

It should be pointed out that each specimen is a different object that has own prestressing status. That is the main reason why it is necessary to do difference preliminary assumptions despite maintaining a constant production procedure. The following conclusion can be drawn: that when the dynamic behaviors of the tested structures are very similar, changes in prestressing forces do not influence the phase velocity. Phase velocity cannot be used as a tool for inferring that stress exists in a structure. Nor is any measurable change in group velocity observed in measurements.

Based on those results, additional tests were carried at the Research Institute of Roads and Bridges in Warsaw. Tests were performed on selected beams (prestressing force 0, 3, 5, and 10 kN) from a set of specimens tested earlier in Poznań. The vibration response to an impulse excitation of each beam was registered during this measurement session. Specimens (1300 × 100 × 140 mm) were loaded, until cracks arose, in a three-bending-point condition with a distance between supports of 1060 mm. The bending force was changed from 0 kN to 19 kN in 1-kN steps.

Analysis of the signals yields very interesting results. Similar to results of measurements made in Poznań, no change in phase velocity was observed either as a function of prestressing or as a function of bending force. Signal spectra from a beam with a 10-kN prestressing force and different loads are shown in Fig. 12.66. It can be seen that natural frequencies, which are related to phase velocity stability, do not shift, in contrast to the sidebands around them, which are related to

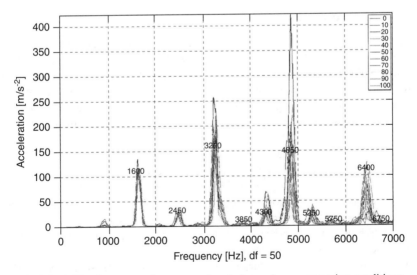

FIGURE 12.65   Spectrum of response signals for various prestressing conditions.

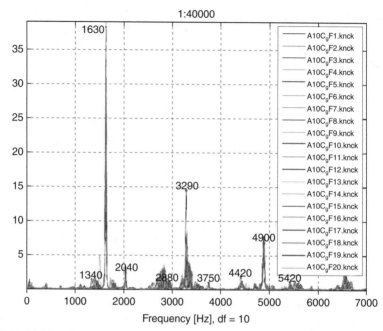

FIGURE 12.66   Signal spectra for a beam with a 10-kN prestressing force and different loads.

group velocity. This phenomenon is caused by dispersion and is quite visible on sidebands around a carrier frequency of 1630 Hz (natural frequency) in Fig. 12.67.

12.8.4   Diagnostic Algorithm

As reported by Radkowski and Szczurowski [96,99–101], dispersion phenomena cause a change in wave group velocity. This change appears as an effect of the amplitude modulation of carrier frequency. It can also be seen that the sidebands can shift, with stress increasing toward higher frequencies. The main conclusion based on such results is that changes in the phase velocity carry information about failure development, and group velocity may be a very useful tool for stress distribution estimation in a structure.

The analysis of the modulation phenomenon of a vibroacoustic signal, especially the amplitude and the frequency structure of the envelope, which is associated directly with the phase and velocity group, may prove to be an effective tool in assessment of the condition of prestressed structures. The amplitude quantity change due to the formation of structure cracks could be considered as an application example for use of the probability distribution of the acceleration envelope as a diagnostic parameter (Fig. 12.68).

On the one hand, there occurs a relationship between phase velocity and frequency; on the other, the group velocity changes lead to the phenomenon of

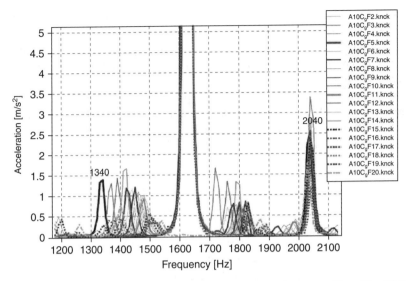

FIGURE 12.67   Zoom of spectra around a 1630-Hz carrier frequency (from Fig. 12.66).

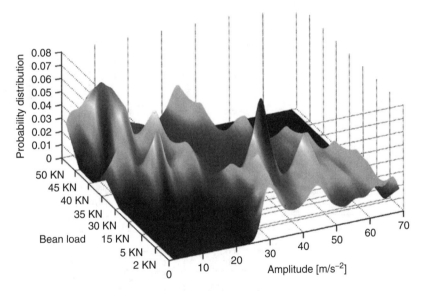

FIGURE 12.68   Probability distribution of an acceleration envelope.

amplitude modulation. The existence of a direct relation between the stress in the concrete and that in the prestressing bars, on one hand, and the values of the phase and group velocities, on the other, creates the possibility of building inverse diagnostic models and thus determining the quantitative changes of parameters, such as Young's modulus or stress in the concrete and prestresseing steel bars.

Descriptions of the phenomena mentioned above, with the relevant mathematical and diagnostic relations, have been presented by Radkowski and Szczurowski [96,99–101]. They developed an algorithm for measuring and analysis diagnostic inference to estimate the health of the prestressed structure described in Fig. 12.69 [97].

In the first step, signal registration and preliminary analysis of the correctness of registered signals are done. With the use of a signal coming from a force sensor, which is placed in a modal hammer used for excitation, the signals are rescaled and averaged by assuming a linear relationship with the excitation force. Such

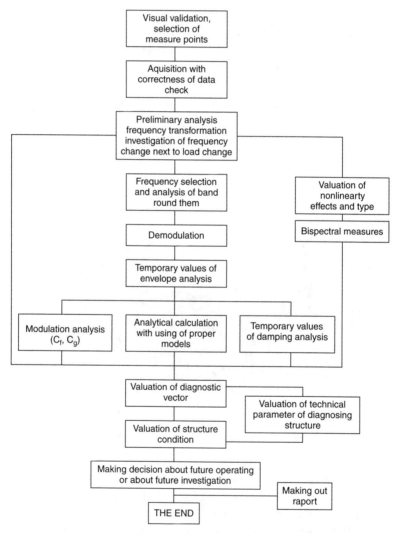

FIGURE 12.69   Diagnostic algorithm.

a relationship between the excitation force and the amplitude response can be examined in connection with prestressed concrete structures. Since the amplitude depends on the force linearly, its scale can be changed in many ways.

In the second step, a Fourier transform procedure is applied to the averaged value. At this stage we can observe the changes in the vibration frequency, which is subjected to substantial changes along with the crack emergence. Defining the frequency values of subsequent forms of vibration enables the determination of phase velocity, while its relation to frequency enables the initial estimation of stressing forces occurring in the structure being examined.

In the final part of the task, the sidebands of the vibration frequency, where we can expect the emergence of modulated bandwidths containing relevantly oriented diagnostic information, can be determined. After filtering in the bandwidths and demodulating the results, we obtain information on the instantaneous values of the envelope. At this point, attention should be drawn to the need to select relevant frequency bands, due to the possibility of omitting information in a case of improper selection. Here the phenomenon of asymmetrical modulation, discussed in earlier papers and reports, is of essential importance [101]. While observing the changes taking place on the envelope of the signal, at the beginning one can define the state of stress by asking that if it is in the area of load where stressing or stretching loads occur. It is also possible initially to determine the occurrence of defects in the object examined.

When using demodulation, it is possible to examine the phenomena associated with group velocity, which along with phase velocity enables us to define the stress in the cross section of the structure examined. While relying on the same signal, it becomes possible to define the damping of the signal examined by the structure. In accordance with the examples presented here, while relying on changes in the damping phenomenon, one can determine the occurrence of unfavorable events, such as defects in the prestressing steel or significant local moisture having influence on the strength of a structure.

The algorithm analyses tasks resulting from the description in a relevant sequence. It is worth noting the need to maintain constant control of the correctness of registered measurement results. It is an outcome of extensive requirements related to measurements, in particular registration at the sampling frequency, which enables determination of the instantaneous phase velocity of wave propagation. Respectively, in the next step, one should select the parameters of transformation into the frequency domain. The selection of the frequency bandwidths should be supported by an analysis of the dynamic model of a structure (in this case it is a beam) and by the analysis of the sensitivity of relevant bandwidth changes of group velocity as a function of its dependence on the properties of materials and the state of stress or the magnitude of prestressing forces.

Signal demodulation is performed as the next step in bandwidths selected earlier to enable estimation of the frequency structure of the signal (its envelope, in this particular case). The demodulation can be performed while using various suitable methods, such as Hilbert transform, which in the task of determining the envelope enables us to obtain highly-precise and sufficiently correct results.

The results obtained in this way allow us to define the parameters of the envelope (especially their instantaneous values), which are required for further analysis. Then, in the next step of the algorithm, these envelope parameters are used to determine the value of phase velocity and group velocity, based on the relationships mentioned above. Similarly, while using a dynamic model that describes a transient motion, it becomes possible to determine the damping value. It is noted that the algorithm essentially yields the possibility of conducting the relevant bispectral analysis, for the purpose of defining the degree of the influence and type of nonlinearity occurring or dominating in a structure.

After conducting the activities presented, a multidimensional vector of diagnostic parameters, which is the next step of the algorithm, can be used to calculate the technical parameters of the structure. Optionally, the proposed procedure offers the possibility of drafting reports in a selected format.

12.9 APPLICATION OF NUCLEAR TECHNIQUES FOR CIVIL ENGINEERING

12.9.1 Introduction and Density Gauges

Many types of manufactured materials used in civil engineering are produced with the aid of nuclear techniques. This is especially true in the construction of buildings that use sheet materials such as roofing felt, glass, linoleum, and sheet metal whose quality is controlled with radioisotopic thickness gauges. In the manufacture of cement and stone products, the most common routine applications of radioisotope instruments occur in measurement of the density of cement and asbestos slurries. Radioisotopic instruments are also used to determine the moisture content of concrete.

Gamma-transmission density gauges are ideal for measuring the density of slurries and consequently are widely used in the manufacture of nonmetallic mineral products. Applications occur in the hydraulic transport of sand, cement slurries used in making pipes, and clay slurries used in cement manufacture [102].

Radioisotopic moisture and density gauges are used extensively in the construction of roads, airfields, and earth dams. Similar instruments are used to measure the density and moisture content of concrete structures, to find defects, and to locate the position of steel reinforcing bars. When electrical power supplies are not available, gamma radiography is frequently used to inspect welds in pipelines and other structures. In suction dredging, density gauges are used to keep the solids content at the maximum concentration for efficient operation.

12.9.2 Stability of Civil Engineering Constructions and Building Foundations

The physical and mechanical properties of soils are primarily dependent on three factors: grain size, compaction, and water content. From these basic properties it is possible to derive values for plasticity, deformability, shear resistance, and permeability. These data are necessary to ensure the correct design for structural stability. Radioisotopic density–moisture gauges are generally used to acquire knowledge about bulk density and moisture content in the construction of roads, airfields, and earth dams [103]. The techniques used in these measurements can be described as an radioisotopic gauge of the surface density of soils and grounds. A device is designed to give technical control and inspection services for quick and objective estimation of the quality of fulfilled construction, hydrotechnical, and road works [103]. A schematic diagram of this type of system can be seen in Fig. 12.70, which shows density meter is a portable device with a power supply from rechargeable battery. The principle of operation is based on the dependence on the medium density of the backscattered gamma-radiation intensity from the tested medium.

The preparation of foundations in many types of civil engineering construction now require a specified soil compaction that has to be checked before the work is accepted. The standard sand-replacement methods of measuring bulk density consist of cutting a cylindrical hole in the soil approximately 10 cm diameter and 15 cm deep. The weight of soil removed is measured, and the volume of the hole is found by backfilling with sand of known density. Although the method is capable of being operated to an accuracy within ±1%, it is time consuming and the results generally vary according to the operator. With modern earth-moving and compacting equipment, rapid, accurate, on-the-spot measurements are essential if machinery is to be kept operating continuously. Overcompaction is time wasting and may cause shear surfaces to develop. It has been shown by experience that with a surface gamma backscatter probe, between 5 and 15 times as many tests can be carried out per labor hour as can be carried out with the sand-replacement method [104]. Measurements can be made on steep embankments and even on vertical surfaces and on materials such as dry sand and coarse gravel, where conventional methods are inaccurate. In many countries, radioisotopic density gauges

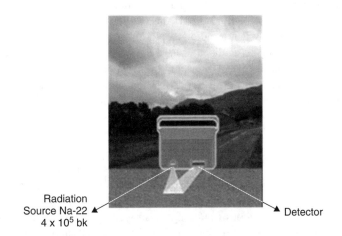

Radiation Source Na-22 4 x 10^5 bk

Detector

FIGURE 12.70 Radioactive density meter working on the backscattering principle.

are specified in the constructor's contract for checking compaction, and it should only be a matter of time before these instruments are universally accepted for this purpose.

Water penetration and migration are important factors in soil stability, and radioisotopic density-moisture gauges are now used widely to evaluate the condition of subsurface soils during the construction of roads and airfields and to study slow variations in the moisture content and density of dams, embankments, dikes, and river banks. By using access tubes, repetitive measurements of the properties of soils in a selected region can be carried out [105]. In the maintenance of some airfield landing strips, access tubes are permanently installed for continuous evaluation of subsurface conditions, in an attempt to avoid sudden breakage or subsidence following heavy loading by aircraft.

Radioactive transmission instruments are used in building construction to control concrete density and to inspect prefabricated reinforced concrete structures. As building components become smaller and lighter (to decrease the consumption of material and to lower costs), density and thickness specifications are made more precise. Classical measurements of concrete density are tedious and difficult, and gamma transmission or scatter methods provide a higher accuracy. In many countries, nondestructive testing of concrete structures by means of gamma radiography is a standard method of investigation. Gamma radiography can give us an idea of the compactness and homogeneity of concrete.

Radioisotopic instruments incorporating low-activity sources and counters are used as an alternative to gamma radiography to investigate the structure of concrete and other prefabricated materials. Transmission methods are employed for thicknesses up to 30 to 40 cm (about 80 g/cm^2). Gamma-backscatter techniques are used at greater thicknesses or when a sample is accessible from one side only.

Surface probes are also used to determine the moisture of asphalt, cement-stabilized soils, and other compacted materials. Analyses are so quick and simple that good control of the compaction of secondary roads and parking areas has become economically feasible. The seasonal migration of water in concrete is also being checked using these methods [104]. The causes of cracking and subsidence in roads have been found by using radioisotopic instruments and have also been used to provide information to settle disputes between contractors. These devices are also used to locate potential landslide areas and to investigate the efficiency of remedial drainage schemes. When refilling trenches after laying pipes and cables, a γ-ray transmission gauge has been used to control the density of the backfilling [104].

12.9.3 Localization of Reinforcing Bars

Gamma-transmission scanning systems incorporating scintillation counter detectors are now being used to inspect objects such as centrifuged reinforced hollow concrete columns used to support electric cables and telephone wires [106]. In the rebuilding of old structures or of buildings destroyed by fire, it is frequently necessary to recalculate the strength of the main load-bearing walls and supporting pillars, particularly if the original drawings have been lost. To do this, the cross-sectional dimensions of the supporting beams must be known accurately and the occurrence of hollows and channels in the beam must be identified. Gamma-transmission density gauges [107] have been used very successfully for this purpose, both to measure the density of the material of the beam and to locate steel reinforcing rods—as an alternative to chipping away concrete. Gamma-radiography is also used to locate reinforcing rods [108].

12.9.4 Density of Spoil in Suction Dredging Using Silt Density Gauges

In the construction and maintenance of navigation channels by suction dredging, the concentration of solids in the slurry must be maintained at an optimum value. If the proportion of solids is too high, the pump becomes clogged and fails. If the solids content is too low, the operation is inefficient. Since the cost of transport and dumping is many times the cost of dredging, particularly with a dry-hopper dredger, it is important to maintain the solids content at the maximum concentration. Gamma-transmission density gauges are now used extensively on suction dredgers. The first instrument for this purpose used a 20 mCi ^{137}Cs source and a Geiger counter, but nowadays ^{60}Co is also used and ionization chambers are preferred, as they can better withstand the arduous environmental conditions at the point of measurement [108].

12.9.5 Neutron Method of Moisture Measurement

The determination of moisture content is based on the slowing effect of fast neutrons during their interaction with the nuclei of the hydrogen atoms in the water contained in soil or asphalt. A measurement system for measuring moisture of the asphalt is based on a backscatter principle in which a fast neutron source and a slow neutron detector (BF_3) are on the surface of the asphalt or soil. The slow neutron detector measures backscattered slow neutrons due to the hydrogen in water. Fast neutrons have excellent penetration properties, and thus measurements through thick asphalt and across substantial distances are possible. Figure 12.71 shows the moisture measurement of soil. Soil contains mineral compounds of differing particle shapes and size, air, and water.

This water is part of the soil, and therefore an extra amount of water should be measured. This moisture component may vary with soil density. When water is added to the soil, air is replaced by water as well as some of the soil (within the given volume of soil). If the soil contains organic matter (containing hydrogen), another aspect is introduced. The soil could also contain traces of elements such as boron, cadmium, salts containing chlorine not precipitated, and other elements causing neutron absorption. Because of this, a calibration method is required to determine a relationship between probe readings and required moisture content. For any neutron probe there exists a unique calibration curve between 0 and 100% by volume of water if the calibration is done in a medium that is otherwise inert or completely devoid of any of the influences described above. This curve, known as the *calibration curve*, has no relationship to any soil, or any other compound,

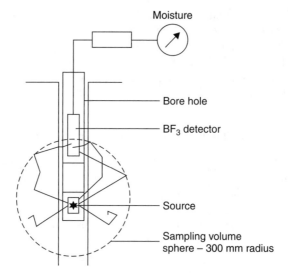

FIGURE 12.71   Neutron moisture gauge.

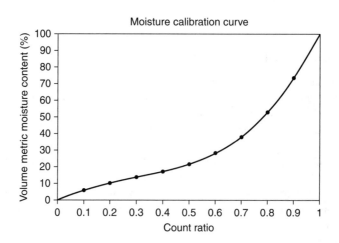

FIGURE 12.72   Typical moisture calibration curve. (From [106].)

except water, the primary measurement objective. Figure 12.72 shows the typical moisture calibration curve of soil. A calibration curve is designed such that the count ratio is for (count-air count)/(water count-air count) and the count is BF_3 slow neutron probe counts per second. The air count is the count taken in air (zero moisture), and the water count is the count taken in clean water. Real counts on this calibration curve and the real moisture content of this soil can be measured.

12.9.6   Other Applications

In the manufacture of concrete for the purpose of making panels for use in the building industry, γ-ray density gauges are used to control the amount of cement in furnaces and γ-ray-level gauges are used on ore-crushing machinery and on storage bins. When it is necessary to control a bed profile—for example, when a container is filled from several inlets—a central source inside the container and several detectors outside are used [108]. To check the strength of concrete used in constructing roads, strain gauges are sometimes buried under the surface and it is necessary to know their inclination to the road direction, as this has a strong influence on the interpretation of the measurements. By fixing two low-activity radioactive sources to the strain gauge and by using a scintillation counter on the surface, the inclination can be measured. Radioisotope ^{60}Co can be used as a marker shot into strata from a borehole and its subsequent movement measured using a scintillation counter to investigate the shrinkage of strata from which water was being pumped. In addition to these, a ^{90}Sr/^{90}Y beta backscatter gauge can be used to measure the rate at which paint applied to road surfaces for the purposes of traffic control is removed by abrasion.

REFERENCES

1. Evans, D. 2006. Engineers call for sustained infrastructure funding. *Assoc. Prof. Eng. Prince Edward Island Newsl.*, Oct.–Nov.

2. ACI Committee 201. 2008. *Guide for Conducting a Visual Inspection of Concrete in Service*. ACI 201.1R-08. American Concrete Institute, Detroit, MI.

3. ACI Committee 201. 2008. *Guide to Durable Concrete*. ACI 202.2R-08. American Concrete Institute, Detroit, MI.

4. ACI Committee 222. 2001. *Corrosion of Prestressing Steels*. ACI 222.2R-01. American Concrete Institute, Detroit, MI.

5. ACI Committee 423. 2001. *Specification for Unbounded Single-Strand Tendons and Commentary*. ACI 423.6-01/423.6R-01. American Concrete Institute, Detroit, MI.

6. Kusenberger, F. N., Barton, J. R. 1991. *Detection of Flaws in Reinforcement Steels in Prestressed Concrete Bridges*. Final Report FH-WA/RD-81/087. Federal Highway Administration, Washington, DC.

7. Muszynski, L. C., Chini, A. R., Andary, E. G. 2003. *Nondestructive Testing Methods to Detect Voids in Bonded Post-tensioned Ducts*. Report DOT HS 809 412. Florida Department of Transportation, Tallahassee, FL.

8. Sawade, G. 2001. *Mobiles SQUID-Meßsystem zur Bauwerksinspektion, Teilvorhaben Magnetisierungsvorrichtung und Signalverarbeitung*. Forschungsbericht 13N 7249/3. Bundesministerium für Bildung Wissenschaft, Forschung und Technologie, Berlin.

9. ASM. 2002. *ASM Metals Handbook*, Vol. 13, *Corrosion*. ASM International, Metals Park, OH.

10. ASM. 2002. *ASM Metals Handbook*, Vol. 11, *Failure Analysis and Prevention*. ASM International, Metals Park, OH.

11. Wichmann, H.-J., Holst, A., Hariri, K., Budelmann, H. 2004. Detection and localization of fractures in tendons by means of electromagnetic resonance measurement. In *Proceedings of the 2nd European Workshop on SHM*, Munich, July 7–9, pp. 269–276.

12. Turk, A. S. Hocaoglu. A. K. 2005. Buried object detection. In *Encyclopedia of RF and Microwave Engineering*, Vol. 1. Wiley-Interscience, Hoboken, NJ, pp. 540–559.

13. Chazelas, A. J. L., Queffelec, G. P., et al. 2007. *EM characterization of bituminous concretes using a quadratic experimental design*. In *Proceedings of the 4th International Workshop on Advanced Ground Penetrating Radar*, Naples, Italy, June 27–29.

14. Benedetto, A. 2007. Prediction of structural damages of road pavement using GPR. In *Proceedings of the 4th International Workshop on Advanced Ground Penetrating Radar*, Naples, Italy, June 27–29.

15. Costa, A., Correia, A. G. 2008. Preliminary results of a 2GHz horn antenna GPR on a pavement section in Portugal. In *Proceedings of the 3rd European Conference on Pavement and Asset Management*, Coimbra, Portugal, July 7–9.

16. Leone, G., Soldovieri, F. 2003. Analysis of the distorted Born approximation for subsurface reconstruction: truncation and uncertainties effect. IEEE *Trans. Geosci. Remote Sens.*, 41(1):66–74.

17. Persico, R., Bernini, R., Soldovieri, F. 2005. The role of the measurement configuration in inverse scattering from buried objects under the Born approximation. *IEEE Trans. Antennas Propag.*, 53(6):1875–1887.

18. Soldovieri, F., Hugenschmidt, J., Persico, R., Leone, G. 2007. A linear inverse scattering algorithm for realistic GPR applications. *Near Surf. Geophys.*, 5(1):29–42.

19. Hugenschmidt, J. 2002. Concrete bridge inspection with a mobile GPR system. *Constr. Build. Mater.*, 16:147–154.

20. Hugenschmidt, J., Partl, M. N., de Witte, H. 1998. GPR inspection of a mountain motorway in Switzerland. *J. Appl. Geophys.*, 40:95–104.

21. Utsi, V., Utsi, E. 2004. Measurement of reinforcement bar depths and diameters in concrete. In *Proceedings of the 10th International Conference on Ground Penetrating Radar*, E. Slob, A. Yarovoy, J. Rhebergen, Eds., Delft, The Netherlands, June, pp. 658–662.

22. Soldovieri, F., Persico, R., Utsi, E., Utsi, V. 2006. The application of inverse scattering techniques with ground penetrating radar to the problem of rebar location in concrete. *Non-Destr. Test. Eval. Int.*, 39(7):602–607.

23. Dodd, C. V., Deeds, W. E. 1968. Analytical solutions to eddy-current probe-coil problems. *J. Appl. Phys.*, 39(6):2829–2838.

24. Ishibashi, K. 1998. Eddy current analysis by integral equation method utilizing loop electric and surface magnetic currents as unknowns. *IEEE Trans. Magn.*, 34(5):2585–2588.

25. Sabbagh, H. A., Sabbagh, L. D. 1986. An eddy-current model for three-dimensional inversion. *IEEE Trans. Magn.*, 22(4):282–291.

26. Bohbot, R. O., Lesselier, D., Duchene, B. 1994. A diffraction tomographic algorithm for eddy current imaging from anomalous fields at fictitious imaginary frequencies. *Inverse Prob.*, 10:109–127.

27. Vertiy, A.A., Gavrilov, S.P. 2003. The eddy current tomography application for the subsurface imaging. In *Proceedings of the International Conference LEOTEST' 2003 (VIII International Scientific Technical Conference and Exhibition on Physical Methods*

and Means for Media, Materials and Product Testing), Slavske (Lviv region), Ukraine, February 17–22, pp. 90–93.

28. Vertiy, A. A., Gavrilov, S. P., Stepanyuk, V. N., Voynovskyy, I. V., Uchanin, V. N. 2004. Subsurface imaging by deep penetrating eddy current tomography. In *Proceedings the Tenth International Workshop on Electomagnetic Nondestructive Evaluation(ENDE')*, Michigan State University, East Lansing, Michigan, June 1–2, pp. 91–92.

29. Vertiy, A. A., Gavrilov, S. P., Voynovskyy, I. V., Stepanyuk, V. N., Ozbek, S. 2002. The millimeter wave tomography application for the subsurface imaging. *Int. J. Infrared Millimeter Waves*, 23(10):1413–1444.

30. BS1881. 1986. *Testing Concrete: Recommendations for Measurement of Ultrasonic Pulses in Concrete*. BS1881: Part 203. British Standards Institution, London.

31. Elvery, R. H. 1971. Nondestructive testing of concrete and its application to specifications. *Concrete*, May, pp. 137–141.

32. Jahnsohn, R., Kroggel, O., Ratman, M. 1995. Detection of tendon ducts with the ultrasonic impulse echo method: an interlaboratory test at BAM, Berlin. *Damstadt Concrete* 10:305–316.

33. Andrews, D. R., Hughes, A. M. 1991. Novel ultrasonic transducers for inspecting concrete. In *IEEE Symposium on Ultrasonics and Ferroelectrics*, pp. 349–352.

34. Andrews, D. R. 2001. Ultrasonics and acoustics. In *Encyclopedia of Physical Science and Technology*, 3rd ed. Academic Press, San Diago, CA.

35. Jones, R., Gatfield, E. N. 1955. *Testing Concrete by Ultrasonic Pulse Technique*. DSIR Road Research Technical Paper 34. HMSO, London.

36. Shevaldykin, V., Samakrutov, A., Kozlov, V. 2003. Ultrasonic low-frequency short-pulse transducers with dry point contact: development and application. In *Proceedings of the International Symposium on Non-destructive Testing in Civil Engineering*, Berlin, Germany, September 16–19.

37. Mayer, K., Marklein, R., Langenburg, K. J., Kreuter, T. 1990. Three dimensional imaging systems based on Fourier transformation synthetic aperture focusing technique. *Ultrasonics*, 28:241–255.

38. Schickert, M., Krause, M., Muller, W. 2003. Ultrasonic imaging of concrete elements using reconstruction by synthetic aperture focusing technique. *ASCE J. Mater.*, 15(3):235–246.

39. Wiggenhauser, H., Streicher, D., Algernon, D., Wostmann, J., Behrens, M. 2007. Automated application and combination of non-destructive echo methods for the investigation of post-tensioned concrete bridges. In *Concrete Platform'07, International Conference*, Belfast, UK, Apr., pp. 261–270.

40. Kohl, C., Streicher, D. 2006. Results of reconstructed and fused NDT-data measured in the laboratory and on-site at bridges. *Cem. Concr. Compos.*, 28:402–413.

41. Kohl, C., Krause, M., Maierhofer, C., Wostmann, J. 2005. 2-D and 3-D visualization of NDT data using data fusion technique. *Mater. Struct.*, 38:283.

42. Andrews, D. R. 1993. Future prospects for ultrasonic inspection of concrete. *Proc. Inst. Civ. Eng. Struct. Build.*, Feb., 99:71–73.

43. Tang, L., Andrews, D. R. 2006. Evaluation of a novel technique RSIM for monitoring the conditions of concrete structures. Presented at the Nondestructive Evaluation in Civil Engineering Conference, St. Loius, MO.

44. Sansalone, M., Carino, N. J. 1989. Detecting delaminations in concrete slabs with and without overlays using the impact-echo method. *ACI Mater. J.*, 86(2):175–184.

45. Sansalone, M., Lin, J. M., Street, W. B. 1998. Determining the depth of surface-opening cracks using impact-generated stress waves and time-of-flight techniques. *ACI Mater. J.*, 95(2):168–177.

46. Kesner, K., Sansalone, M. J., Poston, R. W. 2004. Detection and quantification of distributed damage in concrete using transient stress waves. *ACI Mater. J.*, 101(4):318–326.

47. Tawhed, W. F., Gassman, S. L. 2002. Damage assessment of concrete bridge decks using impact-echo method. *ACI Mater. J.*, 99(3):273–281.

48. Sansalone, M., Streett, W. B. 1997. *Impact-Echo: Nondestructive Evaluation of Concrete and Masonry*. Bullbrier Press, Ithaca, NY.

49. Lin, Y., Sansalone, M. 1992. Detecting flaws in concrete beams and columns using the impact-echo method. *ACI Mater. J.*, 89(4):394–405.

50. Abraham, O., Côte, P. 2002. Impact-echo thickness frequency profiles for detection of voids in tendon ducts. *ACI Struct. J.*, 99(3):239–247.

51. Kim, D. S., Seo, W. S., Lee, K. M. 2006. IE–SASW method for non-destructive evaluation of concrete structure. *Non-Destr. Test. Eval. Int.*, 39:143–154.

52. Ghorbanpoor, A., Benish, N. 2003. *Non-destructive Testing of Wisconsin Highway Bridge*. Wisconsin Highway Research Program Final Report. Wisconsin Department of Transportation, Madison, WI.

53. Gibson, A., Popovics, J. S. 2005. Lamb wave basis for impact-echo method analysis. *J. Eng. Mech.*, 131(4):438–443.

54. Abraham, O., Léonard, Ch., Côte, Ph., Piwakowski, B. 2000. Time frequency analysis of impact echo signals: numerical modeling and experimental validation. *ACI Mater. J.*, 97(6):612–624.

55. Huang, N. E. 1998. The empirical mode decomposition and the Hilbert spectrum for nonlinear and nonstationary time series analysis. *Proc. R. Soc. London*, A454:903–995.

56. Algernon, D., Wiggenhauser, H. 2007. *Impact-Echo Data Analysis Based on the Hilbert–Huang Transform*. TRB 2007 Annual Meeting CD-ROM. National Research Council, Washington, DC.

57. Shokouhi, P., Gucunski, N., Maher, A. 2006. Time-frequency technique for the impact echo data analysis and interpretations. In *Proceedings of the 9th European NDT Conference (ECNDT 2006)*, Berlin.

58. Kijewski-Correa, T., Kareem, A. 2006. Efficacy of Hilbert and wavelet transforms for time–frequency analysis. *J. Eng. Mech.*, 132(10):1037–1049.

59. Reda Taha, M. M., Noureldin, A., Lucero, J. L., Baca, T. J. 2006. Wavelet transform for structural health monitoring: a compendium of uses and features. *Struct. Health Monit.* 5(3):267–295.

60. Goupillaud, P., Grossmann, A., Morlet, J. 1984. Cycle-octave and related transforms in seismic signal analysis. *Geoexploration*, 4(23):85–102.

61. Morlet, J., Arens, G., Fourgeau, E., Glard, D. 1982. Wave propagation and sampling theory: II. Sampling theory and complex waves. *Geophysics*, 47(2):222–236.

62. Takemoto, M., Nishino, H., Ono, K. 2000. Wavelet transform: applications to AE signal analysis. In *Acoustic Emission: Beyond the Millennium*, Elsevier, New York, pp. 35–56.

63. Chiang, C. H., Cheng, C. C., Liu, T. C. 2004. Improving signal processing of the impact-echo method using continuous wavelet transform. In *Proceedings of the 16th World Conference on NDT (WCNDT 2004)*, Montreal, Quebec, Canada.

64. Mallat, S. 1999. *A Wavelet Tour of Signal Processing*, 2nd ed. Academic Press, San Diego, CA.

65. Mertins, A. 1999. *Signal Analysis: Wavelets, Filter Banks, Time-Frequency Transforms and Applications*. Wiley, New York.

66. Yang, W.-X., Barry Hull, J., Seymour, M. D. 2004. A contribution to the applicability of complex wavelet analysis of ultrasonic signals. *Non-Destr. Test. Eval. Int.* 37(6):497–504.

67. Grossmann, A., Kronland-Martinet, R., Morlet, J. 1989. Reading and understanding continuous wavelet transforms, In J. M. Combes, A Grossman and P. Tchamitchain (eds.), *Wavelets: Time-Frequency Methods and Phase Space*, pp. 2-20, Springer-Verlag, Berlin, Germany.

68. Epasto, G., Proverbio, E., Venturi, V., 2010. Evaluation of fire-damaged concrete using impact-echo method, Materials and Structures, 43:235–245.

69. Robertson, A. N., Farrar, C. R., Sohn, H. 2003. Singularity detection for structural health monitoring using Holder exponents. *Mech. Syst. Signal Process.* 17(6):1163–1184.

70. ODOT. 2008. *Acoustic Emission Testing of Reinforced Concrete Bridges*. Research Report. Oregon Department of Transportation, Salem, OR.

71. COST-534. 2007. *New Materials, Systems, Methods and Concepts for Durable Prestressed Concrete Structures*: III. *New Methods for Assessment and Monitoring of Prestressed Concrete Structures*. Final Report. COST, Toulouse, France.

72. Golaski, L., Swit, G., Ono, K. 2006. Acoustic emission behavior of prestressed concrete girder during proof loading. *Prog. Acoust. Emission*, 13:145–151.

73. Ohtsu, M., Yoshiara, T., Uchida, M., Saeki, H., Iwata, S. 2003. Estimation of concrete properties by elastic-wave method. In *Proceedings of the International Symposium on Non-Destructive Testing in Civil Engineering*. Berlin, Germany, September 16–19.

74. Ohtsu, M., Ohno, K., Tokai, M. 2007. Visualized NDT for concrete cracking by SiGMA-AE and SIBIE, NDT for safety. In *Proceedings of the 4th NDT in Progress*, Prague, Czech Republic, November 5–7.

75. ASTM. 2007. *Standard Practice for Acoustic Emission Monitoring of Structures During Controlled Stimulation*. ASTM E569. ASTM, West Conshohoeken, PA.

76. RILEM. 2007. *Recommendation 1: Acoustic Emission and Related NDE Techniques for Crack Detection and Damage Evaluation in Concrete. Measurement Method for Acoustic Emission Signals in Concrete*. RILEM TC 212-ACD. RILEM, Bagneaux, France.

77. Lovejoy, S. C. 2006. Development of Acoustic Emissions Testing Procedures Applicable to Conventionally Reinforced Concrete Deck Girder Bridges Subjected to Diagonal Tension Cracking. Ph.D. Dissertation, Oregon State University.

78. Chotickai, P. 2001. Acoustic Emission Monitoring of Prestressed Bridge Girders with Premature Concrete Deterioration. M.Sc. thesis, University of Texas–Austin.

79. RILEM. 2007. *Recommendation 2: Acoustic Emission and Related NDE Techniques for Crack Detection and Damage Evaluation in Concrete. Test Method for Damage Qualification of Reinforced Concrete Beams by Acoustic Emission.* RILEM TC 212-ACD. RILEM, Bagneaux, France.

80. RILEM. 2007. *Recommendation 3: Acoustic Emission and Related NDE Techniques for Crack Detection and Damage Evaluation in Concrete. Test Method for Classification of Active Cracks in Concrete Structures by Acoustic Emission.* RILEM TC 212-ACD. RILEM, Bagneaux, France.

81. Fowler, T. J., Blessing, J. A., Conlisk, P. J. 1989. New directions in testing. In *Proceedings of the 3rd International Symposium on Acoustic Emissions from Composite Materials*, Paris.

82. Golaski, L., Gebski, P., Ono, K. 2002. Diagnostics of reinforced concrete bridges by acoustic emission. *J. Acoust. Emission*, 20:83–98.

83. Nair, A. 2006. Acoustic Emission Monitoring and Quantitative Evaluation of Damage in Reinforced Concrete Members and Bridges. M.Sc. thesis, Louisiana State University.

84. Gutenberg, B., Richter, F. 1949. *Seismicity of the Earth.* Princeton University Press, Princeton, NJ.

85. Colombo, S., Main, I. G., Forde, M. C. 2003. Assessing damage of reinforced concrete beam using "*b*-value" analysis of acoustic emission signals. *J. Mater. Civ. Eng.*, May–June, pp. 280–286.

86. Colombo, S., Forde, M. C., Main, I. G., Shigeishi, M. 2005. Predicting the ultimate bending capacity of concrete beams from the "relaxation ratio" analysis of AE signals. *Constr. Build. Mater.*, 19:746–754.

87. Shigeishi, M., Ohtsu, M., Shimazaki, J. 2003. Three-dimensional visualization of acoustic emission moment tensor solutions by VRML. In *Proceedings of the international Symposium on Non-Destructive Testing in Civil Engineering*, Berlin, Germany, September 16–19.

88. Masri, S. F., Miller, R. K., Saud, A. F., Caughey, T. K. 1987. Identification of nonlinear vibrating structures: I. Formulation. *J. Appl. Mech.*, 54:923–929.

89. Natke, H. G., Yao, J. T. P. 1990 System identification methods for fault detection and diagnosis. In *International Conference on Structural Safety and Reliability*, American Society of Civil Engineers, New York, pp. 1387–1393.

90. Flesch, R. G., Kernichler, K. 1988. Bridge inspection by dynamic tests and calculations dynamic investigations of Lavent bridge. In *Workshop on Structural Safety Evaluation Based on System Identification Approaches*, H. G. Natke, J. T. P. Yao, Eds. Vieweg, Lambrecht/Pfalz, Germany, pp. 433–459.

91. Yang, J. C. S., Chen, J., Dagalakis, N. G. 1984. Damage detection in offshore structures by the random decrement technique. *ASME J. Energy Resour. Technol.*, 106:38–42.

92. Kurowski, P. 2001. Identification of Modal Models of Mechanical Structures Based on Operational Measurements [in Polish]. Ph.D. dissertation, University of Mining and Metallurgy, Cracow, Poland.

93. Cristides, S., Barrs, A. D. S. 1984. One-dimensional theory of cracked Bernoulli–Euler beams. *Int. J. Mech. Sci.*, 26:639–648.

94. Gudmunson, P. 1983. The dynamic behaviour of slender structures with cross-sectional cracks. *J. Mech. Phys. Solids*, 31:329–345.

95. Chen, J., Garba, J. A. 1988. On-orbit damage assessment for large space structures. *Am. Inst. Aeronaut. Astronaut. J.*, 26:1119–1126.

96. Radkowski, S., Szczurowski, K. 2004. Badanie wpływu stanu naprężeń na proces propagacji naprężeniowej w strukturach sprężonychç [Examining the influence of status of stress on the propagation process in prestressed structures]. *Diagnostyka*, 31:89–94.

97. Radkowski, S., Szczurowski, K. 2007. Amplitude modulation phenomena as the source of diagnostic information about technical state of prestressed structures. Presented at the 6th International Seminar on Technical System Degradation Problems, Liptowski Mikulasz, Czechoslovakia, Apr. 11–14.

98. Galęzia, A., Radkowski, S., Szczurowski, K. 2006. Using shock excitation in condition monitoring of prestressed structures. Presented at, the Thirteenth International Congress of Sound and Vibration, Vienna, Austria, July 2–6.

99. Radkowski, S., Szczurowski, K. 2006. Hilbert transform of vibroacoustic signal of prestressed structure as the basis of damage detection technique. In *Proceedings of the Conference on Bridges*, Dubrovnik, Croatia, May 21–24, pp. 1075–1082.

100. Radkowski, S., Szczurowski, K. 2005. The influence of stress in a reinforced and prestressed beam on the natural frequency. COST Action 534, New materials and systems for prestressed concrete structures. In *Second Workshop of COST on NTD Assessment and New Systems in Prestressed Concrete Structures*, Kielce, Poland, Sept. 19–21, pp. 160–170.

101. Radkowski, S., Szczurowski, K. 2005. Wykorzystanie demodulacji sygnału wibroakustycznego w diagnozowaniu stanu struktur sprężonych [Use of vibroacoustic signal demodulation in diagnosis of prestressed structures]. *Diagnostyka*, 36:25–32.

102. Honeywell. *Water Control in Cement Manufacture*. Instrumentation Data Sheet 1. Honeywell Brown Instruments, Minneapolis, MN.

103. EcoPhysPribor. 2005. http://www.ecophyspribor.com/.

104. Cameron, J. F., Clayton, C. G. 1971. *Radioisotope Instruments*, Vol. 1. Pergamon Press, Elmsford, NY.

105. Morozov, A. A., Motovilov, E. A., Sheinin, V. I., 1998. *Soil Mech. Found. Eng.*, 35:61–64.

106. Sowacs 2001. htttp://www.sowacs.com/.

107. Mikheev, G. E., Postnikov, U. I. 1972. *The Effectiveness of the Use of Radioactive Isotopes in the National Economy*. Field of Atomic Science and Technology, Moscow.

108. Bilge, A. N. 1991. *Nuclear Techniques in Industry*. Report 1445. Istanbul Technical University, Istanbul, Turkey.

Index

Subsurface Sensing, First Edition. Edited by Ahmet S. Turk, A. Koksal Hocaoglu, and Alexey A. Vertiy.
© 2011 John Wiley & Sons, Inc. Published 2011 by John Wiley & Sons, Inc.

WILEY SERIES IN MICROWAVE AND OPTICAL ENGINEERING

KAI CHANG, Editor
Texas A&M University